5th International Conference on Mechanical and Aeronautical Engineering (ICMAE 2019)

IOP Conference Series: Materials Science and Engineering
Volume 751

Sanya, China
12 - 15 December 2019

ISBN: 978-1-7138-4700-7
ISSN: 1757-8981

Printed from e-media with permission by:

Curran Associates, Inc.
57 Morehouse Lane
Red Hook, NY 12571

proceedings.com

Some format issues inherent in the e-media version may also appear in this print version.

Institute of Physics
Dirac House, Temple Back
Bristol BS1 6BE UK

Phone: 44 1 17 929 7481
Fax: 44 1 17 920 0979

techtracking@iop.org

Additional copies of this publication are available from:

Curran Associates, Inc.
57 Morehouse Lane
Red Hook, NY 12571 USA
Phone: 845-758-0400
Fax: 845-758-2633
Email: curran@proceedings.com
Web: www.proceedings.com

TABLE OF CONTENTS

IOP Conf. Series: Materials Science and Engineering **751** (2020) 011001 doi:10.1088/1757-899X/751/1/011001

PREFACE

2019 5th International Conference on Mechanical and Aeronautical Engineering (ICMAE 2019) was held in Sanya, China during Dec 12-15, 2019. ICMAE 2019 was organized by Hong Kong Society of Mechanical Engineers (HKSME). The conference provides a useful and wide platform both for display the latest research and for exchange of research results and thoughts in Mechanical and Aeronautical Engineering, Aircraft Design and Control Engineering and other topics. The participants of the conference were from almost every part of the world, with background of either academia or industry, even well-known enterprise. The success and prosperity of the conference is reflected high level of the papers received.

The proceedings are a compilation of the accepted papers and represent an interesting outcome of the conference. This book covers 8 chapters: Aerodynamics, Propulsion and Control; Spacecraft; Aerospace Design, Testing, and Certifications; Dynamics, Measurement and Control; Design, Modeling and Analysis; Thermal-fluids Engineering; Monitoring, diagnosis and prognosis; Advanced Manufacturing.

We would like to acknowledge all of those who supported ICMAE 2019. Each individual and institutional help were very important for the success of this conference. Especially we would like to thank the organizing committee for their valuable advices in the organization and helpful peer review of the papers.

We sincerely hope that ICMAE 2019 will be a forum for excellent discussions that will put forward new ideas and promote collaborative researches. We are sure that the proceedings will serve as an important research source of references and the knowledge, which will lead to not only scientific and engineering progress but also other new products and processes.

Prof. Y. LAWRENCE YAO, Columbia University, USA

IOP Conf. Series: Materials Science and Engineering **751** (2020) 011002 doi:10.1088/1757-899X/751/1/011002

Committees

General Chair

Prof. Y. Lawrence Yao, Columbia University, USA

Conference Committee Co-Chairs

Prof. Dan Zhang,York University, Canada

Prof. Weijun Pan, Civil Aviation Flight University of China, China

Scientific Committee

Prof. Andrew Rae, University of the Highlands and Islands, UK

Prof. ERASMO CARRERA,Politecnico di Torino, Italy

Prof. Ian McAndrew, Embry Riddle Aeronautical University, UK

Prof. Yu Zhang, Zhejiang Univerisity, China

Prof. Yong Zhao, Nazarbayev University, Kazakhstan

Organizing Committee

Lec. Chenglong Li,, Civil Aviation Flight University of China, China

Prof. Jianping Xiang, Changsha University of Science and Technology, China

Prof. Zhao Hui, Zhengzhou University of Aeronautics, China

International Technical Committees

Prof. Musa bin Mailah, Universiti Teknologi Malaysia, Malaysia

Prof. Craig Law, University of theWitwatersrand, South Africa

Assoc. Prof. Kodakkal Kannan Viswanathan, Universiti Teknoligi Malaysia,Malaysia

Dr. Wenxian Yang, Newcastle University, UK

Dr. Vineet Jain, Amity University Haryana, Gurgaon

Dr. Redouane Zitoune, Clément Ader Institute of Toulouse University, Toulouse, France

Dr. Peng Wei, Iowa State University, USA

Dr. Michele Ferlauto, Politecnico di Torino, Italy

Dr. Max BAKAEV, Novosibirsk State Technical University , Russia

IOP Conf. Series: Materials Science and Engineering **751** (2020) 011002 doi:10.1088/1757-899X/751/1/011002

Dr. Eltayeb Eljack، Mechanical Engineering Department at University of Khartoum, Sudan

Dr. Bhupendra Khandelwal, The University of Sheffield, UK

Dr. Anis Hor, Clément Ader Institute of ISAE - SUPAERO, Toulouse, France

Dr. Andrew Carruthers, University of Bradford, UK

Dr. Krishna Lok Singh,National Aerospace Laboratories, India

Prof. Osman ADIGUZEL, Firat Universit, Turkey

Dr. Vinod Kushvaha，Structures, Indian Institute of Technology Jammu (IIT Jammu) | Jammu, J&K, India

IOP Conf. Series: Materials Science and Engineering **751** (2020) 011003 doi:10.1088/1757-899X/751/1/011003

Peer review statement

All papers published in this volume of *IOP Conference Series: Materials Science and Engineering* have been peer reviewed through processes administered by the proceedings Editors. Reviews were conducted by expert referees to the professional and scientific standards expected of a proceedings journal published by IOP Publishing.

IOP Conf. Series: Materials Science and Engineering **751** (2020) 012001 doi:10.1088/1757-899X/751/1/012001

Numerical Study on Aerodynamic Characteristics of Variable-sweep Morphing Aircraft at Transonic Speeds

Yuchang Lei, Dengcheng Zhang*, Yanhua Zhang and Guangxu Su

School of Aeronautics, Air Force Engineering University, Xi' an 710038, China

*Corresponding e-mail: 1634365359@qq.com

Abstract. The variable-sweep morphing aircraft (VMA) has both high speed and low speed flight performance, and can be competent for combat tasks and practical requirements in different flight environments. Aerodynamic characteristics of VMA have a great impact on flight performance. The transonic stage is particularly important in the whole variation process. It is also one of the most difficult problems in aerodynamic layout design of aircraft. In order to explore the aerodynamic characteristics of the VMA at transonic speeds, this paper adopts an efficient engineering algorithm. The aerodynamic characteristics of VMA at different Sweepback Angle, Angle of attack, Mach number and Altitude are analysed. The result demonstrates that: Sweepback Angle has great influence on aerodynamic characteristics of aircraft. For the lift coefficient, at subsonic speeds, small sweep has great influence. When the Angle of attack (AOA) is 12°, the sweepback changes from 45° to 25°, and the lift coefficient increases by 48.9%. At transonic speeds, large sweepback has great influence. When the AOA is 12°, the sweepback changes from 75° to 55°, and the lift coefficient increases by 85.7%. For the drag coefficient, at subsonic speeds, small sweepback has great influence. When the AOA is 12°, the sweepback changes from 45° to 25°, and the drag coefficient increases by 45.5%.All the changes of sweepback have great influence on drag coefficient at transonic speeds. The influence of sweepback on lift-drag ratio is complex. In general, when the Mach number increases, the sweepback corresponding to the maximum lift-drag ratio increases. The altitude factor only has some influence on the drag coefficient at subsonic speeds. These aerodynamic change rules can provide a basis for the shape design and dynamic analysis of VMA.

1. Introduction

A variant is an object that changes its situation, shape, or character. Therefore, variant aircraft is a new type of aircraft that can change the aerodynamic layout so as to achieve good aerodynamic characteristics in different flight environments and conditions [1]. At the beginning of the 20th century, there were studies on variant aircraft in Britain, the United States and the Soviet union. Compared with fixed-wing aircraft, variant aircraft can take into account maneuverability, take-off and landing performance, penetration and assault performance and so on. In low speed flight, small swept wing is used to obtain better lift characteristics, while in hypersonic flight, large swept wing can achieve better lift and drag characteristics due to shock wave resistance. Therefore, it is difficult for a fixed aircraft configuration to meet all requirements.

At present, the key fundamental problems in the research of variant aircraft focus on the aerodynamic characteristics of variant aircraft, driving material and structure research, layout optimization design, stability and maneuverability, etc. The research on aerodynamic characteristics of variant aircraft is one of the most important fundamental researches.[2]

Around 2000, two far-reaching research projects on variant aircraft were launched at about the same time. NASA's variant project research is mainly to variant aircraft transformation mode and variant mechanism realization form exploration.[3] The DARPA is exploring the application of variant aircraft with folding and telescopic wings.[4-5] The new generation company's sliding skin wing solution USES a special flexible skin to deform the wing, changing its area by 70%[6], The SmartWing program USES a continuous, smooth, non-hinged, high-deflection high-speed control surface to improve roll and pitch performance[7], Many researchers have studied the aerodynamic characteristics and dynamic modeling of the variant aircraft. Yao et al.[8] studied the unsteady aerodynamic characteristics of the variable swept-back aircraft, and proved that the aerodynamic force caused by different swept-back speeds is not very different. Wang et al.[2] conducted the simulation of unsteady aerodynamic characteristics of a variable forward sweeping-wing aircraft, and summarized the variation rules of aerodynamic characteristics in the variation process. Neal et al.[9] studied the effects of variable elongation and variable sweepback on aerodynamic performance in low-speed flight through experiments. Secanell et al.[10] studied airfoil optimization in different flight states based on S-A turbulence model and CFD method. Liang et al.[11] proposed a switching control law based on the mode-dependent dwell time(MDDT), which ensured the stability of variant aircraft in the transformation switching process. Literature [12-13] constructed a switched linear parameter variable (LPV) model, transforming the complex system of variant aircraft into a linear switched model. Yuan et al.[14] conducted aerodynamic modeling and dynamic modeling of variant UAVS. He et al.[15] proposed a switching linear parameter varying based on the average dwell time method. According to the mission requirements of variant aircraft, this method divides the range of backward sweep into different regions.

The above researches on variant aircraft mainly focus on the subsonic stage, while few researches on transonic and supersonic stage. But variant aircraft make a lot of sense in transonic and supersonic designs. In flight, when the airflow speed is close to the speed of sound, the airflow speed on the upper surface of the wing will exceed the speed of sound, and the shock wave will rapidly increase the resistance. An aircraft with a greater forward reach would delay the shock wave, allowing the aircraft to travel at higher speeds. It is necessary to study variant aircraft in transonic and supersonic stages. However, according to the existing research situation, the flight resistance of the aircraft in the supersonic stage is large and the flow field is more complex, so the variation process usually ends before the supersonic stage. Therefore, this paper mainly studies transonic and subsonic phases. Based on the current research status, numerical simulation [16] and wind tunnel experiment [9] are mainly used for research. These two methods are time-consuming and costly in the early stage of aircraft design. This paper takes the VMA as the research object and adopts an engineering algorithm to analyze the aerodynamic characteristics at transonic speeds. The results can provide reference for the refined design and flight control system of VMA.

2. Shape design and aerodynamic characteristics calculation of VMA

2.1. Aerodynamic layout design of aircraft

The research object of this paper is the VMA, which adopts the normal layout. In order to simplify the calculation, based on the MiG-19 swept-back wing layout, some parameters were redesigned, and the original fixed wing design was changed to the variable wing design with "changing backward sweeping Angle". The fuselage is streamlined, with a length of 9.5m. When both the inside and outside sweepbacks are 45°, the wingspan is $4.1m$ and the airfoil area is $22.6m^2$. The wing has an area of $22.6m^2$, a tip chord length of $2.14m$, a wing root chord length of $4.27m$, and a shaft chord length of $2.56m$. Airfoil NACA-W-6-66-012 was used for the wing, vertical tail and flat tail. Figure 1 shows the aircraft models corresponding to different sweepbacks.

The results show that the rotor position of the wing of a VMA has great influence on aerodynamic characteristics. If the rotation axis is more outward, the aerodynamic center movement will be significantly reduced, but the wing area that can be changed will be significantly smaller, unable to bring sufficient aerodynamic benefits. If the rotation axis is located close to the wing root, the area of the wing that can be changed is larger, but this causes the shaft structure to gain weight, limit overload capacity, and make it difficult to rotate. Therefore, according to the current research on the VMA, we choose to set the position of rotation axis at 30% of the wingspan [17]. During the whole variation process, the inside sweepback remains 45°, while the outside sweepback changes from 25° to 75°.(χ=25°-75°)

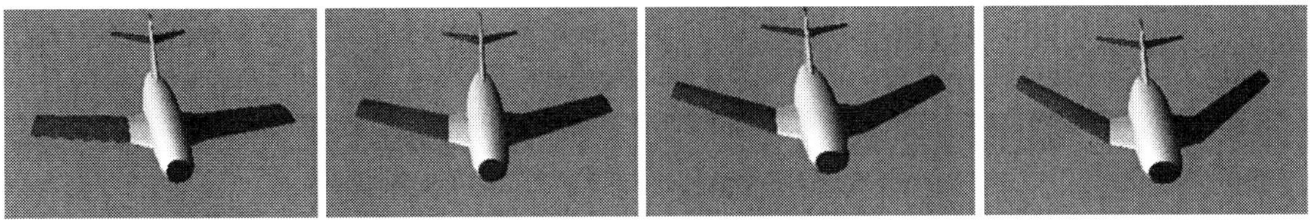

(a) χ=30° (b) χ=45° (c) χ=55° (d) χ=65°

Figure 1. Aircraft models corresponding to different sweepbacks

2.2. Calculation method

After the shape of the aircraft is determined, the designer needs to know the aerodynamic force and torque acting on the aircraft. These data are determined by the geometric shape of the aircraft and directly affect the performance of the aircraft. Generally speaking, the general method is to conduct CFD calculation or wind tunnel experiment. However, wind tunnel experiments are expensive and CFD calculation takes a long time. Especially for the VMA, the sweepback, AOA and other parameters change during the calculation process. Neither approach is suitable for use in the early stages of aircraft design. The optimal situation is that the designer already has an aircraft that can meet the performance requirements. On this basis, the number of wind tunnel experiments and flight tests can be reduced to reduce the design cost.

The engineering algorithm can quickly estimate aerodynamic characteristics by means of analytic expressions. One of the better known is digital DATCOM software designed by the U.S. air force.[18] This exposed software takes airplane geometry as input. The input file should define the geometry of the fuselage, wings, flat tail, and vertical tail. A detailed description of the format of the DATCOM input file can be found in the digital DATCOM manual [19]. Once the input file is created, the DATCOM can be run to estimate the aerodynamic coefficients and derivatives of the selected aircraft shape and given flight conditions.

3. Results Analysis

3.1. Lift characteristics of different sweepback Angles

3.1.1. The influence of AOA on lift coefficient under different sweepbacks Angles. Figure 2 shows the curve diagram of the lift coefficient changing with the AOA under different sweepbacks whenM_a=1.0. All the backward sweep models (BSMs) are not stalling. However, under small sweepback, the increase trend of lift starts to slow down when the AOA is 12°, while under large sweepback, the increase trend of lift does not slow down when the AOA is 14° Therefore, it can be inferred that the stalling Angle of attack of small BSM is smaller than that of large. This is because the velocity component perpendicular to the wing is smaller under a large sweepback.

3.1.2. The influence of Mach number on lift coefficient under different sweepback Angles. Figure 3 shows the curve diagram of the lift slop with Mach number at different sweepbacks. Under the same Mach number, the maximum lift of the small BSM is higher than that of the large. At the same inlet velocity, for the small BSM, the inlet velocity perpendicular to the wing is larger. In addition, there is also a shortage of leading edge lift in the large BSM. Therefore, the lift of small BSM is higher than that of large.

Under subsonic conditions (0.4, 0.6), the influence of small sweeping Angle on lift is greater than that of large sweepback on lift. In transonic condition (0.8, 1.0, 1.2), the influence of small sweepback on lift is smaller than that of large sweepback on lift. The increase of Mach number changes the flow field and reduces the impact of shock wave under small sweepback

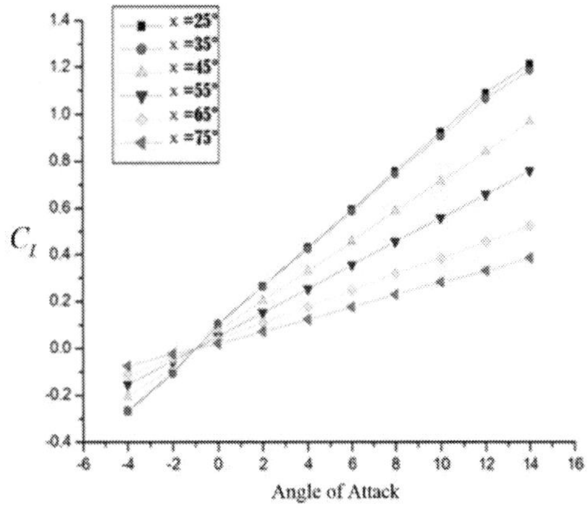

Figure 2. Lift coefficient under different sweepbacks, M_a=1.0

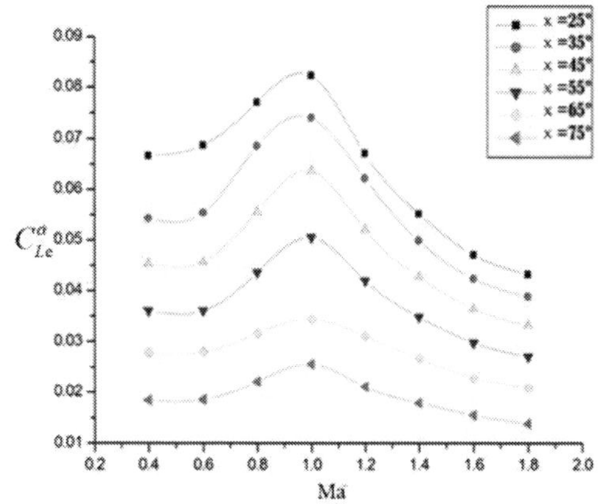

Figure 3. Lift slop under different sweepbacks

As the Mach number increases, C_{Le}^{α} first increases and then decreases, reaching the maximum value near $M_a = 1.0$, remaining basically unchanged at subsonic speeds, changing rapidly at transonic speeds, decreasing slowly at supersonic speed, and the lift also shows the same trend.

3.2. Drag characteristics of different sweepback Angles

3.2.1. The influence of AOA on drag coefficient under different sweepbacks Angles. Figure 4 shows the curve diagram of the change of drag coefficient with the AOA under different sweepbacks when M_a=1.0. In the range from -4° to 0°, as the AOA increases, the drag of each model gradually decreases. Under the same AOA, the drag of small sweepback model is larger. At a small AOA, the drag characteristics are greatly affected by the change of sweepback at subsonic speeds. At high AOA, the drag characteristics are greatly affected by the change of sweepback at transonic speeds.

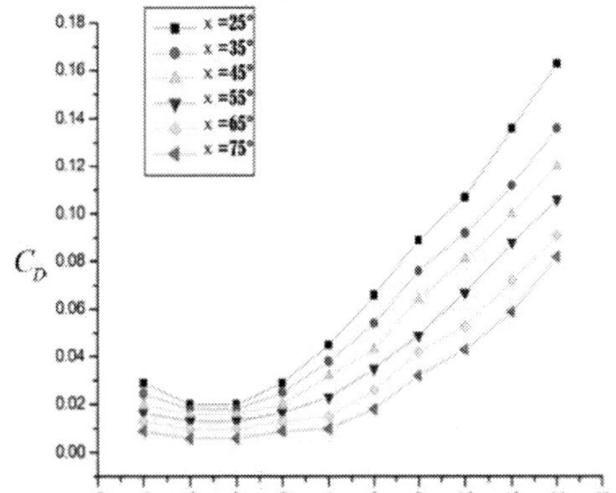

Figure 4. Drag coefficient under different sweepbacks, M_a=*1.0*

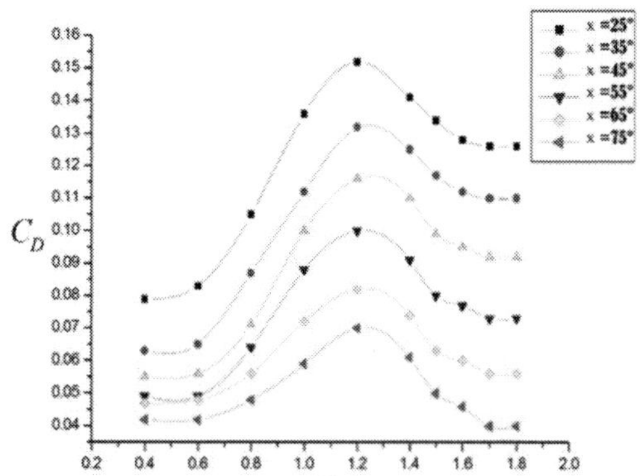

Figure 5. Drag coefficient with mach number, AOA=12

3.2.2. The influence of Mach number on drag coefficient under different sweepback Angles. Figure 5 shows the curve diagram of the change of drag coefficient with Mach number under different sweepbacks when the AOA is 12°. Under different Mach Numbers, the drag of the small BSM is larger than that of the large. The reason is that at the same inflow velocity, the velocity component of the small back sweep model decomposed to the direction parallel to the wing is larger.

Under subsonic conditions (0.4, 0.6), the influence of small sweeping Angle on drag is greater than that of large sweeping Angle on drag. Under the transonic condition (0.8, 1.0, 1.2), all the changes of sweeping Angle have a great influence on the drag.

When the Mach number is lower than 0.6, the drag coefficient of each BSM increases slowly with the Mach number. Then the drag increases sharply with the increase of Mach number and then decreases and tends to be stable in the supersonic section. When the Mach number increases, the local shock wave is generated, the intensity increases and shock wave separation occurs, the drag increases rapidly. When the Mach number continues to increase, the drag coefficient decreases.

In the process of Mach number increasing from 0.8 to 1.0, the drag under a small sweepback increases sharply, while the drag under a large sweepback increases gently. The Mach number corresponding to the maximum drag coefficient increases with the increase of the sweepback. This is because the increase of the sweepback slows down the generation of shock waves. The sweepback is large, the effective partial velocity is small, the local shock wave is late, and the corresponding Mach number is large.

3.3. Lift-drag ratio characteristics of different sweepback Angles

Figure 6 shows the curve diagram of the lift-drag ratio with the AOA when the Mach number is 0.6, 1 and 1.4 respectively. When the Mach number is smaller (M_a=0.6), the lift-drag ratio of the small BSM is larger than that of the large BSM. This is because the lift of the small BSM is much larger than that of the large BSM, while the drag coefficient is less affected by the sweepback. This is particularly evident at small angles of attack.

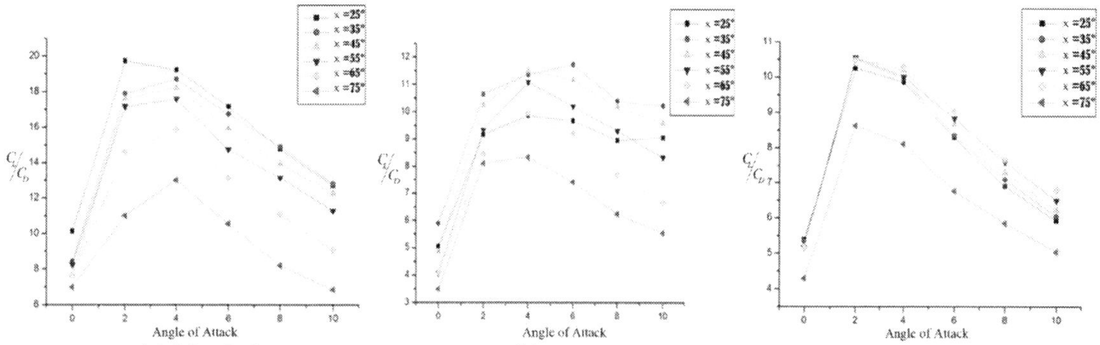

(a) M_a=0.6 (b) M_a=1.0 (c) M_a=1.4

Figure 6. Lift-drag ratio with the AOA

When the Mach number enters the transonic stage ($M_a>0.8$) and continues to increase, the lift-to-drag ratio advantage of the small BSM began to decrease. This is because the increase of Mach number will cause local shock waves on the wing surface, leading to a sharp increase in drag coefficient. A large sweepback can delay the generation of shock waves on the wing surface and reduce the drag coefficient. The lift-drag characteristics of the large BSM begin to show superiority.

Through the analysis of the lift-drag ratio, it can be found that the lift-drag ratio of the model with a sweepback of 75° is always low. Compared with the model with a sweepback of 65°, the lift coefficient of the model with 75° decreases more, but the drag coefficient is similar. Too large sweepback will not improve the lift-drag characteristics, but will make the lift-drag characteristics worse.

3.4. Effect of altitude

After a long period of exploration and summary, in the design process of aircraft, it is generally considered that height has little influence on lift force and pitching moment, but has a great influence on drag coefficient. Such treatment method is very suitable for fighter aircraft [20]. Figure. 7 shows the curve diagram of drag coefficient with height at different sweeping Angles when the AOA is 12°.

At different flight altitudes, the change trend of drag with the change of sweeping Angle is the same, and different sweeping Angle will increase the same drag coefficient with the change of altitude. With the increase of Mach number, the influence of altitude on drag decreases gradually, and becomes very weak at supersonic speed. When the AOA is 12°, the influence of altitude on the drag is nonlinear. When M_a=0.6 and the flight altitude changes from 17km to 22km, the increment of drag coefficient can reach 0.008 ~ 0.012. When M_a=1.4, the altitude changes by 20km, and the drag coefficient increases by 0.001 ~ 0.002 only.

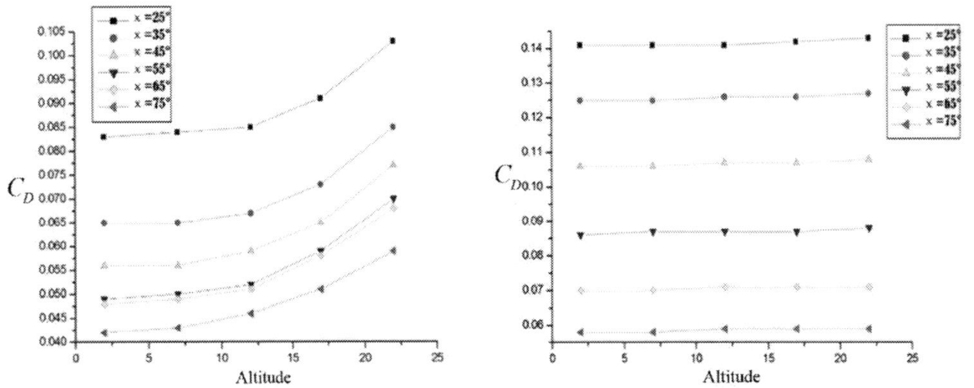

(a) M_a=0.6 (b) M_a=1.4

Figure 7. Drag coefficient with height, AOA=12

4. Conclusion

Based on the above data, the following conclusions can be drawn about the aerodynamic characteristics of the VMA at different sweepbacks:

(1) For a given Mach number, when the sweepback increases, the slope of lift coefficient decreases, but the stalling AOA increases. For a given sweepback, when the Mach number increases, the lift first increases and then decreases. Under the subsonic condition (0.4, 0.6), a small sweepback has a greater impact on the lift. When the AOA is 12°, the sweepback changes from 45° to 25°, and the lift coefficient increases by 48.9%. Under the transonic condition (0.8, 1.0, 1.2), a large sweepback has a greater impact on the lift. When the AOA is 12°, the sweepback changes from 75° to 55°, and the lift coefficient increases by 85.7%.

(2) For a given Mach number, the drag coefficient decreases when the sweepback increases. For a given sweepback, when the Mach number increases, the drag first increases and then decreases. Under the subsonic condition (0.4, 0.6), the drag is almost constant. Under transonic conditions (0.8, 1.0, 1.2), the drag coefficient under small sweepback increases sharply with Mach number, while the drag coefficient under large sweepback increases gently.

(3) The changes of aerodynamic parameters under different BSMs are complex. When the lift-drag ratio is selected as the index to measure aerodynamic benefits, the optimal sweepback increases with the increase of Mach number, but the optimal sweepback is not a linear relationship with the Mach number. There is a maximum sweepback of 75°, under which the aerodynamic effect is not satisfactory.

(4) Altitude has a low influence on lift coefficient, but a large influence on drag coefficient at subsonic speeds. When studying the variable swept-back rule, the influence of height on aerodynamic characteristics under subsonic conditions should be considered to modify the variable swept-back rule.

References

[1] Lu Y P, He Z and Lyu Y, Variable-sweep Morphing Aircraft Technology 2008 *Aeronautical Manufacturing Technology.* **22** p26-29.

[2] Wang X, Su X B and Feng H Y, Simulation Research on Unsteady Aerodynamic Characteristics of Variable Airfoil of VFSW Aircraft 2015 *Computer Simulation.* **32** p54-58.

[3] Love M H, Stroud R L, Bye D R, et al. Demonstration of Morphing Technology Through Ground and Wind Tunnel Tests 2007 *AIAA.* 2007-1729.

IOP Conf. Series: Materials Science and Engineering **751** (2020) 012001 doi:10.1088/1757-899X/751/1/012001

[4] Rodriguez A R, Morphing aircraft technology Survey 2007 *AIAA.* 2007-1258.

[5] Ivanco T G, Scott R C, Love M H, et al. Validation of the Lockheed martin morphing concept with wind tunnel testing 2007 *AIAA.* 2007-2235.

[6] Flanagan J S, Strutzenberg R C, Myers R B, et al. Development and flight testing of a morphing aircraft, the next gen MFX-1 2007 *AIAA.* 2007-1707.

[7] Kudva J, Overview of the DARPA smart wing project 2004 *Journal of Intelligent Material System and Structure.* **15** p261-267.

[8] Yao J K, Qiu J H, Ji H L, et al. Numerical investigation on the unsteady aerodynamic characteristics of flexible variable-sweep morphing aircraft 2015 *Acta Aerodynamica Sinica.* **33(3)** p406-413.

[9] Neal D A, et al. Design and wind tunnel analysis of fully adaptive aircraft configuration 2004 *AIAA.* 2004-1727.

[10] Secanell M, Suleman M, Gamboa P, Design of a morphing aircraft for a light unmanned aerial vehicle using high-fidelity aerodynamic shape optimization 2005 *AIAA.* 2005-1891.

[11] LIANG X H, WANG Q, DONG C Y, Robust adaptive control for morphing aircraft based on switching system 2019 *Journal of Beijing University of Aeronautics and Astronautics.* **45(3)** p538-545(in Chinese).

[12] HE X, ZHAO J, Multiple Lyapunov functions with blending for induced 12-norm control of switched LPV systems and its application to an F16 aircraft model 2014 *Asian Journal of Control.* **16(1)** p149-161.

[13] JIANG W L, DONG C Y, WANG T, et al. Smooth switching LPV robust control for morphing aircraft 2016 *Control and Decision.* **31(1)** 66-72(in Chinese).

[14] Yuan L, He Z, Wang Y, Modeling and trajectory optimization of perching maneuvers for morphing UAV 2018 *Journal of Nanjing University of Aeronauticss & Astronautics.* **50(2)** p266-275.

[15] He Y, Zhang W G, Wang M W, et al. Switching LPV Robust H_∞ Controller for Morphing Aircraft 2016 *Journal of Northwestern Polytechnical University.* 34(6) p1045-1049(in Chinese).

[16] Bowman J, Sanders M B, Weisshaar D T, Evaluating the impact of morphing technologies on aircraft performance 2002 *AIAA.* 2002-1631.

[17] Zhao S Z, Li Z K, Zhang X G, Study of a Variable Wing Sweep Mechanism Design for Unmanned Air Vehicle 2009 *Aeronautical Computing.* **39** p21-24.

[18] Eric J A, Melissa A M. High Angle of Attack Aerodynamic Predictions Using Missile Datcom 2005 *AIAA.* 2005-5086.

[19] Surhone L M, Tennoe M T, Henssonow S F. 2010 *USAF Stability and Control DATCOM* Saarbrücken: Betascript Publishing.

[20] Zhu Z Q, Wu Z C, 2005 *Modern aircraft design aerodynamics.* 2007-1258. Beijing University of Aeronautics and Astronautics Publishing.

IOP Conf. Series: Materials Science and Engineering **751** (2020) 012002 doi:10.1088/1757-899X/751/1/012002

A Simple Modification of HLLEM Approximate Riemann Solver Applied to the Compressible Euler System at Low Mach Number

Hang Yu [1,*], Hua Li [1], Jianqi Lai [1] and Ye Zhang[1]

[1] College of Aerospace Science and Engineering, National University of Defense Technology, Changsha, Hunan, China

*yuhang18@nudt.edu.cn

Abstract. Due to its attractive properties in computing compressible flows, such as sharp capturing of discontinuities, satisfying entropy condition and positivity preservation, the HLLEM approximate Riemann solver has been widely applied for simulations of many compressible flow problems. However, when it comes to weakly compressible or incompressible flows, the HLLEM scheme cannot give physically correct solutions. In the current study, a simple low Mach number fix is applied to improve the accuracy and stability of HLLEM approximate Riemann solver in the low Mach limit. As a result, a modified HLLEM scheme called LM-HLLEM scheme is proposed. Numerical results demonstrate that the proposed LM-HLLEM scheme is able to compute various flow problems accurately and robustly ranging from compressible to low Mach incompressible flows.
Keywords. Low Mach number; Computational fluid dynamics; HLLEM scheme; Euler equations

1. Introduction

Godunov-type Riemann solvers have been widely used in computational fluid dynamics for their clear physical interpretation and robustness. However, many Godunov-type Riemann solvers fail to guarantee physically correct solution in most weakly compressible and incompressible cases[1]. In practical applications, most problems contain both incompressible flow regions with very low Mach number and compressible flow regions. Take the flow around a blunt body as example, there will be a shock in front of the blunt body while in the boundary layer region near the wall, the flow should be considered as incompressible. Therefore, improving Godunov-type schemes' performance in incompressible flows is of great significance.

In order to overcome the low Mach limit of Godunov-type Riemann solvers, many scholars have done a series of researches. Turkel[2] proposed a preconditioning method, improved the accuracy and efficiency of solving the low Mach flows[3]. However, quite a number of test cases show that a cut-off problem is unavoidably for preconditioning method within the region where low Mach flows and high Mach flows both exist. And this problem limits its capability of accurate simulating in transition flow regions and low speed/high speed mixed flow regions. To cure the cut-off problem of Turkel's preconditioning method, Li et al[4] developed a so called All-Speed-Roe scheme. Compared with other previous preconditioned schemes, Li's new scheme is more robust and accurate. And asymptotic analysis is conducted in their subsequent works[5] to study the low Mach number behavior of this

IOP Conf. Series: Materials Science and Engineering **751** (2020) 012002 doi:10.1088/1757-899X/751/1/012002

scheme theoretically. Chen et al[6] proposed a low Mach fixed scheme, i.e. the EC-RoeM scheme. In their opinion, the left/right-moving acoustic waves can be rescaled into the scaling of convective velocity through a scaling Mach number function. Dellacherie[7] also has done a series of works to solve the low Mach number problems and introduced a low Mach correction. Based on this low Mach correction, Dellacherie proposed a low Mach Godunov scheme for the linear wave equation, and further extended it to nonlinear cases. Furthermore, he exhibited that the proposed method can be applied to other schemes and can be used to justify other existing low Mach schemes, like Rieper's LMRoe[8] scheme, K. Oßwald's L2Roe scheme[9], and other Godunov-type methods[10]. Based on the work of Dellacherie, Xie[11] proposed a simple and accurate all Mach HLLC-type scheme, i.e. AM-HLLC-P scheme. This new scheme is of high level of robustness against shock instability and retain the ability of sharp capturing of different kinds of discontinuities. With the introduction of an all Mach correction method, this scheme is further extended to the computation of flows at all Mach numbers.

In 1983, Harten, Lax and van Leer[12] proposed a uniform framework for constructing approximate Riemann solutions, namely the HLL scheme. Based on HLL scheme, the approximate Riemann solutions with different resolution for various discontinuity can be easily obtained. In [13], the authors indicated that under certain wavespeeds estimate, the HLL-type schemes are positively conservative and automatically satisfy the entropy condition. Furthermore, some researchers[13, 14] pointed out that there is a close connection among many upwind schemes, such as Roe scheme[15] and Osher scheme[16]. And these schemes can be represented uniformly by HLL-type scheme framework, therefore, the research on HLL-type schemes is of great significance. Based on HLL scheme, Einfeldt[13] obtained an approximate Riemann solver which can resolve various discontinuities through a method of linear distribution to modify the intermediate state, i.e. the HLLEM scheme. Since HLLEM scheme is based on HLL scheme, it also satisfies the entropy condition and is positivity. HLLEM scheme inherits many attractive properties of HLL scheme, and can resolve linear waves with minimal diffusion. It has been extended to many other applications like multi-phase flows[17], chemically reacting flows[18] and MHD equations[19]. However, when comes to low Mach flows, the HLLEM scheme also cannot guarantee a physically correct solution, like other Godunov-type schemes such as HLLC[20] scheme or Roe[15] scheme. And this low Mach number problem limits HLLEM's performance for accurate simulation of compressible flows.

In this paper, a low Mach correction is introduced for HLLEM scheme to improve its ability for simulating weakly compressible and incompressible flows, and a LM-HLLEM scheme is developed. Inspired by the work of Dellacherie[7], this new scheme preserves the original form of HLLEM scheme to the maximum, thus it will be straightforward to introduce this low Mach correction to existing codes. Besides, LM-HLLEM scheme is more robust and accurate when compressible and weakly compressible phenomenon are coexisting within a flow region.

The paper is organized in the following way. In the second section, the Euler equations and the related discrete method is reviewed, the HLLEM scheme is also introduced in this section. In the third section, a low Mach correction is applied to HLLEM scheme. A few numerical tests are conducted in the fourth section. Finally, concluding remarks are made in the last section.

2. Euler equations and discrete method

2.1. Euler equations

Euler equations are the basic governing equations for inviscid flows, it can be written in the following integral form as,

$$\frac{\partial}{\partial t}\int_{\Omega}\mathbf{U}\mathrm{d}\Omega+\oint_{\partial\Omega}\mathbf{F}\mathrm{d}S=0, \tag{1}$$

where $\mathbf{U}$ and $\mathbf{F}$ are conservative variable vector and inviscid flux vector respectively. Ω is the control volume, $\partial\Omega$ denotes the boundaries of the control volume, $\mathrm{d}S$ denotes the surface of the interface. The conservative variable vector and inviscid flux vector are defined as,

$$\mathbf{U}=\begin{pmatrix}\rho\\ \rho u\\ \rho v\\ \rho e\end{pmatrix},\mathbf{F}=\begin{pmatrix}\rho q\\ \rho uq+pn_x\\ \rho vq+pn_y\\ \rho Hq\end{pmatrix}, \tag{2}$$

where ρ is density, p is static pressure, e is specific total energy, u and v are velocity vector components. $q=un_x+vn_y$, represents the directed velocity, where $\mathbf{n}=\left(n_x,n_y\right)^T$ denotes the outward unit vector normal to the interface.

In this paper, a cell-centered finite-volume method is applied to system (1), assuming that flow variables are stored at the center of grid cell, therefore the time derivative of the conserved variable $\mathbf{U}$ can be transformed as

$$\frac{\partial}{\partial t}\int_\Omega \mathbf{U}\mathrm{d}\Omega=\Omega\frac{\partial \mathbf{U}}{\partial t}. \tag{3}$$

Combine equation (1) and equation (3), we can get the following expression,

$$\frac{\partial \mathbf{U}}{\partial t}=-\frac{1}{\Omega}\oint_{\partial\Omega}\mathbf{F}\mathrm{d}S. \tag{4}$$

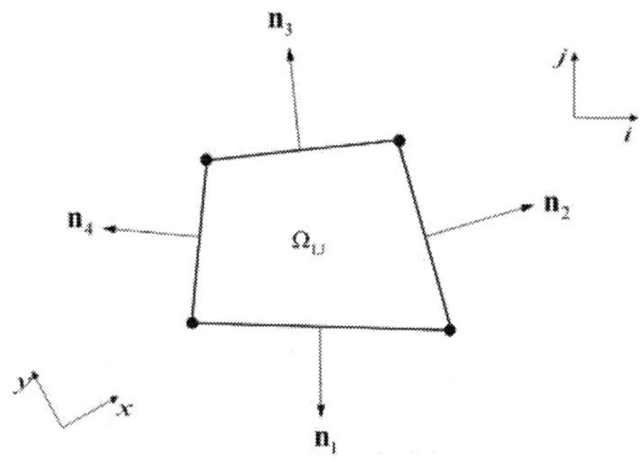

Figure 1. Schematic diagram of control volume in a two-dimensional structured grid.

The semidiscrete finite volume scheme over a control volume Ω_i, as shown in figure 1, can be written as

$$\frac{\mathrm{d}\mathbf{U}_i}{\mathrm{d}t}=-\frac{1}{\left|\Omega_i\right|}\sum_{\Gamma_{ij}\subset\partial\Omega_i}\left|\Gamma_{ij}\right|\mathbf{F}_{ij}, \tag{5}$$

where $\mathbf{U}_i$ is cell average of $\mathbf{U}_i$ on Ω_i, $\left|\Omega_i\right|$ is the volume of Ω_i, Γ_{ij} denotes the common edge of two adjacent cell Ω_i and Ω_j, $\left|\Gamma_{ij}\right|$ is the length of this surface element, $\mathbf{F}_{ij}$ is the numerical flux.

2.2. The HLLEM scheme

The HLLEM scheme, proposed by Einfeldt[13], is a popular method to solve Riemann problem approximately, and the corresponding interface flux can be defined as,

$$\mathbf{F}_{\mathrm{HLLEM}}=\frac{S_R^+\mathbf{F}\left(\mathbf{U}_L\right)-S_L^-\mathbf{F}\left(\mathbf{U}_R\right)}{S_R^+-S_L^-}+\frac{S_L^-S_R^+}{S_R^+-S_L^-}\left(\mathbf{U}_R-\mathbf{U}_L-\hat{\delta}_2\hat{\alpha}_2\hat{\mathbf{R}}_2-\hat{\delta}_3\hat{\alpha}_3\hat{\mathbf{R}}_3\right), \tag{6}$$

with

$$S_L^- \equiv \min\left(0, S_L\right), S_R^+ \equiv \max\left(0, S_R\right), \tag{7}$$

the symbol "^" represents the Roe's averaged values at the interface, S_L and S_R are the left and right wave speeds, and can be calculated from

$$S_L = \min\left(q_L - a_L, \hat{q} - \hat{a}\right), S_R = \max\left(q_R + a_R, \hat{q} + \hat{a}\right), \tag{8}$$

a represents the sound speed.

$$\hat{\mathbf{R}}_2 = \begin{pmatrix} 1 \\ \hat{u} \\ \hat{v} \\ \dfrac{\hat{u}^2 + \hat{v}^2}{2} \end{pmatrix}, \hat{\mathbf{R}}_3 = \begin{pmatrix} 0 \\ -n_y \\ n_x \\ -\hat{u}n_y + \hat{v}n_x \end{pmatrix}, \tag{9}$$

and

$$\hat{\alpha}_2 = \Delta\rho - \frac{\Delta p}{\hat{a}^2}, \hat{\alpha}_3 = \hat{\rho}\left(-n_y \Delta u + n_x \Delta v\right), \tag{10}$$

$$\hat{\delta}_2 = \hat{\delta}_3 = \frac{\hat{a}}{\hat{a} + \left|\hat{q}\right|}, \tag{11}$$

where $\hat{\mathbf{R}}_2$ and $\hat{\mathbf{R}}_3$ are the second and third right eigenvectors of Jacobian matrix $\hat{\mathbf{A}}$, $\hat{\alpha}_2$ and $\hat{\alpha}_3$ are the wave strengths. $\hat{\delta}_2$ and $\hat{\delta}_3$ are the anti-dissipation terms, which control the dissipation terms of linearly degenerate waves.

It is worth noting that when the anti-dissipation terms $\hat{\delta}_2$ and $\hat{\delta}_3$ are equal to 0, the HLLEM scheme will degenerate into HLL scheme.

3. Modification on the HLLEM scheme at low Mach number

As mentioned in the first section, the Godunov-type Riemann solvers cannot guarantee physically correct solutions at low Mach numbers. Dellacherie[7] has done a series of works, and proposed a low Mach correction for Godunov-type schemes. In this part, the low Mach correction will be briefly reviewed, and will be applied to HLLEM scheme.

3.1. An all Mach correction

Here we solve the compressible Euler equations through the finite-volume method, and the spatial convection term is discretized by a Godunov-type scheme, so the low Mach Godunov scheme can be constructed as

$$\frac{\mathrm{d}\mathbf{U}_i}{\mathrm{d}t} + \frac{1}{\left|\Omega_i\right|} \sum_{\Gamma_{ij} \subset \partial\Omega_i} \left|\Gamma_{ij}\right| \mathbf{F}_{ij}^{\mathrm{LM,\,X}} = 0. \tag{12}$$

X represents a Godunov-type scheme, the numerical flux $\mathbf{F}_{ij}^{\mathrm{LM,\,X}}$ can be expressed as

$$\mathbf{F}_{ij}^{\mathrm{LM,\,X}} = \mathbf{F}_{ij}^{\mathrm{X}} + \left(1 - \theta_{ij}\right) \frac{\rho_{ij} a_{ij}}{2} \begin{pmatrix} 0 \\ \Delta q n_x \\ \Delta q n_y \\ 0 \end{pmatrix}_{ij}, \tag{13}$$

where $\mathbf{F}_{ij}^{\mathrm{X}}$ is the original flux of a unmodified scheme X, θ_{ij} is scaling function, and θ_{ij} is related to the local Mach number M_{ij},

$$\theta_{ij} = \theta\left(M_{ij}\right), \theta\left(M\right) = \min\left(M, 1\right), \tag{14}$$

M_{ij}, ρ_{ij} and a_{ij} denote the calculate Mach number, density and sound speed at the interface of i and j respectively, which are calculated from the Roe's averaged values.

3.2. Modification for HLLEM scheme

In this section, the low Mach correction in equation (13) is applied to HLLEM scheme to improve its performance in the simulation of low Mach number flows. Here X scheme is HLLEM scheme. Following the low Mach correction in equation (13), the modified numerical fluxes can be written as

$$\mathbf{F}_{ij}^{\text{LM-HLLEM}} = \mathbf{F}_{ij}^{\text{HLLEM}} + \left(1-\theta_{ij}\right)\frac{\rho_{ij}a_{ij}}{2}\begin{pmatrix} 0 \\ \Delta q n_x \\ \Delta q n_y \\ 0 \end{pmatrix}_{ij}, \tag{15}$$

where θ_{ij} is defined in equation (14), and $(\ )_{ij}$ represents Roe's averaged values.

Besides, we shall introduce a new definition of the local Mach number M_{ij}, namely

$$M_{ij} = \max\left(\frac{\sqrt{u_i^2+v_i^2}}{a_i}, \frac{\sqrt{u_j^2+v_j^2}}{a_j}\right), \tag{16}$$

and this modification is proposed by Thornber et al[21].

Compared with other alternatives, Thornber's modification of local Mach number has little influence on stability of numerical simulation near the sonic point or at high speed flows[22].

4. Numerical experiments

In this section, several numerical test cases will be conducted to verify the effectiveness of Dellacherie's low Mach number correction applied in HLLEM scheme.

4.1. Modified Sod shock tube

This test case is a modified version of the classical Sod shock tube problem[23], which is very useful to evaluate the entropy satisfaction property of numerical schemes. The solution of this problem consists of a right shock wave, a right travelling contact wave and a left sonic rarefaction wave. In this paper, this case is used to verify whether the LM-HLLEM scheme satisfies the entropy condition. The initial condition is set as

$$\left(\rho_L, u_L, p_L\right) = \left(1.0, 0.75, 1.0\right) \quad \left(\rho_R, u_R, p_R\right) = \left(0.125, 0.0, 0.1\right),$$

besides, the position of the discontinuity is set as $x = 0.3$. The computational domain is $x \in [0,1]$, and is evenly discretized with 400 cells. Nonreflecting boundary conditions are used at the boundaries. The time step is determined by local CFL condition, and is set as $\text{CFL} = 0.9$.

The numerical solutions of different solvers at $t = 0.2s$ are shown in figure 2. The results from the LM-HLLEM solver are almost identical to that of the HLLEM scheme, and agree well with the exact solution. As shown in figure 2, this newly developed all Mach number scheme can capture shock wave, rarefaction wave and contact discontinuity sharply. In the left rarefaction wave region, no so-called rarefaction shock generated near the sonic point, which indicates that this LM-HLLEM scheme satisfies the entropy condition. On the contrary, the Roe scheme does not satisfy the entropy condition, so there is an obvious nonphysical rarefaction shock near the sonic point.

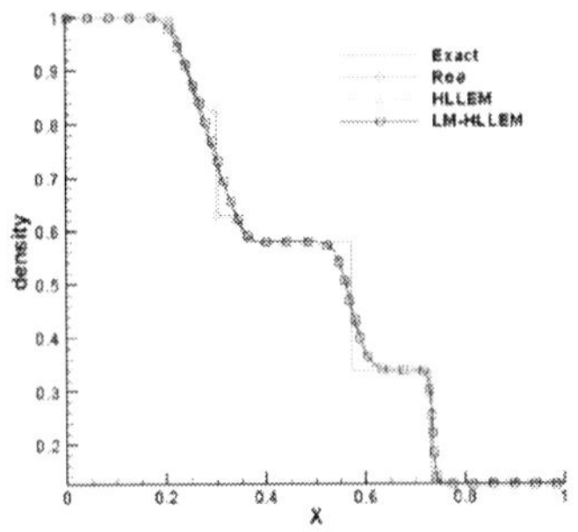

(a)Density profiles

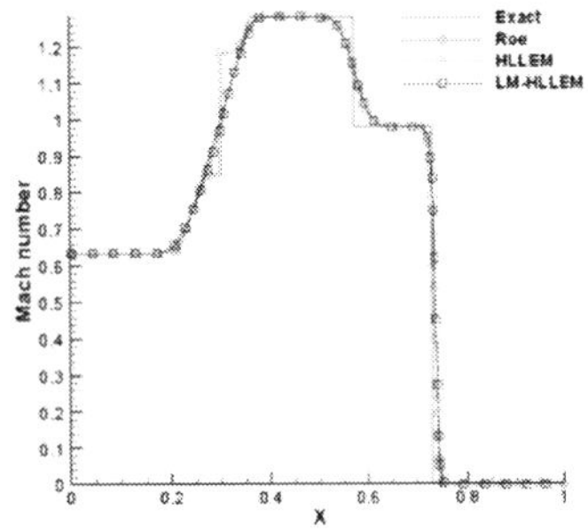

(b)Mach number profiles

Figure 2. Modified Sod shock tube at $t = 0.2s$. (a) Density profiles; (b) Mach number profiles.

4.2. Gresho vortex

The Gresho vortex test is used to demonstrate the accuracy of LM-HLLEM scheme in low Mach incompressible flow regions. Gresho vortex is a rotating flow in which centrifugal force is balanced by pressure gradient. The simulation region is set as $\Omega = [0,1]\times[0,1]$, the initial background state set as

$$\rho_0 = 1.0 \ ,\mathbf{u}_0 = (u_0, 0)^T \ , p_0 = 1.0 \, , a_0 = \sqrt{\frac{\gamma p_0}{\rho_0}} = \sqrt{\gamma} \, , u_0 = M_0 a_0,$$

where M_0 is the global Mach number. A vortex of $R = 0.4$ is set at position $(x_0, y_0) = (0.5, 0.5)$ at the initial time. The initial condition is defined as

$$u_r(r) = u_0 \begin{cases} \dfrac{2r}{R} & 0 \le r < \dfrac{R}{2} \\ 2\left(1 - \dfrac{r}{R}\right) & \dfrac{R}{2} \le r < R. \\ 0 & R \le r \end{cases} \tag{17}$$

And the initial pressure field is given by

$$p(r) = p_0 + u_0^2 \begin{cases} \dfrac{2r^2}{R^2} + 2 - \log 16 & 0 \le r < \dfrac{R}{2} \\ \dfrac{2r^2}{R^2} - \dfrac{8r}{R} + 4\log\left(\dfrac{r}{R}\right) + 6 & \dfrac{R}{2} \le r < R, \\ 0 & R \le r \end{cases} \tag{18}$$

where the radius can be expressed as

$$r = \sqrt{(x - x_0)^2 + (y - y_0)^2}. \tag{19}$$

Periodic boundary conditions are applied to the left and right boundaries of simulating region. The far field boundary conditions are applied to upper and lower boundaries. The computational region is evenly dispersed into a 100 × 100 Cartesian grid, this test case is run for the Mach number $M_0 = 10^{-2}$, and the CFL-number is 0.9.

A dimensionless pressure field is defined to characterize the order of pressure fluctuation in incompressible limit,

IOP Conf. Series: Materials Science and Engineering **751** (2020) 012002 doi:10.1088/1757-899X/751/1/012002

$$p_N(x,y)=\frac{p(x,y)-p_{\min}}{p_{\max}-p_{\min}}. \tag{20}$$

Since Gresho vortex is a steady solution of the incompressible Euler equation, the dimensionless pressure field should not change over time, therefore this test case is very suitable for evaluating the performance of numerical schemes in low Mach flow fields.

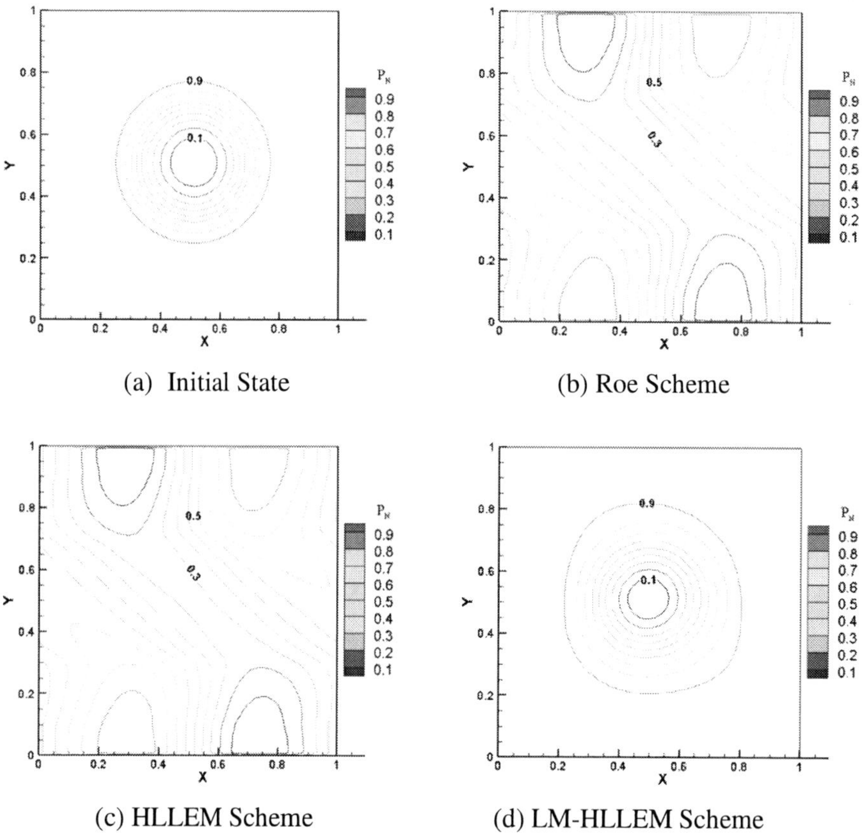

Figure 3. The normalized pressure contour map of the Gresho vortex after one period. (a) Initial State; (b) Roe Scheme; (c) HLLEM Scheme; (d) LM-HLLEM Scheme.

Figure 3 shows the dimensionless pressure field of Gresho vortex after one period under the condition $M_0 = 10^{-2}$, it can be clearly observed that the original Riemann solver, like Roe scheme and HLLEM scheme, contains excessive numerical dissipation in the low Mach number region, so the structure of vortex is completely dissipated after one period, which is greatly different from the initial solution. On the contrary, this newly developed low Mach number scheme maintains the vortex structure well and is in good agreement with the initial solution, showing its low dissipation property. Figure 4 shows the dimensionless pressure distribution curves along the center line of Gresho vortex calculated in different schemes. As shown, the LM-HLLEM scheme is more accurate than original HLLEM scheme, and shows good agreement with the exact solution. This test case verifies the accuracy of LM-HLLEM scheme in low Mach number cases.

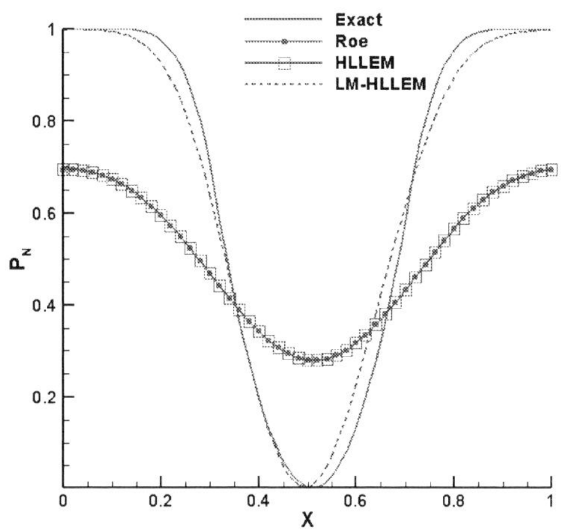

Figure 4. A normalized pressure distribution curve along the midline of the Gresho vortex after one period

4.3. Low Mach number inviscid flow around the NACA0012 airfoil

The inviscid steady flow around the NACA 0012 airfoil is a classical test case to evaluate the performance of numerical schemes in low Mach number flows. An O-type mesh is applied, which contains 241 nodes in the circumferential direction and 121 nodes in the normal direction. The angle of attack is set as 0° , and the inlet Mach number is set as $M_0 = 0.1, 0.01, 0.001$. Each case is executed for 50000 time steps with LU-SGS approach, and the CFL-number is 200. The residuals (L2-norm of density) dropped by at least five orders of magnitude. A normalized pressure field is defined by equation (20) to represent the order of pressure fluctuations in the incompressible limit.

As shown in figure 5, the unmodified HLLEM scheme cannot guarantee a physically correct solution in low Mach flow, while the LM-HLLEM scheme manages to converge to a solution that approaches incompressible flow.

5. Concluding remarks

In this paper, we proposed a low Mach number scheme called LM-HLLEM scheme. Three test cases were conducted to evaluate its performance in compressible and incompressible flows. The results of test cases shown that the LM-HLLEM scheme can capture shocks as sharply as the original HLLEM scheme, and it can be applied to the simulations of weakly compressible and incompressible flows.

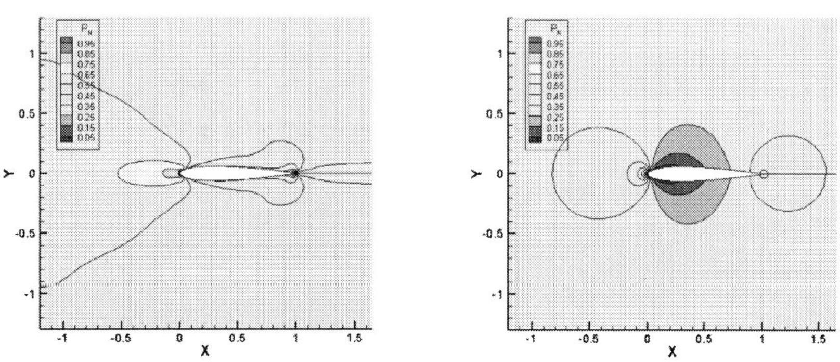

(a) HLLEM, $M_0 = 0.001$ (b) LM-HLLEM, $M_0 = 0.001$

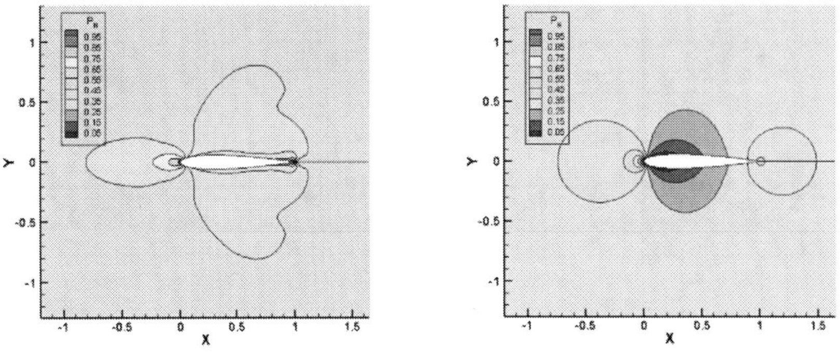

(c) HLLEM, $M_0 = 0.01$ (d) LM-HLLEM, $M_0 = 0.01$

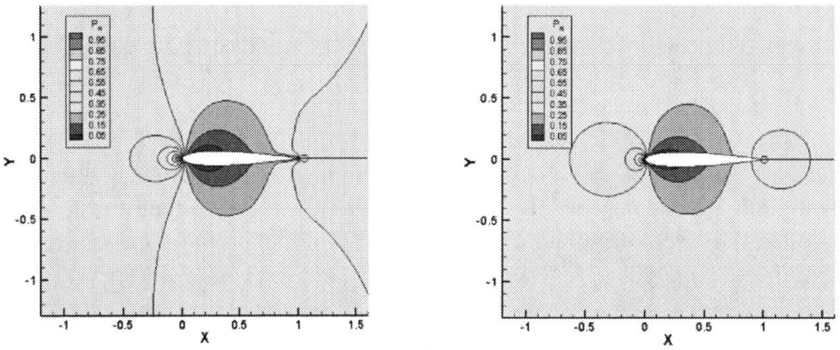

(e) HLLEM, $M_0 = 0.1$ (f) LM-HLLEM, $M_0 = 0.1$

Figure 5. The normalized pressure fields for inviscid flows around NACA0012 airfoil.

References

[1] Volpe G 1993 Performance of compressible flow codes at low Mach numbers *AIAA journal* **31** 49-56
[2] Turkel E 1987 Preconditioned methods for solving the incompressible and low speed compressible equations *Journal of computational physics* **72** 277-98
[3] Edwards JR, Liou M-S 1998 Low-diffusion flux-splitting methods for flows at all speeds *AIAA journal* **36** 1610-7
[4] Li X, Gu C, Xu J 2007 Research of preconditioned roe scheme for all speed flows in turbomachinery *ISABE Conference of AIAA*
[5] Li X-s, Gu C-w 2008 An all-speed Roe-type scheme and its asymptotic analysis of low Mach number behaviour *Journal of Computational Physics* **227** 5144-59
[6] Chen S-s, Yan C, Lou S, Lin B-x 2018 An improved entropy-consistent Euler flux in low Mach number *Journal of computational science* **27** 271-83
[7] Dellacherie S 2010 Analysis of Godunov type schemes applied to the compressible Euler system at low Mach number *Journal of Computational Physics* **229** 978-1016
[8] Rieper F 2011 A low-Mach number fix for Roe's approximate Riemann solver *Journal of Computational Physics* **230** 5263-87
[9] Oßwald K, Siegmund A, Birken P, Hannemann V, Meister A 2016 L2Roe: a low dissipation version of Roe's approximate Riemann solver for low Mach numbers *International Journal for Numerical Methods in Fluids* **81** 71-86
[10] Liou M-S 2006 A sequel to AUSM, Part II: AUSM+-up for all speeds *Journal of computational physics* **214** 137-70

IOP Conf. Series: Materials Science and Engineering **751** (2020) 012002 doi:10.1088/1757-899X/751/1/012002

[11] Xie W, Zhang R, Lai J, Li H 2019 An accurate and robust HLLC - type Riemann solver for the compressible Euler system at various Mach numbers *International Journal for Numerical Methods in Fluids* **89** 430-63
[12] Harten A, Lax PD, Leer Bv 1983 On upstream differencing and Godunov-type schemes for hyperbolic conservation laws *SIAM review* **25** 35-61
[13] Einfeldt B 1988 On Godunov-type methods for gas dynamics *SIAM Journal on Numerical Analysis* **25** 294-318
[14] Van Leer B 1984 On the relation between the upwind-differencing schemes of Godunov, Engquist–Osher and Roe *SIAM Journal on Scientific and Statistical Computing* **5** 1-20
[15] Roe PL 1997 Approximate Riemann solvers, parameter vectors, and difference schemes *Journal of computational Physics* **135** 250-8
[16] Osher S, Solomon F 1982 Upwind difference schemes for hyperbolic systems of conservation laws *Mathematics of computation* **38** 339-74
[17] Nguyen NT, Dumbser M 2015 A new HLLEM-type Riemann solver for compressible multi-phase flows with surface tension *2015 International Conference on Advanced Computing and Applications (ACOMP)* 112-7
[18] Wada Y 1992 Approximate Riemann solvers: Improvement of the HLLEM scheme and its generalization for chemically reacting flows *NASA STI/Recon Technical Report N* **94**
[19] Dumbser M, Balsara DS 2016 A new efficient formulation of the HLLEM Riemann solver for general conservative and non-conservative hyperbolic systems *Journal of Computational Physics* **304** 275-319
[20] Toro EF, Spruce M, Speares W 1994 Restoration of the contact surface in the HLL-Riemann solver *Shock waves* **4** 25-34
[21] Thornber B, Mosedale A, Drikakis D, Youngs D, Williams RJ 2008 An improved reconstruction method for compressible flows with low Mach number features *Journal of computational Physics* **227** 4873-94
[22] Chen S-s, Yan C, Xiang X-h 2018 Effective low-Mach number improvement for upwind schemes *Computers & Mathematics with Applications* **75** 3737-55
[23] Sod GA 1978 A survey of several finite difference methods for systems of nonlinear hyperbolic conservation laws *Journal of computational physics* **27** 1-31

IOP Conf. Series: Materials Science and Engineering **751** (2020) 012003 doi:10.1088/1757-899X/751/1/012003

Effectiveness of Variable Rotor Blades Technique&Variable Stator Vanes Technique in AdjustingAxial-flow Compressor

Long Bingxiang[1,*], Chen Qin[1], Ren Zebin[1]and Liao Daxiong[1]

[1]Facility Design and Instrumentation Institute of Chinese Aerodynamic Research and Development Centre, No. 6, South Section, Second Ring Road, Mianyang City, Sichuan Province, P.R. China

[*]Email: bingxiang2011@126.com

Abstract. Multi-stage Axial-flow compressor is the main driver of continuous transonic wind tunnel and it usually has ultra-wide operating range. High performance wind tunnel generally requires its multi-stage axial-flow compressor has high economic performance and stability in its ultra-wide operating range. The combination of variable compressor rotational speed and variable geometry technique, variable rotor blades technique (VRBT) and variable stator vanes technique (VSVT), has been widely adopted in extending the stable working range with high efficiency of multi-stage axial-flow wind tunnel compressor. Adjusting Effectiveness was defined in this paper to quantify the effectiveness of VRBT and VSVT in extending stable and high aerodynamic efficiency operating range of wind tunnel axial-flow compressor under various operating conditions. The results show that the adjusting effectiveness of VRBT and VSVTis a function of reaction degree, flow coefficient, loading coefficient and the amount of loading coefficient change.

1. Introduction

Multi-stage axial-flow compressor isextensivelyadopted in continuous transonic wind tunnel applications to provide air flow through the tunnel circuit ducting for establishing test conditions, and it has a direct and vital influence on the aerodynamic and economic performance of wind tunnel. Testing Mach number range of continuous transonic wind tunnels is wide and is generally range from 0.2 to 1.3 or even higher, which leads to a ultra-wide loading line for axial-flow compressor (Figure 1). The general aerodynamic design requirements for wind tunnel compressorcan be stated that multi-stage axial-flow compressor with high adiabatic efficiency and sufficient surge margin in its ultra-wide operating range. Thus, how to extend working range with high adiabatic efficiency and satisfactory surge margin of wind tunnel compressor with such a wide operating range is the main challengethat compressor designer has to deal with.

In general, variable geometry method, variable rotor blades technique (VRBT) or variable stator vanes technique (VSVT), is used together with variable speed technique to extend the stable high efficiency working range of wind tunnel compressor.

A three-stage axial-flow compressor with variable stator vanes was used in the Arnold Engineering Development Complex (AEDC) 16-foot transonic wind tunnel[1]and the National Wind Tunnel Complex (NWTC) project[2].R.A.E. considered the VRBT was not realistic in wind tunnel compressor applications in 1950s[3].In the early 1960s, however, German adopted the manual variable rotor stagger technique in the multi-stage axial-flow compressor for the AVA supersonic wind tunnel[4].

Although variable geometry technique has long been used in various wind tunnel compressor applications, few public literature can be found in the area of comparative analysis of these two kinds of variable geometry method. In this paper, a variable -Adjusting Effectiveness- was defined and a mathematical model was developed for the purpose of quantifying the effectiveness of these two kinds of variable geometry method in adjusting wind tunnel compressor. This article is organized as follows. The first section describes the derivation of the mathematical model. A subsequent section presents the comparative analysis results and analysis. This article ends with some conclusions.

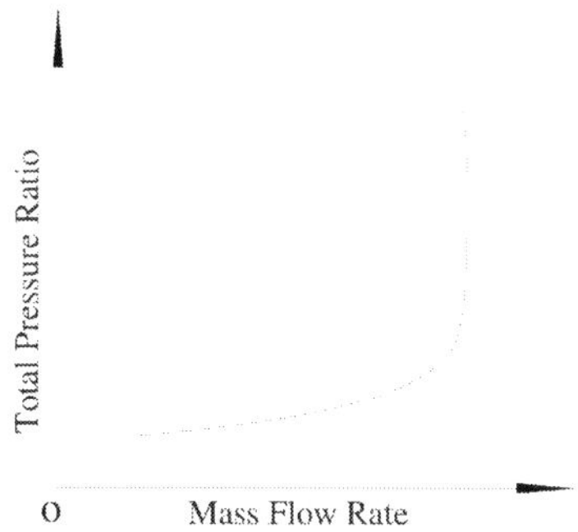

Figure 1. Typical Working Line of Wind Tunnel Compressor

2. Mathematical Model

According to the basic theory of compressor aerodynamics, flow coefficient (ϕ) and loading coefficient (ψ) were defined as follows:

$$\psi = \frac{V1 \times \sin(\alpha 1) - V2 \times \sin(\alpha 2)}{U}, \phi = \frac{V_a}{U} \tag{1}$$

Flow coefficient determines the incidence as well as the mass/volume flow rate into blade row. Loading coefficient is a direct reflection of total pressure ratio. Thus, a typical constant rotational speed performance lineof axial-flow compressor, as it was show in Figure 2, can be presented using flow coefficient and loading coefficient. Working condition A, B and C in Figure 3 represent the optimum incidence condition, negative incidence condition and positive condition respectively. For a fixed compressor rotational speed, regulating the stagger angle of IGV/stator vanes or rotor blades can change the working condition of compressor and achieve the purpose of extending the axial-flow compressor operating range.

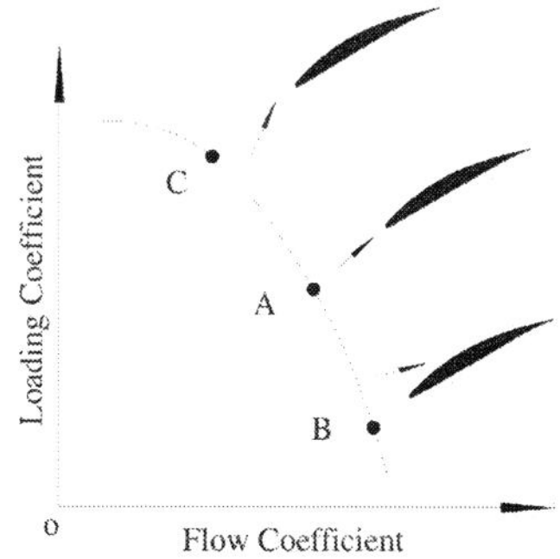

Figure 2. Typical Axial-flow Compressor Performance Line

Adjusting Effectiveness ($\eta_{adjusting}$) was defined to quantify the effectiveness of the VRBT and VSVT in adjusting wind-tunnel compressor. As can be seen from equation 2, Adjusting Effectiveness is the ratio of the amount of loading coefficient change ($\Delta\psi$) to the amount of stagger angle change ($\Delta\gamma$).

$$\eta_{adjusting} = \frac{\Delta\psi}{\Delta\gamma} \tag{2}$$

Two dimensional cascade model and velocity triangle, which were presented in Figure 3, were used in the process of developing the mathematical model.

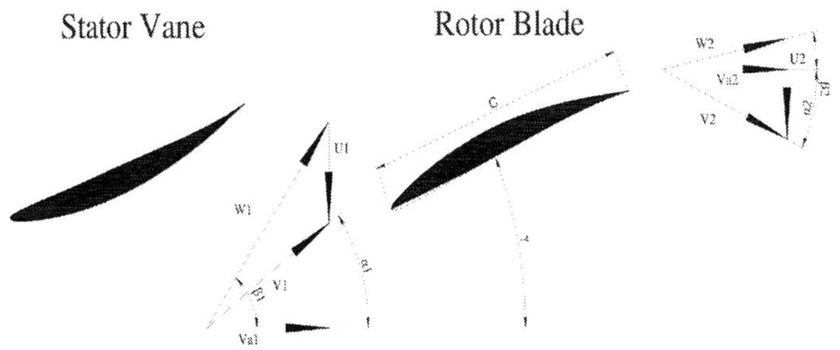

Figure 3. Cascade geometry

In axial-flow compressor aerodynamic design, the inlet/outlet axial velocity ratio is given as a aerodynamic design input and it generally equal or close to 1.0. Thus,in order to simplify the derivation of the analysis model, the axial velocity at the inlet of the cascade was assumed to equal to that at the outlet. Re-plot the inlet and outlet velocity triangle as Figure 4.For a given rotational speed (U), a given loading coefficient and a given flow coefficient, $\triangle P_1AB$ and $\triangle P_2AB$ denote the inlet and outlet velocity triangle respectively.

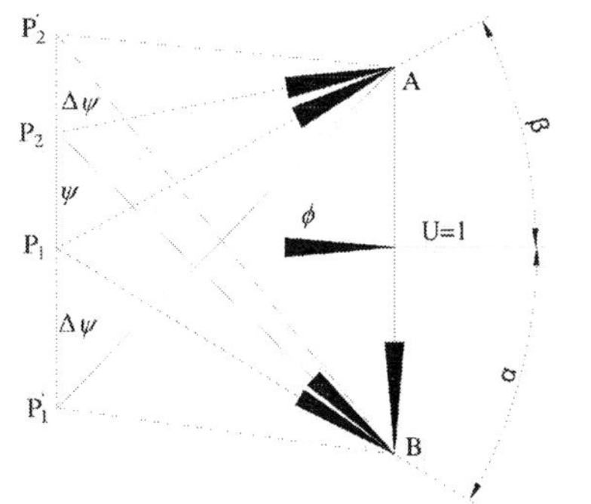

Figure 4. Velocity Triangle

For a givenchange in loading coefficient is ($\Delta\psi$).If variable IGV/stator vanes method was used, the inlet velocity triangle will change from $\triangle$P1AB to $\triangle$P'1AB, the variation of the IGV/stator stagger angle is $\angle$P1BP'1. If variable rotor blade technique was adopted, the outlet velocity triangle will change from $\triangle$P2AB to $\triangle$P'2AB, the variation of the rotor stagger angle is $\angle$P2AP'2.The edge length of each velocity triangle is as follows:

$$P_1B = \sqrt{1+(\tan\alpha)^2}\phi,\ P_1'B = \sqrt{\phi^2+(\Delta\psi+\phi\tan\alpha)^2},\ P_1'P_1 = \Delta\psi \tag{3}$$

$$P_2A = \sqrt{(1+\phi\tan\alpha-\psi)^2+\phi^2},\ P_2'A = \sqrt{(1+\phi\tan\alpha-\psi-\Delta\psi)^2+\phi^2},\ P_2'P_2 = \Delta\psi \tag{4}$$

The absolute flow angle (α) was eliminated so that the Adjusting Effectiveness can be presented as a function of more commonly used aerodynamic design inputs of axial-flow compressors. According to the definition of the reactiondegree (R)[5], equation 5 and 6 can be derived.

$$\cos\beta = \phi\left\{\phi^2+(R+\psi/2)^2\right\}^{-0.5},\ \tan\alpha = \frac{\phi\tan\beta-1}{\phi},\ -90^\circ<\alpha,\beta<90^\circ \tag{5}$$

$$\alpha = \arctan\left(\frac{R+0.5\psi-1}{\phi}\right) \tag{6}$$

Substitute equation 6 into equation 3 and 4, rewrite quation 3 and 4 as follows:

$$P_1B=\sqrt{\phi^2+(1-R-0.5\psi)^2},\ P_1'B=\sqrt{\phi^2+(-1+R+0.5\psi+\Delta\psi)^2},\ P_1'P_1=\Delta\psi \tag{7}$$

$$P_2A=\sqrt{(R-0.5\psi)^2+\phi^2},\ P_2'A=\sqrt{(R-0.5\psi-\Delta\psi)^2+\phi^2},\ P_2'P_2=\Delta\psi \tag{8}$$

The variation of rotor stagger angle and IGV/stator stagger angle can be calculated using equation 9 and 10 respectively.

$$\Delta\gamma_{stator} = \arccos\left(\frac{(P_1B)^2+(P_1'B)^2-(P_1'P_1)^2}{2(P_1B)(P_1'B)}\right) \tag{9}$$

$$\Delta\gamma_{rotor} = \arccos\left(\frac{(P_1A)^2+(P_1'A)^2-(P_1'P_1)^2}{2(P_1A)(P_1'A)}\right) \tag{10}$$

It is clear, from the definition of adjusting effectiveness as well as from equation 7 to 10, that adjusting effectiveness is a function of reaction degree, flow coefficient, loading coefficient and the amount of loading coefficient change and can be expressed using the flowing mathematical function:

$$\eta_{adjusting} = f(\psi,\phi,\Delta\psi,R) \tag{11}$$

3. Results and Analysis

In the following analysis, the range of reaction degree, loading coefficient and flow coefficient were chosen base on the Lewis Chart and are as follows:

$$R\in[0.5\ \ 0.9],\phi\in[0.2\ \ 0.7],\psi\in[0.15\ \ 0.45] \tag{12}$$

The adjustable range was chosen so that the loading coefficient of compressor after adjusting still lies within the range from 0.15 to 0.45

3.1. Adjusting Effectiveness of VRBT

Figure 5 to Figure 7 represent the change of the Adjusting Effectiveness of VRBT with the variation of reaction degree, flow coefficient, loading coefficient and the amount of loading coefficient change.

As can be seen from Figure 5, for compressor with a reaction degree of 0.5, the Adjusting Effectiveness of VRBT decreases with the increase of the amount of loading coefficient change in conditions with flow coefficient and loading coefficient lie in the range defined in equation 12.

For working condition whose flow coefficient range from 0.4 to 0.7, it is clear from Figure 5 that the Adjusting Effectiveness of VRBT decrease with the decrease of flow coefficient. Figure 5 also demonstrates that, for a givenamount of loading coefficientchange and a given flow coefficient, the Adjusting Effectiveness of VRBT increase with the decrease of loading coefficient.

From Figure 5 to Figure 7, it is easy to infer that the trend of the variation of Adjusting Effectiveness with the variation of the amount of loading coefficient change is similar for compressor with reaction degree in the range of 0.5 to 0.9.

Figure 5 to Figure 7also show that the Adjusting Effectiveness of VRBT increase with the increase of reaction degree, and the amplitude of the Adjusting Effectiveness change increase with the decrease of flow coefficient for given amount of loading coefficient change and a given loading coefficient.

As can be seen from Figure 7, for compressor with a reaction degree of 0.9, the Adjusting Effectiveness of VRBT increases with the decrease of flow coefficient.

3.2. Adjusting Effectiveness of VSVT

Figure 8 to Figure 10 show the change of the Adjusting Effectiveness of VSVT with the variation of reaction degree, flow coefficient, loading coefficient and the amount of loading coefficient change.

Figure 8 clearly demonstrates that, for compressor with reaction degree of 0.5, the Adjusting Effectiveness of VSVT increases with the increase of the amount of loading coefficient change in conditions with flow coefficient and loading coefficient lie in the range defined in equation 12. As the last chart in Figure 10 shows, for high-reaction highly loaded compressor, the Adjusting effectiveness of VSVT decreases with the increase of the amount of loading coefficient change.

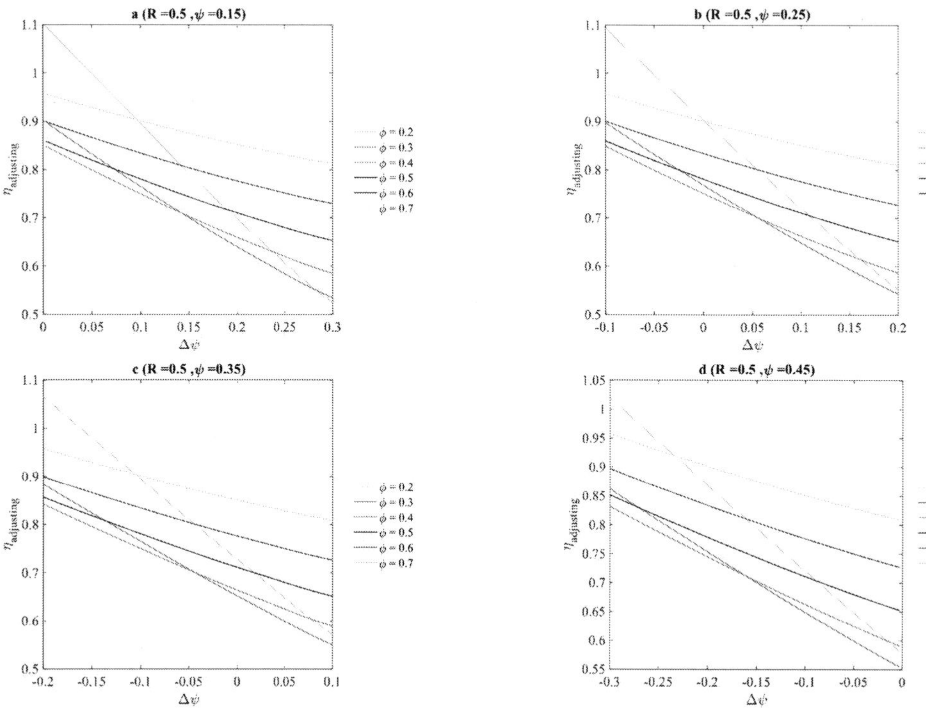

Figure 5. Adjusting Effectiveness via VRBT (R=0.5)

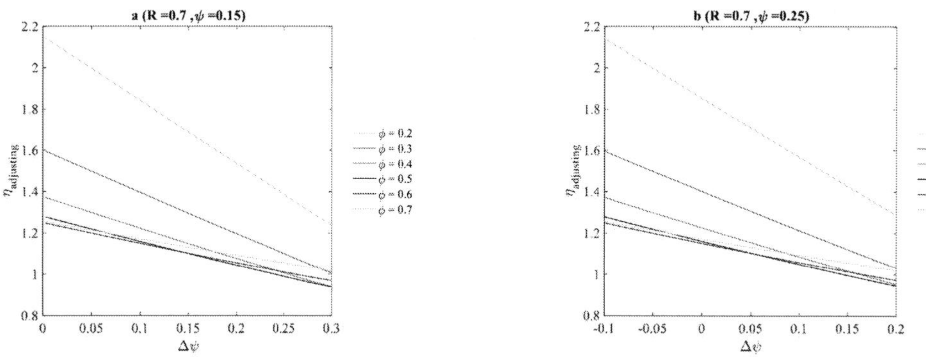

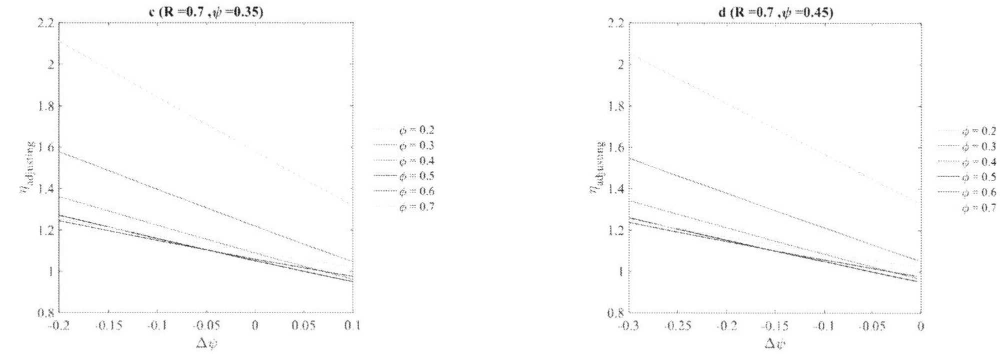

Figure 6. Adjusting Effectiveness via VRBT (R=0.7)

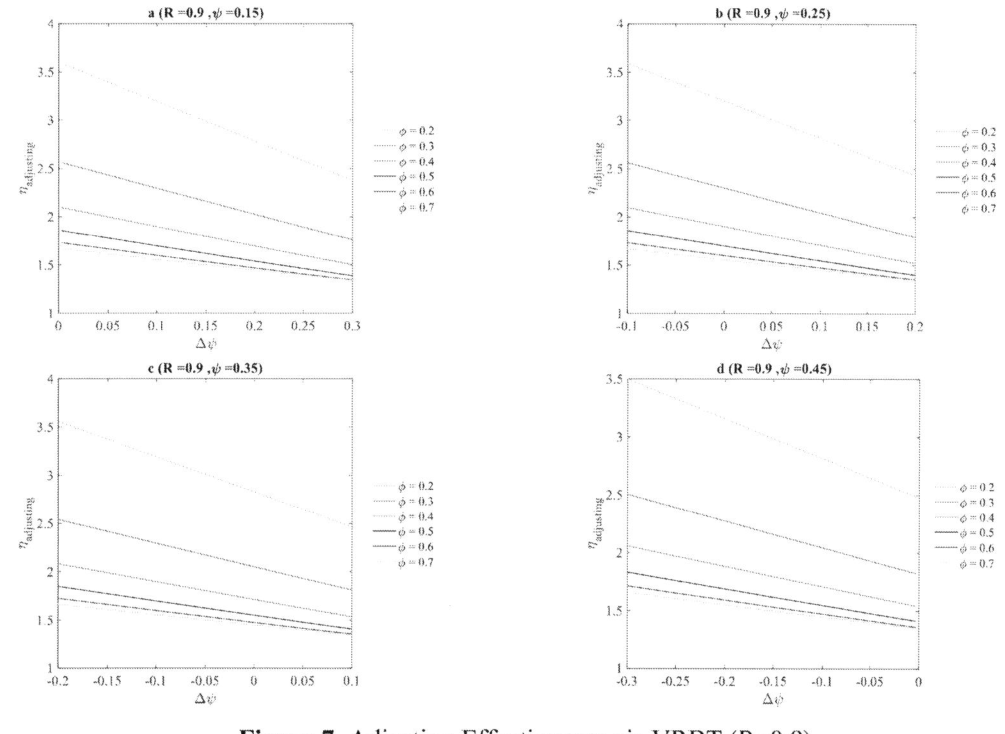

Figure 7. Adjusting Effectiveness via VRBT (R=0.9)

The Adjusting Effectivenessof VSVT,as can be seen from Figure 8 to Figure 9, decreases with the increase of loading coefficient for the same amount of loading coefficient change and the same flow coefficient condition. Figure 10, however, presents a contrary trend.

Figure 8 to Figure 10 demonstrates that for low-reaction low-loaded compressor, VSVT has larger Adjusting Effectiveness in small flow coefficient conditions. For high-reaction highly-loaded compressor, however, VSVT has greater Adjusting Effectiveness under large flow coefficient conditions.Figure 8 to Figure 10 also shows that the Adjusting Effectiveness of VSVT decreases with the increase ofreaction degree.

3.3. Adjusting Effectiveness Difference Analysis

Figure 11 to Figure 13 show the change of the Adjusting Effectiveness differencebetweenVRBT and VSVTwith the variation of reaction degree, flow coefficient, loading coefficient and the amount of loading coefficient change.

It is clear from Figure 11 to Figure 13 that the difference between the adjusting effectiveness of VRBT and VSVT change linearly with the variation of the amount of loading coefficient change.

As can be seen from Figure 11, for compressor with reaction degree of 0.5, VSVT is more effective than VRBT in any working conditions except when the amount of loading coefficient change is smaller than 0.0. When the amount of loading coefficient change is larger than 0.0, Figure 11 demonstrates that the smaller the flow coefficient the greater superiority the VSVT has.

For any working condition with flow coefficient in the range of 0.2 to 0.7 and loading coefficient in the range of 0.15 to 0.45, it is easy to infer from Figure 12 and Figure 13 that, within the adjustable range, VRBT is more effective than VSVT in adjusting compressor with reaction degree of 0.7 to 0.9. Figure 12 and Figure 13 also demonstrate that the smaller the flow coefficient, the larger superiority margin the VRBT has.

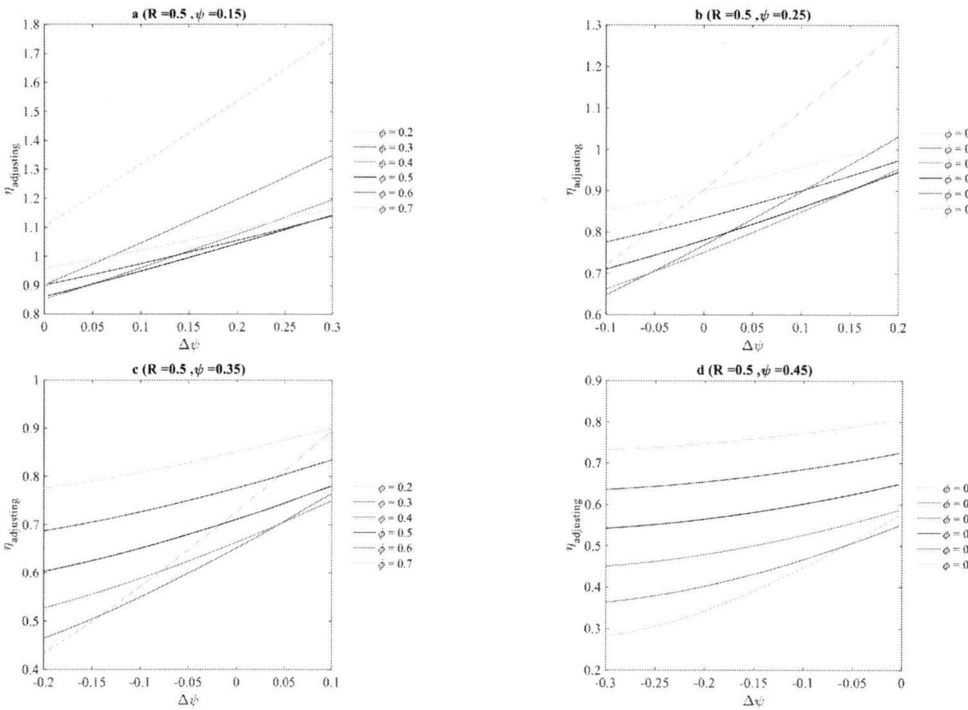

Figure 8. Adjusting Effectiveness via VSVT (R=0.5)

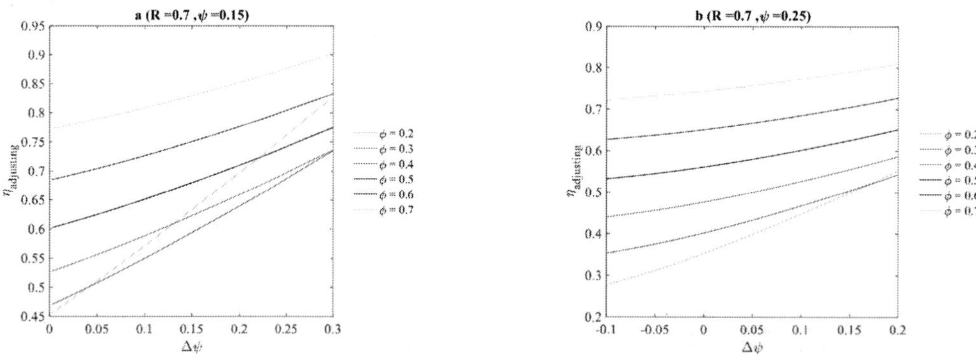

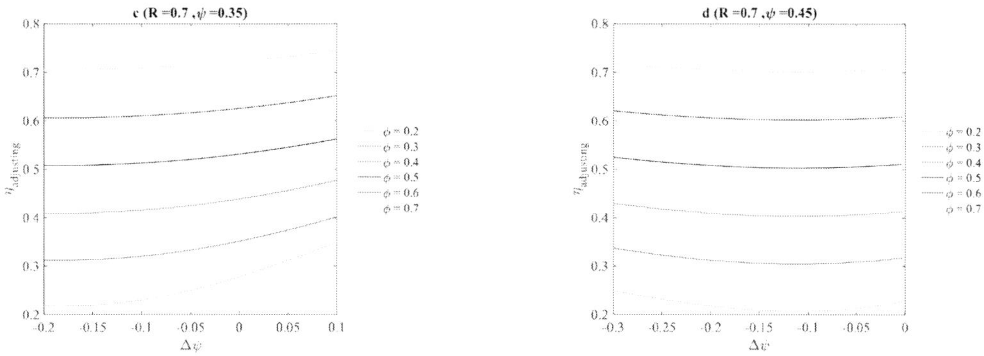

Figure 9. Adjusting Effectiveness via VSVT (R=0.7)

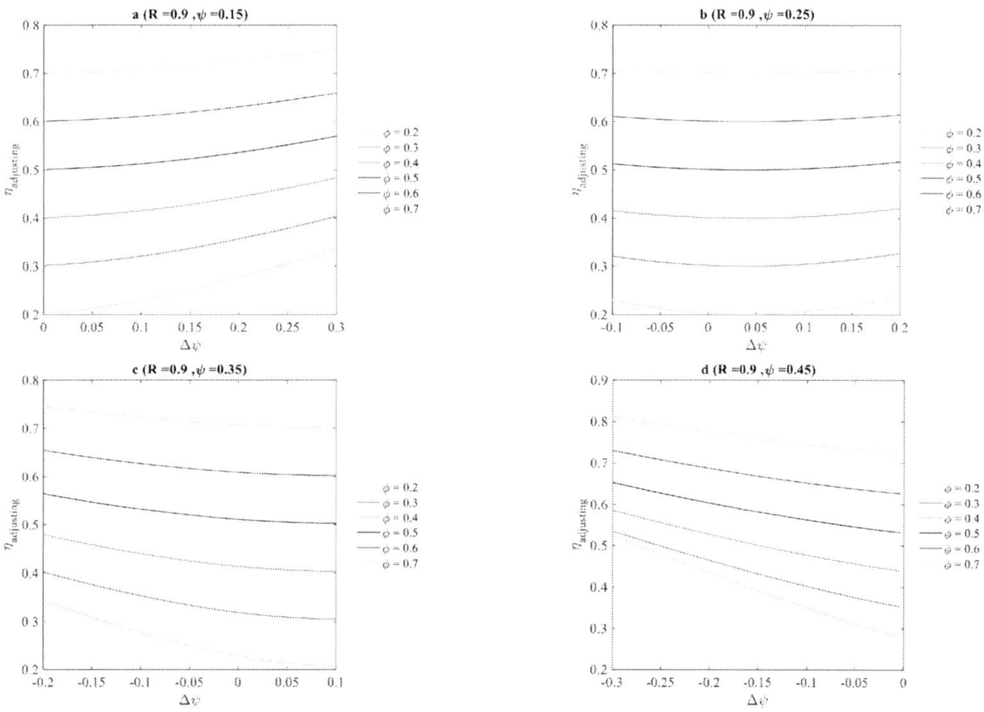

Figure 10. Adjusting Effectiveness via VSVT (R=0.9)

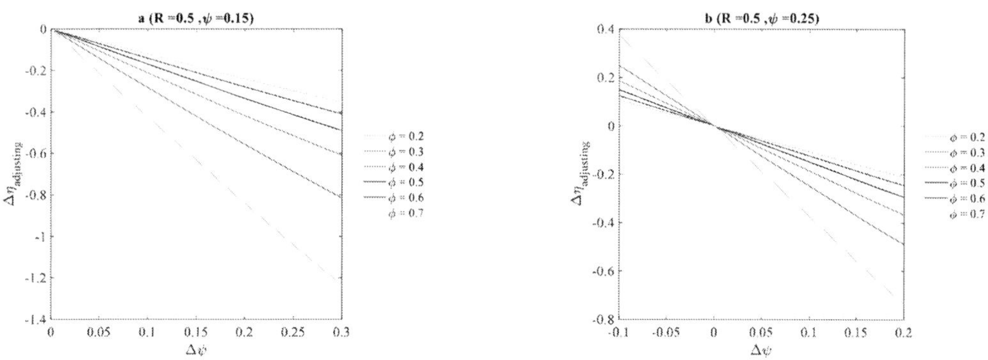

IOP Conf. Series: Materials Science and Engineering **751** (2020) 012003 doi:10.1088/1757-899X/751/1/012003

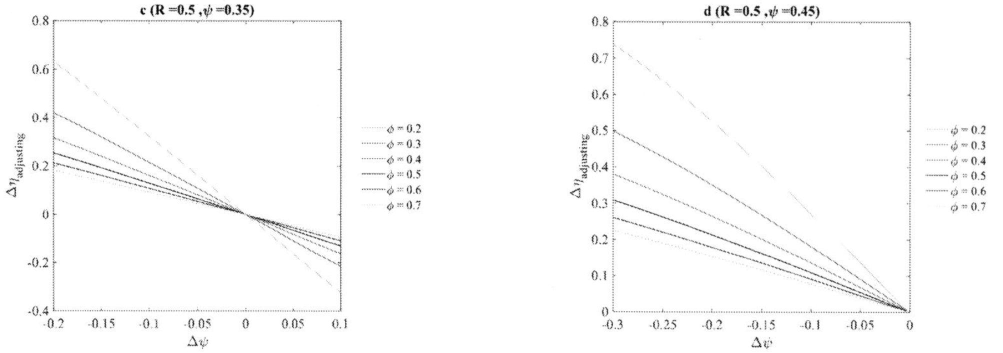

Figure 11. Adjusting Effectiveness Difference (R=0.5)

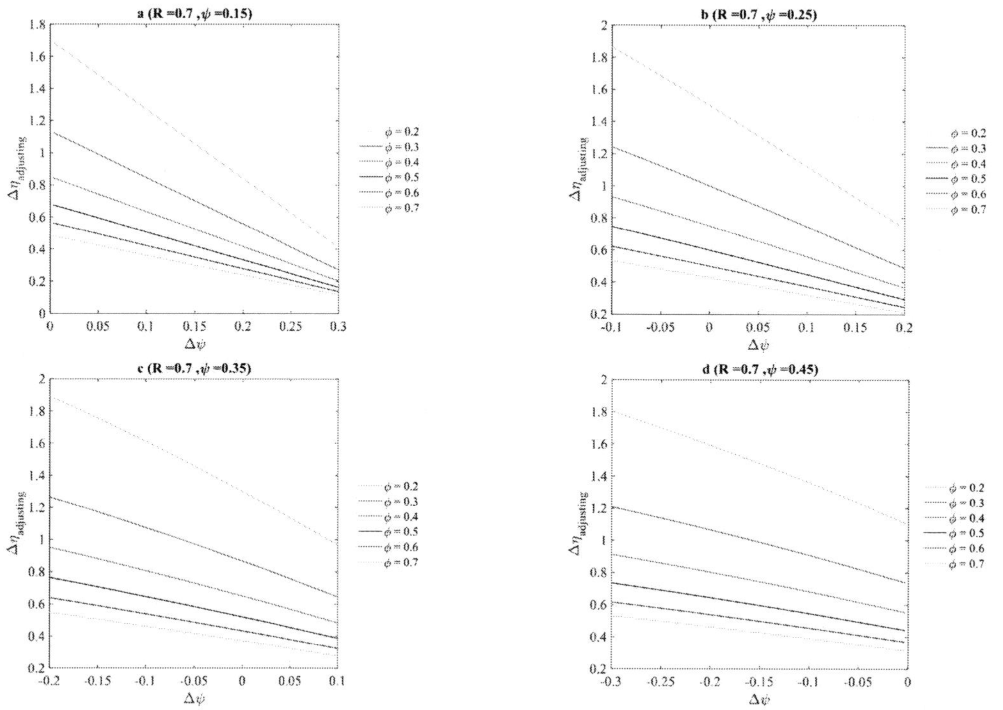

Figure 12. Adjusting Effectiveness Difference (R=0.7)

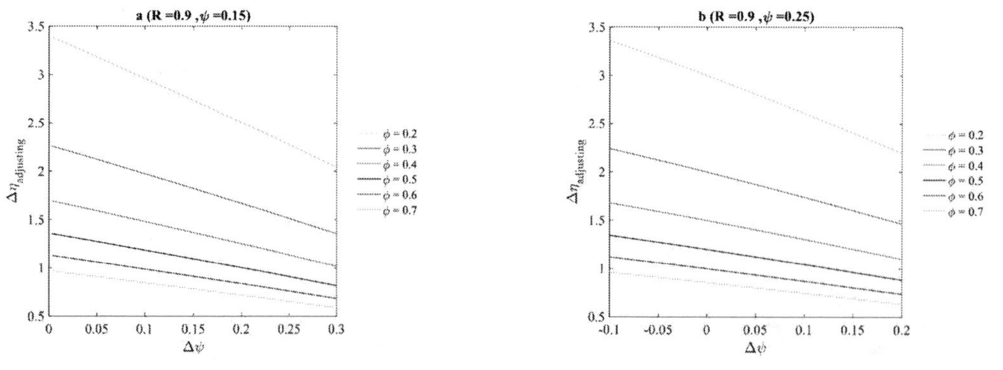

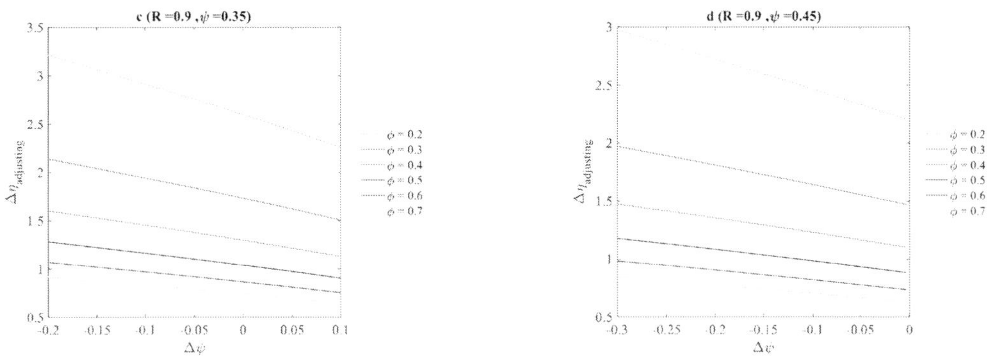

Figure 13. Adjusting Effectiveness Difference (R=0.9)

4. Conclusion

Adjusting Effectiveness was defined to quantify the effectiveness of the VRBT and the VSVT in adjusting wind tunnel compressor. It is found that the adjusting effectiveness if a function of reaction degree, flow coefficient, loading coefficient and the amount of loading coefficient change;

For low-reaction low-loaded wind tunnel compressor, VRBThas higher Adjusting Effectiveness in conditions with large flow coefficient; while for high-reaction highly-loaded wind tunnel compressor, VRBThas higher Adjusting Effectiveness in conditions with small flow coefficient;

For low-reaction low-loaded wind tunnel compressor, VSVThas higher Adjusting Effectiveness in conditions with small flow coefficient; while for high-reaction highly-loaded wind tunnel compressor, VSVThas higher Adjusting Effectiveness in conditions with large flow coefficient;

The adjusting effectiveness of VRBT increase with the increase of reaction degree, while the Adjusting Effectiveness of VSVT decrease with the increase of reaction degree; For high-reaction compressor adjusting, VRBT is more effective than VSVT from the aerodynamic point of view.

References

[1] Estabrooks B B, Robinson R A, and Jones M H 1961 Aerodynamic Performance of A Third Stage Plenum Evacuation Compressor of The ARNOLD CENTER Propulsion Wind Tunnel. AEDC-TN-61-90.

[2] Jim Young, John Dison, Mike Chaney, et al. 1996 Compressor & Drive System Segment Specification. NWT-SS-B-300-0000-01.

[3] Cheshire L J, Evans J Y G, Goodsell W A, *et al.* 1958 The design and construction of the compressor for the 8ft by 8ft high-speed wind tunnel at R.A.E. Bedford *PROC. INST. MECH. E.***172** 549-583

[4] Hans Hofmann and Heidenheim. 1967 Constructive details of a 8-stage axial compressor for the transonic wind tunnel of aerodynamic testing facility Gottingen, *Voith Research and construction.*Issue 17, Item 5 (In German)

[5] N.A. Cumpsty 2004 *Compressor Aerodynamics*(KRIEGER PUBLISHING COMPANY, Malabar Florida)

IOP Conf. Series: Materials Science and Engineering **751** (2020) 012004 doi:10.1088/1757-899X/751/1/012004

Modeling and Numerical Simulation of Stratospheric Aircraft Propulsion System

XingjunYuan[1,*], Jiming Wang[1,*], Xuan Tong[1] andFengmei Wei[2]

1.Systems Engineering Research Institute, Beijing, 100094, China
2.CASIC Space Engineering Development Co.Ltd, Beijing, 100143, China

[*]E-mail: yuanxingjun123@163.com; 953186879@qq.com

Abstract. Aiming at propulsion system of stratospheric, a high altitude propeller was designed. Avoiding the singularity of equations in the iterative process, the interference angle equation was improved and the designed propeller was calculated by the strip theory. The three-dimensionalNavier-Stokes equations were used to simulate the complex flow around the designed propeller and its aerodynamic characteristics were analyzed, in which the k-ω turbulence model was used. The accuracy of the numerical simulation was determined by the calculation of different grid numbers. The results show that the designed propeller can satisfy the requirement of propulsion system and the strip theory and numerical simulation can provide the reference for the design of stratospheric propeller.

Keywords: stratospheric; propulsion system; high altitude propeller; strip theory; numerical simulation

1. Introduction

In recent years, stratospheric aircrafts have received the great attention and many countries around the world have carried out some relevant researches. The stratospheric aircraft has important military applications, which flies between aeronautics and astronautics. Influenced by the high altitude environment, the propulsion system of the stratospheric aircraft has its unique composition.[1-2] Since the propeller is an important part of the propulsion system, it is necessary to study. The efficiency of the high altitude propeller directly affects the efficiency of the whole propulsion system, so the design and analysis of the high altitude propeller becomes the primary task of the stratospheric aircraft propulsion system.

At present, some scholars have done a lot of researches in designing and analyzing the high altitude propeller. L.Danielle Koch used the advanced ducted propfan analysis code (ADPAC) and numerical calculation to calculate the aerodynamic performance of the high altitude propeller.[3] Anthony Colozza had introduced in detail the design method of the high altitude propeller and proposed the consistent design method of the stratospheric propeller.[2] Wang Haojie had carried out the aerodynamic design of the propeller for the unmanned aerial vehicle using the blade theory and obtained good results.[4] Liu Yuanqiang used the strip theory to predict the propeller performance and verified the feasibility of the strip theory to calculate the high altitude propeller by the wind tunnel test.[5] Deng Zhiwei calculated the interference angle by the secant line method and established the propeller model by the strip theory.[6] Liu Peiqing used the numerical method to calculate the performance parameters of the high altitude propeller, compared with the wind tunnel test results, and proved the feasibility of the numerical simulation to calculate the aerodynamic characteristics of the propeller.[7] Wang Yufu used the sliding grid method to calculate the high altitude propeller, compared with the wind tunnel test results, and proved the feasibility of the numerical simulation to calculate the complicated flow around the propeller.[8] In this paper, the high altitude propeller is designed by analyzing the requirements of the stratosphere aircraft propulsion system and the accuracy of the design propeller is verified by the strip theory and numerical simulation method. At the same time, the efficiency of the high altitude propeller is relatively high at the design point.

2. Strip theory of high altitude propeller

The blade theory of the propeller is mainly based on the vortex theory and Prandlt's finite wings theory. The relationship between the blade geometry and aerodynamic force is established.[9]

If the number of propeller blades is N_B and the radius of propeller hub is r_0, the propeller thrust is

$$T = \frac{1}{2}\rho V_0^2 N_B \int_{r_0}^{R} T_c dr \tag{1}$$

The circumference force of the propeller is

$$F = \frac{1}{2}\rho V_0^2 N_B \int_{r_0}^{R} \frac{Q_c}{r} dr \tag{2}$$

The torque of the propeller is

$$M = \frac{1}{2}\rho V_0^2 N_B \int_{r_0}^{R} Q_c dr \tag{3}$$

The efficiency of the propeller is

$$\eta = \frac{TV_0}{2\pi n_s M} = \frac{V_0}{2\pi n_s} \frac{\int_{r_0}^{R} T_c dr}{\int_{r_0}^{R} Q_c dr} \tag{4}$$

When the aerodynamic characteristics of propeller are calculated, The known variables are as follows: the radius of the blade element r, the chord of the blade element b, the number of blade N_B, the incidence angle θ, the lift coefficient C_L, the drag coefficient C_D.

The axial velocity interference factor is

$$a = \frac{\tan\varphi[1+\tan\varphi_0 \tan(\varphi+\gamma)]}{\tan\varphi_0[1+\tan\varphi \tan(\varphi+\gamma)]} - 1 \tag{5}$$

The circumference velocity interference factor is

$$a' = a\tan\varphi_0 \tan(\varphi+\gamma) \tag{6}$$

The interference angle β is determined by the Newton iteration method.

$$C_L\sigma = \frac{4\sin(\varphi_0+\beta)\tan\beta}{1-\tan\gamma\tan\beta} \tag{7}$$

$$K = \frac{C_L b(1+a)^2}{\sin^2\varphi\cos\gamma} \tag{8}$$

$$T_c = K\cos(\varphi+\gamma) \tag{9}$$

$$Q_c = Kr\sin(\varphi+\gamma) \tag{10}$$

IOP Conf. Series: Materials Science and Engineering **751** (2020) 012004 doi:10.1088/1757-899X/751/1/012004

The strip theory of the propeller is derived from the assumption that the number of the blade is infinite. Actually, the number of the blade is finite. Prandlt proposed the momentum loss factor to modify the strip theory.[9] The equation is as follows.

$$F = \frac{2}{\pi}\arccos e^{-f} \tag{11}$$

$$f = \frac{N_B}{2}\frac{R-r}{R\sin\varphi_1} \tag{12}$$

In general, $R\sin\varphi_1$ of the equation (12) can be modified with $R\sin\varphi_0$ and the error can meet the requirements of the engineering precision. So the thrust and torque need multiply by the equation (11).

3. Calculation on the high altitude propeller by the strip theory

It is very important to ensure the interference angle, when using the strip theory to calculate the aerodynamic characteristics of the propeller. We can use the Newton iteration method and secant method to obtain the interference angle. The solution equation of the interference angle is as follows.

$$f(\beta) = C_L\sigma - \frac{4\sin(\varphi_0+\beta)\tan\beta}{1-\tan\gamma\tan\beta} = 0 \tag{13}$$

Changing the equation (13) to

$$f(\beta) = \frac{C_L\sigma(1-\tan\gamma\tan\beta) - 4\sin(\varphi_0+\beta)\tan\beta}{1-\tan\gamma\tan\beta} = 0 \tag{14}$$

The equation (14) can be changed as

$$F(\beta) = C_L\sigma(1-\tan\gamma\tan\beta) - 4\sin(\varphi_0+\beta)\tan\beta = 0 \tag{15}$$

The lift coefficient CL can be obtained by the XFOIL program or the Fluent software. The XFOIL program, developed by Mark Drela, is the turbulent transport model which is based on the surface element method, viscous boundary layer and the free boundary layer.[10] The angle of attack of the blade element changes constantly, so CL and CD also change. In this paper, XFOIL program and Fluent software are used to calculate the aerodynamic of S1223 airfoil. Then these data are called by the spline interpolation functions of Matlab software.

Fluent is used to calculate the aerodynamic characteristics on condition of the low Reynolds number, because the results of Fluent are more accurate. In this paper, we calculate the aerodynamic data on condition of Re=10000, 20000, 30000, 40000, 50000 and α=1, 2, 3, 4, 5, 6, 7, 8, 9, 10, 11, 12. The resulting curves of the lift and drag coefficients are shown in Figure 1.

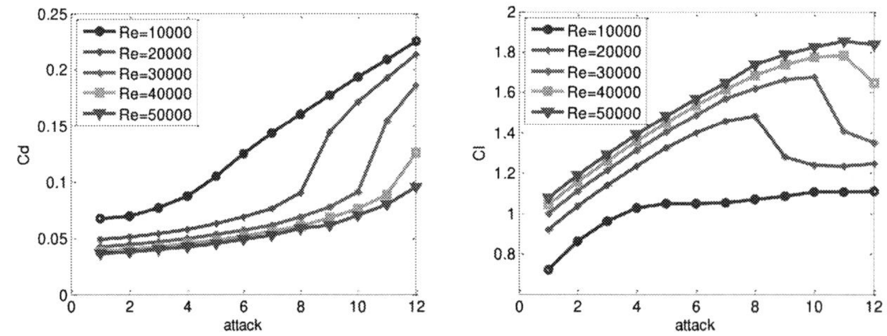

Figure 1.the lift and drag coefficients of S1223 airfoil

The lower Re is, the smaller the Angle of attack at the separation point of S1223 laminar flow is from figure 1. When Re is 40000, the Angle of attack (AOA) at the separation point is 11 degrees. When Re is 30000, AOA at the separation point is 10 degrees. When Re is 20000, AOA at the separation point is 8 degrees.

4. Numerical simulation of high altitude propeller

4.1. Calculation model

Fluent software is used to calculate the aerodynamic characteristics of high altitude propeller. The propeller data is selected from Reference [11]. CATIA software is used to establish the propeller solid model. Gambit software is used for meshing. The entire flow field area of the propeller is set as a large cylinder. The upstream length is 10 times diameters, the downstream is 20 times diameters, and the span is 10 times diameters. The flow field around the propeller is a small cylinder, which can be used as the encryption grid.

4.2. Calculation method

The pressure based solver is adopted to calculate the aerodynamic characteristics of the propeller in this paper. The governing equations are based on Reynolds mean Navier-Stokes equations. The governing equation is discretized by the finite volume method. The coupled pressure and velocity adopts the Coupled algorithm. The Shear Stress Transport k-ω turbulence model is used to calculate the propeller.

4.3. Boundary condition

The boundary conditions are the velocity inlet boundary and the pressure outlet boundary. The atmospheric pressure at 20km is 5474.9pa, the pressure outlet pressure is 0Pa, and the velocity is 20m/s. The rotating coordinate system is adopted in the flow field area. The propeller WALL is set as a stationary relative coordinate system, and the propeller rotates at a certain speed relative to the flow field. So the real flow field of the propeller can be simulated in the high altitude environment.

4.4. Calculation results

The comparison curves of the numerical calculation to the thrust coefficient, power coefficient and efficiency of the propeller, strip theory and wind tunnel experiment results are shown in Figure 2.

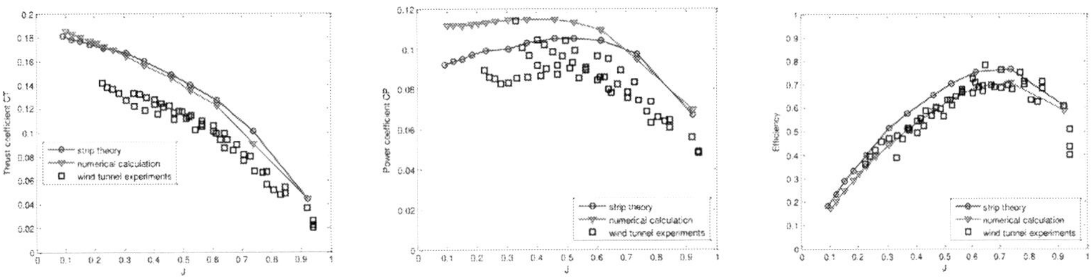

Figure 2.thrust coefficient, power coefficient and efficiency curves

The thrust coefficients calculated by the numerical simulation and strip theory are basically consistent from Figure. 2. The power coefficient calculated by the numerical simulation is larger than the strip theory and the wind tunnel experiment. The efficiency of the numerical simulation is more consistent with the experimental data. So the numerical simulation can be used to predict the performance of the high altitude propeller.

When the propeller speed is 20m/s, the aerodynamic characteristics of the high altitude propeller at different advance ratios are shown in Table 1.

Table 1 aerodynamic characteristics of the high altitude propeller

λ	T	M	C_T	C_P	η
0.7385	246.6	266.5	0.0904	0.0945	70.69
0.6154	480.3	443.4	0.1223	0.1091	68.96
0.5275	724	624.8	0.1354	0.1130	63.23
0.4615	1017	826.3	0.1457	0.1144	58.77

0.3692	1709	1293.5	0.1567	0.1146	50.47
0.3077	2576	1858.1	0.1640	0.1143	44.13
0.2637	3620.3	2512.5	0.1693	0.1136	39.31
0.2308	4816.9	3261.6	0.1725	0.1129	35.26
0.2051	6181.5	4113.6	0.1749	0.1125	31.89
0.1846	7724.6	5061.8	0.1770	0.1121	29.15
0.1538	11295.7	7249.4	0.1798	0.1115	24.80
0.1231	17967.7	11355.4	0.1830	0.1118	20.15
0.1026	26136.9	16303.7	0.1849	0.1115	17.01

5. Design of the high altitude propeller

The design parameters of the high altitude propeller mainly include the propeller speed V_0, the rotation speed n_s, diameter D, incidence angle θ, chord b, section airfoil, and blade number N_B.

(1) The tip Mach number of the propeller generally is less than 0.8. So the rotation speed of the propeller can be ensured.

$$Ma=\frac{\sqrt{V_0^2+(\pi D n_s)^2}}{c}\le 0.8 \tag{16}$$

(2) There is a consistent relationship between the incidence angle θ and chord b for the design of the high altitude propeller.

$$\begin{aligned} b=(&0.084241-0.85789r+4.7176r^2-\\ &9.6225r^3+8.5004r^4-2.7959r^5)D \end{aligned} \tag{17}$$

$$\begin{aligned} \theta=\varphi_{0.7}&+0.4387+0.3040r-3.9616r^2\\ &+5.1180r^3-1.6284r^4-0.3244r^5 \end{aligned} \tag{18}$$

(3) Eppler387 and S1223 airfoils are both suitable for the section airfoil of the high altitude propeller. But the aerodynamic performance of S1223 at the low Re is better than Eppler387. So S1223 is selected to design the high altitude propeller.

According to the requirements of the stratosphere propulsion system, the design parameters of the propeller: D=2.4m, V_0=28m/s, n_s=1000rpm, T=98N, R_h=0.1R, $\theta_{0.7R}$=20.5°.

6. Results of high altitude propeller

The propeller thrust and torque are calculated under different mesh numbers, when the rotation speed is 960rpm. The calculation results are shown in Figure 3.

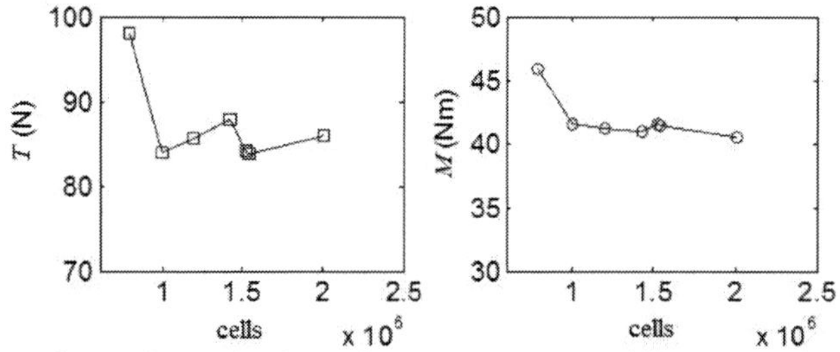

Figure 3.aerodynamic data of propeller under different mesh numbers

From Figure 3 different cells have influence on the propeller thrust and torque and the results tend to be stable with the increase of mesh. Considering the calculation time and cost, 152 million cells are used to calculate the high altitude propeller.

The aerodynamic characteristics of the high altitude propeller are obtained by the strip theory and numerical simulation under different advance ratios and the results are shown in Figure 4. From Figure 4 the errors of the strip theory and numerical simulation are less than 5%. The thrust coefficient, power coefficient and efficiency can satisfy the engineering precision. So the strip theory and numerical simulation can be used to calculate the aerodynamic characteristics of the high altitude propeller. When λ=0.7, the thrust obtained by the CFD method is 98.7N, the torque is 44.8Nm, the efficiency is 63.1%. The design high altitude propeller can meet the requirement of the stratosphere propulsion system.

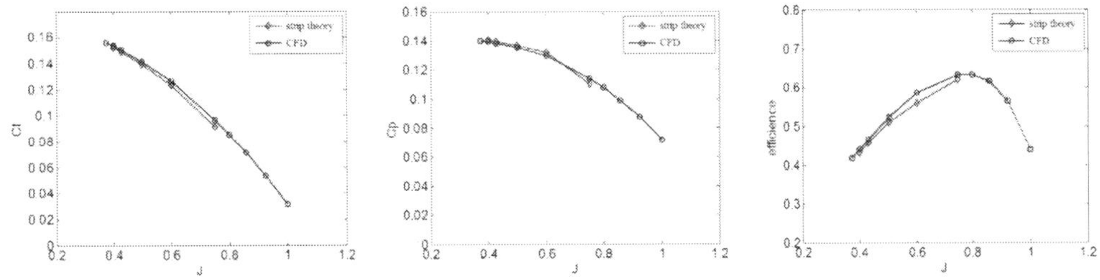

Figure. 4 thrust coefficient, power coefficient and efficiency curves

7. Conclusions

In this paper the aerodynamic characteristics of the high altitude propeller are calculated by the numerical simulation method. The feasibility of the numerical calculation to the high altitude propeller is verified by comparison with the strip theory and the wind tunnel experiment data. Avoiding the singularity of equations in the iterative process, the interference angle equation was improved.

Aiming at the stratospheric propulsion system, a high altitude propeller is designed and the design high altitude propeller is calculated by the strip theory and numerical simulation method. The results show that the design propeller can satisfy the requirement of propulsion system and the strip theory and numerical simulation can provide the reference for the design of stratospheric propeller.

References

[1] Pancotti A. An overview of advanced concepts for near-space systems[J]. 2009.

[2] Colozza A, Dolce J. Initial feasibility assessment of a high altitude long endurance airship[J]. 2003.

[3] Koch L D. Design and performance calculations of a propeller for very high altitude flight[J]. NASA, 1998 (19980017535).

[4] Wang Haojie, Li Jie, Fu Weijia, and Zhou Zhou. Aerodynamic design of propeller for unmanned aerial vehicles[J]. Chinese Journal of Applied mechanics, 2012, 04:380-385.

[5] Liu YuanQiang, Guo Jinsuo, Xiang Song. Aerodynamic performances calculation of propellers based on standard strip analysis[J]. Journal of Shenyang Aerospace University, 2013, 01:43-46.

[6] Deng Zhiwei, Huang Xianghua, Tian Chao. Real-time modeling of propeller in a turboprop engine[J]. Journal of Aerospace Power, 2014, 02:434-440.

[7] Liu Peiqing, Ma Lichuan, Duan Zhongzhe. Study and verification on similarity theory for propellers of stratospheric airships[J]. Journal of Beijing University of Aeronautics and Astronautics, 2012, 07:957-961.

[8] Wang Yufu, Liu Zhenguo, Tao Guoquan. Numerical simulation of high altitude propeller's aerodynamic characteristics and wind tunnel test[J]. Journal of Beijing University of Aeronautics and Astronautics, 2013, 08: 1102-1105.

[9] Dorfling J, Rokhsaz K. Constrained and unconstrained propeller blade optimization[C]//52nd Aerospace Sciences Meeting. 2014: 0563.

[10] Drela M. XFoil: An analysis and design system for low Reynolds number airfoils[C]//Conference on low Reynolds number airfoil aerodynamics. Notre Dame: University of Notre Dame, 1989: 1-12.

[11] Liu P, Duan Z, Ma L, et al. Aerodynamics properties and design method of high efficiency-light propeller of stratospheric airships[C]//Remote Sensing, Environment and Transportation Engineering (RSETE), 2011 International Conference on. IEEE, 2011: 8041-8044.

IOP Conf. Series: Materials Science and Engineering **751** (2020) 012005 doi:10.1088/1757-899X/751/1/012005

Study on Rotor Eddy Current Losses in High-speed Permanent Magnet Brushless Machines for Aircraft Applications

Yaohua Hu[1,*] and Shushu Zhu[2]

[1]Aviation Key Laboratory of Science and Technology on Aero Electromechanical System Integration, Nanjing Engineering Institute of Aircraft Systems, Nanjing 211106, China
[2]Center for More-Electric-Aircraft Power System, Nanjing University of Aeronautics and Astronautics, Nanjing 211106, China

[*]Email: huyaohua413@126.com

Abstract. To improve the power density of aviation electric drive system, the high-speed permanent magnet machines has more and more popular in the air compressor, radial compressor, and electro-hydrostatic actuator applications. Two methods have studied to reduce the rotor eddy current losses in high-speed permanent magnet machines in this paper. Firstly, rotor eddy current losses at different stator slot openings are investigated, the stator core loss are considered simultaneously. Secondly, to reduce the magnet eddy current loss in carbon fiber retaining sleeve, the copper shield is added between permanent and carbon fiber, and the rotor eddy current losses at different copper shield width are studied.

1. Introduction

The high-speed permanent magnet brushless machines have more and more popular in the air compressor, radial compressor, and electro-hydrostatic actuator for aircraft applications [1]. The rotor eddy current loss is generated when the permanent magnet brushless machines operations, the rotor eddy current loss will induce the temperature rise especially on high-speed operation [2]. Due to the weak cooling situation on rotor, the high temperature will cause the magnets performance reduction and irreversible demagnetization [3]. Therefore, the low rotor eddy current losses structure is a key factor in the high-speed permanent magnet brushless machines design stage for aircraft applications.

For the high-speed permanent magnet brushless machines, the magnets should be protected by add high-strength retaining sleeve against the centrifugal force [4]. The sleeve can be made of metal such as stainless steel, titanium or Inconel, or made of fiber such as glass fiber or carbon fiber. Many scholars have studied the rotor eddy current loss in high-speed permanent magnet brushless machines. The influence of stator slot openings for the rotor eddy current losses has studied in [5], the larger the slot opening is, the more significant the rotor eddy current loss. Only considerate the influence of rotor eddy current losses, not consider the stator core losses. The rotor eddy current losses at different retaining sleeve material are analyzed in [6], it has found the rotor eddy current losses is the minimum when the retaining sleeve material is carbon fiber, but not segregation of the rotor eddy current losses. A high conductivity materiel added between the magnets and the carbon fiber has been studied in [7] to reduce the magnets eddy current loss, but the reason is not clearly.

This paper studied on rotor eddy current losses in high-speed permanent magnet brushless machines. The methods of reduction rotor eddy current losses are investigated. The influence of stator slot openings width in the stator core losses and rotor eddy current losses are analyzed. A copper shield has added between the magnets and the carbon fiber. The conclusion of optimal width of copper shield has been analyzed.

2. Machine Specification

The high-speed permanent magnet brushless machine consists of the stator, rotor, air-cool enclosure, and air bearing. The structural diagram of the high-speed permanent magnet brushless machine is shown in Figure 1. The stator has the iron core and three phase windings, and the rotor has the carbon fiber retaining sleeve, magnets, and the shaft. The rated power of the high-speed permanent magnet brushless machine in this paper is 45 kW, and the rated speed is 90000 r/min.

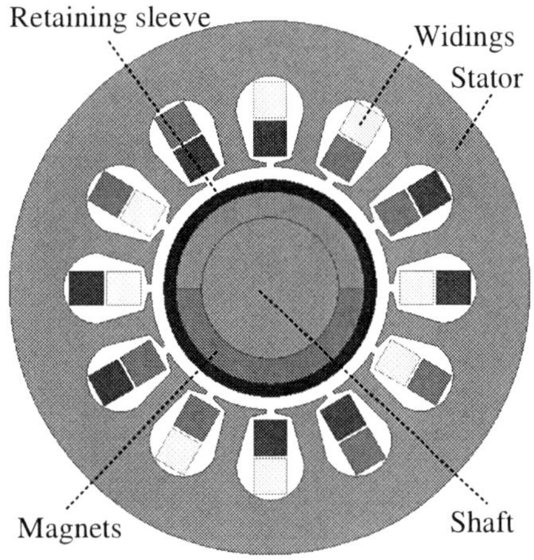

Figure 1. High-speed permanent machine.

The specification of the investigated high-speed permanent magnet brushless machines is shown in Table 1.

Table 1. Machine Specification.

Design Parameters	Value
Phase, poles, and slots	3 phases, 2 poles, and 12 slots
Rated speed	90000 r/min
Rated current density	10 A/mm^2
Rated torque	5 Nm
Continuous rated power	45 kW
Type of cooling	Air cooling
Diameter of stator and stack length	125 mm, and 100 mm
Width of retaining sleeve, magnets, and airgap	3mm, 6mm, and 3mm
SmCo (20°C)	B_r = 1.13 T, H_c = -830000 A/m
Thickness of electrical steel sheet	0.1 mm

3. Study on Rotor Eddy Current Loss

3.1. Influence of stator slot openings

IOP Conf. Series: Materials Science and Engineering **751** (2020) 012005 doi:10.1088/1757-899X/751/1/012005

Figure 2 shows the stator core losses and rotor eddy current losses density distribution at different stator slot openings of high-speed permanent magnet brushless machines. The material of retaining sleeve is the carbon fiber, B_{s0} is the width of stator slot openings, the current I is 160 A, and the frequency is 1500 Hz. It can be found that the rotor eddy current losses are focus on the retaining sleeve and magnets, the shaft is very small. The rotor eddy current losses of the enclosure slot are 53.8 W, B_{s0} = 2 mm are 62.9 W, B_{s0} = 4 mm are 76.4 W, B_{s0} = 6 mm are 124.2 W. From the region 2 in Figure 2, it should be pointed out that, with the increase of stator slot openings widths, the rotor eddy current losses are increased. The cause is the larger the slot opening is, the severer the permeance variation, and the more significant the rotor eddy current losses.

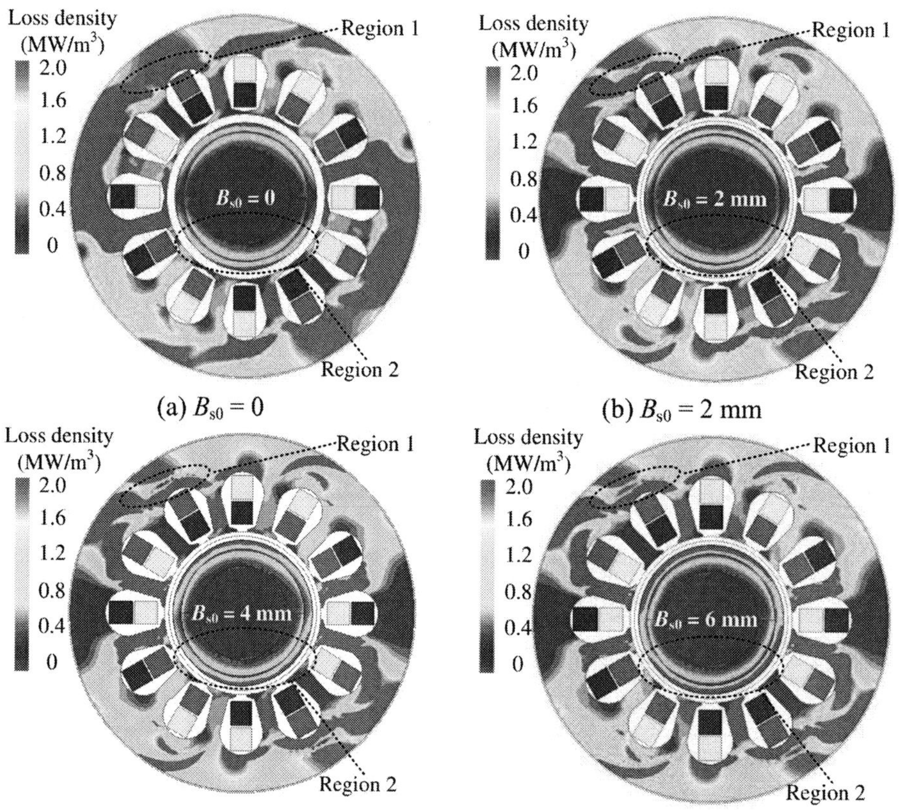

(a) $B_{s0} = 0$ (b) $B_{s0} = 2$ mm

(c) $B_{s0} = 4$ mm (d) $B_{s0} = 6$ mm

Figure 2. Stator core losses and rotor eddy current losses density distribution at different stator slot openings. (I = 100 A, speed = 90000 r/min)

It can be found that, the rotor eddy current losses are the minimum when the stator slot openings selected as the enclosure slot. However, the stator core losses should be considerate when we calculated the whole power losses, because the stator core losses are the main component of power loss in the permanent magnet brushless machines under high-speed operation. They depend essentially on the flux-density waveform of stator core and rotor core, the magnetic field in the machines includes much time harmonics. The core loss with the Fourier transformation of the flux-density waveform at each point can be expressed as

$$W_c = W_e + W_h = \sum_{n=1}^{n_{max}} k_{en}(nf_1, B_n) \cdot (nf_1)^2 + \sum_{n=1}^{n_{max}} k_{hn}(nf_1) \cdot (nf_1) \cdot B_n^{a(nf_1, B_n)} \quad (1)$$

where W_e is eddy current losses, W_h is hysteresis losses, f_1 is the fundamental frequency, k_{en} is an eddy current component coefficients, k_{hn} and a are hysteresis component coefficient, n is the nth harmonics components.

The stator core loss and rotor eddy current losses at different slot openings width are shown in Figure 3. Combine the region 1 in Figure 2, with the increase of stator slot openings widths, and the stator core losses are decreased. The stator core losses are the maximum when the stator slot openings selected as the enclosure slot, which 1820.6 W. The cause is that, the enclosure slots parts are short circuit severely, it will reduce the distortion of the stator core losses. Therefore, tradeoff design the stator core losses, rotor eddy current losses, and winding process, the stator slot openings width is selected as 3 mm. The rotor eddy current losses are 67 W, and the stator core losses are 1385.8 W.

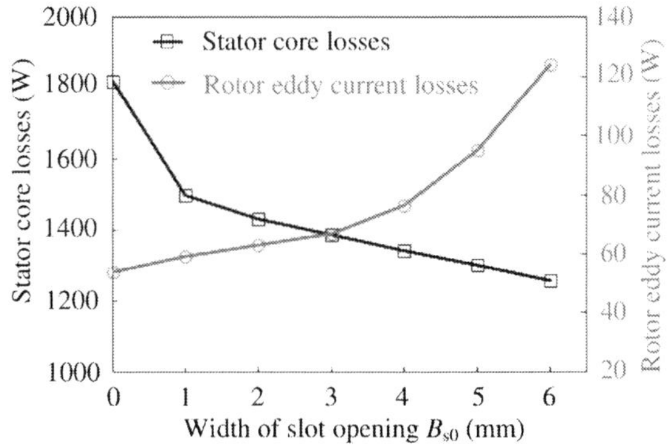

Figure 3. Stator core loss and rotor eddy current losses at different slot openings width. (I = 100 A, speed = 90000 r/min)

3.2. Influence of conductive shield

The rotor eddy current losses are focus on the retaining sleeve and magnets when the material of retaining sleeve is the carbon fiber. To reduce the magnet eddy current loss, the method is preventing the harmonics eddy current permeate into the magnets. A high conductivity materiel can be added between the magnets and the carbon fiber. Thus, the harmonics eddy current is focus on the shield not to the magnets. Although the eddy current leads to an extra loss in the shield itself, the loss is not high because the resistance of the shield is small. Therefore, the overall rotor eddy current loss can be reduced by employing the conductive shield. Figure 4. shows the copper shield is added between the permanent and the carbon fiber.

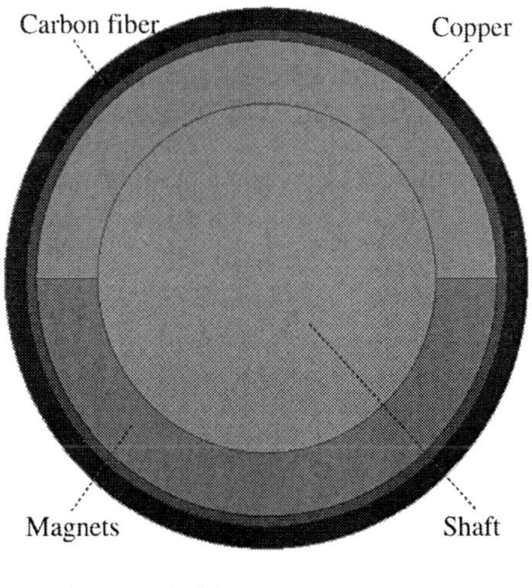

Figure 4. Copper shield is added between the magnets and the carbon fiber.

The added copper shield will influence the rotor eddy current losses. The width of copper shield should be analysed. The rotor eddy current losses density distribution at different copper shield width is shown in Figure 5, where the *d* is the width of copper shield. It can be found that, the magnet eddy current losses density is extremely reduced when the copper shield width is 0.6 mm. The eddy current losses density of copper shield is higher than the magnets.

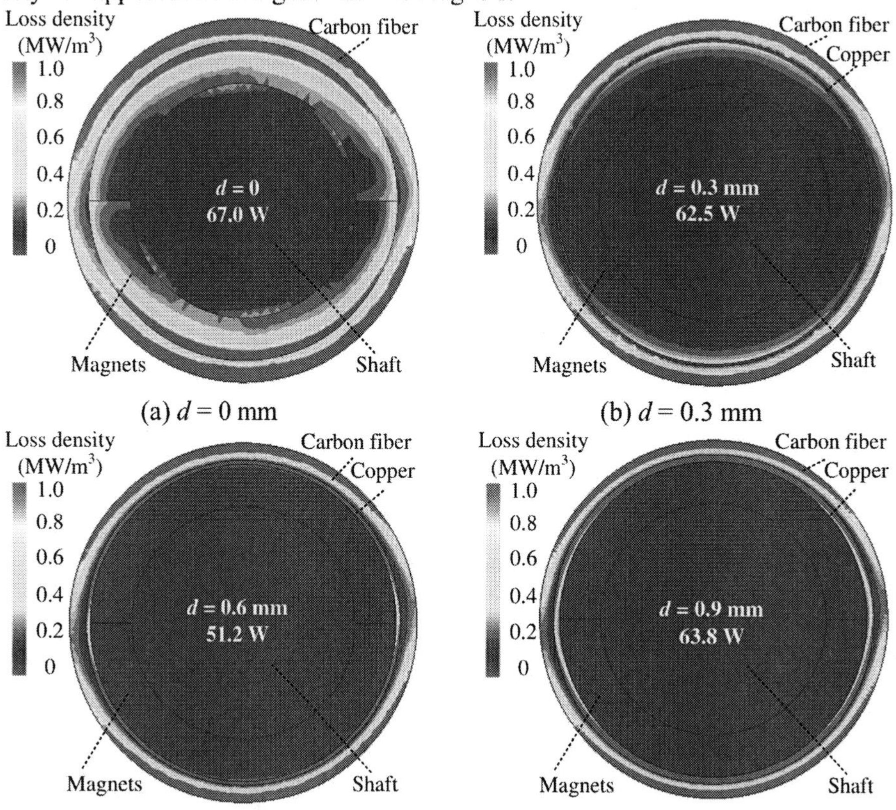

Figure 5. Rotor eddy current losses density distribution at different copper shield width. (I = 100 A, speed = 90000 r/min)

To clearly show the proportion of rotor eddy current losses, the segregation of rotor eddy current losses at different copper shield width are shown in Figure 6.

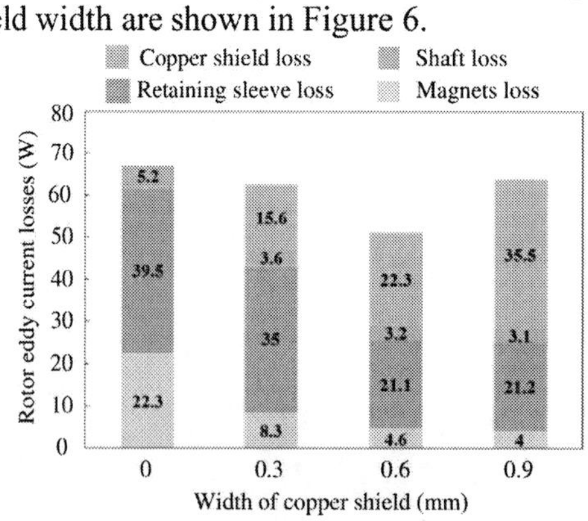

Figure 6. Segregation of rotor eddy current losses at different copper shield width slot openings width. (I = 100 A, speed = 90000 r/min)

It can be found that, magnet eddy current losses can be reduced extremely by the copper shield added. The total rotor eddy current loss is the minimum when the copper shields width is 0.7 mm, which equals the skin depth of 6th fundamental frequency in copper shield (the skin depth of 6th fundamental frequency in copper shield is 0.696 mm). The rotor eddy current loss is 50.5 W, reduction of 25% compared with the non-copper shield situation.

4. Conclusion

The rotor eddy current losses in high-speed permanent magnet brushless machines with carbon fiber retaining sleeve for aircraft applications are studied in this paper. It can be found that, with the increase of stator slot openings widths, the rotor eddy current losses are increased, but the stator core losses are decreased. Therefore, the stator slot openings width should be tradeoff design between in the stator core losses, rotor eddy current losses, and winding process. The magnet eddy current losses can be extremely reduced by the copper shield added between in the magnets and the carbon fiber. When the copper shields width equals to the skin depth of 6th fundamental frequency in copper shield, the rotor eddy current loss is the minimum.

Acknowledgments
This work was supported by the Aeronautical Science Foundation of China under Project 2018ZC09002, National Natural Science Foundation of China under Project 51807090, and Natural Science Foundation of Jiangsu Province under Project SBK2018040827.

References

[1] Jang S, Cho H, Choi S, "Design and analysis of a high-speed brushless DC motor for centrifugal compressor,"*IEEE Trans. Magn.*, vol. 43, no. 6, pp. 2573-2575, Jan. 2007.
[2] Hyun-Woo Jun, Jong-Wook Lee, et al., "Optimal Design of the PMSM Retaining Plate With 3-D Barrier Structure and Eddy-Current Loss-Reduction Effect," *IEEE Trans. Ind. Electron.*, vol. 65, no. 2, pp. 1808-1818, Feb. 2018.
[3] Ming Cheng, and Sa Zhu, "Calculation of PM Eddy Current Loss in IPM Machine Under PWM VSI Supply With Combined 2-D FE and Analytical Method," *IEEE Trans. Magn.*, vol. 53, no. 1, pp. 6300112, Jan. 2017.
[4] D. Gerada, A. Mebarki, N. L. Brown, C. Gerada, A. Cavagnino, and A. Boglietti, "High-speed electrical machines: Technologies, trends, and developments," *IEEE Trans. Ind. Electron*, vol. 61, no. 6, pp. 2946-2959, Jun. 2014.
[5] S. S. Nair, J. Wang, R. Chin, L. Chen, and T. Sun, "Analytical prediction of 3-D magnet eddy current losses in surface mounted PM machines accounting slotting effect," *IEEE Trans. Energy Convers.*, vol. 32, no. 2, pp. 414-423, Jun. 2017.
[6] F. Zhang, G. Du, T. Wang, G. Liu, and W. Cao, "Rotor retaining sleeve design for a 1.12-MW high-speed PM machine," *IEEE Trans. Ind. Appl.*, vol. 51, no. 5, pp. 3675-3685, Sep./Oct. 2015.
[7] Manoj R. Shah, and Ayman M. EL-Refaie, "Eddy-Current Loss Minimization in Conducting Sleeves of Surface PM Machine Rotors With Fractional-Slot Concentrated Armature Windings by Optimal Axial Segmentation and Copper Cladding," *IEEE Trans. Ind. Appl.*, vol. 45, no. 2, pp. 720-728, Mar/Apr. 2009.

IOP Conf. Series: Materials Science and Engineering **751** (2020) 012006 doi:10.1088/1757-899X/751/1/012006

Study on Magnet Eddy Current Losses in Interior Permanent Magnet Machines for Aircraft Electric Green Taxiing Systems

Yaohua Hu* **and Pengfei Zhang**

Aviation Key Laboratory of Science and Technology on Aero Electromechanical System Integration, Nanjing Engineering Institute of Aircraft Systems, Nanjing 211106, China

*E-mail: huyaohua413@126.com

Abstract. The magnet eddy current of multi-layered interior permanent magnet (IPM) machines is studied in the paper for the aircraft electric green taxiing systems. The armature reaction flux density distribution, eddy current density and harmonic spectrum at the end of permanent magnet are analysed. A method to reduce the magnet eddy current of multi-layered IPM machines is presented. That is to increase the distance between outer permanent magnet and rotor surface and the distance between two layers of permanent magnet so as to provide a larger core path for armature reactive flux. This method can effectively reduce the magnet eddy-current loss of the multi-layered IPM machines and improve the operation efficiency.

1. Introduction

Using electric machines for the aircraft traction during taxi phase is a current technology [1]-[2]. Because of the advantages of high torque density, wide operation range of constant power and low rotor eddy current loss, the interior permanent magnet (IPM) synchronous machine can be used for the aircraft electric green taxiing systems [3]. When the permanent magnet synchronous machine (PMSM) is running at high speed, the spatial harmonic of the machine magnetic field caused by stator grooving, the spatial harmonic of the machine magnetic field caused by winding current and the temporal harmonic of the machine magnetic field will produce a lot of eddy current loss in the rotor permanent magnet [4]. Excessive eddy current loss of permanent magnet will lead to irreversible demagnetization at high temperature. Therefore, it is of great significance to study eddy current loss of permanent magnet of high-speed IPM machines.

Literature [5]-[6] optimized the stator tooth and rotor separation magnetic bridge of the built-in permanent magnet synchronous machine. After optimization of the structure, the eddy current loss of the permanent magnet is reduced by about 60%. The literature [7] studies the eddy current loss of the fractional slot centralized winding built in machine. The research results show that the eddy current loss on the fixed rotor core can be reduced by adding auxiliary magnetic barrier to the straight axis. By adding auxiliary magnetic barrier to the quadrature axis, the eddy current loss on the permanent magnet can be reduced.

In this paper, the magnet eddy current loss of multi-layered IPM machine is studied. The distribution of magnetic density, the eddy current density at the end of the permanent magnet and the harmonic spectrum of the pure armature reaction of the multi-layered IPM machine are studied, and a method to reduce the rotor eddy current loss of the multi-layered IPM machine is presented.

IOP Conf. Series: Materials Science and Engineering **751** (2020) 012006 doi:10.1088/1757-899X/751/1/012006

2. Machine Specification

Multi-layered IPM machine is composed of stator, rotor, wind-cooled housing and bearing. The rotor structural diagram of the high-speed multi-layered IPM machine is shown in Figure 1. PM1-PM3 are permanent magnets, bridge1-bridge2 are bridges, rib1 is the reinforcement of permanent magnet, and barrier1-barrier3 are permanent magnetic barriers, a_1 is the permanent magnet angle. The first layer of permanent magnet is composed of permanent magnet PM1 and PM2, and its shape is V-shaped. The second layer of permanent magnet is composed of permanent magnet PM3. When the two layers are superimposed, they can be seen as an inverted triangle, which is initial structure. The power rating is 30 kW continuous, the rated torque is 100 Nm, the over load torque is 200 Nm, the rated speed is 3000 r/min, the maximum speed is 9000 r/min, the reduction ratio of the gears is 36.

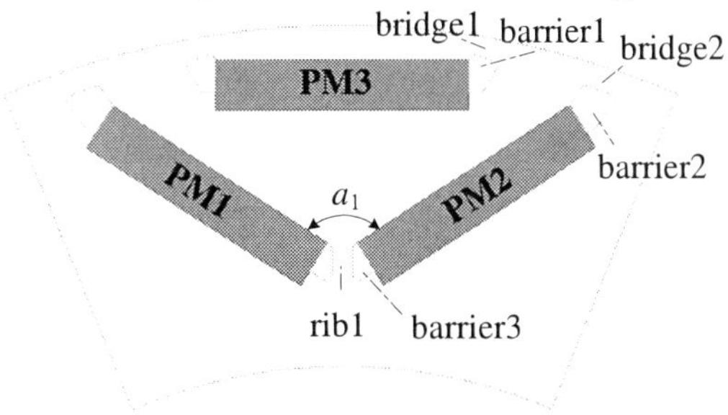

Figure 1. Rotor of high-speed multi-layered interior permanent machine.

3. Study on Magnet Eddy Current Loss

3.1. Magnetic density distribution of pure armature reaction

Figure 2 shows the magnetic density distribution of pure armature reaction at 9000 r/min with inverted triangle structure. For the IPM structure, the permanent magnet is buried in the core. The armature reaction flux returns to the air gap and stator mainly through the core and the bridge. The bridge and reinforcement on the rotor provide an additional path for the armature reaction flux. For the initial structure, the second layer of permanent magnet PM3 is located at the outer edge of the rotor surface, which has a great hindrance to the armature reaction flux. Armature reaction after regional 1 magnetic flux, the part through the magnetic separation of magnetic bridge back to the air gap and stator, partly through regional 2 and through the first layer of the permanent magnet reinforcement back to another extreme.

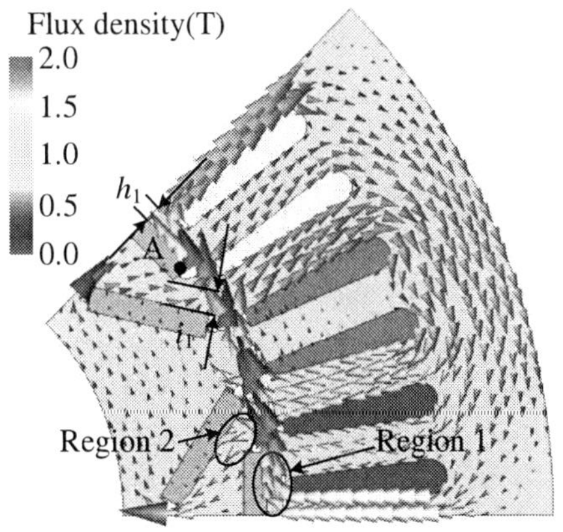

Figure 2. Magnetic density distribution of pure armature reaction at 9000 r/min with initial structure.

3.2. Eddy current loss density cloud of the permanent magnet

Figure 3 shows the eddy current loss density cloud of the permanent magnet with an initial structure at 9000 r/min, h_1- h_2 is the depth of two layers of permanent magnets, i_1 is the distance between two layers of permanent magnets. It can be seen from the figure that the eddy current loss of the permanent magnet is mainly concentrated in the end of the permanent magnet. The magnet eddy current loss of the initial structure is 86.5 W, which is relatively large and needs to be optimized.

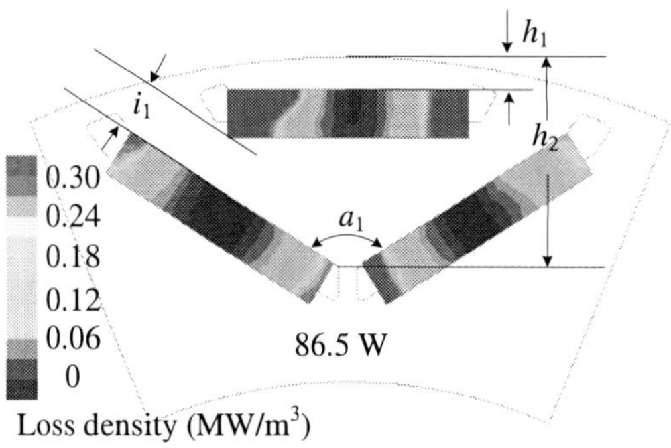

Figure 3. Eddy current loss density cloud of the permanent magnet with an initial structure at 9000 r/min.

4. Optimization of Eddy Current Loss of Permanent Magnets

According to the above analysis of eddy current loss of the permanent magnet with initial structure, the second layer of permanent magnet PM3 is located at the outer edge of the rotor surface, which has a great obstacle effect on the armature reaction flux. Therefore, the rotor core region can be increased to provide a path for the armature reaction flux, so as to reduce the eddy current loss of the permanent magnet. Figure 4 shows the rotor structure diagram of the optimized high-speed multi-layered IPM machine.

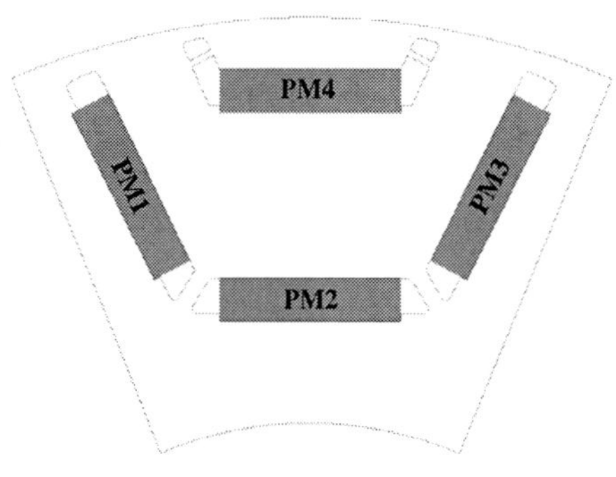

Figure 4. The optimized rotor structure of high-speed multi-layered interior permanent magnet machine.

Figure 5 shows the magnetic density distribution of pure armature reaction at 9000 r/min for the optimized structure. For the optimized structure, the second permanent magnet layer depth h_1 is bigger than the initial structure, the optimized structure of iron core area of 1 is larger than the initial structure. The second permanent magnet and the first layer of the permanent magnet distance i_1 is bigger than the initial structure, the optimized structure of the core area of 2 is larger than the initial structure core area. The added core region of the optimized structure can provide a larger path for the armature reaction flux.

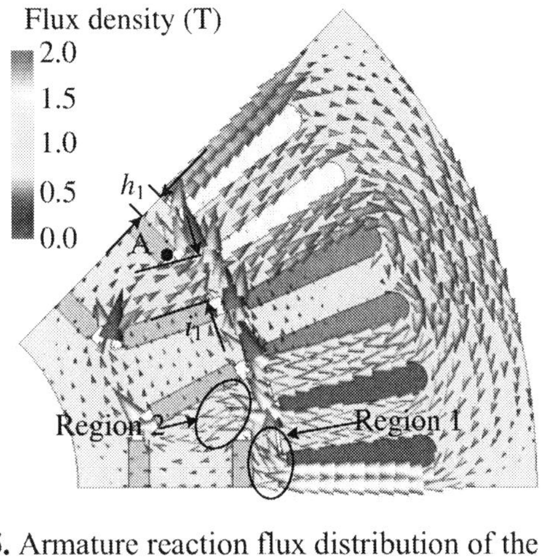

Figure 5. Armature reaction flux distribution of the optimized structure at 9000 r/min.

In order to analyse the eddy current density on the permanent magnets of the two structures, the rotor permanent magnets in Figure 2. and Figure 5. are added with point A, which is located at the end of the outermost permanent magnet. Figure. 6. shows the eddy current density and harmonic spectrum analysis of the ends of the permanent magnets of the two structures at 9000 r/min. The eddy current density of the permanent magnet with the optimized structure is smaller than that of the initial structure. The eddy current loss of the permanent magnet is directly proportional to the square of the eddy current density of the permanent magnet. Therefore, the eddy current loss of the permanent magnet with initial structure is smaller than that of the initial structure.

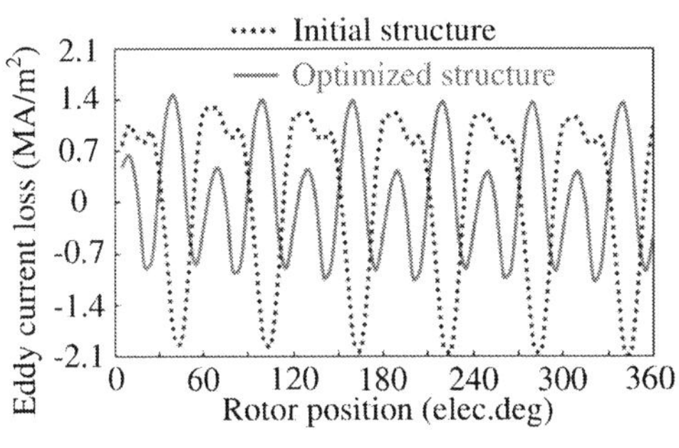

(a)

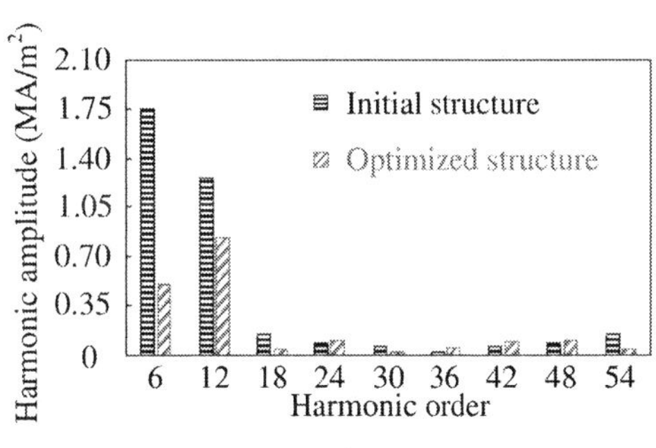

(b)

Figure 6. Eddy current density and harmonic spectrum analysis of the ends of the permanent magnets of the two structures at 9000 r/min. (a) Eddy current density profile. (b) Harmonic spectrum analysis results.

The eddy current density harmonic component of two IPM machines at 9000 r/min are shown in Table 1. As can be seen from the Table 1, the harmonics of eddy current density are mainly concentrated on the 6th and 12th. The 6th and 12th harmonic vortices of the optimized structure are much less dense than those of the initial structure.

Figure 7. shows the eddy current loss density cloud of the permanent magnet with the optimized structure at 9000 r/min. The magnet eddy current loss of the optimized structure is 45.2 W, which is 47.74% lower than that of the initial structure. Therefore, for double-layer IPM structure, increasing the distance between the depth of the outer permanent magnet and the number of layers of the permanent magnet can effectively reduce the eddy current loss of the permanent magnet.

IOP Conf. Series: Materials Science and Engineering **751** (2020) 012006 doi:10.1088/1757-899X/751/1/012006

Table 1. Eddy current density harmonic component of two structures at 9000 r/min.

Harmonic component	Initial structure	Optimized structure
6th (A/mm^2)	1750000	497057.75
12th (A/mm^2)	1260000.07	826004.35
18th (A/mm^2)	143909.78	34479.58
24th (A/mm^2)	82253.85	101363.71
30th (A/mm^2)	60278.35	11424.03
36th (A/mm^2)	3953.63	42035.04
42th (A/mm^2)	60278.35	84161.06
48th (A/mm^2)	82253.85	98016.17
54th (A/mm^2)	143909.78	39200.19

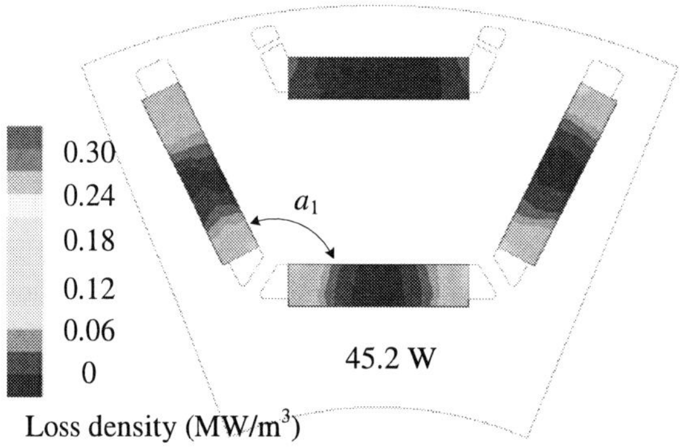

Figure 7. Eddy current loss density cloud of the permanent magnet with the optimized structure at 9000 r/min.

Table 2 shows the design parameters of two structures. It includes the depth of the permanent magnet, the distance between the layers of the permanent magnet, the Angle of the permanent magnet, the length of the permanent magnet PML and the width PMW.

Table 2. Design parameters list of two structures.

Harmonic component	Initial structure	Optimized structure
h_1 (mm)	3	5.2
h_2 (mm)	19.5	25.9
i_1 (mm)	4.6	8.8
a_1 (°)	116	117
PML (mm)	23	18.5
PMW (mm)	4.5	4.2

5. Conclusion

In this paper, the rotor eddy current loss of multi-layered high speed IPM machine is studied for the aircraft electric green taxiing systems. The distribution of magnetic density in pure armature reaction is studied. The eddy current density and harmonic spectrum at the end of permanent magnet are analysed. The results show that for the structure of two-layer permanent magnet, increasing the distance between the outer permanent magnet and the rotor surface and the distance between the two layers of permanent magnet, so as to provide a larger core path for the armature reaction flux, can effectively reduce the eddy current loss of the permanent magnet.

Acknowledgments
This work was supported by the Aeronautical Science Foundation of China under Project 2018ZC09002.

References

[1] F. Re, "Assessing environmental benefits of electric aircraft taxiing through object-oriented simulation," *Soc. Automotive Engineers Int. J. Aerosp.*, vol. 5, no. 2, pp. 503-512, 2012.
[2] H. Oyori and N. Morioka, "Power management system for the electric taxiing system incorporating the more electric architecture," *The SAE Aero Tech Congress. Exhibition*, Montréal, Canada, 2013.
[3] E. D. Ganev, "Electric Drives for Electric Green Taxiing Systems: Examining and Evaluating the Electric Drive System," *IEEE Electrification Magazine*, vol. 5, no. 4, pp. 10-24, Dec. 2017.
[4] D. Gerada, A. Mebarki, N. L. Brown, C. Gerada, A. Cavagnino, and A. Boglietti, "High-speed electrical machines: Technologies, trends, and developments," *IEEE Trans. Ind. Electron*, vol. 61, no. 6, pp. 2946-2959, Jun. 2014.
[5] Katsumi Yamazaki, and Atsushi Abe. "Loss Investigation of Interior Permanent-Magnet Machines Considering Carrier Harmonics and Magnet Eddy Currents," *IEEE Transactions on Industry Applications*, 2009, 45(2):659-665.
[6] Katsumi Yamazaki, Yuji Kanou, Yu Fukushima, Shunji Ohki, Akira Nezu, Takeshi Ikemi, and Ryoichi Mizokami. "Reduction of Magnet Eddy-Current Loss in Interior Permanent-Magnet Machines With Concentrated Windings." *IEEE Transactions on Industry Applications*, 2010, 46(6):2434-2441.
[7] Gilsu Choi and Thomas M. Jahns. "Reduction of Eddy-Current Losses in Fractional-Slot Concentrated-Winding Synchronous PM Machines." *IEEE Transactions on Magnetics*, 2016, 52(7): 8105904.

IOP Conf. Series: Materials Science and Engineering **751** (2020) 012007 doi:10.1088/1757-899X/751/1/012007

Study on Lumped Magnetic Circuit in Interior PM Machines for Aircraft Electric Green Taxiing Systems

Pengfei Zhang* **and Yaohua Hu**

Aviation Key Laboratory of Science and Technology on Aero Electromechanical System Integration, Nanjing Engineering Institute of Aircraft Systems, Nanjing 211106, China

*Email: zhangpf@neias.cn

Abstract. This paper considers the possibility of adopting a V-shaped interior permanent magnet (IPM) machine for Electric Green Taxiing (EGT) application. This rotor topology of the IPM machine has been optimized to reduce the torque ripple and iron loss. An analytical lumped magnetic circuit model was adopted to minimize the total harmonic distortion (THD) of air-gap flux density for the V-shaped IPM machine. The results of analytical lumped magnetic circuit model were compared with finite-element analysis (FEA). With the optimized rotor structure parameters, the torque ripple at the maximum torque per ampere and iron loss at the flux-weakening control were investigated.

1. Introduction

The concept of using electric machines for the aircraft traction during taxi phase is a current technology challenge driven by the global effort for environmentally responsible air transportation [1-3]. There are several types of electric machines that could be used for electric taxi applications. Interior permanent magnet (IPM) machines have many merits compare with surface-mounted permanent magnet machines, such as wide constant-power speed range, low EMF and flux-weakening ability, so it has more popular for the aircraft electric green taxiing systems [4].

In order to calculate the harmonics in the air-gap flux density distribution for the IPM machines, a lumped magnetic circuit model usually is a good compromise between simplicity and accuracy [5]. A parametric design for rotor geometry was applied to find the best shape for the rotor spatial MMF distribution that can minimize the harmonic distortion or torque ripple [6].

This paper presents a lumped magnetic circuit model to compute the flux density distribution in the air gap, minimize THD of the air-gap flux density for the V-shaped IPM machines. The results of analytical lumped magnetic circuit model were compared with FEA, it proves to be useful to confirm the rotor geometry. Using the optimized rotor structure parameters, iron losses under the flux-weakening control were investigated.

2. Machine Specification

The V-shaped IPM machine and gears are installed on main land wheel. The structural diagram of the V-shaped IPM machine is shown in Figure 1. The stator has the iron core and three phase windings, and the rotor has the magnets, ribs, bridges, barriers, and the shaft. The power rating is 30 kW continuous, the rated torque is 100 Nm, the over load torque is 200 Nm, the rated speed is 3000 r/min, the maximum speed is 9000 r/min, the reduction ratio of the gears is 36.

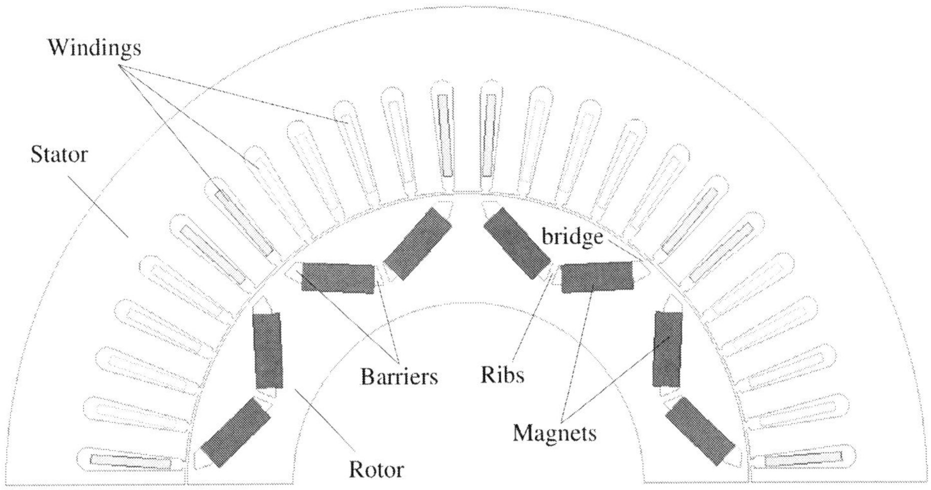

Figure 1. Proposed IPM machine for electric green taxiing systems.

3. Study on Lumped Magnetic Circuit

3.1. Flux line distribution of the V-shaped IPM machines

In order to highlight the influence of rotor design parameters and provide a simple and useful means at early design stage, analytical models will be developed for the V-shaped IPM machine in this section, with the open-circuit and assuming a smooth stator, i.e., neglecting the slotting effect. Figure 2 shows the Flux line distribution of the V-shaped IPM machines. The flux line 1 represents the leakage flux through the *bridge*1, flux line 2 represents the leakage flux through the *barrier*1, flux line 3 represents the leakage flux through two *bridges*, flux line 4 represents the leakage flux through the *barrier*2, and flux line 5 represents the leakage flux through the *rib*1.

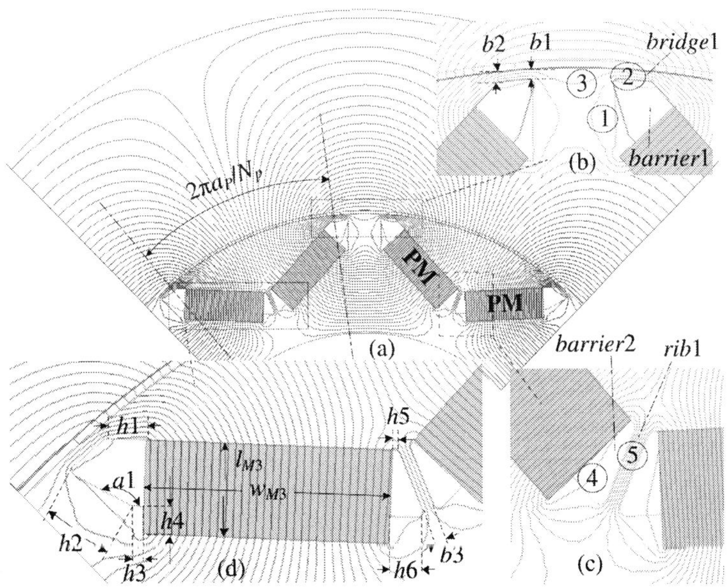

Figure 2. Flux line distribution of the V-shaped IPM machines.

FEA predicted no-load flux lines of the V-shaped IPM machine are shown in Figure 3. The no-load air-gap flux density distribution is curve 1, which can be simplified to curve 2, where the N_p is the number of magnet poles, and the a_p is the pole-arc to pole-pitch ratio.

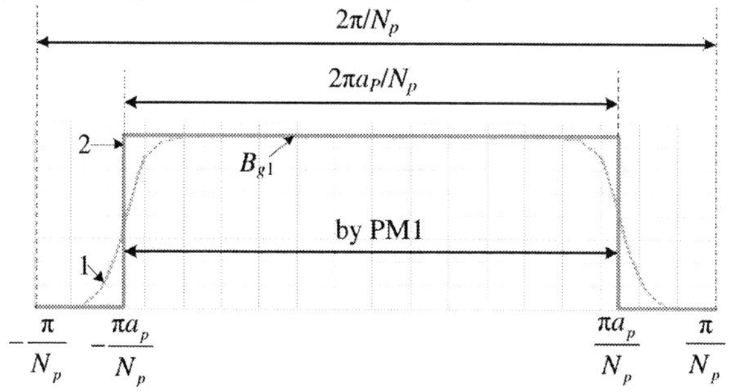

Figure 3. Equivalent no-load air-gap flux density distribution of the V-shaped IPM machine.

3.2. Analysis of lumped magnetic circuit

Figure 4(a) shows the lumped magnetic circuit associated with the fluxes excited by magnets. φ_{g1} is the air-gap fluxes excited by PM over one magnet pole, while the corresponding reluctances is R_{g1}. φ_{r1} andφ_{mo} are the flux sources and the leakage fluxes of PM over one magnet pole, the corresponding leakage flux reluctances is R_{mo}. φ_{ml2} andφ_{ml4} are the leakage fluxes of PM over *barrier*1 and *barrier*2, the corresponding leakage flux reluctances is R_{ml2} and R_{ml4}. φ_{ml1}, φ_{ml3} andφ_{ml5} are the leakage fluxes of PM over *bridge*1, two *bridges* and *rib*1, the corresponding leakage flux reluctances is R_{ml1}, R_{ml3} and R_{ml5}. R_r and R_s are the reluctances of the rotor yoke and the stator yoke. In general, there is no significant magnetic saturation in the yokes. Therefore, and may be neglected in comparison with R_{g1}. φ_{ml1} andφ_{ml3} can be simplifyφ_{ml13}, the corresponding leakage flux reluctances is R_{ml13}, so Figure 4 (a) can be simplified of Figure 4(b). Figure 4(b) also can be simplified circuit of Figure 3(c) due to symmetry.

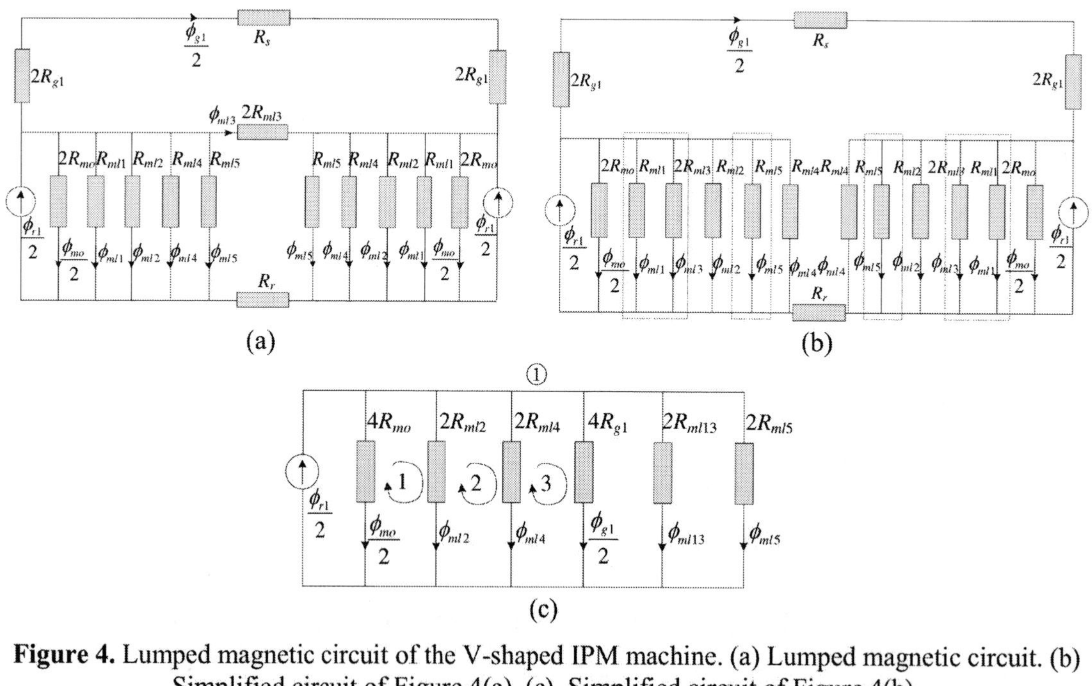

Figure 4. Lumped magnetic circuit of the V-shaped IPM machine. (a) Lumped magnetic circuit. (b) Simplified circuit of Figure 4(a). (c). Simplified circuit of Figure 4(b).

With reference to Figure 2, the following expressions can be obtained:

$$\phi_{r1} = B_r A_{m1} = 2B_r w_{M1} l_a \tag{1}$$

$$R_{g1} = \frac{g}{\mu_0 A_{g1}} \tag{2}$$

$$A_{g1} = \frac{2\pi\alpha_p (l_{SID} - g/2) l_a}{N_p} \tag{3}$$

$$R_{mo} = \frac{l_{M1}}{\mu_0 \mu_r A_{mo}} = \frac{l_{M1}}{2\mu_0 \mu_r w_{M1} l_a} \tag{4}$$

$$R_{ml2} = \frac{l_{M1} - h_4}{\mu_0 A_{ml2}} = \frac{4(l_{M1} - h_4)}{\mu_0 (2h_2 \sin(\alpha_1 - 90^o) + 3h_3 + h_1) l_a} \tag{5}$$

$$R_{ml4} = \frac{l_{M1} - h_4}{\mu_0 A_{ml4}} = \frac{2(l_{M1} - h_4)}{\mu_0 (h_5 + h_6)) l_a} \tag{6}$$

Where μ_0 is the permeability of air, μ_r is the magnet relative recoil permeability, B_r is the magnet remanence, l_a is the lamination stack length, g is the air-gap length, l_{SID} and is the stator bore radius. l_{M1} is magnet length, while w_{M1} is magnet width, h_1–h_6 are barrier width, a_1 is the angle of h_2 and h_3. Because of the saturation in the bridge, R_{ml13} and R_{ml5} are nonlinear. But the leakage flux through the bridge can be approximated as:

$$\phi_{ml13} \approx B_{st} A_{ml13} = B_{st} l_a (b_1 + b_2)/2 \tag{7}$$

$$\phi_{ml5} \approx B_{st} A_{ml5} = B_{st} l_a b_3 / 2 \tag{8}$$

where A_{ml13} and A_{ml5} denotes the cross-sectional area of the *bridge* and *rib*, b_1 and b_2 are the bridge width, b_3 are the *rib* width. B_{st} is the saturation level on the B-H curve of the lamination, which is 2 T. This simplification may cause some errors because of the variation in magnetic saturation level in the *bridges* and *ribs* under different loading conditions. However, since the *bridges* and *ribs* are usually designed to be highly saturated even at open-circuit, the accuracy of lumped magnetic circuit models will still be acceptable, as will be illustrated later.

From Figure. 4(c), the Kirchhoff's law is applied to node ①, and loop 1-3:

$$\begin{cases} \dfrac{\phi_{r1}}{2} = \dfrac{\phi_{mo}}{2} + \phi_{ml2} + \phi_{ml4} + \dfrac{\phi_{g1}}{2} + \phi_{ml13} + \phi_{ml5} \\ 4R_{mo}\dfrac{\phi_{mo}}{2} = 2R_{ml2}\phi_{ml2} = 2R_{ml4}\phi_{ml4} = 4R_{g1}\dfrac{\phi_{g1}}{2} \end{cases} \tag{9}$$

$$\begin{pmatrix} 1 & 2 & 2 & 1 \\ R_{mo} & -R_{ml2} & 0 & 0 \\ 0 & R_{ml2} & -R_{ml4} & 0 \\ 0 & 0 & R_{ml4} & -R_{g1} \end{pmatrix} \begin{pmatrix} \phi_{mo} \\ \phi_{ml2} \\ \phi_{ml4} \\ R_{g1} \end{pmatrix} = \begin{pmatrix} \phi_{r1} - 2\phi_{ml13} - 2\phi_{ml5} \\ 0 \\ 0 \\ 0 \end{pmatrix} \tag{10}$$

Therefore, the average air-gap flux density excited by PM1 is

$$B_{g1} = \frac{\phi_{g1}}{A_{g1}} \tag{11}$$

Table 1 lists the design parameters of the V-shaped IPM machine.

Table 1. Design parameters of the V-shaped IPM machine.

Design Parameters	Value	Design Parameters	Value	Design Parameters	Value
N_p	8	$h1$ (mm)	3.39	l_{SID} (mm)	80.95
a_p	0.75	$h2$ (mm)	5.0	l_a (mm)	45
$b1$ (mm)	1.6	$h3$ (mm)	1.0	B_r (T)	1.22
$b2$ (mm)	2.0	$h4$ (mm)	2.2	B_{sat} (T)	2.0
$b3$ (mm)	1.2	$h5$ (mm)	0.5	μ_0	$4\pi\times10^{-7}$
l_{M1} (mm)	7.6	$h6$ (mm)	3.0	μ_r	1.08
w_{M1} (mm)	20.5	a_1 (deg)	155	H_c (A/m)	-899000
		g (mm)	0.73		

Figure 5 shows the air-gap flux density distribution of the V-shaped IPM machine. Figure 5(a) compares the analytical lumped magnetic circuit results and the finite-element analysis (FEA) results when the a_p is 0.75. The airgap flux density result of the analytical method is 0.762 T, the FEA result is 0.742 T, they have very good agreement. Figure 5(b) shows the analytical lumped magnetic circuit results when a_p from 0.75 to 0.9. Figure 5(c) shows the amplitude of the fundamental airgap flux density and THD with different a_p. The lowest THD of the a_p is located at the 0.75, therefore, the optimized rotor structure parameters $a_p = 0.75$ is selected.

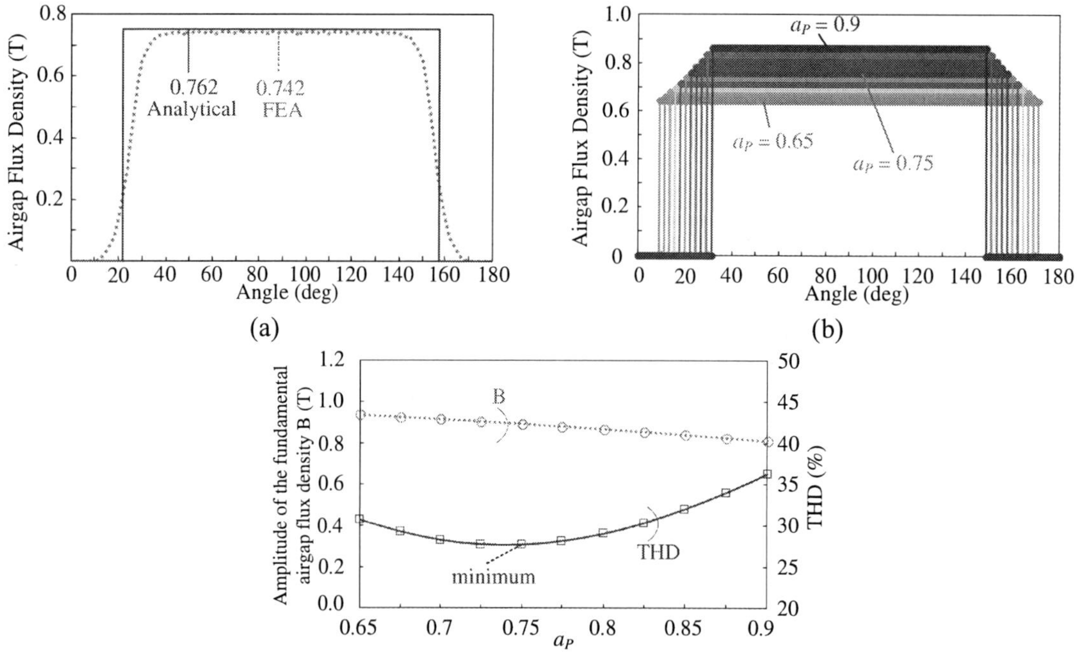

Figure 5. Air-gap flux density distribution of the V-shaped IPM machine. (a) $a_p = 0.75$. (b) a_p from 0.75 to 0.9. (c) Amplitude of the fundamental airgap flux density and THD with different a_p.

3.3. The average torque, torque ripple, and iron loss

Figure 6(a). shows the torque waveforms of the optimized rotor at the maximum torque per ampere (MTPA) control. The average torque of the model II is 106.59 Nm. The torque ripple of the model I is 10.49%. Figure 6(b). shows the iron loss under flux-weakening control (FWC) at 9000 r/min, the amplitude of phase voltage is 150 V, the amplitude of phase current is 200 A, and the current angle is 80 deg. Iron losses consist of core loss and magnet eddy current loss. It is observed that the eddy current loss of stator and rotor is higher than hysteresis loss, the iron loss of the machine is 1239.61 W.

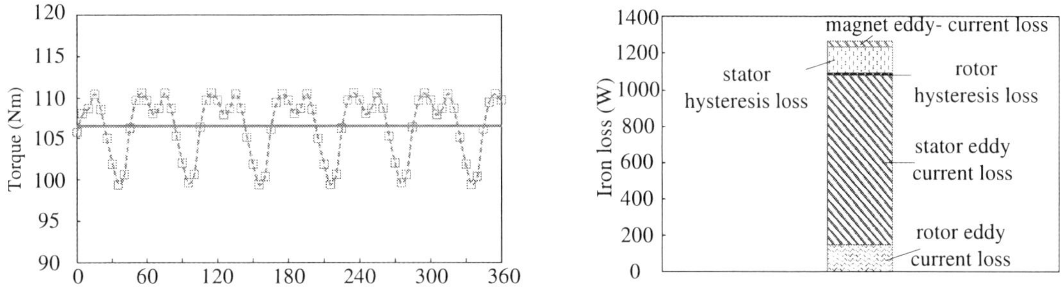

Figure 6. The torque ripples and core loss density of the V-shaped IPM machines when the a_p is 0.75. (a) Torque waveform. (b) Iron loss at 9000 r/min.

4. Conclusion

This paper considers the possibility of adopting a V-shaped IPM machine for EGT systems. An analytical lumped magnetic circuit model was adopted to minimize the total harmonic distortion (THD) of air-gap flux density for the V-shaped IPM machines. The results of analytical lumped magnetic circuit model were compared with finite-element analysis (FEA), they have very good agreement. Using the optimized rotor structure parameters, the average torque of the machine is 106.59 Nm, the torque ripple of the machine is 10.49%, and the iron loss of the machine is 1239.61 W at 9000 r/min.

Acknowledgments

This work was supported by the Aeronautical Science Foundation of China under Project 2018ZC09002.

References

[1] A. Teo, K. Rajashekara, J. Hill, and B. Simmers, "Examination of aircraft electric wheel drive taxiing concept," *The Soc. Automotive Engineers Power Syst. Conf.*, Seattle, WA, 2008.
[2] F. Re, "Assessing environmental benefits of electric aircraft taxiing through object-oriented simulation," *Soc. Automotive Engineers Int. J. Aerosp.*, vol. 5, no. 2, pp. 503-512, 2012.
[3] H. Oyori and N. Morioka, "Power management system for the electric taxiing system incorporating the more electric architecture," *The SAE Aero Tech Congress. Exhibition*, Montréal, Canada, 2013.
[4] E. D. Ganev, "Electric Drives for Electric Green Taxiing Systems: Examining and Evaluating the Electric Drive System," IEEE Electrification Magazine, vol. 5, no. 4, pp. 10-24, Dec. 2017.
[5] Dong-Kuk Lim, Kyung-Pyo Yi, Dong-Kyun Woo, et al., "Analysis and Design of a Multi-Layered and Multi-Segmented Interior Permanent Magnet Motor by Using an Analytic Method," IEEE Trans. Magn., vol. 50, no. 6, pp. 8201308, June. 2014.
[6] Seok-Hee Han, Thomas M. Jahns and Z. Q. Zhu, "Design Tradeoffs Between Stator Core Loss and Torque Ripple in IPM Machines," *IEEE Trans. Ind. Appl.*, vol. 46, no. 1, pp. 187-195, Jan./Febr. 2010.

Experimental Study on the Performance of One Micro Pulsed Plasma Thruster

Xinru Du*, Jianjun Wu, Sheng Tan, Yu Zhang, Yang Ou and Zhaofu He

College of Aerospace Science and Engineering, National University of Defense Technology, 410073, Changsha, China

*Corresponding author:1289253951@qq.com

Abstract. In the vacuum environment, the experimental platforms of discharge waveform measurement, impulse bit measurement and electromagnetic interference testing were set up for one designed micro pulsed plasma thruster (μPPT) to study the effect of the capacitance capacity, discharge voltage and electrode pads' spacing on the discharge waveforms and impulse bit of it, as well as the electromagnetic interference. The experimental results show that the oscillation of discharge waveforms can be reduced by increasing the capacitance capacity and the distance between the two electrodes or decreasing the discharge voltage. With the increasing of the capacitance capacity, the discharge voltage and the distance between two electrodes, the impulse bit of μPPT rises. In the frequency range of 0.8~2.6 GHz, the work of μPPT will not cause electromagnetic interference to spaceborne equipment and its signal transmission.

1. Introduction

Pulse plasma thruster (PPT) is an electromagnetic thruster which produces electromagnetic field by pulse discharge of the capacitor, produces plasma by ablation and ionization of the working medium, and then accelerates plasma by electromagnetic force. Usually, PPT, the mass of which is less than 1 kg and the discharge energy of which is less than 5 J is called μPPT, whose volume, mass and power consumption are at least one order of magnitude lower than that of PPT [1]. Because μPPT is light in mass, small in volume, simple in structure, long in life and high in specific impulse and control precision, it has a broad application prospect in orbit maneuver, attitude control, resistance compensation and formation flying of modern NanoSats. Although the structure of μPPT is simple, the pulse working time is short, the working process is complex, and the coupling of electric, magnetic, optical and thermal fields is involved [2]. Therefore, it is extremely necessary to study the real working conditions of μPPT through experiments.

In the experiment, the effect of the capacitance capacity, discharge voltage and electrode pads' spacing on the discharge waveforms and impulse bit of μPPT was studied [3-7]. At the same time, the electromagnetic interference of μPPT to the spaceborne equipment and its signal transmission were also tested.

2. Experimental devices

2.1. The μPPT prototype

The main parameters of the μPPT experimental prototype (as shown in figure 1) are shown in table 1.

Table 1. Parameters of the μPPT prototype

parameters	values
Sparkplug size（screw thread diameter）	M10
Sparkplug mass（g）	8/15
Electrode pads size（length*width*thickness/mm）	12*10*3.5
Electrode pads spacing（mm）	0-33（adjustable）
Circuitry system size (mm）	100*100*30
Circuitry system mass（g）	＜200
Prototype body mass（g）	＜150
Single capacitance capacity（μF）	0.1
Charging voltage（V）	＜1700
Main discharge energy（J）	＜5（adjustable）
Ignition frequency（Hz）	＜2（adjustable）
Total size（length*width*thickness /mm）	35*50*40
Total mass（g）	＜350

Figure 1. The μPPT experimental prototype

2.2. Vacuum environment

The vacuum chamber used in the experiment, the final atmospheric pressure of which is 5×10^{-4} Pa, is 3 m in length, 2.4 m in diameter. In the vacuum system, the pre-pumping subsystem is mainly composed of one ZJP-600 pump with Roots relief valve and two 2X-70A oil-seal rotary vane pumps. The working subsystem is mainly composed of one 2X-30A oil-seal rotary vane pump, one Z-300 oil diffusion booster pump and two K-900C oil diffusion pumps of high vacuum and wide area. When the vacuum system was working, the vacuum degree of the vacuum chamber was pumped below 5 Pa by the pre-pumping subsystem, then it was pumped below 3×10^{-3} Pa by the working subsystem to meet the requirements of the working environment of μPPT.

2.3. Discharge waveform measuring equipment

The voltage of the capacitor was measured by the P5100 high voltage probe produced by Tektronix Company. Before using the high voltage probe for data acquisition, the DPO4034 four-channel oscilloscope produced by Tektronix Company was used to compensate and correct it. The attenuation coefficient of the probe is 100, the bandwidth is 250 MHz, and the allowable peak voltage is 2500 V. The current of the discharge loop was measured by the CWT150 rogowski coil produced by PEM Company and its corresponding integrator. The conversion coefficient of the CWT150 rogowski coil is 5000 A/V, the sensitivity is 0.2 mV/A, and the allowable peak current is up to 30 kA.

2.4. Impulse bit measuring system

Impulse bit is one of the most important performance parameters of μPPT. In this paper, the C-type torsion pendulum designed for micro thrust measurement in our laboratory was used to measure the impulse bit of μPPT. The specific structural composition is shown in figure 2.

1-thruster, 2-torsion arm, 3-C-type torsion rod, 4-reflecting lens, 5-He-Ne laser, 6-calibration system, 7-counterweight, 8-Position Sensitive Detector (PSD), 9-optical path magnifying glass group, 10-wiring device

Figure 2. The C-type torsion pendulum

The impulse bit of the thruster can be calculated by processing calibration data and measuring data. The distance between the calibration device and the torsion rod, namely the torque of the calibrated impulse, is L. The distance between the thruster and the torsion rod, that is the torque of the impulse to be measured, is R. The relationship between the calibrated impulse and the impulse to be measured leading to the same swing amplitude is as follows:

$$I_{bit} = \frac{L}{R} I \tag{1}$$

In the experiment, a non-contact electromagnetic damping method was used. It is that basing on the real-time motion state of the torsion pendulum, the adjustable current is input into the calibration coil, thus the corresponding ampere force is produced to hinder the motion of the torsion pendulum. In this way, the attenuation can be accelerated and the measurement efficiency can be improved.

2.5. Electromagnetic interference testing facility

The μPPT with 3 μF ultrahigh-voltage pulsed film capacitor was selected for the electromagnetic interference testing. The testing facilities include a AV4051F spectrum analyzer, a HPE4421B signal generator and a number of transmitting and receiving antennas. The AV4051F spectrum analyzer can display 300 MHz frequency spectrum at both sides of the setted center frequency. The signal generator and the spectrum analyzer are connected to antennas respectively. The μPPT, the distance from which to the spectrum analyzer antenna is 0.5 m and to the signal generator is 1 m, is placed between them. And its distance from the ground is 1.2 m.

3. Experimental measurement

3.1. Experimental results and analysis of μPPT performance testing

3.1.1. Discharge characteristics. Figure 3(a) and figure 3(b) are the discharge voltage curve and the current curve respectively with different capacitance capacities. From figure 3, we can see that when the capacitance capacity increases, the discharge period of the capacitor increases, the peak discharge current increases, the reverse voltage and reverse current decrease, so the impact of the discharge process on the capacitor is reduced.

Figure 4(a) and figure 4(b) are the discharge voltage curve and the current curve respectively with different discharge voltages. From figure 4, it can be seen that when the initial discharge voltage of the capacitor increases, the discharge period of it is basically unchanged and the peak discharge current

IOP Conf. Series: Materials Science and Engineering **751** (2020) 012008 doi:10.1088/1757-899X/751/1/012008

increases. The reverse voltage and reverse current increase, and the oscillation of discharge waveforms intensifies.

Figure 5(a) and figure 5(b) show the discharge voltage curve and the current curve with different plate spacing. From figure 5, we know that when the plate spacing increases, the electric resistance increases while the oscillation intensity of discharge waveforms weakens.

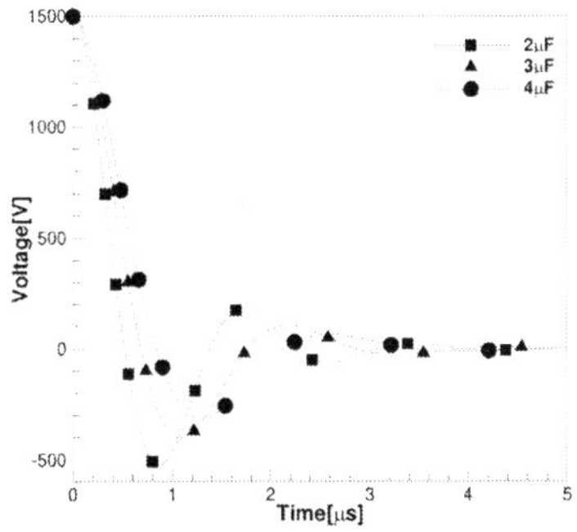

Figure 3(a). Discharge voltage curve with different capacitance capacities

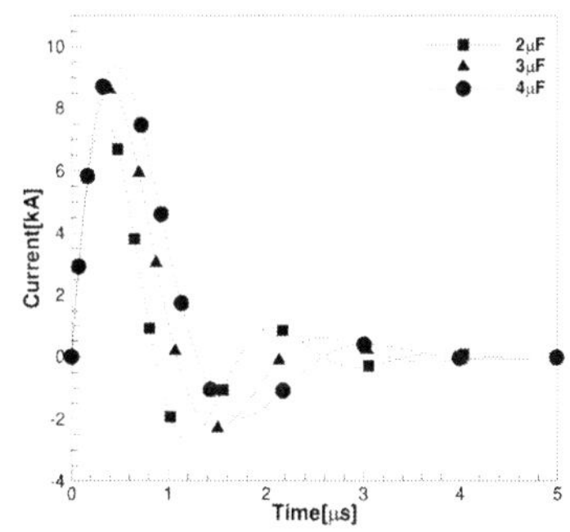

Figure 3(b). Discharge current curve with different capacitance capacities

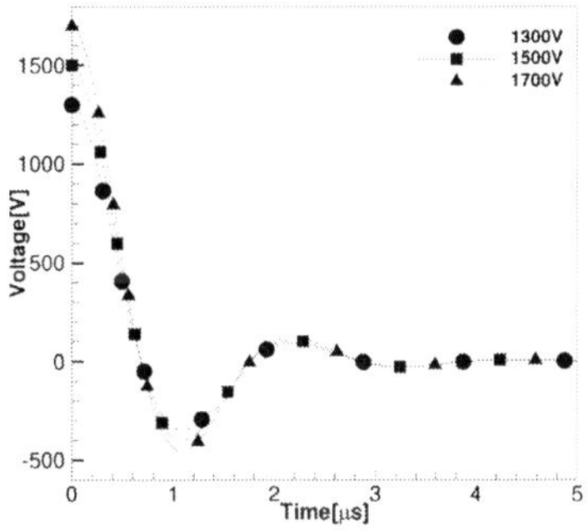

Figure 4(a). Discharge voltage curve with different discharge voltages

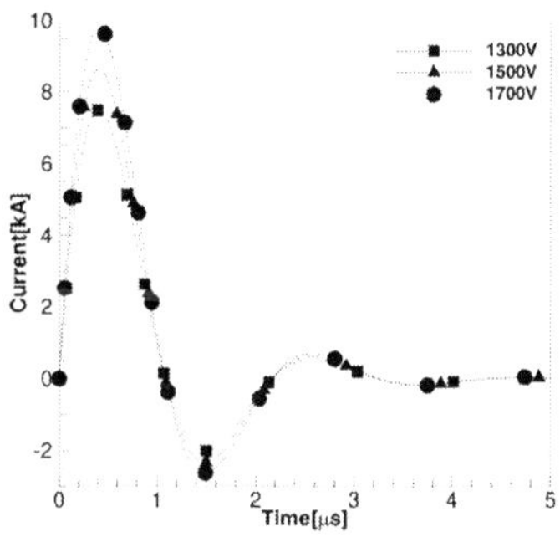

Figure 4(b). Discharge current curve with different discharge voltages

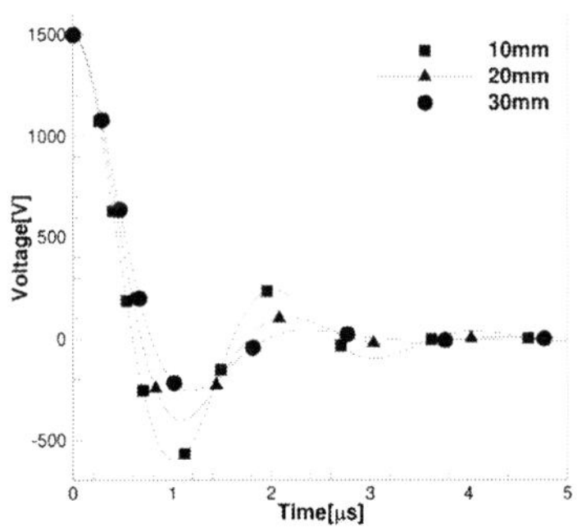

Figure 5(a). Discharge voltage curve with different plate spacing

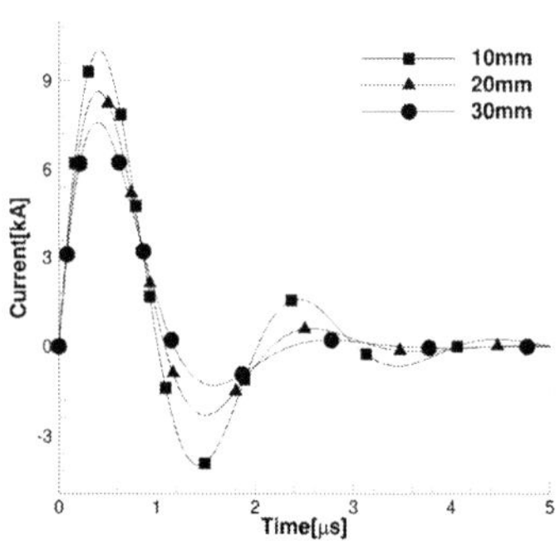

Figure 5(b). Discharge current curve with different plate spacing

To sum up, the oscillation of discharge waveforms can be slowed down by increasing capacitance capacity and plate spacing or decreasing discharge voltage, which is very beneficial to the performance improvement of μPPT.

IOP Conf. Series: Materials Science and Engineering **751** (2020) 012008 doi:10.1088/1757-899X/751/1/012008

3.1.2. Impulse bit measurement. Figure 6 shows the impulse bit curve with different capacitance capacities. From figure 6, it can be seen that when the capacitor capacity increases, the impulse bit of μPPT increases and the increasing rate is almost unchanged.

Figure 7 shows the impulse bit curve with different discharge voltages. From figure 7, we know that when the initial discharge voltage of capacitor increases, the impulse bit of μPPT increases, and the increasing rate rises slightly with the increase of initial discharge voltage.

Figure 8 shows the impulse bit curve with different plate spacing. From figure 8, we can see that when the capacitor capacity increases, the aspect ratio of the electrode plate of μPPT increases, the corresponding impulse bit also increases with the increasing rate almost unchanged.

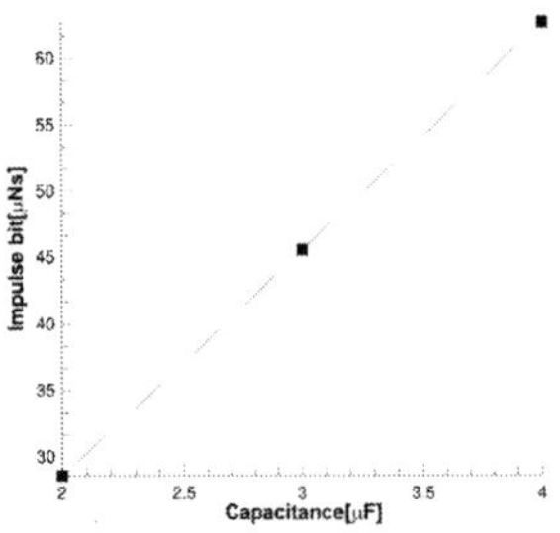

Figure 6. Impulse bit curve with different capacitance capacities

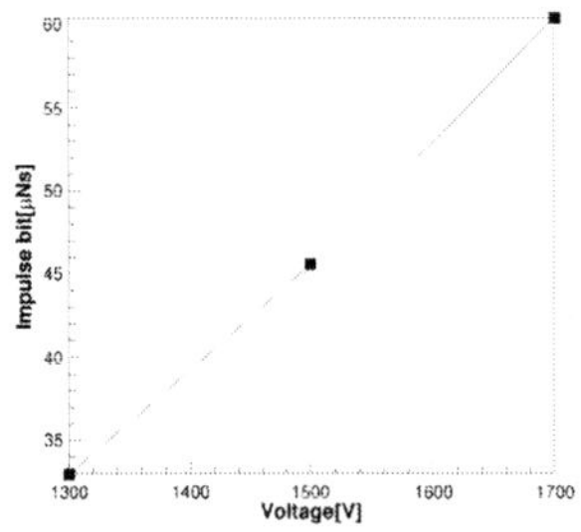

Figure 7. Impulse bit curve with different discharge voltages

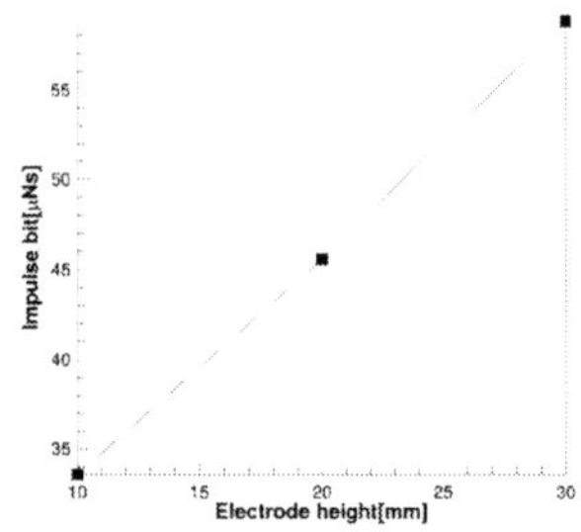

Figure 8. Impulse bit curve with different plate spacing

In conclusion, the increase of discharge voltage, capacitance capacity and plate spacing is beneficial to increasing the impulse bit of μPPT when the discharge energy is low (< 5J).

3.2. Electromagnetic interference testing

With the operation of μPPT ignited at a frequency of 2 Hz, four frequency bands (0.5~1.1 GHz, 1.1~1.7 GHz, 1.7~2.3 GHz, 2.3~2.9 GHz) were selected for electromagnetic interference testing. The spectrum analyzer was adjusted to send out signals with the center frequency of 0.8 GHz, 1.4 GHz, 2.0 GHz and 2.6 GHz respectively. Observing the electromagnetic wave spectrums received by the spectrum analyzer during the operation of μPPT, it was found that the received spectrums when μPPT was in operation are basically the same as those when μPPT was not working. It illustrates that there is no interference in the transmission and reception of signals in the neighboring frequency ranges of 0.8 GHz, 1.4 GHz, 2.0 GHz and 2.6 GHz during the discharge of μPPT. The testing results of signals whose center frequency is 2.0 GHz are shown in figure 9 and figure 10.

Figure 9. Spectrum image without μPPT 's operation when the center frequency is 2.0 GHz

Figure 10. Spectrum image with μPPT 's operation when the center frequency is 2.0 GHz

To conclude, in the frequency range of 0.8~2.6 GHz, the work of μPPT will not cause electromagnetic interference to the related instruments and their signal transmission.

4. Conclusions

In this paper, the performance of a specific μPPT prototype, mainly the discharge characteristics and impulse bit variation with different capacitance capacities, discharge voltages and electrode pads'

IOP Conf. Series: Materials Science and Engineering **751** (2020) 012008 doi:10.1088/1757-899X/751/1/012008

spacing, has been studied. In addition, the electromagnetic interference experiment of μPPT has been completed, either. The main conclusions are as follows:

- When the capacitance capacity is increased, the single pulse discharge energy of μPPT increases and the oscillation degree of discharge waveforms weakens, making the impulse bit of μPPT increase. Therefore, when the quality and volume requirements of μPPT are satisfied, its system performance can be optimized by increasing the capacitance capacity.
- When the discharge voltage increases, the peak value of discharge waveforms and the impulse bit of μPPT increase. However, the high discharge voltage will cause the intense oscillation of discharge waveforms, shortening the life of the capacitor and possibly leading to the "secondary ablation" of the working medium which is very harmful to the performance of μPPT. Therefore, we can't blindly increase the discharge voltage.
- With the increase of plate spacing, the oscillation degree of discharge waveforms weakens, the releasing time of capacitor energy decreases, and the impulse bit of μPPT increases. However, in the design of μPPT, we still need to consider the space size of thruster when selecting the distance between two electrodes.
- The electromagnetic interference testing of different frequency bands has been carried out. The results indicate that the working medium of μPPT will not cause electromagnetic interference to the spaceborne equipment and its signal transmission in the frequency range of 0.8~2.6 GHz.

Acknowledgments
The authors would like to thank National Natural Science Foundation of China for the financial assistance provided under the grant number 11772354 for this work.

References

[1] Rayburn C, Campbell M, Hoskins W, et al. Development of a micro pulsed plasma thruster for the Dawgstar nanosatellite[R]. Seattle, WA: University of Washington, AIAA 2000-3256, 2000.
[2] Simon D H, Landiii H B. Micro pulsed plasma thruster technology development[R]. Laurel, MD: The Johns Hopkins University Applied Physics Laboratory, AIAA 2004-3622, 2004.
[3] Coletti M, Guarducci F, Gabriel S B. A micro PPT for cubesat application: design and preliminary experimental results[J]. *Acta Astronautica*, 2011, **69**: 200-208.
[4] Ziran Li. Theoretical and experimental study on the design and performance of pulsed plasma thruster[D]. Changsha: National University of Defense Technology, 2008.
[5] Le Yang. Theoretical and experimental study on the working process of pulsed plasma thruster[D]. Changsha: National University of Defense Technology, 2007.
[6] Guman W J, Peko P E. Solid propellant pulsed plasma micro-thruster studies[J]. *Journal of Spacecraft*, 1968, **5(6)**: 732-733.
[7] Mcleavey C. Measurement and control of current waveform in an ablative z-pinch pulsed plasma thruster[D]. 2000.

IOP Conf. Series: Materials Science and Engineering **751** (2020) 012009 doi:10.1088/1757-899X/751/1/012009

Theoretical Analysis of Supersonic Low Reynolds Number Flow Control Technique Based on Flow Mixing

Xu Dachuan*, Ren Zebin, Chen Zhiqiang, Wang Haifeng and Guo Longde

Facility Design and Instrument Institute, China Aerodynamics Research and Development Center, Mianyang 621000, China

*E-mail: xudc_ustc@hotmail.com

Abstract. The thickness of boundary layer increases rapidly in supersonic low Reynolds number flows. As a result, there will be shock wave-boundary layer interference. The disturbances of downstream can spread upstream through in boundary layer because the flow in boundary layer is subsonic, which may decrease flow field uniformity and stability. The pressure of this type of flow is always very low. For the sake of flow continuity, the increase of flow pressure is necessary. A kind of supersonic low Reynolds number flow control technique was proposed. The high pressure supersonic flow jet-ted from plates which located in the flow channel blew away boundary layer, cut off the disturbances spread from downstream to upstream, and eliminated the shock wave-boundary layer interference. The pressure of primary flow was also increased by mixing with the high pressure jetting flow. The calculation model of this flow control technique was established. A group of standard inputs were selected to calculate the parameters of flow and the preliminary law of flow was also obtained.

1. Introduction

In some supersonic low Reynolds number flows, such as flow with chemical reaction or combustion, the thickness of boundary layer increases rapidly, which will cause Shock wave/Boundary Layer Interation. The major feature of Shock wave/Boundary Layer Interation is strong adverse pressure gradient created near surface. Separation will be prompt by strong adverse pressure gradient. The separation shock wave in leading edge and reattachment shock wave in trailing edge will make the flow more complex. Unsteady fluctuation will be generated as unsteady characteristics of separation. Disturbances created downstream could spread upstream through boundary layer, as flow in boundary layer is subsonic, which will destroy flow field structure, decline the quality of chemical reaction or combustion. Besides, the exhaust needs to be pressurized for continuity of chemical reaction or combustion, as the pressure of flow is usually very low [1-2].

The flow control technique proposed in this paper is a type of active control. High pressure supersonic air flow streamed through pylons which are arranged in flow passage can blow away boundary layer, eliminate the shock waves interference, and improve the character of flow field, meanwhile increase pressure of exhaust.

2. Theoretical analysis model

2.1. Technique principle

IOP Conf. Series: Materials Science and Engineering **751** (2020) 012009 doi:10.1088/1757-899X/751/1/012009

As shown in figure 1, pylons, through which high pressure supersonic air flow jets out, are arranged in reaction cavity flow passage. As the name implies, central pylons are arranged in the central area of flow passage, periphery pylons are arranged around flow passage. Major flow and jetting flow enter into mixing section after preliminary mixing. It is assumed that after interaction and interblending, the major flow and jetting flow become a uniform flow in mixing section. The passage area of mixing section is not constant. The sectional area of front part is contractive, and constant in the middle part. It becomes a diffuser in the latter part. The purpose of this type of mixing section is decreasing pressure loss. The contractive area can lower flow Mach number, which will decrease the loss of shock loss. The uniform flow becomes subsonic from supersonic through several oblique shock wave near the inlet of diffuser. The pressure of subsonic flow in diffuser is increasing with area diffusing, and deceleration of velocity. In the calculation proposed in this paper, subscript 1 presents the parameters of major flow, subscript 2 presents the parameters of jetting flow, subscript 3 presents the parameters of mixing flow.

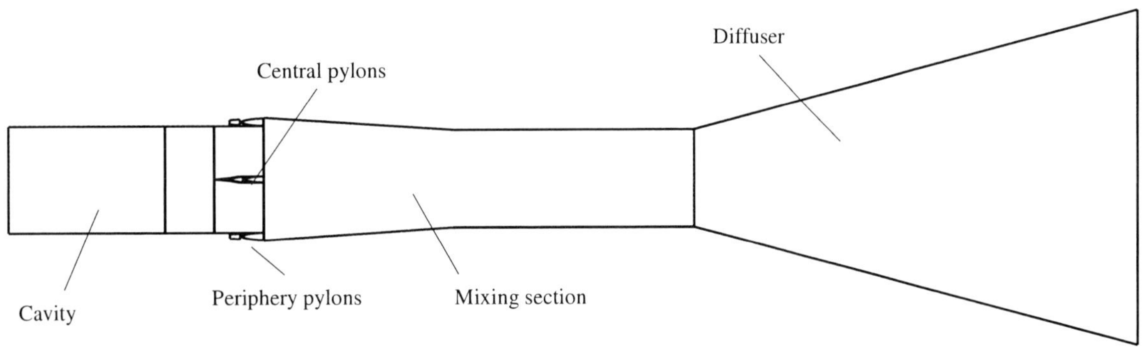

Figure 1. Technique sketch diagram

The main functions of periphery pylons are blowing away the boundary layer of major flow in reaction cavity, weakening or even eliminating the Shock wave/Boundary Layer Interation by jetting high pressure supersonic air flow. The main functions of central pylons are eliminating the wave structures, improving the flow field in reaction cavity. Meanwhile, as the pressure of jetting flow is higher than major flow, the pressure of major flow can be increased by the mixing with jetting flow. The increase of flow pressure is beneficial for the design of downstream system.

Before deriving the theoretical analysis model, several hypotheses are listed as follows.

1) The flow is steady.
2) Both major flow and jetting flow are uniform, and the mixing flow is also uniform.
3) The fluid of major flow, jetting flow and mixing flow is perfect gas.
4) The shear stress of wall is ignored.
5) The flow is adiabatic.

2.2. Area ratio and flux ratio

Area ratio is the ratio of total area of jetting flow passage and total area of flow passage in cross section of pylons outlet, and can be written as:

$$F = \frac{A_2}{A_1+A_2} \tag{1}$$

F is area ratio, A_2 is the area of jetting flow passage, A_1 is the area of major flow passage, and the position is in the cross section of pylons outlet.

Flux ratio is the ratio of major flow rate and jetting flow rate, and can be written as:

$$n = \frac{\dot{m}_1}{\dot{m}_2} \tag{2}$$

n is flux ratio, $\dot{m}_1$ is major flow rate, $\dot{m}_2$ is jetting flow rate.

Both $\dot{m}_1$ and $\dot{m}_2$ are expanded by flow formula, and the equation of area ratio and flux ratio can be obtained.

$$n = \frac{C_1 q_1 P_{01}(1-F)}{C_2 q_2 P_{02} F}\sqrt{\frac{T_{02}}{T_{01}}} \tag{3}$$

C_1 is aerodynamics parameter which is related to the gas physical property, q is the mass flow coefficient, P_0 is the total pressure of flow, T_0 is total temperature of flow.

In the theoretical analysis model, the flux ratio is confirmed firstly, then the area ratio can be obtained by calculation. Once the geometrical parameters of reaction cavity are confirmed, both the major flow passage area and the jetting flow passage area can be obtained.

2.3. Momentum equation in mixing section

The flow process prediction and control of major flow and jetting flow in mixing section is one of the key problems in the theoretical analysis model establishment [3]. It is assumed that the major flow and jetting flow become a mixing uniform flow in mixing section. The variation rule of main flow parameters in mixing section can be described by momentum equation, which can be written as:

$$\dot{m}_3 V_3 - (\dot{m}_1 V_1 + \dot{m}_2 V_2) = (p_1 A_1 + p_2 A_2) - p_3 A_3 + \int_{wall} p(l)dl \tag{4}$$

$\dot{m}_3$, V_3, p_3 presents the mass flow rate, velocity and pressure of uniform flow at the outlet of contractive part in mixing section respectively. A_3 is the flow passage area at the outlet of contractive part. p(l) is the pressure distribution in mixing section, which reflects the interaction and interblending of major flow and jetting flow in mixing section. The pressure distribution is also related to the geometrical parameters of mixing section. The rule of pressure distribution is the key to solve the momentum equation. If there is significant difference between the pressure distribution adopted in solving momentum equation and real pressure distribution, the correct theoretical analysis model can not be obtained [4-5].

Some special pressure distribution in mixing section can be adopted in order to simplify the calculation, for example, the linear pressure distribution as follows:

$$\int_{wall} p(l)dl = -\frac{p_2+p_3}{2}(A_1 + A_2 - A_3) \tag{5}$$

Equation (4) and (5) give:

$$E_3\lambda_3 = E_2\lambda_2 + n\sqrt{\theta}E_1\lambda_1 + \frac{\pi_2}{C_2 q_2}\left(\frac{2-k+k\Phi}{2kF} + \frac{k-1}{k}\right) - D\frac{\pi_3(1+\Phi)}{2C_3 q_3 \Phi} \tag{6}$$

Φ is contraction ratio, c is ratio of specific heat at constant pressure of major flow and jetting flow, k is ratio of static pressure. D, E, π is the aerodynamics parameter related to gas physical property, λ is velocity coefficient.

$$\Phi = \frac{A_3}{A_1 + A_2}, k = \frac{p_2}{p_1}, c = \frac{Cp_1}{Cp_2}, E = \sqrt{\frac{2\gamma R}{\gamma+1}}, D = (n+1)\sqrt{\frac{nc\theta+1}{nc+1}}, \pi = \left(1 - \frac{\gamma-1}{\gamma+1}\lambda^2\right)^{\frac{\gamma}{\gamma-1}}$$

The relationship of velocity coefficient λ_3 at mixing section outlet and contraction ratio Φ can be obtained once the parameters of major flow and jetting flow are confirmed.

2.4. Calculation of parameters at outlet

Each velocity coefficient λ_3 at mixing section outlet corresponding a contraction ratio Φ. Through λ_3 and parameters of diffuser, the parameters at outlet can be obtained by calculation. In real calculation, a reasonable range of λ_3 can be selected firstly, then each corresponding contraction ratio and diffuser outlet parameters can be obtained by calculation. If the main objective is high pressure at diffuser outlet, the contraction ratio which corresponding the highest outlet pressure is selected as optimal contraction ratio.

The total pressure ratio at mixing outlet is defined as:

$$P_m = \frac{P_{03}}{P_{01}} \tag{7}$$

After derivation, P_m can be rewritten as:

$$P_m = \frac{P_{02}C_2q_2DF}{P_{01}C_1q_1\Phi} \tag{8}$$

The uniform flow becomes subsonic from supersonic through several oblique shock wave near the inlet of diffuser. Experiment results indicate: the pressure of flow which becomes subsonic from supersonic through several oblique shock wave approximately equals to the flow which through a normal shock wave. Therefore, the normal shock wave relation can be used to estimate the static pressure recovery of flow.

The relationship of normal shock wave pressure recovery coefficient σ and wavefront Mach number Ma can be written as follows:

$$\sigma = \left(\frac{2\gamma}{\gamma+1}Ma^2 - \frac{\gamma-1}{\gamma+1}\right)^{-\frac{1}{\gamma-1}} \left[\frac{(\gamma+1)Ma^2}{(\gamma-1)Ma^2+2}\right]^{\frac{\gamma}{\gamma-1}} \tag{9}$$

The total pressure recovery coefficient is defined as:

$$P_t = P_m\sigma \tag{10}$$

The contraction ratio which corresponding the highest total pressure coefficient is the optimal contraction ratio. Therefore, the calculation of theoretical analysis model can be accomplished once the parameters of major flow and jetting flow are confirmed.

3. Theoretical analysis

3.1. Standard input parameters

In order to summarize the effects of key parameters on flow performance, a group of standard input parameters is selected in theoretical analysis, which are listed in table 1 and table 2.

Table 1 Parameters of major flow

parameter	value	unit
Ma_1	0.3	
T_{01}	500	K
γ_1	1.4	
P_{01}	592	Pa
P_1	400	Pa
R_1	287	J/（kg•K）
$\dot{m}_1$	200	g/s

Table 2 Parameters of jetting flow

parameter	value	unit
Ma_2	4.0	
T_{02}	300	K
γ_2	1.4	
P_{02}	84.4	kPa
P_2	400	Pa
R_2	287	J/（kg•K）
$\dot{m}_2$	1.33	kg/s

3.2. Optimal contraction ratio

The variation curves of mixing section contraction ratio Φ, pressure recovery coefficient σ, total pressure ratio P_m with velocity coefficient λ_3 are shown in figure 2. The curves indicate that in the case of confirmed inlet flow parameters of mixing, the optimal contraction ratio exist which the corresponding total pressure recovery coefficient P_t is maximum. As shown in figure 2, with increase of velocity coefficient λ_3, the pressure recovery coefficient σ decreases gradually and total pressure ratio P_m increase. The product of σ and P_m is maximum at optimal contraction ratio, which means the total pressure recovery coefficient P_t is maximum.

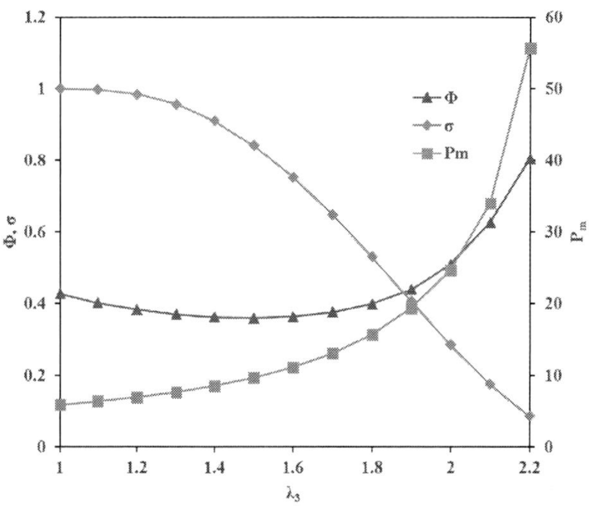

Figure 2. Variation curves of key design parameters with λ_3

3.3. Flux ratio

The variation curve of total pressure recovery coefficient P_t with flux ratio n is shown in figure 3. The diagram indicates that the total pressure recovery coefficient P_t decrease gradually with increase of flux ratio n. The reason is that with increase of flux ratio n, the proportion of low energy major flow is increase, and the energy of mixing uniform flow is also decrease. In the case of constant major flow Mach number Ma_1 and jetting flow Mach number Ma_2, it result in the decreasing of total pressure recovery coefficient Pt. Meanwhile, in order to keep the static pressure ratio of major flow and jetting flow constant, the area ratio decrease as the static pressure of major flow decrease.

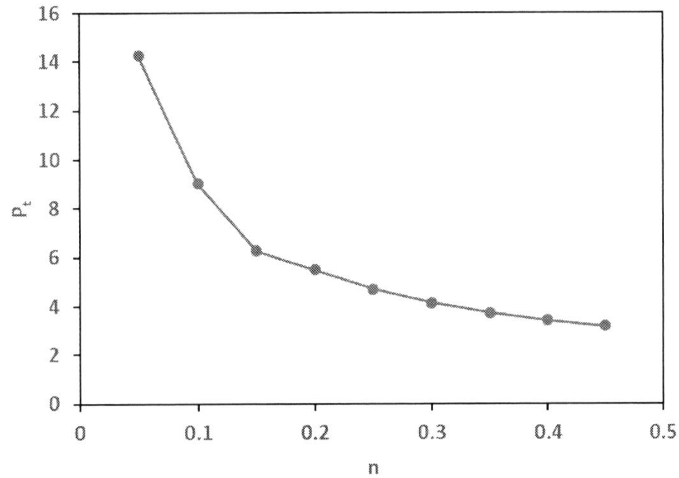

Figure 3. Variation curve of P_t with n

3.4. Jetting flow Mach number

The variation curve of total pressure recovery coefficient P_t with jetting flow Mach number Ma_2 is shown in figure 4. The result indicates that the total pressure recovery coefficient P_t increase with the increase of jetting flow Mach number Ma_2. The reason is that in the case of constant flux ratio n, increase of jetting Mach number Ma_2 means the pressure of jetting flow is higher, the area of pylons outlet decrease but kinetic energy increase. As a result, the total pressure recovery coefficient P_t increases with the increase of jetting flow Mach number Ma_2.

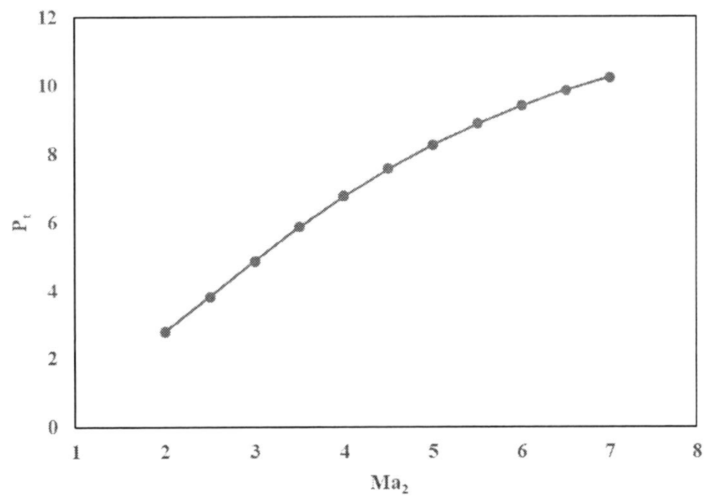

Figure 4. Variation curve of P_t with Ma_2

4. Conclusion

The technique principle of supersonic low Reynolds number flow control technique based on flow mixing is proposed in this paper. The basic flow process is analysed, and the theoretical analysis model is obtained. The effects of key parameters on flow performance are summarized by calculation through a group of standard input parameters. There is an optimal contraction ratio to achieve the highest total pressure recovery coefficient. The lower flux ratio is beneficial for improving total pressure recovery coefficient. Increase jetting flow Mach number is also beneficial for improving total pressure recovery coefficient.

References

[1] Xu H 1987 *Fundamentals of aerodynamics* (Beijing: Beijing Aeronautics College Press)
[2] Liu Z 2003 *Aerodynamic and structure design of high speed and low speed wind tunnel* (Beijing: National Defence Industry Press)
[3] Liao D and Ren Z 2006 *High Power Laser and Particle Beams,* **18** 728
[4] Liao D 2003 *An investigation of optimum performance and mixing enhancement for gaseous ejectors* (Xi'an: Northwestern Polytechnical University)
[5] Emanuel G 1974 *AIAA Journal,* **14** 1292

IOP Conf. Series: Materials Science and Engineering **751** (2020) 012010 doi:10.1088/1757-899X/751/1/012010

Fluid-structure Interactions on Steerable Cruciform Parachute Inflation Dynamics

Xinglong Gao[1,2]; Qingbin Zhang[3]; Qin Chen[1,2,*] and Wenkai Wang[4]

[1] State Key Laboratory of Aerodynamics, China Aerodynamics Research and Development Center, Mianyang, China
[2]Facility Design and Instrumentation Institute, China Aerodynamics Research and Development Center, Mianyang, China
[3]School of Aerospace Science and Engineering, National University of Defense Technology, Changsha, China
[4]Technology Center, AVIC Aerospace Life-Support Industries. Ltd., Xiangyang, China

[*]E-mail: qchen2016@hotmail.com

Abstract. To develop the decelerating and rigging potential performances of cruciform parachute during to its relatively large stand-off with lower cost production in precision aerial delivery operations, the fluid structure interaction behaviours of steerable cruciform parachute is studied. The concept of suspension lines control of cruciform parachute is firstly proposed. Based on the multi-material Arbitrary Lagrange-Euler method, the coupling dynamic model between a viscous incompressible fluid and a flexible large deformation structure of the cruciform parachute is solved. The inflation performance of a cruciform parachute under different steerable conditions is analysed. The decelerating parameters of the parachute, including drag area, and opening loads are obtained from FSI simulation. Meanwhile, the evolution of the three-dimensional shape of the cruciform parachute inflation is predicted.

1. Introduction

As the huge demands in military and humanitarian resupply, the precision aerial delivery systems experience a significant development during the last 20 years. Accordingly, the steerable ability of parachute during the decelerating process become mature and make great efforts to the precision aerial delivery operations. The ram air parafoil is most familiar and know to all for its accurate steerability, while the control mechanism of parafoil is relatively complex and even require some intelligence of payload to keep system stable in some extreme environments[1, 2]. Nowadays, people are still making efforts to seek a low-cost, minimally complex, and accurate precision delivery system[3, 4].

As an alternative to ram-air parafoils, round parachute control technology has been extensively developed by U.S. Army Natick Soldier Research Development and Engineering Center (NSRDEC)[5-7], such famous accurate delivery system of round parachute is the Affordable Guided Airdrop System (AGAS)[8]. Based on the extensively studied in the past from the standpoint of structural integrity[9] and aerodynamic effects of canopy geometry[10], including intentionally introduced asymmetry[11], as well as the evaluation of control line reefing techniques for high controllability of parachute with minimal actuator power requirements[12]. Yakimenko and Fields et al. are seeking the feasibility of steerable ability of cruciform by implementing the actuator on the suspension lines[13, 14]. The construction of cruciform parachute is simple and economical, of which the canopies are composed by two identical fabric rectangles, crossed and joined to each other at the square intersection to form a flat surface having four equal arms[13]. Fagley *et. al.* examined the bleed air spoiler (BAS) control technique on cruciform parachute canopy by computational study method[15], Potvin *et. al.* performed an extensive field testing regimen to explore the potential gliding capabilities of cruciform and hybrid cruciform canopies[16]. Herrington *et. al.* presents an experimental approach for testing a steerable cruciform parachute system using a vertical wind tunnel[17].

To develop the dynamic response performance of a cruciform parachute during inflation stages at line-reefing steering conditions, the ALE code in LS-DYNA software is utilized in this study to perform the FSI simulation of parachute. The steerable parachute concept is firstly designed, then the finite element model of parachute and fluid domain are built and the penalty coupling method is applied for FSI simulation.

2. Problem Statement and concept design

The concept of steerable cruciform parachute origin from the low-cost demands in precision airdrop, delivery and recovery tasks. The cruciform parachute addressed in this paper is a 16 suspension lines cross parachute consisting of a cruciform canopy, which can be steered by adjusting the length of a dynamic line through the single actuator on it during flight, and one static line can be shortened beforehand with a constant length during flight. Thus the asymmetry of initial cruciform parachute can be broken and the spin control can be achieved by intentionally introducing a difference between the steerable and static lines, $\Delta l = \delta_d - \delta_s$,and L=2.8m is the arm length, W=0.72m is the arm width and l=2.82m is the suspension length of cruciform parachute (Figure 1).

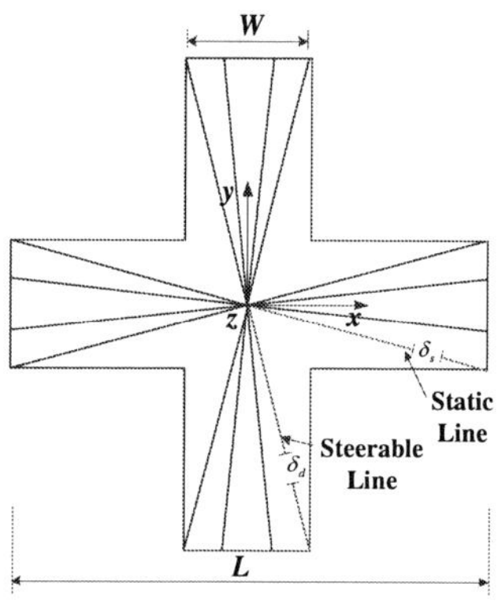

Figure 1. Layout of steerable cruciform parachute

3. Fluid Structure Interaction Model

3.1. Structure and fluid dynamics

Parachute components are flexible and continuous media. The governing equation of structure system is

$$\rho_s \frac{\partial \boldsymbol{u}}{\partial t} = \boldsymbol{\sigma}_s(\boldsymbol{u}) + \rho_s \cdot f_s \text{ on } \Omega^s \tag{1}$$

where ρ_s is material density, u is the velocity vector of structure media, σ_s is Cauchy stress tensor, and f_s is the external body forces acting on structure. For canopy structure formulation, a special membrane element like traditional airbag fabric constitutive material model is better suited to it. For the thin fabrics, buckling (wrinkling) can occur with the associated inability of the structure to support compressive stresses. The membrane is a two-dimensional (2D) shell suited for a 3- or 4-node element.

Besides on, air permeability of parachute fabric is one of the important properties must be considered, here it's determined by the Ergun equation.

$$\frac{dP}{dr} = a(\mu,\beta)\cdot v_{rel} + b(\mu,\beta)\cdot v_{rel}^2 \tag{2}$$

The fluid field during parachute inflation is time-variant spatial domains, let Ω^f be the spatial domain, the N-S equations for incompressible flows are [18]

$$\rho_f\left(\frac{\boldsymbol{u}_f}{\partial t} + \boldsymbol{u}_f\cdot\nabla\boldsymbol{u}_f + \boldsymbol{f}_f\right) - \nabla\cdot\boldsymbol{\sigma}_f = 0 \text{ on } \Omega^f \tag{3}$$

$$\nabla\cdot\boldsymbol{u} = 0 \text{ on } \Omega^f \tag{4}$$

ρ_f, $\boldsymbol{u}_f$, $\boldsymbol{f}_f$ and σ_f are the density, velocity vector, external body force and stress tensor respectively. By introducing the ALE formulation, the finite mesh can be freely moved. The fluid particle coordinates are $\mathbf{X}_i(t)$, $(i = x, y, z)$, where $t \in (0,T)$.

3.2. Coupling method

We use the ALE penalty-coupling algorithm implemented in the LS-DYNA nonlinear dynamic software to solve the FSI equations for steerable cruciform parachute inflation process. Both the fluid and structure models can be independently meshed, and the time step of explicit dynamic integral method in ALE algorithm is in proportion to the mesh size. Besides on, the negative volume and an hourglass effect should also be avoided by introducing the scale factor of time step. At each time step, the velocity and pressure should be decoupled by a penalty method, and the fluid and structure equations would be associated, solved, and iterated once.

4. Numerical Simulation

In this study, the infinite mass inflation is simulated, the cruciform the parachute is attached to a fixed point in space and flow can pass through the simulation domain, like a wind tunnel configuration, as shown in figure 2. The parachute's folding and deployment process are also not considered in this paper, and we focus on the fabric inflation behaviours during steerable process, thus the initial state of cruciform parachute model is simplified as illustrated in figure 2. The fabric was meshed by 2D tetrahedral shell and the line was meshed by 1D discrete beam elements. The fluid domain was meshed by 3D solid elements.

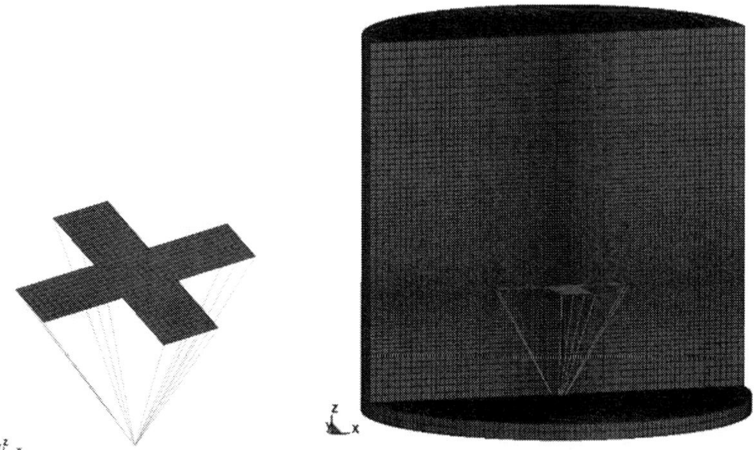

Figure 2. Finite element model of cruciform parachute and fluid domain for FSI simulation

IOP Conf. Series: Materials Science and Engineering **751** (2020) 012010 doi:10.1088/1757-899X/751/1/012010

5. Results and Discussions

The inflation process of cruciform parachute without steering by line control are presented in figure 3, the canopy is asymmetrically inflated, and the steady shape is hemisphere. Figure 4 illustrates the FSI phenomenon between canopy structure and surrounding flow, the cruciform parachute begins to swing and rotate after the canopy inflated. It is clearly to know that the numerical simulation of FSI has captured parachute's opening process.

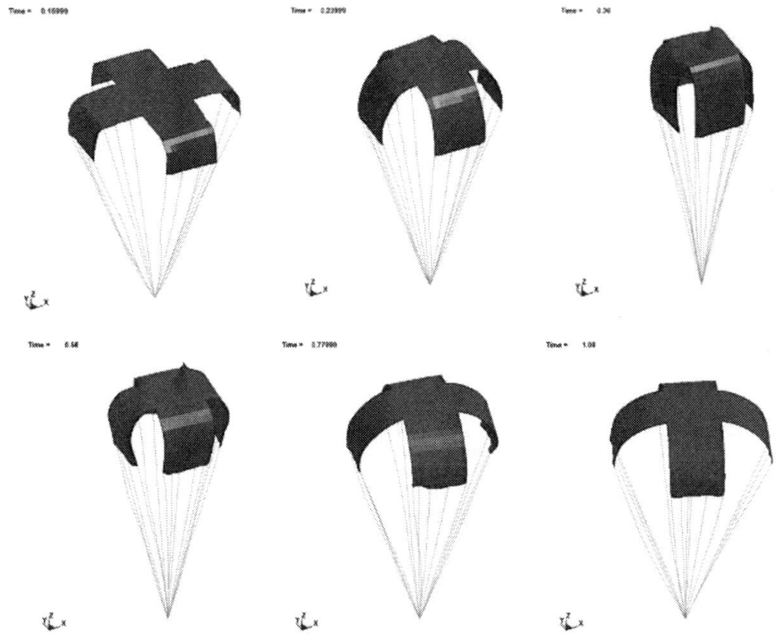

Figure 3. Inflation geometry of cruciform parachute

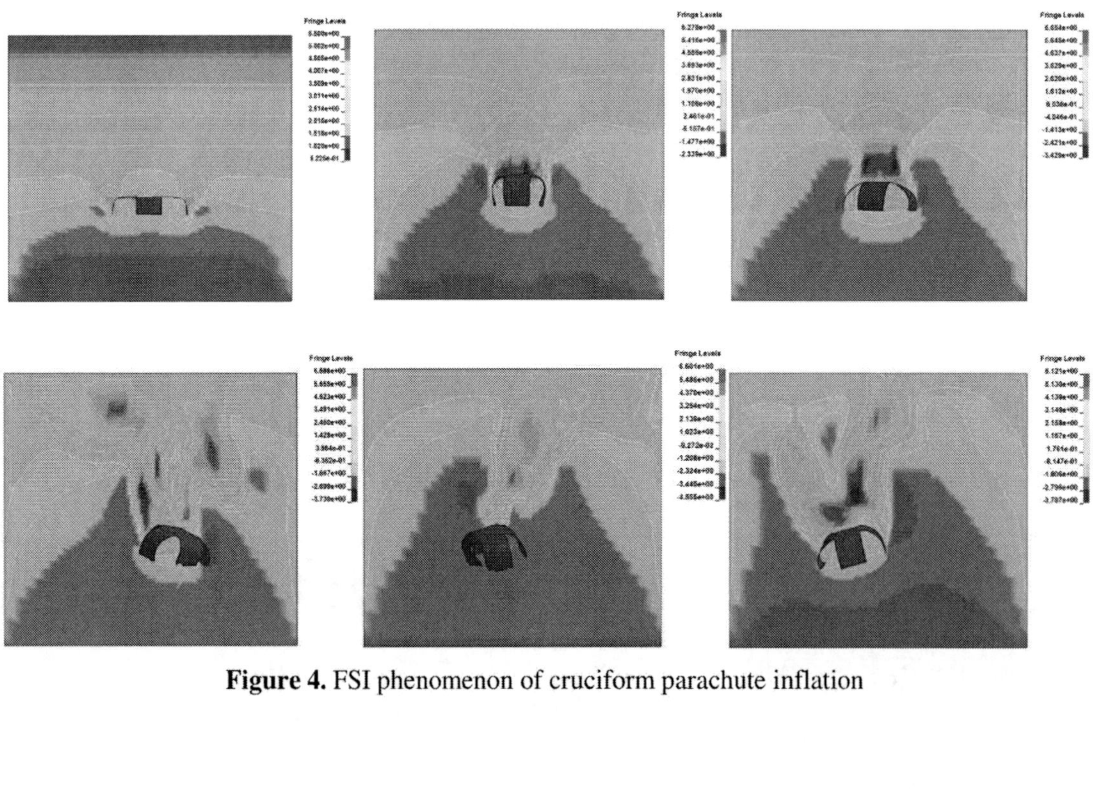

Figure 4. FSI phenomenon of cruciform parachute inflation

Figure 5 shows the drag area of parachute's projection under 3 cases. It is shown that the canopies reach to a maximum area at about 1.1s for all 3 cases after a slight shrinkage occur. After the canopies are fully inflated, the drag area keeps at a steady value in case 1. On the contrary, the canopies regularly fluctuate after fully inflated both for the other two steering cases, which indicate that the aerodynamic shape of canopy can be controlled and steered by the suspension lines retraction control.

Figure 6 predicts the opening forces evolution of canopy for all 3 cases, it can be seen from the figure 5 and 6 that the peak forces appear at the shrinkage of canopy at about 0.4s before the fully inflate state. The peak force values are same, and the evolution rules are similar for all 3 cases.

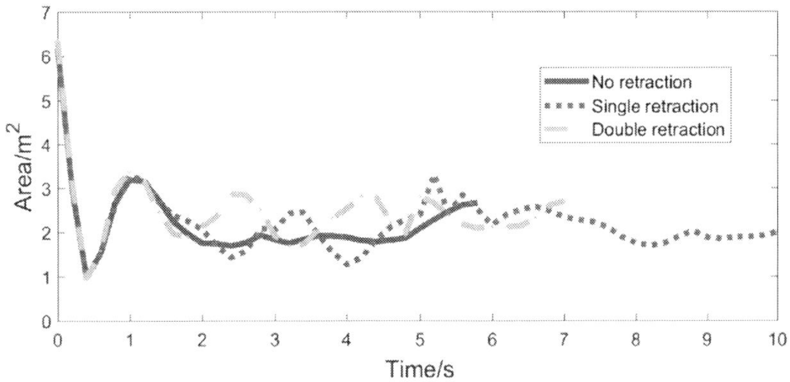

Figure 5. Evolution of cruciform parachute's drag area

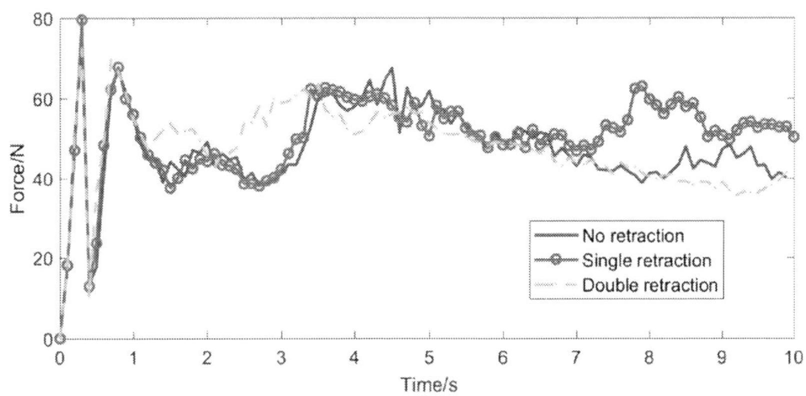

Figure 6. Opening force of canopy

6. Conclusions

This paper addressed the steering concept and FSI simulation of cruciform parachute under infinite mass inflation. The steering conditions were designed and the numerical models for different steering conditions were built and solved, and curves from different work conditions of the numerical models were also compared and investigated. The results show that the aerodynamic shape and decelerating performance can be regulated and steered by the line retraction control, the method used in this paper can also provide a simplified approach for FSI dynamics estimation of steering parachute inflation, as well as the indication of fluid and structure dynamics. Future study should consider much more steering conditions and comparing with the airdrop and wind tunnel experimental results.

Acknowledgments

This study was supported by Research Project of National Natural Science Foundation of China (No. 11902331).

IOP Conf. Series: Materials Science and Engineering **751** (2020) 012010 doi:10.1088/1757-899X/751/1/012010

References

[1] Benney, R., et al., *The Joint Precision Airdrop System Advanced Concept Technology Demonstration*, in *Proceedings of the 18th AIAA Aerodynamic Decelerator Systems Technology Conference and Seminar*. 2005: Munich, Germany.

[2] Fogell, N., L. Iannucci, and K. Bergeron. *Fluid-Structure Interaction Simulations of the Inflated Shape and Associated Flowfield of the MC4/5 Parafoil During Steady Gliding flight*. in *24th AIAA Aerodynamic Decelerator Systems Technology Conference*. 2017. Denver, Colorado.

[3] Yakimenko, O.A., *Precision Aerial Delivery Systems: Modeling, Dynamics, and Control*. 2015, Arlington, VA: AIAA Progress in Astronautics and Aeronautics.

[4] Gilles, B., M. Hickey, and W. Krainski. *Flight Testing of a Low-Cost Precision Aerial Delivery System*. in *Proceedings of the 18th AIAA Aerodynamic Decelerator Systems Technology Conference and Seminar*. 2005. Munich, Germany.

[5] Fields, T.D., J.C. LaCombe, and E.L. Wang, *Autonomous Guidance of a Circular Parachute Using Descent Rate Control*. Journal of Guidance, Control, and Dynamics, 2012. **35**(4): p. 1367-1370.

[6] Fields, T.D., J.C. LaCombe, and E.L. Wang, *Time Varying Descent Rate Control Strategy for Circular Parachutes*. Journal of Guidance, Control and Dynamics, 2015. **38**(8): p. 1468–1477.

[7] Yakimenko, O.A., V.N. Dobrokhodov, and I.I. Kaminer, *Synthesis of Optimal Control and Flight Testing of an Autonomous Circular Parachute*. Journal of Guidance, Control, and Dynamics, 2004. **27**(1): p. 29–40.

[8] Brown, G., et al. *The Affordable Guided Airdrop System (AGAS)*. 1999.

[9] LaFarge, R.A., J.M. Nelsen, and K.W. Gwinn. *A Novel CFD/Structural Analysis of a Cross Parachute*. in *Proceedings of the 32nd AIAA Aerospace Sciences Meeting and Exhibit*. 1994. Reno, NV: AIAA Paper.

[10] Levin, D. and Z. Shpund, *Canopy Geometry Effect on the Aerodynamic Behavior of Cross-Type Parachutes*. Journal of Aircraft, 1997. **34**(5): p. 648-652.

[11] Shpund, Z. and D. Levin. *Dynamic Investigation of Asymmetric Cross-Type Parachutes*. in *Proceedings of the 14th AIAA Aerodynamic Decelerator Systems Technology Conference and Seminar*. 1997. San Francisco, CA: AIAA Paper

[12] Fields, T.D., *Evaluation of Control Line Reefing Systems for Circular Parachutes*. Journal of Aircraft, 2016. **53**(3): p. 855-859.

[13] Fields, T.D. and O.A. Yakimenko, *Feasibility of a Steerable Single-Actuator Cruciform Parachute for Targeted Aerial Delivery*, in *24th AIAA Aerodynamic Decelerator Systems Technology Conference*. 2017: Denver, Colorado.

[14] Haller, J.S., T.D. Fields, and O.A. Yakimenko, *Precision Aerial Delivery with a Steerable Cruciform Parachute*, in *24th AIAA Aerodynamic Decelerator Systems Technology Conference*. 2017: Denver, Colorado.

[15] Casey Fagley, et al., *Computational Study of Air Drop Control Mechanisms for Cruciform Parachutes*, in *24th AIAA Aerodynamic Decelerator Systems Technology Conference*. 2017: Denver, Colorado.

[16] Potvin, J., et al. *Glide Performance Study of Standard and Hybrid Cruciform Parachutes*. in *Proceedings of the 17th AIAA Aerodynamic Decelerator Systems Technology Conference and Seminar*. 2003. Monterey, CA.

[17] Herrington, S.M., et al., *Modeling and Control of a Steerable Cruciform Parachute System Through Experimental Testing*, in *AIAA Scitech 2019 Forum*. 2019: San Diego, California.

[18] Stein, K., Benney R., Kalro, V., Tezduyar, T. E., Leonard, J. and Accorsi, M., *Parachute fluid-structure interactions: 3-D computation*. Comput. Methods Appl. Mech. Engrg., 2000. **190**: p. 373-386.

IOP Conf. Series: Materials Science and Engineering **751** (2020) 012011 doi:10.1088/1757-899X/751/1/012011

Rapid Aerodynamic Calculation Method for Hypersonic Gliding Vehicle

Bin Zhang*, Zhiwei Feng, Boting Xu, Tao Yang and Wuyu Peng

College of Aerospace Science and Engineering, National University of Defence Technology, Changsha, China

*Corresponding author e-mail: 15084904287@163.com

Abstract. The accuracy and efficiency are always conflict in the process of aerodynamic shape design, especially for hypersonic gliding vehicle (HGV). In this work, a rapid aerodynamic calculation method for hypersonic gliding vehicle is conducted, including shape parametric modelling, automatic mesh generation and rapid computational fluid dynamics (CFD) method. In details, the aerodynamic shapes of HGV are modelled by multi-section solids, which parameterized by class function/ shape function transformation (CST) method and exponential function. Then, the computational meshes are generated automatically in two steps: one is generating surface mesh by Delaunay method, and another is generating volume mesh by adaptive Cartesian method. In order to minor the computational cost, the hypersonic aerodynamic are analysed by combining the Euler equation and the viscosity force. We have tested the accuracy and efficiency of this rapid aerodynamic calculation method in two flight states for a typical HGV. Compared with Navier Stokes (NS) equations, the results shown the errors of axial aerodynamic force are all less than 6%, but time costs are reduced in hundreds of times.

1. Introduction

At present, the HGV has become a research hotspot for its high speed and high mobility. The main flight altitude of HGV is in near-space of 20km-100km and the Mach number is greater than 5 [1]. Due to the special flight environment with high-altitude and high-speed, the aerodynamic simulation of HGV is difficult and costly, especially in time-consuming. In addition, the configurations of HGV are unique and complex, such as lifting body and wave-rider vehicle [2].

An accurate shape parametric model is the basis for the aerodynamic shape design of HGV, which plays a crucial role in its performance. When calculating the aerodynamic force in design of a lifting body, the accuracy and efficiency are always conflicted. With multiple iterations in the aerodynamic shape design works, it is important to minimize the computational consumption.

Consideration above challenges, this paper developed a rapid hypersonic aerodynamic calculation method for aerodynamic design of HGV, which contains three key points: shape parametric modeling, automatic mesh generation and rapid CFD method. Based on this rapid aerodynamic calculation method, the designer can predict the aerodynamic results for HGV with high efficiency and credible accuracy.

2. Shape parametric model

According to the characteristics of aerodynamic shape of HGV, the class function/shape function transformation (CST) method [3] is suitable to generate the section curve. Combined with exponential function generates the profile curve, the shape parametric model of HGV can be completed.

2.1. Section curve based on CST method

The CST method is a commonly used parametric geometry representation technique, which can be developed to generate the complicated three dimensional configurations of HGV. Before using the CST method to describe the section curve, the actual geometry coordinates (*x*, *y*) should be normalized with parameterized coordinates (η,ψ). The normalized function of section curve based on CST method can be written as:

$$\eta(\psi)=C(\psi)S(\psi) \tag{1}$$

where $\eta\in[0,1]$ and $\psi\in[0,1]$ are the normalized coordinates of section curve, $C(\psi)$ is the class function and $S(\psi)$ is the shape function.

In the expression of CST method, the class function of $C(\psi)$ represents the classification and geometric properties of section curve, and the shape function of $S(\psi)$ defines the figure of the curve more accurately.

2.2. Profile curve based on exponential function

Generally, the profile curve in the top view of HGV is a smoothed curve, which extending smoothly from the top to the bottom. Based on above consideration, we can use the exponential function to generate the profile curve, defined as:

$$y=\frac{W}{2L^n}x^n \tag{2}$$

where *W* is the maximum width of HGV, *L* is the total length of HGV, and *n* is an index that controls the curvature change of the profile curve.

2.3. Shape parametric model based on "Section- Profile" curve

Combined the "Section" curve based on CST method with the "Profile" curve based on exponential function, we can complete the shape parametric model for HGV based on "Section- Profile" curve. In details, the execution steps are as follows:

a) Provide the maximum width *W* and total length *L* of the HGV to define the outer outline;

b) Ascertain the number of section curves to express the geometry characteristics of HGV;

c) Ascertain the functions of $C(\psi)$ and $S(\psi)$ in each section, then generate the section curves based on the CST method;

d) Ascertain the value of index *n* to generate a profile curve based on exponential function;

e) Generate the shape parametric model of HGV based on above "section-contour" curves.

3. Automatic grid generation

When the shape model of HGV has been completed by above parametric method, the next work is to generate the computational grids, including the surface grid and volume grid. At present, there are three mainly used mesh types, such as structured grid, unstructured grid and Cartesian grid.

On one hand, the unstructured grid is easy to generate surface grid without the constraints of geometry boundaries. On the other hand, the adaptive Cartesian grid can provide the automatic generation of volume grid, which based on the surface grid. Therefore, for the shape model HGV, the surface grid is generated based on the unstructured mesh, and the volume grid is generated based on the adaptive Cartesian method.

3.1. Surface grid based on Delaunay method

When generating the unstructured meshes, the Delaunay method is a commonly used and reliable method. The basis theory of Delaunay method is dividing a plane into some convex polygons based on the point sets we have known. The Delaunay method has some advantages: good quality, high efficiency, strict basis of mathematical theoretical, and can generate equilateral triangles as much as possible.

3.2. Volume grid based on adaptive Cartesian method

The data structure of Cartesian grid is based on the dynamic fork tree, such as a dynamic quadtree structure for two-dimensional grid and a dynamic octree structure for three -dimensional grid. On the basis of geometric shape and flow field characteristics, the adaptive Cartesian method is able to continuously generate mesh adaptive refinement in the local area.

The adaptive Cartesian mesh method has some advantages: high efficiency, no division of mesh topology, automation mesh generation, etc. Based on the adaptive Cartesian mesh method, a high-quality volume grid can be generated without human intervention.

4. Rapid CFD method

In the flight environment with high-altitude and high-speed, the viscous effect is the mainly factor interfering the aerodynamic force of HGV [4]. In order to balance the calculation efficiency and precision, a rapid CFD method for HGV is developed.

First, an Euler equation is used to calculate the non-viscous aerodynamic force. Then, the viscosity force is computed based on the friction coefficient. Combined the non-viscous with viscosity force, the final aerodynamic force of HGV can be predicted.

4.1. Euler equation

Generally, the Euler equation can calculate the three-dimensional compressible and non-viscous aerodynamic force. When the incoming fluid is considered as the perfect gas, the Euler equation can be written as:

$$\oiiint_{\Omega} \frac{\partial Q}{\partial t} dV = -\oiint_{\partial\Omega} \left(\vec{F} \cdot \hat{n} \right) ds \tag{3}$$

where, $\hat{n}$ is a unit vector, $\vec{F}$ is the flux function, and Q is the state vector, which consisting of flow variables, such as the density ρ and the velocity vectors of u, v, w.

4.2. Viscosity force

The viscous correction is calculated by using the axial force viscous component based on the friction coefficient. First, the friction coefficient of the aircraft is calculated, and then the viscous force of the surface is integrated to obtain the viscous force component of the axial force.

For the turbulent flow, the definition of friction coefficient can be written as

$$C_f = \frac{0.37}{(\log \mathrm{Re}^*)^{2.584}} \left(\frac{\rho^*}{\rho_\infty} \right) \tag{4}$$

where, C_f is the friction coefficient, Re^* is the referenced Reynolds number at the edge of the boundary layer, ρ^* is the referenced density, and ρ_∞ is the density of incoming fluid.

The referenced values of Re^*, ρ^* and μ^* can be calculated by the reference enthalpy method, which developed by Eckert [5]. In this paper, we take the Meador's reference enthalpy method [6] to calculate the viscosity force:

$$\frac{H^*}{H_e} = 1 + 0.5\left(\frac{H_w}{H_e} - 1 \right) + 0.16r\frac{\gamma - 1}{2} Ma_e^2 \tag{5}$$

where, H^* is the reference enthalpy, H_w is the enthalpy on the wall, H_e is the enthalpy at the edge of the boundary layer, Ma_e is the Mach number at the edge of the boundary layer, r is the recovery factor that depends on the Prandtl number, and γ is the ratio of specific heats.

5. Results

Integrated above shape parametric model, automatic grid generation and rapid CFD method, we can set up the rapid aerodynamic calculation method for hypersonic gliding vehicle. In order to verify the accuracy and efficiency of the proposed method, we demonstrate a typical example.

5.1. Shape parametric Model

Based on the shape parametric model based on "Section- Profile" curves, we take three section curves to shape the parametric model as a typical example of HGV, as shown in Fig. 1.

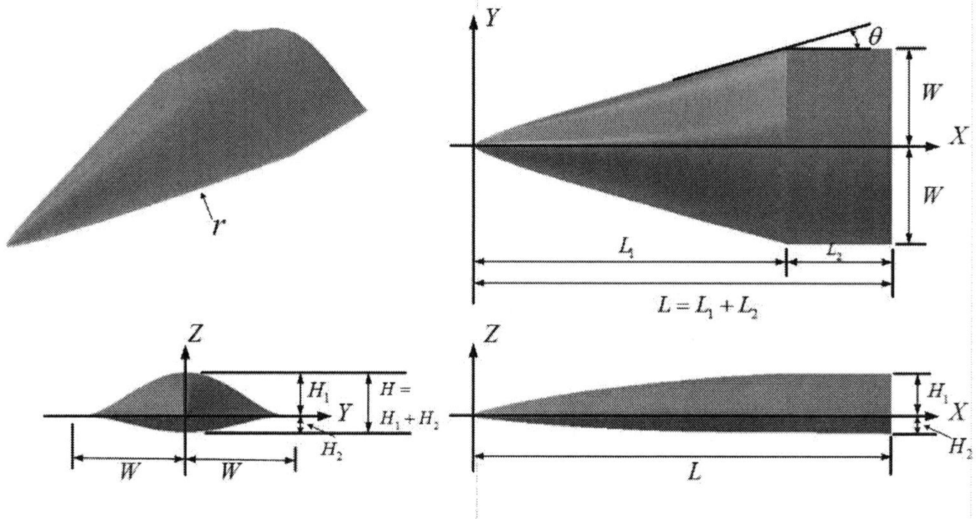

Figure 1. The shape parametric model of HGV.

Table 1. Main values of shape parametric model in Fig. 1.

Shape parameter	L	L_1	L_2	W	H	H_1	H_2	r	θ
Main values（mm）	4000	3000	1000	1000	560	400	160	10	18°

5.2. Automatic mesh generation

First, we generate the surface mesh for above shape parametric model of HGV based on Delaunay method, as show in Fig. 2.

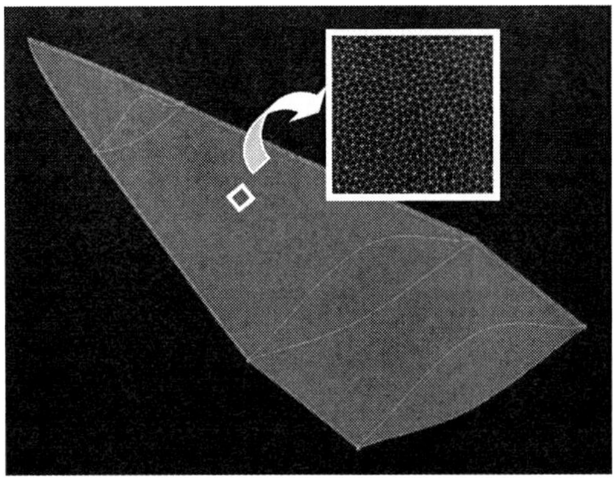

Figure 2. The automatic generated surface grid of HGV.

Then, we generate the volume mesh for surface mesh of HGV based on adaptive Cartesian method, as show in Fig. 3.

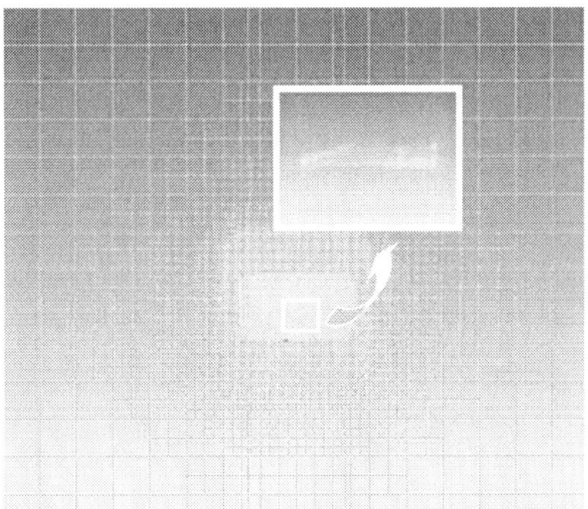

Figure 3. The automatic generated volume grid of HGV.

5.3. CFD results

For the above typical HGV, we compute its axial aerodynamic force results under the flight height of 55km with two commonly used computational states of Ma=8, α =8° and Ma=15, α = 10°. Taking the results of NS equations as the reference, we compare the calculation accuracy and efficiency between the rapid aerodynamic calculation method proposed in this paper and the NS equations.

The results between rapid aerodynamic calculation method and NS equations are shown in Table 1 and in detail, *Rapid* represents the results of the rapid aerodynamic calculation method, *NS* represents the results of the NS equations, and the *Delta* represents the relative errors between the results of rapid aerodynamic calculation method and the NS equation, which can be written as:

$$Delta = \frac{Rapid - NS}{NS} \times 100\% \qquad (6)$$

As Table 2 shows, the results of *Delta* in aerodynamic force are all within 6% for the two computational states, which means the calculation accuracy of this rapid aerodynamic calculation method is credible. On the other hand, the computation time of NS equations is as 160 times as the rapid aerodynamic calculation method. What's worse, the rapid aerodynamic calculation method just takes one central processing unit (CPU), but the *NS* equations need four CPUs. Otherwise the difference of time consumption is larger. In conclusion, the rapid aerodynamic calculation method not only equips with credible calculation accuracy, but also have higher efficiency.

Table 2. Results between rapid aerodynamic calculation method and NS equation.

H=40km		Ma=6, α =8°	Ma=10, α = 8°
Aerodynamic force	*NS*	0.05795	0.03981
	Rapid	0.04627	0.04032
	Delta	4.42%	1.29%
H=55km		Ma=8, α =8°	Ma=15, α = 10°
Aerodynamic force	*NS*	0.07211	0.05867
	Rapid	0.06933	0.06218
	Delta	-3.86%	5.98%
Time	*NS* (4 CPU)	80h	80h
	Rapid (1 CPU)	0.5h	0.5h

6. Conclusion

This paper introduces a rapid aerodynamic calculation method for hypersonic gliding vehicle, which contains shape parametric modeling, automatic mesh generation and rapid CFD method.

Based on this rapid aerodynamic calculation method, we have tested its accuracy and efficiency for a typical HGV. Compared the computational results of NS equations, the errors of axial aerodynamic force are all less than 6%, but time costs are reduced in hundreds of times.

Acknowledgments

This work was partially financially supported by the National Natural Science Foundation of China (No.11772353).

References

[1] Ye Youda, Study on aerodynamic characteristics and design optimization for high speed near space vehicles [J]. Advances in Mechanics, 2009, 39(6): 683-694.

[2] Zhao Guilin, Hu liang, Wen Jie, Peng Hui and Zhang Mianchun. An overview of the research on waveriders and waverider. Derived hypersonic vehicles[J], Advances in Mechanics, 2003, 33(3): 358-374.

[3] Yang Ma, Tao Yang, Zhiwei Feng and Qingbin Zhang. Hypersonic lifting body aerodynamic shape optimization based on the multiobjective evolutionary algorithm based on decomposition [J]. Journal of Aerospace Engineering, 2015, 229 (7): 1246–1266.

[4] Hanqiao Han, Chenan Zhang and Famin Wang. Study on multi-physical effects of hypersonic gliding waverider vehicles [J]. Acta Aerodynamica Sinica, 2014, 32 (1): 101-108.

[5] E. E. R. C. Engineering Relations for Heat Transfer and Friction in High-Velocity Laminar and Turbulent Boundary Layer Flows over Surfaces with Constant Pressure and Temperature [J]. Transactions of the ASME, 1956 ,78, pp: 1273-1283.

[6] W. E. Meador and M. K. Smart. Reference Enthalpy Method Developed from Solutions of the Boundary-Layer Equations [J]. AIAA Journal, 2005, 43: 135-139.

Study on LabVIEW-based Aviation BLDCM Control System

Linhai Zhang[1, 2,a], Yeming Yao[1,2, b] and Lilin Xu[1, 2,c]

[1] Aviation Key Laboratory of Science and Technology on Electromechanical System Integration;

[2] Nanjing Engineering Institute of Aircraft Systems, AVIC Nanjing, Jiangsu Province, China, 211100;

[a] 646058138@qq.com, [b] 178348423@qq.com, [c] xulilin@126.com;

Abstract. with the increasing market demand and rapid development of aircraft technology, more electric concept has become an important trend for the future development of aircraft. Therefore, motors are widely applied in the aviation field increasingly. This paper mainly introduces the hardware environment and LabVIEW implementation process in the development of aviation high speed and high voltage BLDCM control system based on NI hardware and software platform. The test shows the control system is able to realize effective control over aviation BLDCM. This system is highly flexible and can lay a foundation for the algorithms used in subsequent related research efforts.
Keywords: LabVIEW FPGA; Brushless DC Motor (BLDCM); BLDC control; Aviation motor;

1. Introduction

With the development of more-electric and all-electric aircraft technology, motors are more and more widely applied in the aviation field, among which the high speed and high voltage BLDCM has become popular due to its high power density, high reliability, high power-weight ratio and other advantages[1-2]. Motor control technology constitutes a vital part of motor development. An excellent motor control technology can unleash the motor potential to the utmost extent, and motor performance test conducted under the technology can promote the design of motor itself [3-4]. The topological structure and principles of BLDCM drive are shown in Figure 1. The paper uses a three-phase full-bridge drive circuit. The rotor position is indicated by the three-phase Hall position signal to control the on-off time sequence of 6 switching tubes of the drive circuit and to further control the electrification time sequence of the stator coil of the motor to drive motor rotation.

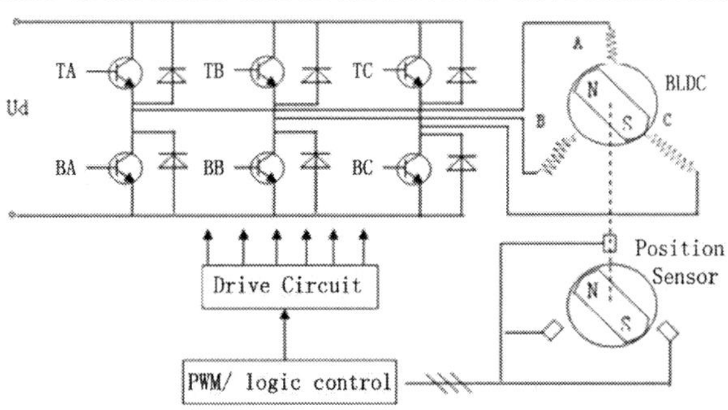

Figure 1 Schematic Diagram of BLDCM Drive

IOP Conf. Series: Materials Science and Engineering **751** (2020) 012012 doi:10.1088/1757-899X/751/1/012012

The paper uses LabVIEW to develop the high speed and high voltage BLDCM control system. LabVIEW is a graphical programming language characterized by simple and direct programming, high development efficiency, usability and high learning property, etc. Development of motor control system based on LabVIEW is favorable for in-depth study and verification of motor control strategy and control algorithm in the future, which saves repetitive development and shorten iteration cycle [5].

2. Development of control system

2.1. Introduction to the architecture

The motor control system architecture is shown in Figure 2. The NI-based RCP semi-physical motor testing platform can be used to verify control strategy and control methods.

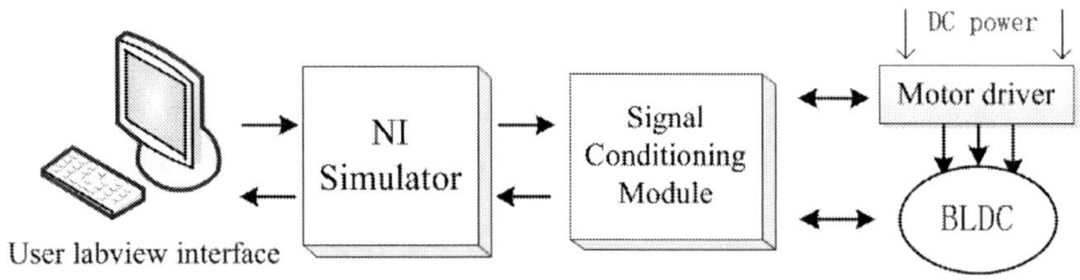

Figure 2 NI-based Motor Control System Architecture

Main hardware involved in the paper is shown in Figure 3, mainly including NI PXIe-1071 case, NI PXIe-8880 controller, NI PXIe-7858R board card, Infineon HybridPACK™1 6ED100HP1-FA drive module, optocoupler isolator, certain aviation BLDCM and multiple sensors (voltage, current and temperature), etc., among which NI PXIe-7858R is the FPGA board card of NI. It is characterized by rich resources and high running speed, etc. The default single-cycle frequency is 40MHz.

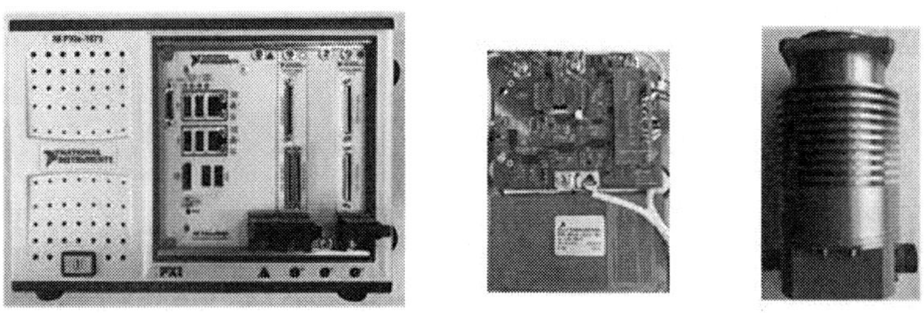

Figure 3 Main Hardware on NI-based RCP Semi-Physical Motor Testing Platform

2.2. LabVIEW Program Implementation

The LabVIEW-based motor control architecture is shown in Figure 4. The first is the acquisition of Hall signals and anti-shake treatment to obtain the on-off time sequence of 6 switching tubes with dead zone control. Meanwhile, the actual speed of motor is demodulated. Later the speed command and the actual speed command will be compared to output current command via the speed controller. It is then compared with the collected current to output the duty ratio command via the current controller. Finally the voltage is put to the motor to drive motor rotation.

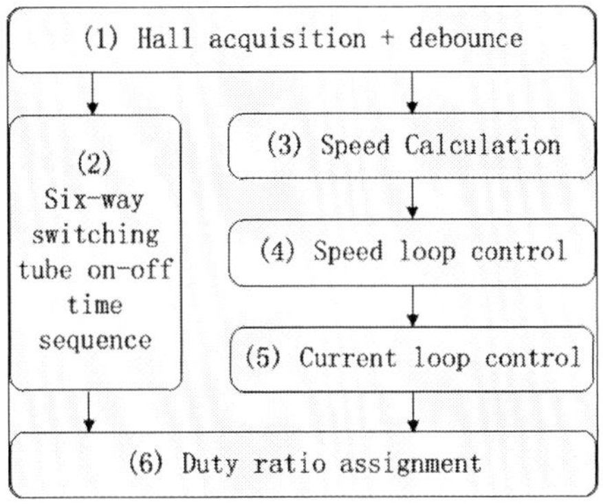

Figure 4 LabVIEW-based BLDCM Control Architecture

2.2.1. Hall sensor acquisition and debounce treatment. Hall acquisition and debounce treatment program is shown in Figure 5. After the motor Hall position signal passes through the optocoupler isolator, it is acquired by the three-phase DI channel of NI PXIe-7858R. To reduce lag, the acquisition frequency is set to be 40MHz; meanwhile, debounce treatment is performed on Hall signal to eliminate the impact of interference signal. "Debounce" principle: when the signal acquired changes, the counting starts. If the signal remains the same in a certain period, the signal is recognized as correct signal. The signal is assigned variables HA/HB/HC, or variables should follow the previous values. The debounce time set in the program is 1us.

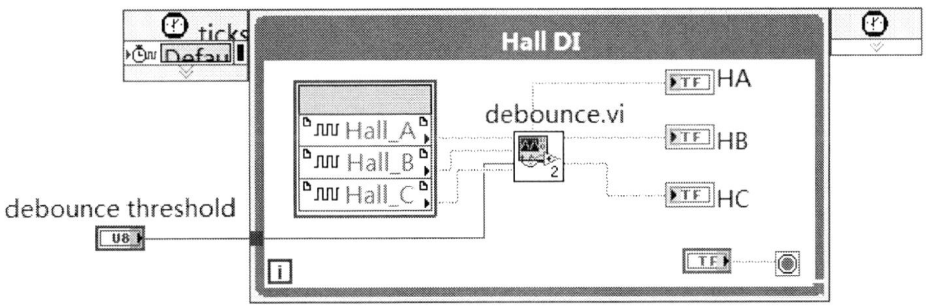

Figure 5 Hall acquisition and debounce treatment

Other paragraphs are indented (BodytextIndented style)

2.2.2. Six-way IGBT on-off time sequence control. The common breakover method for three-phase full-bridge drive circuit can be divided into two-two breakover and three-three breakover. The paper chooses the two-two breakover method, which refers to breakover of two phases on the motor every moment. The third phase is suspended. The breakover sequence and time of each phase are determined by rotor position signal. Under the method, when the motor is working normally, the upper and lower bridge arms are respectively equipped with one conductive power device only every moment. Even during commutation, the upper and lower bridge arms on the same bridge are difficult to be conductive simultaneously. Each time, the motor goes through a commutation, the direction of the resultant torque turns 60 electrical degrees. In a cycle, the torque has to go through six direction changes [6-7]. The signal logic of two-tow breakover is as follows:

IOP Conf. Series: Materials Science and Engineering **751** (2020) 012012 doi:10.1088/1757-899X/751/1/012012

$$\text{Forward:}\quad HAg\overline{HB} = TA \qquad \text{Reverse:}\quad HAg\overline{HB} = BA$$
$$HBg\overline{HC} = TB \qquad HBg\overline{HC} = BB$$
$$HCg\overline{HA} = TC \qquad HCg\overline{HA} = BC$$
$$\overline{HA}gHB = BA \qquad \overline{HA}gHB = TA$$
$$\overline{HB}gHC = BB \qquad \overline{HB}gHC = TB$$
$$\overline{HC}gHA = BC \qquad \overline{HC}gHA = TC$$

Wherein, HA, HB and HC represent the three-phase Hall signal acquired; TA, TB and TC represent three top tubes; BA, BB and BC represent three bottom tubes. The control program is shown in Figure 6, dead zone is set in the program. The principle of a dead zone is similar to that of "debounce". As the commutation under two-two breakover is difficult to have conductive top and bottom tubes on the same bridge arm at the same time, the dead zone time may be set at a small value or even not set at all. The set value in the program is 0.5us.

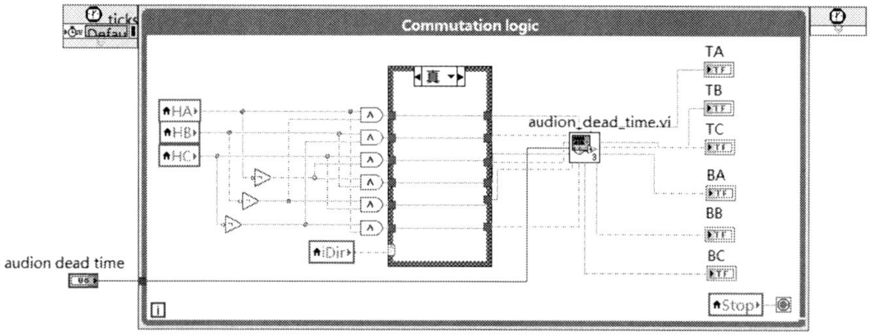

Figure 6 Six-way IGBT on-off time sequence control

2.2.3. Duty ratio assignment control. The duty ratio assignment control program is presented in the Figure 7. Nested loop is adopted on the output part of the duty ratio. The inner loop is the loop of duty ratio assignment, which is to assign the duty ratio signal to six-way IGBT via DO Interface. The loop frequency is 40MHz (single cycle timed loop). The stop condition for the loop is "once every PWM cycle"; the external loop is the conventional while loop, and it is to guarantee the internal loop in cycle.

The default PWM frequency for the program is 10kHz, which means each PWM cycle contains 40,000 ticks. Duty ratio control may be effectively realized via ticks counting. Besides, by logical judgment, the breakover time of IGBT in each PWM cycle happens to be the middle position of the whole cycle to guarantee the uniformity of current acquisition timing. Current acquisition trigger command is also set in the program.

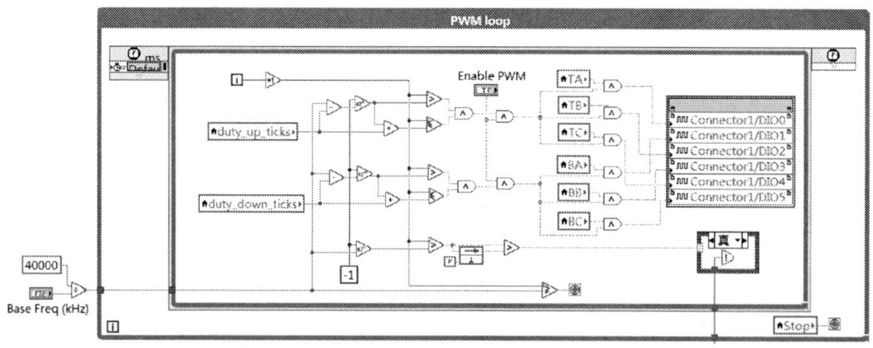

Figure 7 Duty ratio assignment control

2.2.4. Other Parts of the Program. The logic of speed calculation of Hall signal is relatively complicated. The main principle is to record the number of ticks between two groups of Hall signals,

IOP Conf. Series: Materials Science and Engineering **751** (2020) 012012 doi:10.1088/1757-899X/751/1/012012

which is T-method speed testing. Besides, the signal change problem, speed fluctuation in single Hall calculation and speed lag in six Hall calculation, etc. upon motor reversing and initial start shall be taken into consideration.

Compound control strategy of feedforward and feedback is adopted on the speed loop and current loop, which is suited to the filtering control method and variable parameter four-quadrant drive algorithm. Meanwhile, anti-integral windup PI controller is used to solve the excess integral problem. To guarantee the flexibility of algorithm, the paper has designed the anti-integral windup PI controller based on LabVIEW, which is able to realize the variable parameter control demand of the algorithm presented in the Figure 8. The basic idea is: when the controller output reaches its limit, the symbol of error signal is judged to determine whether to stop the integral action or not.

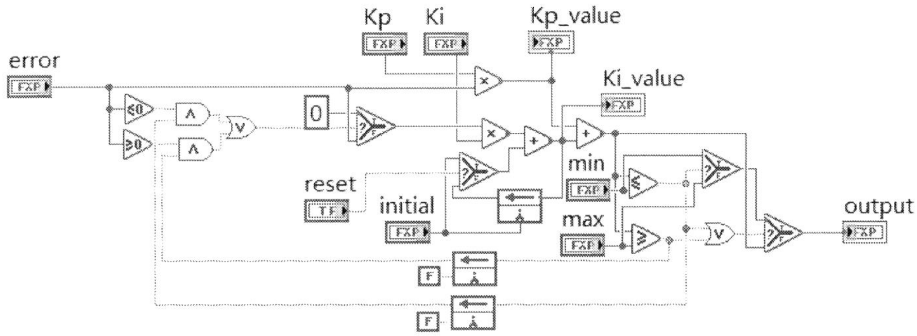

Figure 8 LabVIEW-based Anti-Integral Windup PI Controller

3. The first section in your paper

The paper tests and verifies the above control system via a certain aviation high speed and high voltage BLDCM. In the experiment, the current is acquired by Lime LA55-P SP50 current sensor. Both current sampling frequency and PWM frequency are 10KHz. The main parameters of the motor used in the experiment are presented in the table 1.

Table 1 Main Parameters of Test Motors

Name	Numerical value
Rated power/W	500
Rated voltage/V	310
Rated speed/rpm	20000
Rotational Inertia/(Kg·m2)	1.8×10-5
Winding phase inductance/mH	2.3
Winding phase resistance/Ω	2.6
Torque Coefficient/(Nm·A-1)	0.13

In the experiment, the motor is respectively supplied with phase step command and sinusoidal command to observe its response. The left figure in Figure 9 shows the phase step response curve. The phase step command is ±2000r/min. The right figure in Figure 9 shows the sinusoidal response curve. The amplitude and frequency of the sinusoidal command are 11000r/min and 5Hz respectively. Wherein, the red line represents the speed command signal and the white line the speed response signal. It can be seen from the figure that the phase step response time is approximately 0.13s, and the overshoot is around 3.7%; the speed change of sinusoidal response near zero speed is not smooth. On one hand, it is because that BLDCM gives feedback of position signal based on Hall sensor and therefore the speed fluctuation in low-speed segment is large; on the other hand, it is caused by the switchover between the speed of single Hall calculation and the speed of six Hall calculation. The left and the right in Figure 10 respectively represent the speed of single Hall calculation and the speed of six Hall calculation under the command of 2000r/min. It can be seen from the figure that the speed of single Hall calculation and the speed of six Hall calculation both fluctuates around 2000r/min. The maximum deviation is around 400r/min for the former and around 40r/min for the latter.

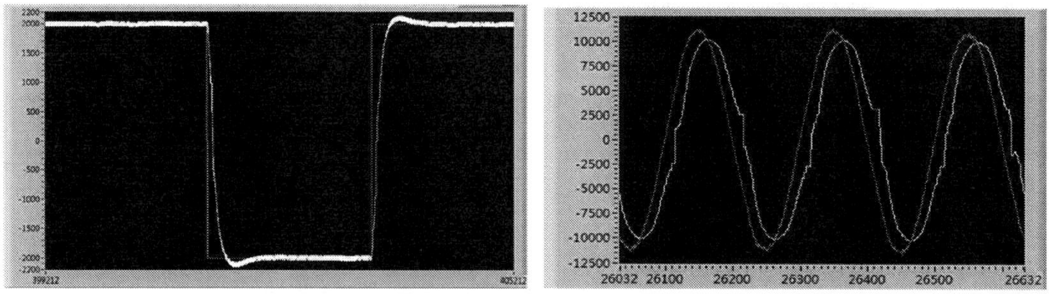

Figure 9 Motor Phase Step Response and Sinusoidal Response Curve

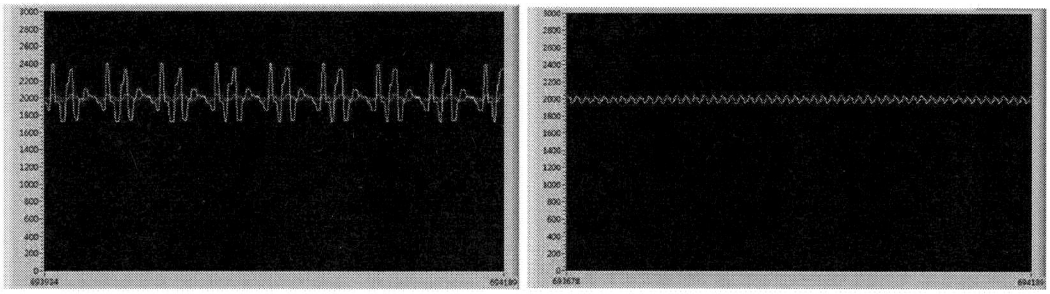

Figure 10 Comparison between Speeds of Single Hall Calculation and Six Hall Calculation

4. Conclusions

This paper introduces in detail the implementation process from Hall acquisition to PWM dispatch in the development of aviation high speed and high voltage BLDCM control system using LabVIEW based on NI RCP semi-physical simulation platform. The test result shows that the control system designed in the paper can effectively implement the control over aviation high speed and high voltage BLDCM. The control system is highly flexible and extendable, laying firm foundation for later studies on relevant algorithms.

Acknowledgments

This work was supported by the Aeronautical Science Foundation of China under Project 2017ZC09004.

References

[1] Tan Jiancheng, PM BLDCM Technology [M], Beijing: China Machine Press, 2011
[2] Xia Changliang, Zhang Maohua, Wang Yingfa, Liu Dan. Direct Torque Control of PM BLDCM [J].Proceedings of the CSEE,2008(06):104-109.
[3] Xia Changliang. BLDCM Control System [M]. Beijing: Science Press, 2009
[4] Xia Changliang, Li Zhengjun, Yang Rong, Qi Wenya, Xiu Jie, ADRC-based BLDCM Control System [J], Proceedings of the CSEE, 2005(02):85-89
[5] Gao Jin, Xu Qiuxia, Dong Zhaoqiang, BLDCM Semi-Physical Real-Time Simulation and Testing [J], Electric Machines and Control, 2018,22(04):8-15.
[6] Xia Changliang, Fang Hongwei. PM BLDCM and Its Control[J].Transactions of China Electrotechnical Society,2012,27(03):25-34.
[7] Zhang Xiangjun, Chen Boshi, Impact of PWM Modulation Method on Commutation Torque Ripple in BLDCM Control System [J], Electric Machines and Control, 2003(02):87-91.

IOP Conf. Series: Materials Science and Engineering **751** (2020) 012013 doi:10.1088/1757-899X/751/1/012013

Investigation of the Influence of Blocking Effect on the Aerodynamic and Hydrodynamic Characteristics of a Powered Aircraft Model

Zhijian Xiao[1,2], Bin Wu[1,2], Mingzhen Wang[1,2] ,Qi Hu[1,2] and Wenjing Yang[3*]

[1] Key Aviation Scientific and Technological Laboratory of High-Speed Hydrodynamic, Jingmen 448035, China
[2] China Special Vehicle Research Institute, Jingmen 448035, China
[3] Huazhong University of Science and Techonolgy, Wuhan 430074, China

[*]E-mail address: yangwenjing10@hust.edu.cn

Abstract. This paper investigated the influence of blocking effect on the aerodynamic characteristics of a powered aircraft model by experimental and numerical methods. Firstly, the towing test were carried out in the bottom and front of the trailer. The experimental results showed that the darg and trim of two conditions were different, indicating that the blocking effect was obvious. RANS method, slidding mesh and overset grid were adopted in numerical simulation. The simulation of the wind field in the bottom and front of the trailer were carried out. And the numerical results were compared with the experimental results to verify the accuracy of numerical method adopted in this paper. The results showed that the wind speed in the bottom of the trailer was larger than that in the front of the trailer on average and there was stable speed increment interval when the wind filed simulation was conducted in the front and the bottom of the trailer without model. Then the hydrodynamic performance of the powered aircraft model in infinite domain, in the bottom of the trailer and in front of the trailer were carried out respectively. The aerodynamic downward moment was larger when the model was in the bottom, but the overall aerodynamic lift force of the model was similar under the two conditions. The heave amplitude was also similar under the two conditions, but the trim angle and the total drag was smaller when the model was in the bottom of the trailer.

Keywords: Powered aircraft; RANS; Overset mesh; Blocking effect

1. Introduction

According to the theory of fluid mechanics, the bottom and tank wall limit the flow of water around the ship model which would increase the velocity of the reflow, resulting in a change in viscous drag, called the blocking effect[1]. At present, the research on the blocking effect of the towing tank was mainly focused on the low and medium speed ship, and mainly focused on the blocking effect of water. Sheng Zhenbang pointed out that the shallow water effect and the wall effect of the ship model in the towing tank were usually negligible, and the blocking effect was the main part of the tank boundary effect[2]. The drag of ship model obtained by the towing test conducted in the towing tank was usually greater than the that of the same speed in the infinite domain, where the increase of the drag mainly caused by the tank boundary[3][4]. According to the results of five experiments in the tank of Zhejiang Ocean University and related theoretical results, Li Guangnian pointed out that the blocking effect can be ignored when the blocking ratio was less than 0.2%[5], indicating that the extent of water blocking effect of the ship model was related to the blocking ratio. At present, the research results on the blocking effect of high-speed surface vehicles were rarely published. And for the towing test of high-speed vehicles such as the aircraft and ground-effect aircraft, the change of wind speed also had a great impact on the test results[6] [7]. In the ideal state, the wind speed in the bottom of the high-speed trailer was equal to the speed of the high-speed trailer, but when the high-speed trailer traveled quickly, the trailer chassis, the tank wall and the water surface limit the air flow and cause the blocking effect, resulting the increase of the wind speed of the air field in that area. Then the aerodynamic characteristics of the aircraft, such as the aerodynamic forces and moments of propellers, elevators and wings were changed, affecting the test result such as drag, heave and pitch of aircraft.

Therefore, this paper took a powered aircraft model as the research object to study the influence of blocking effect on its aerodynamic and hydrodynamic characteristics by experimental and numerical methods. The model towing test was carried out in the two test areas, in the front and bottom of the trailer. The numerical simulation mainly included two parts. Firstly, the wind field simulation of the front and bottom test areas of the trailer were carried out, and the extent of wind field blocking effect in the two areas was studied. And then the simulation when the model in infinite domain, front of and bottom of the trailer were carried out to study the effects of the blocking effect on the aerodynamic and hydrodynamic characteristics of the model.

2. Geometry and Gird

2.1 Geometry

We conducted the experiment and numerical simulation of the powered aircraft mode shown in figure 1. The angles of flaps in this model was 0 degree and the angle of elevators was 16 degree. The size, rotation direction and installation angle of the four six-blade propellers of the aircraft were all the same. When the towing test was carried out in the bottom and front of the trailer, there was no different in the size and other physical conditions except the position of the model. The layout of towing tank was shown in figure 2. And the simulation was carried out in the finite and infinite domain. Two conditions were employed in the finite domain simulation, which were the same as the experiment.

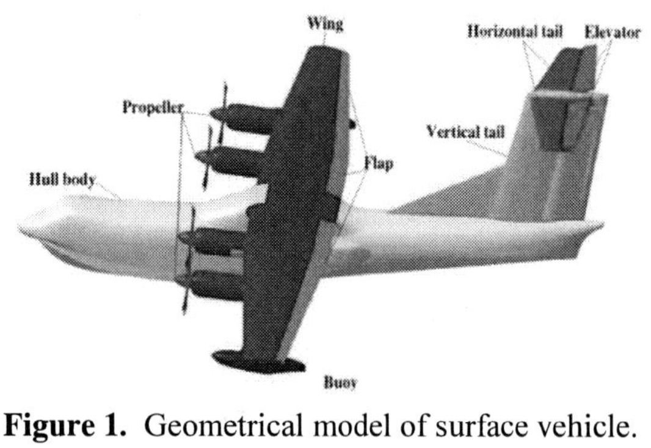

Figure 1. Geometrical model of surface vehicle.

Figure 2. The layout of towing tank.

2.2. Gird generation and boundary condition

The hex-unstructured mesh was adopted in this paper. The mesh in the area with serious flow separation or complex surface was refined by volumetric controls. The overset mesh was used to solve the problem of model motion. The surface mesh in the overset area was refined to capture near-wall flow accurately. The sildding mesh and the "interface" were adopted to achieve the rotation of the propeller and the data exchange between the aircraft and the propeller. The size of finite domain was the same as the towing tank, while the infinite domain were set as -5.0L to 1.5L in x direction, 0L to 2.5L in y direction and -2.0L to 2.0L in z direction.

Figure 3. Mesh distributions of aircraft in the bottom of trailer.

Figure 4. Mesh distributions of aircraft in the infinite domain.

Because the model and the trailer were both geometrically symmetric, the half model was adopted in the simulation, and the trailer model was simplified that only the trailer chassis was kept. The model and the trailer surface were defined as no-sliding wall boundary. In the infinite domain, inlet, top, side and bottom were set as velocity inlets and the outlet was set as pressure-outlet. The boundary conditions in the finite domain were the same as the finite domain except that the side and bottom should be set as wall. The grid distribution of finite domain and infinite domain was shown in figure 3 and figure 4. Since the model position was the only different of the two simulation in finite domain, so the grid distribution of the model in front of the trailer was set as the example to be shown in figure 3. In this paper, RANS method, SST $k-\omega$ turbulence model and two phase volume of fluid (VOF) technology were used in all simulations [8].

3. Experimental methods

In order to verify the blocking effect in the bottom of the trailer, the tests with the model in the front and bottom of the trailer were carried out respectively. It was a routine test to carry out the test in the bottom of trailer, which was shown in Figure 5. There was some description for the test with the model in the front of the trailer because it was the first time to do such a test. The device of the test that in front of the trailer shown in figure 6 mainly included the limiting device, the moving device, the former stretch device and the trailer. In order to extend the model position to the front of the trailer, the trailer should be consolidated with the former stretch device. There were a small pulley, a lifting rod and Barycentric connecting rod in the moving device, which was used to provide the aircraft with the DOF(degree of freedom),such as the translation in the heading direction , heave and pitch. In order to avoid the craft of the model and the tank wall, the limit device was used to limit the yaw movement of the model. The limit function was realized by the navigation bar.

Figure 5. The experimental device in the bottom of trailer.

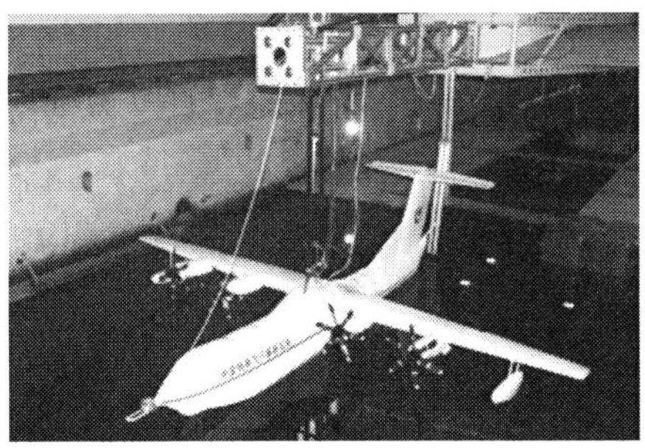

Figure 6. The experimental device in the front of trailer.

4. Results and Discussion

4.1. The experimental results analysis

Figure 7-10 showed the results of the model test. The relative Error was defined as Error(%)=(extension-bottom)/bottom*100%. It's important to note that Δ (displacement) and L (the length of aircraft) were constant value which was used to normalize the simulation results in this article. Because the research object belonged to the surface high-speed vehicle, the experimental and numerical simulation results under medium and high speed conditions were studied in this paper.

Figure 7-10 showed that the drag when the model was in the bottom of the trailer were significantly greater than that when the model was in the front of the trailer. As can be seen from figure 8, the test results of the model heave in the two conditions were basically consistent, with the error within 2%. As can be seen from figure 9, when the model was in the front of the trailer, the trim angle was larger than that when the model was in the bottom of the trailer, and the error increased with the speed. As can be seen from figure 10, the difference of heave amplitude under the two conditions changed stably with the increase of speed and fluctuated around 0. While the difference of drag and trim angle increased with the increase of speed. When the speed was greater than 10m/s, the tail of the model in the bottom of the trailer started to leave the water, and with the increase of the speed, the trim angle decreased. While the tail of the model in the front of the trailer still did not leave the water, and the trim angle gradually increased. The heave amplitudes of the model under the two test conditions were basically the same. Compared with the model in the front of the trailer, when the model was in the bottom of the trailer, the drag was larger but the trim angle was smaller. Therefore, the experimental results showed that the model test of powered aircraft in the bottom of the trailer was affected by blocking effect obviously. Considering that the simulation object was a powered aircraft model, the influence of blocking effect on the aerodynamic and hydrodynamic characteristics of the model need to be studied further.

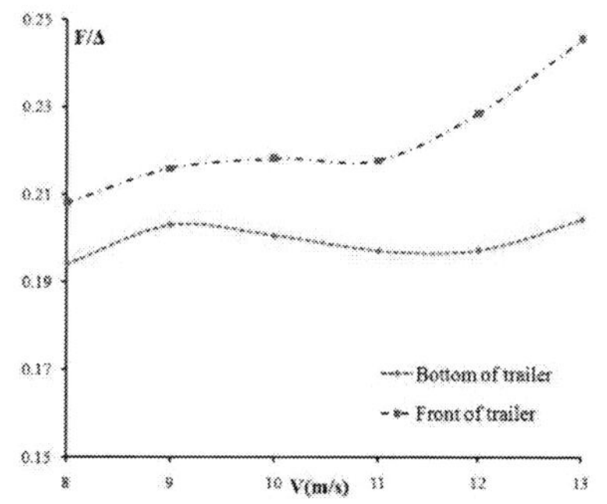

Figure 7. Experimental results of drag.

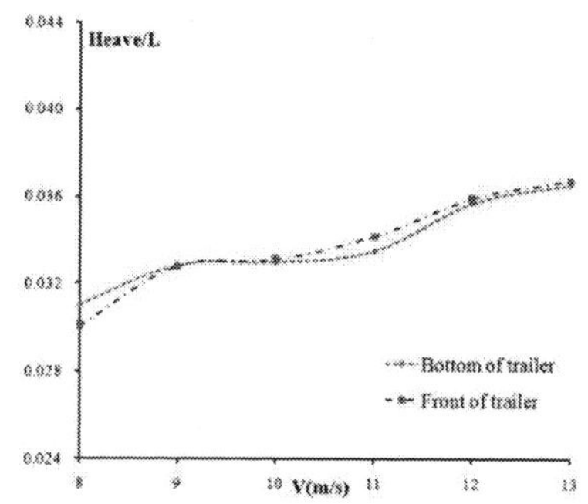

Figure 8. Experimental results of heave.

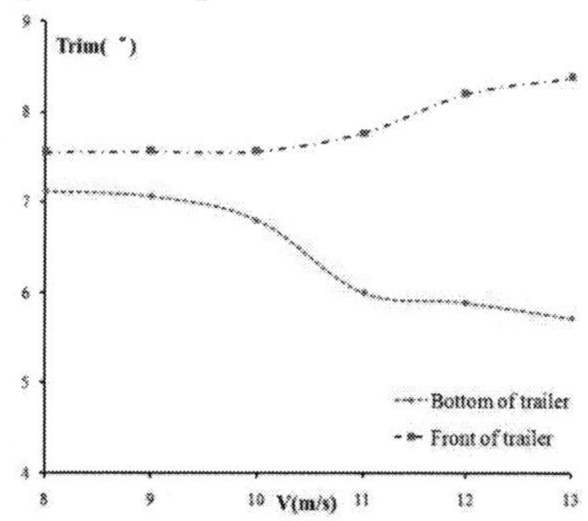

Figure 9. Experimental results of trim.

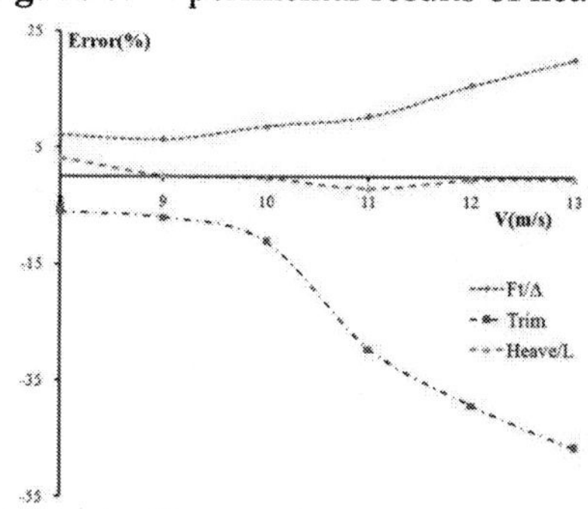

Figure 10. Errors of experimental results

4.2. The numerical simulation results analysis

4.2.1 Numerical method verification. Taking the working condition of the model in the bottom of the trailer as an example, the numerical method verification was carried out by comparing the test result with the simulation results. The comparison results of total drag, pitch angle and heave amplitude were shown in figure 11. The drag and attitude results obtained by experiment and simulation were in good agreement, indicating that the simulation method used in this paper was accurate and reasonable.

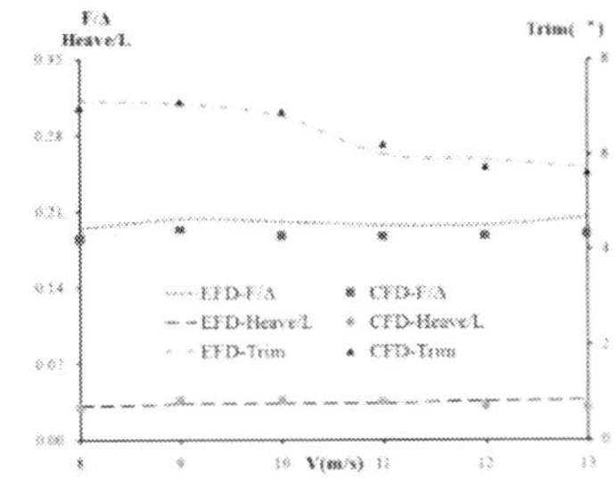

Figure 11. Comparing simulation and experiment

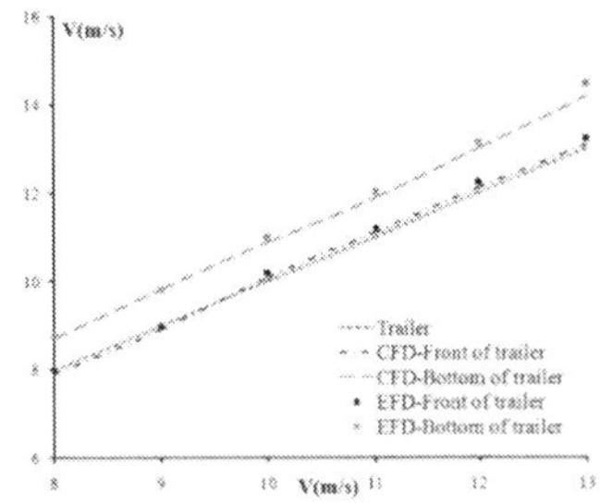

Figure 12. Computed and experimental results of wind speed.

4.2.2 The blocking effect of the wind in the bottom of trailer. In an ideal state, the wind speed in the bottom of the trailer was equal to the speed of the high speed trailer. However, due to the blocking effect, when the high speed trailer was fast moving in the pool, the wind speed in the bottom of the high speed trailer increased due to the existence of the bottom of the trailer, water surface and tank wall. To research the blocking effect of wind in the bottom of the trailer, the wind field simulation under two conditions was conducted with no model. The simulation result was shown in figure 12 with the test result. The pitot tube and ultrasonic sounder were used in the test. As can be seen from figure 12, the computed and experimental result were in good agreement with each other. The average wind speed in the bottom area of the trailer was 9.1% greater than the trailer speed. While the average wind speed in the front of the trailer was consistent with the trailer speed. The difference between them was less than 1.8%.

The wind direction of each measuring point was parallel to the incoming wind direction, and the error of angle was less than 2.0%, which was less than the error of measurement and equipment installation. It could be seen that the side wall and water surface did not affect the wind field. But the wind speed was increased. Therefore, the wind field in the bottom of the trailer had obvious blocking effect, while the wind field in the front of the trailer had no blocking effect.

4.2.3 the influence of blocking effect on the model. Figure 13 and figure 14 showed the simulation results of drag and attitude when the model was in the front, bottom of the trailer and infinite domain. The simulation result showed that the relative error of the drag, pitch angle and heave was small between the condition of infinite domain and the front of the trailer, and the overall error was within 3%. The average wind speed in the front of trailer area was the same as the speed of trailer, indicating that there was not blocking effect in the front of the trailer. When the simulation was carried out in the bottom and front of the trailer, there was no different in the size of model and other physical conditions except the position of the model. The relative error between the condition of infinite domain and the bottom of the trailer was obvious and consistent with the experimental result. Considering the average wind speed in the bottom of the trailer 9.1% larger than the trailer speed, we can conclude that the influence of the blocking effect in the bottom of the trailer was dominate when conducting a towing test.

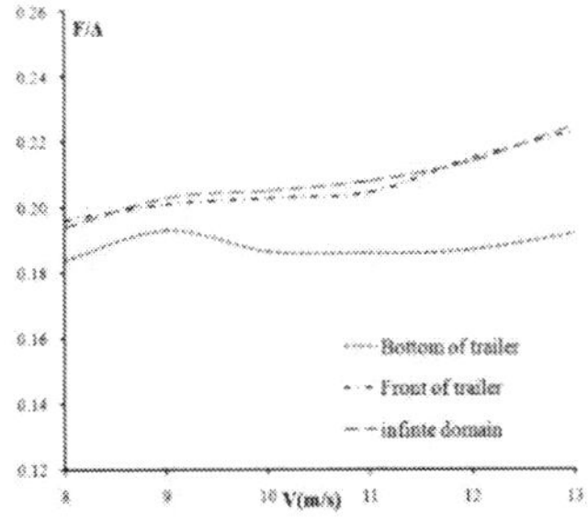

Figure 13. Computed drag at the three conditions.

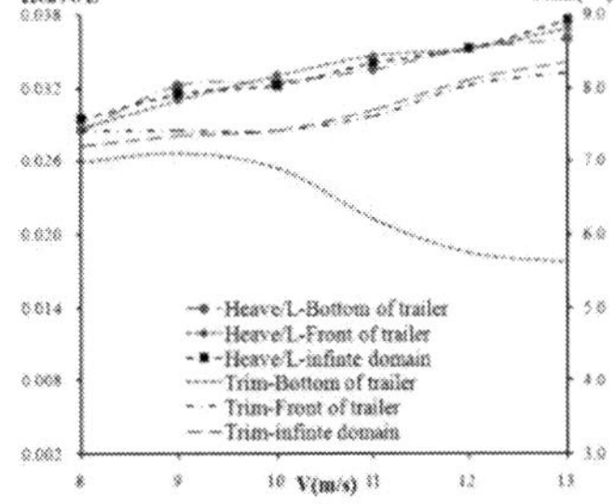

Figure 14. Computed attitudes at the three conditions.

IOP Conf. Series: Materials Science and Engineering **751** (2020) 012013 doi:10.1088/1757-899X/751/1/012013

To study the effect of blocking effect on the aerodynamic characteristics of the model appendage, the aerodynamic lift and longitudinal moments on the propeller, wing + flap and tail + elevator under the two calculated conditions were shown in figure 15 and figure 16. In these figure, lift was positive when z was greater than 0 and moment was positive when there was trim by bow. These figure showed that the aerodynamic lift and upward moment of flat tail + elevator and wings + flap in the bottom of the trailer were significantly greater than that in the front of the trailer due to the average in the bottom of the trail was 9.5% larger than that in the front of the trailer. Analyzed the simulation of propeller. On the one hand, the heave amplitude was very close under two condition. It explained by that the influence of the larger downward pull of propeller when it was in the bottom of trailer counteracted the influence of the greater wind speed. On the other hand, the downward moment when the propeller in the bottom of the trailer was greater than that in the front of the trailer and the difference with the increment of the downward moment was significantly greater than the increment of the upward moment of the flat tail + elevator and the wing + flap under the same condition. Therefore, the trim angle of the model in the bottom of the trailer was less than that in the front of the trailer. In summary, affected by the blocking effect of the wind in the bottom of the trailer, the lifting force in z direction of the powered aircraft model was similar and the downward moment was larger when the model was in the bottom of the trailer compared with that in front of the trailer, which explained that the heave amplitude of the model in two conditions was similar and the trim angle of model in the bottom of trailer was smaller. Thus the drag of the model in the bottom of trailer was smaller.

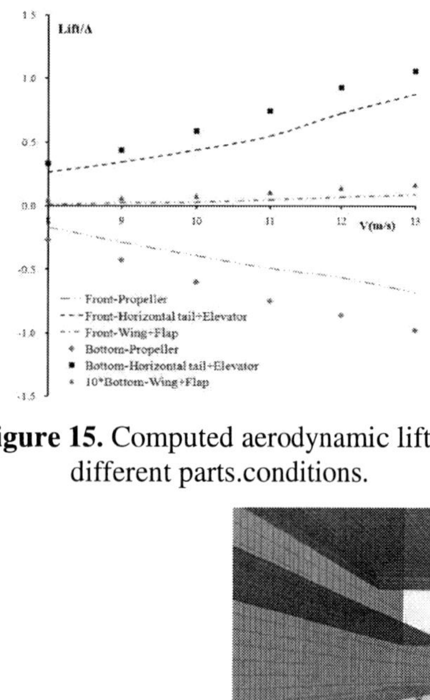

Figure 15. Computed aerodynamic lift of different parts.conditions.

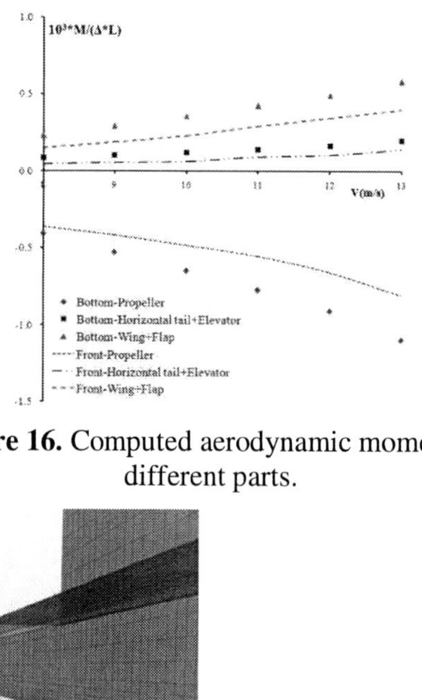

Figure 16. Computed aerodynamic moment of different parts.

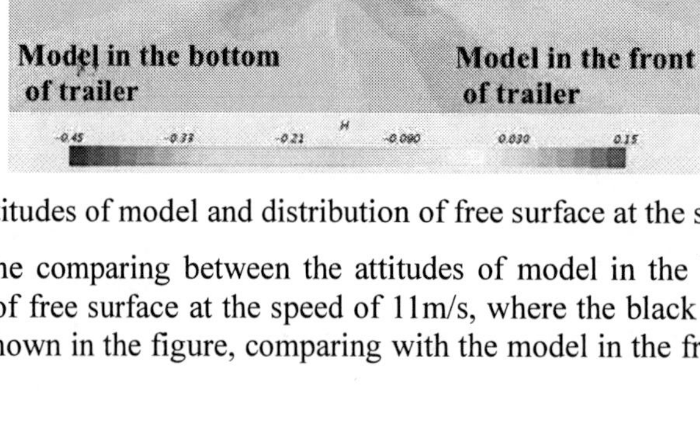

Figure 17. Attitudes of model and distribution of free surface at the speed of 11m/s.

Figure 17 showed the comparing between the attitudes of model in the bottom and front of the trailer and distribution of free surface at the speed of 11m/s, where the black grid area was the trailer and the pool wall. As shown in the figure, comparing with the model in the front of the trailer, model, the trim angle and height of the wave system around model was smaller. Wave energy was provided by the body, so the wave amplitude small means less drag. From the height of the fault terrace, the heave of model in the two conditions were similar.

5. Conclusion

In this paper, the effect of blocking effect on the aerodynamic characteristics of a powered aircraft model was studied by experiment and numerical simulation. The following conclusions can be drawn from the above work:

1) The results of the test in the front and bottom of the trailer show that the drag and the trim angle were smaller when the model was in the bottom of the trailer, but the heave amplitude under the two conditions was similar.

2) The speed of the wind field in the bottom of the trailer was 9.1% higher than that in the front of the trailer on average, indicating that there was obvious blocking effect. While the wind speed in the front part of the trailer was consistent with the theoretical value which meant that there was not blocking effect.

3) The computed results of the model in the infinite domain and the front of the trailer were similar indicating that the blocking effect of water on the model can be negligible. The total drag and the trim angle of the model were significantly different when the model was in the front and the bottom of the trailer respectively, indicating that the blocking effect of the wind on the aerodynamic characteristics of the model was obvious.

4) By analyzing the aerodynamic components of each control surface of the model, compared with the model in the front of the trailer, there were similar aerodynamic lift force but larger downward moment when model in the bottom of the trailer. Therefore the heave amplitude was similar, the trim angle decreased, which meant the windward area of the ship was smaller and then the drag decreased.

Reference

[1] Sheng Zhengbang. Principle of Ship[M]. Shanghai: Shanghai Jiao Tong University Press,2003.
[2] Zhou Guangli. Uncertainty analysis of ship model towing drag experiment [D]. Harbin: Harbin Engineering University,2002.
[3] Liu Yanwei. Numeircal simulation of the effect of blockage ratio on drag measurement in water tunnel test[J].Experimental technology and management,2007,24(12):44-47.
[4] GROSS A. On blockage correction [C]. Proceedings of the 13th ITTC. Hamburg & Berlin, Germany, 1972: 209-240.
[5] Li Guangnian. Design of towing tank[J]. Ship Building of China, 2011,52 (3): 109-114.
[6] Xie K Z 1978 The experiment study of blockage effect of the tank. *Journal of Shanghai ship and shipping research institute*, 1(2) ,1-27.
[7] Yang G. On coupling effect of two vortex systems of chinese aircraft with turbo-propellers[J]. Journal of Northwestern Polytechnical University, 1997,15 (4): 511-516.
[8] Zhang Lang. Coupled hydrodynamic and aerodynamic performance analysis of seaplane take-off process in calm water[J]. Science Technology and Engineering, 2018, 18(11): 190—195

IOP Conf. Series: Materials Science and Engineering **751** (2020) 012014 doi:10.1088/1757-899X/751/1/012014

Research and Design of the Control Law of Passive Side-Stick System with Human Factor Fault

Yao Ziyu*, **Yang Xizhong and Qiu Xuyi**

Key laboratory of avionics system integration technologies. Shanghai, 200233.

*Email:yzy4028@gmail.com

Abstract. For modern commercial civil aircraft, failure tolerant control (FTC) method is the basic strategy. Human factor fault is becoming more and more important for the design of the active control technology for the electrical flight control system or fly-by-wire system (EFCS or FBW). This paper focuses on the problem caused by the human factor with passive side-stick equipment. A new design concept was proposed based on the Boeing and Airbus experience and a control law of feedforward structure with alarm functions was designed in MATLAB\simulink. The functions were tested in a flight simulation platform in a desktop simulator based on the open-loop dynamic model. The results showed that each part of the control law worked well and the functions were effective. The human factor failure problem of dual inputs was well solved. The work provided a reference for the designer in the civil aviation field especially for those of the developing modern commercial aircrafts.

1. Introduction

The fly-by-wire (FBW) control system equipped with the passive side-stick [1] has been generally applied to the modern large commercial civil aircraft. Compared with the conventional controls, the passive side-stick FBW enjoys the advantages of light weight, wide vision, large space, comfortable feel, etc. However, because of the lack of back-drive mechanism, the loss of human-machine or PF\PM information interaction is very likely to occur. And it may cause serious danger especially in case of emergency, e.g. In 2011, July the 22nd, in NO 34 runway of Narita International Airport of Tokyo, an accident occurred because the fuselage of the Airbus A-380 civil aircraft which carried 168 passengers slanted to right, and it led to a friction between the wing and the ground; It was not a unique instance, but had its counterpart. In 2009, October 28th, the tail of an Airbus A-320 civil aircraft which carried 147 passengers was destroyed because of the accident of friction. The two accidents above both occurred in the Airbus A series civil aircrafts which used the FBW equipped with passive side-stick as the control system. And it was concluded by analysis that the human error control caused by the inherent defect of the passive side-stick control system was the main reason of the air crash [2]. So the research of the passive side-stick control law is important for the civil aircraft safety. It is highly significant study work for Chinese civil aviation development.

This paper is organized as follows: in section 2, the design concepts of Boeing and Airbus experience about the control system of human-machine and PF\PM relationship is discussed. The control law structure is discussed in section 3. The section 4 shows the detailed design of our control law and simulation system. The simulation results are shown in section 5. And the section 6 is the conclusion.

2. Airbus and Boeing experience

In order to do our own design, a survey was made to get information of the recent passive side-stick system development and a scientific analysis of the design concept was made.

2.1. Man\machine authority distribution

According to the Airbus civil aircraft design concept [3,4], there exists a 'hard limit'. When the pilots do some unconventional handling, the Primary Flight Computer (FPC) will interfere with the pilots' inputs, so that the aircraft can remain in a preset orbit. Especially in case of emergency, the preset control law will rule. In contrast, the Boeing design concept is a 'soft limit'. The 'soft' means just suggesting but not interfering. In every case, the pilots of Boeing civil aircraft can hold the highest authority and the FPC provides the warning information by the cockpit (e.g. Flight Control Display and Primary Flight Display, PCD and PFD), the stick force gradient or something else. Sometimes, the plane can fly across the flight envelope.

The two concepts are both aimed at safety. The reason of the difference in concept is due to the difference of their cockpit control device, which is the difference between the passive side-stick of Airbus and the central stick with mechanical connections of Boeing. For the pilots, they certainly do not want their authority to be limited, but the lack of feedback information of the passive side-stick is very likely to let the pilots make mistakes, so that the protection measures and control laws such as pitch attitude protection, high angel-of-attack (AOA) protection and high speed protection are necessary. According to the Boeing design concept, there exists an override button, whose function is to get the highest control authority. For our design, to combine the advantages of the two design concepts was the basic idea. The key was the design balanced point between the override control and the effective protection. Furthermore it helped to deal with the man\machine authority distribution.

2.2. PF\PM authority distribution

PF and PM are the two pilots of the civil aircraft. It is easy to realize the information interaction by the mechanical linkages of the Boeing aircrafts [5]. But for the passive side-stick system, the PF or PM sends the command to the flight control computer independently. It is very likely to occur that the PF and PM handled the side-stick at the same time which leads to an error dual input and that is the human factor fault. The designers of Airbus divide the dual handling inputs into three types, 'unintentional' 'intentional' and 'instinctive' [6,7]. Unintentional dual inputs always occur during the cruising flight, it is caused by the PM's touch without intention. The signal is always short in time and of small amplitude. The intentional dual inputs are always caused by the PM too. If the PM feels uncomfortable during the flight, he will handle the side-stick to adjust to suit the passengers. Owing to the handling of the PF, although the dual inputs last a long time, it will not lead to serious consequences. The instinctive dual inputs always occur in the take-off and the load phases. Especially in case of emergency, the PF and PM probably handle the side-stick fast with large amplitude. It is always the main reason of the friction accident.

In order to solve the problem of the passive side-stick to prevent the accident, the PF/PM authority must be well distributed. For our design of this part, the experience of Airbus is used and an advanced and reasonable design was made.

3. The structure of control law

There are two forms of the control law, feedback and feedforward. The feedback control laws are always based on the deviation adjustment. The feedback control laws always have time lag. In case of emergency, the lag is deathful. And the feedback control laws have closed control loop, so it relates to a problem of stabilization. In contrast, the feedforward control laws have the advantages of open loop stabilization, advanced control, less deviation, etc. In our designed simulation system, the control variable is measurable and the flight dynamics model is clear. It meets the requirement of the feedforward control design. So the feedforward structure was chosen in our design. And in the control close-loop, the control law is the feedforward part of the command channel.

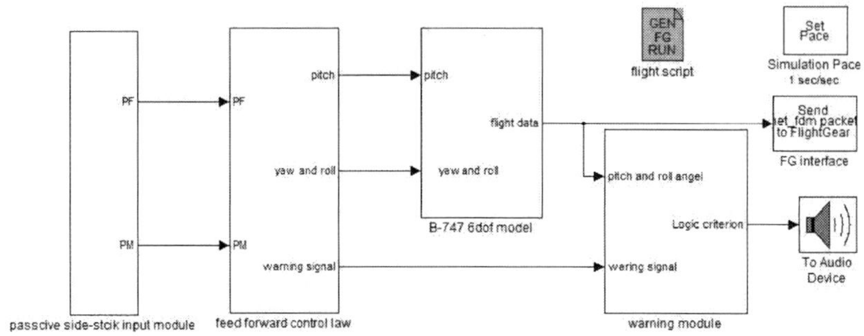

Figure 1. Synopsis of the software program

4. Detailed design of the control law and the simulation system

In order to solve the problem well, the control law provided a lot of functions [8,9]. And the function helped deal with the man\machine and the PF\PM authority distribution to prevent the accident. The modularization designed control law had clear process and well defined functions [3]. In the design, the reasonable hardware and software were chosen to build a full simulation system. And the functions could be tested by the simulation system which is shown in figure 1. The process is consistent with the airbus golden rules [6,7].

4.1. The hardware input module

The function of this module was to collect the signal from the hardware and to send the digital signal to the after module [3]. Saitek X52 pro Joystick was chosen as the input hardware. The Saitek X52 pro Joystick is a kind of displacement joystick. People could handle the stick easily and the feeling was much closed to the passive side-stick [10,11]. The signal from the joystick was sent through the USB interface. The USB transmission speed is very quick and exact. And it sent the signal to the program development software—matlab\simulink. The joystick input module of Aerospace Blockset tools was used to complete the process. The signal consists of two parts, the three axis signal and the button signal. The function of passive side-stick hardware input was realized.

4.2. The feedforward control law

This module is the core of the whole control law. This part received the handling signal from the PF and PM as shown in figure 1. After the actuator model control, the module sent the processed data to the flight dynamic model and provided the warning signal to the warning module. In order to achieve the goal of finding the balance point between the Boeing and Airbus design concept, the control law we designed contained the functions below: 1) The short in time and small in amplitude signals will not be warned; 2) The authority of a single pilot will not be limited.; 3) The PF and PM can switch the authority through the priority logic module easily; 4) When the override button is pressed, the pilots will have the highest control authority. The structure is shown in figure 2. It is simple and clear and consistent with the FTC strategy trend of simplification.

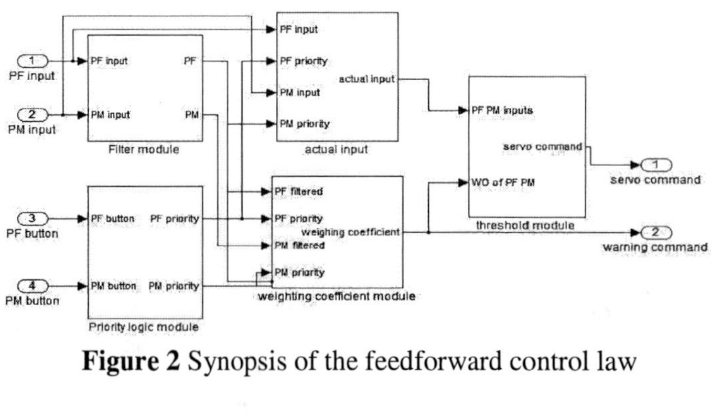

Figure 2 Synopsis of the feedforward control law

4.2.1. The filter module. For the unintentional and intentional dual inputs, the Airbus set an initial stick force or a stick displacement-signal graph shown in Figure 3 in the passive side-stick system[1]. Although it could solve the two dual inputs situations, actually it limited the pilots' authority and the handling sensitivity and quality was not good.

The function of our filter module is to erase the small signals caused by the unintentional touch to prevent frequent alarm and send the result to the warning module. According to our design, the signals were not limited in this part, and criterion was provided to decide whether the warning module worked. After the filter, if the signal was zero, it meant it need not be warned. So the time and amplitude threshold in this part was important and according to the Airbus Flight system specification, if the value was less than 2degree and 500ms, the signal would not impact the flight much. And our initial filter threshold value was set 2degree and 500ms.

4.2.2. The threshold module. The design of the Airbus passive side-stick did not contain this module. So that the inputs sometimes could be so large that the transient response of the aircraft caused the accidents. And that is the reason why the module was set before the servo part. The threshold itself is a kind of limit, however, the dual inputs are wrong controls and the limits are reasonable. And in order to guarantee a single pilot's authority, the threshold value set initially was the max value of one side-stick. The value of Saitek X52 pro joystick is from -1 to 1 for the roll, yaw and pitch axis, 0 to 1 for the throttle. So in the design, the threshold values were set based on the hardware data. As the threshold module is also applied in the servo-loop, the servo-loop threshold value could be set double value of this module for the override function described in 4.2.5.

4.2.3. The passive side-stick priority logic module. According to the Airbus control, the take-over button could help the PF and PM switch the priority. When a pilot kept pressing the take-over button, the pilot had the input priority and the other pilot's input signal was shielded. After 40 seconds continuously pressing, the pilot would keep the priority. There were two disadvantages of the Airbus design. Long time keeping pressing made the pilots feel tired and in addition, once the priority was locked, it took time to switch the priority, and in case of emergency, the design was quite unfriendly.

In our design, the button actions are all short pressing. So the switching process is quick and convenient. Two buttons were set called 'take-over button' and 'lock button' on the joystick. The function of 'Take-over button' is to switch the handling priority in normal flight time. And the function of 'lock button' is to lock the priority to deal with the emergencies. The digital input of the button is 0, 1 for 'take-over button' and 0, 2 for 'lock button'. The priority level of 'lock button' is higher than the 'take-over button'. That means when the priority was locked, the 'take-over button' would lose its function. In order to achieve the button function of take-over and lock, the design of the whole logic was completed using some Simulink blocks. Besides, when this module worked, any dual inputs could be avoided. The output of the passive side-stick priority logic module was the Boolean number 0 and 1 which showed the priority of the PF and PM, 1 for yes and 0 for no. The module made the PF and PM switch the priority quickly, conveniently and comfortably, it helped deal with the PF\PM authority distribution.

4.2.4. The weighting coefficient module. This module worked when the 'instinctive' dual inputs occurred. If the PF and PM handled the passive side-stick at the same time and the signal was not zero after the filter, meanwhile the priority outputs of PF and PM were both 1, the module would set a weighting coefficient and multiply the signal by it. The weighting coefficient could help reduce the impact of the dual inputs and it made the pilots' responsibility and authority clear in case of dual inputs. In our design, the initial number was. That meant when the dual inputs happened only the inputs of PF was effective and the PF would take the full responsibility. The coefficient value would be set by the PF and PM before the flight. In order to avoid large inputs, it is suggested that, so that the responsibility and authority were clear and the PF/PM authority distribution was well dealt with.

4.2.5. The override button module. The design concept of override button came from the Boeing aircraft. The function of the override button was to provide the highest control authority to the pilots. In our design, when the override button was pressing, the two pilots' inputs will be sent to the dynamic model directly without any limit and dual inputs were allowed. The input amplitude value was 2 times of the original, and the pilots could give full play to the flight performance.

In our design, the function of the override button was merely to disconnect the feedforward control law, so we provided enough data interface for the further study to improve the function of this module.

4.3. 6 DOF dynamic model. We chose the B-747 parameters to build the 6 DOF dynamic model. B-747 is a classical commercial civil aircraft. The aerodynamic and geometry parameters have been published. In the design, we made some assumptions to simplify the equations: the influence of the earth rotation was ignored; the plane was seen as a rigid body; the change of the plane was ignored during the process.

According to the survey of the passive side-stick, the side-sticks were always handled in the take-off phase and the landing phase. So we chose the aerodynamic data in such phase. And then we simplified the 12 dynamic equations and got the small perturbation linear equation set. The form of the equation set is as follows:

$$\dot{X} = AX + BU \tag{1}$$

According to the initial conditions, the linear equation set was divided into 2 equation sets, which we called longitudinal dynamic model and lateral dynamic model. The form of the two models is as follows:

$$\dot{X}_\theta = A_\theta X_\theta + B_\theta \delta_e \tag{2}$$

$$\dot{X}_{\psi\phi} = A_{\psi\phi} X_{\psi\phi} + B_{\psi\phi} \begin{bmatrix} \delta_a \\ \delta_r \end{bmatrix} \tag{3}$$

$[\delta_e, \delta_a, \delta_r]$ is the deflection of the elevator, aileron and the rudder. We call it controlled variable. $X_\theta = [V, \alpha, q, \theta, z, x]^T$,$X_\psi = [\beta, p, r, \phi, \psi, y]^T$ they are the 12 quantities of state. $[x, y, z]$ stands for the displacement; $[\psi, \theta, \phi]$ stands for the 3 attitude angle; $[p, q, r]$ stands for the 3 attitude rate; $[\alpha, \beta, V]$ is the angle of attack, angle of side slip and air speed. The matrix A and B are the state space matrix and input matrix. The numerical value of A and B were from the B-747 sea level aerodynamic data.

The dynamic model module received the signal of the deflection of the control surface from the joystick signal. Then the 12 linear equations were solved and we got the 12 quantities of state. The process is shown in figure 1. The output of this module was the flight quantities of state, and they were sent to the visual flight module and the warning module.

4.4. Warning module

The function of warning module is to provide the alarm information. There were 3 situations we thought should be warned.

4.4.1. Dual inputs warning. The criterion of the dual inputs came from the priority logic module and filter module. If none of the four output values were zero, we provided the warning signal. In this part, a logical AND module was set to achieve the goal.

4.4.2. Tail warning. The criterion whether the warning signal was provided was based on the result of calculation. In the body coordinate frame, we assume the tail coordinate is $\xi_t = [-\delta_t 0,0]$. The transition matrix from the body coordinate frame to the earth-surface inertial reference frame is L:

$$L=\begin{bmatrix}\cos\theta\cos\psi & \sin\theta\sin\phi\cos\psi-\cos\phi\sin\psi & \sin\theta\cos\phi\cos\psi+\sin\phi\sin\psi\\ \cos\theta\sin\psi & \sin\theta\sin\phi\sin\psi+\cos\phi\cos\psi & \sin\theta\cos\phi\sin\psi-\sin\phi\cos\psi\\ -\sin\theta & \sin\phi\cos\theta & \cos\phi\cos\theta\end{bmatrix} \quad (4)$$

So in the earth-surface inertial reference frame we have:

$$(\xi_t)_g=\begin{bmatrix}x\\ y\\ z\end{bmatrix}_g+L(\xi_t)_b \quad (5)$$

We get:

$$\begin{bmatrix}0\\ 0\\ z_g\end{bmatrix}=\begin{bmatrix}x\\ y\\ z\end{bmatrix}+L\begin{bmatrix}-\xi_x\\ 0\\ 0\end{bmatrix} \quad (6)$$

And in the two frames we have $h=-z$ and $h_g=-z_g$ where h is the height.

After simplifying, we get:

$$h=h_{\xi_t}+\sin\theta\xi_x \quad (7)$$

Equation (7) shows the relationship between the center of mass height and the tail height. If $h_{\xi_t}<0$, then the accident would take place. We fitted a curve with cubic function based on the equation (7) to find the safety limits as shown in figure 4.

$$\theta=0.0010h^3-0.0367h^2+1.8902h-0.7063 \quad (8)$$

4.4.3. Wingtip warning

The wingtip warning criterion had the similar solution procedure with the tail warning. In this paper, the height result is as follows:

$$h=h_{\xi_r}+\sin\theta\xi_x-\left|\cos\theta\sin\phi\xi_y\right| \quad (9)$$

Where the wingtip coordinate is $\xi\_l=(-\xi\_x,-\xi\_y,0)$ and $\xi\_r=(-\xi\_x,\xi\_y,0.)$.

Finally, the warning signals were sent to an interface to the computer sound card and the visual software. So the pilots could get the visual and auditory warning information and the safety of the flight would improve.

4.5. Visual flight module

The visual flight module was designed to turn the simulation result into animation [12,13]. And it provided the direct visual feelings. The free open source software FlightGear was chosen to provide the visual window. The FlightGear had the global terrain database and a variety of aircraft picture. Its internal data interface made it easy to connect to the matlab\Simulink.

The visual flight module consists of four Simulink modules. The four modules were all from the aerospace block set tool. When the simulation experiment was made, the simulation system first got the input signal from PF and PM through the passive side-stick interface module. Then the feedforward control law module process the signal and provided the deflection of each control surface and the priority signal of the PF and PM. The 6 DOF dynamic model sent the flight quantities of state to the visual software. The warning module sent the alarm information to the computer sound card. At last, the PF and PM could receive the visual information from the computer window and sound information from the sound card. And they make the next decision. The flight data were saved to the workspace and the data curve could be seen at any time. Figure 5 shows the pilots' first-person spatial perspective.

IOP Conf. Series: Materials Science and Engineering **751** (2020) 012014 doi:10.1088/1757-899X/751/1/012014

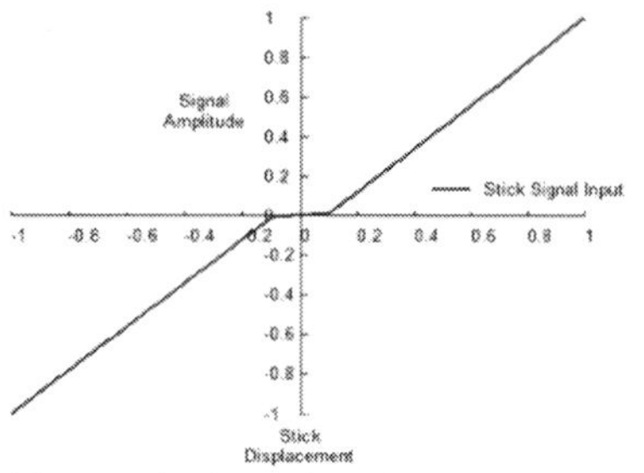

Figure 3 Side stick displacement

Figure 4 Safety Limits

Figure 5 Flight Simulation Station

5. Simulation Results

In order to test and verify the functions of each module accurately, the inputs of each module were from the Simulink source modules instead of the passive side-stick hardware. However, some results were appeared as the sound and animation, e.g. the warning module and the visual flight simulation module and the function of the override button module is to disconnect the control law. The simulation results of them are difficult to show in this paper. Actually, those modules had been all tested in the desktop simulator and the functions were all effective.

5.1. filter module

The signal with perturbations was sent to the filter module in the test and the result is shown in figure 6. The curve shows that the filter module can filter the small-amplitude and short-time signals and prevent the frequent alarm. The result was the same as expected.

5.2. weighting coefficient module

The test signals of this part are sinusoidal signals of different initial conditions. Figure 7 and figure 8 show the two results of the weighting parameter module. Figure 7 shows that the PF and PM handle the passive side-stick at the same time in the same direction. Figure 8 shows the situation that the PM's operation is wrong and just opposite the PM. The results show that the module has good correction capability. And without side-stick logic, the PM would always have the priority and be responsibility for the flight process. It was quite recommended that the weighting coefficient of the pilot who has more experience is set 1. So the human factor error could be reduced greatly.

5.3. threshold module

Figure 9 shows the result of the threshold module. Although, the output of the threshold module does not quite tally with the expected, it increased the input amplitude. The expected function of the module was to keep the input amplitude from being too large, so the module achieved the goal.

5.4. priority logic module

Figure 10 is the button inputs of the PF and PM. Figure 11 is the priority outputs of them. In order to test the 'take over' and 'lock' function of the module, a special input signal is set as shown in figure 10. During the first two seconds, the inputs of the button were both zero, so the priority of the PF and PM were both 1. From the 2nd second to the fifth second, the inputs were only 1 and 0, and the figure 11 shows the 'take-over' priority switch function. From the fifth second to the tenth second, the inputs were 2 and 0, and the figure 11 shows the 'lock' priority switch function. We observed that during the ninth second, the input signal of PM is 1, but it didn't work. So the function of 'take-over' and 'lock' was effective. Actually, there are 63 possible situations in this logic module, when figure 10 and 11 are typical results. It had been confirmed that every tested result could meet the function requirement.

5.5. The result of the angle of pitch

If a command signal is send to the dynamic model to change the parameter, the longitudinal states will change. Figure 12 and figure 13 show the results of the angle of pitch with and without control by

IOP Conf. Series: Materials Science and Engineering **751** (2020) 012014 doi:10.1088/1757-899X/751/1/012014

some dual inputs. The value reduces from 58.7 degree to 25.4 degree in weighting coefficient control and from 58 .7 degree to 26.1 degree duo to the threshold. The results indicated that the feedforward control law was effective. The two modules both reduced the open loop responses and the flight safety and flight quality were improved. Actually, if no dual inputs happened, there was no difference between the two result curves and it met the requirement in section 4.2. Consequently, the control law solved the problem of dual inputs.

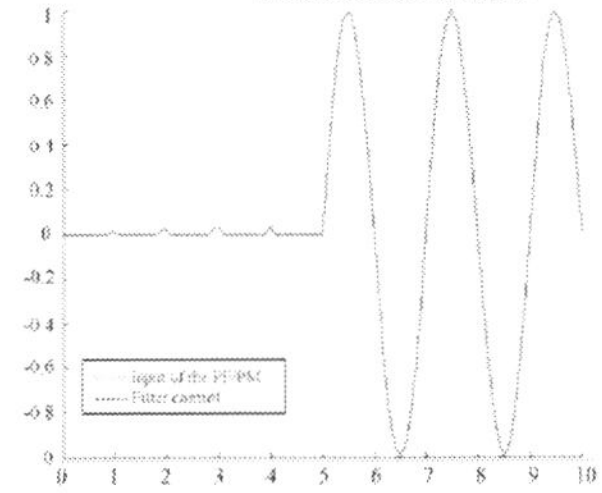

Figure 6 Filter Simulation Result

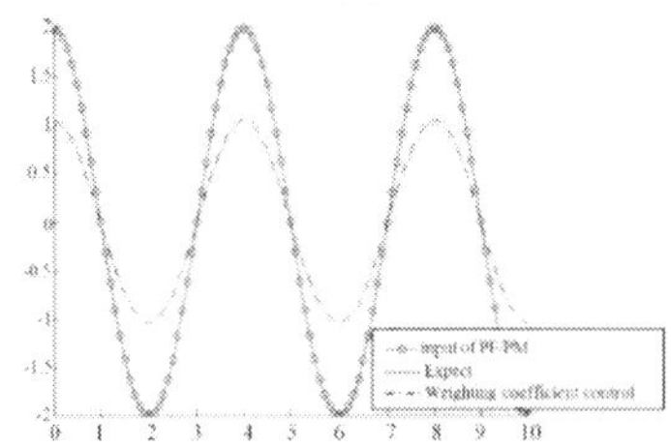

Figure 7 the Same Direction

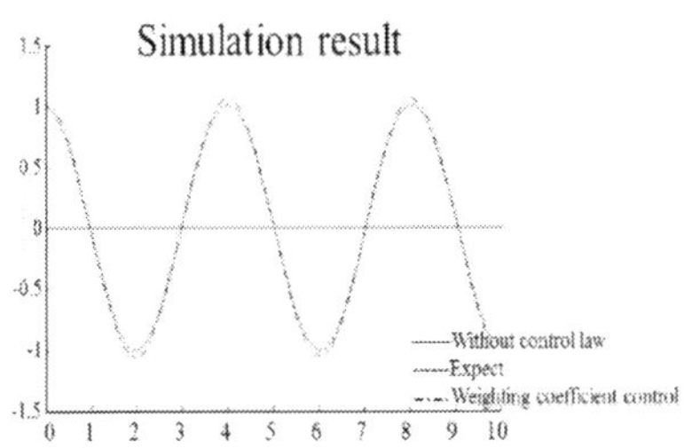

Figure 8 Opposite Direction

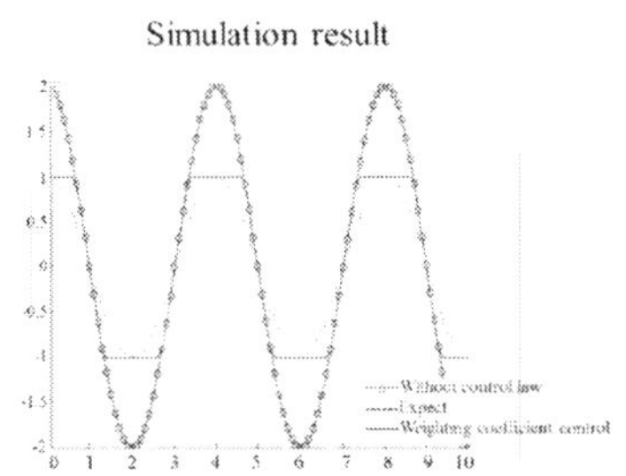

Figure 9 Threshold Result

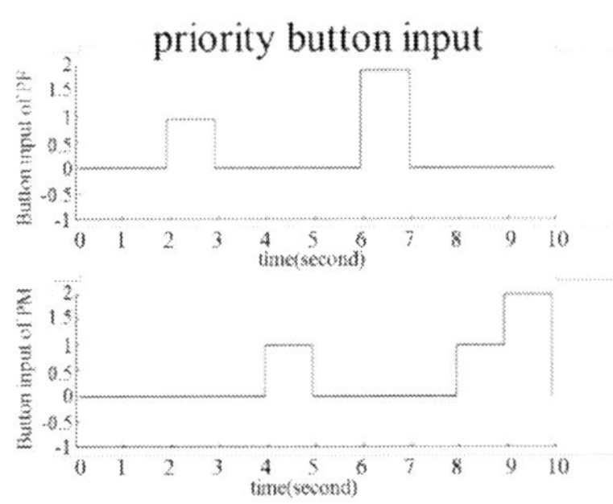

Figure 10 Priority of PF and PM

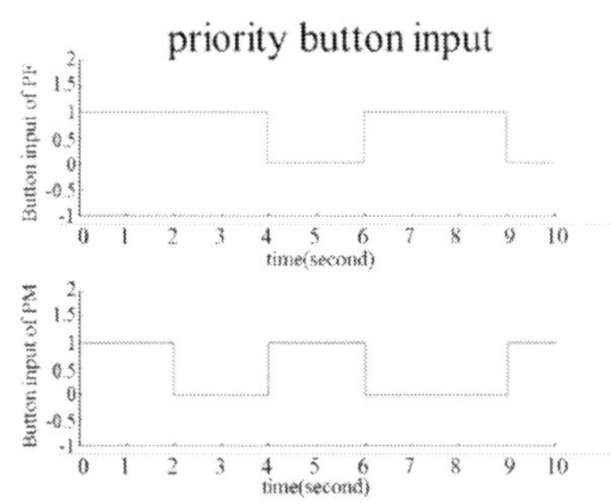

Figure 11 Logic button inputs of PF and PM

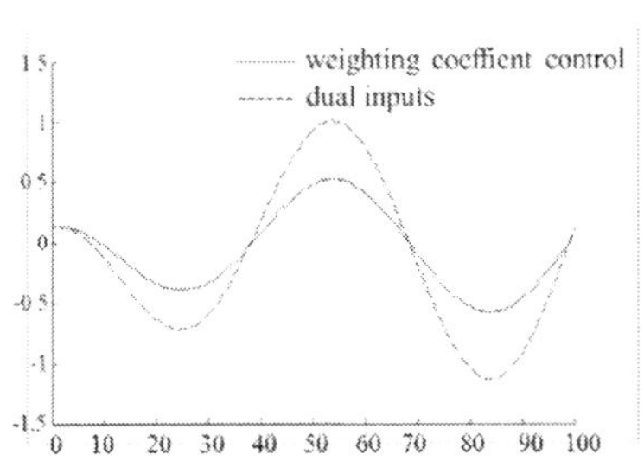

Figure 12 PM=1 and PF=0

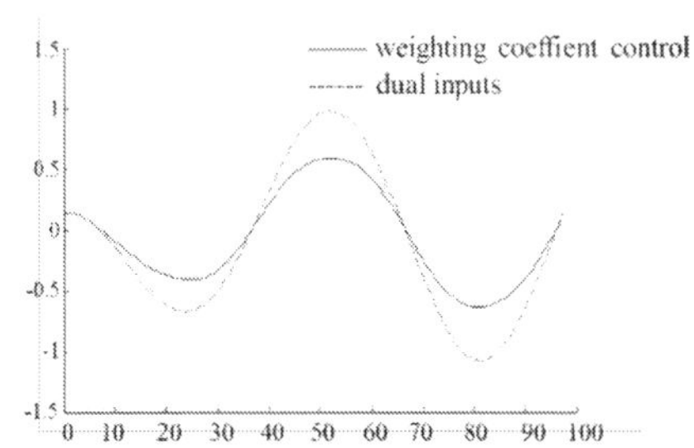

Figure 13 Threshold value=1

IOP Conf. Series: Materials Science and Engineering **751** (2020) 012014 doi:10.1088/1757-899X/751/1/012014

6. Conclusion

This paper studied the passive side-stick control system in the civil aircraft and found the disadvantage and design defect with human factor of it. In order to solve the problem, a new design concept was put forward for the passive side-stick control system from the Airbus and Boeing's design. We improved the design of the Airbus and combined the advantage of the Boeing design and the possible human factors were considered in the function design of the control law with feedforward structure. The visual flight simulation system could help test and verify the functions of the control law and also it was a good platform for the aerospace researchers. The simulation results showed that the functions of each module were effective, and the design could help reduce the human factor error operations. So it would surely reduce the number of the air crashes. The flight safety and quality were both improved especially in the Load\Take-Off phase.

The research and design have great significance for the aircraft safety. Above all, the design is beneficial to the development of our country's aviation industry especially the design development of the commercial large civil aircraft.

Acknowledge

I owe great gratitude to all the partners in our research group for their valuable ideas. Thank the National Program on Key Basic Research Project (973 program: NO.2010CB734102) for funding the research.

References

[1] "A319/A320/A321 Flight deck and systems briefing for pilots" published by Airbus, Sep. 1998.

[2] Li Lin, Human Factor in the Passive Side-stick Design for Civil Airplanes [J]. Civil Aircraft Design and Research, 2012(2):67-69.

[3] Favre C. Fly-by-wire for commercial aircraft: the Airbus experience [J]. International Journal of Control, 1994, 59(1): 139-157.

[4] Lombaerts T J J, Chu Q P, Mulder J A, et al. Modular flight control reconfiguration design and simulation[J]. Control Engineering Practice, 2011, 19(6): 540-554.

[5] Dietrich H, Christian H. Active side stick technology – a means for improving situational awareness [J]. Aerospace Science and Technology, 1999(3):525 – 532.

[6] Goupil P. AIRBUS state of the art and practices on FDI and FTC in flight control system [J]. Control Engineering Practice, 2011, 19(6): 524-539.

[7] Goupil P. Oscillatory failure case detection in the A380 electrical flight control system by analytical redundancy [J]. Control Engineering Practice, 2010, 18(9): 1110-1119.

[8] Wiegmann D A, Shappell S A. A human error approach to aviation accident analysis: The human factors analysis and classification system[M]. Ashgate Publishing, 2012.

[9] Gopal P. Analysis of factors leading to pilot error accidents in civil aviation [J].

[10] Fielding C. The design of fly-by-wire flight control systems [J]. Aeronautical Journal, 2001, 105(1051): 543-549.

[11] Traverse P, Lacaze I, Souyris J. Airbus fly-by-wire: A total approach to dependability [M]//Building the Information Society. Springer US, 2004: 191-212.

[12] Côté M È. Design characteristics to reduce inadvertent cross-axis coupling during side stick handling of aircraft pitch and roll axis control [D]. École Polytechnique de Montréal, 2012.

[13] HUANG H, XU Y, DENG Z. Real—time visual flight simulation system based on FlightGear simulator [J]. Journal of System Simulation, 2007, 19(19): 72-74.

IOP Conf. Series: Materials Science and Engineering **751** (2020) 012015 doi:10.1088/1757-899X/751/1/012015

Experimental Analysis on Combined Vibration and Acoustics Test of a Satellite

Yan Tingfei[1,*], Gao Haiyang[1] and Xiang Shuhong[1]

[1]Beijing Institute of Spacecraft Environment Engineering, Beijing100094, China

[*] Corresponding author: 1951294335@qq.com

Abstract: Random vibration tests and acoustics tests are typical environmental testing approaches for verifying the dynamic environmental adaptability of a satellite, which should be selectively carried out on the ground before launch. Existing test criteria suggests that large spacecraft should be excited in a reverberation chamber, while a shaker should be used for a small spacecraft. With the development of experimental technique of environment simulation, the combined vibration and acoustics (vibro-acoustic) test are increasingly investigated for reproducing a more realistic environment and avoiding an over test or an under test. However, few studies have been published on coupling effect of combined vibration and acoustic test of a full scale prototype satellite. This paper discusses the equivalence between random vibration and acoustics excitation for satellite ground testing. The experiment results have shown advantages of the combined test in reproducing flight environment. And the coupling effect and distributions of high frequency response attenuation during the combined vertical vibro-acoustic test is observed and analyzed in detail.

1. Introduction

As we all know that the realistic launch environment should be reconstructed in ground experimental verifications of spacecraft. In fact, satellites would be generally subjected to vibration and acoustic loads simultaneously during launch and ascent.

In the early stage of space development, satellites were designed to be small and compact. Single random vibration test was known to be adequate. With the development of space technology, more spacecraft with a large surface area, such as solar wings or antenna arrays, are increasingly designed for long-time tasks. To investigate the equivalence between random vibration tests and acoustic tests on an assembled satellite, some assessment methods[1-6] based on experimental results and simulations have been proposed. In the recent years, the combined vibro-acoustic testing technique has been drawing attentions in dynamic environment simulation. The combined vibro-acoustic tests for subsystems or even an assembled system were performed [7-9].

It is known that the dynamic environment testing level of a satellite is generally determined by the envelope lines of spacecraft-vehicle coupling analysis in a separate environment. However, this approach ignores the influence of the combined vibro-acoustic coupling effect in reality. Frendi and Robinson[10] showed that the coupling effect is important for plate response predication at high excitation levels. And they also indicated that the acoustic coupling may attenuate plate response at high frequency bandwidth. Consequently, it seems to be inaccurate to deduce the component testing levels based on complex environment decomposition, which may lead to an overestimation or an underestimation, and even result in an inappropriate strength and stiffness design.

In this paper, a combined vertical vibro-acoustic test for a prototype satellite is executed.

2. Combined vibro-acoustic test of a prototype satellite

2.1. Equipment and tested satellite

A full scale, dynamically equivalent, prototype satellite is designed for vibration and acoustic qualification test before the launching the satellite. The size of the prototype satellite in Z-direction is about 2560mm, and the mass is about 990kg.

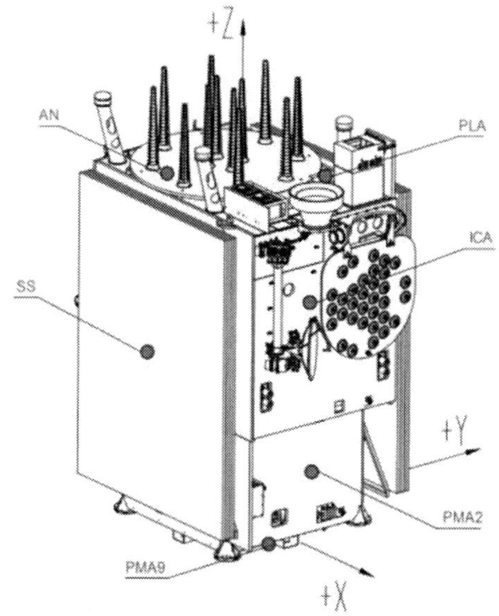

Figure 1 Coordinates system of the prototype satellite and sensor placement

The combined vibro-acoustic test system consists of a reverberation chamber subsystem and an electromagnetic shaker subsystem, in which each subsystem has a controller to achieve the reference test level. During the tests, the satellite was mounted on a special fixture by using four M12 screws. There are five acoustic modulators in reverberation chamber with the ability to provide 156dB acoustic pressure level. To acquire the acceleration responses on the satellite, a total of thirty-eight three-directional acceleration sensors are mounted on the structure, spreading over the propelling module, instrument capsule, payload capsule, solar wings and antenna of the tested satellite. Some representative measurement points, as shown in Fig.1.

2.2. Control spectrum

The loading levels of random vibration test and acoustic test for vertical direction are taken from China Long March series launching condition. In consideration of subsequent use, half-acceptance level tests were adopted in this work. Table 1 and Table 2 show the control spectrums of the separate random vibration and acoustics test. In the combined vibro-acoustic test, these two kinds of loads are simultaneously acting on the prototype satellite.

Table 1 The control acceleration spectrum of half-acceptance level random test in vertical direction

Frequency range (Hz)	20~200	200~800	800~1000	1000~2000	Total RMS
Magnitude	+6dB/Oct	0.01g2/Hz	-6dB/Oct	-6dB/Oct	3.38grms
Tolerance (dB)		±1.5		±3.0	±1.0

Table 2 The control sound spectrum of half-acceptance level acoustic test

Central frequencies (Hz)	31.5	63	125	250	500	1000	2000	4000	8000	OSPL
Magnitude (dB)	121	126	131	135	130	126	125	124	119	138.5
Tolerance (dB)				-2~+4				-5~+4	-5~+4	-1~+3

2.3. Loading and data acquiring method

The loading method of the combined vibro-acoustic test is similar with single vibration or acoustic test. In this work, four vibration accelerators are mounted on the interface between the satellite and the fixture for average peak value control strategy. For the acoustic test, four sound sensors were symmetrically located about 0.5m away from the satellite with a height of 1200mm which is close to satellite COG.

It should be noted that reverberation would influence the random vibration self-check before loop control start. Therefore, random vibration self-check should be carried out firstly and up to -12dB of the full vibration level, then wait for the acoustic control system till it starting up to -12dB of the full acoustic level, and then random vibration and acoustic excitation would be acting simultaneously to full test level (0dB).

Acceleration responses of random vibration and acoustic tests are acquired in steady state by using statistical average method. According to empirical knowledge, the response acquisition approach of random vibration test is essentially identical to that of acoustic test. The only difference lies in the frequency ranges. To conveniently compare the responses, the same frequency range in the interval of 20Hz to 2000Hz is shown and discussed.

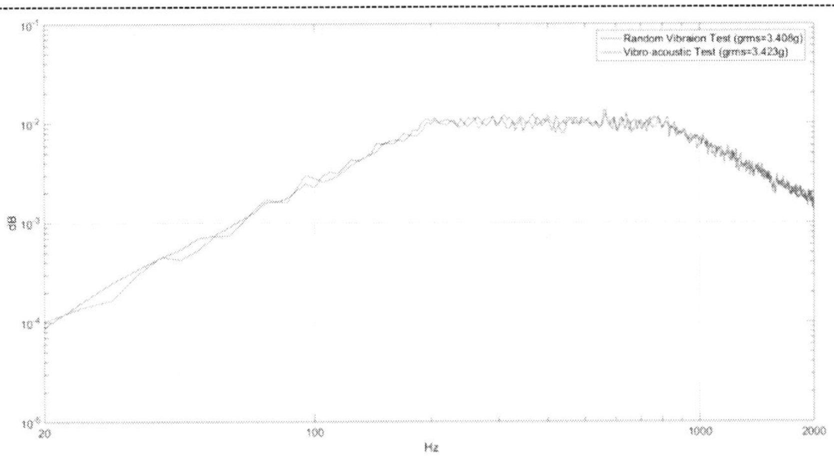

Figure 2 Control PSDs of random-only test and combined test

The acceleration control PSDs of the single random test and the combined test are shown in Fig.2. It can be seen that the RMS (root mean square) values of both curves are approximately equal to each other, which indicates that the random acceleration control PSD of the combined vibro-acoustic test has not been strongly affected by acoustic excitation. As shown in Fig.3, the sound control spectrums of the acoustic-only test and the combined test are also compared, in which negligible differences appear at each center frequencies.

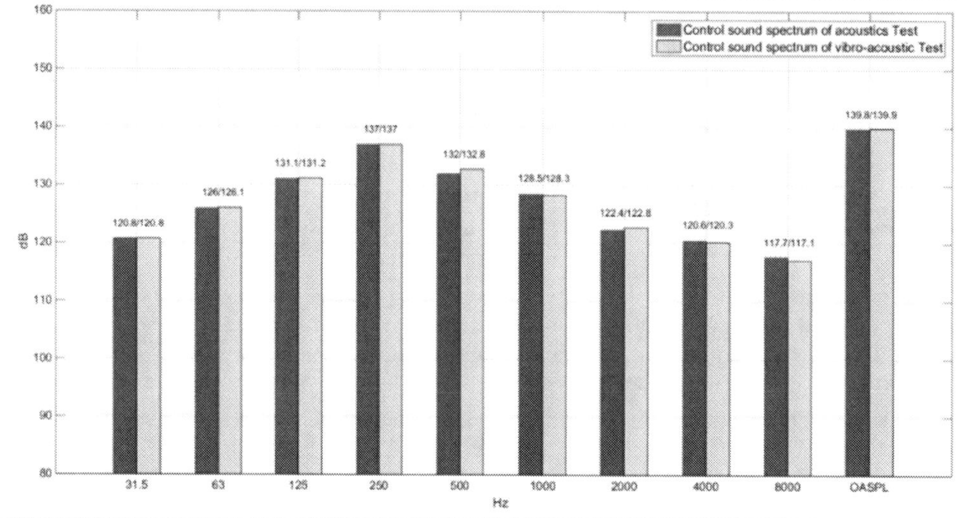

Figure 3 Averaged control sound spectrums of acoustic-only test and combined test

3. Test results and analysis

The main difference between random-only and acoustic-only test is input energy transition and acting position. In the acoustic test, the satellite was located at the center of high level reverberation chamber, and a number of modulators with horns provided acoustic environment around it. While for a random vibration test, the interface frame of the satellite is connected on a shaker that provides mechanical excitation from the joint interface. It means that random vibration excitation is transmitted from shakers to the satellite, which essentially is point or surface excitation. And acoustic loads are directly acting on the surface of the satellite, considered as field excitation from all the directions. From the view of structural responses, acoustic excitation has higher impact on plate-like structures than random vibration.

The response RMS values resulted from random vibration are mostly greater than that of acoustic test. For a typical measurement point PMA2, the RMS of random response is 5.083grms, and that of acoustics test is 3.1grms. As shown in Fig.4, the responses of PMA2 are enhanced by shaker excitation in the whole frequency bandwidth. It can be seen that the combined vibro-acoustics test response is similar with that of single random vibration test, and basically envelopes the response of acoustics test. Significant disparity appears between single acoustic and random vibration test for PMA9.

The response curves of antenna subsystem during the three individual tests are compared in Fig.5, two resonate peaks in low frequency range appear in the curves of random excitation and combined vibro-acoustic excitation, and none for acoustic test. It means that the antenna is excited adequately by mechanical random excitation. Above 300 Hz, the response of the combined test and the single acoustic test are in good agreement. The solar wings exposed outside the main structure frame is far away from source of mechanical vibration. It can be obviously seen from Fig.6 that the response at the center of solar wing from the combined vibro-acoustic test is almost identical to that of single random vibration test below 150Hz, and close to that of the acoustic-only test above 150Hz. It can be concluded that the combined vibro-acoustic test can give a less conservative test, compared with single vibration and acoustic test, for plate-like structures exposed out of the main structure.

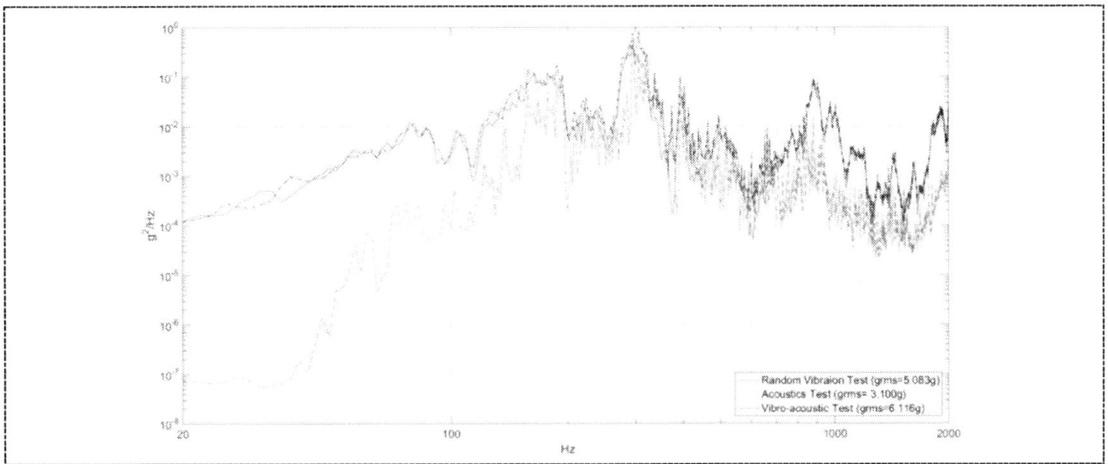

Figure 4 PSDs of response at PMA2

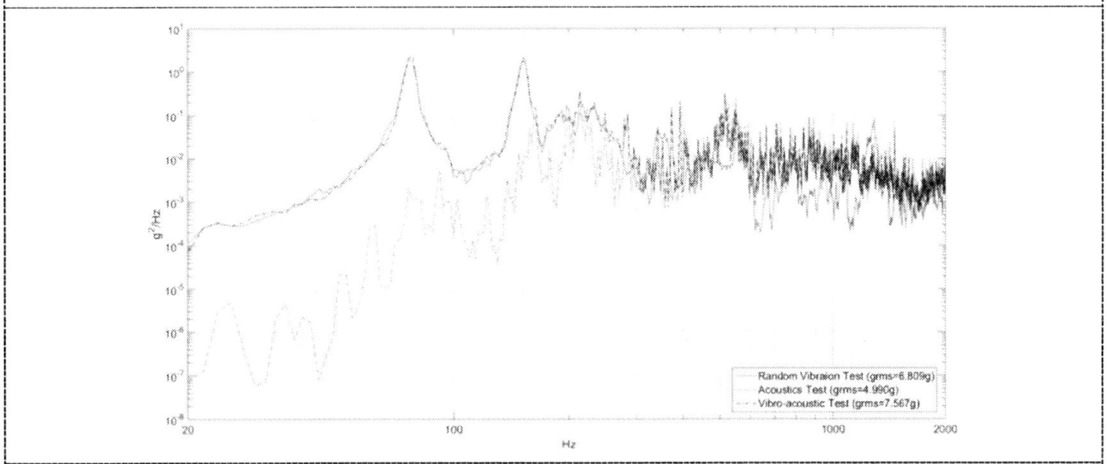

Figure 5 PSDs of response at AN

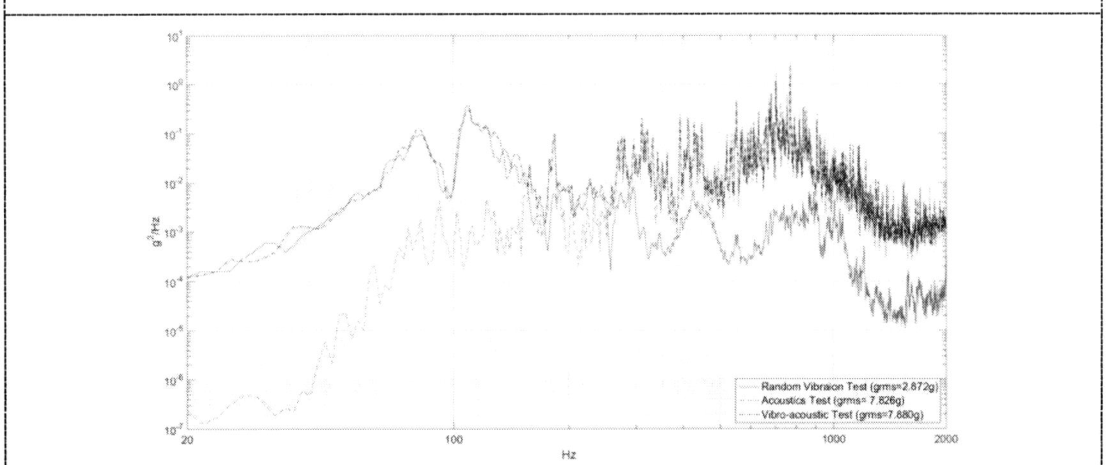

Figure 6 PSDs of response at SS

4. Conclusion

IOP Conf. Series: Materials Science and Engineering **751** (2020) 012015 doi:10.1088/1757-899X/751/1/012015

This paper focuses on a combined vertical vibro-acoustic test of a prototype satellite. The coupling effect of combined test in high frequency response attenuation and its distributions on a prototype satellite are experimentally investigated. Some appealing results can be summarized.

In most representative measuring points, the responses are enhanced by random vibration in low frequency range, while dominated by reverberation field loads in high frequency range. Compared with single random vibration or acoustic test, the combined vibro-acoustic test not only makes up the deficiency of acoustic test in low frequency, also improves the attenuation of vibration transmission far away from excitation source, which might be close to true flight environment intuitively.

References

[1] Salzberg A and Haughton J 1990 Combined high level acoustic and mechanical vibration testing and analysis *8th International Modal Analysis Conference Kissimmee* FL 933-939

[2] Forgrave J.C, Man K.F and Newell J.M 1998 Acoustic and random vibration test for low-cost missions *The 44th Annual Institute of Environmental Sciences Technical Meeting* Phoenix

[3] Scharton T.D 2002 Vibration and acoustic testing of spacecraft *Sound and Vibration* 36: 14-19

[4] Shigemasa A, Keiichi Y and Qinzhong S 2009 The test effectiveness factor for random & acoustic tests on spacecraft subsystem *Institute of Environmental Sciences and Technology 25th Aerospace Testing Seminar* Manhattan Beach, California

[6] Zou Y.J, Han Z.Y, Liu S.K, Wang Z.Y and Zhu W.H 2016 Comparison of acoustic and random vibration testing for large spacecraft *23th International Congress on Sound and Vibration* Athens, Greece

[7] Larko J.M and Cotoni V 2008 Vibro-acousitc response of the NASA ACTS spacecraft antenna to launch acoustic excitation *NASA/TM-2008-215168*

[8] Yan T.F, Zhang J.G, Fang G.Q and Shen Z.Q 2014 A comparison between vibro-acoustic environment test and single environment test for a spacecraft antenna. *Spacecraft Environment Engineering* 31(2), 154-157

[9] Shen Z.Q, Yan T.F, Zhang J.Q and Fang G.Q 2015 The vibro-acoustic test technology for satellite *Spacecraft Environment Engineering* 32(4), 395-399

[10] Frendi A and Robinson J 1993 Effect of acoustic coupling on random and harmonic plate vibrations *AIAA Journal* 31(11) 1992-1997 (1993)

IOP Conf. Series: Materials Science and Engineering **751** (2020) 012016 doi:10.1088/1757-899X/751/1/012016

Spacecraft's Partially Invariant Stabilization System

Nickolay Zosimovych

Intelligent Manufacturing Key Laboratory of Ministry of Education, Shantou University, 243, Daxue Road, Shantou, Guangdong, 515063, China

nzosimovych@stu.edu.cn

Abstract. In most of the known spacecraft stabilization systems the control signal in the centre of mass movement channel is generated according to proportional plus integral control law, based on the measurements of tangential velocity of the centre of mass and its integral-linear drift. In the angular stabilization channel, the control signal shall be generated in proportion to the spacecraft deviation angle in the transverse plane and the angular velocity of the spacecraft rotation in this plane. In practice, a spacecraft's stabilization system, which is partially invariant to the disturbing moment, is the easiest to implement. The velocity performance limitation of the control actuator has the most influence upon stability of the system among all the nonlinearities considered in the mathematical modelling. Therefore, we shall consider a model of an invariant stabilization system with due regard to nonlinearity. For the system under consideration, it is possible to construct lines of equal values of the auto oscillation amplitude in a two-parameter plane. The study of the stability of the proposed partially invariant stabilization system revealed that it is possible to ensure sufficient stability margins in the system under consideration by choosing parameters for the stabilization controller. At the same time, it allows to provide high quality of the transition process.

Keywords: accuracy, spacecraft, stabilization, controller, nonlinearity.

1. Introduction

The flight control profile of the spacecraft, its power performance, dynamic and accuracy characteristics largely determine the type of tasks to be solving and the quality of their solution by a specific spacecraft. In some cases, when using a control system built according to the principle of program control (the "robust trajectories" method) the efficiency of task solution is much influenced by the accuracy of the spacecraft stabilization system in the powered portion of flight.

Some publications [1-6] included synthesis of an invariant centre of mass motion stabilization system of a spacecraft and a comparative analysis of accuracy and quality of transition processes when conventional stabilization system is used. They demonstrated advantages of the proposed algorithms in terms of accuracy improvement.

An essential criterion determining practical application of the control system in addition to the control accuracy is a possibility to provide proper stability margins of the system. Consequently, the challenge is to select appropriate parameters for a stabilization controller. Therefore, a detailed study of the stability of the partially invariant algorithm proposed in [4-6] as well as selection of the proper parameters for the stabilization controller based on conducted studies, which shall provide sufficient stability margins for the system and the required quality characteristics for the transition process shall be the task to be solving herein.

The selection of stabilization controller parameters shall be more convenient and more informative if we build the system stability area in the stabilization controller parameter plane: for the angular stabilization channel and for the centre of mass motion stabilization channel. In practice, sufficient stability margins of the system shall be 1.5 – 2.0 time decrease (increase) margin of the system parameters with respect to their boundary stability conditions [7]. Therefore, ensuring of the specified stability margins is an essential criterion in the selection of parameters for automatic system in this research.

2. Problem formulation

In order to ensure invariance of the coordinate $\dot{y}$ under influence F it is necessary that the condition $\Delta\dot{y}_F = 0$ is met, or

$$s^2\left(K_{FB} + \frac{1}{K_{CA}}s\right) + W_{AS}C_{\vartheta\delta} = 0.$$

Fig. 1 [1, 3-6] shows two channels transferring an impact from the origin of the disturbing moment M to the controlled condition $\dot{y}$ while there is only one channel between the origin of destabilizing force F and the controlled condition $\dot{y}$. This explains the earlier conclusion about the physical inability to implement a system which would be invariant both for M, and F.

Therefore, an analysis of the possibility of an invariant stabilization system shows that such a system can be implementing only under one of the influences, i.e. disturbing moment [1, 4]. As this impact is a determining one, it is useful to consider a possibility to build such a system [3].

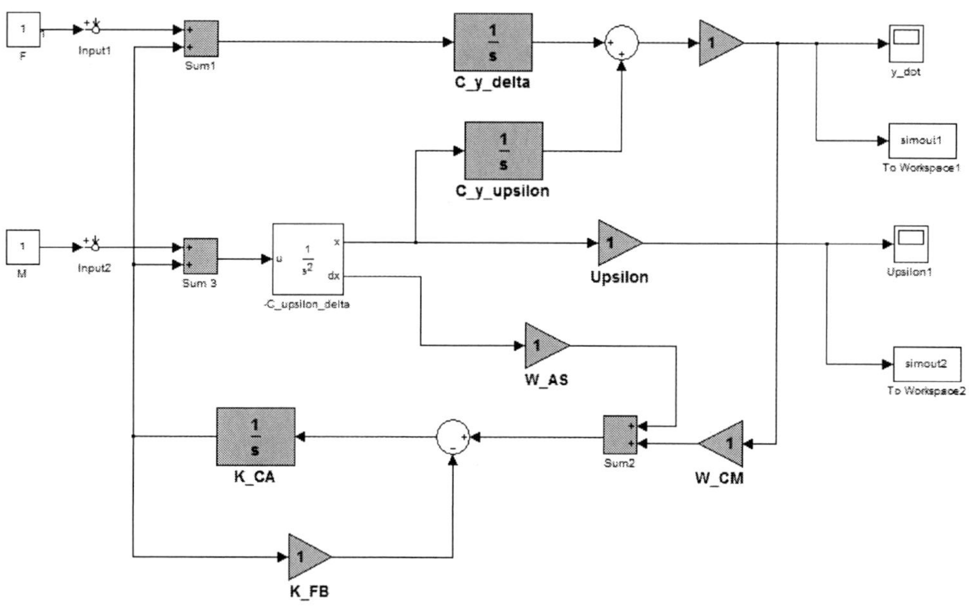

Figure 1. Block diagram of the spacecraft centre of mass motion stabilization system under study.

Let us examine the stability of the proposed partially invariant stabilization system considering the influence of nonlinear elements of the system. As regards stability, the velocity performance limitation of the control actuator [8] has the most influence upon stability of the system among all the nonlinearities considered in the mathematical modelling [9, 10]. Therefore, we shall consider a model of an invariant stabilization system with due regard to nonlinearity. A block diagram corresponding to a model of the system allowing for nonlinearity is showing in Fig.2.

Objectives: to solve the task of significant increase in stabilization accuracy of centre of mass tangential velocities during the trajectory correction phases when using the "rigid" trajectory control principle.

Subject of research: The centre of mass movement stabilization system in the transverse plane, which is using during the trajectory correction phases.

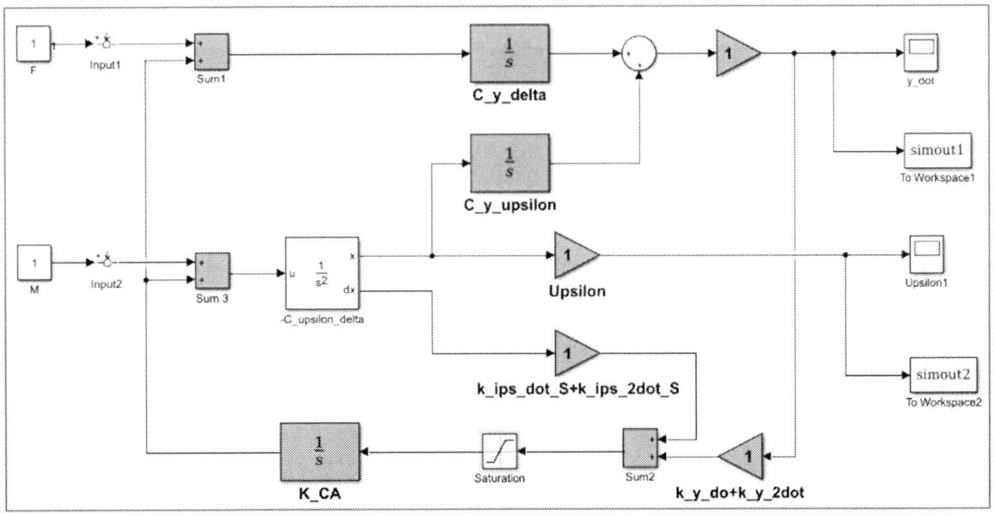

Figure 2. Block diagram of an invariant centre of mass motion stabilization system considering nonlinear velocity performance of the control actuator [6].

We shall define the transfer functions for the linear and nonlinear parts of the system. According to the block diagram, the transfer function of the linear part of the system is follows:

$$W_L(s)=\frac{K_{OD}}{s^4}\left[s^3\left(C_{y\delta}k_{\ddot{y}}-C_{\vartheta\delta}k_{\ddot{\vartheta}}\right)+s^2\left(C_{y\delta}k_{\dot{y}}-C_{\vartheta\delta}k_{\dot{\vartheta}}\right)-sC_{\vartheta\delta}C_{y\delta}k_{\ddot{y}}-C_{\vartheta\delta}C_{y\delta}k_{\dot{y}}\right]. \quad (1)$$

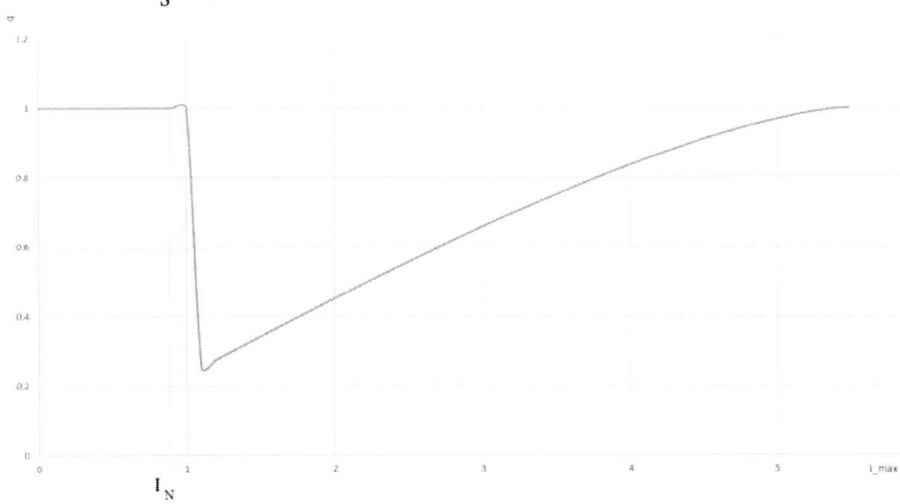

Figure 3. Dependence of the coefficient of harmonic linearization q from the signal amplitude i_{max}.

For nonlinearity with a saturation zone, the expression for the coefficient of harmonic linearization is [9]:

$$\begin{cases} q = \frac{2}{\pi}\left(\arcsin\frac{I_N}{i_{\max}} + \frac{I_N}{i_{\max}}\left(1-\frac{I_N^2}{i_{\max}^2}\right)^{1/2}\right), when: i_{\max} > I_N; \\ q = 1, when: i_{\max} \le I_N. \end{cases} \quad (2)$$

where $i_{\max}$ – the amplitude of the input is signal for the control actuator; I_N – is velocity performance saturation current for the control actuator.

According to the above formulas, the dependence of the coefficient of harmonic linearization q from the amplitude of the input signal $i_{\max}$ is showing in Fig. 3.

3. Synthesis of a stabilization system partially invariant under destabilizing force and disturbing moment

We used the same principle for research and synthesis in the previous paragraphs of this work: we checked the criterion for physical implementation of the invariant stabilization system described above under disturbing moment and destabilizing force. As a result, we revealed a possibility to synthesize an invariant system only under disturbing moment. That is, the synthesis resulted in a partial invariant stabilization system under disturbing moment [4, 8].

As noted above, disturbances influencing the spacecraft over the active phase are slowly changing, so it is necessary to fulfil the invariance conditions for drift velocity under disturbances themselves rather than under their derivatives [4].

As may be seen from the analysis of the characteristic equation of the system in question, the dynamic delay of the stabilization controller, in this case, does not have a fundamental impact on the structure and stability of the system, so the dynamic delay of the stabilization controller will not be taken into account hereinafter. Thus, the block diagram for the stabilization system under consideration will be as follows (Fig. 2) [5].

Mathematical modeling shows that the simultaneous use of invariant algorithms and the initial adjustment of the operating device give an increase in the accuracy of the drift velocity twice or more as compared to the model stabilization system using similar additional algorithms [1, 5].

The mathematical modeling shows that simultaneous use of initial adjustment of operating device and the body of a spacecraft with invariant stabilization algorithms improves the accuracy of regulation approximately by one order as compared with a typical stabilization system using the same configuration elements [8].

4. Modeling of transition processes in a partially invariant centre of mass motion stabilization system

The following values of the characteristics were selecting for modeling: $K_{AVS} = 1\frac{V \cdot s}{grad}$; $T_F = 0.01s$; $T_1 = 0.01s$; $k_{\dot{\vartheta}} = 14s; k_{\ddot{\vartheta}} = 6s; k_{\dot{y}} = 40\frac{V \cdot s}{m}; k_{\ddot{y}} = 80\frac{V \cdot s^2}{m}$.

The angular velocity sensor (AVS) signal error $\xi_{\dot{\vartheta}}$ is considered Gaussian uncorrelated random value (Fig. 4), with a zero mathematical expectation and mean square deviation (MSD) $\sigma_{\xi_{\dot{\vartheta}}} = 0.01\frac{grad}{s}$. The MSD value was chosen from the condition that the accuracy of modern AVS is about one hundredths degree per second. The values of the stabilization controller coefficients $k_{\dot{\vartheta}}, k_{\ddot{\vartheta}}, k_{\dot{y}}, k_{\ddot{y}}$ were selected based on the author's algorithm.

The following values of the disturbance characteristics were selected for modeling: $T_k = 0.1s; \sigma_F = 0.1^0; \sigma_M = 0.3^0$; $0.1^0 \le m^H F \le 0.35^0; 0.3^0 \le m^H M \le 3.5^0$.

Thus, a structural scheme of a partially invariant stabilization system used in the modeling shall be as follows (Fig. 4). The response functions of generating filters W_{ff}^{M}, W_{ff}^{F} according [8] shall be as follows: $W_{ff}^{F(M)}(S) = \frac{K_{ff}}{T_{ff}S+1}$, where $K_{ff} = \frac{b}{a}, T_{ff} = \frac{1}{a}$.

The modeling was doing by numerical integration with application of Runge-Kutt 4th order method, with a controlled step size. Random values $\xi, \xi_{\dot{\vartheta}}$ were modeling with application of Gaussian pseudo-random number sensors.

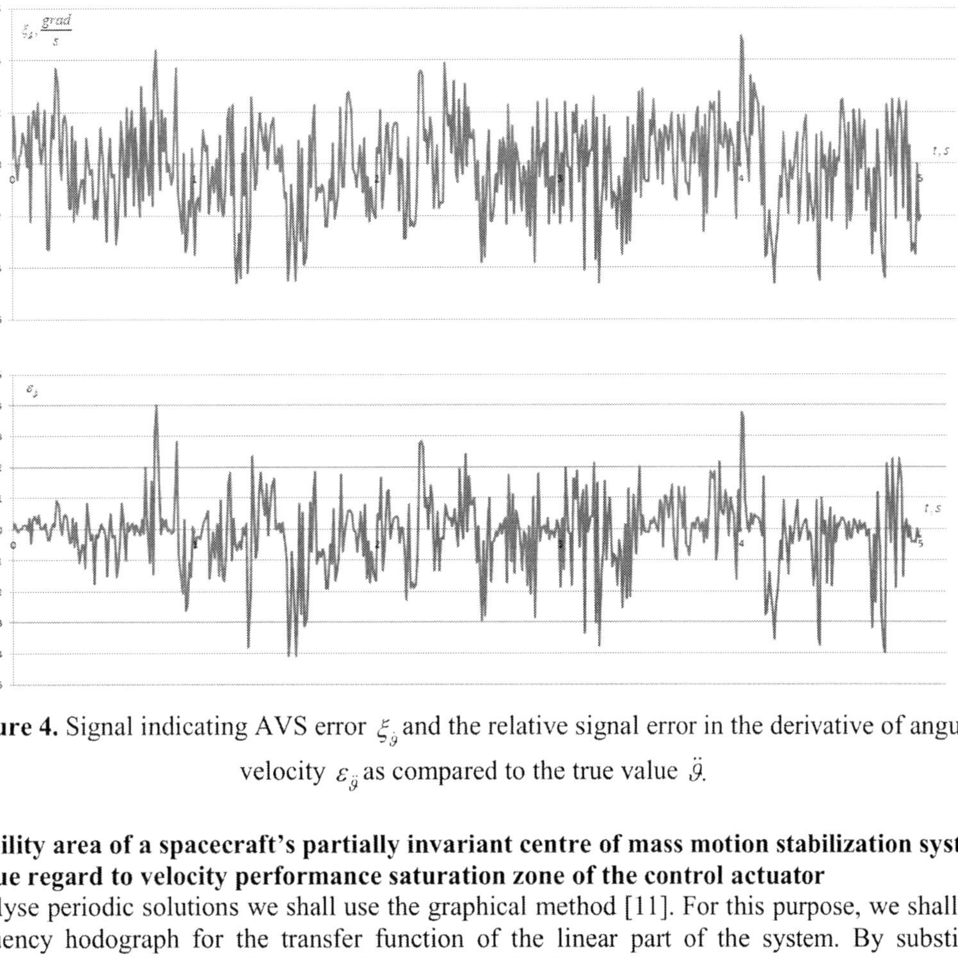

Figure 4. Signal indicating AVS error $\xi_{\dot{\vartheta}}$ and the relative signal error in the derivative of angular velocity $\varepsilon_{\ddot{\vartheta}}$ as compared to the true value $\ddot{\vartheta}$.

5. Stability area of a spacecraft's partially invariant centre of mass motion stabilization system with due regard to velocity performance saturation zone of the control actuator

To analyse periodic solutions we shall use the graphical method [11]. For this purpose, we shall build a frequency hodograph for the transfer function of the linear part of the system. By substituting Laplace operator s for $j\omega$, in the expression $W_L(s)$ we obtain:

$$W_L(j\omega) = \frac{K_{OD}}{\omega^4}\left[j\omega^3\left(-C_{y\delta}k_{\ddot{y}} + C_{\vartheta\delta}k_{\ddot{\vartheta}}\right) + \omega^2\left(-C_{y\delta}k_{\dot{y}} + C_{\vartheta\delta}k_{\dot{\vartheta}}\right) - j\omega C_{\vartheta\delta}C_{y\delta}k_{\ddot{y}} - C_{\vartheta\delta}C_{y\delta}k_{\dot{y}} \right]. \quad (3)$$

Let us introduce the following notations:

$$\alpha_1 = \left(C_{\vartheta\delta}k_{\ddot{\vartheta}} - C_{y\vartheta}k_{\ddot{y}}\right);\ \alpha_2 = \left(C_{\vartheta\delta}k_{\dot{\vartheta}} - C_{y\vartheta}k_{\dot{y}}\right);\ \alpha_3 = C_{y\delta}C_{\vartheta\delta}k_{\ddot{y}};\ \alpha_4 = C_{y\vartheta}C_{\vartheta\delta}k_{\dot{y}}. \quad (4)$$

Given the above notations, the expression (1) will be as follows:

$$W_L(j\omega) = \frac{K_{OD}}{\omega^4}\left[j\omega^3\alpha_1 + \omega^2\alpha_2 - j\omega\alpha_3 - \alpha_4 \right]. \quad (5)$$

We shall separate the real and imaginary parts from the expression (5):

$$\mathrm{Re}(W_L) = \frac{K_{OD}}{\omega^4}\left(\omega^2\alpha_2 - \alpha_4\right) \quad (6)$$

$$\operatorname{Im}(W_L) = j\frac{K_{OD}}{\omega^4}\left(\omega^3\alpha_1 - \omega\alpha_3\right). \tag{7}$$

The characteristic equation for the linear part of the open-loop system has four zero roots, so when the frequency changes between -0 and +0 the hodograph of the transfer function $W_L(j\omega)$ will move along a circle with infinitely large radius with $\omega_1 = +180^0$ till $\omega_2 = -180^0$.

If $\omega = 0$ the expressions (6) and (7) will take on values $-\infty: \operatorname{Re}(W_L)_{\omega=0} = -\infty; \operatorname{Im}(W_L)_{\omega=0} = -\infty$. In this case, the phase shift of frequency hodograph will be:

$$tg\varphi_{\omega=0} = \frac{\operatorname{Im}(W_L)_{\omega=0}}{\operatorname{Re}(W_L)_{\omega=0}} = \left.\frac{\omega\left(\omega^2\alpha_1 - \alpha_3\right)}{\omega^2\alpha_2 - \alpha_4}\right|_{\omega=0} = 0; \varphi_{\omega=0} = -180^0. \tag{8}$$

If $\omega = +\infty$ the expressions (6) and (7) will take on values $0: \operatorname{Re}(W_L)_{\omega=+\infty} = 0; \operatorname{Im}(W_L)_{\omega=+\infty} = 0.$

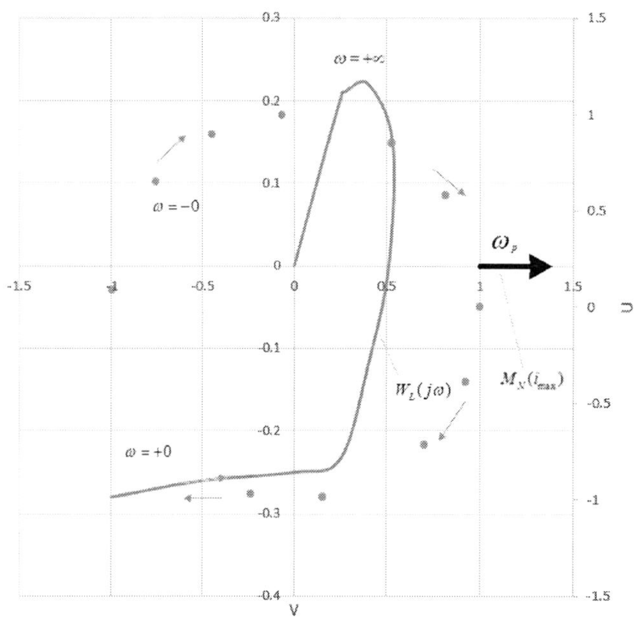

Figure 5. The amplitude-frequency characteristics of the partially invariant spacecraft centre of mass stabilization system.

Thus, if frequency ω changes from 0 to $+\infty$, the transfer function hodograph $W_L(j\omega)$ beginning in the third quadrant with the phase shift $\varphi = -180^0$, while moving through the 4th and 1st quadrants reaches the origin of coordinates. In this case, since closure of the system was made through +1 and not through -1, it is necessary that the branch of the phase and gain performance corresponding to the positive frequencies, together with a fragment of the circle beginning with a positive real semi axis doesn't cover the point $(+1;0j)$ to provide stability of the closed-loop system [12].

The existence of periodic solutions in the system can be determined from the condition $W_L(j\omega)W_N(i_{\max}) = 1$, where $W_N(i_{\max})$ a transfer function of the nonlinear element is, or otherwise:

$$W_L(j\omega) = \frac{1}{q(i_{\max})}. \tag{9}$$

According to (6) and (7), condition (8) shall be split into two conditions:

$$\operatorname{Im}(W_L) = 0 \tag{10}$$

$$\mathrm{Re}(W_L) = \frac{1}{q(i_{\max})}. \quad (11)$$

By substituting the expression for the imaginary part of the transfer function of the linear part of the system into the condition (8), we get:

$$\frac{K_{OD}}{\omega^4}\left(\omega^3\alpha_1 - \omega\alpha_3\right) = 0, \quad (12)$$

where from $\omega^2\alpha_1 - \alpha_3 = 0$, or

$$\omega = \left(\frac{\alpha_3}{\alpha_1}\right)^{1/2}. \quad (13)$$

The determined frequency is the frequency of a periodic solution for the given nonlinear system. Further, we will denote it as ω_p. The self-oscillation amplitude in the system under consideration $i_{\max\,p}$ shall be found from the expression (13):

$$\frac{K_{OD}}{\omega_p^4}\left(\omega_p^2\alpha_2 - \alpha_4\right) = \frac{1}{q(i_{\max\,p})}, \quad q(i_{\max\,p}) = \frac{\omega_p^4}{K_{OD}\left(\omega_p^2\alpha_2 - \alpha_4\right)}. \quad (14)$$

By substituting the expression (2) for $q(i_{\max\,p})$ into the last formula, we finally obtain:

$$\frac{2}{\pi}\arcsin\left(\frac{I_N}{i_{\max\,p}}\right) + \frac{I_N}{i_{\max\,p}}\left(1 - \frac{I_N^2}{i_{\max\,p}^2}\right)^{1/2} = \frac{\omega_p^4}{K_{OD}\left(\omega_p^2\alpha_2 - \alpha_4\right)}. \quad (15)$$

The expression (15) is transcendental relative $i_{\max\,p}$, therefore, it can be solved only numerically. The value $i_{\max\,p}$ can be approximately determined from the diagram $q(i_{\max})$ (Fig. 5).

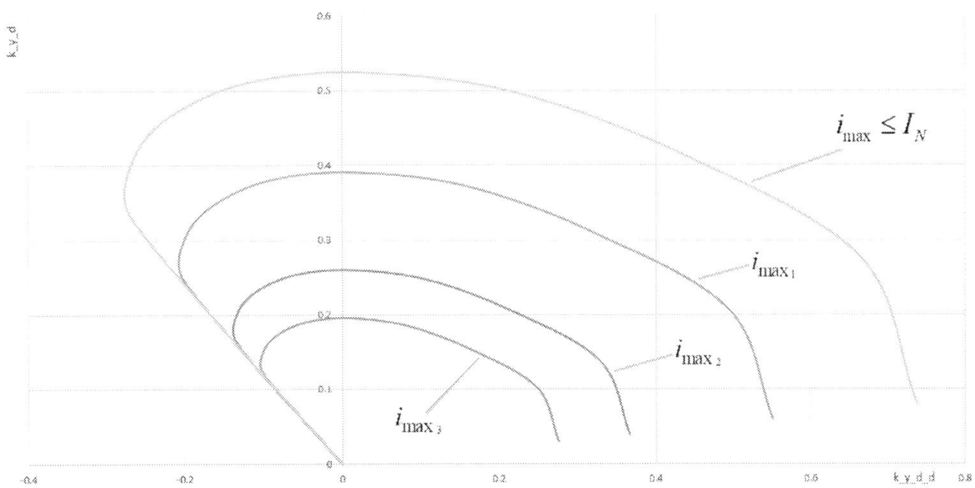

Figure 6. Stability area limits of a partially invariant stabilization system in the parameter plane $k_{\dot{y}}, k_{\ddot{y}}$

Now we shall consider stability of periodic solution. To this end, we shall build a hodograph of the function $M_N = \dfrac{1}{q(i_{\max})}$ in the plane UV (Fig. 6). According to (2), when $0 \le i_{\max} \le I_N, M_N = 1$, if $i_{\max} > I_N, M_N$ monotonically increases to $+\infty$. Since hodograph M_N crosses the amplitude-frequency characteristic of the linear part of system "from the outside to the inside" [13], then the periodic solution of the system under study $(i_{\max_p}, \omega_p)$ will be unstable. Therefore, when $i_{\max} < i_{\max_p}$ the system solution will converge to 0, that is, the system will be stable, and if $i_{\max} > i_{\max_p}$ we will observe a divergent transient process, that is, the system will be unstable.

Thus, the stability condition for the system in general shall be determined by the initial conditions of the transition process, at which the amplitude $i_{\max}$ at the input of the nonlinear element is less than the value $i_{\max_p}$:

$$i_{\max} < i_{\max_p}. \tag{16}$$

For the system under consideration, it is possible to construct lines of equal values of the auto oscillation amplitude in a two-parameter plane. Such parameters will be the gain of the angular velocity and angular acceleration stabilization controller $k_{\dot\vartheta}, k_{\ddot\vartheta}$. The following can be obtained from (15) and (16):

$$\omega_p = \left(\frac{\alpha_3}{\alpha_1}\right)^{1/2} = \left(\frac{C_{\vartheta\delta} C_{y\delta} k_{\ddot y}}{C_{\vartheta\delta} k_{\ddot y} - C_{y\delta} k_{\ddot y}}\right)^{1/2} = \left(\frac{C_{\vartheta\delta}}{\frac{C_{\vartheta\delta}}{C_{y\delta}} - 1}\right)^{1/2}. \tag{17}$$

Let us set $c_1 = \left(\frac{C_{\vartheta\delta}}{C_{y\delta} k_{\ddot y}}\right)^{1/2}$, then $\omega_p = \left(\frac{C_{\vartheta\delta}}{c_1 k_{\ddot\vartheta} - 1}\right)^{1/2}$. From $\frac{K_{OD}}{\omega_p^4}\left(\omega_p^2 \alpha_2 - \omega \alpha_4\right) = \frac{1}{q(i_{\max_p})}$, the expression (11):

$$\frac{K_{OD}}{\omega_p^4}\left(\omega_p^2 \alpha_2 - \omega \alpha_4\right) = \frac{1}{q(i_{\max_p})}, \tag{18}$$

from which

$$\alpha_2 = \left[\frac{\omega_p^4}{K_{OD} q(i_{\max_p})} + \alpha_4\right] \frac{1}{\omega_p^2}. \tag{19}$$

By expanding the expression for α_2, we get after transformation:

$$k_{\dot\vartheta} = \frac{\frac{1}{\omega_p^2}\left[\frac{\omega_p^4}{K_{OD} q(i_{\max_p})} + \alpha_4\right] + C_{y\vartheta} k_{\dot y}}{C_{\vartheta\delta}}. \tag{20}$$

After substituting in the expression for ω_p:

$$k_{\dot\vartheta} = \frac{\frac{C_1 k_{\ddot\vartheta} - 1}{C_{\vartheta\delta}}\left[\frac{C_{\vartheta\delta}^2}{K_{OD} q(i_{\max_p})(C_1 k_{\ddot\vartheta} - 1)^2} + \alpha_4\right] + C_{y\vartheta} k_{\dot y}}{C_{\vartheta\delta}}. \tag{21}$$

After transformation, we get:

$$k_{\dot\vartheta} = \frac{1}{K_{OD} q(i_{\max_p})(C_1 k_{\ddot\vartheta} - 1)} + \alpha_4 \frac{C_1 k_{\ddot\vartheta} - 1}{C_{\vartheta\delta}^2} + \frac{C_{y\vartheta}}{C_{\vartheta\delta}} k_{\ddot\vartheta}. \tag{22}$$

By expanding the expressions for α_4 and c_1, we finally have:

$$k_{\dot\vartheta} = \frac{C_{y\vartheta} k_{\ddot\vartheta}}{K_{OD} q(i_{\max_p})(C_{\vartheta\delta} k_{\ddot\vartheta} - C_{y\vartheta} k_{\ddot\vartheta})} + \frac{k_{\dot y}}{k_{\ddot y}} k_{\ddot\vartheta}. \tag{23}$$

It should be noting that the right-hand side of expression (23) is different from the right side of the condition (10), which determines the stability limit for the linear systems only by the presence of harmonic linearization coefficient in the denominator of the first term. Since $q(i_{max}) \leq 1$, then the presence of nonlinearity $I_N < i_{max\,1} < i_{max\,2} < i_{max\,3}$ will lead to narrowing of the stability area according to the parameters $k_{\dot{\vartheta}}, k_{\ddot{\vartheta}}$.

Accordingly, the expression for the stability limits according to the parameters $k_{\dot{y}}, k_{\ddot{y}}$ will be similar to the right side of inequality (16) taking into account the coefficient of harmonic linearization $q(i_{max})$:

$$k_{\dot{y}} = -\frac{C_{y\vartheta} k_{\ddot{y}}^2}{K_{OD} q(i_{max\,p})(C_{\vartheta\delta} k_{\ddot{\vartheta}} - C_{y\vartheta} k_{\ddot{y}}) k_{\ddot{\vartheta}}} + \frac{k_{\dot{\vartheta}}}{k_{\ddot{\vartheta}}} k_{\ddot{y}}. \quad (24)$$

A general view of the stability area in the parameter plane $k_{\dot{y}}, k_{\ddot{y}}$ for different amplitude values at the input of the nonlinear element is shown in Fig. 6.

As can be seen from the expressions (13)-(14) and Fig. 6, stability of the invariant stabilization system considering nonlinearity and having a saturation zone, will directly depend on the value of the input signal in the nonlinear element, and ultimately on the disturbances acting on the spacecraft [14]. Therefore, in order to use such a stabilization system it is necessary to choose in practice such parameter values of the control action under which the system will be stable under possible perturbations. On the other hand, we will need to determine for the system under study conditions for its absolute stability regardless of perturbation.

According to [10] the absolute stability area shall be determined as an area of system parameters, which obviously escapes periodic solutions, and in which any transition process fades under any initial conditions to the value $x = 0$.

To determine an absolute stability area we can apply any linear stability criterion to the harmonically linearized system.

Let us apply the Hurwitz criterion to the system. According to the block diagram (Fig. 2) and the expression for the transfer function of the linear part of the system (1), a characteristic equation for a harmonically linearized system will be:

$$s^4 + K_{OD} = \left[s^3 (C_{\vartheta\delta} k_{\ddot{y}} - C_{y\vartheta} k_{\ddot{\vartheta}}) + s^2 (C_{\vartheta\delta} k_{\dot{\vartheta}} - C_{y\vartheta} k_{\dot{y}}) + s C_{\vartheta\delta} C_{y\vartheta} k_{\ddot{y}} + + C_{\vartheta\delta} C_{y\vartheta} k_{\dot{y}} \right] q(i_{max}) = 0. \quad (25)$$

Let us introduce the following notations:

$$\beta_1 = (C_{\vartheta\delta} k_{\ddot{\vartheta}} - C_{y\vartheta} k_{\ddot{y}}) K_{OD};\ \beta_2 = (C_{\vartheta\delta} k_{\dot{\vartheta}} - C_{y\vartheta} k_{\dot{y}}) K_{OD};\ \beta_3 = C_{y\vartheta} C_{\vartheta\delta} k_{\ddot{y}} K_{OD};\ \beta_4 = C_{y\vartheta} C_{\vartheta\delta} k_{\dot{y}} K_{OD}. \quad (26)$$

Then the characteristic equation for the linearized system can be written as:

$$s^4 + s^3 \beta_1 q + s^2 \beta_2 q + s \beta_3 q + \beta_4 q = 0. \quad (27)$$

The Hurwitz conditions for the characteristic equation of the fourth order [8]:

$$q\beta_1 > 0; q\beta_2 > 0; q\beta_3 > 0; q\beta_4 > 0; \beta_3 q(\beta_1 \beta_2 q^2 - \beta_3 q) - \beta_4 \beta_1^2 q^3 > 0. \quad (28)$$

To ensure absolute stability of the system it is necessary that the above conditions be satisfied for any admissible values of the coefficient of harmonic linearization q. For the system under study $0 \leq q \leq 1$. Obviously, with the boundary condition $q = 0$ of the Hurwitz conditions for this system shall be not performed. Therefore, this system escapes absolute stability with any parameters. This is also confirmed by the graphic method of periodic solutions investigation. Let us refer to Fig. 5. To ensure stability of a linear system it is necessary that the frequency hodograph for the transfer function of the linear part of the system $W_L(j\omega)$ should not cover point $(+1; 0j)$. When frequency changes from 0 to $+\infty$. On the other hand, the hodograph of the function $\frac{1}{q(i_{max})}$ when q changes from 0 to 1 goes through a real positive semi axis from $+\infty$ to $+1$. According to the above studies, the point of intersection of the frequency hodograph $W_L(j\omega)$ and the hodograph of the function $\frac{1}{q(i_{\max})}$ is a point of unstable periodic solutions and provided the conditions for the stability of the nonlinear system fulfilled (when the hodograph does not cover point $(+1;0j)$, there will be always an area of unstable periodic solutions.

6. Conclusion

Based upon research results we can conclude the following:

1. It is impossible to implement a centre of mass stabilization system, which is invariant regarding both the disturbing force and disturbing moment.
2. In practice, a stabilization system, which is partially invariant to the disturbing moment, is the easiest to implement. In order to comply with the invariance conditions, there must be a positive control actuator feedback with a gain equal to the object's angular deflection gain of in the angular stabilization channel. Stability of the system shall be ensuring by introduction of an additional second derivative action from the object's deflection angle into the action as well as by introduction of an equivalent delay loop in the feedback of control actuator in order to compensate for the dynamic delay of the stabilization controller.
3. It is also possible to synthesize a stabilization system, which shall be partially invariant fewer than two disturbances simultaneously. Open feedback of the control actuator and exclusion of control according to object's deflection angle and of the spacecraft centre of mass drift coordinate from the angular stabilization channel are the invariance conditions in this case. Such a stabilization system has obvious advantages over a system, which is invariant under disturbing moment M, and therefore it is more suitable for practical implementation.
4. A partially invariant under two disturbances stabilization system provides a significant increase (several times) in the accuracy of the centre of mass tangential stabilization velocities as compared to known stabilization systems.
5. The tangential velocity transition process in a partially invariant stabilization system has a significantly shorter (several times) decay time as compared to known stabilization systems.
6. Employment of additional self-regulation elements in a partially invariant stabilization system reveals the significant advantages of such a system in terms of greater accuracy when compared to known stabilization systems.
7. The study of the stability of the proposed partially invariant stabilization system revealed that it is possible to ensure sufficient stability margins in the system under consideration by choosing parameters for the stabilization controller. At the same time, it allows to provide high quality of the transition process.
8. The system under consideration has an unstable limit cycle due to "saturation zone" nonlinearity, which cannot be excluding by system parameter adjustment. It means the system is missing the absolute stability area.

References

[1] Zosimovych N, Modeling of Spacecraft Centre Mass Motion Stabilization System 2017 *Int. Refereed Journ. of Eng. and Science (IRJES)* Vol. 6, Issue 4 **34** 41

[2] Zosimovych N, Increasing the Accuracy of the Center of Mass Stabilization of Space Probe 2017 *"Новината за Напреднали Наука - 2017", XIII Int. Scientific and Pract. Conf.,* May, 15-22, Techn. Science, Vol. 10 София, «Бял ГРАД-БГ» ООД, **31** 37

[3] Zosimovych N, Improving the Spacecraft Center of Mass Stabilization Accuracy 2017 *IOSR Journ. of Eng. (IOSRJEN)* Vol. 7, Issue 6 **7** 14

[4] Zosimovych N, Increasing the Accuracy of the Center of Mass Stabilization of Space Probe with Partially Invariant System 2017 *Science and Education, a New Dimension. Natural and Techn. Science*, Vol. 14, Issue 132 **105** 08

[5] Zosimovych N, Raj A N J. Synthesis of Stabilization Algorithms in the System Controlling Rotations of the Operating Device 2017 *«Prospects of World Science -2017», XIII Int. Scientific and Pract. Conf.,* Jul. 30 – Aug. 7, Vol. 4, Sheffield Science and Education ltd, Yorkshire, England, S1 4LR **14** 20

[6] Zosimovych N, Invariant Stabilization Algorithms in a Control System with Rotating Operating Device 2017 *American Journ. of Eng. Research (AJER)*, Vol. 6, No. 10 **297** 11

[7] Sineglazov V M 2003 *Theory of automatic control*. In 2 Vol.: Manual for students of all specialities, Kyiv, NAU

[8] Malyshev V V, Krasilshikov M N, Bobronnikov V T, Dishel V D 1996 *Aerospace vehicle control*, Moscow, MAI

[9] Andreev V D 1967 *Theory of Inertial Navigation (Aided Systems)*, Izdatelstvo 'Nauka', Moscow, (Available from the U.S. Department of Commerce, Springfield, Va. 22151), **409**

[10] Dorf R C, Robert Bishop R H 1973 *Modern Control Systems*, Pearson, 12th edition, **807**

[11] Popov E P 1973 *Prikladnaya teoriya processov upravleniya v nelineynyh sistemah*, Moscow: Mashinostroeniye

[12] Greensite A L 1970 *Elements of Modern Control Theory*, New York, Spartan Books

[13] Furman T T 1981 *Approximate Methods in Engineering Design*, Vol. 155: Math. in Science and Eng., Edited by Richard Bellman

[14] Zosimovych N, Xiaodong N, Simeone A, Peng X 2018 SmallSats Rational Design 6th Int. *Conf. and Exhibition on Mech. & Aerospace Eng.*, Nov, 07-08, Atlanta, USA. *Journ. of Aeronautics & Aerospace Eng.*, Nov., Vol. 7, Conf. Series LLC Ltd: 47 Churchfield Road, W3 6AY, London, UK

A Method of On-line Monitoring for Vibration Table Bearings Based on VMD

Du Xiaozhou*, **Xin Qi, Gao Haiyang and Zhu Zihong**

Beijing Insititute of Spacecraft Environment Engineering, Beijing

*E-mail: 171269766@qq.com

Abstract. Aiming at the problems that the bearing inside the sliding table of the shaker used for the dynamic test of spacecraft cannot be observed and the failure cannot be known at the initial stage, an online prediction method for bearing failure is proposed. In this paper, the acceleration sensor is selected as the monitoring medium of bearing motion state response, and the VMD (variational mode decomposition) is used as the analysis method of bearing acceleration response data to calculate the harmonic components of the bearing except the fundamental wave. Through the calculation results, the failure bearing can be intuitively interpreted.

1. Introduction

In aerospace, aviation, vehicle and other engineering fields, vibration test, as an effective means to test product reliability and dynamic strength, has been widely used in product performance assessment and dynamic strength appraisal. Vibration test equipment generally includes shaker, control system, measuring instrument and auxiliary equipment. As an important auxiliary equipment, sliding table is indispensable in vibration test, and its performance state also affects the test control effect and product response. If the sliding table suddenly breaks down during the test, the validity of product verification will be affected and the product will be damaged seriously. During A satellite vibration test in the United States, the bearing of the sliding table was damaged, resulting in severe structural damage. Three of the four solars were damaged, which was classified as A class A accident [1]. To avoid similar situations, it is essential to evaluate the condition of the slide before testing. In view of the problem of online monitoring and slider bearing state, this paper puts forward a kind of based on VMD (variational mode decomposition) of the embedded bearing acceleration state monitoring method, the purpose is to get the bearing status indicators, bearing danger ahead of time, to avoid bearing failure in the testing product loss, fill the shaker bearing more accurate state monitoring of blank.

2. Embedded bearing acceleration testing technology

2.1.Sliding table t-type bearing structure

The two key components of the sliding table are the slide plate and the hydrostatic bearing. The slide plate has high strength and stiffness. The upper surface is used for the installation of test pieces, while the lower surface is threaded with hydrostatic bearing lugs to generate horizontal vibration under the thrust action of the shaker. As shown in figure 1, under the normal working condition of the sliding table, the sliding plate can slide freely, the working curve is smooth and without burr, and the bearing capacity can reach the maximum.

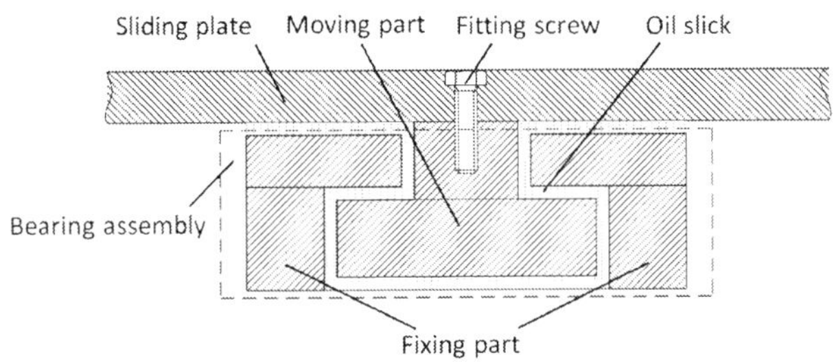

Figure 1. Schematic diagram of relative position between slide and bearing

2.2. *Embedded installation of acceleration sensor*

Since the bearings are located inside the sliding table and there are a large number of them, each bearing state cannot be checked by visual inspection, so sensors are used to monitor the working state of the bearings. The acceleration sensor should meet the following requirements:

- The size is smaller than the bearing clearance inside the sliding table;
- Protection meets oil and corrosion resistance;
- The range is more than 200g;
- High impact resistance;
- Type is IEPE piezoelectric sensor;
- The upper frequency response range is 10kHz.

The shaker bearings have two types: tile motion and shaft motion. To obtain dynamic signals, sensors should be installed on the moving parts of the bearing, and try to be close to the easily worn parts of the bearing to facilitate data acquisition. Considering that the vibration sensor will be soaked in lubricating oil for a long time, it is necessary to drill holes in the bearing and choose the screw connection mode. See the figure 2 for details.

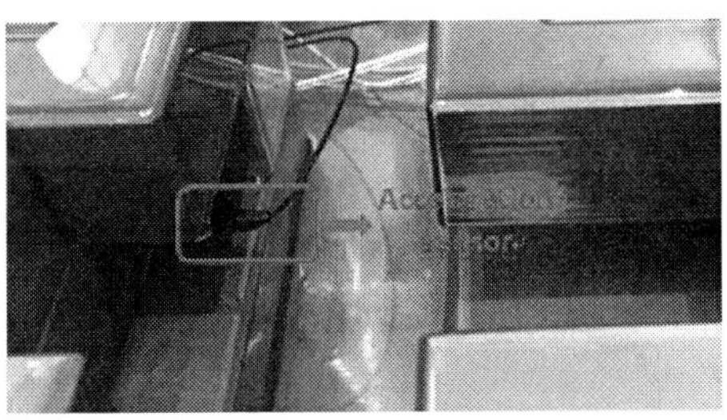

Figure 2. Physical picture of acceleration sensor installation on site

3. Data processing method

3.1. *Variational mode decomposition(VMD)*

Variational modal decomposition (VMD) is a multi-scale non-recursive signal decomposition method which is suitable for the analysis of unsteady signals. This method transforms modal estimation into a variational problem, and determines the frequency center and bandwidth of each component by iteratively searching for the optimal solution of the variational model in the process of obtaining the decomposed components, so as to realize the effective separation of the frequency domain parts and components of the signal adaptively. The variational constraint problem can be expressed as

$$\min_{\{u_k\},\{\omega_k\}}\left\{\sum_{k=1}^{K}\left\|\partial_t\left\{\left[\delta(t)+\frac{j}{\pi t}\right]*u_k(t)\right\}e^{-j\omega_k t}\right\|_2^2\right\} \tag{1}$$

In the formula, u_k is the set of modal functions, ω_k u_k is the set of central frequency, ∂_t is the heap function takes the partial derivative of "*t*", $\delta(t)$ is unit impulse function, "*j*" is imaginary unit, "*"expresses convolution.

In order to solve the constrained optimization problem, the advantages of the second penalty term and Lagrange multiplier method are used in the implementation of the algorithm. Introduced the augmented Lagrangian function ζ, kicking direction multiplier method, a series of iterative optimization, find the ζ minimum value:

$$\zeta(\{u_k\},\{\omega_k\},\lambda)=a\sum_{k=1}^{K}\left\|\partial_t\left\{\left[\delta(t)+\frac{j}{\pi t}\right]*u_k(t)\right\}e^{-j\omega_k t}\right\|_2^2+\left\|f(t)-\sum_{k=1}^{K}u_k(t)\right\|_2^2+\left\langle\lambda(t),f(t)-\sum_{k=1}^{K}u_k(t)\right\rangle \tag{2}$$

In the formula, "*a*"is the bandwidth parameter, $\lambda(t)$ is the lagrangian multiplier.

The modal components are iterated directly in the frequency domain, and finally the inverse Fourier transform to the time domain. The VMD method can overcome the shortcomings of modal aliasing and frequency effect that appear in the recursive decomposition algorithm such as empirical modal decomposition, and has good noise robustness and noise reduction effect. It has been applied in signal noise reduction, and its detailed theory is shown in references [2, 3]. Bearing damage may lead to motion stagnation, distortion of periodic acceleration signals, or superposition of high-frequency signals. Therefore, the VMD method can effectively separate the high frequency noise from the signal.

3.2.Application of VMD method in the field of bearing failure

VMD method is widely used in fault diagnosis of rolling bearings, fault features appear periodically in random signals in the form of pulses. It is difficult to obtain the fault characteristics of rolling bearing directly by the modal decomposition of the original vibration signal. Research has introduced the concept of spectral kurtosis, using VMD decomposed each order modal component (the intrinsic mode function, the IMF), and then calculate the corresponding component of the kurtosis value for its adaptive reconfiguration, then rapid spectral kurtosis analysis was carried out on the reconstructed signal, and band-pass filter design, according to the reconstructed signal is demodulated resonance after the line judge bearing failure[4]. For example, VMD parameters are optimized by means of the principle of minimum information entropy. Meanwhile, IMF where the minimum information entropy is located is obtained, and the envelope demodulation analysis is carried out on it to extract the characteristic frequency of bearing fault[5]. Or according to a kind of mean square error - Euclidean distance index selecting the IMF component containing abundant fault information for signal reconstruction, the reconstructed signal singular value decomposition for diagonal matrix singular values, combining the information entropy theory to calculate the diagonal matrix singular value of the entropy, using the singular value of entropy to distinguish between the size of the rolling bearing working state and fault type[6].

In the analysis of rolling bearing faults by VMD method, due to the large amount of original signal frequency information, the fault frequency will often be overlapped in the original signal, and too many or too few decomposition layers will lead to under decomposition or over-decomposition, resulting in modal aliasing. The application of spectrum kurtosis and information entropy is to obtain fault characteristic frequency accurately on the basis of VMD method.

3.3. Bearing state monitoring index

In the field of vibration test of aerospace products, the internal bearing of the sliding table is sliding bearing, and the resistance of the bearing movement after the formation of oil film comes from the liquid friction, so there is almost no noise generated when the healthy bearing moves. In addition, the vibration test of space products incudes sine sweep and steady-state random two kinds, in view of the limitations of VMD random vibration signal analysis, this paper uses the sine sweep test as sample data analysis, the bearing vibration response of oneness to ensure that the original signal fundamental frequency, is advantageous to the VMD process fault component extraction, and no need to make optimization method of VMD.

In the sine sweep frequency test, the bearing vibration response at the initial low frequency stage

IOP Conf. Series: Materials Science and Engineering **751** (2020) 012017 doi:10.1088/1757-899X/751/1/012017

contains fault characteristics. At this time, the fault characteristic frequency of friction caused by the fault of sliding bearing is much higher than the fundamental frequency of the normal vibration response of the bearing, while the vibration response signal of healthy bearing only exists after filtering out the noise interference. Therefore, the fault bearing can be accurately located only by screening out the response of other harmonic components besides the fundamental frequency signal. The decomposition layer number of 4 is sufficient in the VMD process below 10Hz.

The acceleration signals $x_i(t)$ of each bearing in a certain period of low frequency range are collected and intercepted. The data are decomposed into many modal sequences by variational modal decomposition like $u_1^i(t)$, $u_2^i(t)$, ..., $u_n^i(t)$. $u_1^i(t)$ is the time domain signal composed of the fundamental frequency of the original signal, other signals are the time domain signals corresponding to the residual harmonic frequency except the fundamental frequency. To facilitate the interpretation of the amplitude of harmonic components, the amplitude square of the original signal is separated from the amplitude square of $u_1^i(t)$:

$$R_i(t) = x_i(t)^2 - u_1^i(t)^2 \tag{3}$$

"t" is the time range of intercept data, and the absolute value of variance of squares is taken to obtain the monitoring index of bearing status:

$$ENGR_i(t) = |R_i| \tag{4}$$

4. Test analysis

4.1. Test set

During the horizontal vibration test of a product with a mass of nearly 1 ton, the acceleration sensor monitoring the bearing state collected the response signal of abnormal bearing state. The equipment used in the test is the domestic 20-ton thrust electric vibration table, the size of the slide plate is 1.5m×1.5m, and a total of 16 bearings under the slide plate are distributed 4×4 uniformly. The test condition is 4~100Hz sine sweep frequency test, in which 4~8Hz is equal displacement input stage.

In the frequency sweep process, burrs appear in the response curve of the bearing at the low frequency stage, which is most obvious when the vibration starts at 4Hz. Through preliminary interpretation, it is found that all the bearing responses have burrs, which is because the dry friction of the faulty bearing leads to the obstructed running of the sliding plate, and the other abnormal bearing responses are indirectly caused by the sliding plate. After further processing and analysis of the response data of each group, the fault bearing can be accurately located.

4.2. Data processing process

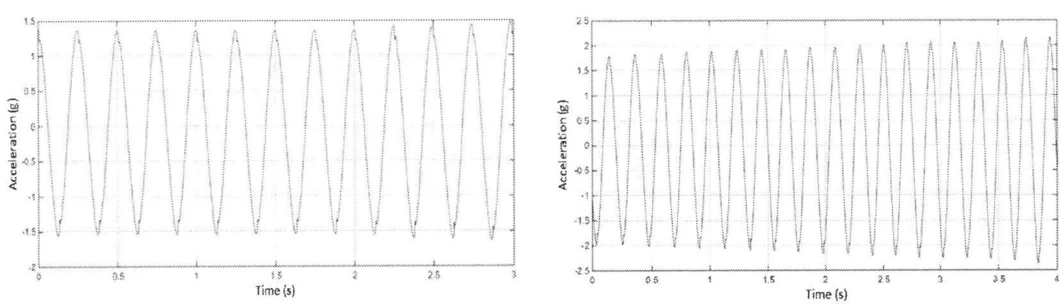

(a) (b)

Figure 3. Typical acceleration time domain signal of bearing

Figure 3 shows the time-domain curve of typical bearing acceleration. Figure 3 (a) is the data collected within a period of time after the sweep test reaches the initial position of the control spectrum; Figure 3 (b) is the data with certain changes in frequency and amplitude. The VMD method is suitable for the decomposition of unsteady signal, so the frequency of signal is allowed to have certain frequency conversion characteristics. Each bearing signal is decomposed, and the number of decomposition layer is 4. Figure 4 shows the curves of each layer after data decomposition of group a and group b. The first layer of data is the basic frequency waveform of the signal, and the second layer to the fourth layer are all high order frequency waveform. It can be seen that higher-order frequency waveform will appear "burr" at the corresponding position of peak and trough of the original signal.

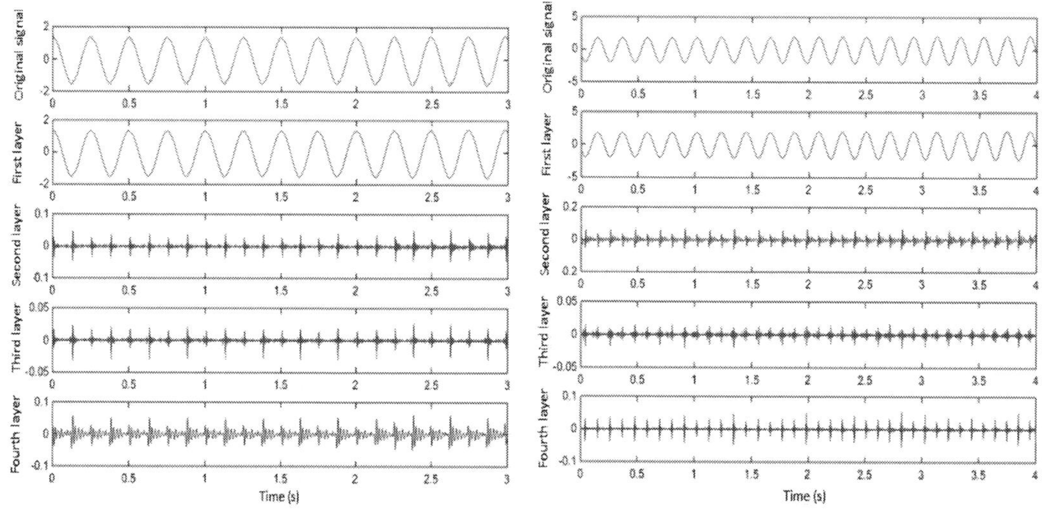

(a)Data decomposition results of group a (b)Data decomposition results of group b

Figure 4. Data decomposition results of two groups

Calculate according to formula 3 and formula 4, and draw the index waterfall diagram (Figure 5.).It can be seen from the figure that the bearing No. 4 shows significant spikes in the intercept time range. It can be seen that the method proposed in this paper can effectively identify the bearing numbers in abnormal states.

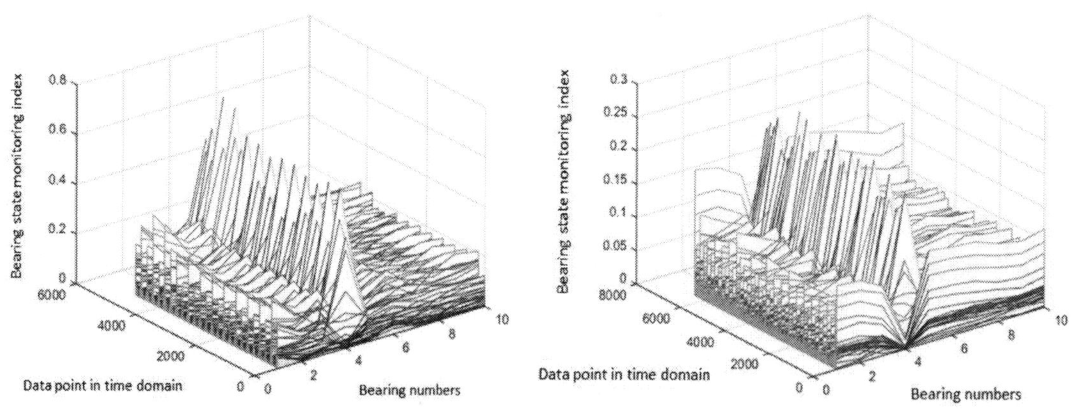

(a)Test results of group a (b)Test results of group b

Figure 5. Bearing condition test results

4.3. Analysis instructions

After locating No. 4 bearing as a failure bearing through variational mode decomposition method, the test personnel removed the horizontal slide plates of the vibration table and checked each bearing state one by one. Only No. 4 bearing showed wear, See the figure 6 for details.

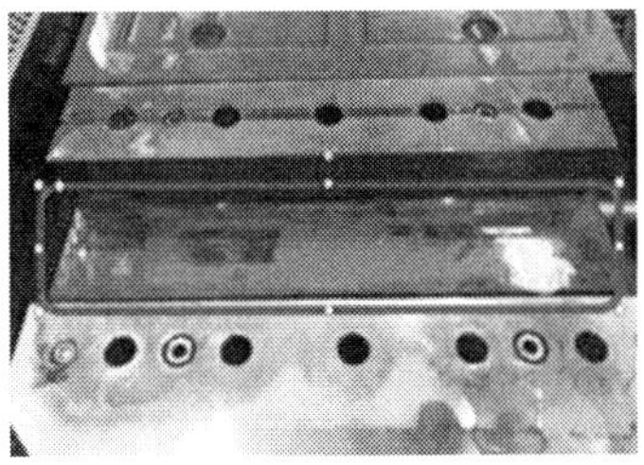

Figure 6. Physical drawing of bearing wear

Due to the role of the bearing monitoring system, the fault bearing was found and positioned at the early stage of failure. The monitoring system effectively avoided the possibility of bearing problems worsening or even locking, and ensured the safety of test products.

5. Conclusion

As an effective evaluation method, the embedded acceleration monitoring method is suitable for fault diagnosis of multi-guide rail or multi-bearing sliding platform. A monitoring sensor is installed inside the sliding plate for each bearing in an embedded way, which can ensure the accuracy of bearing status indicator positioning. Status indicators evaluation to realize online synchronous calculation, after the own law, has the good real-time and intuitive, in the early stages of the abnormal state of bearing, effectively avoid shaker fault caused by damage to the product structure seriously, research is of great significance to the success of space products, and the production development of shaker has very good reference value.

References

[1] Jin Xushu. A satellite vibration test accident in the United States 2002. *Engineering of Spacecraft environmental*, 01.

[2] Dragomiretskiy, K., Zosso, D. Variational mode decomposition 2014. *IEEE Transactions on Signal Processing*, 62 (3), 531-544.

[3] Ming, Z., Jiang, Z., Feng, K. Research on variational mode decomposition in rolling bearings fault diagnosis of the multistage centrifugal pump 2017. *Mechanical Syatems and Signal Processing*, 93(460), 460-493 .

[4] Ma Zengqiang. Li Yachao. Ruan Wanying. Rolling bearing fault feature extraction based on VMD and spectral kurtosis 2017. *Journal of Electronic Measurement and Instrumentation*, 31(11), 1782-1787.

[5] LI Hua, WU Xing, LIU Tao. Bearing fault feature extraction based on VMD optimized with information entropy 2018. *Journal of Vibration and Shock*, 37(23), 219-225.

[6] ZHANG Chen, ZHAO Rongzhen, DENG Linfeng. Weak fault identification of rolling bearings based on VMD singular value entropy 2018. *Journal of Vibration and Shock*, 37(21), 87-91.

IOP Conf. Series: Materials Science and Engineering **751** (2020) 012018 doi:10.1088/1757-899X/751/1/012018

Comparative Research on Early Warning Determination Methods of Spacecraft Collision

Sun Hongqiang[1,*], Zhang Zhanyue[2] and Hu Ewen[1]

[1] Space Engineering University, Graduate School, Huairou, 101416, China
[2] Space Engineering University, Huairou, 101416, China

[*]Email: sjtushq@163.com

Abstract. With the development of space science, more and more spacecraft have entered into the space, and the operating environment of spacecraft in space has become more complex. Various signs show that the future space field will face congestion, collision and other safety problems, so the collision warning in space is becoming more and more important. It is the primary guarantee for the spacecraft to carry out its mission in space in the future to obtain the early warning information and complete the avoidance within the early time to ensure the environmental safety. In this paper, based on the perspective of collision warning and determination methods, two kinds of warning and determination methods are introduced, including BOX region method and probability method. In addition, by comparing the warning efficiency of these two methods, the results show that, under the same constraint, the probability method is more accurate and saves more space resources than the BOX region method. Considering that the spacecraft is limited in fuel and other resources when working in orbit, it is more scientific to adopt to avoid unnecessary maneuver and waste of resources.

1. Introduction.

Since most space missions are concentrated in low-earth and Geostationary Orbit regions, the vast majority of space debris is currently scattered in Low Earth Orbit (LEO) and Geostationary Earth Orbit (GEO). Figure 1 shows the distribution of space targets. LEO debris, its speed is close to the first cosmic speed, about 7 ~ 8 km/s, and spacecraft collision speed can be as high as 10 km/s, even could be up to 16 km/s[1]. The greater the mass of space debris, and the grater the damage is.

Due to the space debris which is more than 10 cm are the biggest threat to the safety of the astronauts, space missions, evasive maneuvers[2], so take the initiative to avoid spacecraft and space debris collision method, is currently the most practical value, the method of space debris collision warning is a prerequisite for successful active evasive maneuvers.

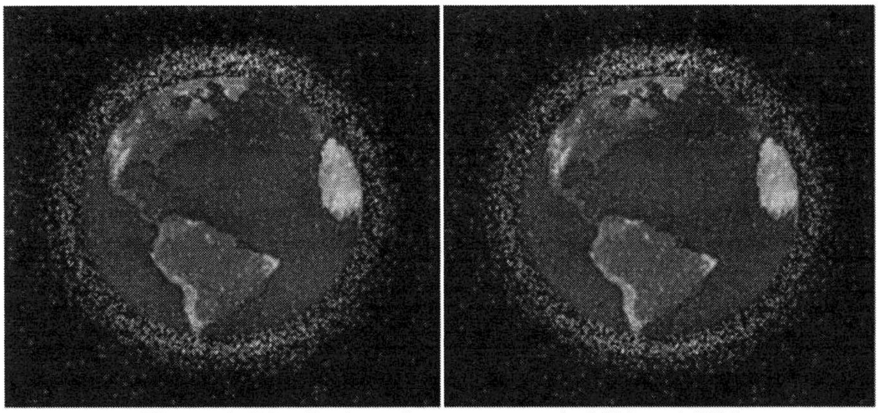

Figure 1. Distribution of space targets in LEO and GEO

2. Evaluation method of collision warning

In space environment observation, there are certain errors in observation, determination and prediction of spacecraft and space debris in orbit, as shown in figure 2. These errors make it become more uncertain for the collision events. it is necessary to evaluate the risk of collision with corresponding criteria[3]. The main criterions are minimum distance and collision probability.2.1.2. research status of formation satellites.

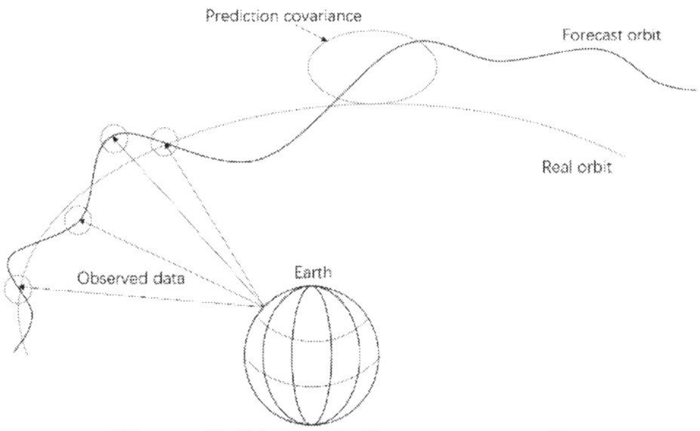

Figure 2. Diagram of error propagation

The BOX region method uses the minimum distance as the criterion for collision warning. The space area defined by the US Space Shuttle Collision Warning System is a rectangular box with the space shuttle as the center, track direction of ±25km, orbital plane normal direction and vertical track direction of ±5km. When the target enters the area, the system will give warning and provide more detailed orbital forecast data[4]. As shown in figure 3, when space debris enters this BOX area, the system will regard it as a dangerous target and conduct collision warning to provide evasive basis for subsequent spacecraft maneuvers.

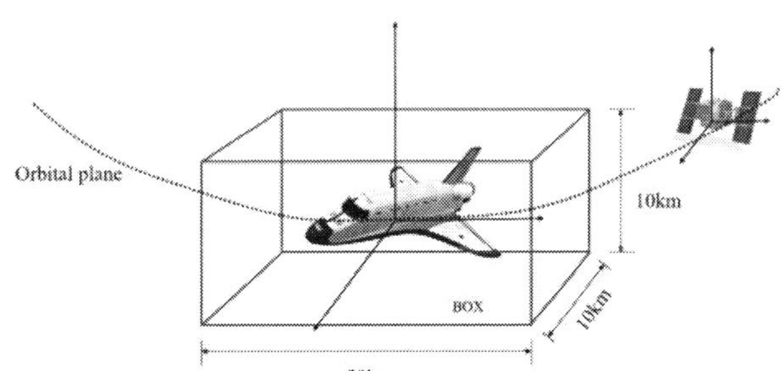

Figure 3. BOX region method

Collision probability method using the near distance, time of approaching and the information such as track error covariance, with the basis of the calculated probability to determine the collision probability between spacecraft and debris, when the collision probability is greater than the given threshold, early warning system will send out warning signals, the spacecraft perform evasive maneuvers. NASA divided the probability Threshold of evasive maneuvers into Yellow Threshold with a value of 10^-5 and Red Threshold with a value of 10^-4. When the collision probability is greater than the red threshold, evasive maneuver is carried out. When the collision probability is higher than the yellow threshold, evasive maneuvers will be carried out without harming the target of the mission.

When the distance between the two targets is less than the sum of their equivalent radius, then it can be judged as the occurrence of collision. The calculation of collision probability by analytic method can be simplified into the following steps[5]:

1, Obtain the information (including time of approaching, the position and the speed of the spacecraft and debris, the minimum distance between them) and the error covariance matrix;

2. Establish the encounter plane and the encounter coordinate system, and project the motion state vectors of spacecraft and space debris onto the encounter coordinate system;

3. According to the errors and radius of the two targets, the joint error ellipsoid and union are formed and projected to the meeting coordinate system;

4. Calculate the collision probability based on the above data.

3. Collision warning simulation

The above two collision warning methods are simulated and verified as follows. Take the spacecraft in low earth orbit as an example. The scene is set in a space with an orbital altitude of H=1000km. The position of spacecraft orbits is shown in table 1. Simulation environment: STK11+Matlab2015b.

Table 1. Position of main space spacecraft

	Semi-major Axis (km)	Eccen tricity	Inclination (deg)	RAAN (deg)	Arg of Perigee (deg)	True Anomaly (deg)
Main Target	7478.14	0	50	100	0	40

3.1. Early warning simulation based on BOX region

The BOX region was defined as a rectangular BOX centered on the spacecraft, with the track direction being ±25km, the orbital plane normal ±5km, and the vertical direction being ±5km. Randomly generate several debris in the space near the cuboid, including the location information of randomly generated debris.

As shown in figure 4, the central blue ball is the main spacecraft, black dashed cuboid around the spacecraft is divided by the BOX area. Circles are 100 randomly generated debris around the spacecraft space, among them, the green circles are outside the BOX cuboid region, the red circles are in the BOX the cuboid region.

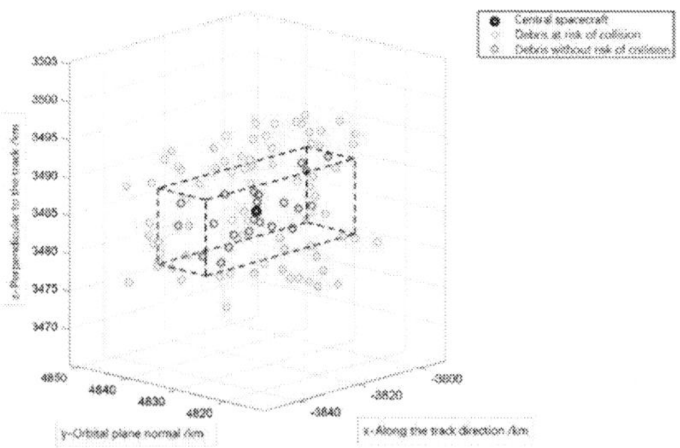

Figure 4. Randomly generate space debris around the BOX area

According to the minimum distance principle of the BOX region method for collision warning of spacecraft, debris entering the BOX region is deemed as debris with collision risk, that is, all space debris marked with red circles in the figure 4. Collision warning is issued for these debris to remind spacecraft of collision avoidance. Among the 100 randomly generated space debris, 22 are distributed in the BOX region. Therefore, the proportion of warning for this random scene is 22÷100=22%.

To reduce the influence caused by the accidental event, do above random space debris of scene simulation in many times to get average number of warning and warning probability, as shown in figure 5, although each time the randomly generated 100 space debris, the number of dangerous debris is variable, but it showed a trend of the distribution of the number is in the 20 or so. After processing these 1000 sets of data, the median of dangerous debris is 21 and the average is 20.8830.

Figure 6. shows the relationship between the number of randomly generated debris and the proportion of dangerous debris. It can be seen that when the number of randomly generated debris is small, the proportion of dangerous debris tends to be unstable and divergent. When the number of randomly generated debris reaches a certain level (thousand degrees), the proportion of dangerous debris shows a convergence trend, and the convergence trend becomes more stable with the increase of the total number of debris. According to figure 6, the proportion of dangerous debris converges at y=0.20874, so the average warning rate in this scene is 20.874%.

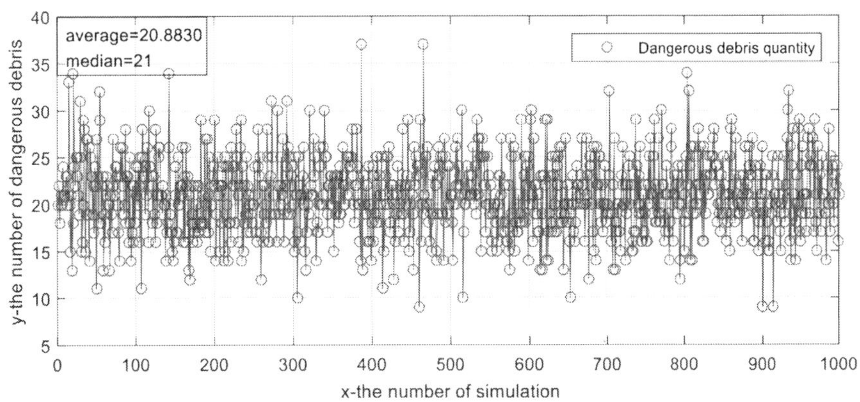

Figure 5. Number of dangerous debris in multiple scenarios (BOX region method)

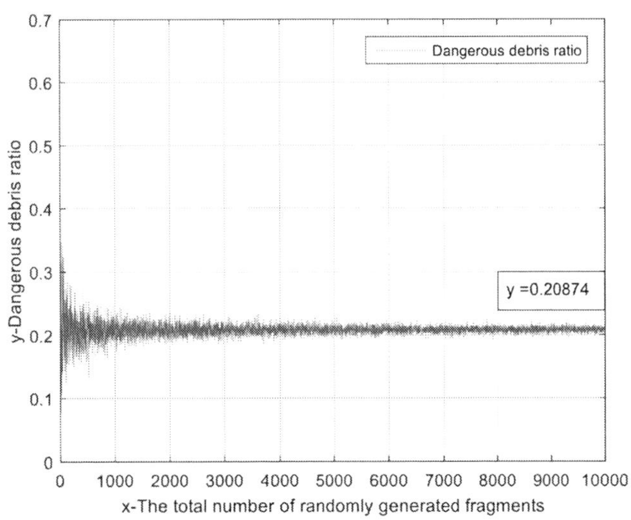

Figure 6. The number of debris and the dangerous ratio (BOX region method)

IOP Conf. Series: Materials Science and Engineering **751** (2020) 012018 doi:10.1088/1757-899X/751/1/012018

3.2. Early warning simulation based on collision probability

In A number of randomly generated debris are generated in the space area centered by the main spacecraft, including the location and velocity information of randomly generated debris as the initial conditions.

As shown in figure 7, the blue ball in the center is the main spacecraft, and the circle represents the generated space debris randomly around the spacecraft. After the calculation of collision probability, the space debris with collision risk is marked as red, and the debris without collision risk is marked as green. The BOX area for the spacecraft is represented by a black dotted line.

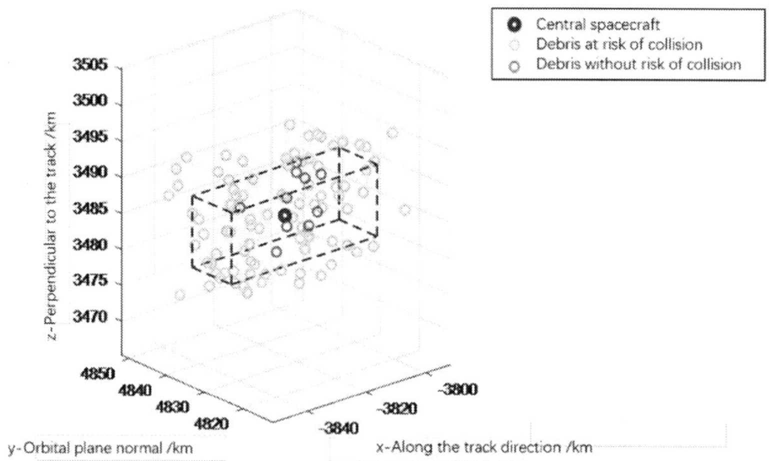

Figure 7. Randomly generate space debris around the main spacecraft

Through comparison, the following two points can be drawn:

1) Not all the debris in the BOX area exceeds the collision risk of the safe threshold;

2) The collision probability of debris outside the BOX area is not always below the safety threshold.

Among these 100 randomly generated space debris, there are 10 of them have collision risk. Therefore, the proportion of warning for this scenario is 10÷100=10%.

To reduce the influence caused by the accidental event, do above random space debris of scene simulation in many times to get average number of warning and warning probability, as shown in figure 8, although each time the randomly generated 100 space debris, the number of dangerous debris is variable, but it showed a trend of the distribution of the number is in the 10 or so. After processing these 100 sets of data, the median of dangerous debris is 11 and the average is 10.7050.

Figure. 9 shows the relationship between the number of randomly generated debris and the proportion of dangerous debris. It can be seen that when the number of randomly generated debris is small, the proportion of dangerous debris tends to be unstable and divergent. When the number of randomly generated debris reaches a certain level (thousand degrees), the proportion of dangerous debris shows a convergence trend, and the convergence trend becomes more stable with the increase of the total number of debris. According to figure 9, the proportion of dangerous debris converges at y=0.10677, the average warning rate in this scene is 10.677%.

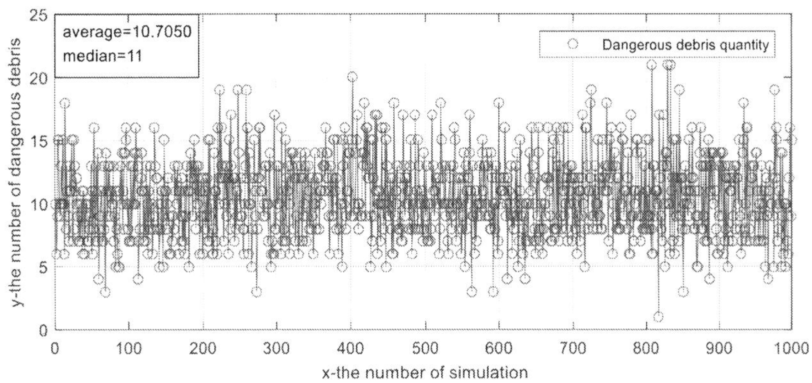

Figure 8. Number of dangerous debris in multiple scenarios (collision probability method)

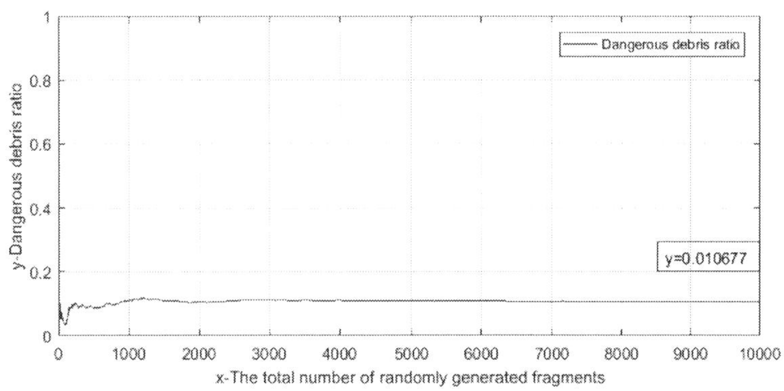

Figure 9. The number of debris and the dangerous ratio (collision probability method)

3.3. Compare and analysis

For this scenario, the probability of collision warning and collision warning BOX area method are compared, as shown in figure 10, the collision probability method used to determine the number of dangerous debris is around 11, and the BOX area method used to determine the number of dangerous debris is around 21, significantly higher than the early warning method number.

Therefore, the use of BOX area collision warning will produce more than a few "false alarms", which will cause waste to the spacecraft within fuel and other resources, not conducive to the continuous work in space.

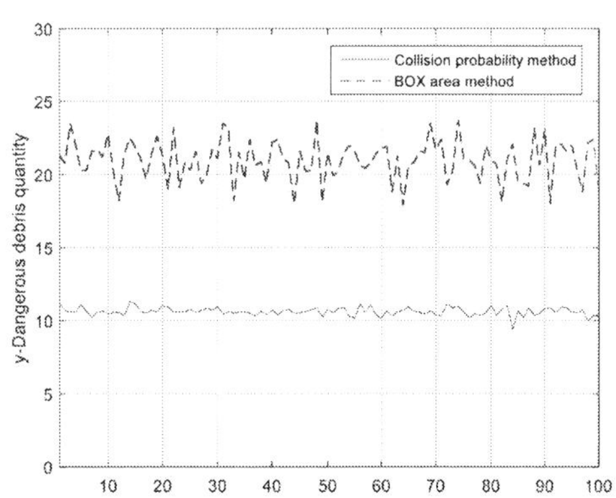

Figure 10. Comparison of the number of danger warning

IOP Conf. Series: Materials Science and Engineering **751** (2020) 012018 doi:10.1088/1757-899X/751/1/012018

4. Conclusion

The space environment of spacecraft in orbit is complex, so it is very important to avoid dangerous space debris. In this paper, two common collision warning methods, namely BOX region collision warning method and collision probability warning method, are compared. These two methods respectively calculate collision problems from geometric and analytical perspectives. After comparing the two methods, which can be found that the warning number of BOX area method is more than collision probability method, this because the BOX area method just takes the distance between the spacecraft and debris into consideration, and neglects the collision velocity and other important judgment information. To a certain extent, it creates a "false alarm" phenomenon, mobile the spacecraft with no risk of collision, wasting fuel of spacecraft, also caused the waste of space resources. Two conclusions can be drawn:

1) Not all the debris in the BOX area exceeds the collision risk of the safe threshold;

2) The collision probability of debris outside the BOX area is not always below the safety threshold.

Therefore, the collision probability calculated by using the distance, velocity and error covariance information between space spacecraft and debris is more accurate, and the number of dangerous debris screened by threshold value is relatively small, thus reducing the number of maneuver. Considering that the spacecraft is limited in fuel and other resources when working in orbit, it is more scientific to adopt collision probability calculation method to avoid unnecessary maneuver and waste of resources.

References

[1] Li Yiyong, Shen Huairong, Li Zhi. *Environmental hazards of space debris and countermeasures* [J] Missile and space delivery technology, 2008 (6): 31-35

[2] Sun Hao, Dong Jie, Luo Chao. *Enlightenment from research and development of space debris monitoring technology* [J]. Chinese Radio, 2016(3):63-64.

[3] Bai Xianzong. *Study on prediction error and collision probability of space target orbit* [D]. National university of defense technology, 2013.

[4] Cheng Tao, Liu Jing, Wang Ronglan. *Research on collision probability method in space debris warning* [J] Journal of space science, 2006, 26(6):452-458

[5] Huo Yurong. *Research on error analysis method of space debris collision warning* [J].Science and technology of China, 2016(7).

IOP Conf. Series: Materials Science and Engineering **751** (2020) 012019 doi:10.1088/1757-899X/751/1/012019

Research on FMEA Application in Assembly Process of Complex Space Probes

Chenxing Wang, Wei Feng, Linzheng Wang and Song Song

Beijing Institute of Spacecraft Environment Engineering,Beijing 100094, China

E-mail: artemisstars@qq.com

Abstract. The process of developing new spacecraft is time-consuming and task-intensive. This paper analyses various problems existing in the application of spacecraft FMEA, and points out that it is necessary to it is necessary to assure the validity of FMEA work, including normative FMEA work carried out in time, comprehensive identification of failure modes and their causes, correct expression of failure modes and their causes and effects, and the pertinence and maneuverability of measures to be taken. Taking the development of Mars rover as an example and considering the actual situation, the assembly process and FMEA agreement are proposed and formulated. According to the development process, the failure modes, causes and effects (severity) of the main processes are analyzed. The risk sequence number (RPN) is obtained by combining the frequency of occurrence and the difficulty of detection. Finally, through sorting out the assembly work and key points in each stage of AIT, 60 processes and 95 failure modes are analyzed, the possible weaknesses are found out, and the control methods and measures are put forward, so that the RPN values are below 12. Improve the quality and efficiency of product development by strengthening FMEA management, and ensure the smooth development of Mars probe on.

1. Introduction

FMEA(Failure Mode Effects Analysis) is an analysis technology that analyses every potential failure mode in a product and determines its impact on the product and classifies each potential failure mode according to severity level[1].Since the first application of this method in the Grumman Aircraft Company(USA) development of aircraft control system in the 1950s, many international aerospace companies have taken compulsory measures to vigorously implement the FMEA method. For example, various types of satellites in the United States have stipulated that FMEA must be widely used. FMEA runs through the entire satellite product. The process, including strategy research and implementation, system management and operation, product design, process design, and so on, has achieved good results. FMEA has been gradually implemented in spacecraft design in China, which has played an important role in improving the reliability of spacecraft[2]. However, the reliability analysis of spacecraft assembly process is still in the exploratory stage in China, and there is a big gap compared with the mature reliability management system of spacecraft assembly abroad. Therefore, the research and application of process FMEA should be strengthened.

Functional FMEA, hardware FMEA and process FMEA are involved in spacecraft development. Functional FMEA functional FMEA and hardware FMEA hardware FMEA can be completed in the design stage, eliminating the weak links. Potential failure modes of process FMEA are eliminated iteratively along with product implementation process, mainly including processing, assembly, assembly, testing and so on[3].

Aiming at the actual situation of Mars Explorer development process, the standards are formulated and followed. Through sorting out the assembly work and key points in each stage of AIT, the possible weaknesses are found out, and the control methods and measures are put forward to improve the quality and efficiency of product development in management.

2. FMEA Workflow and Conditions

2.1. FMEA Workflow

Spacecraft companies pay more attention to and mostly carry out FMEA work than before, but process FMEA is seldom carried out, and there are various problems at the same time[4].

- FMEA management is not in place, and FMEA work has no closed-loop;
- FMEA lacks systematicness and low effectiveness.
- There are phenomena of "replacing reliability professionals" and "making up afterwards" in FMEA work, and the effect of FMEA is difficult to guarantee.

The FMEA of spacecraft assembly process should be standardized, especially the new process. Process FMEA should be implemented at the beginning of process design and finished before the process is formally applied. Assuming that the product design meets the requirements, identify the possible failure modes, causes and their effects on the product or process in each step of the process, identify the weak links in the process, and propose the possible improvement measures. The FMEA workflow of spacecraft product process is shown in Figure 1.

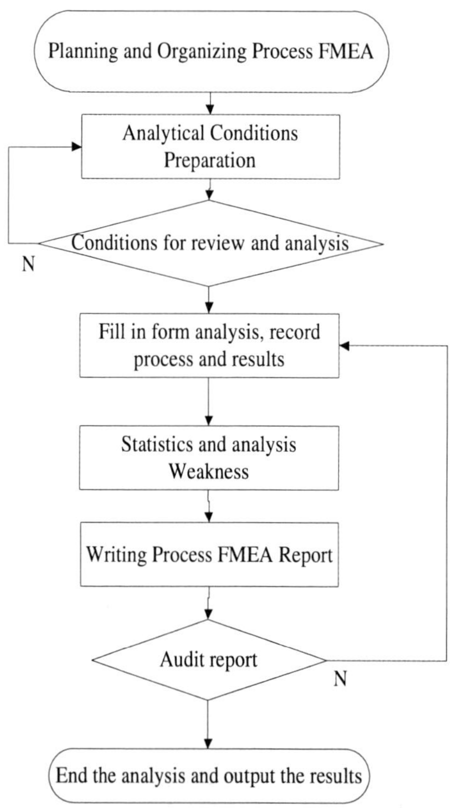

Figure 1. The FMEA workflow of spacecraft product

2.2. Conditions
According to the assembly technology process, the main processes are analyzed according to the sequence of stages, and the following agreements are made:

- The assembly process FMEA only deals with the important links in the assembly process at all stages starting from the delivery of the product, excluding the production process of the stand-alone product.
- Process FMEA analysis does not involve the installation of separation devices and pyrotechnics, which is reflected in special documents.
- The repetitive work, such as docking, decomposition, deck opening and closing, product hoisting, installation and disassembly of equipment, and cable insertion, is analyzed only once.
- It does not involve obvious and non-technical human factors, such as the instability of the grasp during the transmission of instruments and equipment.
- FMEA analysis does not involve potential hazards caused by force majeure.

3. Technical Scheme
According to the technical status and process of the assembly, the failure modes, causes and effects (severity) of the main processes are analyzed. The risk sequence number (RPN) is obtained by combining the frequency of occurrence and the difficulty of detection. The total risk sequence number (TRPN) is taken as the basis for determining the control measures[5].

3.1. Severity of failure effects
A severity category classification shall be assigned to each identified failure mode analysed. Severity categories are assigned without consideration of existing compensating provisions to provide a qualitative measure of the worst potential consequences resulting from item failure. The agreement on severity is shown in table 1.

Table 1. Severity numbers applied at the different severity categories

Definition	Severity category	**SN**
catastrophic	If the product is scrapped and the economic loss is serious, it will have a serious impact on the function of the system product, and will cause harm to people, such as loss of life or permanent disability injury or occupational disease.	4
critical	It causes product damage or major overshoot and cannot be repaired; tasks can not be completed; temporary disability but no life-threatening injury or temporary occupational disease.	3
major	It causes difficulties in the processing of this or subsequent processes, product overshoot, difficulty in rework and repair, task degradation or process degradation.	2
negligible	Excess tolerance is acceptable by concession; any other effect.	1

3.2. Probability of occurrence
The probability of process FMEA failure mode occurrence is called process FMEA failure mode occurrence degree, which is a qualitative measure of the possibility of process FMEA failure mode occurrence caused by a certain reason. It is usually determined by experience or process statistics.

IOP Conf. Series: Materials Science and Engineering **751** (2020) 012019 doi:10.1088/1757-899X/751/1/012019

Only through process design to prevent or control the causes of process failure can the occurrence of process failure modes be reduced. Identify the probability of occurrence of Very likely, Likely, Unlikely, Extremely unlikely are defined as 4, 3, 2, 1.

3.3. probability of detection
Process FMEA fault mode detection degree refers to the qualitative measurement of process fault mode detected, which is divided into four levels. The agreement on the detection level is shown in Table 2.

Table 2. Detection numbers (DN) for probability of detection

Definition	**Instructions for Degree of Detection**	DN
Undetectable	There is no known detection method and means to detect the failure mode.	4
Indirect detection or visual measurement	The current detection methods and means are difficult to detect fault modes, or can only be detected by high-level assembly or testing	3
Direct inspection after completion	In the follow-up process or product, the fault mode can be detected by the current detection methods and means	2
Direct Inspection of Process	The fault mode can be detected by the current detection methods and means in the process	1

3.4. Determination of the criticality number (CN)
Risk Priority Number (RPN) is a measure used when assessing risk to help identify critical failure modes associated with your design or process. The criticality number (CN) shall be defined as the product of the numbers assigned to failure mode severity, probability of occurrence, and probability of detection according to:

RPN= （SN） × （PN） × （DN）

The RPN values range from 1 (absolute best) to 64 (absolute worst). The FMEA RPN is commonly used in the automotive industry and it is somewhat similar to the criticality numbers used in Mil-Std-1629A.

3.5. Criticality acceptance criteria
The risk of a potential weak point is regarded as unacceptable and a recommendation for additional preventive or compensatory provisions shall be given if:

- the severity number $SN \geq 3$
- the probability number $PN = 4$
- the detection number $DN = 4$
- the criticality number $CN \geq 12$

Critcality acceptance criteria may be tailored to suit individual projects.

4. Detector assembly process and FMEA results

4.1. Detector assembly process
The general development process of spacecraft is shown in Figure 2. In this paper, process FMEA analysis is carried out according to the flow chart and assembly characteristics of Mars probe development.

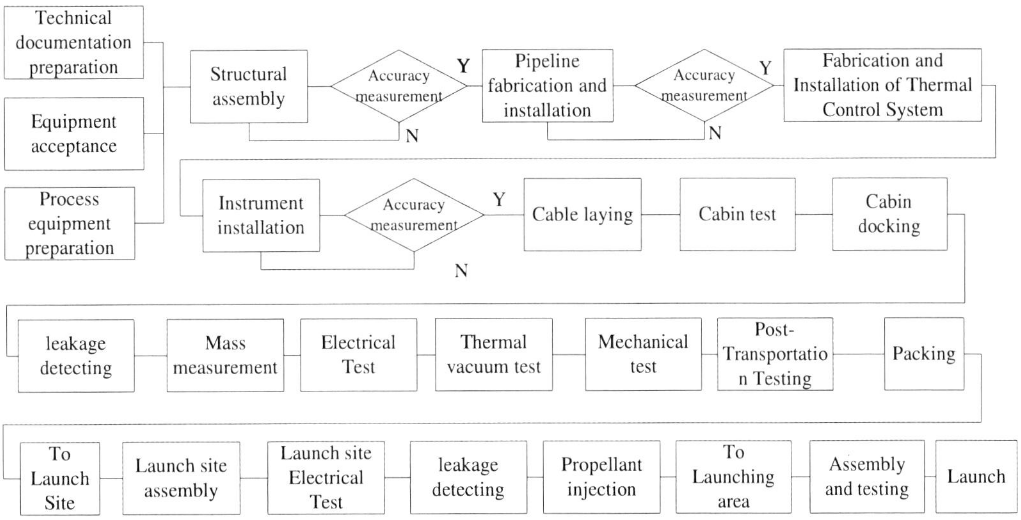

Figure 2. The FMEA workflow of spacecraft product

4.2. FMEA results

FMEA of MARS-1 detector assembly process has analyzed 60 processes and 95 failure modes. The detailed analysis results are shown in Table 3.

Table 3. The detailed analysis results

Process Function Requirements	Potential Effects Of mode	Potential Effects of Failure			Potential Causes of Mechanisms of Failure	Detection means	Before improvement				Process Control Prevention and Inspection	Improving measures after implementation			
		Follow-up process	The whole process	Product (spacecraft)			SN	ON	DN	RPN		SN	ON	DN	RPN
Probe packing	Structural Collision Damage	Follow-up process delay	delay	Decreased structural strength or loss of function	1.Uncontrollable sloshing of suspension; 2.Hanging point connectors are damaged during storage and previous use.	Visual	4	2	2	16	Strictly control the number of personnel on site; Personnel command hoisting; Strict control of crane speed; Before use, check and confirm that there is no mistake, then connect and install the hanger.	2	1	2	4

According to the FMEA analysis results of the assembly process, considering the severity, occurrence, detection and RPN values, the key control links are refined assembly requirements, quantitative indicators are proposed, mandatory checkpoints are set, forms are formed, and tracked in the actual process.

The completed assembly work proves that after the implementation of the improvement measures, the risk priority of the assembly process mainly concentrates in the lower numerical range, and most of the faults are detectable, the ground state can be solved, the severity is low, and the impact on the

IOP Conf. Series: Materials Science and Engineering **751** (2020) 012019 doi:10.1088/1757-899X/751/1/012019

detector flight is small. The follow-up assembly work, due to its long cycle, complex tasks and many interface features, is mainly to clarify the overall requirements, improve the quality of personnel, refine process regulations, and strengthen site management is an effective means to prevent and control failures.

5. Conclusions

In the end, taking the development of Mars probe as an example and considering the actual situation, the assembly process and FMEA agreement are proposed and formulated.According to the development process, the failure modes, causes and effects (severity) of the main processes are analyzed, and the risk sequence number (RPN) is obtained by combining the frequency of occurrence and the difficulty of detection as the basis for determining the control measures. By sorting out the assembly work and key points in each stage of AIT, 60 processes and 95 failure modes are analyzed, and possible weaknesses are identified. Process control measures to eliminate these failure modes and their causes or to reduce the risk to an acceptable range (RPN below 12) are proposed.And improve the quality and efficiency of product development by strengthening the management of FMEA.

References

[1] Department of the NAVY Commanding Officer. MILSTD-1629A Failure mode effect analysis[S].Department of Defense,Washington D.C.,1980
[2] YI Wangmin,SUN Gang,LIU Hongyang.Reliability of Spacecraft Assembly,Space Environment Engineering, 2006,23(3):161-164
[3] Guidelines for failure modes and effects analysis of spacecraft products,Q/W 1364-2012
[4] LIU Zhiquan,GONG Ying. Consideration about the Validity of Aerospace Product FMEA,Space craft Engineering,2011,20(1):142-146
[5] ESA. Space Product Assurance Failure modes, effects and criticality analysis (FMECA),ECSS-Q-30-02A-2001

IOP Conf. Series: Materials Science and Engineering **751** (2020) 012020 doi:10.1088/1757-899X/751/1/012020

Assembly Method for Satellite Device Installation based on HoloLens

Wei Zhang[1,*], Qiang Zhang[1], Zhe Liu[1], Chengli Zhang[1], Wei Wang[1], and Chengshuang Yin[1]

[1] Beijing Institute of Spacecraft Environment Engineering, Beijing Engineering Research Center of the Intelligent Assembly Technology and Equipment for Aerospace Product, Beijing, China

[*]E-mail: zhangwei308308@163.com

Abstract. As the satellite design changes from the traditional 2D mode to the 3D digital mode, the satellite assembly also enters the 3D assembly mode. An effective and accurate device installation process information transmission method from designers to operators is not formed, and accidents with incorrect polarity in device installation often occur. In order to improve efficiency and quality of satellite device installation, modelling method of R point of device is regulated and a system based on HoloLens is designed. With the help of the system, the code and R point are clear at a glance, so an ordinary operator can verify the polarity correctness of the device installation and the efficiency can be doubled.

1. Introduction

Device installation is a frequent operation in AIT process of satellite, installation errors of important device can lead to accidents of final assembly, such as equipment bumping and cable squeezing, telemetry signal errors, etc., and even more directly affects the functionality of the satellite. The devices which are numerous are divided into several subsystems, and the devices have a complicated layout on the satellite. Before the installation of devices, the operator must confirm the installation location, the installation direction, and the installation process information through the 2D drawings and process files. After the installation of devices, the inspector must check the polarity of the devices, and other installation processes such as the thermal control implementation of the device.

Model based definition (MBD) technology is a method of 3D digital product definition, the core issue of the design is ensure that the MBD technology is perfectly integrated into the model to make it work better[1-2]. As the satellite design changes from the traditional 2D mode to the 3D digital mode, research on how to create effective MBD process information model is one of the major problems. Most satellites belong to one-piece customized production, the device layouts of different satellites have large differences, the final assembly space is small, the operation conditions are complex, the various layouts of small devices are scattered and coded, and therefore, the final assembly of the satellite device is heavily dependent on manual work. In order to complete the installation, the worker needs to obtain the required information from the drawing or 3D model before installation and confirm installation polarity and fastener mounting specifications after installation in the satellite cabin. The disadvantage of this type of method is that the relevant design information is manually matched to the real physical environment, and it is inefficient and error-prone. Work efficiency is reduced significantly because workers have to bear a lot of additional cognitive burden to confirm the correctness of assembly.

The device installation is error-prone, in recent years, virtual assembly technology is widely used [3-9]. In order to improve efficiency and quality of satellite device assembly, a device assembly system based on AR technology is designed. This method automatically matches the device installation information with the real physical environment, and guides the workers to complete the assembly. In the field of complex mechanical assembly, enterprises have researched the assembly guidance method based on augmented reality, but cannot meet the needs of satellite single-piece production and unable to systematically solve the problem of automatic conversion of device assembly 3D models to visual process information.

2. Device Installation in Traditional AIT Process of Satellite

2.1. Process of Satellite Device Installation

Device installation is a frequent operation in AIT process of satellite, and the process of device installation in AIT process of satellite is showed in Figure 1. Before the installation of devices, the operator must confirm the installation location through the code of the device using 2D drawings, confirm the installation direction through the R hole using 2D drawings, and confirm the installation process information using process files. The basis of the process documentation includes 2D drawings and technical requirements of mechanical designer, technical requirements of thermal control designer, etc. After the installation of devices, the inspector must check whether the position and polarity of the devices installation are correct, and whether other implementation processes such as the thermal control implementation of the device are correct. In short, it takes a lot of time to find the response drawings and process documentation before the operation.

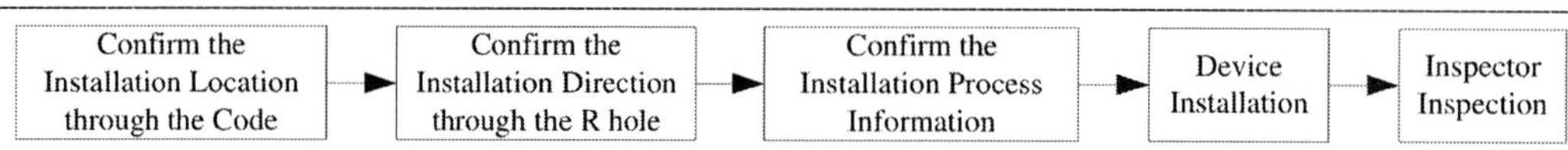

Figure 1. Process of Satellite Device Installation

2.2. The traditional 2D drawings of Satellite Device Installation

In addition to the process documentation, the operator needs to look at the drawings before installing the device. The 2D drawings of satellite device Installation is showed in Figure 2, for example, the code of the temperature controller is R01, the number of fasteners is four, the specification is M4*14, the position of the device is on the +Y side of the deck 21-03, the R Point Feature which is labelled on the device is on the +X-Z direction of the deck. The polarity of the devices and the mechanical installation process information are clear, but the thermal control installation process information which can only be seen in process documentation is not included. The operator needs a lot of prior knowledge to confirm the position of the devices.

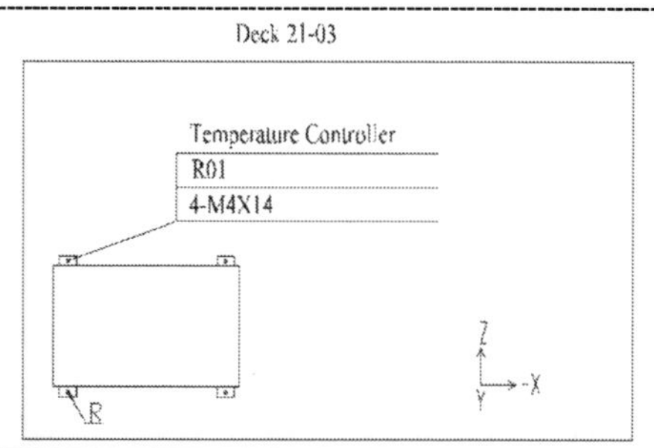

Figure 2. 2D Drawings of Satellite Device Installation

3. R Point Modelling of Device in 3D design model

R Point Feature is an important process parameter in AIT process of satellite, and it guarantees the correct polarity of the installed instrument. In designing 3D models, different designers use different expressions, mainly in three forms. As is show in Figure 3 and Figure 4, the stretch feature and the sketch feature are suitable for operators to use when viewing model entities on a computer, not suitable for machine identification. The point Feature which is show in Figure 5 is suitable for machine identification, but not suitable for operators to use when viewing model entities on a computer, the large amount of annotation information may lead to the hedgehog phenomenon, which will impact on the quick and correct expression.

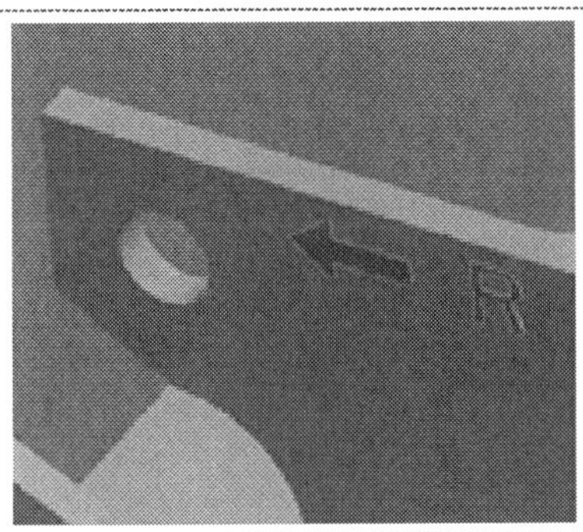

Figure 3. Stretch Feature.

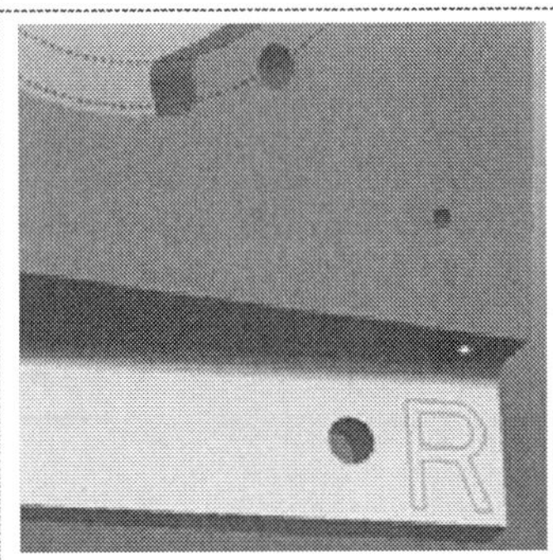

Figure 4. Sketch Feature

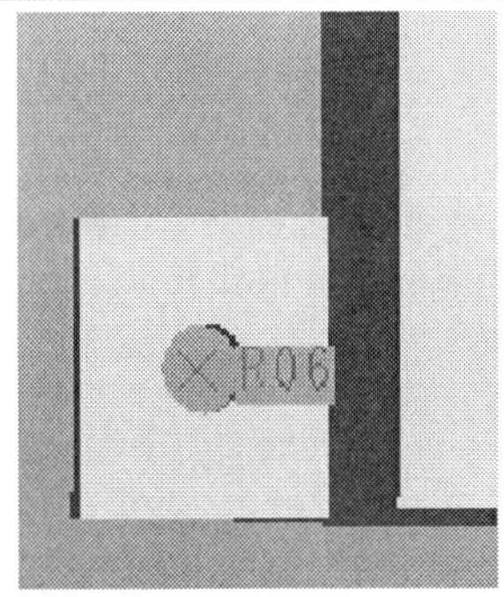

Figure 5. Point Feature

Suggested modelling method of R point of device which is show in Figure 6 can be easily accepted by the operator through subsequent technical means, R point which is labelled on the device is also created by point feature. For example, the device has 6 mounting holes, and the A06 is the R point.

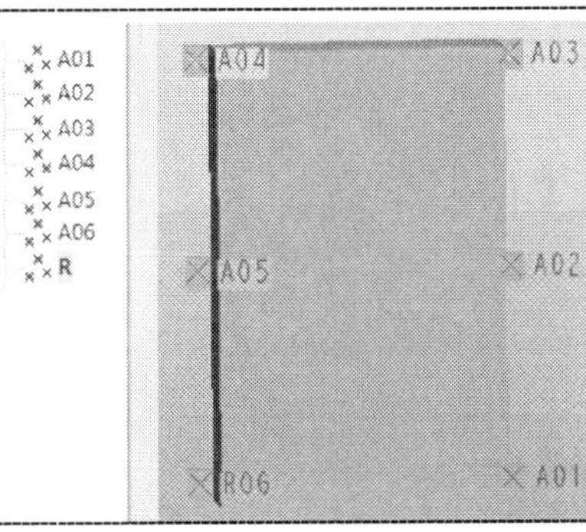

Figure 6. Suggested Modelling Method of R Point Feature of Device

4. Augmented Reality and the Microsoft HoloLens

As the satellite design changes from the traditional 2D mode to the 3D digital mode, the satellite device installation also enters the 3D assembly mode. Because the 3D design input is inconvenient to use, and an effective and accurate device information transmission method from designers to operators is not formed, accidents with incorrect polarity in device installation often occur. The current 3D model based on desktop computer has poor human-computer interaction, In the process of device installation, operators need to frequently confirm status back and forth between satellite and computer stations, and therefore, new technology is needed to improve assembly efficiency. Compared with Virtual Reality technology, Augmented Reality technology can not only see virtual scenes, but also superimposes virtual scenes in real scenes by calculating operator position and perspective in real time. The Microsoft HoloLens was released as a development edition in 2016 and is now available as a consumer version. It is the ideal choice for investigating an AR assembly application on a commercially available device.

5. The Process of Assembly Method for Satellite Device Installation based on HoloLens

A system based on AR technology is designed to improve efficiency and quality of satellite device installation, the flow chart of the method is showed in Figure 7. Compared with the traditional method, this method automatically matches the device installation information with the real physical environment, and guides the workers to complete the assembly.

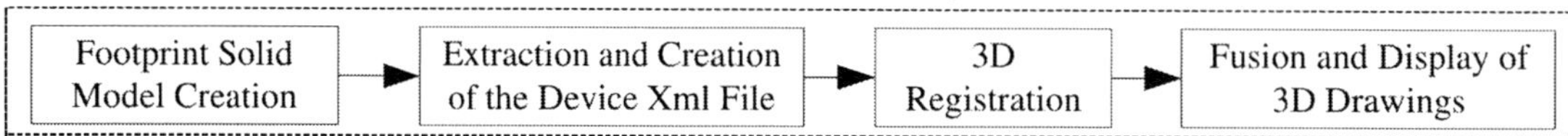

Figure 7. Process of Assembly Method for Satellite Device Installation based on HoloLens

5.1. Footprint Solid Model Creation

The 3D device model which is designed using CAD modelling tools such as PROE is large, and the required hardware performance is relatively high, so the 3D design model is not suitable for application in mobile terminals. The 3D design device model contains much modelling data, such as cables and electrical connectors. In order to improve the subsequent display rendering effect, the unnecessary model information should be deleted first. The most important indicator that affects Unity rendering efficiency or whether the picture is stuck is the Tris and Verts in the states panel, Tris are the total number of triangular faces rendered within the camera's field of view, and Verts are the total number of vertices rendered within the camera's field of view. In order to reduce the Tris of the model, footprint surface model is created, but footprint solid model is used because the footprint surface is coincident with the deck and the flickering occurs when displayed. The footprint solid model is created by stretching 1 mm of footprint surface. The footprint solid model has the advantages of small file size, easy operation and smooth display.

5.2. Extraction and Creation of the Device Xml File

While installing device, the operator needs to know the code, the position, the fastener and the related process information, so it is necessary to extract some information in design 3D model through secondary development of software. The installation process information of the device is showed in Table 1, and they will be added to the corresponding position of the lightweight and Footprint Solid Model. The code of the device, the centre position, R point Position and the Normal Vector are extracted through 3D model; the thermal control, the Fastener and the name are created from technical requirement document of mechanical designer and thermal control designer. All the installation process information of the device is written in an xml file named by the device code as the xml file provides a way to describe a cross-platform language.

Table 1. The Installation Process information of the Device.

Type	Device
Code	K405f
Name	Line Box F
Model Name of Footprint	K405f_Fp.prt
Fastener	6*M5×16
Thermal Control	Coated thermal Grease
Ground Wire	Installation
Centre Position(X,Y,Z) (mm)	450,-862,451
Normal Vector(X,Y,Z)	-1,0,0
R Position(X,Y,Z) (mm)	450,-965,373

5.3. 3D Registration

The 3D registration technology mainly realizes the correct fusion of virtual components and real assembly scenes through the conversion between coordinate systems. Accurate 3D registration of virtual and real scenes is a key technical problem to be solved in the real-time fusion, to accurately register the virtual assembly object in the corresponding position in the real environment, the conversion relationship between the virtual space and the real space coordinate system must be established in the virtual fusion assembly system. Registration error reaches a certain value will result in erroneous results. In this system based on AR technology, 2D codes which can be recognized very quickly are used in 3D registration, and it is pasted on satellite deck structure.

5.4. Fusion and Display of 3D Drawings

The system based on AR technology was developed using Unity3D, Vuforia and the Microsoft HoloLens. Unity3D is an excellent development tool for the Microsoft HoloLens. To achieve higher precision, the embedded Vuforia system was used to perform marker based tracking. After 3D registration, the HoloLens can properly track position as the user moves around in the assembly area using its IMU and environment processing cameras. An assembly application with a user friendly UI has been built to improve the user experience.

6. The Typical Case of Application in Satellite Device Installation

The assembly method for satellite device based on HoloLens is tested during the AIT process of a remote sensing satellite in the paper. The 3D model of device installation is showed in Figure 8. The footprint surface model of device is showed in Figure 9, and the footprint solid model of device is showed in Figure 10, the footprint solid model is better than footprint surface model in terms of display. As is shown in Figure 11, a 2D QR code is printed on paper and pasted on the satellite deck structure where is easy to determine the location. The operator must first wear Microsoft HoloLens before device installation and use the gesture to enter the corresponding module of the remote sensing satellite, secondly, scan 2D QR code in the satellite deck structure to complete 3D registration, and then select device to be installed from the UI interface. The process of this system is very simple, and the assembly information is clear at a glance, thereby greatly improving the visibility of the 3D model and guiding principle. Figure 12 shows a typical case of application in a satellite device installation, the code and the R point of the device is clear, if needed, other installation process information is also can be shown in this system. With the help of the assembly system based on AR technology, an ordinary operator can verify the polarity correctness of the device and the efficiency can be doubled.

Figure 8. The 3D Design Model of Device Installation.

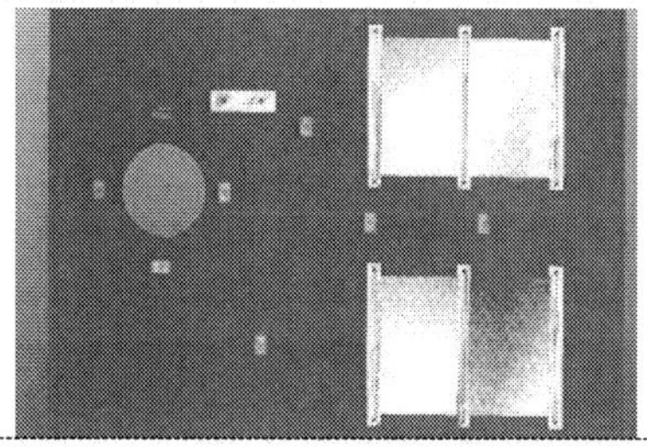

Figure 9. Footprint Surface Model.

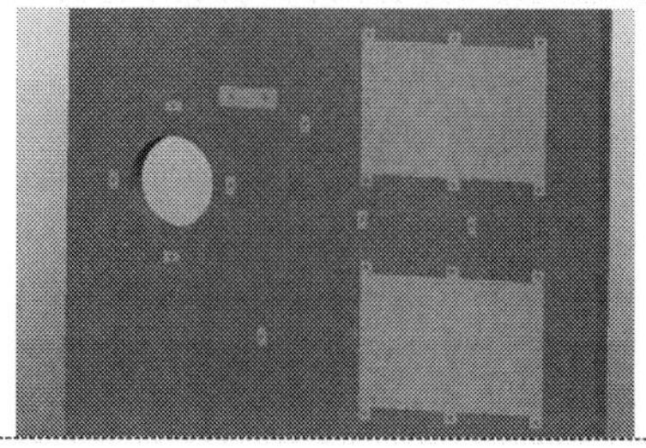

Figure 10. Footprint Solid Model.

Figure 11. The 2D QR Code Pasted on the Satellite Deck.

Figure 12. Fusion and Display of Device Installation.

7. Conclusion

This paper reviews the traditional device installation in AIT process of satellite. As the satellite design changes from the traditional 2D mode to the 3D digital mode, the satellite assembly also enters the 3D assembly mode. In order to improve efficiency and quality of satellite device installation, modelling method of R point of device is regulated and a system based on HoloLens is designed. The process of this system is very simple, and the assembly information is clear at a glance, the successful application of this system provides a new solution for satellite device installation and marks a new step in the digital assembly capacity.

References

[1] ZHANG Yu 2018 Research and Application of Design Method of MBD Technique with Simplified Dimension Annotation *Aeroengine* 44(1):22-26

[2] Ye Sheng 2015 MBD-based three-dimensional assembly process modeling and application Journal of Drainage and Irrigation Machinery Engineering 33(2):179-184

[3] ZHANG Qiuyue 2017 Application of Virtual Reality and Augment Reality in Aircraft Assembly *Aeronautical Manufacturing Technology* 530(11):40-45

[4] LI Shiqi 2009 Mixed Reality-based Interactive Technology for Aircraft Cabin Assembly *CHINESE JOURNAL OF MECHANICAL ENGINEERING* 22(3):403-409

[5] YIN Xuyue 2018 Augmented Reality Training System for Aerospace Product Assembly Process Guidance and Its Application *Aeronautical Manufacturing Technology* 61(1):48-53

[6] Gao Xiang 2018 A Survey on Mobile Augmented Reality Visualization *Journal of Computer-Aided Design & Computer Graphics* 30(1):1-8

[7] YANG Qing 2017 Development of the intelligent assembly assistant system for complex product based on augmented reality *MACHINE DESIGN AND MANUFACTURING ENGINEERING* 46(11):33-37

[8] Hou Ying 2017 A Survey of Augmented Reality Technology *Computer Measurement & Control* 25(2):1-7

[9] G Evans 2017 Evaluating the Microsoft HoloLens through an augmented reality assembly application *Proc.of SPIE* Vol.10197

IOP Conf. Series: Materials Science and Engineering **751** (2020) 012021 doi:10.1088/1757-899X/751/1/012021

Application and Research of Air Cushion Transportation in Flexible Assembly for Spacecraft

Zhu Yahong[12*] Zhang Chengli[1] Xing Shuai[1] and Li Xiaohuan[1]

[1] Beijing Institute of Spacecraft Environment Engineering, Beijing 100094, China
[2]Beijing Engineering Research Center of the Intelligent Assembly Techinology and Equipment for Aerospace Product, Beijing 100094, China

[1*]zhyhzymail@163.com

Abstract: With the development of spacecraft, the problem of flexible transfer between large weight and large size spacecraft stations needs to be solved urgently. Air-cushion transporter is a special transporting platform for heavy load. It can effectively solve the transfer problem of large spacecraft. This paper introduces a parallel air-cushion vehicle for large cargo transportation. In this way, there may be a certain height difference on the road surface where the air cushion vehicle moves, which will affect the transfer. The air volume adjustment of the air cushion vehicle is introduced to ensure the smooth transportation method. Through test, it summarizes the key indicators of air-cushion transporting, which can provide reference for the research of large spacecraft transporting methods.

1. Introduction

With the development of a series of super-large spacecraft, the spacecraft cabin and spare parts are very large and heavy. In order to meet the different transshipment requirements of large spacecraft in the assembly process, maneuverable and flexible posture adjustment needs to be solved urgently. Air cushion transporter is a special transporting platform applied to heavy load[1] [2]. It is used in equipment manufacturing and heavy industry as well as special handling in special environment. Compared with omnidirectional[3] transshipment and Mecanum[4]transportation, air cushion transporter has the advantages of high load and small size.In recent years, AGV[5] technology has also been applied in air cushion transporter based on the demand of accurate mobile positioning.

2. Principle introduction

The working process is shown in Figure 1. The air cushion module is the core component of the air cushion vehicle. It consists of an aluminium alloy flat disc and a special ring-shaped reinforced rubber airbag bonded to the flat disc. When the compressed air enters the gasbag from the upper intake pipe, the gasbag expands into a circular shape. At this time, gasbag joins the ground, and the air in the air chamber is sealed, the pressure rises, when the air pressure is enough to support the load on the aluminium alloy, the aluminium alloy plate will be lifted off the ground. Finally, the compressed air escapes from the air chamber at the bottom of the gasbag and lifts the load off the ground. At this time, the air entering and escaping the air chamber reaches balance, and the air pressure is stable, and the whole system is in balance. The air cushion suspension transportation system has flat-panel type, parallel type and AGV type. Different forms of air cushion vehicles can meet the needs of different industries. Figure 2 is a parallel type of air cushion vehicle and Figure 3 is an AGV guided air cushion vehicle.

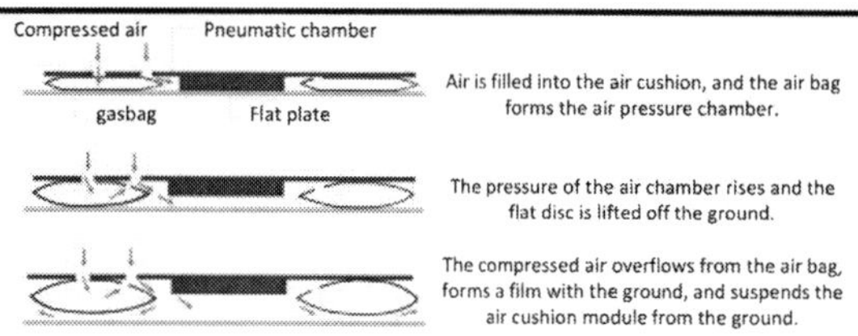

Figure1. Principle

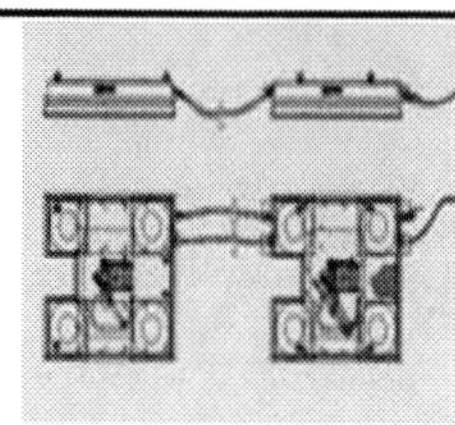

Figure 2. Parallel mode

Figure 3. AGV mode

3. Composition of Air Cushion Transfer Vehicle

The number of air cushion vehile used for transport goods should consider not only the load, but also the size of the object to be transported. The typical composition of a single vehile is shown in Fig. 4, Fig. 5. In order to ensure the stability of transportation, the parallel transportation of large-sized spacecraft is usually carried by several vehicle. Take the case of spacecraft's packing container transportation as an example. Transportation weight: 60t, size: 13m×6m×6m (length×width×height), Two air cushion trucks are used to lift the packing box at the front and back positions, and the two air cushions are connected in parallel for spacecraft transportation.Parallel transportation is shown in Fig. 6 and mode of transportation is shown in Fig. 7.

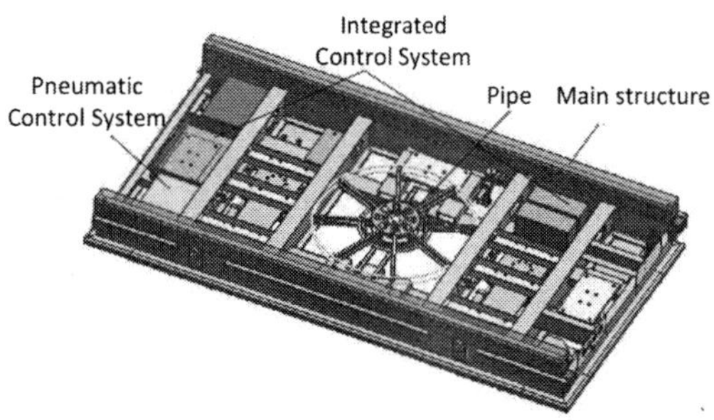

Figure 4. Air cushion transfer vehicle(a)

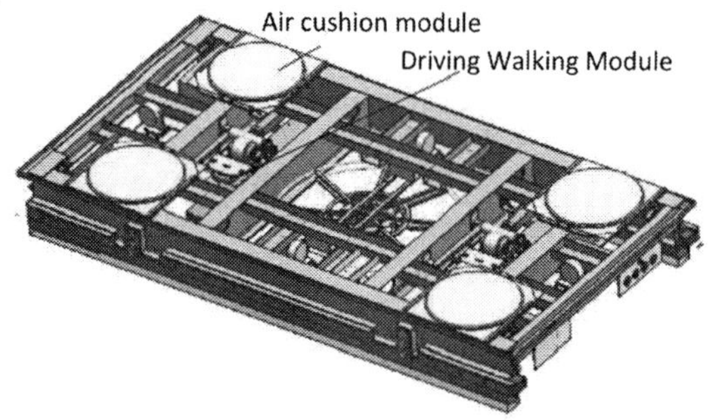

Figure 5. Air cushion transfer vehicle(b)

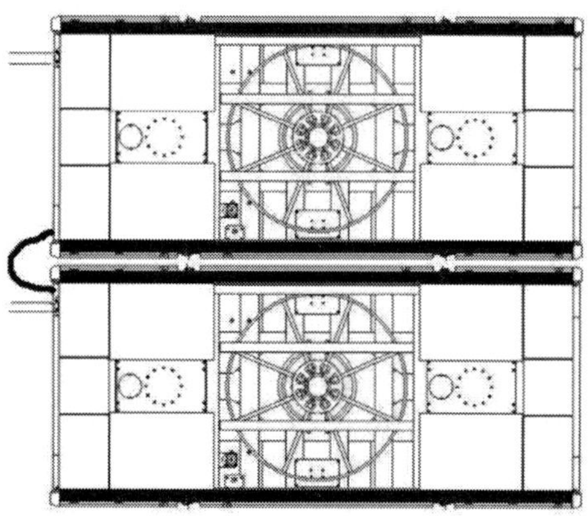

Figure 6. Parallel air cushion vehicle

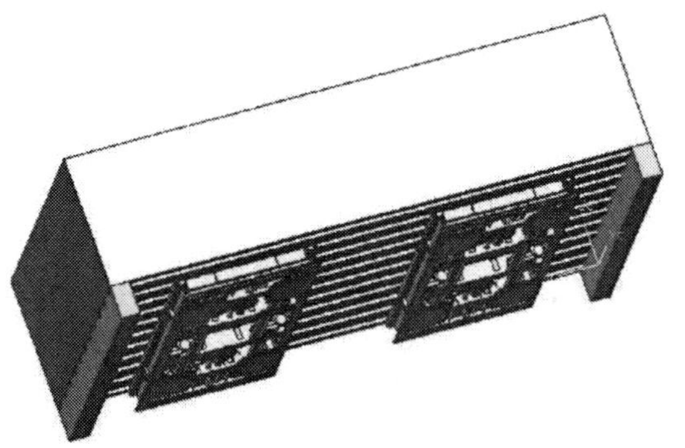

Figure 7. Parallel transshipment

3.1.Main structure

The vehicle body is a rigid carrier. The loads and tools to be carried are placed on the vehicle body. The air cushion module is installed under the vehicle body. The balance module and the driving module are installed in the vehicle body. According to different uses and requirements, the size, shape and structure of the vehicle body are different. When in use, the load can be transported as long as the vehicle body is driven under the object being carried.

3.2. Air cushion module

The air cushion module is the core component of the air cushion suspension transportation system. It consists of an aluminium alloy flat disc and a special ring reinforced rubber airbag attached to the flat disc.

3.3. Driving Walking Module

The electric drive device provides the power needed for the operation of the whole system, as shown in figs. 8 and 9. The electric driving device is mainly composed of driving wheel, walking motor, steering motor and down-pressure device. The drive has the function of slow start, that is, when the whole vehicle starts or stops, it will not overload the heavy objects due to suddenly acceleration or deceleration. Drive can complete the following three directions: downward pressure, walking, steering.

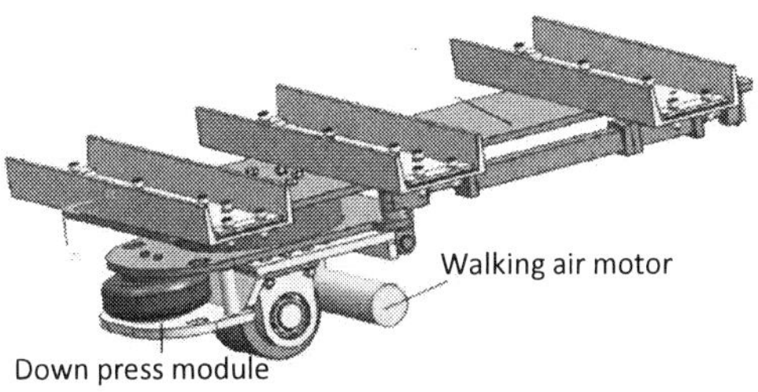

Figure 8. Driving Walking equipment(a)

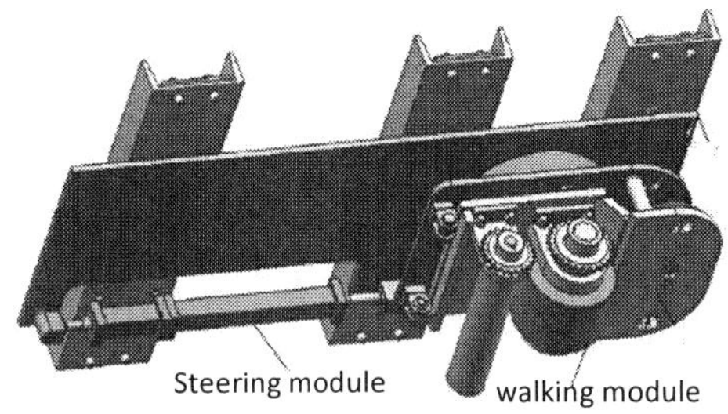

Figure 9.1 Driving Walking equipment(b)

Downward pressure: After the cylinder is inflated, the downward pressure provided overcomes the spring force and friction force, the driving wheel will move downward along the guide rod and land, and the gasbag pressure will gradually increase, and the adhesion of the driving wheel will also increase. On the contrary, when the system stops, the gasbag deflates and the driving wheel leaves the ground under the force of the spring and returns to its position.

Walking: The walking motor provides power, which is transmitted to the driving wheel through sprockets and chains, and the driving wheel rotates. Because the airbag provides enough adhesion, the drive can drive the load to move.

Steering: The steering motor provides power to transfer the force to the outer ring of the slewing bearing after gear deceleration. The outer ring of the slewing bearing drives the driving wheel installed on the outer ring to rotate. The angle of driving rotation is acquired by the encoder of the motor. The driving wheel can stop accurately at 0° , 90° .

3.4. Balanced module

Balanced module can eliminate the bias effect during walking by changing the air pressure in each air cushion module. It has the function of automatic adjustment and balance for different pavement conditions and different loads of weight and center of gravity. Balancing module is composed of sensor, PLC, valve, buffer tank, pressure gauge, pipeline and so on. The sensor is used to detect the distance between each air cushion module and the PLC receives the feedback signal of the sensor to control the valve to adjust the air pressure in the gasbag. Buffer tank can balance the uniformity of compressed air flow.

3.5. Integrated Control System

The integrated control system is mainly composed of three parts: power supply system, motion control system and human-computer interaction system.The power supply system is the power supply for the whole vehicle, which provides the power for the vehicle electrical equipment, mainly including the storage battery group and the power conversion management module. Motion control system is the core of electrical integrated control system, which mainly completes the functions of vehicle motion control, safety monitoring and self-diagnosis of running state.Human-computer interaction system provides the main control interface for user operation, including vehicle status display system and remote control system.For parallel air cushion truck transportation, its motion control processing system works by changing the conduction relationship of the driving steering control valve group to change the air flow direction of the steering cylinder, thus completing the lateral/vertical switching of the driving wheel. The working mode of the motion control and processing system is as follows: by changing the conduction relationship of the drive steering control valve group, the direction of air flow in the steering cylinder is changed to complete the lateral/vertical switching of the driving wheel. The driving wheel moving control device includes a driving wheel moving direction control valve group and a driving wheel moving speed control device. The driving wheel moving direction control valve group is used to change the direction of air intake of the walking air motor, realize the positive and negative rotation of the air motor, and change the moving direction of the air cushion vehicle. he driving wheel moving speed control driver is used to change the opening of the driving wheel moving speed control valve group and adjust the air intake of the air motor, so as to realize the control of the driving wheel moving speed.

3.6. Pneumatic control system

The pneumatic control system consists of air cushion control valve, air cushion, external air pipe reel, air source treatment valve group, driving down pressure gasbag control valve and air pipe. Pneumatic control system is mainly used for driving and steering control, tracheal recovery control and air cushion control. The air cushion vehicle can monitor the total air pressure in real time. When the air supply pressure can not meet the requirements of the air cushion vehicle, the system gives a pressure alarm. The air cushion vehicle is equipped with air source processing device to ensure the working environment of the pneumatic system.

4. Transportation Index test

In order to verify the transshipment capability of the developed parallel air cushion vehicle, the test load is 80t, and the two air cushion vehicles are transported in parallel.

(1) Walking test. The functions of translation, in-situ rotation, acceleration and deceleration were tested to observe whether the moving process was smooth and to measure the parking acceleration.

(2) Travel acceleration: When parking at maximum speed(8m/min), measure its acceleration.

(3) Measure the braking distance when parking. The braking test is carried out during the moving process of the mobile platform, and the braking distance is measured.

(4) Point test. Measure the minimum displacement of the air cushion transfer platform under one control command, that is, the pointing accuracy.

(5) Cooperative performance. Whether parallel two vehicles can be transported smoothly in the process of transshipment.

The load test is shown in Fig. 10. The results of load test of mobile platform are shown in Table 1. Load test results show that it can meet the transshipment requirements of spacecraft from the test results.

Figure 10. Load test

Table 1. Test result

No.	Item	data						Average value	Result
		1	2	3	4	5	6		
1	In-situ rotation migration	0.2	0.4	0.3	0.4	0.2	0.7	0.37	<5mm
2	Parking acceleration	1.06	1.08	1.12	1.05	1.07	1.11	1.08	<1.3g
3	Stopping distance（8m/min）	82	63	91	65	77	87	77.5	<100mm
4	Point distance	5	2	1	2	3	5	3	<5mm
5	Stationarity(Parall el connection)	√	√	√	√	√	√	√	stable

5. Typical application

Air cushion vehicle has been used in aerospace field. Fig. 11 is an air cushion truck for transshipment spacecraft packing container used in domestic space transportation field. Fig. 12 is for transporting large spacecraft and brackets. Fig. 13 shows the application of air cushion transfer platform in docking of foreign aircraft cabins. Fig. 14 shows that in the engine assembly building of the French Guiana Space Center, the solid rocket engine section is transported by air cushion vehicle with a load of 220 tons.

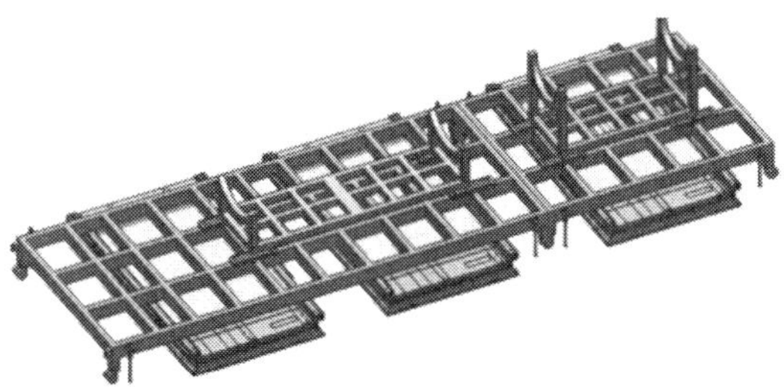

Figure 11. Heavy packing container transportation **Figure 12.** Spacecraft and Bracket Transportation

Figure 13. Fighter transport

Figure 14. Rocket Engine Transportation

In production practice, such as Airbus, Nokia, ABB, Siemens, Toshiba, Bombardier, General Electric, ALSTOM, BMW, MERCEDES BENZ, Volvo, etc. use air cushion suspension transport equipment to solve the transportation problems of equipment on their production lines. Fig. 15 is the application of Air cushion vehicle in locomotive manufacturing industry. Fig. 16 shows the application of generators in transportation.

Figure 15. Locomotive body transport

Figure 16. Generator Transportation

6. Concluding remarks

Through the cooperative operation of two or more hovercraft, the large-scale and heavy product handling and moving of spacecraft can be accomplished[6]. Air cushion vehicle transshipment has the advantages of small size and heavy load. It has no advantage in lean control compared with electronic control transshipment truck. It can provide reference for aerospace transshipment logistics field.

Reference

[1] LI Zhenhua. Application of DELU aircushion technology in new generation aircraft factory[J]. Aeronautical ManufacturingTechnology, 2009(12): 104—105.
[2] Xu weimin The Requirements of the Air Cushion Suspension Transportation System in the Engineenng Design and the Application in the Heavy Equipment Manufacturing Industry[J] Construction & Design for Project, 2011(12): 70-71.
[3] YAO Ding, SHE Jing, HUANG Xiang, et a1. Large aircraft component transport based on omnidirectionalmobile and muhi-point supporting[J]. Journal of Nanjing University of Aeronautics & Astronautics, 2012 (2): 73-78.
[4] Mecanum Wheel Technology and its Application in Ship Wheel Devices[J] Ship Engineering201133(3): 84-87.
[5] Assembling Overloading Omni-directional Mobile AGV Control System Techniques of Automation and Applications201837(2)64-68.
[6] FENG Huashan, QIN Xiansheng, WANGRunxiao. Developing trend of industrialrobot in aerospace manufacturing industry[J]. Aeronautical Manufacturing Technology, 2013(1 91: 32-37.

IOP Conf. Series: Materials Science and Engineering **751** (2020) 012022 doi:10.1088/1757-899X/751/1/012022

Simulation of Space Flexible webs on Capture Process Based on Nonlinear Finite Element Method

Boting Xu[1], Yueneng Yang[1,2,*], Bin Zhang[1] and Ye Yan[3]

[1] College of Aerospace Science and Engineering, National University of Defense Technology, Changsha 410073, China
[2] College of Intelligence Science and Engineering, National University of Defense Technology, Changsha 410073, China
[3] Innovation Research Institute of Defense Technology, Academy of Military Sciences, Beijing, 100071, China

*Corresponding author: yangyueneng_nudt@163.com

Abstract. Aiming at the problem of the space flexible webs collision during the space capture process, the general method of collision detection was given based on the nonlinear finite element method, and the collision dynamic model of the space flexible webs and target was established. In order to verify the versatility of the space flexible webs structure in different capture scenarios, the target capture position and attitude were changed respectively, and different capture scenarios were established to simulate and analyze the working conditions by using the finite element software. The results showed that the target position and attitude during the collision impact process is the key of the results of the flexible webs capture mission, which can provide some reference for the structural design and on-orbit task planning of the space flexible webs system.

1. Introduction

As of May 30, 2019, the total number of on-orbit space targets (diameter greater than 10 cm) catalogued by the Space Surveillance Network (SSN) has reached 19,643, of which 5,134 are payloads (there are 2,315 still in normal working conditions), there are14509 abandoned rocket body and space debris, accounting for 77% of the total number of targets in the orbital space. Space debris is the potential "perpetrator" of the space collision accident, in the low earth orbit, the space debris runs fast, and the relative velocity of collision between the debris and spacecraft can reach more than 10km/s, which will seriously damage the spacecraft and even lead to catastrophic consequences such as structural disintegration[1,2]. At present, the active debris removal technologies mainly include manipulators capture, flexible webs capture, laser removal and so on. The flexible webs capture technology proceed the debris removal mission by developing a flexible webs to wrap space debris in some corresponding deployment mode. This method has become a research hotspot in the field of space debris removal which can be applied to capture space debris of different shapes and sizes and has great fault-tolerant advantage.

The space flexible webs system is a typical nonlinear dynamic system, which is characterized by variability, relaxation and winding. The capture process of the space flexible webs mainly includes two stages: flexible webs deployment and target capture. Mankala studied the dynamic modeling and simulation of the entire mission process of flexible network systems[3]. Zhai Guang studied the attitude dynamics during the capture process of flexible network systems[4]. Yu Yang developed the finite element model of the flexible webs projectile deployment process, implemented in the software THUsolver, in order to study the the mechanical properties of the webs under different environments[5,6]. Gardsback Tibert developed the finite element model, implemented in the commercial software LS-DYNA, in order to imitate a spinning deployment process of the space webs and verify a robust control method for the spinning deployment[7,8]. Chen Qin established the lumped-mass model and analysed the dynamics problems of the webs deployment process. The above research mainly focuses on the deployment stage of flexible webs, and there are relatively few researches on the collision and wrapping process in the capture stage of flexible webs[9]. During the process of target capture, the collision between flexible webs and target will produce rapid stress and strain, which may lead to the breakage of rope segment or the instability of web configuration, thus causing the target to be escaped from the webs. Therefore, the capture process of space flexible webs also needs to be studied in depth. Yu Yang et al. analyzed the stress and strain of flexible webs with different mesh sizes under impact based on the finite element model[10]. Benvenuto used the penalty function method to simulate the target capture process of the flexible webs[11]. Xu Boting used finite element software to analyze the capture ability of the bionic flexible webs[12].

In this paper, the dynamic modeling and simulation of the collision wrapping process between the flexible webs and the target will be carried out. The general method of collision detection is given based on the nonlinear finite element method, at the same time, the collision dynamic model of the space flexible webs and target is established. In order to verify the universality of the flexible webs in different capture scenarios, we focus on the variation of the collision force and energy of the webs when capture different position/attitude targets, which can provide some reference for for the structural design and on-orbit task planning of the space flexible webs system.

2. Dynamic model

2.1. Collision detection

The flexible webs capture process is generally a collision of irregular objects. In this paper, we focus on the simulation analysis of the flexible webs collision process. In order to save the simulation time of the collision detection process, the shape of target is simplified which consider to be a rigid cube.

For the collision detection between the webs and target, firstly judge the spatial position of the checkpoint and the target, if detected that there is intrusion between the moving bodies, then the quantity of intrusion will be calculated. The contact model calculates the collision force and update the state of collision motion, repeating the aboved process in the next calculation. The detection process is shown in Figure 1. The point-plane detection methol is applied in the collision detection process, which by calculate the distance of the checkpoint and detection suface to determine whether the intrusion into the detection surface. Because of the grid of the web unit is sparse, it is necessary to arrange several checkpoints in the unit, and these checkpoint are distributed in the unit evenly.

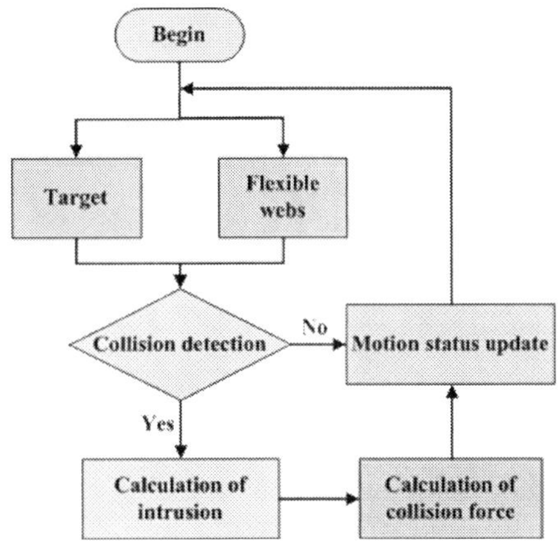

Figure 1. The flow of the collision detection process.

2.2. Dynamics model of collision

The mechanical process of collision can be described as Lagrange impulse method:

$$\begin{bmatrix} M & \Phi^{\mathrm{T}} & \partial s/\partial q \\ \Phi & 0 & 0 \\ (\partial s/\partial q)^T & 0 & 0 \end{bmatrix} \begin{bmatrix} \Delta\dot{q} \\ \lambda \\ -p \end{bmatrix} = \begin{bmatrix} 0 \\ 0 \\ -(1+e)(\partial s/\partial q)^T \dot{q}^- \end{bmatrix} \tag{1}$$

where M is the mass matrix, Φ is the Jacobian matrix, s is the distance between the contact points, q is the location coordinate, $\dot{q}^-$ is the velocity before collision, e is the coefficient of restitution, λ is the counter-force of impact force, p is the impact force.

2.3. Dynamics model of contact

The contact in the flexible webs system can be divided into two categories: the contact between the flexible webs and the rigid body, and the contact between the rigid body and the rigid body, which regardless of the self-collision between the rope segments of the webs. For a rigid body or particle in contact, $u_n^i\,(i=1,2)$ is the distance to the vertical contact surface (or point), $u_n^i\,(i=1,2)$ is the displacement in the contact surface. The contact can generally be divided into several types:

$$\begin{aligned} &u_n^2 - u_n^1 + a_0 > 0, P_n = 0 && \textit{no touch with webs} \\ &u_n^2 - u_n^1 + a_0 = 0, P_n < 0 && \textit{in touch with webs} \\ &|P_\tau| < -\mu P_n, u_\tau^2 - u_\tau^1 = 0 && \textit{in touch with webs, no slide} \\ &|P_\tau| = -\mu P_n, \operatorname{sgn} P_\tau\left(u_\tau^2 - u_\tau^1\right) \le 0 && \textit{in touch with webs, slide} \end{aligned} \tag{2}$$

where a_0 is the safe distance, that is the threshold of distance threshold detection, P_n is the contact force, which is perpendicular to the contact surface, P_τ is the force of friction of contact surface.

3. Numerical simulation and analysis

In order to verify the versatility of the flexible webs system for different capture scenarios, the target capture position and attitude will be changed respectively, and different capture scenarios will be simulated and analyzed by using the finite element method. In the part of the attitude change of target, three typical collision situations of flexible webs in capture process are simulated by designing initial contact states of surface, line and point respectively. In the numerical calculation of different working conditions, only the position and attitude of the target are changed, while the other simulation parameters remain unchanged. The collision scenarios of initial surface is detailed in the article [12] and will not be repeated here. Simulation parameters are listed in Table 1.

Table 1. Simulation parameters of flexible webs

Parameter	Value
Size of webs/m	5×5
Size of target/m	1×1
Size of grids/m	0.25×0.25
Mass of masses/kg	0.5
Sectional area of rope units/m^2	2.73e-5

Simulation assumptions:

(1) Prior to contact and collision, the flexible webs posture remain stable, the speed of each point is identical, and the webs move at a constant speed;

(2) The flexible webs and target experience straight-line motion; the influence of orbital motion is ignored.

(3) The flexible webs are assumed to be stationary while the target moves relative to the webs.

3.1. Scenarios 1: Capture position change of the target

As shown in Fig. 2, the capture position is not in the center of the flexible webs, but on the left side of the center, the target is 2m away from the surface of webs, and the relative velocity between the target and webs is 100m/s.

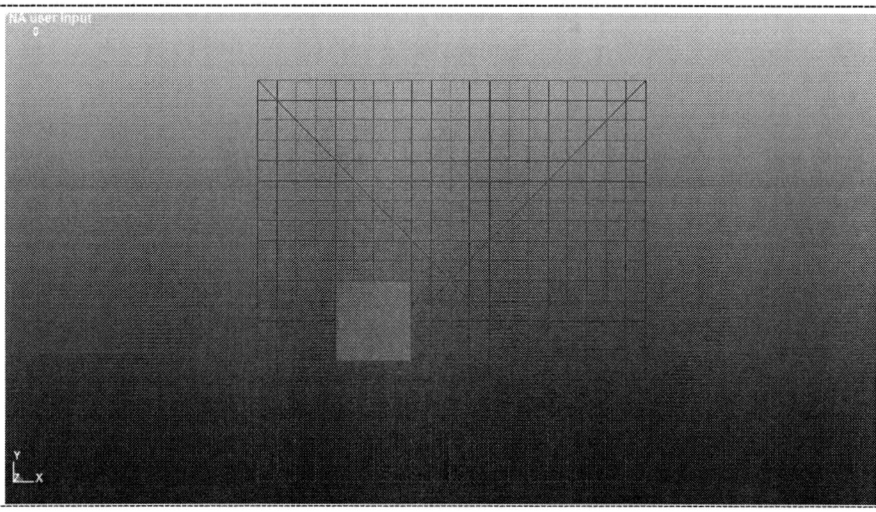

Figure 2. Capture position change of the target.

Fig. 3 shows the movement of the flexible webs and the target at different times during the collision process. The target is wrapped by the flexible webs after collision, and the two will form a rigid-flexible coupling body to move together. Fig. 4-Fig. 6 are the changes of deformation, strain and stress of flexible rope segment during collision when the initial collision position is changed. It can be seen from the simulation results that although the flexible webs can successfully capture the target, the target is not completely wrapped by the network, which increases the difficulty of the next step of the off-track control process of the complex, and there is a risk of failure of the capture task. Therefore, in order to ensure the success of the capture task, the initial devolopment angle of the webs is designed to ensure that the acquisition position is the center area of the flexible network.

(a) t=0.019 (b) t=0.023 (c) t=0.036

Figure 3. The deformation of the flexible webs.

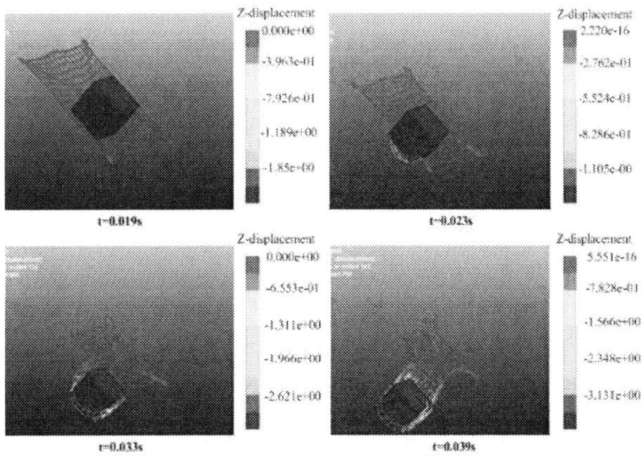

Figure 4. The variations in the deformation of the flexible-webs rope segment.

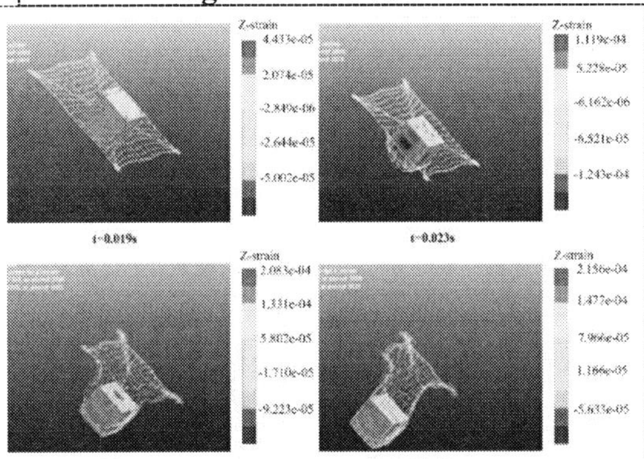

Figure 5. The strain in the deformation of the flexible-webs rope segment.

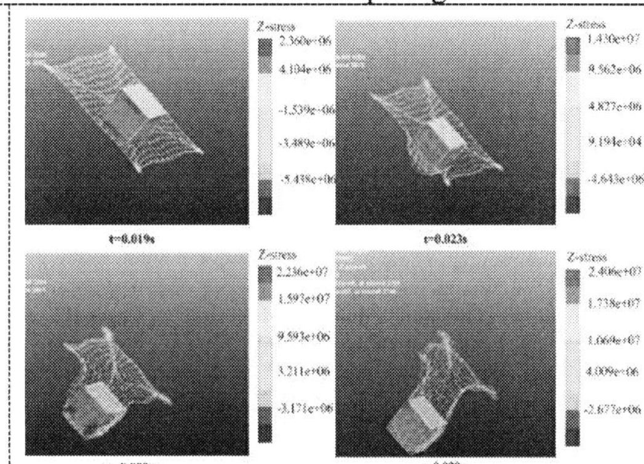

Figure 6. The stress in the deformation of the flexible-webs rope segment.

3.2. Scenarios 2: Capture attitude change of the target (Initial line collision)

Fig. 7 shows the movement of the flexible webs and the target at different times during the collision process. The target is wrapped by the flexible webs after collision, and the two will form a rigid-flexible coupling body to move together. Fig. 8-Fig. 10 are the changes of deformation, strain and stress of flexible rope segment during initial line collision. It can be seen from the simulation results that the target is completely wrapped by the flexible network, and the initial line collision can also ensure the success of the capture task.

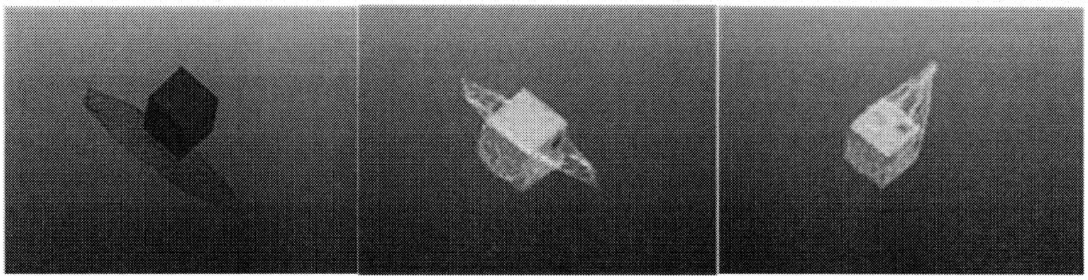

(b) t=0.019 (b) t=0.029 (c) t=0.043

Figure 7. The deformation of the flexible webs(Initial line collision).

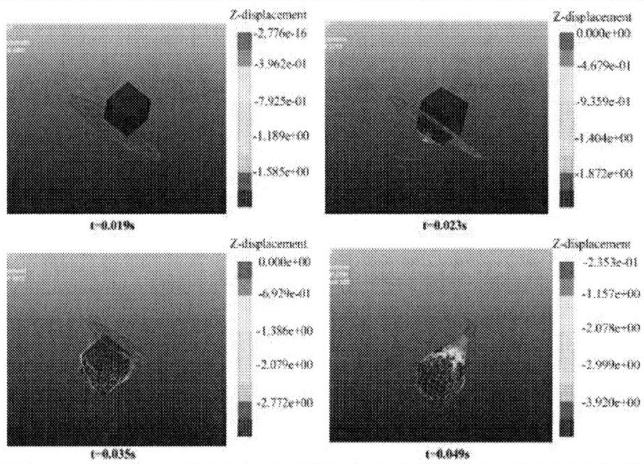

Figure 8. The variations in the deformation of the flexible-webs rope segment(Initial line collision).

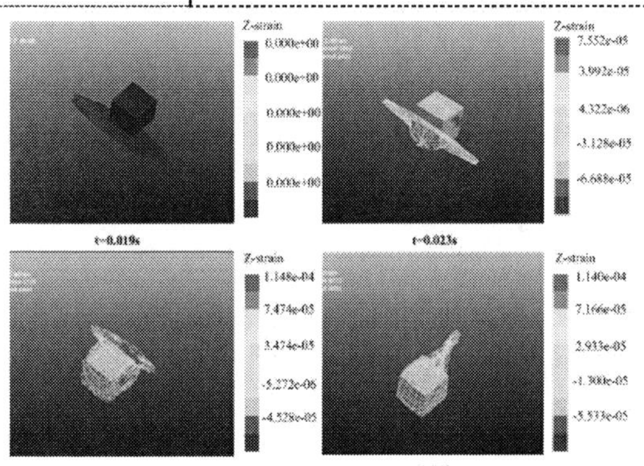

Figure 9. The strain in the deformation of the flexible-webs rope segment (Initial line collision).

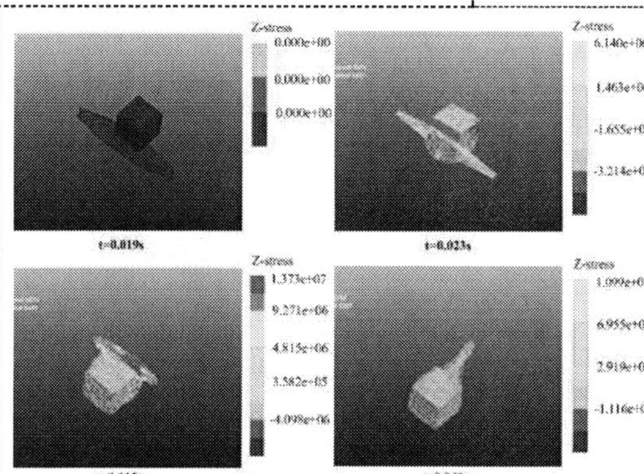

Figure 10. The stress in the deformation of the flexible-webs rope segment (Initial line collision).

3.3. Scenarios 3: Capture attitude change of the target (Initial point collision)

Figure 11 shows the movement of the flexible net and the target at different times during the collision process. In the case of initial point collision, the simulation results are not ideal, the flexible webs does not capture the target and the capture task fails. Fig. 12-Fig. 14 are the changes of deformation, strain and stress of flexible rope segment during initial line collision. It can be seen from the simulation results that in the initial stage of the collision between the target and the flexible webs, the excessive stress on the flexible webs is the main reason for the failure of the capture task.

(c) t=0.019 (b) t=0.029 (c) t=0.043

Figure 11 The deformation of the flexible webs(Initial point collision).

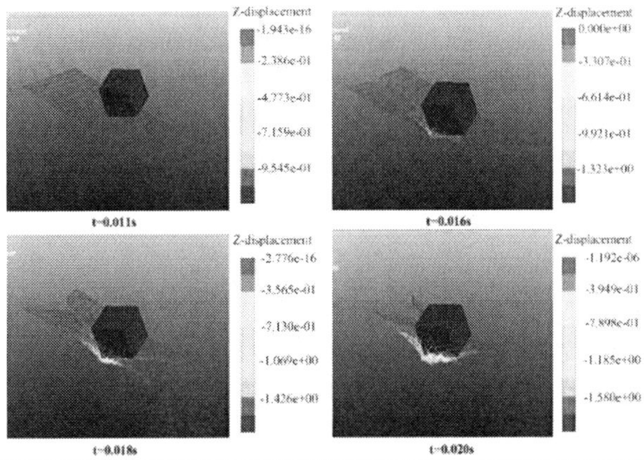

Figure 12 The variations in the deformation of the flexible-webs rope segment (Initial point collision).

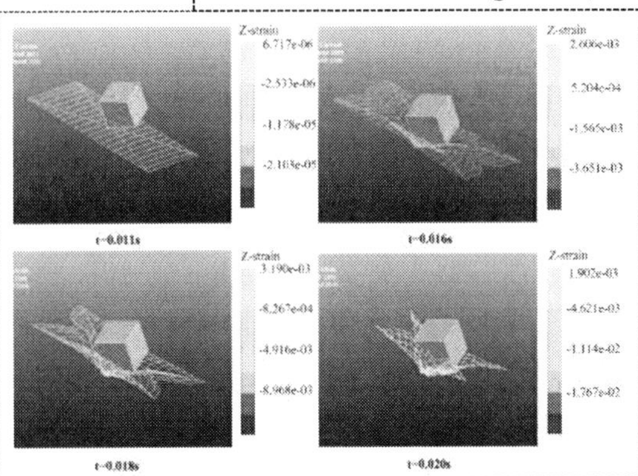

Figure 13 The strain in the deformation of the flexible-webs rope segment(Initial point collision).

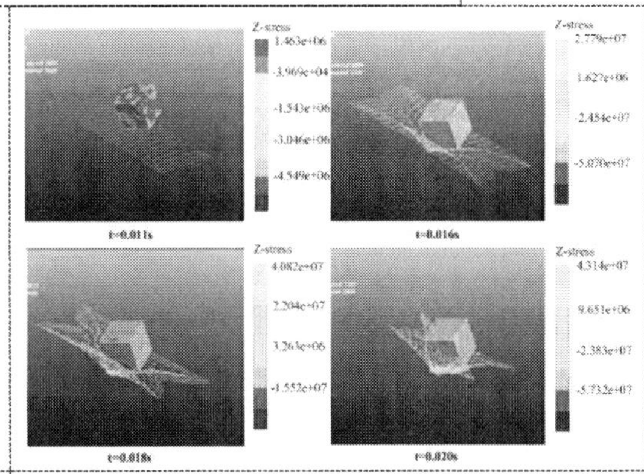

Figure 14 The stress in the deformation of the flexible-webs rope segment(Initial point collision).

4. Conclusion

Based on the non-linear finite element method, the dynamic modeling and simulation of the capture process of the space flexible webs are carried out. The influence of the position/attitude change of the target on the capture results of the flexible webs is analyzed. The results show that under the condition of the same relative collision speed, the initial surface collision and the initial line collision can ensure the success of the capture mission. In the case of initial point collision, the capture failure occurs due to the excessive stress on the flexible webs at the initial moment of the collision. Therefore, in the structural design process of space flexible webs system, the configuration analysis and verification of the energy dissipation of the flexible webs should be emphasized. In the process of space on-orbit mission design, the deployment angle of flexible webs structure can be adjusted by the attitude of the target, and the appropriate initial contact and collision position should be selected to ensure the success of the capture mission.

It should be pointed out that in this paper, the capture process of space flexible webs is qualitatively analyzed, but the accuracy of the model in this paper still needs to be corrected and modified by means of ground test. For the analysis of the capture process of space flexible webs, further verification and improvement are needed.

Acknowledgement
The work reported in this paper was partially sponsored by the Science and Technology Committee of China (No. 02-ZT-005-021-01), National Science Key Lab Project of the Technology of Space Intelligent Control(No. HTKJ2019KL5022016).

References

[1] J.-C. Liou, NicholasL.Johnson 2009 Asensitivity study of the effectiveness of active debris removal in LEO, Acta Astronautica, 236–243.
[2] J.-C. Liou, NicholasL.Johnson, N.M. Hill 2010 Controlling the growth of future LEO debris populations with active debris removal Acta Astronautica, (66) 648-653.
[3] Mankala K K, Agrawal S K 2004 Dynamic modeling and simulation of impact in tether net/gripper systems, Multibody System Dynamics, (11) 235-250.
[4] Zhai G, Qiu Y, Liang B 2009 On-orbit capture with flexible tether–net system, Acta Astronautica, (65) 613-623.
[5] Yu Y, Baoyin H X, Li J F 2010 Modelling and simulation of space webs projecting dynamics, Journal of Astronautics, 1289–1296.
[6] Yu Y, Baoyin H X, Li J F 2011 Dynamic modelling and analysis of space webs, Science China Physics, Mechanics and Astronomy, (54) 783-791.
[7] Gärdsback M, Tibert G 2009 Deployment control of spinning space webs, Journal of guidance, control, and dynamics, (32) 40-50.
[8] Gärdsback M, Tibert G, Izzo D 2007 Design considerations and deployment simulations of spinning space webs, in: AIAA (Ed.) Proc. of 48th AIAA/ASME/ASCE/AHS/ASC Structures, Structural Dynamics, and Materials Conference.
[9] Chen Q, Yang L P 2008 Dynamic modeling and simulation of orbital net-capture systems, in: The 14th CASI As-tronautics Conference, Montreal,Quebec Canada.
[10] Yu Y, Baoyin H X, Li J F 2011 Dynamic modelling and analysis of space webs. SCIENCE CHINA Physics, Mechanics & Astronomy, 54(4): 783-791.
[11] Benvenuto R, Salvi S, Lavagna M 2015 Dynamics analysis and GNC design of flexible systems for space debris active removal. Acta Asronautica, 110: 247-265.
[12] Boting Xu, Yueneng Yang, Ye Yan, Bin Zhang 2019 Bionics design and dynamics analysis of space webs based on spider predation[J]. Acta Astronautic, 159: 294-307.

IOP Conf. Series: Materials Science and Engineering **751** (2020) 012023 doi:10.1088/1757-899X/751/1/012023

Numerical Simulation of Air Resistance on the Mars Probe Cabin Separation

Qiwei Chen*, **Wei Feng, Laiying Tang, Linzheng Wang, Shengyu Zheng, Wei Sun and Wei Wang**

Beijing Institute of Spacecraft Environment Engineering, Beijing Engineering Research Center of the Intelligent Assembly Technology and Equipment for Aerospace Product, Beijing, China

*wherechen@163.com

Abstract. In order to verify the separation requirements of Mars probe's cabins, the Flunet analysis models of the heatshield and the aeroshell of Mars probe are established. The relationships between air resistance and the velocity, deflection angle of the cabins are obtained by numerical calculation method. The relationships between air resistance moment and the velocity, deflection angle of the cabins are also gained. According to simulation and analysis, the air resistance of the cabins obviously with velocity, but has little change with the increase of deflection angle. The air resistance moment of the cabins increases obviously with velocity and deflection angle.The dynamic simulation results of air resistance during the courses of the cabins separations provide effective data support for the simulation and ground validation of Mars probe cabin separation.

1. Introduction

During the exploration of Mars, the EDL (Entry, Descent and Landing) process is fully autonomously controlled, called "black 7 minutes". To verify the separation scheme design of the heatshield, which is facing on the surface of Mars and ablated in the process of falling, and the aeroshell, which is back facing on the surface of Mars and separated by parachute, it is necessary to carry out ground test verification. Mars's atmospheric density is less than 1% of the Earth's atmosphere, which is one of the main factors affecting the ground test of Mars rover in the process of EDL cabin separation. At present, the main research methods of air resistance are numerical simulation [1] [2] [3] [4] and experimental research [5] [6].In this paper, the Flunet analysis models of the heatshield and the aeroshell are established by numerical calculation. Through numerical simulation, the relationships between air resistance and the velocity, deflection angle and the relationships between air resistance moment and the velocity, deflection angle are obtained.

2. Test technical scheme

The stress realization scheme of the cabins is shown in figure 1. The cabins are subjected to axial force, normal force and torque. The axial force is controlled by servo motor, while the normal force and torque are mainly controlled by springs.

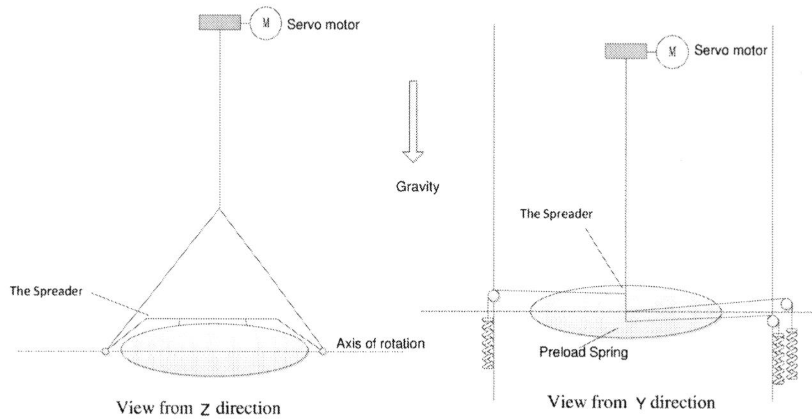

Figure 1. Test technical scheme

3. Model Building

3.1. Geometric Model

The moving direction of the cabins defined as X axis. The deflection angle is defined as the angle between the deflected X axis and the initial X axis. The maximum diameter of the cabins is 3400mm, so the models of the cabins are carried into the calculation domain of 12000mm*10000mm in the analysis module of ANSYS. In order to simplify the calculations, the symmetrical surfaces are set up and half of the models are taken for analysis. The top surface of the cuboid is an air inlet and the base surface is an outlet. The calculation domains of heatshield and aeroshell are shown in figure 2.

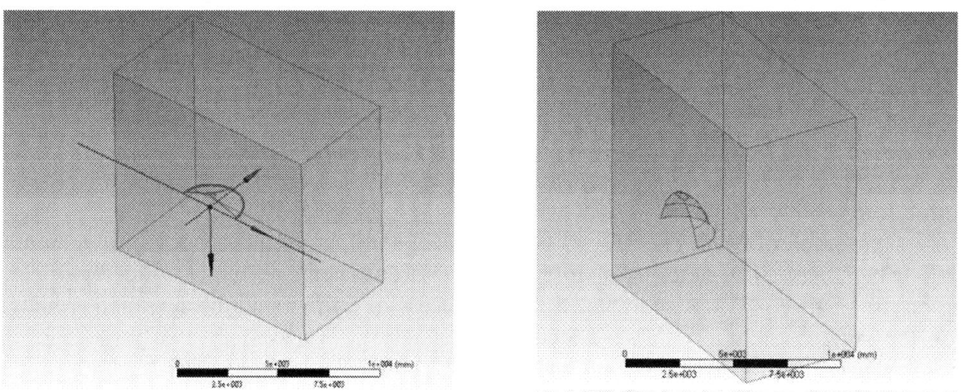

(a) Heatshield (b) Aeroshell

Figure 2. The computational domain of the heatshield and the aeroshell

3.2. Mathematical Model and Boundary Conditions

Considering the small volume of the heatshield and the aeroshell, the change of air density surrounding them should be neglected. Make the following assumptions.

a. The isotropic assumption is introduced for turbulent motion. The standard k-ε turbulence model is used to solve the three-dimensional N-S equation, and the standard wall function is used near the shell.

b. The flow is incompressible.

c. There is no heat exchange in the air flow.

The continuum equation and momentum equations are as follows:

$$\frac{\partial u_i}{\partial x_i} = 0 \tag{1}$$

$$\frac{\partial (u_i u_j)}{\partial x_i} + \frac{1}{\rho}\frac{\partial p}{\partial x_j} = \frac{\partial}{\partial x_j} \nu \left(\frac{\partial u_i}{\partial x_j} + \frac{\partial u_j}{\partial x_i} \right) - \overline{u_i' u_j'} \tag{2}$$

where, $u_i, u_j (i, j = 1,2,3)$ are the average velocity component;

$x_i = (i, j = 1,2,3)$ are the coordinate component;

p is the average pressure of fluid;

ν is the kinematic viscous coefficient of fluid;

ρ is the fluid density;

$\overline{u_i' u_j'}$ are unknown Reynolds stress component.

To simulate the working conditions as accurately as possible, the inlet and outlet channels of air are extended during modeling. The inlet boundary is located at 5000mm away from the center of the heatshield, where the wind speed is given. The outlet boundary is located at 5000mm away from the axle centre of the heatshield, where the pressure is atmospheric pressure. According to the simulation results of ADAMS in previous period, the velocity of the heatshield changes from 3.5m/s to 8.5m/s with 0.5m/s intervals, and the deflection angle is set to 0°, 1°, 2°, 3°, 5°. Meanwhile, the velocity of the aeroshell changes from 3m/s to 8m/s with 0.5m/s intervals, and the deflection angle is set to 0°, 1°, 2°, 3°, 5°.

4. Simulation results

4.1. Heatshield

4.1.1. Resistance. The relationship between the resistance on the heatshield and velocity, deflection angle of the heatshield is shown in table 1 and figure 3.

Table 1. The relationship between resistance and velocity, deflection angle of heatshield

Resistance (N) / Velocity (ms-1) / Deflection Angle(°)	3.5	4	4.5	5	5.5	6	6.5	7	7.5	8	8.5
0	84	102.7	123.6	146.8	172.8	200.4	230	262	295.2	330.8	368.8
1	82.4	101.6	122.8	145.8	171.4	199.0	229.4	260.6	294.7	330.4	368.4
2	80.7	100.8	121.2	145.1	171	199.2	228.4	263.1	294	332.7	368.8
3	79.8	99.6	120.4	146.2	170.6	200.5	229.2	264.2	295.6	333.5	371.2
5	78.6	98.4	120.6	145	171.6	200.4	231.4	264.4	300	337.4	376.0

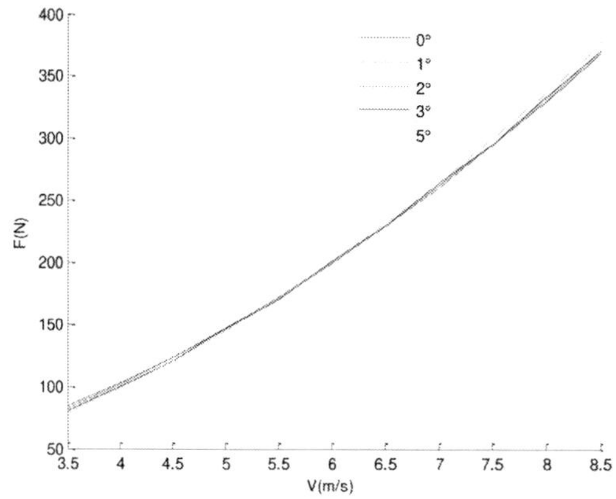

Figure 3. Graph of the relationship between resistance and velocity, deflection angle of heatshield

When the heatshield velocity increases from 3.5m/s to 8.5m/s and the deflection angle increases from 0° to 5°, the air resistance increases from 78N to 376N. The air resistance of the heatshield increases obviously with velocity, but has little change with the increase of deflection angle.

4.1.2. Resistance moment. The relationship between resistance moment on the heatshield and velocity, deflection angle of the heatshield is shown in table 2 and figure 4.

Table 2. The relationship between resistance moment and velocity, deflection angle of heatshield

Moment (N.m) / Velocity (ms-1) / Deflection Angle(°)	3.5	4	4.5	5	5.5	6	6.5	7	7.5	8	8.5
0	0	0	0	0	0	0	0	0	0	0	0
1	0.08	0.29	0.58	0.95	1.24	1.54	1.82	2.12	2.46	2.82	3.2
2	0.26	0.58	1.16	1.56	2.16	2.59	3.2	3.8	4.2	4.8	5.38
3	0.56	1.21	1.9	2.78	3.26	3.51	4.72	5.21	6.4	7.2	8.38
5	1.4	2.28	3.12	3.99	4.92	5.9	7.16	8.38	9.72	11.12	12.6

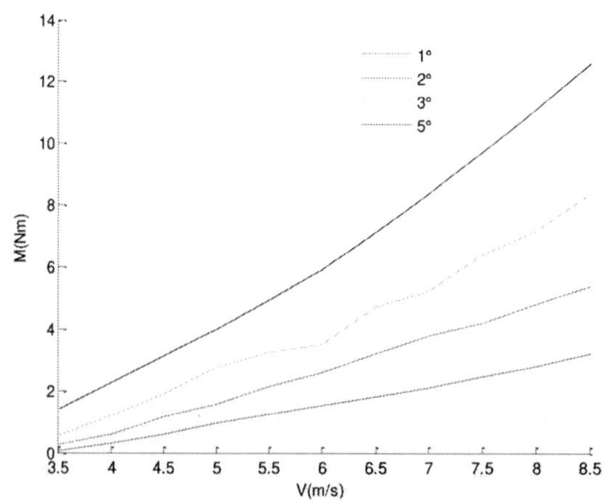

Figure 4. Graph of the relationship between resistance moment and velocity, deflection angle of heatshield

When the heatshield velocity increases from 3.5m/s to 8.5m/s and the deflection angle increases from 0° to 5°, the air resistance moment increases from 0 to 12.6 N.m. The air resistance moment of the heatshield increases obviously with velocity and deflection angle.

4.2. Aeroshell

4.2.1. Resistance. The relationship between the resistance on the aeroshell and velocity, deflection angle of the aeroshell is obtained by simulation calculation, as shown in table 3 and figure 5.

Table 3. The relationship between resistance and velocity, deflection angle of aeroshell

3	3.5	4	4.5	5	5.5	6	6.5	7	7.5	8
46.2	53.2	67	83.5	94.3	107.9	125	140.1	164.2	185.1	209.8
47.4	53.8	67.4	83.8	94.9	108.1	125.8	141.8	165.6	181.7	210.2
45.2	52.2	66.8	84.4	93.6	106.9	124.7	140.2	163.1	182.1	208.1
46.8	54.2	67.5	85.5	95.3	108.6	127.9	143.1	165.2	181.3	207.9
46.9	51.3	67.7	87.1	94.8	108.3	128.5	141.1	167.2	184.1	209.2

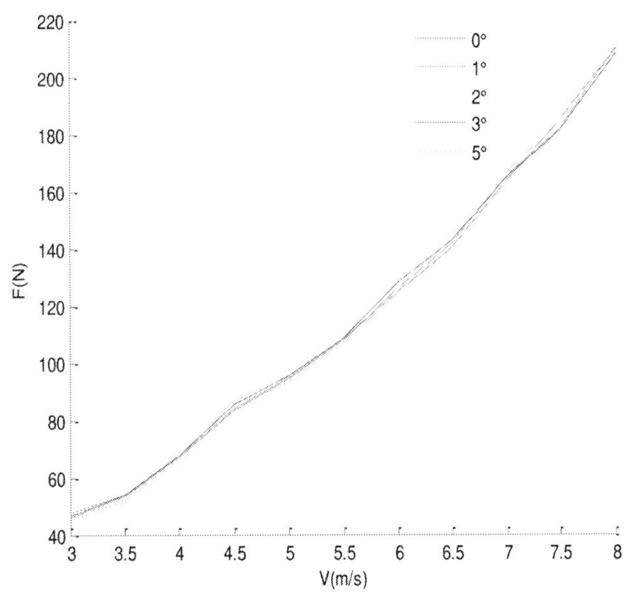

Figure 5. Graph of the relationship between resistance and velocity, deflection angle of aeroshell

When the aeroshell velocity increases from 3m/s to 8m/s and the deflection angle increases from 0° to 5°, the air resistance increases from 45N to 210N. The air resistance of the aeroshell increases obviously with velocity, but has little change with the increase of deflection angle.

4.2.2. Resistance moment. The relationship between resistance moment on the aeroshell and velocity, deflection angle of the aeroshell is obtained by simulation calculation, as shown in table 4 and figure 6.

Table 4. The relation between resistance moment and velocity, deflection angle of aeroshell

Moment (Nm) / Velocity (ms-1) / Deflection Angle(°)	3	3.5	4	4.5	5	5.5	6	6.5	7	7.5	8
0	0	0	0	0	0	0	0	0	0	0	0
1	0.04	0.15	0.29	0.47	0.62	0.77	0.91	1.06	1.23	1.41	1.6
2	0.13	0.29	0.58	0.78	1.08	1.29	1.6	1.9	2.1	2.4	2.69
3	0.28	0.61	0.95	1.39	1.63	1.75	2.36	2.61	3.2	3.6	4.19
5	0.7	1.14	1.56	1.99	2.46	2.95	3.58	4.19	4.86	5.56	6.3

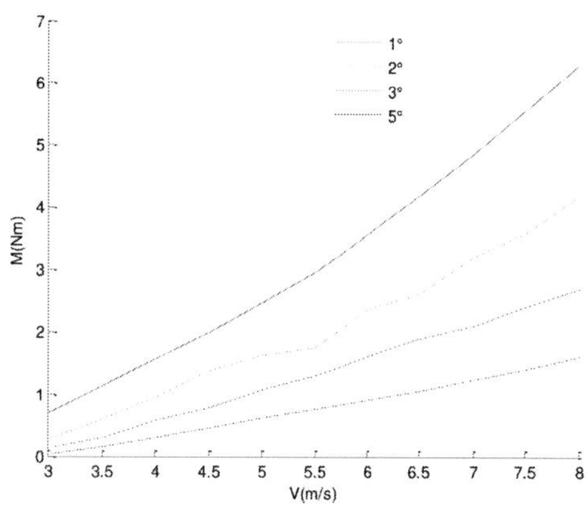

Figure 6. Graph of the relationship between resistance moment and velocity, deflection angle of aeroshell

When the aeroshell velocity increases from 3m/s to 8m/s and the deflection angle increases from 0° to 5°, the air resistance moment increases from 0 to 6.3 N.m. The air resistance moment of the aeroshell increases obviously with velocity and deflection angle.

5. Conclusion

In this paper, the Flunet analysis models of the cabins (heatshield and aeroshell) are established by numerical calculation. According to simulation and analysis, the air resistance of the cabins obviously with velocity, but has little change with the increase of deflection angle. The air resistance moment of the cabins increases obviously with velocity and deflection angle. The results of this paper have been used as input conditions for system simulation and ground test.

References

[1] Zheng Xun-hao, Zhang Ji-ye, Zhang Wei-hua. Numerical simulation of aerodynamic drag for high-speed train bogie[J]. Journal of Traffic and Transportation Engineering. 2011, 11(2): 45-51.

[2] Hu Xiaonan, Zhang Jianbin, Zhao Guowei. Dynamic simulation of a Flexible Rope with Air Resistance[J]. Mechanical Science and Technology for Aerospace Engineering. 2010, 29(10): 1334-1337.

[3] Chen Feng, Li Qiliang, Yang Zhigang. Numerical analysis on aerodynamic drag of generic rear view mirror[J]. Computer Aided Engineering. 2009, 18(4): 24-28.

[4] Zhou Xiao, Zhang Yaoping, Yao Yingfeng. Numerical Simulation on the Aerodynamic Drag of High-Speed Train in Evacuated Tube[J]. Science Technology and Engineering. 2008, 8(6): 1626-1628.

[5] Tian Hongqi. Study on the Characteristics of Train Air Resistance under Wind Environment[J]. China Railway Science. 2008, 29(5): 108-112.

[6] Huang Zhi-xiang, Chen Li, Jiang Ke-lin. Wind Tunnel Test of Air-drag Reduction Schemes of High-Speed Trains[J]. Journal of The China Railway Society. 2012, 34(4): 16-21.

IOP Conf. Series: Materials Science and Engineering **751** (2020) 012024 doi:10.1088/1757-899X/751/1/012024

Quick Trajectory Optimization Method for Mars Airplane Based on Adaptive Gauss Pseudospectral Method

Z Y Sun[1,*], Z Y Liu[1] and P Zhang[1]

[1] Beijing Institute of Control Engineering, Beijing, China, 100194

[*]Email: biceszy@sina.com

Abstract. This paper gives a new method based on adaptive Gauss pseudospectral method for quick trajectory optimization problem of Mars airplane. The method converts the trajectory optimization problem to a non-linear programming problem, taking into account for the dynamic constrains, the boundary constrains, the path constrains and the control constrains during the cruise phase of the Mars airplane. And the sequential quadratic programming is adapted to solve the non-linear programming problem. In combination with the adaptive Gauss pseudospectral method and sequential quadratic programming in Matlab, numerical results of the optimal trajectory with 267 collocations could be obtained in 800s with an approximation accuracy of 10^{-6}.

1. Introduction

Recently, NASA is carrying out a research on a new Mars exploration vehicle named Daedalon, which is a kind of unmanned airplane and can enlarge its platform area in low speed to gain enough lift [1]. This kind of designed Mars exploration vehicle is not only satisfied with the requirement of Mars entry, cruise and landing phases, but also could realize large-range glide to gain more scientific data than traditional Mars exploration vehicles.

The Daedalon flight profile consists of three phases after Mars entry, high-speed flight descent phase, low-speed cruise phase and lading phase The high-speed flight descent phase begins from Mars atmosphere entry, and then the vehicle discards the heat shield TPS. In this phase, the vehicle's speed decreases from 5km/s to about 0.5km/s and the height from 120km to about 3km. And in low-speed cruise phase, the vehicle performs a continuous morphing of the wing at about 500m altitude, such that the final low-sweep, high-aspect-ratio configuration to realize a large range cruise. In this phase, the vehicle also surveys the ground in search of the ideal landing site. In landing phase, the vehicle initiates a deep stall at 10m altitude and the thrusters allow it to land gently on its wheels.

Different from traditional Mars exploration vehicles, this kind of vehicle such as Daedalon has the property of nonlinear, parameter-varying and model-varying during the Mars entry, cruise and landing phases. In addition, the dynamic pressure constrains and the control output constrain must be taken into account in the trajectory design process. All of these constrains result in difficulties for the trajectory design. However, the trajectory programming theory, which could give an optimal trajectory of some specified mission under the consideration of the control constrains, the path constrains and boundary constrains, is a good choice to solve the problem.

The trajectory programming problem is an optimization problem essentially, and many fruits about optimization method could be borrowed. The method to solve these optimization problem consists of direct method and indirect method [2]. The indirect method solve the problem to get analytic result

IOP Conf. Series: Materials Science and Engineering **751** (2020) 012024 doi:10.1088/1757-899X/751/1/012024

using maximum principle, but some problem with complicate constrains couldn't get the analytic result usually. In contrast, direct method is of fast convergence velocity and broad convergence region, which converts the optimization problem to a nonlinear programming problem and then solves the problem with numerical method. Recently, as a kind of direct method, pseudospectral method gained considerable attention in solving complicated optimization problem. In addition, as a kind of pseudospectral method, Gauss pseudospectral method (GPM) has been applied widely in ascent and reentry trajectory programming and lunar exact-landing trajectory optimization with its advantages of high accuracy and fast convergence velocity [3][4][5]. The GPM, like all other pseudospectral methods, approximates the state using a basis of global interpolating polynomials [6]. These global polynomials are based on a set of discrete points across the interval. Adding discrete points is the only method to increase the approximate accuracy, which increases the optimization variable number in nonlinear programming and thus the optimization time. In order to increase the optimization accuracy and reduce the optimization time, Darby introduced an adaptive GPM [7]. In the low accuracy region, the adaptive GPM could increase the segment or colocation points, and in the high accuracy region, it will not make any changes, which could raise the computation efficiency enormously and has been applied on robot trajectory programming and ascent Phase trajectory optimization [8] [9].

This paper gives a quick trajectory programming method of Mars airplane as Daedalon in cruise phase adapting the adaptive GPM.

2. Mathematical model

Under in no consideration of the Mars rotation and wind, the Mars airplane dynamic equation can be described as [10],

$$\dot{r} = v * sin\theta \tag{1}$$

$$\dot{x} = v * cos\theta * cos\psi \tag{2}$$

$$\dot{y} = -v * cos\theta * sin\psi \tag{3}$$

$$\dot{v} = -D - g * sin\theta \tag{4}$$

$$\dot{\theta} = L * \frac{cos\sigma}{v} - g * \frac{cos\theta}{v} \tag{5}$$

$$\dot{\psi} = L * \frac{sin\sigma}{v * cos\theta} \tag{6}$$

where r is the distance between the airplane and the Mars centre, v the airplane's velocity, θ the flight path angle, x、y the distance to the destination, σ the bank angle, ψ trajectory deflection angle, g the gravitational acceleration at the Mars surface. And L and D are the lift and drag respectively, which can be expressed as,

$$D = \frac{1}{2}\rho v^2 C_D \frac{S}{m} \tag{7}$$

$$L = \frac{1}{2}\rho v^2 C_L \frac{S}{m} \tag{8}$$

where C_D, C_L are the drag and lift coefficient related to the Mach number and the attack angle [11], S the vehicle's area, m the mass of the vehicle. And ρ is the density of the Mars atmosphere, which can be expressed as [12],

$$\rho = exp\left(\sum_{i=0}^{4}\beta_i h^i\right) \tag{9}$$

where h is the altitude, β_i is the constant coefficient.

According to the attitude dynamic equation, the velocity and acceleration of the bank angle σ and the attack angle α can be expressed as,

$$\dot{\sigma} = \omega_1,\ \ddot{\sigma} = u_1,\ \dot{\alpha} = \omega_2,\ \ddot{\alpha} = u_2 \tag{10}$$

where ω_1, ω_2 are the angular velocity of the bank angle σ and the attack angle α and u_1, u_2 the acceleration of the bank angle σ and the attack angle α produced by the control torque from the attitude control system on board.

3. Principle of adaptive Gauss pseudospectral method

3.1. Continuous Bolza problem

The trajectory programming problem for Mars airplane could be considered to be a general Bolza optimal control problem that is to find the control variable $\boldsymbol{u}(t) \in \mathbf{R}^m$, minimizing the Bolza index, under all constrains in practice. The Bolza index can be expressed as,

$$J = \Phi\left(\boldsymbol{x}(t_0), t_0, \boldsymbol{x}(t_f), t_f\right) + \int_{t_0}^{t_f} g\left(\boldsymbol{x}(t), \boldsymbol{u}(t), t\right) \mathrm{d}t, \tag{11}$$

where $\boldsymbol{x}(t) \in \mathbf{R}^n$, the initial time t_0 and the final time t_f are satisfied with the following constrains,

(1) Dynamic constrains

$$\dot{\mathbf{x}}(t) = \mathbf{f}\left(\mathbf{x}(t), \mathbf{u}(t), t\right), t \in \left[t_0 \quad t_f\right], \tag{12}$$

The dynamic constrains of the Mars plane consist of (1) ~ (6) and (10).

(2) Path constrain

$$\boldsymbol{C}\left(\boldsymbol{x}(t), \boldsymbol{u}(t), t\right) \le 0, t \in \left[t_0 \quad t_f\right], \tag{13}$$

The path constrain is just the dynamic pressure constrain, which can be expressed as,

$$q = \frac{1}{2}\rho v^2 \le q_{max} \tag{14}$$

where q_{max} is the maximum dynamic pressure in the duration.

(3) Boundary conditions

$$\phi\left(\boldsymbol{x}(t_0), t_0, \boldsymbol{x}(t_f), t_f\right) = 0, \tag{15}$$

The boundary constrain consists of the initial state constrain and the final state constrain, and can be expressed as,

$$r(t_0) = r_0,\ v(t_0) = v_0,\ \theta(t_0) = \theta_0,\ x(t_0) = x_0,\ y(t_0) = y_0,\ \psi(t_0) = \psi_0 \tag{16}$$

$$r(t_f) = r_f,\ v(t_f) = v_f,\ \theta(t_f) = \theta_f,\ x(t_f) = x_f,\ y(t_f) = y_f,\ \psi(t_f) = \psi_f \tag{17}$$

The optimal trajectory of the state variable and control variable could be gained by solving the equation set inclusive of Bolza index, dynamic constrains, path constrain and boundary constrains.

3.2. Adaptive Gauss pseudospectral method

GPM converts the above continuous Bolza problem to a discrete optimal problem by approximating the state variable and control variable using Lagrange interpolation in a series of separate points. The time derivative of the state variable can be expressed by the time derivative of global interpolation polynomial, the integration item in Bolza index by Gauss integration equation and the final state constrain by integration during the total process from initial state. Under the above numerical approximation, the continual optimal control problem could be converted to be a discrete and nonlinear programming problem with some constrains.

In GPM, the Lagrange polynomial is constructed at the Legendre-Gauss points, initial point and final point. And the state variable and control variable could be approximated as [7],

$$x(\tau) \approx X(\tau) = \sum_{i=0}^{K} X_i l_i(\tau), \tag{18}$$

$$\mathrm{u}(\tau) \approx \mathrm{U}(\tau) = \sum_{i=1}^{K} \mathrm{U}_i \tilde{l}_i(\tau), \tag{19}$$

where, $X_i = X(\tau_i)$, $U_i = U(\tau_i)$, and $l_i(\tau)$ is the Lagrange basis function.

In combination with equation (18), the dynamic constrains in equation (12) can be expressed as,

$$DX - \frac{t_0 - t_f}{2} F\left(X^{\mathrm{LG}}, U^{\mathrm{LG}}, \tau^{\mathrm{LG}}, t_0, t_f\right) = 0, \tag{20}$$

where the differential matrix $D \in \mathbf{R}^{k\times(k+1)}$.

The final state X_f should meet the dynamic constrains,

$$x(\tau_f) = x(\tau_0) + \int_{-1}^{1} f\left(x(\tau), u(\tau), \tau\right) d\tau, \tag{21}$$

According to the Gauss-Lobatto formula, equation (21) could be expressed as,

$$X_{N+1} = X(\tau_{N+1}) = X_0 + \frac{t_f - t_0}{2} w^{\mathrm{T}} F, \tag{22}$$

where τ_i are LG points, w^{T} is the Gauss integration weight.

And thus, the Bolza index can be expressed as,

$$J = \Phi\left(X_0, X_f, t_0, t_f\right) + \frac{t_0 - t_f}{2} \sum_{i=1}^{K} \omega_i g\left(X_i, U_i, \tau_i, t_0, t_f\right), \tag{23}$$

The boundary constrain and path constrain can be described at LG points as,

$$\phi\left(X_0, \tau_0, X_{N+1}, \tau_f\right) = 0, \tag{24}$$

$$C\left(X_k, U_k, \tau_k, t_0, t_f\right) \leq 0, \tag{25}$$

After the above numerical approximation, the continuous Bolza problem can be converted to a discrete and nonlinear programming problem, to find the discrete state variables $\left(X_1, X_2, \cdots, X_K\right)$ and control variables $\left(U_1, U_2, \cdots, U_K\right)$ at LG points, which both are under the constrains of (20), (24) and (25) and minimize the index expressed in (23).

The above described global GPM can't heighten precision except for adding more LG points, and this in turn prolong the computation time extensively. In order to taking into account both the precision and the computation time, a new adaptive GPM is introduced in [7], which could decide whether to add LG points or segments according to the approximating precision. In contrast to the above global GPM, the adaptive GPM divides the NLP into some segments, and applies the GPM in each segment. Assuming there are *S* segments and *N* LG points.

And in this condition, the dynamics in (20) can be expressed as,

$$\begin{bmatrix} D_1 & 0 & \mathrm{L} & 0 \\ 0 & D_2 & \mathrm{L} & 0 \\ \mathrm{M} & & \mathrm{O} & \mathrm{M} \\ 0 & \ldots & 0 & D_s \end{bmatrix} X = \frac{1}{2} \begin{bmatrix} t_1 - t_0 & 0 & \mathrm{L} & 0 \\ 0 & t_2 - t_1 & \mathrm{L} & 0 \\ \mathrm{M} & & \mathrm{O} & \mathrm{M} \\ 0 & \ldots & 0 & t_s - t_{s-1} \end{bmatrix} F, \tag{26}$$

Obviously, the derivative matrix in (27) is sparser than the one in (24), and the sparse matrix helps to heighten computation efficiency and decrease computation time. The adaptive GPM adds segments or LG points in low precision area and does nothing in the area where the precision is enough. And hence, this method can heighten the computational efficiency at the same time assuring the precision.

At the low precision area, the adaptive GPM decides whether to add segments or LG points according to the midpoint residuals matrix. Each middle time $\bar{t}_i$ can be expressed as,

$$\bar{t}_i = \frac{t_i + t_{i+1}}{2}, (i = 1, ..., N_s - 1), \tag{27}$$

Define $\bar{\boldsymbol{X}} \in \mathbf{R}^{(N_s-1)\times n}$, $\bar{\boldsymbol{U}} \in \mathbf{R}^{(N_s-1)\times m}$ as the state variable and control variable at $\bar{t}_i$, and then the midpoint residuals matrix $\boldsymbol{R}$ can be expressed as,

$$\boldsymbol{R} = \left| \bar{\boldsymbol{D}}\bar{X} - \frac{t_s - t_{s-1}}{2} \boldsymbol{F}\left(\bar{X}, \bar{U}, \bar{\boldsymbol{\tau}}, t_s, t_{s-1}\right) \right| \in \mathbf{R}^{(N_s-1)\times n}, \tag{28}$$

where $\bar{\boldsymbol{D}} \in \mathbf{R}^{(N_s-1)\times(N_s-1)}$ is the derivative matrix at $\bar{\boldsymbol{\tau}} = \left(\bar{\tau}_1, ..., \bar{\tau}_{N_s-1}\right)$.

Define the scaled midpoint residual vector $\boldsymbol{\beta}$ as,

$$\boldsymbol{\beta} = \left[\beta(\bar{t}_1), ..., \beta\left(\bar{t}_{N_s-1}\right)\right] = \boldsymbol{r} / \bar{r}, \tag{29}$$

where $\boldsymbol{r} \in \mathbf{R}^{(N_s-1)\times 1}$ is the column matrix with biggest residuals in matrix $\boldsymbol{R}$ and $\bar{r}$ can be expressed as,

$$\bar{r} = \frac{\sum_{i=1}^{N_s-1} r(t_i)}{N_s - 1}, \tag{30}$$

Define ε as the tolerance and consider the condition where the maximum residual is bigger than ε. When the element of uniform-type errors in $\boldsymbol{\beta}$ is bigger than the threshold ρ ($\rho > 1$), we add LG points to heighten the approximated precision, that is, $N_s^{'} = N_s + L$, where L is the LG points to be added.

The adaptive GPM adds LG points or segments at low precision region to heighten the precision. And it performs well in both approximation precision and computation efficiency for non-smooth optimization problems.

4. Numerical results

Taking into consideration the limited fuel of the Mars airplane, the Bolza index of the trajectory programming is selected to be the fuel consumed during the cruise phase, which can be expressed as followed,

$$J = \int_{t_0}^{t_f} \boldsymbol{u}\boldsymbol{u}^T dt, \tag{31}$$

where $\boldsymbol{u}$ is the control variable, satisfied with the equation $\boldsymbol{u} = \left[\ddot{\sigma}, \ddot{\alpha}\right]$.

The control constrain consists mainly of the maneuver capability of the vehicle, inclusive of the bank acceleration constrain and the attack angle constrain, which can be expressed as,

$$\left|\ddot{\sigma}\right| \le \ddot{\sigma}_{max}, \left|\ddot{\alpha}\right| \le \ddot{\alpha}_{max} \tag{32}$$

The state variable is selected to be $\boldsymbol{x} = \left[r, \theta, x, y, v, \psi, \sigma, \dot{\sigma}, \alpha, \dot{\alpha}\right]$.

The dynamic equation is listed in (1) ~ (6) and (8). The path constrains and the boundary constrains are listed in (14), (16) ~ (17) respectively.

The parameters in the trajectory programming are listed in Table 1~3.

Table 1 Physical parameters in simulation

Parameter	Value
Radius of Mars (km)	3397
Gravitational constant	4.283×10^{13}
Mass of the vehicle (kg)	600
Heat flux coefficient	8.49×10^{-8}

Table 2 Boundary parameters in simulation

Parameter	Initial Value	Final Value
r (km)	3504. 1	3400.08
v (m)	500	160.5
x (m)	-20000	-4282.35
y (m)	70000	98.23
θ (°)	0.192	0.03
ψ (°)	90	0.006
σ (°)	0	0
$\dot{\sigma}$ (°/s)	0	0
α (°)	0	0
$\dot{\alpha}$ (°/s)	0	0

Table 3 Path and control constrain parameters in simulation

Parameter	Value
Maximum bank angle σ_{max} (°)	90
Maximum angular velocity of the bank angle $\dot{\sigma}_{max}$ (°/s)	5
Maximum angular acceleration of the bank angle $\ddot{\sigma}_{max}$ (°/s^2)	3.36
Maximum attack angle α_{max} (°)	15
Maximum angular velocity of the attack angle $\dot{\alpha}_{max}$ (°/s)	5
Maximum angular acceleration of the attack angle $\ddot{\alpha}_{max}$ (°/s^2)	3.36
Maximum dynamic pressure q_{max} (Pa)	1000

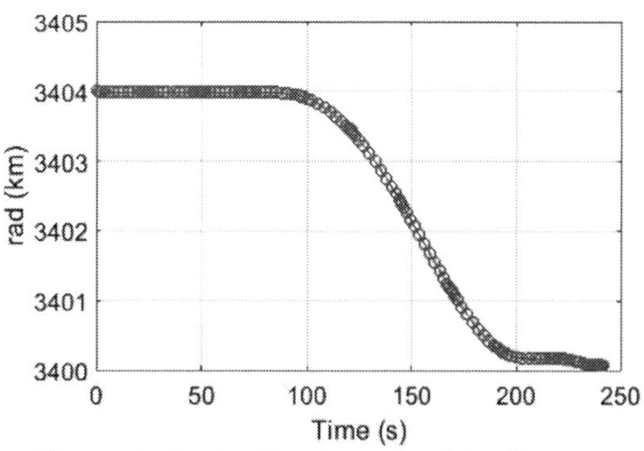

Figure 1. Optimal trajectory of the distance

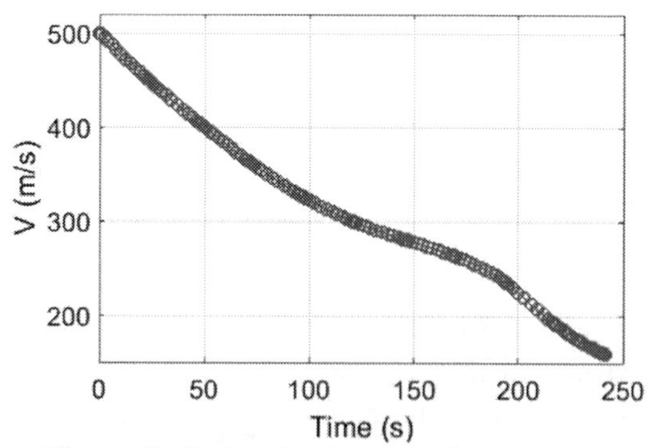

Figure 2. Optimal trajectory of the velocity

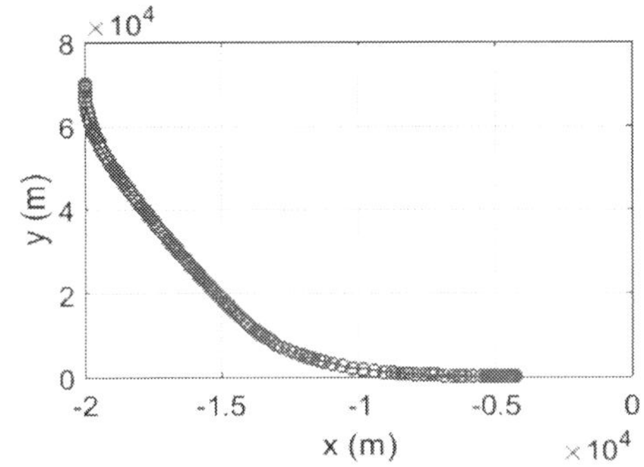

Figure 3. Optimal trajectory of the x and y

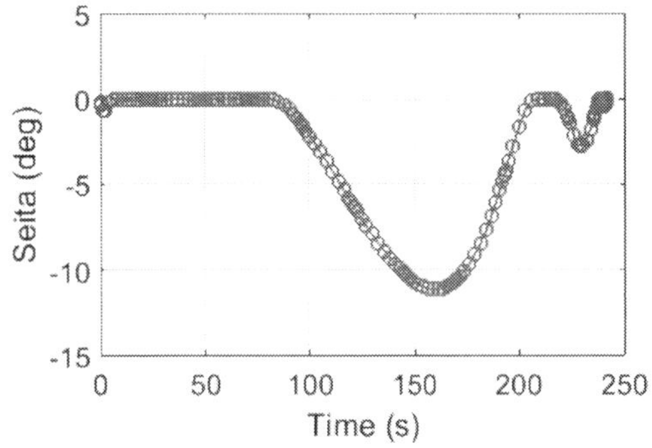

Figure 4. The flight path angle curve

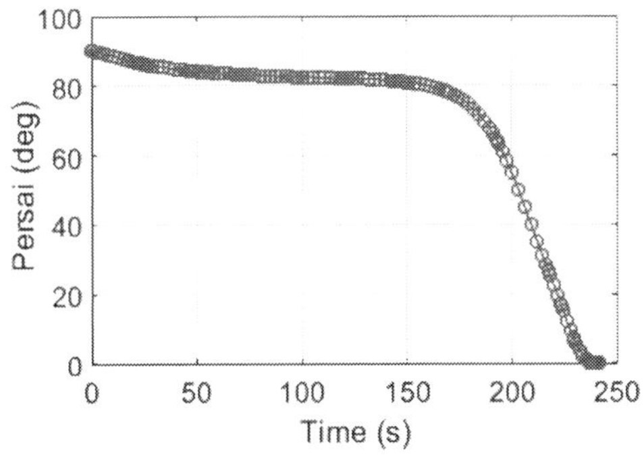

Figure 5. The trajectory deflection angle

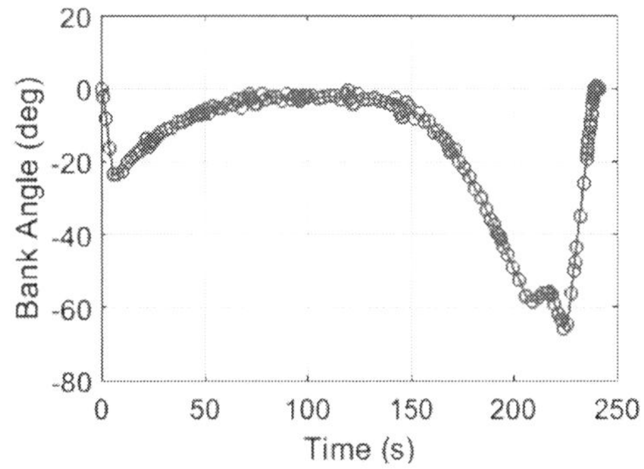

Figure 6. The bank angle curve

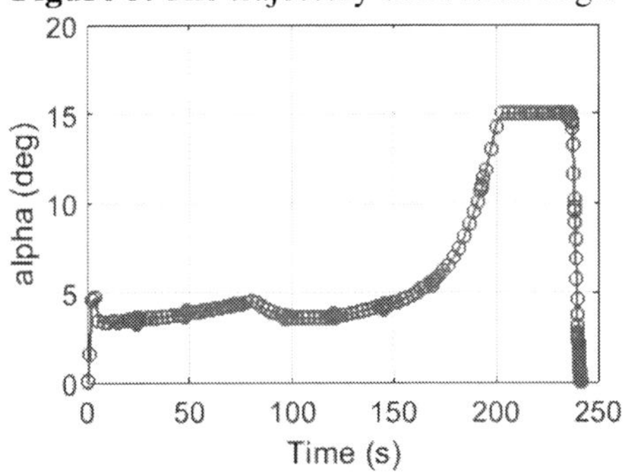

Figure 7. Attack angle curve

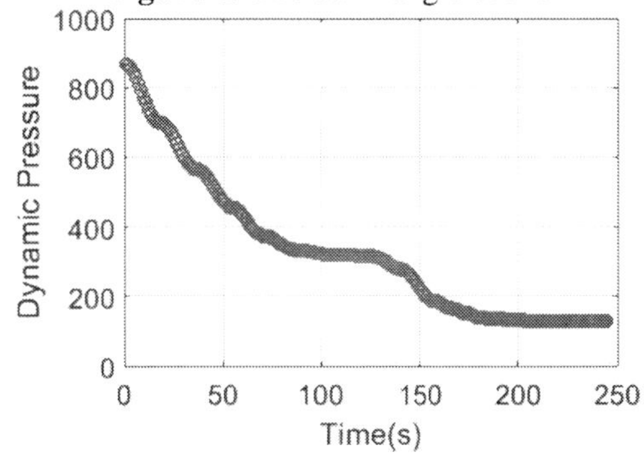

Figure 8. The dynamic pressure curve

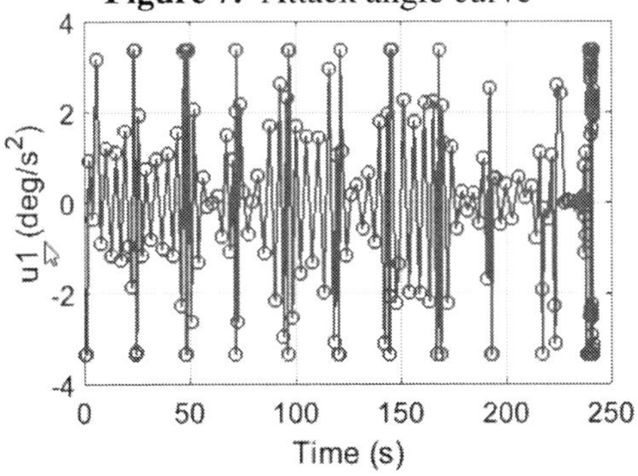

Figure 9. Bank angle acceleration curve

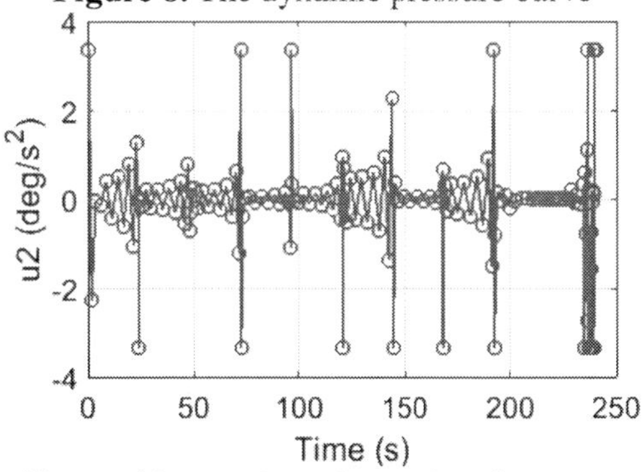

Figure 10. Attack angle acceleration curve

The trajectory programming problem is run on a computer with Matlab R2017b and GPOPS II (V1.0) software [13]. The minimum number of the collocation point in each segment is 4 and the maximum error tolerance is 10^{-6}. The GPOPS software adopted sequential quadratic programming to solve the nonlinear programming problem, and gave the optimal reference trajectory with 257 collocation points in about 800 seconds after 21 iterations. And the trajectory programming results are shown in Figure 1~Figure 10. Figure 1~Figure 5 are the optimal trajectories of some state variables, and suggest that the Mars airplane could fly from the initial point to the destination accurately by tuning the bank angle and the attack angle as shown in Figure 6~figure 7 in spite of all the constrains. And the cruise phase lasts about 240s. Fig. 8 is the dynamic pressure trajectory, which is satisfied with the path constrains. Figure 9~figure 10 are the angular acceleration of the bank and attack angle, which is also under the control constrains. In conclusion, the adaptive Gauss pseudospectral method solves the trajectory programming problem with complicated constrains for Mars airplane perfectly.

5. Conclusions

In combination with the adaptive Gauss pseudospectral method and sequential quadratic programming, this paper gives a new method to solve the trajectory programming problem of the Mars airplane quickly. This method takes into account all constrains inclusive of path constrains, boundary constrains and control constrains, and gives the optimal trajectory of the Mars airplane in cruise phase. Numerical results suggest that the high accuracy and satisfying all constrains trajectory could be produced in Matlab quickly, and thus the method introduced in this paper is of great practical application.

Acknowledgement

This work is supported by National Nature Science Foundation under Grant 61803029.

References

[1] Jarret M L, John R O, Robert D B. Daedalon: A Revolutionary Morphing Spacecraft Design for Planetary Exploration [C]. 1st Space Exploration Conference: Continuing the Voyage of Discovery 30 January -1 February, 2005, Orlando, Florida.

[2] Yong En-mi, Chen Lei, Tang Guo-jin. A survey of numerical methods for trajectory optimization of spacecraft [J]. Journal of Astronautics, 2008, 29(2): 397-406.

[3] Yang Xi-xiang, Zhang Wie-hua. Rapid optimization of ascent trajectory for solid launch vehicles based on GPM [J]. Journal of Astronautics, 2011, 32(1): 15-21.

[4] Zhou Wenya, Yang Di, Li Shunli. Using Gauss Pseudospectral Method to Solve Optimal Reentry Trajectory for Lifting Vehicle [J]. Journal of Nanjing University of Science and Technology, 2010, 34(1): 85-90.

[5] Peng Qi-bo, Li Hai-yang, Shen Hong-xin. Rapid lunar exact-landing trajectory optimization via GPM [J]. Journal of Astronautics, 2010, 31(4): 1012-1016.

[6] Benson D A, Huntington G T. Direct trajectory optimization and costate estimation via an orthogonal collocation method [J]. Journal of Guidance Control, and Dynamics, 2006, 29(6): 1435-1440.

[7] Darby C L, Hager W W, Rao A V. An hp-adaptive pseudospectral method for solving optimal control problems [J]. Optimal Control Applications and Methods, 2011, 32(4): 476-502.

[8] Zeng Xiangxin, Cui Naigang, Guo Jifeng, Path planning of space robot based on hp-adaptive pseudospectral method [J], Robot, 2018, 40(3): 385-342.

[9] Liu Ruifan, Yu Yunfeng, Yan Binbin, Ascent Phase Trajectory Optimization for Hypersonic Vehicle Based on hp-Adaptive Pseudospectral Method [J], Journal of Northwestern Polytechnic University, 2016, 10(5): 790-797.

[10] Vinh N X. Optimal trajectories in atmospheric flight [M], New York: Elsevier Scientific Publishing Co., 1981.

[11] Ming Yin, Coordinate control of deformation and flight for morphing aircraft [D], Nanjing University of Aeronautics and Astronautics, 2015.

[12] Benito Manrique J. Advances in spacecraft atmospheric entry guidance [D]. University of California, Irvine, 2010.

[13] www.gpops2.com

VISHKIRA, A Gallinaceous Aircraft

K Taneja[1, a], R Jaiswal[2,b] and S K Kalwala[1,c]

[1]Student, University of Petroleum and Energy Studies, INDIA (2016-2020).
[2]Assistant professor, Aerospace department, University of Petroleum and Energy Studies, INDIA.

[a]kashishtaneja1997@gmail.com; [b]rjaiswal@ddn.upes.ac.in; [c]sreekavyakalwala@gmail.com

Abstract: A Vishkira is an idea of a gallinaceous aircraft useful for military purposes. It is the modest version of morphing of wings and Ornithopter. This aircraft can morph itself by decreasing its span length by folding mechanism and locked at the folded position by electro-magnetism. The folded mechanism in wings uses electro hydraulics. This provides the flexibility to pass through narrow passages between mountains avoiding the aircraft to tilt and accomplish the military mission and also have an advantage of flying like a bird by flapping its wing as in Ornithopter producing a huge amount of thrust in a single flap of the wing. This provides a convenience of having a shorter runway which is suitable for military aircrafts when they are in a need of emergency take-off.

1. Introduction

Vishkira is the modest form of morphing of wing of an aeroplane with an ancient ideology of birds. Morphing of wing of an aeroplane is to change the wing configuration that is shape, size, and dimensions for better flight performance which is classified on what changes are made in the conventional wing. There is platform morphing which constitute the change in sweep, span and chord. And there are out-of-plane and aerofoil morphing in the classification. Vishkira (the proposed wing) utilizes platform and morphing by changing its span and chord length of the wing as its morphing characteristic furthermore is unidirectional which reduces the complexity.

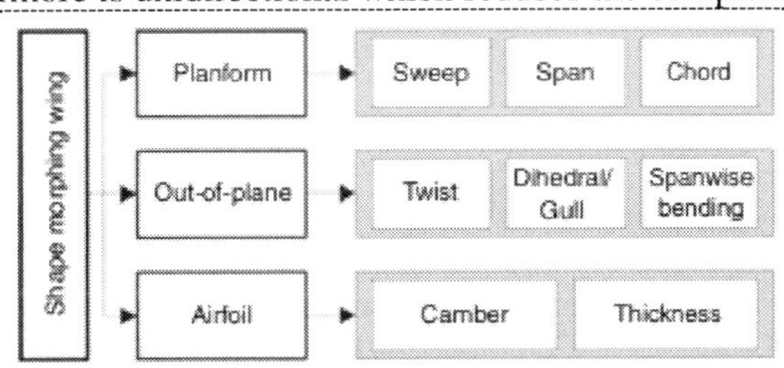

Figure 1. Types of shape morphing wing

Till date, the morphing techniques are commonly applied to the infinite wing [1] which are categorized in figure 1 but there is less number of experiments going on the finite wing as there are a lot of complications. Nevertheless, vishkira is an aircraft which contains a wing of unidirectional infinite wing morphing system.

In the Wing of Vishkira, the whole length of wing folds at two positions and changes its aerofoil in the morphed section; one is at the joint where the wing is attached to the fuselage as it experiences a maximum of Bending Moment at that point, so to reduce this, the wing has a folding at this position. The other folding position is at almost the mid span of the wing. If one position folds clock-wise, the other will fold anti-clock-wise to balance the moment depending on the configuration of the wing that is, if the aircraft has top wing configuration, the first fold is folded in clock-wise and the other in vice versa. If the aircraft has low wing configuration, the first fold is anti-clock-wise folded and other in vice versa. But if the aircraft is mid-wing configured then it has an advantage of folding in any direction.

This type of mechanism is specially designed for military aircrafts as it needs to fly through narrow passages and in between of valleys for certain missions[1]. The old conventional method to overcome this situation is by tilting the horizontal axis of the aircraft which is very complicated and increases the burden on pilot or by surrounding the whole valley by changing path which consumes time and fuel. The Wing of Vishkira has an advantage of reducing its span length in flight condition by folding mechanism. It also has and add on advantage for taking less space at the ramp. Wings having this type of configuration cannot place their hard points i.e., weapon stations on the wing.

2. Ideology

The idea for this concept is truly inspired from the birds which morph their wings according to the conditions. And moreover, the name Vishkira is derived from a gallinaceous heavily bodied terrestrial bird which is capable of reacting to any situation hastily.

3. Literature survey:

After a long study on Morphing and the techniques used, a detailed literature survey is shown in Table-1 which helped in designing the morphing techniques. The Table-1 also shows the way of proceedings tried for the successfulness of the paper.

Table 1. Literature review

Author	Research name	Description	inference
A.Y.N Sofla, S.A.Meguid, K.T.Tan, W.K Yeo [2]	Shape morphing of aircraft wing: Status and challenges	A research on evaluation and classification of shape morphing wing and the techniques used to morph the wing with the help of smart materials.	Aerofoil profile adjustment, span change morphing using techniques.
Kazuya Satio, Fabio Agnese, Fabrizio Scarpa. [3]	A cellular Kirigami Morphing Wingbox Concept	Use of Japanese old art techniques of paper folding for the morphing of wing box using sheet metal and heat.	Use of finite element approach in the model and the manufacturing.
Rajesh Purohit, Divya Gosain and Gauri Gupta[4]	Metal Rubber: the new age nanomaterial	An introduction to the new age nanomaterial called Rubber Metal	How can metal rubber change its shape according to the

		and its fabrication which continues with its advantages and scope in the future industry.	properties and fabrication and use of flexible actuators.
Jennifer H. Lalli, William Harrison, Theodore Distler, Michael Bortner, Keith Hill, Jake Dennis, Bradley Davis, and Richard O. Claus [5]	Shape Memory-Metal Rubber Morphing Aircraft Skins.	A whole experimental research of morphing of aircraft wing using the technique of folding by use of metal rubber.	Properties and advantages of metal rubber in aerospace industry.

4. Methodology

This method of morphing includes two techniques; firstly, span morphing [2] which makes the aircraft fold at two places and the other is changing the NACA series of the morphed wing. Changing the aerofoil configuration decreases the vortices and hence therefore drag created is less. This type of morphing is can be done using Metal Rubber [4].

4.1. Metal rubber;

Metal rubber is the new age super material which is formed by plastic polymers and metal ions [4]. This poly-material needs only one per cent of metal ions to maintain its conductive properties through when electrical or thermal means these metal ions are conducted and they tend to change their shape according to the actuators. This composite is super flexible and can maintain its properties at temperatures from 338°F to -76°F and can resist through stress [5]. Also it can strain itself up to 1000 per cent of its original dimensions which brings out the advantage for using to morph. Metal rubber is prepared by a nanotechnology process called Electro Static Molecular Self Assembly (ESA).

For the use of metal rubber in morphing [5], divide the wing into two sections, the section which is attached to the fuselage is section-A and the outer part is Section-B. The joint between fuselage and section-A is made up of Metal Rubber under which flexible actuators are fixed. And also from the joint between section-A and Section-B to the end the wing is made up of metal rubber with same actuators. Using actuators, thermal heat is produced electrically which changes the shape of metal rubber. The wing is folded at the desired sections for morphing [3]. So by this, the span length of the wing is reduced which allows the aircraft to fly at narrow paths. As the span length decreases, the span area of the wing also decreases which results in lower lift.to overcome this condition, that is, to increase the span area we can morph the section-B of the wing by changing its aerofoil.

4.2. Designing;

The model designing is done in a software named CATIA, a multi-platform software suite for computer aided design. The software designed by Dassault system of version V5-R20 is used. Firstly, a simple wing is designed using sketch based tools and multi surface volume command of NACA series 2412 as shown in figures 2 and 3.Moreover, the wing is 60% tapered wing an aspect ratio of 3.75 with the chord being 1000mm and the span of the wing being 3000mm.After morphing at 200 mm from the fuselage, the aspect ratio of the wing becomes 3.5593.

IOP Conf. Series: Materials Science and Engineering **751** (2020) 012025 doi:10.1088/1757-899X/751/1/012025

CATIA model of NACA 2412 wing

Figure 2. Top view

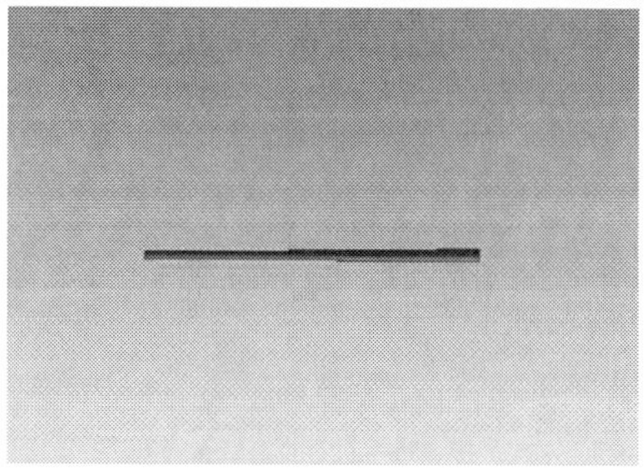

Figure 3. Front view

Now, another model of the same NACA series is made by morphing its structure **[1]**. The wing is morphed in this way such that it folds at two positions as said before. Consider it as a low wing configuration, so that it bends upwards. This type of morphing is done so that the span length of the wing decreases. The models are shown below

NACA 2412 MORPHED WING

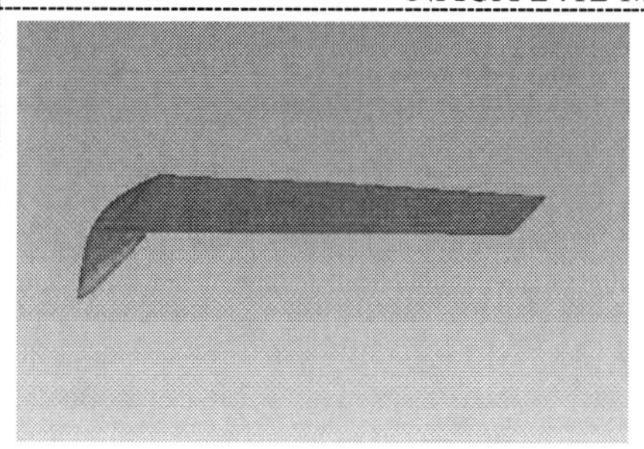

Figure 4. Sideways view

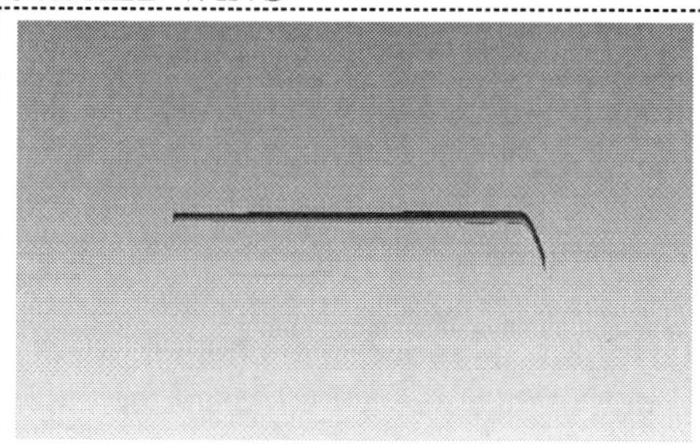

Figure 5. Front view

Due to the above wing design, the wing experiences wing tip vortices, which create pressure drag. Due to this pressure drag, the overall lift of the wing is decreased. And also by decreasing the span length of the wing, the span area is also decreased, so the lift produced will be less. So to overcome this condition, the cl of the wing is changed by changing the cross-section i.e. by changing the NACA series of the morphed part. So, from NACA 2412 as shown in figures 4 and 5 the cross-section is changed to NACA 4412 as depicted in figures 6 and 7.

NACA 4412 MORPHED WING

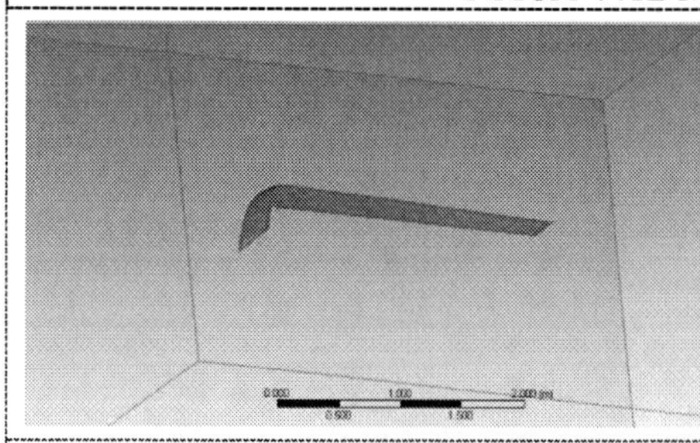

Figure 6. Side view

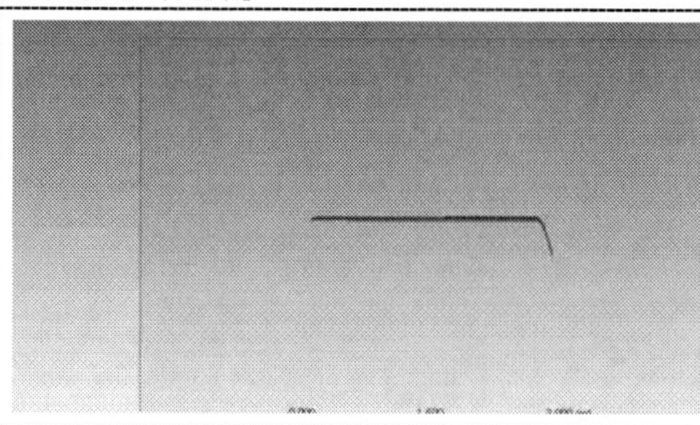

Figure 7. Front view

The whole analysis is done using CATIA and ANSYS. The model is made in CATIA and the flow analysis is done in ANSYS. Firstly, a straight wing is designed in CATIA as shown in figures, then flow analysis is done on that. Next, the straight wing is morphed by folding as shown in figures. Later on, the NACA series is changed on the morphed wing and the analysis is done.

Pressure Distribution over NACA 2412 Wing

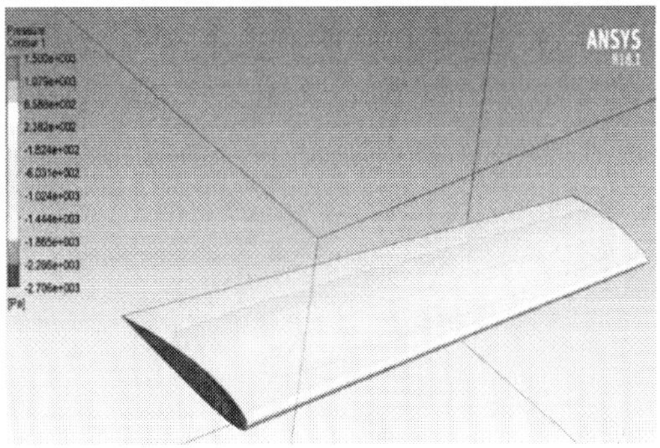

Figure 8. Isometric view

Figure 9. Top view

Pressure distribution over NACA 2412 morphed wing

Figure 10. Isometric view

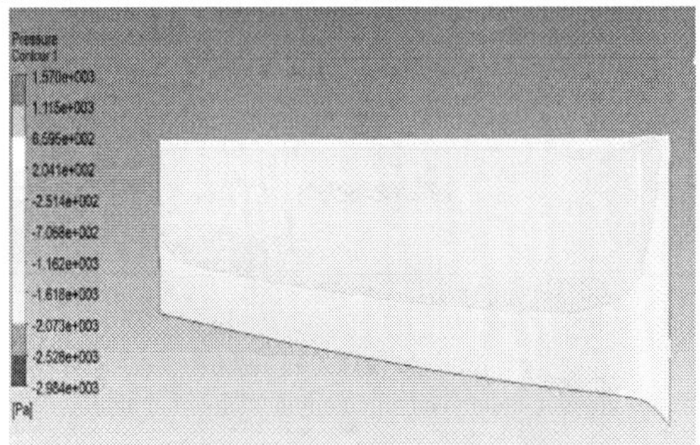

Figure 11. Front view.

The model used for analysis of the wing is a standard k-epsilon viscous model with enhanced wall treatment model. On comparing the cl and lift and pressure forces of the straight wing with the morphed wing as shown in figures 8 to 13, we found a slight decrease from the former to the later. We can conclude that the aforementioned idea for the morphed wing is feasible.

Pressure distribution over 4412 morphed wing

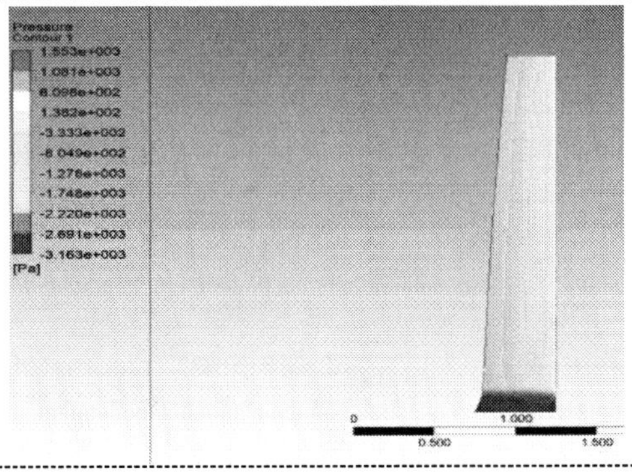

Figure 12. Top view

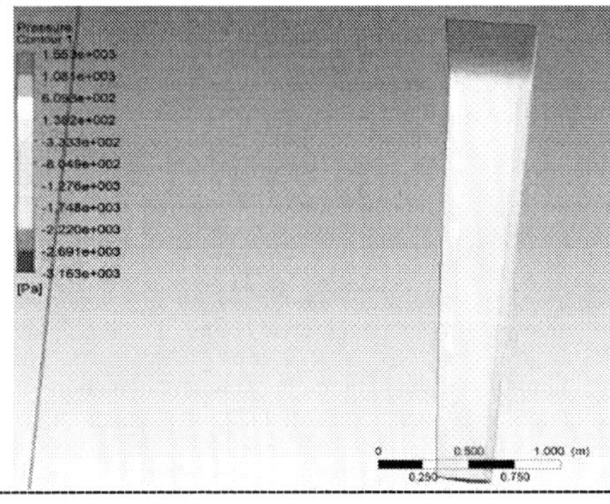

Figure 13. Bottom view

IOP Conf. Series: Materials Science and Engineering **751** (2020) 012025 doi:10.1088/1757-899X/751/1/012025

Lift coefficient

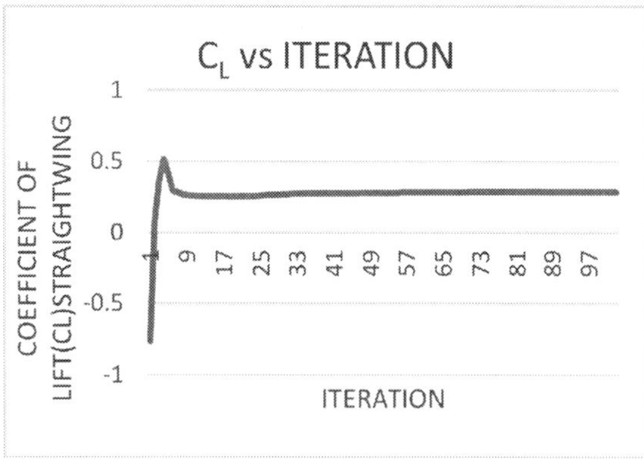

Figure 14. Lift coefficient of NACA 2412 straight wing.

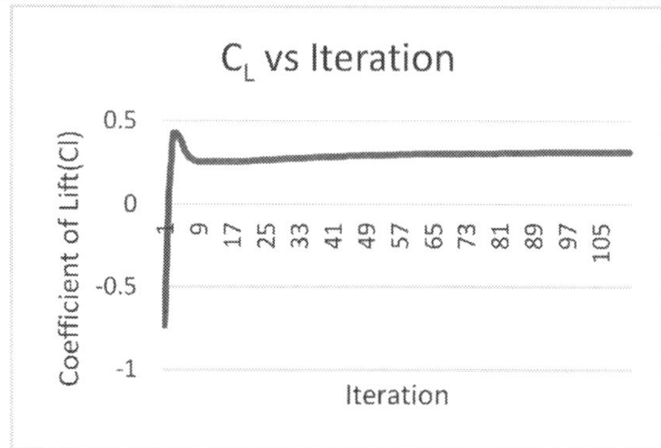

Figure 15. Lift Coefficient of NACA 4412 morphed wing.

Drag Coefficient

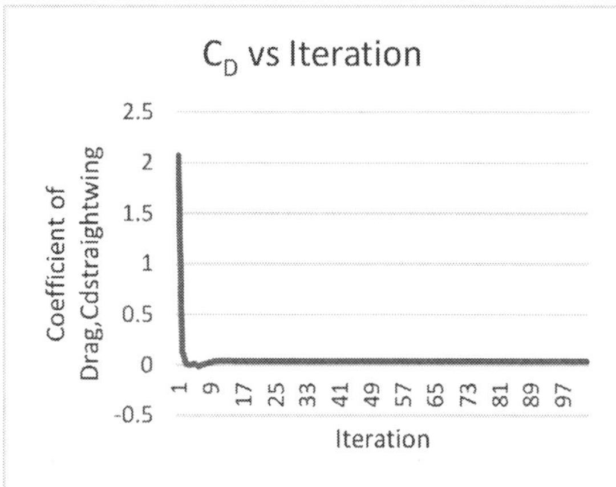

Figure 16. drag coefficient of NACA 2412 straight wing.

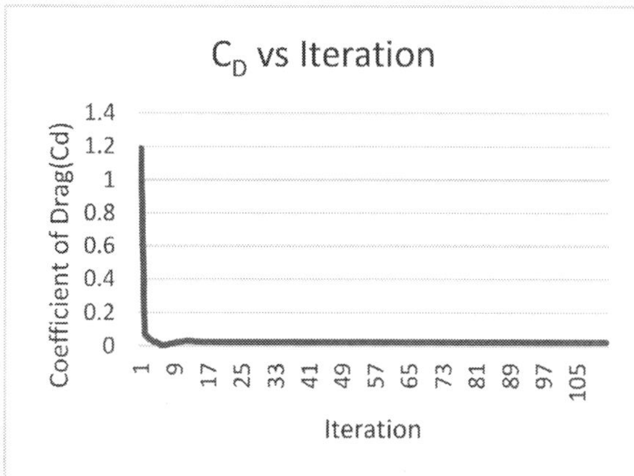

Figure 17. drag coefficient of NACA 4412 morphed wing.

On Comparing the analysis reports,we notice that, the minimal difference in coefficient of lift and coefficient of drag for straight wing (fig.14 and fig.16) and morphed wing (fig15 and fig. 17) is proportional to minimal difference in aspect ratio of the straight wing and tapered wing. the lift force of NACA 2412 straight wing and NACA 4412 morphed wing are almost same. From this we can infer that we can produce the same amount of lift decreasing its span length.

After the aircraft is morphed, to keep the wings in the locked position, electromagnetic locks are placed at the folded area and fuselage. Whenever the pilot requires to lock the system he will actuate the mechanism which will produce the magnetism using electricity.

5. Future Scope;

With enough research and advancement, wing morphing can help in developing smoother flight experience, it can be utilized in landing and take-off in order to increase and decrease lift instead of flaps which will help in removing more mechanical parts and increasing wing efficiency.

IOP Conf. Series: Materials Science and Engineering **751** (2020) 012025 doi:10.1088/1757-899X/751/1/012025

Reference

[1] Rodriguez A. Morphing Aircraft Technology Survey. *45th AIAA Aerospace Sciences Meeting and Exhibit* 2007 1258-1274.
[2] Sofla A Y N, Meguid S A, Tan K T and Yeo W K 2009 *Materials & Design.* **31** 1284-1292
[3] Saito K, Agnese F, Scarpa F. A Cellular Kirigami Morphing Wingbox Concept. *Journal of Intelligent Material Systems and Structures*. 2011;**22**(9):935-944.
[4] Purohit R, Gosain D and Gupta G 2008 *Int.J,Nnomanufacturing* **2** 659
[5] Lalli J, Harrison W, Distler T, Bortner M, Hill K, Dennis J et al. Shape Memory-Metal Rubber™ Morphing Aircraft Skins. **Volume 2:** 2009 Multifunctional Materials; Enabling Technologies and Integrated System Design; Structural Health Monitoring/NDE; Bio-Inspired Smart Materials and Structures

IOP Conf. Series: Materials Science and Engineering **751** (2020) 012026 doi:10.1088/1757-899X/751/1/012026

Research on Model Correction of Turbofan Engine Based on Quantum-behaved Particle Swarm Optimization

Renjun Qian*, Benwei Li, Siqi Yan, Shufan Zhao and Huailiang Teng

Aviation Foundation College, Naval Aviation University, Yantai, Shandong, China

*18902016296@163.com

Abstract. In order to solve the problem that the calculation result of mathematical model and the measured data of the whole machine deviate greatly due to the conditional simplification in the modeling process and engine component difference, an engine model correction method based on Quantum-behaved Particle Swarm Optimization(QPSO) is proposed. And the engine model calculation results are compared with the measured data of the whole machine performance. The results show that using the QPSO to modify the engine model can significantly improve the accuracy of the model. Before the correction, comparing the performance calculation result of the engine model with the measured data, the maximum error reaches 4.84%. After the correction, the accuracy of the model is greatly improved, and the maximum error is only 0.966%. The correction effect is good.

1. Introduction

The mathematical model of the aero-engine is an important part of the engine development process and is the basis of the engine control system and performance evaluation system. A good simulation model can greatly shorten the development cycle of the engine and save development costs. The accuracy of the model directly determines the accuracy of the model and its practical value [1]. Compared with the actual working process, many simplifications are made in the engine modeling process. In addition to the influence of the manufacturing process level, the gas path performance parameters output by the engine model are often quite different from the actual engine test data. Therefore, it is necessary to correct the mathematical model of the engine so that it can match the real data [2].

In the 1990s, scholars began to study correction method of aero-engine components characteristics and model parameter. Among them, Stamatis A first proposed a method for engine model correction, taking the minimum error of engine measured data and engine model output value as the objective function by iterating component characteristic correction coefficient [3]. Based on the engine simulation model, the appropriate correction factors are selected and added to component characteristics. The minimum error of engine measured data and engine model output value is taken as the objective function. Quantum-behaved Particle Swarm Optimization(QPSO) with high efficiency is to optimize the solution of the objective function to achieve the correction of the engine model.

2. Building an engine model

According to the aerodynamic thermodynamic relationship between the gas path structure and the cross sections of a turbofan engine, the calculated cross sections are divided as shown in Figure 1.

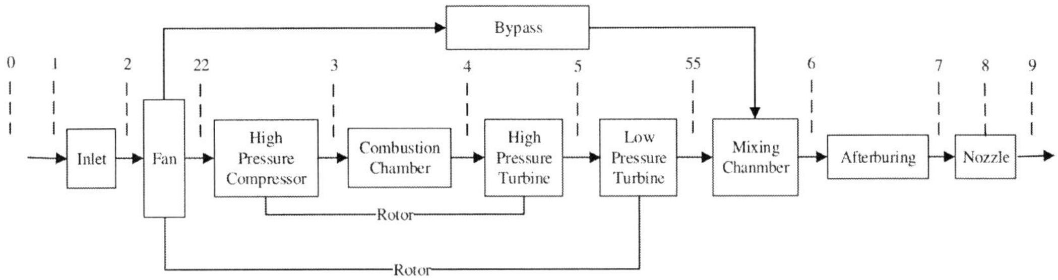

Figure 1. Section diagram of turbofan engine

According to the design point parameters, the engine thermodynamic model is established. The input parameters are atmospheric temperature, pressure, height and Mach number. The output parameters of the model are the total temperature and pressure of each section, thrust, fuel flow and other performance parameters. The initial predicted values of engine model are low-pressure speed, low-pressure ratio function, high-pressure ratio function, turbine front temperature, high-pressure turbine flow and low-pressure turbine flow. The equilibrium equations corresponding to the solution process are high-pressure turbine flow balance, low-pressure turbine flow balance, power balance between high-pressure turbine and compressor, power balance between low-pressure turbine and fan, static pressure balance between inlet duct and outlet duct of mixed chamber entrance and area balance of nozzle, which are:

$$W_{g5}/W_{g5}' - 1 = 0 \tag{1}$$

$$W_{g55}/W_{g55}' - 1 = 0 \tag{2}$$

$$L_{TH}/L_{CH} - 1 = 0 \tag{3}$$

$$L_{TL}/L_{CL} - 1 = 0 \tag{4}$$

$$P_{25}/P_{55} - 1 = 0 \tag{5}$$

$$A_8/A_8' - 1 = 0 \tag{6}$$

Where W_g is the calculated value of gas flow, W_g' is the interpolation result of characteristic data, L_C is the calculated power of fan and compressor, L_T is the calculated power of turbine characteristic data after interpolation, P_{55} and P_{25} are the static pressure of inlet duct and outlet duct of mixed chamber entrance respectively. When the engine is in steady state, the model solving problem is transformed into solving a set of nonlinear equations composed of six common working equations [4]. In this paper, N+1 residual method is used to solve the problem iteratively. If the residual satisfies the requirement, the result is convergent and the model is solved.

3. Engine model correction method based on QPSO

3.1 Correction factor selection

The component characteristic data is the basis of the engine component level model. Whether the characteristic data conforms to the real situation of the engine is directly related to the accuracy of the engine model. Therefore, engine model correction usually improves the accuracy by correcting the component characteristic data. Based on the measured data of the intermediate state of the engine, combined with the initial guess type of the common working equation of the engine model, the selected correction factors are fan flow, fan efficiency, compressor flow, compressor efficiency, high and low pressure turbine efficiency, a total of six correction factors [5], recorded as: $X = [x_i], i = 1,2...6$.

3.2 Objective function construction

The engine steady-state model correction method is to correct the original characteristic data in the model calculation process by selecting an appropriate correction factor X, so that the error between the output parameter and the measured performance parameter of the model is reduced. Therefore, the objective function of the model correction by selecting the measured performance parameters of the engine is:

$$F(x_1,x_2,x_3,x_4,x_5,x_6)=\sum_{i=1}^{m}\left(\frac{y_{i,cal}-y_{i,act}}{y_{i,cal}}\right)^2 \tag{7}$$

Where $y_{i,cal}$ is the calculated value of the model parameters, $y_{i,act}$ is the measured parameter, and m is the number of selected measured performance parameters. In this paper, the thrust *FN*, the fuel flow *WFT*, the turbine rear temperature *T55*, the turbine rear pressure *P55*, the compressor outlet temperature *T3*, the compressor outlet pressure *P3*, and the fan outlet pressure *P22* are selected as objective parameters. According to the objective function, the engine model correction method is to solve $\min F(X)$ by a suitable optimization algorithm.

3.3 QPSO optimization algorithm

The basic idea of the PSO is to have a group of M particles flying at a certain speed to find the optimal position. During the searching process, the flight speed is updated according to the historical optimum point *Pbest* of a single particle and the historical optimum point *Gbest* of a group particle, and the position is updated [6]. QPSO introduces quantum mechanics principle into PSO. By solving the Schrodinger equation, the wave function is obtained, and then the probability density function and probability of particles appearing at a certain point in space are obtained. Finally, the position of particles is measured by random sampling [7].

The specific process of the QPSO is:

(1) Algorithm initialization:

Determining the size of the group M and Randomly generating the initial position of all particles in the group:

$$\boldsymbol{X}^0=\left(\boldsymbol{x}_i^0,\boldsymbol{x}_2^0,\cdots,\boldsymbol{x}_i^0,\cdots\ \boldsymbol{x}_M^0\right),\boldsymbol{x}_i^0\in\left[\boldsymbol{x}_{lb},\boldsymbol{x}_{up}\right] \tag{8}$$

Initializing local optimal position ***Pbest*** and global optimal position ***Gbest***:

$$\begin{cases}\boldsymbol{P}_i^0=\boldsymbol{x}_i^0\\ \boldsymbol{G}^0=\min\left\{f\left(\boldsymbol{P}_i^0\right),\cdots,f\left(\boldsymbol{P}_i^0\right),\cdots,f\left(\boldsymbol{P}_M^0\right)\right\}\end{cases} \tag{9}$$

(2) The *kth* generation evolution process:

Step1: Calculating the fitness from each particle position: $f\left(\boldsymbol{x}_i^k\right)$

Step2: Updating $\boldsymbol{P}_i^k$ and $\boldsymbol{G}^k$:

$$\boldsymbol{P}_i^k=\begin{cases}\boldsymbol{P}_i^{k-1}, & \text{if } f\left(\boldsymbol{x}_i^k\right)\ge f\left(\boldsymbol{P}_i^{k-1}\right)\\ \boldsymbol{x}_i^k, & \text{if } f\left(\boldsymbol{x}_i^k\right)< f\left(\boldsymbol{P}_i^{k-1}\right)\end{cases} \tag{10}$$

$$\boldsymbol{G}^k=\min\left\{f\left(\boldsymbol{P}_i^k\right),\cdots,f\left(\boldsymbol{P}_i^k\right),\cdots,f\left(\boldsymbol{P}_M^k\right)\right\} \tag{11}$$

Step3: Calculating learning tendency point $\boldsymbol{p}^k$ and average best position $\boldsymbol{mbest}^k$:

$$p_{ij}^k=\varphi_{1j}^kP_{ij}^k+\left[1-\varphi_{1j}^k\right]G_j^k \tag{12}$$

$$\boldsymbol{mbest}^k=\left(\boldsymbol{mbest}_1^k,\cdots,\boldsymbol{mbest}_N^k\right)=\left(\sum_{i=1}^{M}\boldsymbol{P}_{i1}^k/M,\sum_{i=1}^{M}\boldsymbol{P}_{i2}^k/M,\cdots,\sum_{i=1}^{M}\boldsymbol{P}_{iN}^k/M\right) \tag{13}$$

Step4: Updating each dimension component x_{ij}^{k+1} of the particle position vector;

$$x_{ij}^{k+1} = p_{ij}^{k} \pm \beta \left| \boldsymbol{mbest}_{j}^{k} - x_{ij}^{k} \right| \ln\left(1 / u_{ij}^{k}\right) \tag{14}$$

$$p_{ij}^{k} = \varphi_{1j}^{k} P_{ij}^{k} + \left[1 - \varphi_{1j}^{k}\right] G_{j}^{k} \tag{15}$$

$$\boldsymbol{mbest}_{j}^{k} = \sum_{i=1}^{M} \boldsymbol{P}_{ij}^{k} / M \tag{16}$$

Where φ and u are random numbers uniformly distributed between $(0,1)$, and the contraction-expansion factor β is an important parameter of QPSO.

Step5: Testing algorithm termination test:

It is checked whether the updated fitness value corresponding to G^k reaches the termination requirement or the search total algebra reaches the maximum algebra limit, and if it is satisfied, it terminates; otherwise, let $k=k+1$ and return to Step1.

Combined with the engine model to calculate the solution process and the real test data, the flow chart of using the QPSO to modify the engine model is shown in Figure 2.

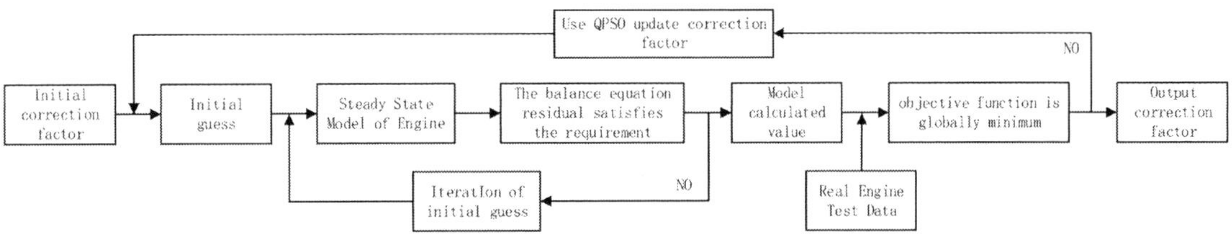

Figure 2. Modification of engine model flow chart using QPSO algorithms

4. Results and analyses of simulation

4.1 Model calculation results before correction

Taking an engine for an example, the general model of the engine is established according to its design point parameters, and its specific performance parameters are calculated under standard atmospheric conditions. Compared with the measured performance of the whole engine. the results are shown in Table 1. It can be seen from the table that there is a significant deviation between the calculated value of the model and the measured value of the engine, and the calculation error of the thrust is the maximum error, which is 4.84%.

Table 1. Comparisons between calculated and measured values of the model before correction

Performance parameter	*T3*	*T55*	*P55*	*FN*	*WFT*	*P22*	*P3*
Relative error	1.73%	-2.11%	1.55%	4.84%	-2.10%	-0.20%	-0.98%
Absolute error	1.73%	2.11%	1.55%	4.84%	2.10%	0.20%	0.98%

4.2 Modified model calculation results

The QPSO described above is used to optimize the objective function of the engine correction model. The particle group is set to 30, and the maximum number of iterations is 500. The optimized correction factor is calculated as *X*=（0.934，1.0314，0.999，1.016，0.934，0.931）. The particle group fitness changing process is shown in the Figure 3. The minimum value is 1.2044e-04.

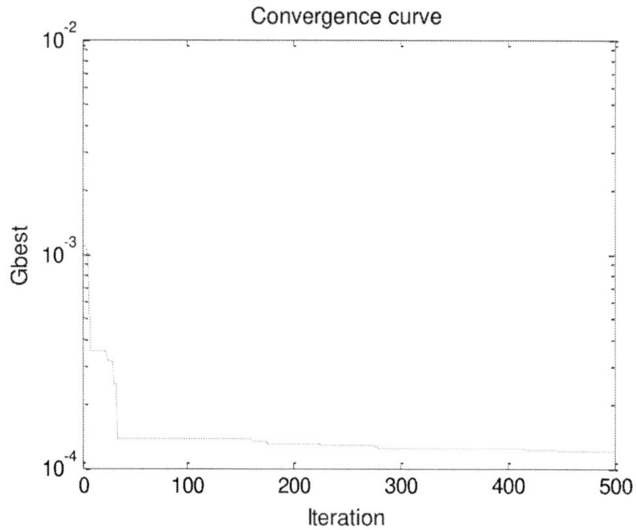

Figure 3. Iterative convergence graph of QPSO algorithms

The correction factor obtained by the QPSO optimization solution is brought into the engine model for calculation, and the calculated value of the intermediate state performance parameter of the modified engine model is obtained. Compared with the measured value of the engine, the results are shown in Table 2. It can be seen from the table that the calculated value of the model has less deviation from the measured value of the engine, and the maximum error is only 0.966%.

Table 2. Comparisons between calculated and measured values of the model before correction

Performance parameter	*T3*	*T55*	*P55*	*FN*	*WFT*	*P22*	*P3*
Relative error	0.034%	0.366%	-0.054%	0.097%	-0.274%	-0.171%	-0.966%
Absolute error	0.034%	0.366%	0.054%	0.097%	0.274%	0.171%	0.966%

The errors between calculated values of engine model and measured data before and after modification are compared. It can be seen that compared with the engine model before correction, the calculated accuracy of the corrected model performance parameters is significantly improved.

5. Conclusion

The aerodynamic thermodynamic simulation model of a turbofan engine is established in this paper, and the engine model correction method based on Quantum-behaved Particle Swarm Optimization is proposed. The engine model calculation results are compared with the measured engine data. Before the correction, the maximum error between the calculated value of the engine model and the measured data reached 4.84%. After the correction, the accuracy of the model is greatly improved, and the maximum error is only 0.966%. The correction effect is good.

References

[1] Dong Zhen, Zhou Wenxiang, and Pan Muxuan, "Modification and Updating Method in Component Characteristics of Turboshaft Engine", *Aeroengine*, vol. 44(6), pp. 11-6, 2018.
[2] Zhang Haoyuan. *Research on Modeling Correction for Turboshaft Engine*. Nanjing University of Aeronautics and Astronautics, 2015.
[3] Stamatis A, Mathioudakis K, and Berbs G, "Jet engine fault detection with discrete operation gas path analysis", *Journal of Propulsion*, vol. 7(6), pp. 1043-1048, 1991.

[4] Sheng Bolin, Li Qiong, and Wu Heng, "Aerothermodynamic Simulation Calculation of a Certain Turboprop Engine", *Journal of Naval Aeronautical and Astronautical University*, vol. 31(2), pp. 121-126, 2016.

[5] Gu Bin, Li Meijin, and Yu Qiuxia, "Investigation on steady-state performance model correction for the engine core of variable cycle engine", *Gas Turbine Experiment and Research*, vol. 32(02) pp. 12-6, 2019.

[6] Yin Dawei. *Algorithms for Solving Aeroengine Nonlinear Mathematical Model and Parameter Estimation in Performance-Seeking Control*. National University of Defense Technology, 2011.

[7] Fang W., Sun J, and Ding Y. R, "A Review of Quantum-behaved Particle Swarm Optimization", *IETE TECHNICAL REVIEW*, vol. 27(4), pp. 336-348, 2010.

IOP Conf. Series: Materials Science and Engineering **751** (2020) 012027 doi:10.1088/1757-899X/751/1/012027

Certification Techniques for the Deformation Compliance of Civil Aviation Passenger Seat

Zhou Jianbin*, Zhang Xinyuan and Wang Xiaozheng

No. 5188 Jinke Road, Pudong District, Shanghai China

*E-mail: zjbnuaa@163.com

Abstract. The deformation requirements for passenger seats of civil aviation are complicated. The author provides the experience of real civil aviation programs for seat deformations' certification, including the significance of seat deformations' airworthiness compliance, relevant regulations and tests and the measuring requirements for passenger seat deformations. FAR25 and CCAR25 both require that the passenger seats shall not yield to the extend they would impede rapid evacuation of the airplane occupants, but the relevant factors are not defined clearly. Seat deformations are the key factors that would impede occupants' rapid evacuation. This article discussed the related requirement and the right way to measure and analyze seat deformations.

1. Introduction

Emergency Evacuation Demonstration Test is a vitally important test for civil transport aircraft, as a practicable way to show the compliance with FAR/CS/CCAR25.803. But the most easily neglected factor is the deformations of PASSENGER SEAT. According to AC25-17A[1], the passenger seats used in Emergency Evacuation Demo Test are assumed that they have no deformations. This strategy provides convenience for Emergency Evacuation Demo Test, but set a high requirement for seat deformations.

If we make a deep research into FAR/CS/CCAR25.561 and 562 regulations[2], it will be found that these two regulations both only required "the passenger seats shall not yield to the extend they would impede rapid evacuation of the airplane occupants", but what deformations, which would impact occupations' evacuation, are not defined clearly in regulations

The truth is the requirements for passenger seat deformations are so highly hidden that it's easy to overlook them before making preparations for Emergency Evacuation Demonstration Test.

Which deformations need concerning and where to collect these deformations, what kind of deformations are acceptable and what aisle width should be used in Emergency Evacuation Demonstration Test will be discussed in this article.

2. Tests Required by Regulation

According to TSO-C127b, a passenger seat has to pass Dynamic Test and Static Test[3]. FAR/CS/CCAR 25.561, 562 and 785 have the similar requirement. The difference is that TSO-C127b's concerning is only the performance of seat itself, but FAR/CS/CCAR concern the interface strength between the seat and the track as well.

Dynamic Test includes 16g Dynamic test, 14g Dynamic test, Head-path test, Front-row HIC test and Row-to-row HIC test.

IOP Conf. Series: Materials Science and Engineering **751** (2020) 012027 doi:10.1088/1757-899X/751/1/012027

Passenger Seat is a TSO part. Its certification can be divided to TSO approval and Installation approval. To reduce cost, it's suggested to combine these two certificate activities.

For 16g Dynamic test, the real aircraft seat track shall be used. Otherwise, a supplemental dynamic test or static test shall be carried out to prove the interface strength between seat and seat track is acceptable[4]. To carry out this supplemental test, the 6 loads (3 forces and 3 moments) shall be measured in 16 dynamic test using steel track (or other aircraft track). And then carry out the supplemental dynamic or static test with these loads.

For 14g Dynamic test, the unreal aircraft seat track (like steel track or other aircraft's track) is acceptable The 14g load can be decomposed to two component loads: one (component A) is along track and the other one (component B) is perpendicular to track. For component A, it must be smaller than 14g and then also smaller than 16g, which means 16g dynamic test can cover this situation. For component B, it's a compressing force not drawing force, which means it will not destroy track since there's structure beams under track.

Head-path test is used to collect the head-path data of passenger seat. This data will be used to analyze head hit area in row-to-row HIC test and front-row setback distance.

Front-row HIC test is usually avoided by using a big front-row setback distance, because bulkhead in aircraft is usually made by honeycomb sandwich panel and will cause HIC value exceed 1000.

Row-to-row HIC test needs to do two tests using min pitch and max pitch. If ATD head only has a glance on the seatback, an additional row-to-row HIC test using a 3 inches smaller pitch than max pitch shall be carried out.

Static test actually includes 5 tests using the loads in different directions: forward, rearward, upward, downward and sideward. The load factors for static tests shall take int account the aircraft Ground/Flight load envelope and the load factors required in FAR/CS/CCAR25.561, even though TSO/ETSO/CTSO-C127 only requires to comply with FAR/CS/CCAR25.561. It's required to select the max load factor in each direction for static test.

In dynamic test, it's usually use high-speed photography to measure the deformations. In static test, optical measurement or linear displacement measurement are commonly used.

3. Deformations That Needs Considering

FAR25.803[2] requires Emergency Evacuation Demonstration Test. But this regulation doesn't define the aisle width for the test mentioned above. FAR25.815 defines a min aisle width of 15 inches for civil aircraft[2], although the actual aisle width used in aircraft operation is usually bigger than this requirement and an aisle width of 17 inches or more is common.

In Emergency Evacuation Demonstration Test, it's not required to use the passenger seat with the deformations caused by dynamic/static tests. But passenger seat deformations have to be in a acceptable range to avoid impacting crew and passengers' egress (required by FAR25.561(d) and 562(c)(8)[2]).

In principle, passenger seat deformations (forward, downward, upward, sideward and rearward) are measured in 16g dynamic test, 14g dynamic test, row-to-row HIC test and static test, and finally the max deformation in each direction is used as seat deformation.

The extreme sideward deformation usually appears in sideward loaded static test.

The extreme forward deformation usually appears in 16g dynamic test. In this test, a special tooling is used to simulate the aircraft floor with 10 degrees yaw, 10 degrees pitching and 10 degrees rolling. The real aircraft seat track is necessary for this test, otherwise the deformation couldn't represent the real deformation in the aircraft, because in dynamic test the seat track also has deformations.

The extreme downward deformation usually appears in 14g dynamic test.

The extreme forward deformation of seatback usually appears in row-to-row HIC test. The head of the ATD would hit the seatback of the front-row seat and cause a big deformation of the seatback.

The extreme rearward deformation appears in rearward loaded static test. It has influence on the row-to-row distance after seats deform. But, according to experience, its influence is smaller than forward deformation of the rear row seat, since the rearward inertia factor (commonly less than 2.0g)

IOP Conf. Series: Materials Science and Engineering **751** (2020) 012027 doi:10.1088/1757-899X/751/1/012027

is usually much smaller than the forward one (commonly 9.0g). The forward deformation in static test usually is smaller than the one in 16g dynamic test. The max sideward deformation can be obtained in static sideward test. But in some less common cases, the 16g dynamic test also can cause the max sideward deformation.

When a passenger seat type is certificated for installation approval, not every test would be carried out, because many tests can be covered by the tests of TSO prototype seat. TSO prototype seat uses the critical configuration of the seat family. At the aspect of strength of seat, this idea is perfect. But it may not at the aspect of deformation of seat. Usually critical seat configuration generates critical deformations. If every aircraft type's aisle width and seat pitch had to to keep the margin for the critical deformations of passenger seat, the economical efficiency would be affected negatively. According to engineering experience, it's suggested to do 16g dynamic test and sideward loaded static test at least, because the forward deformation and the sideward deformation have a great influence on aisle width and passenger seat pitch which have further influence on aircraft economical efficiency and passenger experience.

4. Seat Deformation Requirements

AC25.562-1B defined passenger seat deformations' acceptable range[5], as shown in Table 1:

Table 1. Seat Deformation Requirements

No.	Direction	Source	Requirements
1	Forward Rearward	AC25.562-1B. appendix 2.a	a) Minimum clearance between two deformed seats is 6". See Figure 1. The displacements on Armrest and Bottom Cushion both need concerning. b) The deformation on seatbacks shall comply with the requirement in Figure 2.
2	Sideward	AC25.562-1B. appendix 2.d.(1)	The deformed seat should not encroach more than 1.5 inches (38 mm) into the required longitudinal aisle space at heights up to 25 inches (635 mm) above the floor.
		AC25.562-1B. appendix 2.d.(2)	The deformed seat should not encroach more than 2.0 inches (50 mm) into the longitudinal aisle space at heights 25 inches (635 mm) or more above the floor.
3	Downward	AC25.562-1B. appendix 2.b	There is no limit on downward permanent deformation, provided that the feet or legs of occupants will not be trapped by the deformation.
4	Rotation	AC25.562-1B. appendix 2.c	a) The seat bottom rotational permanent deformation must not result in an angle that exceeds 20 degrees pitch down or 35 degrees pitch up from the horizontal plane. This rotational deformation is measured between the fore and aft extremities of the seat pan at the centerline of each seat bottom. See Figure 3. b) Rotation of the seat pan must not entrap the occupant.

No.	Direction	Source	Requirements
5	Additional requirements	AC25.562-1B. appendix 2.e.(1)	Affect the operation of any emergency exit or encroach into an emergency exit opening for a distance from the exit not less than the width of the narrowest passenger seat installed in the airplane.
		AC25.562-1B. appendix 2.e.(2)	Encroach into any required passageway.
		AC25.562-1B. appendix 2.e.(3)	Encroach more than 1.5 inches (38 mm) into any cross-aisle or flight attendant assist space.

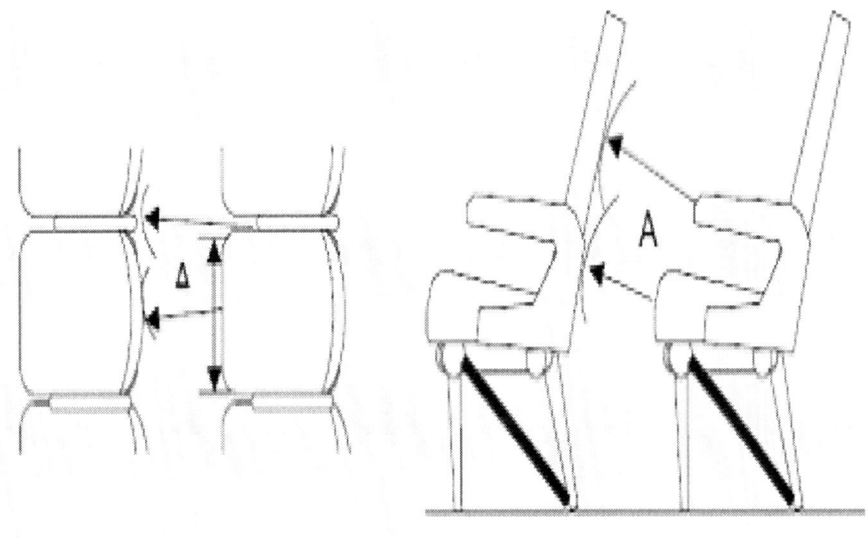

Figure 1. Minimum clearance

The deformation for seatback also needs concerning, as following:

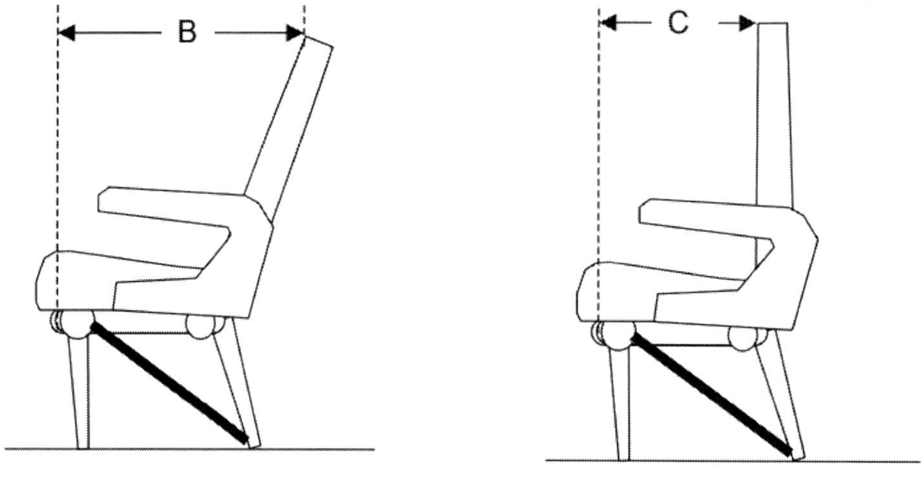

Pre-test Condition Post-test Condition

C/B shall be not smaller than 50%

Figure 2. Deformation requirement for seatback

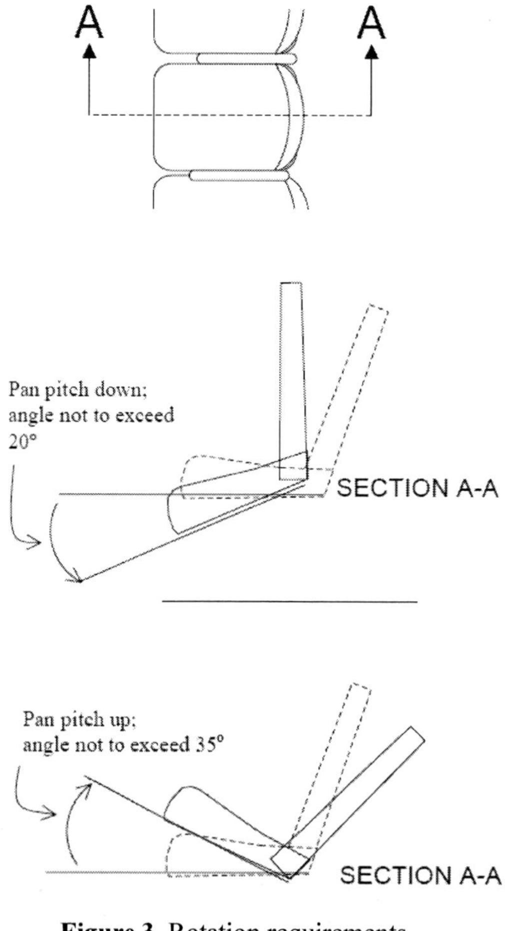

Figure 3. Rotation requirements

5. Conclusion

This article summarized the deformation requirements for passenger seat in dynamic/static tests and are reference to Emergency Evacuation Demonstration Test, passenger seat upgrading and cabin retrofitting. In aircraft certification, a common mistake is carry out Emergency Evacuation Demonstration Test exactly using the aisle width in cabin LOPA (Layout of Passenger Accommodation). That would cause not enough space left for the deformations of passenger seat. It's suggested to use an as narrow aisle width in this test as possible to make sure passenger seat upgrading (like promoting seat width) will have enough space. For making a narrow aisle, foam is commonly used to increase the whole width of passenger seats in Emergency Evacuation Demonstration Test. It should be attached to seats' backrest and armrest to create the aisle width which the Emergency Evacuation Demonstration Test wants to certificate.

References

[1] AC25-17A Transport Airplane Cabin Interiors Crashworthiness handbook, 05/18/2009.
[2] 14 CFR Part 25 - AIRWORTHINESS STANDARDS: TRANSPORT CATEGORY AIRPLANES (FAR25). 11/18/2003.
[3] TSO-C127b Rotorcraft, Transport Airplane, And Small Airplane Seating Systems, 06/06/2014.
[4] SAE AS8049B Performance Standard for Seats in Civil Rotorcraft, Transport Aircraft, and General Aviation Aircraft, 01/2005.
[5] AC25.562-1B Dynamic Evaluation of Seat Restraint Systems and Occupant Protection on Transport Airplanes. 01/20/2006.

Requirements Management Applied in Airworthiness Certification in the Civil Aircraft

Yi Wang[1] and Xinai Zhang[2]

No 5188, Jinke Road, Pudong District, Shanghai, China

E-mail: [1]wangyi5@comac.cc, [2]zhangxinai@comac.cc

Abstract. The civil aircraft certification is the process, which indicates whether aircraft design complies with airworthiness regulations, or not. In recent years, requirements management has been gradually implemented in civil aircraft design, showing a method, which can implement a series of stakeholder, needs or expectation including airworthiness into specific products through requirements definition, validation and verification. Requirements are hierarchical and structured, however, the lack of technical hierarchy of regulations as provisions results in a poor correlation between regulations and requirements. This paper focuses on the research about the application of requirement based engineering method in the development of civil aircraft, and exploring how to implement the regulations in the whole life cycle of the aircraft development to effectively ensure that the regulations are fully implemented into the product requirements, and fully interacted with the system development process. To a certain extent, avoid the disconnection between the airworthiness regulations and the development process, and reduce the risk that the development of civil aircraft does not meet the airworthiness regulations.

1. Introduction

The civil aircraft is a typical complex product system, with the characteristics of high and new technology intensity, strong interdisciplinary, high system integration, long development cycle, huge project investment and complex management[1]. It is necessary to creatively integrate professional technologies and systems in multiple fields to optimize the overall performance of the aircraft. With the further improvement of aircraft's demands for safety, economy, environmental protection and comfort, aircraft's functions are becoming increasingly complex and the information exchange between systems is getting bigger and bigger, which further increases the difficulty of aircraft design. In the face of new challenges, the exploration and application of new design methods and procedures to meet the needs of modern civil aircraft development, to ensure the comprehensive success of civil aircraft project development, airworthiness certification, market and operation, has become a common problem that aviation enterprises around the world urgently need to solve.

In recent years, we have gradually implemented requirement-oriented requirement engineering in the whole life cycle of civil aircraft development. Through the requirements definition, requirements validation and verification, we have implemented the needs or expectations of a series of stakeholders into specific products[2]. There are many stakeholders involved in creating a new airplane design, including the internal and external stakeholders like customer needs or regulations. New practices and processes have been implemented to ensure that every stakeholder's inputs are considered. However, regulations belong to case law, and not only the scope of regulations contains constraints on the technical characteristics of products, but also puts forward demands on the process, method and use of product research and development, so it is difficult to establish the relationship between regulations and requirements.

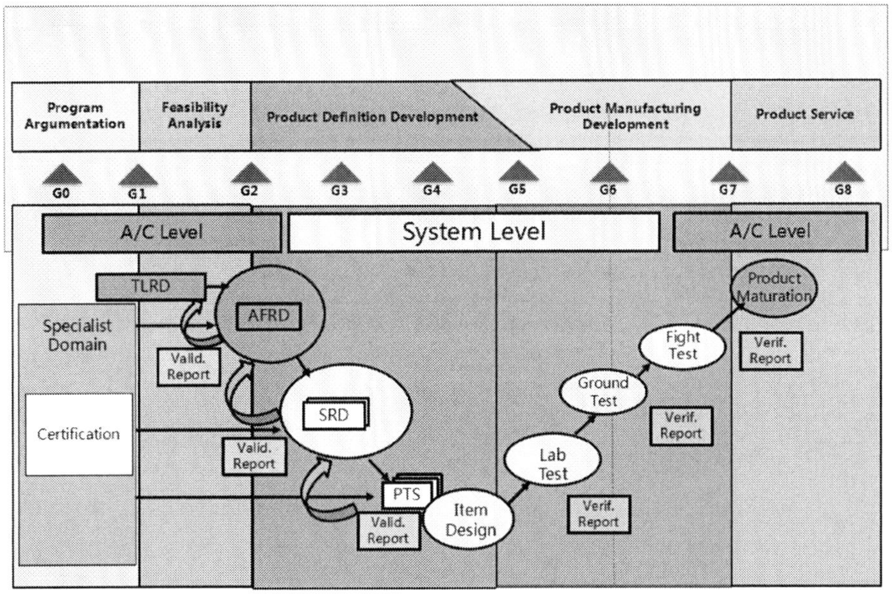

Figure 1. The civil aircraft requirement based development process.

This article is doing the research about the application of requirement based engineering method in the development of civil aircraft, and exploring how to implement the regulations in the whole life cycle of the aircraft development, to manage and control related design activities throughout the project lifecycle in strict compliance with aircraft requirements. Airworthiness demands are considered from the definition stage of requirements and considered as requirements in the whole process of aircraft development, so as to avoid the disconnection between airworthiness demands and research & development process and reduce the risk of civil aircraft development, which cannot meet airworthiness demands.

2. Generic civil aircraft development process and civil aircraft certification

2.1. The civil aircraft requirement based development process

Requirements definition and management should be throughout the whole life cycle of product research and development[3]. That is to say, requirements must be clearly defined for all products firstly, and all research and development management activities including project planning, design activities, evaluation testing, production delivery and trade-off analysis throughout the life cycle are driven by requirements from the beginning of the project to the testing and verification stage of the product. As figure 1 below, the needs of customers were captured through market research and other means in the early stage of design. On this basis, customer needs are analysed to help designers understand the market, customers and other stakeholders that make it can complete the definition of the top-level requirements of the product. To ensure that the defined requirements meet the customer's initial needs, it is necessary to check whether the defined top-level requirements are correct, complete, and fully meet the customer's needs. Then carrying out top-level design activities of products based on requirements, and allocating the requirements to the next level. Upon completion of this work, a review of the defined design schemas is required to verify that the design can meet the requirements at the appropriate level. After the concrete implementation of all the schemes, the corresponding requirements verification work should be carried out to verify that the developed products meet the requirements of the corresponding levels and the final integrated system meets the needs of customers.

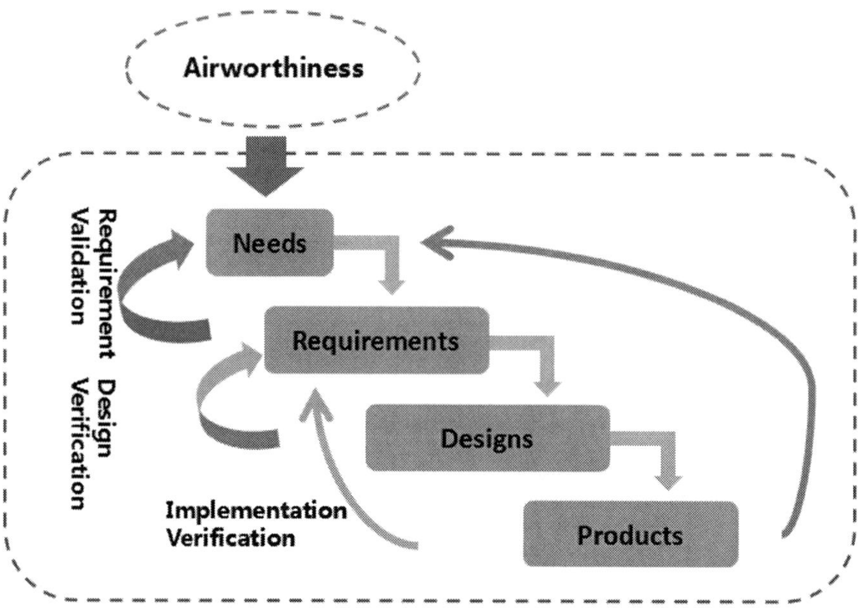

Figure 2. Relationship between regulations and requirements.

2.2. Airworthiness

Airworthiness is an attribute of a civil aircraft, it means that the aircraft or aviation components meet the necessary demands for flying in a safe state within the permitted restrictions. The certification process of modern civil aircraft is a process to show the conformity of design to airworthiness regulations. The development process of modern civil aircraft and the certification process interact and proceed synchronously. From the planning stage, airworthiness certification will affect the final attributes of products at different levels including aircraft, system, equipment, hardware and software through the interaction with relevant technical management process, technical assessment process and design process based on requirements. Therefore, the regulations have multi-levels. At the same time the types of regulations are diverse, there are technical demands, management demands, and demands on the relevant process. According to the regulations demands of the technical attributes of the product, it can directly enter the system development process and participate in the system development as the technical requirements. According to the regulations demands of management activities and technical process, it is necessary to enter the relevant technical and management process. And ensure adequate interaction with the system development process so that the regulations are fully implemented [4].

2.3. Relationship between regulations and requirements

Based on the multi-layered and diverse features of regulations and combined with the development process of the whole life cycle of civil aircraft, the transmission path from regulations to requirements was established as figure 2. First of all, regulations are identified as the needs of stakeholders at each level in the early requirements definition stage. Through the technical path interpretation of the regulations by airworthiness team, identify the responsible parties for the regulations and assign the regulations. The responsible parties of the regulations accept the regulations demands through consensus with airworthiness professionals, and carry out to design after requirements definition and validation. And in the subsequent verification activities to ensure that the design of the product can meet the requirements, so as to meet the regulations.

3. Requirements management

Requirements Management is a process performed throughout a system's life to elicit, identify, develop, manage, and control requirements and associated documentation in a consistent, traceable, correlatable, verifiable manner[5]. It ensures solution compliance with stakeholder needs and expectations using allocation, verification, and adaptation to and control of changes.

After the regulations are transformed into requirements, the requirements transformed by requirements can be taken as the management objects by adopting the requirements management method. Scientific and orderly management and monitoring to the captured requirements from the beginning of the concept phase to the end of the life cycle of the product, which can ensure that all product activities are focused on the regulations demand and the project objectives are achieved accurately and efficiently.

3.1. Requirement validation

The purpose of requirement validation is to ensure that the defined requirements are correct and complete[2], which can meet the needs of customers and stakeholders. In general, requirement validation must be completed before formal product design activities to ensure that all design activities are correct and complete those are driven by requirements. But in the process of complex product system development, requirement validation and design is a dynamic process of mutual iteration. Improving the correctness and completeness of requirements in the initial condition can effectively reduce the design iteration, shorten the development cycle and reduce the development cost.

After the identification of the items transformed into requirements, requirement validation strategy, deliverables, validation activity schedule, executor and other contents shall be formulated according to requirements type, requirements orientation and other characteristics, those are the basis of requirement validation plan to be formed. The requirement validation plan is used to guide the definition of specific requirement validation activities, which can only be carried out after the requirement validation plan has been approved.

3.2. Implementation verification

Implementation verification is the scientific method of showing that the implementation at each level meets the specified requirements[2]. For complex products, the implementation verification should provide a systematic verification approach and a complete evidence system to show stakeholders that the final product meets their functional and performance expectations. The input of the verification work is the validated requirements of the product or system, and the designed product or system is verified to meet the requirements of the corresponding level by means of inspection, review, analysis, test, modelling and other methods.

After the regulations are confirmed into the content of the requirements, it is necessary to define the verification strategy, deliverable, verification activity schedule and roles & responsibilities, taking into account the requirements, which have been validated. This shall be recorded in the implementation verification plan. The implementation verification plan will be established with the support of domain experts since it will guide the definition of specific verification activities. During the development and implementation of the validation plan, the requirements definition team also needs to work closely with the airworthiness team to clarify the work plan of meeting the requirements. Determine how compliance work will be performed, and define certification plan. At the same time, capture the evidence of conformity to satisfy the regulations.

3.3. Requirements management database

The traditional conventional file management method cannot meet the needs of practical use because of the large scale, complex system and big quantity of civil aircraft. Under this background, the application software focusing on requirements management came into being[6]. IBM Rational DOORS is an application that is widely used for requirements management and enables the management of requirements on an itemized basis to facilitate requirements tracking and change analysis, see as figure 3. We can quickly grasp the relationship between regulations and requirements by using this function to establish the link between regulations and requirements in an itemized way, In addition, the verification status can be recorded through the database, and the compliance evidence can be provided to meet the regulations by building a bridge between regulations, requirements and verification evidence.

Information	**2 Subpart B – Flight**	No
Information	**2.1 General**	No
Information	**2.1.1 CCAR 25.21 Proof of compliance.**	No
Technical Index	(a) Each requirement of this subpart must be met at each appropriate combination of weight and center of gravity within the range of loading conditions for which certification is requested. This must be shown—	Yes
Technical Index	(1) By tests upon an airplane of the type for which certification is requested, or by calculations based on, and equal in accuracy to, the results of testing; and	Yes
Technical Index	(2) By systematic investigation of each probable combination of weight and center of gravity, if compliance cannot be reasonably inferred from combinations investigated.	Yes
Information	(b) [Reserved]	No
Technical Index	(c) The controllability, stability, trim, and stalling characteristics of the airplane must be shown for each altitude up to the maximum expected in operation.	Yes
Activity	(d) Parameters critical for the test being conducted, such as weight, loading (center of gravity and inertia), airspeed, power, and wind, must be maintained within acceptable tolerances of the critical values during flight testing.	Yes
Technical Index	(e) If compliance with the flight characteristics requirements is dependent upon a stability augmentation system or upon any other automatic or power-operated system, compliance must be shown with CCAR 25.671 and 25.672.	Yes
Activity	(f) In meeting the requirements of CCAR 25.105(d), 25.125, 25.233, and 25.237, the wind velocity must be measured at a height of 10 meters above the surface, or corrected for the difference between the height at which the wind velocity is measured and the 10-meter height.	Yes
Technical Index	(g) The requirements of this subpart associated with icing conditions apply only if the applicant is seeking certification for flight in icing conditions.	Yes
Technical Index	(1) Each requirement of this subpart, except CCAR 25.121(a), 25.123(c), 25.143(b)(1) and (b)(2), 25.149, 25.201(c)(2), 25.207(c) and (d), 25.239, and	Yes

Figure 3. Airworthiness module in requirement management database.

3.4. Process monitoring

Process monitoring is the process of monitoring one or a series of activities in real time, usually in order to achieve a specific goal or achieve a specific phase state. The product system of civil aircraft is complex and the development period is long. It can effectively help project managers to control project progress, stage results, technical status and other contents for the introduction of process monitoring in the development process of the whole life cycle of civil aircraft. Reduce project risks, labour, time and financial input by Identifying key issues through data.

In the process of requirements management, the process monitoring implemented is carried out on the basis of requirements. By tracking key metrics over time to ensure that requirements are executed according to requirements management processes and methods. Conduct scientific and quantitative inspection and control analysis through different indicators, so as to display to quality, design, test and other departments, check the implementation effect of requirements in each stage, and provide management data and decision basis for project management. These are called KPI (Key Performance Indicator).

During the transformation of regulations into requirements, two KPI indicators can be defined based on the process described above: regulations allocation rate and requirements linking rate to regulations. Regulations allocation rate indicates whether all the regulations that need to be converted into requirements and entered into the design are allocated. Link rate indicates whether all the assigned regulations are linked with corresponding requirements to ensure that the requirements are in line with the terms. And through the form of visual report pictures, clearly show the relation between regulations and requirements, see as figure 4.

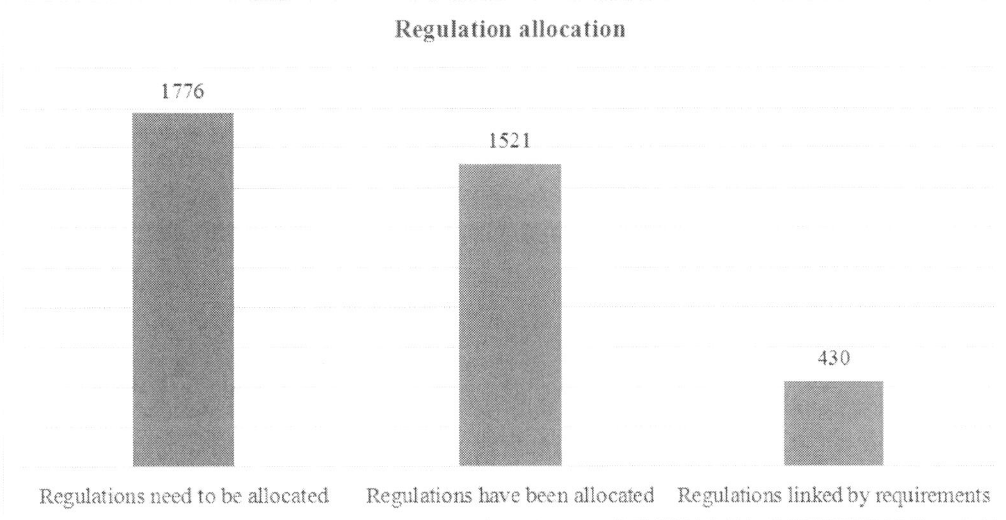

Figure 4. Regulations allocation KPI displaying

4. Applications

Taking CCAR25 899(a) as an example, this paper describes in detail how the regulations are translated into requirements and managed, see as figure 5. CCAR25 899(a) includes the demands of electrical bonding and protection against static electricity. According to the interpretation of the technical path by airworthiness professionals, the discharge device or electric lap should be installed on the aircraft to ensure the minimum of electrostatic accumulation. Thus, identify the structure or E3 team associated with completing the design, which consider the demands of regulations in the requirement definition phase and capture the requirements.

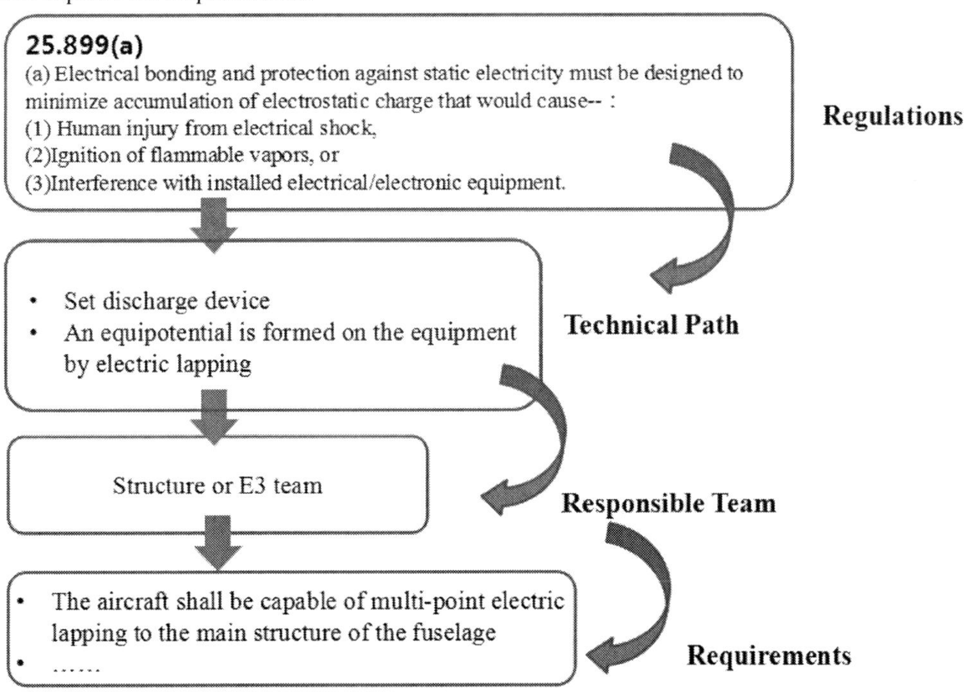

Figure 5. Example of regulation transformation process.

IOP Conf. Series: Materials Science and Engineering **751** (2020) 012028 doi:10.1088/1757-899X/751/1/012028

In order to effectively manage requirements for regulations transformation, store regulations and requirements in the requirements management platform, DOORS. After the above interpretation of the technical path of the article has clarified the relevant technical team, the technical team can be assigned in the platform. At the same time, the technical team stored the requirements captured based on regulations in the requirements module and the validation and verification evidence of the requirements in the matrix module through using the DOORS platform's linking capabilities to establish compliance links for regulations, requirements, and evidence, see as figure 6. In the requirements definition stage, airworthiness research is considered comprehensively and systematically in the whole life cycle of product development, so as to facilitate the definition of conformity ideas and conformity methods in the subsequent certification plan definition stage.

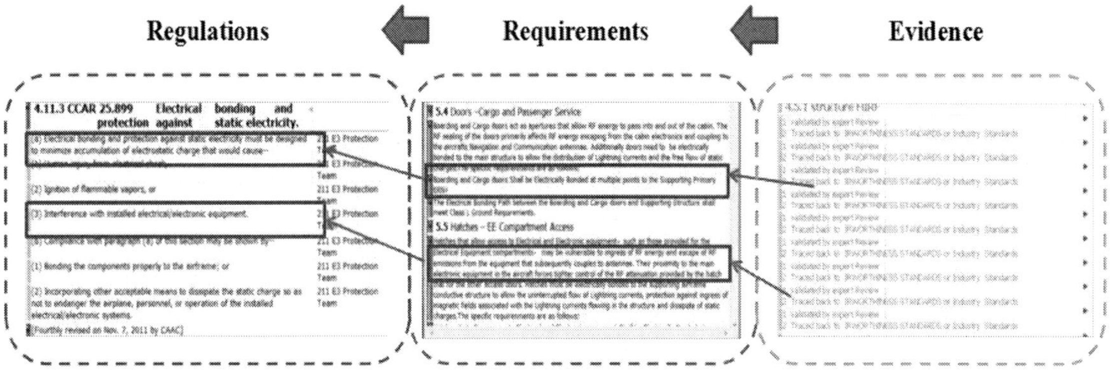

Figure 6. Applications of regulations in DOORS.

5. Conclusions

In this article, according to the requirements based civil aircraft development process and the characteristics of airworthiness regulations, realize the implementation of provisions into requirements in the whole life cycle of civil aircraft development, and ensure that the transformed requirements into the design and product through the means of requirements management. At the stage of requirement definition, airworthiness research is considered comprehensively and systematically in the whole life cycle of product development to reduce the risk of airworthiness research not meeting airworthiness requirements. It is convenient to plan airworthiness compliance verification plan from aircraft level and solve the correlation and reference relationship between different professions and teams in solving airworthiness compliance verification of airworthiness regulations, so as to achieve full coordination, no repetition and no omission in the verification.

References

[1] Guo Bozhi, Li Haomin. The Application of Requirements Management in Civil Aircraft Development[J].2013, No4.

[2] "SAE ARP 4754A Guidelines for Development of Civil Aircraft and Systems," Society of Automotive Engineers. December 2010.

[3] "Development process for civil airplane," HB 8525-2016.

[4] He Jing and etc. Some Considerations for Establishing Civil Aircraft Design Assurance System[J]. Advances in Aeronautical Science and Engineering. 2015,6(4).

[5] FAA System Engineering Manual.

[6] Orlena Gotel, Patrick Mader. How to Select a Requirements Management Tool: Initial Steps. 17th IEEE International Requirements Engineering Conferences[C]. 365-367, 2009.

Research on Civil Aircraft Emergency Evacuation Time for Ditching

Wu Yang

Aircraft Integration Department, Shanghai Aircraft Design and Research Institute

E-mail: wuyang1@comac.cc

Abstract. Emergency evacuation for ditching is an important airworthiness requirement for civil aircraft to ensure safety. However, to demonstrate the emergency evacuation for ditching of a real aircraft is a time-consuming and laborious test with significant safety risks. Based on the analysis of airworthiness clauses, this paper proposes a method for calculating and analysing the emergency evacuation time for ditching. Taking a certain type of aircraft as an example, a typical case is selected for calculation and analysis, and the evacuation time under this typical case is obtained, which provides data support for the airworthiness emergency evacuation airworthiness compliance.
Keyword: Civil aircraft, Airworthiness, Ditching, Emergency Evacuation

1. Introduction

According to CCAR25.801(d) [1], for civil aircraft, “It must be shown that, under reasonably probable water conditions, the flotation time and trim of the aeroplane will allow the occupants to leave the aeroplane and enter the life rafts required by 25.1415.” That is, the flotation time of the aeroplane must be longer than the time of the emergency evacuation of all occupants so that the occupants can safely evacuate the aircraft and enter the life rafts. However, to demonstrate the emergency evacuation for ditching of a real aircraft is a time-consuming and laborious test [2], this paper proposes a method for calculating and analysing the emergency evacuation time for ditching, and provides data support for airworthiness compliance.

2. Airworthiness requirements

According to CCAR25.803, for all newly developed aeroplanes, it must be shown that the maximum seating capacity, including the number of crew members required by the operating rules for which certification is requested, can be evacuated from the aeroplane to the ground under simulated emergency conditions within 90 seconds. To demonstrate emergency evacuation on the ground involves a wide range of personnel, professions and large funds, and for the volunteers participating in the test, it is a test of huge safety risks. It is also one of the key tests for transport forensic aircrafts to obtain the model certificate issued by the airworthiness authority.

There is no relevant requirements for emergency evacuation for ditching in CCAR25.803. All requirements for ditching are reflected in related verification tasks in CCAR25.801 (d). Therefore, the emergency evacuation for ditching does not need the real demonstration test to show conformity as emergency evacuation on the ground, but only by means of computational analysis. The ditching for

IOP Conf. Series: Materials Science and Engineering **751** (2020) 012029 doi:10.1088/1757-899X/751/1/012029

civil aircraft is a problem of water entry impact [3]. Most of the early studies were theoretical [4], but after 1970s, it tended to solve it numerically [5].

Almost none of the transport aircraft designed independently in China has conducted systematic research on ditching [6]. Especially for large transport aircraft, there is a lack of experience in airworthiness forensics [7].

According to CCAR25.801 (d), it must be indicated that the time of the aircraft floating on water is longer than the total time spent by all members evacuating from the aircraft to the life raft, which includes the time spent by the crew taking the raft, opening the hatch, throwing the raft, inflating the raft, and evacuating. This paper decomposes the total evacuation time, measures the decomposed time by experiments, and accumulates them to provide support for measuring the total time of emergency evacuation for ditching.

3. Computational conditions

a) Total number of evacuated passengers

The longest evacuation time is required when the number of passengers is the largest. Therefore, the evacuation time is calculated based on the maximum seating capacity, and the number of pilots and all crew members should be added.

b) Emergency exits arrangement

The arrangement of emergency exits, which contains the number and location of emergency exits, has a significant impact on the emergency evacuation time. The number of emergency exits is closely related to the maximum number of passengers.

According to CCAR25.807 (e), ditching emergency exits must be provided in accordance with the following requirements whether or not certification with ditching provisions is requested. For aeroplanes that have a passenger seating configuration of 10 seats or more, excluding pilots seats, one exit above the waterline in a side of the aeroplane, meeting at least the dimensions of a Type III exit for each unit (or part of a unit) of 35 passenger seats, but no less than two such exits in the passenger cabin, with one on each side of the aeroplane. The passenger seat/exit ratio can be increased through the use of larger exits or other means, provided it can show the evacuation capability during ditching has been improved.

c) Life raft layout

The layout of the life rafts (including the standard capacity of passengers of the life raft and the location of the life rafts on the aircraft) is also critical to the emergency evacuation for ditching.

4. Calculation

4.1. Analysis of typical experiments

According to the calculation results of the aircraft waterline and the location of the emergency exits, as well as the situation that the emergency exit may be unable to open due to the ditching, this paper analyses the unusable conditions that may occur in all the emergency exits of the aircraft, and finds out the most typical case for consideration and analysis. For example, all exits can be used, one emergency exit on the wing cannot be used, and two exits on the rear cannot be used, and so on.

Considering that one or several life rafts are not working properly when ditching, it is necessary to analyse the most typical situation that life rafts cannot be used according to the location and the actual seating capacity of life rafts. For example, the life raft with the maximum seating capacity cannot be used, the life raft nearest to the emergency exit cannot be used, and the life raft most convenient to use cannot be used.

Combine the situation the emergency exit cannot be opened and the life raft cannot be used to find out the most typical situation or the most critical time of emergency evacuation time, and carry out comprehensive experimental analysis.

4.2 Experimentation time

According to the experiment process of the emergency evacuation for ditching and the need of calculation and analysis, the whole experimental process is decomposed step by step and the time of each step is measured by experiment. Specific time decomposition may include:

t1: The time taken by the attendant to unlock the seat belt and stand up from the seat;

t2: The time taken by the attendant to run from the attendant seat to the life raft storage, take out the life raft and get the life raft out to the ready handling state with the help of two designated passengers (for each life raft, there will be a corresponding t2);

t3: The time taken by the two designated passengers to move the life raft packages taken off and ready for handling to the available emergency exit (according to the combination of the life raft and emergency exit, there will be a corresponding t3);

t4: The time taken by the attendant to open the hatch (there will be a corresponding t4 for each hatch);

t5: Total time for n(which represents the number of people participating in the experiment)crew members to jump from emergency exit to the life raft, which is the total evacuation time.

t6: The total time taken by the n crew members to run at a specified distance in the cabin.

4.3. Calculating time

According to the above experimental time, the time of each analysis item can be calculated. Specifically include:

T1: The time taken by the attendant to unfasten the seat belt and stand up. In the calculation analysis, the average t1 of all records is T1.

T2: The time taken by the attendant to run to the life raft, take the life raft out of the storage and move it to the emergency exit with the help of the designated passengers. The combination of the measured t2 and t3 is T2. For different emergency exits and life raft storage locations, there will be a variety of situations of T2.

T3: The time taken by the attendant to run from one emergency exit to another. The corresponding t3 multiply the coefficient C, which is the distance from one emergency exit to another divided by the distance from the life raft to the emergency exit in the experiment, is T3.

T4: The time taken by the attendant to open the cabin door. Different hatch opening times can be obtained by averaging t4 measured at the hatch, taking the maximum value as T4.

T5: The average time taken by the crew to jump from the hatch to the raft. T5=t5/n.

T6: The time taken by the crew to move from one emergency exit to another. The average time of T6 measured multiply the coefficient C, which is got through the distance from one emergency exit to another emergency exit divided by the distance running in the experiment, is T6.

T7: The time taken by the attendant to tie the life raft to a fixed rope, throw the life raft out of the door and pull the inflatable rope.

T8: Inflation time of life raft. Due to the high cost of dumping and retracting life rafts, T7 can be estimated from competing aircraft of the same kind, and T8 can be provided by life raft suppliers.

T9: The time, which is not counted in the above time measurement process, contains the time to judge the internal and external conditions of the aircraft after confirming the aircraft has stopped, and the clearance time of the attendant. The delay time T9 caused by the confusion can be estimated by referring to the experimental data of the competition aircraft.

5. Case analysis

This chapter takes a certain type of aircraft as an example, selects a typical case, performs calculation and analysis, and obtains the evacuation time in this typical case.

a) Total number of evacuees

When a type of aircraft makes emergency evacuation for ditching, the total number of evacuees includes 90 passengers, 2 pilots and 2 crew members, for a total of 94. The detailed cabin layout is shown in Figure 1.

IOP Conf. Series: Materials Science and Engineering **751** (2020) 012029 doi:10.1088/1757-899X/751/1/012029

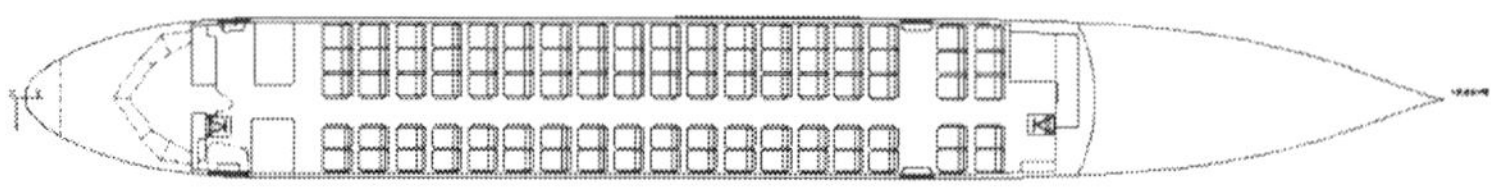

Figure 1. All Economy Class Cabin Layout

b) Emergency exit arrangement

There are 4 emergency exits for a type of aircraft (shown in figure 2). All 4 doors are type I.

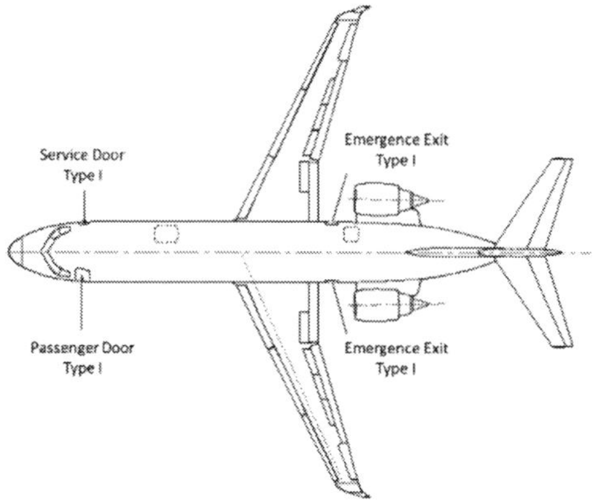

Figure 2. Emergency Exit Arrangement

c) Life raft layout

A type of aircraft is equipped with three 38-seat (standard capacity) life rafts, whose overload capacity is 1.5 times of the standard capacity, that is, 57 seats. The three life rafts are arranged as follows: one is in the first suitcase on the left side of the cabin, another is in the second suitcase on the right side, and the third one is in the storage room behind the aircraft.

d) Test and Data Processing

Through experiments, the measured data are shown in Table 1 below.

Table 1. Time Data Processing Table

step	Test measurement items		Average value of measurement (s)
t1	the attendant unfastens the seat belt and stands up		1.62
t2	time for passengers to pick up rafts at the life raft storage area	raft fetching in front suitcase	3.51
		raft fetching in rear storage room	9.36
t3	time for passengers moving rafts to corresponding emergency exits	From the front suitcase to the front emergency exit	11.03
		From the front suitcase to the rear emergency exit	9.42
		From the rear storage room to the front emergency exit	20.01
		From the rear storage room to the reart emergency exit	7.01
t4	Time for opening the cabin door	The front emergency exit	6.35
		The rear emergency exit	3.01
t5	Total evacuation time of 16 passengers		28.44
t6	Running time of 16 passengers at specified distances in the cabin		17.12

IOP Conf. Series: Materials Science and Engineering **751** (2020) 012029 doi:10.1088/1757-899X/751/1/012029

The experimental process is shown in Figure 3 below.

Figure 3. Experimental process diagram

e) Typical case calculation

Typical cases are selected as follows: if a life raft in the front of a certain type of aircraft is found to be unusable, and two emergency exits in back cannot be used, all the crew members evacuate from the boarding gate and service gate.

In this typical case, the evacuation method will be carried out according to Figure 4.

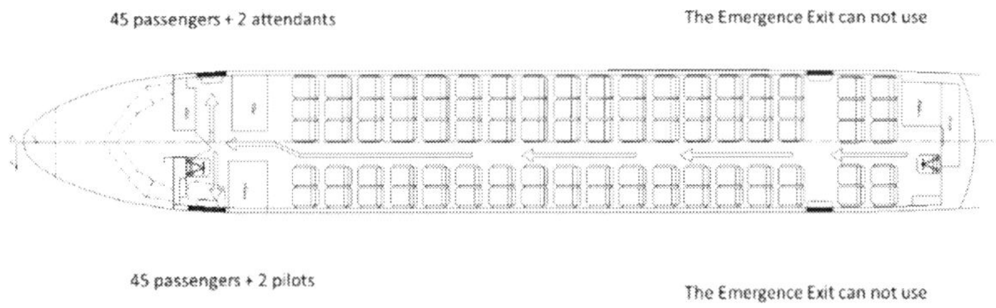

Figure 4. Evacuation Method

The calculation results are shown in Table 2 and Figure 5:

Table 2. Computational Decomposition

step	action	Time item	Time(s)
1	the attendant unfastens the seat belt and stands up	T_1	1.62
2	the attendant runs to the life raft, with the help of the designated passengers takes the life raft out of the storage and moves it to the emergency exit	T_2	14.54
3	the attendant opens the cabin door	T_4	6.36
4	the attendant ties the raft to a fixed rope, throws the raft out of the door and pulls the inflatable rope	T_7	19.50
5	Inflate life raft	T_8	45.00
6	The time taken by 47 crew members to board the life raft and each person costs T0 (the front life raft can contain a maximum of 47 people)	T_{47}	83.90
7	the latter attendant finds one of the front life raft cannot be used and move the latter life raft to the other side of the front emergency exit	T_{10}	25.47
8	47 crew members board the life raft at T0 per person	T_{47}	83.90
9	the delay and confusion time	T_9	52.50
	total evacuation time	T	332.76

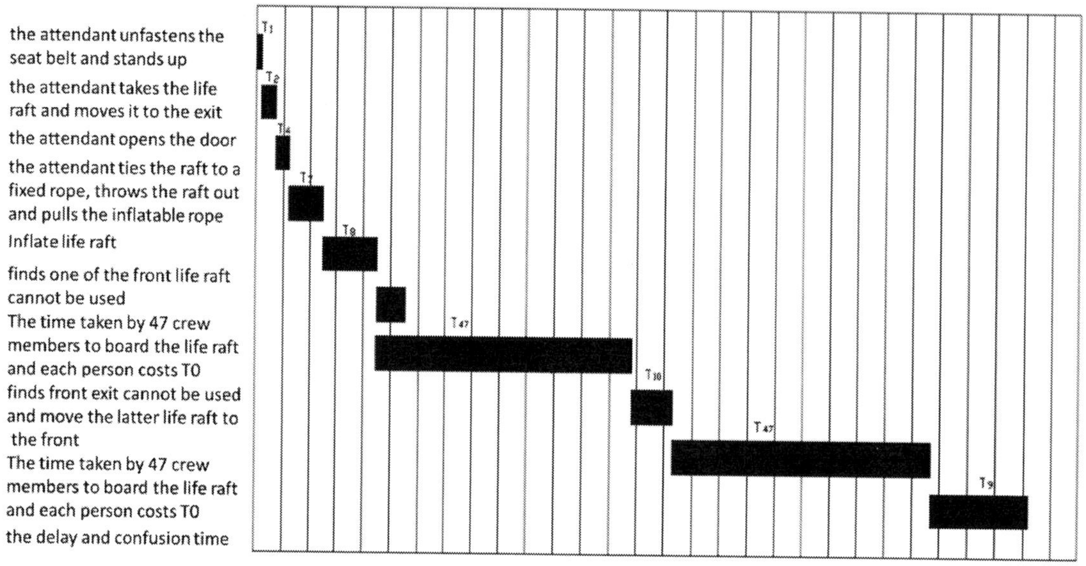

Figure 5. Computational Decomposition

In this case, while the former attendant open the front door of the aircraft, the latter attendant finds that the latter emergency exit cannot be opened. After the evacuation of the front passengers, the latter flight attendants move to the front and the remaining passengers evacuate.

So the total evacuation time: T= T1+ T2+ T4+ T7+ T8+ T47+ T10+ T47+ T9.

T1, T2 and T4 can be measured directly through experiments. T7 can be estimated by reference to similar competing models. T8 can be provided by life raft suppliers.

T47=47×T0.

T10 can be calculated from test data.

T9 is confusion and delay time, and can be estimated by competing models.

The total evacuation time is 332.76 seconds. This time is much shorter than the floating time of a certain type of aircraft.

It can be concluded that the aircraft meets the relevant airworthiness requirements of emergency evacuation for ditching.

6. Conclusion

This paper analyses the airworthiness clauses of emergency evacuation for ditching firstly, and concludes that the emergency evacuation for ditching does not need real demonstration test to show conformity, so as to avoid the real verification test which involves a wide range of personnel, a large number of professions, a large amount of funds and a huge safety risk to the volunteers. On this basis, a method for calculating the emergency evacuation time for ditching is proposed, which is to decompose the total evacuation time for ditching of the aircraft's crew, and measure the decomposition time step by step through experiments, and then accumulate it to get the total evacuation time. Finally, taking a civil aircraft as an example, a typical case is selected for calculation and analysis, and the total evacuation time for ditching under this typical case is obtained, which provides data support for the airworthiness of emergency evacuation for ditching.

References

[1] Civil Aviation Regulation of China (CCAR), *Part 25-Airworthiness Standards: Transport Category Airplanes*. R4.

[2] Li Dan , Civil Aircraft Numerical Simulation Technology Based on CCAR25 Ditching Regulation , 2016.09, *China Science and Technology Information*, 43-44.

[3] Zhang Tao, Li Shu, Jiang Xiang, and Zhao Jinqiang. Analysis model and numerical simulation for civil aircraft ditching. 2010.03. *Journal of Nanjing University of Aeronautics & Astronautics*, 392-394.

[4] Von Karman T. The impact on seaplane floats during landing [R]. *NACA-TN-321*, 1929.

[5] Miloh T. On the oblique water entry problem of a rigid sphere [J]. J Eng Math, 1991, 25(1): 77-92.

[6] Li Bin, Yang Zhichun. Research progress on ditching of large transport aircraft. *High-level Forum on Key Technologies of Large Aircraft and* (Shenzhen Guangdong China: *2007 Annual Academic Conference of China Aviation Society*). 200-209.

[7] Jin Lei, Liu Youdan. Research on ditching Airworthiness Technology of Civil Aircraft. 2012.03. *Aeronautic Standardization & Quality*. 29-32.

IOP Conf. Series: Materials Science and Engineering **751** (2020) 012030 doi:10.1088/1757-899X/751/1/012030

Accuracy Control of Big-scale Space 3D Measurement System in Aircraft Digital Assembly

Yan Cheng*, Yuan-heng Xu, Xue-liang Huang and Lei Wang

Institute of Systems Engineering, CAEP, Mianyang 621999, Sichuan, China

*E-mail:654803236@qq.com

Abstract. Large scale space measurement technology is one of the key technologies of aircraft assembly. The accuracy of measurement system affects the efficiency and quality of aircraft assembly directly. However, the accuracy of the measurement system depends on the process of constructing the measurement system. Therefore, this paper explores three principles of constructing high-precision large-scale space measurement system, through a large number of experiments and simulations.

1. Introduction

The rapid development of large-scale space measurement technology has changed the traditional aircraft design and manufacturing mode [1-2], which meets the requirements of large measurement range, high precision, high efficiency and high flexibility in the process of aircraft digital assembly. Constructing digital measurement system to locate the spatial position and posture of aircraft components accurately is the foundation and premise of modern aircraft digital assembly [3], therefore, the accuracy of measurement system affects the efficiency and quality of aircraft assembly directly. According to the current aircraft manufacturing process, the assembly accuracies at many stages are just above the minimum level, so the cumulative effect may even make that the final quality of aircraft assembly cannot satisfy the standard. Therefore, it is necessary to explore the large-scale space measurement technology, to control the accuracy of the measurement system effectively, and provide better support for the aircraft digital assembly.

2. Survey of large scale space measurement technology

Different from the traditional measurement technologies, large scale space measurement system consists of professional optical measuring equipment and target receiver. According to different principles, large scale measurement equipment can be divided into three types which are widely used in industry:

2.1. IGPS (Indoor GPS)

IGPS [4] is a concept of three-dimensional principle based on regional GPS technology, which is proposed in the early 21st century. The measuring principle of IGPS is similar to GPS, the infrared pulse laser transmitters are used as the satellite, and the indoor receiver receive the laser and time characteristic parameters passed by the transmitters. According to the angle and position characteristics of the receivers, the third-party measurement software calculates and gets the high-precision measurement results [5-6]. The main features of IGPS technology are as follows: the system has no limitation of measurement range, and it can expand the measurement range by increasing the number of transmitters; As long as there is no physical occlusion, there will be no light drop; IGPS can be used indefinitely after the location is calibrated without considering the station moving. However, the cost of deploying IGPS measurement system is high, and its location is inconvenient to move.

2.2. Laser radar

Laser radar is an instrument that can acquire 3D spatial target location data in real time. Combined with airborne system and remote sensing technology, Laser radar can achieve large scale and high-precision space measurement. Laser radar measurement system combines laser technology with photoelectric detection technology, which includes transmission system, receiving system, information processing system and other subsystems. When the measurement system works, the transmitters send a laser beam (detection signal) to the target, the receiving system receives the reflected signal and restores it to a kind of electrical pulse that can be analysed. The information processing system analyses the detected signal and the returned electrical pulse signal, calculating the spatial position information of the target, including distance, orientation, posture and external contour parameters, etc. The main features of the laser radar measurement system are as follows: the measuring speed is fast and the range is wide, which is suitable for the large-scale measured target; the measuring precision is high, the single point repetitive measuring accuracy is less than 0.05 mm, within five meters; but it is difficult to process the measuring data.

2.3. Laser Tracker

Laser tracker is a kind of high-precision and large-size measuring equipment in industrial measurement field, which is mainly used for real-time tracking and measurement of space moving targets. Laser tracker measurement system consists tracker host, controller, upper computer, target ball and some measurement accessories (such as T-probe, T-cam). When measuring, the laser beam is emitted from the main hole of the tracker and irradiated on a reflecting prism which can rotate around the horizontal axis; the main head of the mirror can rotate around the vertical direction. Thus, the laser beam can track all points in the measured space. The target ball is placed at the measured point, receiving the laser beam and reflecting the original laser beam back. The reflected beam is divided into two paths when passing through the spectroscope: one beam of laser light enters the laser interference system, forming the interference fringes, which can measure the linear distance between the target ball and the origin of the laser tracker; another laser light enters into two angle encoders, the optical elements convert and amplify the photoelectric signal, the encoder can measure the horizontal and vertical angle between the target ball and the laser tracker's rotation centre [7], and then calculate the relative spatial position of the target ball. The main features of the laser tracker measurement system are as follows: tracking the moving target automatically; the measuring precision is high and the absolute accuracy can be controlled within (+0.1mm); measuring instruments are easy to carry. And laser tracker can even measure the shading points and scan the contour of targets, with T-probe, T-scan and other accessories. But it's inconvenient to transfer station.Overall, laser tracker measurement system has become the most popular large-scale space measurement method in the field of aircraft digital assembly.

3. Spatial distribution of common reference points and accuracy of measurement System

Measurement field must be constructed to calibrate the position of aircraft components in each coordinate system, and transform it with the aircraft design coordinate system to describe the spatial position and posture of the aircraft components accurately, before aircraft digital assembly. Therefore, the accuracy of the measurement field determines the final accuracy of aircraft assembly directly, while the essence of constructing measurement field is the process of measuring ERS points and transferring station to create an enhanced coordinate system, which accuracy is related to the spatial distribution of the common reference points (ERS points) directly. Therefore, this paper explores the relationship between the spatial position distribution of ERS points and the accuracy of measurement system through a large number of laser tracker measurement experiments, hoping to provide some valuable reference for the accuracy control of large-scale spatial measurement system in aircraft digital assembly.

3.1. Enveloping property of ERS points and accuracy control

Enveloping property refers to whether the measurement field constructed by ERS points envelops all the OTP (Optical Tooling Points).

Experiment process: keep the position of the laser tracker and ERS points fixed, configure four different OTP spatial distribution schemes. Spatial envelope relationship between OTP groups and ERS points are shown in figure 1 to figure 4, the spherical centre is the position of the laser tracker, ERS points envelope a sphere with a radius of 6000 mm, four groups of OTP are scattered irregularly on the spherical surface with radius of 3000mm, 7500mm, 9400mm and 14000mm. As shown in Figure 1, the first group of OTP are located within the enveloping region of the measurement field, while the spatial distribution of the second, the third and the fourth groups of OTP have exceeded the region of measurement field, and the distance is getting farther and farther, as shown in Figure 2 to Figure 4. According to the measured coordinates and the theoretical coordinates in aircraft coordinates of the four groups of OTP to calculate the measurement errors, which represent the accuracy of the measurement field.

The results of the transformation errors between the four groups of OTP and the ERS points are shown in Figure 5. The measurement errors of the first group of OTP enveloped in the inner region of ERS points are almost the same as the transformation errors. Because of the uncertainty of the spatial position, the measurement errors of each optical tooling point have a little difference, but the error values are fluctuating within the maximum transformation error. The measurement errors of the other three groups of OTP, which are distributed outside the region of ERS points, cannot be controlled within the transformation errors. Therefore, when constructing the measurement system, ERS points envelope all the points to be measured is helpful to control the measurement accuracy. If the actual condition is limited, the points to be measured should be put closer to the measurement field.

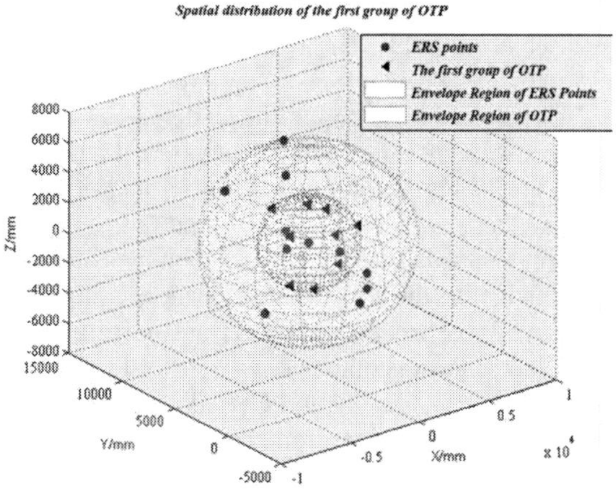

Figure 1. Spatial envelope relation between the first group of OTP and ERS points.

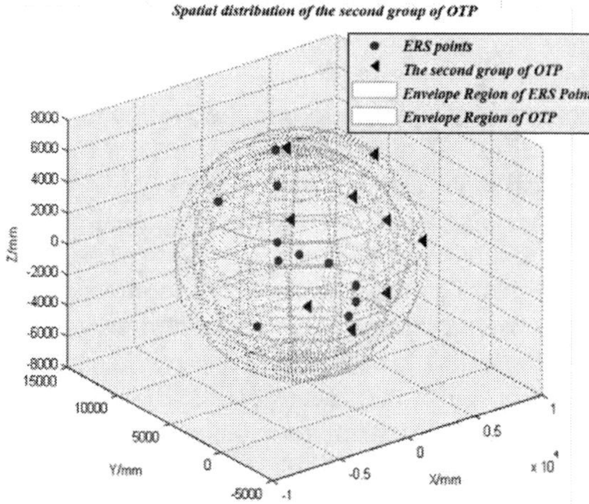

Figure 2. Spatial envelope relation between the second group of OTP and ERS points.

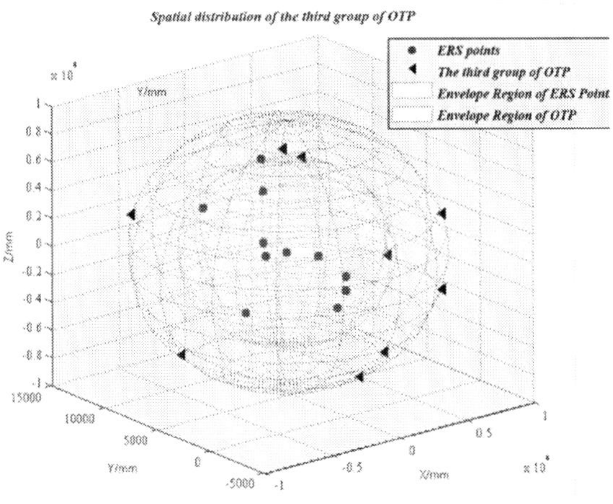

Figure 3. Spatial envelope relation between the third group of OTP and ERS points.

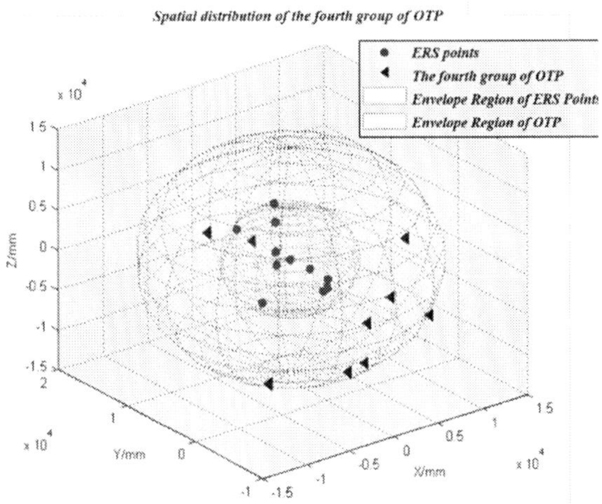

Figure 4. Spatial envelope relation between the fourth group of OTP and ERS points.

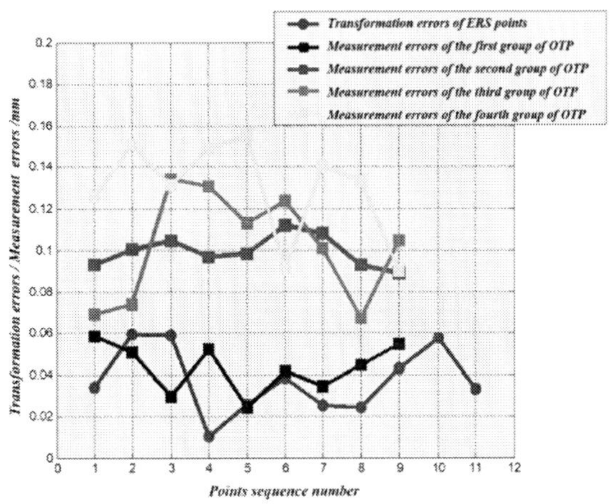

Figure 5. Measurement errors of the four groups of OTP and transformation errors of ERS Points.

3.2. The number of ERS points with accuracy control

Measuring different number of ERS points in the same measurement area (10m × 10m × 3m), to explore the relationship between the number of ERS points and the measurement accuracy.

Experiment process: keep the position of laser tracker and the point P to be measured fixed, measuring ERS points to transfer station, and the number of ERS points participating in transferring station increases one by one. Then measuring the point P and calculating the measurement error of point P at different numbers of ERS Points participating in transferring stations. The spatial distribution of point P, laser tracker and all ERS points is shown in Figure 6. The transformation errors of ERS point with different number are shown in Figure 7, Figure 8 and Figure 9. The measurement errors of point P are shown in Figure 10.

As shown in these figures, when the number of ERS is three or four, the transformation errors are zero, but the measurement error of point P is very large; when the number of ERS points increases from five points, the transformation errors tend to be normal, and the measurement errors of point P decrease with the increase of the number of ERS points; when the number of ERS points is 7 or more than 7, the transformation errors of ERS points are small, and tend to be stable with the increase of the number of ERS points, the measurement errors of point P are stable; when the number of ERS points is more than 11, the transformation errors of ERS points increase slightly, and the measurement errors of point P also increase slightly. Theoretically, the more the ERS points are, the higher the accuracy of the measurement filed will be. However, due to the interference of many external factors in the actual measuring process, ERS points have measurement errors, therefore, the more ERS points are, the more factors affecting the transformation accuracy, so, the number of ERS points increases while the transformation accuracy decreases. The results of experiments show that it is appropriate to keep the number of ERS points at 7~11. If the region of measurement field is large, the number of ERS points can be increased appropriately.

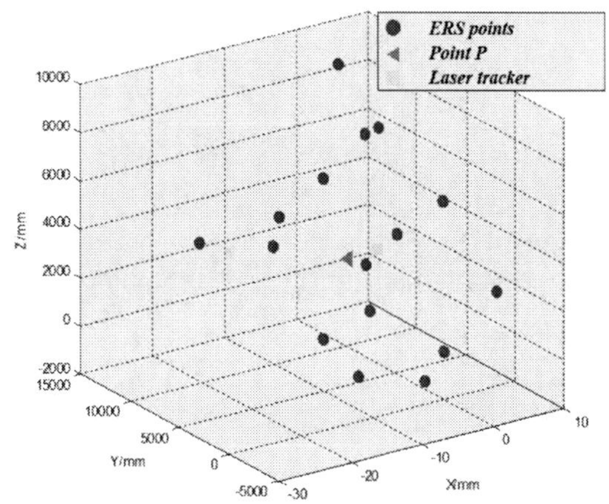

Figure 6. The spatial distribution of point P, laser tracker and all ERS points.

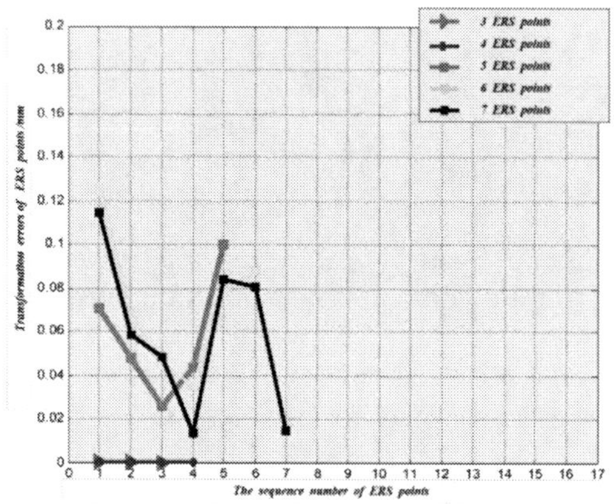

Figure 7. Transformation errors of 3-7 ERS points.

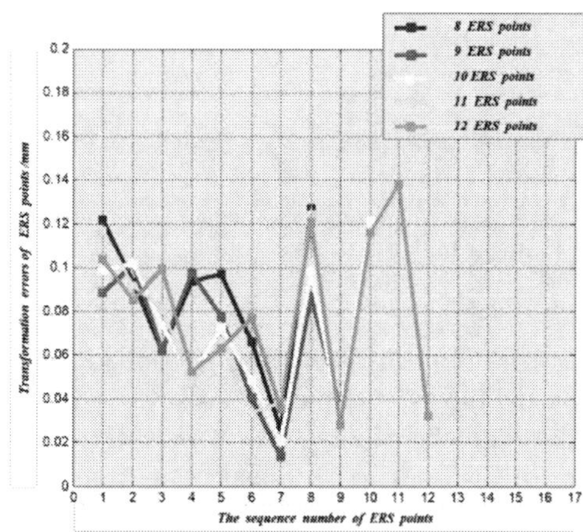

Figure 8. Transformation errors of 8-12 ERS points.

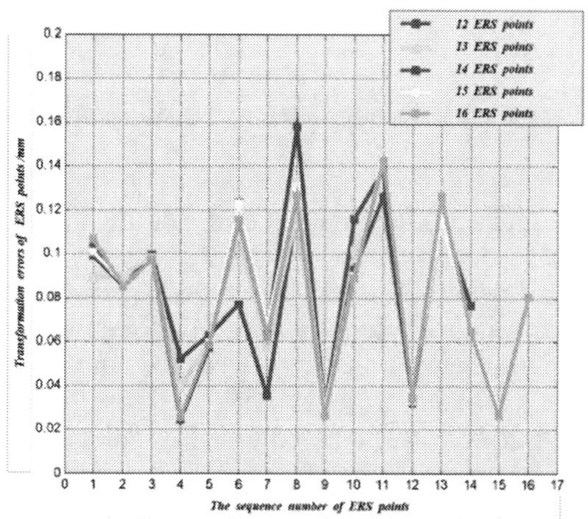

Figure 9. Transformation errors of 12-16 ERS points.

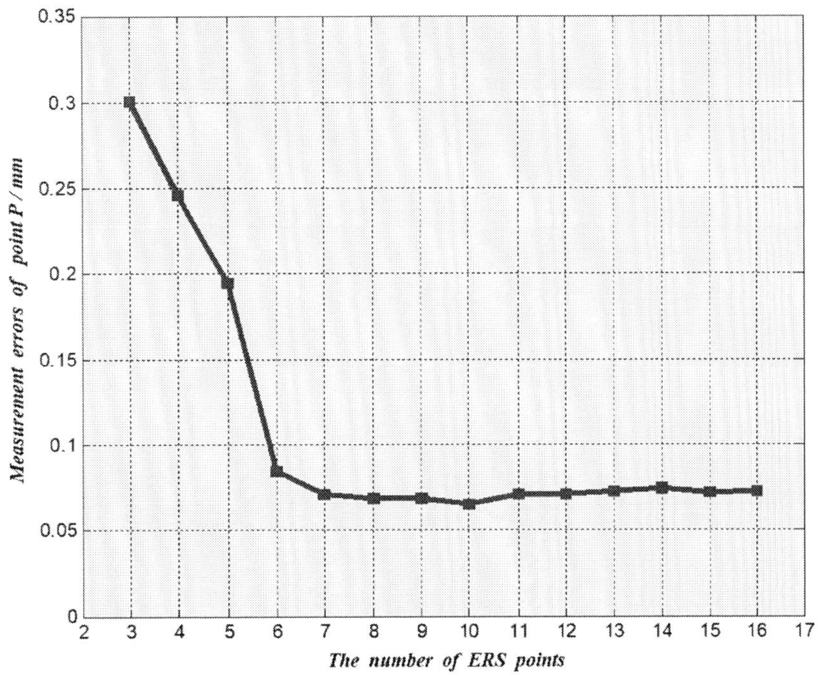

Figure 10. Measurement errors of point P.

3.3. Flat distribution of ERS points with accuracy control

The spatial distribution of ERS points is flat on a certain coordinate, the measurement accuracy will be affected on this certain coordinate direction.

Experiment process: in the same measuring area, first, measuring two groups of ERS points with the same number, then measuring the same point q and analysing the measurement errors of point q in three coordinate directions, after establishing the measurement field. The spatial distribution of the first group of ERS points is very small in the vertical direction, while the spatial distribution of ERS points in the second group is relatively uniform in these three coordinate directions.

As shown in Figure 11, ignoring the single-point transformation errors, transformation errors of these two groups of ERS points are almost the same. However, as shown in Figure 12, comparing the measurement results, the measurement errors of point q in the Y (vertical) direction appear a mutation, in the measurement field established by the first group of ERS points. Because of the consistency of measurement environment, instruments and coverage area, it can be inferred that the measurement errors in single direction (vertical) are caused by the difference of spatial distribution of ERS points. The spatial distribution of the first group of ERS points is relatively flat; the difference in the Y direction is small, while the spatial distribution of the second group of ERS points is relatively stereoscopic, there are obvious differences in the three coordinate directions. This spatial distribution difference is consistent with the performance of the final measurement errors of point q. Therefore, it is considered that ERS points should be distributed evenly in space, avoiding flat in a certain direction, which is helpful to improve the accuracy of the measurement system.

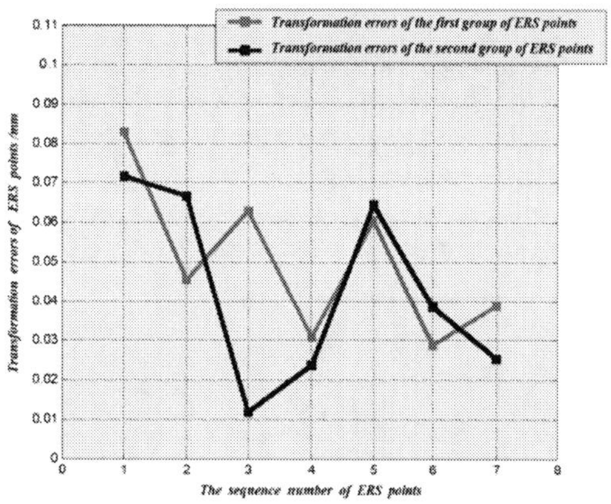

Figure 11. Transformation errors of ERS points.

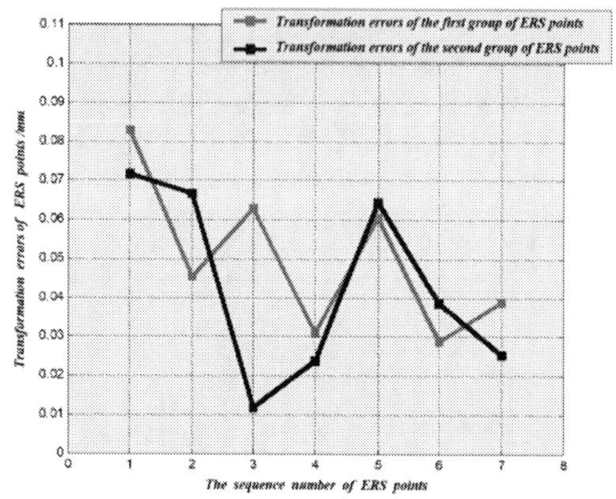

Figure 12. Measurement errors distribution of point q in three coordinate directions.

4. Conclusions

The rapid development of aircraft digital assembly technology has put forward higher demand for accuracy control of digital measurement system. Many factors, such as measuring instruments, measuring environment, coordinate transformation parameter solving algorithm, spatial distribution of ERS points, will affect the measurement accuracy. Based on the digital measurement system of laser tracker, this paper studies the relationship between spatial distribution of ERS points and system accuracy control, hoping to provide some theoretical suggestions for building high-precision digital measurement system to improve the accuracy of aircraft digital assembly. The main conclusions are as follows:

- All the OTP points are enveloped in the region formed by ERS points can help to guarant the measurement accuracy.
- There should be no less than 7 ERS points for the establishing measurement field, and the number of ERS points can be increased appropriately when the measuring area is large.
- The spatial distribution of ERS points should ensure uniformity instead of flatty.

References

[1] Corbett J, Naing S. *Feature based design for jigless assembly*[J]. Cranfield University, 2004.
[2] Weber, A. Aerospace prepares to soar again [J].*Assembly Magazine*,2003,46(6):58-59.
[3] Zhao Lele. *Research on Construction Technology of Digital Measurement Field for Aircraft Large Component Assembly* [D]. Nanjing University of Aeronautics and Astronautics, 2013.
[4] Wang Z, Mastrogiacomo L, Franceschini F, et al. Experimental comparison of dynamic tracking performance of iGPS and laser tracker[J]. *International Journal of Advanced Manufacturing Technology*, 2011, 56(56):205-213.
[5] Wire B. Arc Second, Inc. Announces Technology Partnerships with ROMER CimCore and Other Metrology Industry Leaders; Facility-wide indoor GPS Gains Acceptance[J]. *Business Wire*, 2004,24(1-3): 29-35.
[6] Norman A R, Schönberg A, Gorlach I A, et al. Validation of iGPS as an external measurement system for cooperative robot positioning[J]. *The International Journal of Advanced Manufacturing Technology*, 2013, 64(1-4): 427-446.
[7] Aguado S, Samper D, Santolaria J, et al. Identification strategy of error parameter in volumetric error compensation of machine tool based on laser tracker measurements[J]. *International Journal of Machine Tools and Manufacture*, 2012, 53(1): 160-169.

The Comfort Design for Civil Aircraft Cabin Using Ergonomics Theory

Wu Yang*, **Mu Tengfei and Wang Haiqiang**

Aircraft Integration Dep, Shanghai Aircraft Design and Research Institute

*wuyang1@comac.cc

Abstract. In order to improve the comfort of civil aircraft cabin, the cabin comfort system is built in this research based on the ergonomics theory and considering the influence of cabin environment, facilities, configuration, etc. In this process, filtrating and classification for the indexes which mainly influence the comfort are conducted, weight value setting is completed according to the analytic hierarchy process method and the quantify score for all the indexes is pushed. Finally, the results of cabin comfort evaluation are obtained. The case analysis results verify the effectiveness of the method. To sum up, based on this method proposal in this research, the main factors that can influence the comfort of the cabin are obtained, the comfort factors of the cabin are quantized and analysed, and last, these research results could direct the cabin comfort design.

Keyword: civil aircraft cabin; comfort design; ergonomics theory; evaluation system; analytic hierarchy process

1. Introduction

Nowadays passenger's comfort has become one of the important factors in the cabin design of civil aircraft, since the passenger's requirement for aircraft comfort has been continuously increased with the development of economy and the improvement of people's living [1-3]. Civil aircraft cabin comfort mainly involves cabin cross-section design, cabin layout, interior, lighting effects, cabin noise and vibration, as well as systems and equipment such as seats, trunks and passenger service devices. Systematic cabin evaluation can provide guidance for the design process of civil aircraft cabin as well as data support and design feedback for designers, which is of great significance. However, work in this area still needs to be improved.

At present, the evaluation methods of civil aircraft cabin comfort mainly include subjective evaluation and objective evaluation. Subjective evaluation method refers to the use of various subjective scales, questionnaires or interviews to derive the subjective comfort evaluation of the subjects. Subjective evaluation method needs to set the weight values of each indicator, which has greater subjectivity and varies with each individual, consequently reducing the credibility of the evaluation results [4, 5]. The objective evaluation method is mainly to measure and statistically analyse the physiological data of the subjects, such as pressure distribution, myoelectric signals, EEG signals, etc., which can reflect the comfort or discomfort of the subjects. The objective method establishes an evaluation model through quantifiable indicators. However, the model needs a great number of sample points and there are usually big differences between sample points. It needs to verify the accuracy of the model [6, 7].

IOP Conf. Series: Materials Science and Engineering **751** (2020) 012031 doi:10.1088/1757-899X/751/1/012031

In this paper we present an evaluation system which can quantitatively assesse the comfort of civil air-craft cabins. Firstly, the designers build a comfort index system based on different factors associating with the cabin comfort design. Secondly, through the analytic hierarchy process and considering the incommensurability between the indicators, the weight values of each indicator are set. Finally, we applied this method to the comfort evaluation and compilation of three different designs of dual-channel civil aircraft cabins, which verifies the feasibility and effectiveness of our evaluation system.

2. Cabin comfort assessment system

Cabin comfort design involves multidisciplinary knowledge such as cabin environment, cabin configuration, cabin interior, cabin facilities, cabin entertainment system, seat comfort and service, so the assessment of cabin comfort design is a comprehensive Multi-factor evaluation system. Based on the establishment of a comprehensive multi-factor evaluation system, expert questionnaire survey and scientific data analysis are carried out on the system, and then the effects of various factors on cabin comfort design can be quantified. Finally, through engineering discussion, the cabin design scheme with better comfort can be determined. According to the design process of the civil aircraft cabin, the equipment and systems involved and the criteria such as ergonomics and etc., the main civil aircraft cabin design elements are selected, and the cabin comfort evaluation system can be summarized as shown in Figure 1.

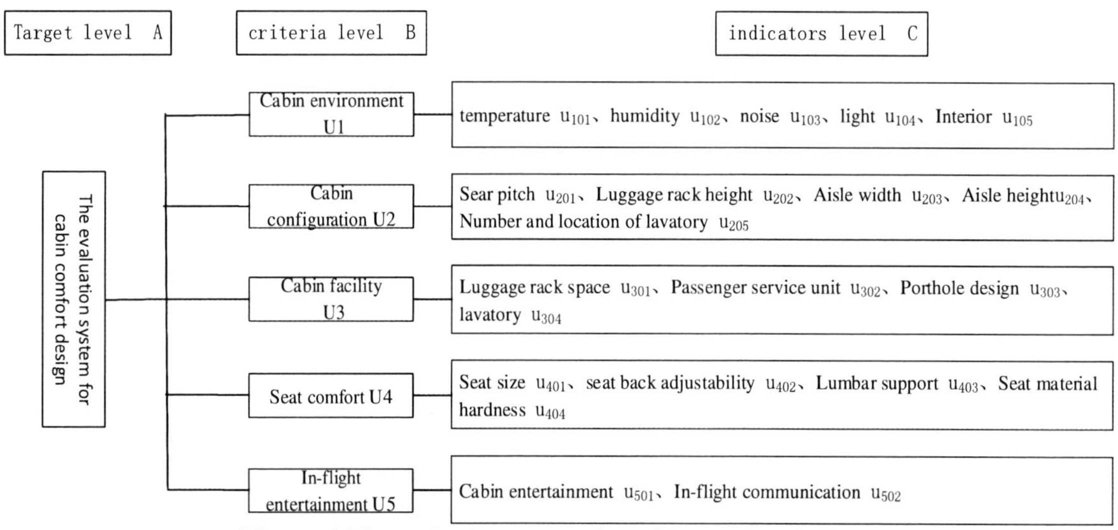

Figure. 1 The evaluation system for cabin comfort design.

In view of the above-mentioned cabin comfort evaluation system, the statistical data of cabin comfort system rating are obtained by means of design expert investigation. These data are the basis of cabin comfort design for civil aircraft. Based on the effective statistical data, the data are analysed by the combined evaluation method adopted in this paper, and the engineering discussion of cabin comfort is carried out. Finally, a multi-objective balanced optimal cabin design scheme can be determined.

3. Combined evaluation method

In Figure 1, there are six criteria and 20 indicators for cabin comfort. Therefore, it is a typical multi-attribute decision-making problem. Among them, the rationality of the weight setting of each comfort index will directly affect the evaluation conclusion of cabin comfort. In this paper, a combination evaluation method is used to analyse data samples [8]. In the weight setting of each indicator system, the analytic hierarchy process is used to compare the importance of each level of the comfort indicator system described in Figure 1 through the expert survey score, and finally derive the weight values of the bottom level indicator layer to the top level target layer. Subjective indicators, such as "access convenience" and "personal privacy space", are transformed into quantitative indicators. Then quantitative indicators are further analysed. For example, quantifiable indicators are divided into six types: cost type, benefit type, fixed type, interval type, deviation type and deviation interval type. Finally, the existing evaluation in document [8] is adopted. The model carries on the numerical analysis to each index.

The cabin comfort index P(x) is composed of subjective evaluation φ(x) and objective evaluation ψ(x). The expressions are as follows:

$$\boldsymbol{P}(x) = \sum_{i=1}^{N1} \boldsymbol{\alpha}_i \, \boldsymbol{\varphi}(\boldsymbol{x}_i) + \sum_{j=1}^{N2} \boldsymbol{\beta}_i \, \boldsymbol{\psi}(\boldsymbol{x}_j) \tag{1}$$

In the formula, N1 is the number of subjective comfort indicators and N2 is the number of objective comfort indicators. $\boldsymbol{\alpha}_i$ and $\boldsymbol{\beta}_j$ are the weights of subjective and objective evaluation results, which are determined by analytic hierarchy process. $\alpha \in [0,1], \beta \in [0,1], \alpha + \beta = 1$.

Generally, when evaluating ergonomics, it is necessary to consider the predicted value of human factor failure probability (mainly considering the effect of human factors on the accuracy of operation, which should be taken into account in the evaluation of human factor or maintainability of cockpit). In the case of cockpit comfort, the predicted value of human factor failure probability should be considered. However, in the case of cabin comfort, passengers have less interaction with cabin equipment during flight. In this paper, the predicted value of human failure probability is not considered.

3.1 Weight Setting of Comfort Index

The weight of the comfort index is set by the analytic hierarchy process. Firstly, the indicator layer in the comfort index system is weighted, and then the indicators under each criterion are weighted. The weight set-tings follow the following process:

Firstly, the judgment matrix is constructed, which divides the comparison results of the two indicators into scales of equally important, slightly important, obviously important, strongly important and extremely important. The cabin design experts only need to give the upper half matrix scale or the lower half matrix scale of the matrix diagonal, and the other half matrix can directly take its reciprocal. In order to quantify the evaluation indexes, the relative weights of the indexes can be measured according to the (1-9) scaling method in document [8]. According to the scaling meaning, the equally, slightly, obviously, strongly and extremely important values of the experts' evaluation can be assigned to 1, 3, 5, 7 and 9 respectively. Each subjective judgment made by the experts can be assigned to a value knot to get a sum and mean value.

Secondly, a consistency check on cabin comfort factors is required. The consistency check is performed on the judgment matrix by using C.R. = C.I./R.I. Among them, C.R. is the consistency ratio of the judgment matrix, C.I. is the consistency index of the judgment matrix and R.I. is the average random consistency index of the judgment matrix. When judging the consistency ratio of the matrix C.R. = C.I./R.I. < 0.1, the judgment matrix has satisfactory consistency. Otherwise, it is necessary to use the expert experience to modify the judgment matrix again until the consistency test is passed. The average random consistency index RI values are shown in Table 1.

Table 1. RI value of average random consistency index

order	1	2	3	4	5	6	7	8	9
RI	0	0	0.58	0.90	1.12	1.24	1.32	1.41	1.45

3.1.1 Weight Setting of Criteria Layer. Evaluation of experts at all levels, CI = 0.0318, indicating that the expert evaluation meets the consistency requirements. The criterion layer construction matrix and the weight values of each criterion are shown in Table 2.

Table 2. The matrix and the weight value of criterion level

	Cabin environment U1	Cabin configuration U2	Cabin facility U3	Seat comfort U4	In-flight entertainment U5	Weights
Cabin environment U1	1.000	1.000	3.000	0.500	3.000	0.2276
Cabin configuration U2	1.000	1.000	3.000	1.000	5.000	0.2815
Cabin facility U3	0.333	0.333	1.000	0.333	3.000	0.1085
Seat comfort U4	2.000	1.000	3.000	1.000	5.000	0.3259
In-flight entertainment U5	0.333	0.200	0.333	0.200	1.000	0.0566

3.1.2 Weight Setting of Index Layer. Similar processes at the criteria level are used to set the weight of each index. The results are shown in table 3.

Table 3. The weight value of comfort index

	Cabin environment	Cabin configuration	Cabin facility	Seat comfort	In-flight entertainment	Weights
temperature	0.3270					0.0744
humidity	0.1009					0.0230
noise	0.4534					0.1032
light	0.0468					0.0107
Interior	0.0719					0.0164
Sear pitch		0.4067				0.1145
Luggage rack height		0.2963				0.0834
Aisle width		0.1451				0.0408
Aisle height		0.0592				0.0167
Number and location of lavatory		0.0927				0.0261
Luggage rack space			0.4875			0.0529
Passenger service unit			0.2234			0.0242
Porthole design			0.1266			0.0137
lavatory			0.1625			0.0176
Seat size				0.6893		0.2246
seat back adjustability				0.1264		0.0412
Lumbar support				0.1264		0.0412
Seat material hardness				0.0579		0.0189
Cabin entertainment					0.8333	0.0472
In-flight communication					0.1667	0.0094

3.2 Evaluation Model of Subjective Indicators

In this paper, Miller's method is used to set the subjective evaluation level of comfort to 5 levels, namely 0, 2.5, 5, 7.5 and 10. 0 means very bad, inappropriate, etc. 2.5 means slightly better, but not suitable for acceptance. 5 means general and acceptable. 7.5 means good. 10 means excellent. Specific quantitative criteria for subjective indicators is shown in tables 4 and 5.

Table 4. The quantification standard of subjective comfort index

Comfort criterion	Comfort Index	evaluation result				
		Very bad	Bad	common	good	Very good
		0	2.5	5	7.5	10
Cabin environment	Cabin color matching		√			
......	......			√		
Seat comfort	Seat material hardness				√	

Table 5. The index list of subjective comfort

Comfort criterion	Comfort Index	Subjective indicator	Objective indicator	Index type	Quantization function
Cabin environment	light		√		
	Interior		√		
Cabin facility	Passenger service unit		√		
	Porthole design		√		
	Lavatory and gallery design		√		
Seat comfort	Lumbar support		√		
	Seat material hardness		√		
In-flight entertainment	Cabin entertainment		√		
	In-flight communication		√		

3.3 Evaluation Model of Objective Indicators

The list of objective comfort indicators is shown in Table 6. Among them, the benefit attribute refers to the attribute whose value is the bigger the better, the cost attribute refers to the attribute whose value is the smaller the better, the fixed attribute refers to the attribute which is better when its value is closer to a fixed value, and the interval attribute refers to the attribute which is better when its value is to a fixed interval (including falling into the interval). However, the deviation attribute refers to the attribute whose value is better when deviating further from a fixed value and the deviation interval attribute refers to attribute whose value is better when deviating further from a certain interval.

Table 6. The index list of objective comfort

Comfort criterion	Comfort Index	Index type	Quantization function
Cabin environment	temperature	Interval	f_{region}
	humidity	Interval	f_{region}
	noise	Cost	f_{cost}
Cabin configuration	Sear pitch	benefit	$f_{benifit}$
	Luggage rack height	Interval	f_{region}
	Aisle width	benefit	$f_{benifit}$
	Aisle hight	benefit	$f_{benifit}$
	Number and location of lavatory	benefit	$f_{benifit}$
Cabin facility	Luggage rack space	benefit	$f_{benifit}$
Seat comfort	Seat size	benefit	$f_{benifit}$
	seat back adjustability	benefit	$f_{benifit}$

$$f_{region} = \begin{cases} \frac{5(x_j-c)}{c-q} + 10, & x_j < c \\ 10, & x_j \in (c,d) \\ \frac{5(x_j-d)}{d-q} + 10, & x_j > c \end{cases} \quad (2)$$

$$f_{cost} = \frac{5(x_j-p)}{q-p} \quad (3)$$

$$f_{benifit} = \frac{5(x_j-q)}{p-q} \quad (4)$$

In the above formula, P and Q are the upper and lower limits of values respectively, and [c, d] is the interval value of interval attributes. In order to ensure consistency with the subjective quantization value range, 5 is used as the boundary point of quantization value. The optimal value is 10, and the worst value is 0.

4. Examples and results analysis

In this study, three typical double-channel passenger cabins are selected for comparison. Aircraft A and B have been delivered to operation, while Aircraft C is in the design stage. The analysis results are shown in Table 7.

Table 7. The results of cabin comfort analysis

Comfort Index	weight	Aircraft A	Aircraft B	Aircraft C
temperature	0.0744	8.4	8.4	8.4
humidity	0.0230	9.3	9.3	9.3
noise	0.1032	7.9	7.9	7.9
light	0.0107	9.0	10.0	9.0
Interior	0.0164	9.0	10.0	9.0
Sear pitch	0.1145	8.5	8.5	8.5
Luggage rack height	0.0834	7.7	8.3	8.3
Aisle width	0.0408	8.5	9.3	9.3
Aisle height	0.0167	7.4	8.6	8.6
Number and location of lavatory	0.0261	8.5	8.7	9.0
Luggage rack space	0.0529	8.3	8.3	8.3
Passenger service unit	0.0242	9.0	9.0	9.0
Porthole design	0.0137	9.1	8.3	8.5
lavatory	0.0176	9.0	10.0	9.0
Seat size	0.2246	7.4	8.5	8.5
seat back adjustability	0.0412	8.1	8.1	8.1
Lumbar support	0.0412	7.5	8.7	8.7
Seat material hardness	0.0189	9.1	9.5	9.3
Cabin entertainment	0.0472	8.5	8.7	8.7
In-flight communication	0.0094	7.7	8.2	8.2
Comfort quantitative score		8.1	8.6	8.5

5. Concluding remarks

From the perspective of ergonomics, this paper establishes an evaluation model of cabin comfort. Through the analysis of factors affecting the comfort, the various evaluation indicators are classified and stratified, and a more reasonable evaluation system of cabin comfort is established. By using the analytic hierarchy process, it's found that the most influential indicators of cabin comfort are seat size, seat pitch, noise, height of luggage rack, temperature and space of luggage rack. It should be pointed out that the reasonableness of setting the weights of comfort indicators depends on the qualification of experts and the reliability of scoring. The method in this paper does not involve factors in this respect. Therefore, based on the improved hierarchy method, the number of scoring experts can be further increased and the accuracy of the method can be improved to guide the design of airplane cabin comfort.

References

[1] AHMADPOUR N, LINDGAARD G, ROBERT J M, et al. The thematic structure of passenger comfort experience and its relationship to the context features in the aircraft cabin [J] *Ergonomics*, 2014, 57(6): 801-815.
[2] ZHANG Y, WANG H P. Some suggestions on evaluating preliminary overall design of Chinese passenger aircraft [J]. *Journal of Northwester Polytechnic University*, 2006, 24(6): 791-794.
[3] GOULART F R, VALLS-SOLÉ J. Patterned electromyography activity in the sit-to-stand movement [J]. *Clinical Neurophysiology*, 1999, 110(9): 1634-1640.
[4] XU J B, YU D H, CHU J J, et al. Study on system and method of workover rigs man-machine ergonomics evaluation [J]. *Machinery Design & Manufacture*, 2016(11):97-100.
[5] QI S J, WANG T, WANG F. Research on ergonomic integrated evaluations of UAV ground control station [J]. *Advances in Aeronautical Science and Engineering*, 2013, 4(4):474-480.
[6] CHEN D K, WANG Y Q, YE C, et al. Research on the method of ergonomics evaluation in manned submersibles cockpit [J]. *Ship Science and Technology*, 2016, 38(7):105-109.
[7] GUO Z, GUO D, Yang J C, et al. Study on evaluation index system of helicopter display interface ergonomics [J]. *Electronics Optics & Control*, 2011, 18(5):67-71.
[8] WANG M H, HU F H, LIU D X, et al. Machine tool optimization selection technology research based on fuzzy analytic hierarchy process with grey relational analysis method [J]. *Machinery Design & Manufacture*, 2015(10): 161-164.

IOP Conf. Series: Materials Science and Engineering **751** (2020) 012032 doi:10.1088/1757-899X/751/1/012032

The Application of Overset Grid in the Analysis of Impact Loads of Amphibious Aircraft during Landing on the Water

HU Qi[1,2,*], WANG Ming-Zhen[1,2], WU Bin [1,2], ZHANG Jia-Xu[1,2] and XIAO Zhi-Jian [1,2]

[1] China Special Vehicle Research Institute, Hubei Jingmen 448035
[2] Key Laboratory of High-Speed Hydrodynamic Aviation Science and Technology, HuBei Jingmen 448035

[*]E-mail: 1252847359@qq.com

Abstract. In view of the complex three-phase coupling problem of gas, liquid and solid when amphibious aircraft landed on the water surface, based on the finite volume method, the overset grid technology was used to simulate the landing load characteristics of amphibious aircraft; firstly, the overset grid technology was used to establish the landing model of amphibious aircraft, and the load characteristics under different landing conditions were obtained, and compared with the model test results. The results show that, firstly, the larger the sinking speed is, the larger the load is, and the more obvious bounce phenomenon occurs; secondly, the transverse distribution of the bottom pressure gradually increases along the keel to bilge; thirdly, the load characteristics obtained by numerical simulation are basically consistent with the model test. Therefore, the overset grid technology can simulate the landing load characteristics of amphibious aircraft and provide an effective method for the landing load design of amphibious aircraft.

1. Introduction

Amphibious aircraft is one of the important equipment in the emergency rescue system of our country. It can put out fire in the forest or carry out search and rescue at sea. It is a strong guarantee to protect the ecological environment of our country and maintain the maritime Silk Road [1]. According to CCAR 25.519, the landing load of amphibious aircraft shall be designed according to the most severe load conditions that may be encountered in actual operation, and the landing load is the most important input condition for aircraft structure design and strength check. Therefore, the water load is closely related to the safety of the aircraft structure and the weight of the fuselage structure, which affects the overall performance of the aircraft. In the design of amphibious aircraft, reasonable and accurate water load design is a very important task.

With the development of high performance computer, numerical simulation technology has become a powerful means to analyze the landing load characteristics of amphibious aircraft. Ma [2] used ALE method to simulate the wave landing process of amphibious aircraft, and analyzed the load characteristics under different sinking speed and different encounter position. Sun [3] used the general coupling algorithm to simulate the landing process of amphibious flight. The simulation results are in good agreement with the experiments. Zeng [4] used ALE method to simulate the landing response of amphibious aircraft under wave conditions, and analyzed the aircraft attitude, overload and ship bottom pressure under different waves. Yao [5] used the cel method to analyze the time-dependent curves of the ship bottom pressure and vertical velocity of amphibious aircraft under different vertical speeds and attitude angles. Qu [6] uses the finite volume method to solve the RANS equation and K - ε , and uses the global dynamic grid method to simulate the dynamic response of the simplified aircraft model landing on water. Zhang [7] used SPH method to simulate the landing process of the aircraft under the sea waves, and obtained the acceleration and attitude angle curves with time under different working conditions.

In this paper, based on CFD numerical simulation method, using overset grid technology, the impact load of amphibious aircraft during landing on the water surface is simulated, and the load characteristics and motion response under different working conditions are obtained. The influence of sinking speed on the load characteristics is analyzed, and the law of pressure distribution at the bottom of the ship with time is summarized.

2. Overset grid technology

Overset grid technology divides each part of the object into separate grids, and then embeds them into another set of grids. There will be overlaps between the grids. After pre-processing such as digging holes, the grids outside the calculation domain (such as the grid cells located in the surface of the object) will be removed and excluded from the calculation, and the interpolation relationship will be established in the remaining overlapped grid areas. Finally, the data exchange between each set of grids can be carried out in the overlapped grid through the interpolation method. Overset grid technology makes the grid of complex shape easier to generate, and the quality of local grid can be guaranteed. By using the characteristic of overset mesh, the unconstrained six freedom motion of the object can be realized.[8][9]

$$\phi_I = \sum_{i=1}^{n} \omega_i \times \phi_i \tag{1}$$

In this equation, ω_i are the interpolation weighting factors, ϕ_I are the values of the dependent variable. ϕ_i at donor cells N_i ($N_1 \sim N_3$) and subscript i runs over all donor nodes of an interpolation element (denoted by the green triangles in the figure). This way, the algebraic equation for the three neighbor cells from the same mesh and three cells from the overset mesh ($N_4 \sim N_6$). As shown in Figure 1.

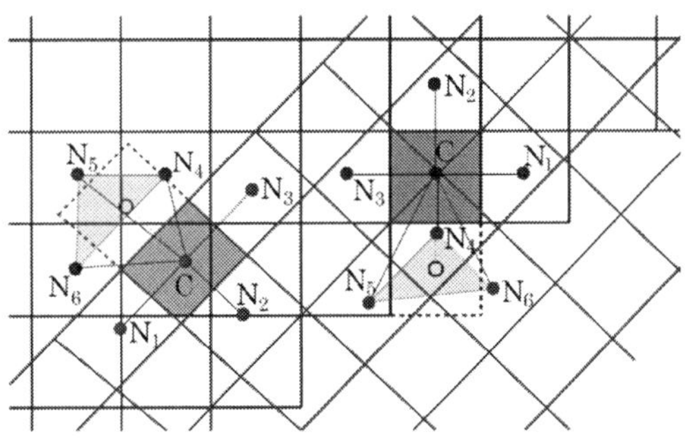

Figure 1. Diagram of overset grid

3. Geometric and model test

3.1 Geometric

Amphibious aircraft includes fuselage, wing, flap, vertical tail, flat tail, elevator and wing pontoon. When it lands on the water surface, the flap has a certain deflection angle, which makes the lift coefficient increase and the horizontal speed decrease when the aircraft lands on the water. At the same time, the elevator has a certain deflection angle, which provides a certain amount of lift moment for the aircraft, and avoids the aircraft's head burying under the action of hydrodynamic moment. The geometry of amphibious aircraft is shown in Figure 2 below:

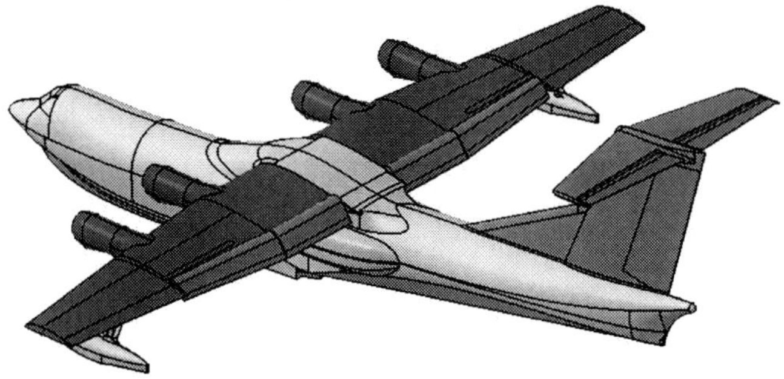

Figure 2. The geometry of amphibious aircraft

3.2 Model test

According to Fu rude's similarity criteria, the scale-up model of amphibious aircraft is designed, the model landing test device is developed, and the landing test is carried out in the high-speed towing pool, as shown in Figure 3:

Before the test, adjust the initial height and attitude angle of the model, and fix it with front and rear, left and right struts. During the test, the high-speed trailer is used to accelerate the model to a certain horizontal speed. When the speed is stable, the model will be released. Under the action of gravity and aerodynamic lift, the model will impact the water surface after falling freely, and the acceleration and bottom pressure during the impact process will be tested.

Figure 3. Test of amphibious model

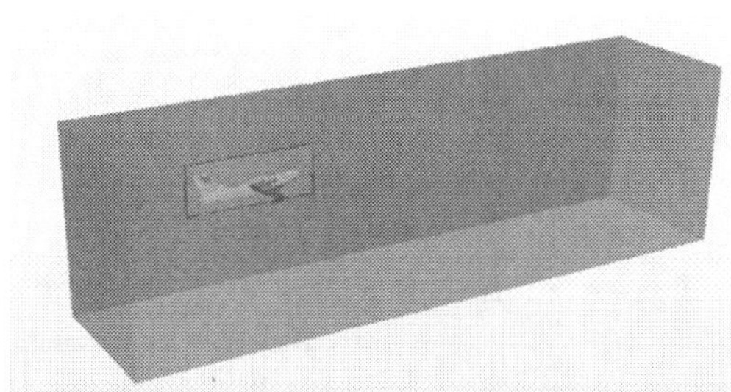

Figure 4. Computational domain

4. Mesh generation and boundary conditions

4.1 Mesh generation

The length of the aircraft is L. the entrance of the calculation area is 4L from the head of the aircraft, the exit is 1.5L from the tail of the aircraft, and the top and bottom are 1L from the aircraft respectively. In order to save computing resources, half of the basin is taken for calculation The trim grid is adopted. The hull surface grid size is 8 ‰ of the aircraft length. The wing and tail are locally encrypted. The grid size is 2 ‰ of the aircraft length. The free surface is locally encrypted, and the total mesh is 6.5 million. The distribution diagram of calculation domain is shown in Figure 4. The multi-phase flow model and VOF model are used to capture the free surface. SST k-omega turbulence model is used in the turbulence model, and the first-order upwind discrete scheme is adopted.

After the calculation domain is established, grid generation is carried out. In mesh generation, all kinds of meshes use cut volume mesh. According to the calculation domain grid, it is divided into background grid and overset grid. The overset grid is divided into three layers as a whole, namely, the surrounding part, the transition part and the outermost part of the aircraft. The outermost grid overlaps the background grid, and the size of this part of the grid is consistent.

4.2 Boundary condition

First of all, the background grid area, the inlet is set as the speed inlet, the outlet is set as the pressure outlet, and the rest, except for the symmetry plane, are all set as the speed inlet. Secondly, the overlapped grid area, the fluid area and the hull surface are set as overlapped grids, and the motion rules of the overlapped grid area are defined as DFBI rotation and translation. Finally, the interface between the background grid area and the overset grid area is calculated.

In DFBI, an aircraft is defined as a 6-DOF body, and its centroid is set. At the same time, set the initial value of the aircraft, such as the moment of inertia, initial speed. In this paper, the initial horizontal velocity of aircraft is defined as 15.7m/s. The time step of the solver is 0.025 s, and the maximum number of iterations is 5.

5. Results and discussion

5.1 Calculation condition

When amphibious aircraft land on the water surface, it usually keeps a small sinking speed close to the water surface. In this paper, five different sinking speeds are selected as calculation conditions. As shown in the following table 1:

Table1. Calculation conditions

Condition	Attitude /°	Horizontal velocity /m/s	Sinking velocity / m/s
1			0.2
2			0.4
3	6	15.7	0.6
4			0.8
5			1.0

When the calculation time is set to 0.8s, the complete landing process of amphibious aircraft is obtained. The vertical acceleration, vertical velocity, attitude angle, bottom pressure and horizontal velocity time history curve of the aircraft are monitored, and the load characteristics, motion response and ship bottom pressure under different sinking speeds are analysed . As shown in Figure 5, the free surface of the amphibious aircraft landing at different times.

Figure 5. Free surface of amphibious aircraft landing

5.2 Load characteristics

The simulation results are compared with the model test, as shown in Figure 6, the vertical acceleration time history curve, showing good consistency, indicating the accuracy of numerical simulation.

Figure 6 shows that the larger the initial sinking speed is, the more obvious the vertical acceleration increases, and the earlier the peak occurs. In a short time, the peak acceleration of the aircraft is up to 2.15g. It can be seen that the landing process of amphibious aircraft is short and the load is very large.

A one-way coupling relationship is established between the fluid area and the structural area to transfer the bottom pressure to the structural area, extract the bottom pressure of each analysis step, select a cross section before the fault step, and analyze the pressure distribution law at different times.

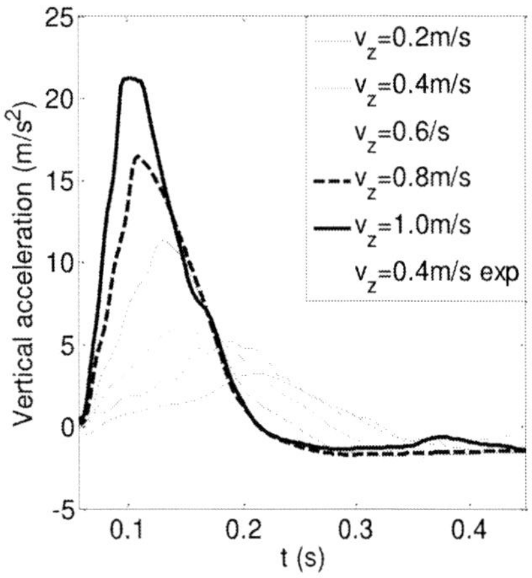

Figure 6. Time history curve of vertical acceleration

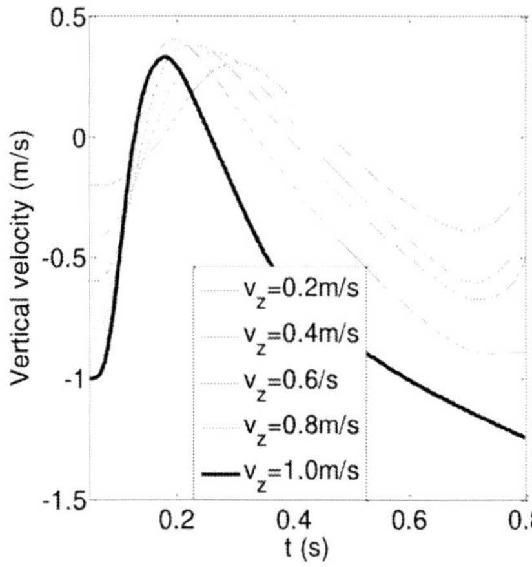

Figure 7. Time history curve of vertical velocity

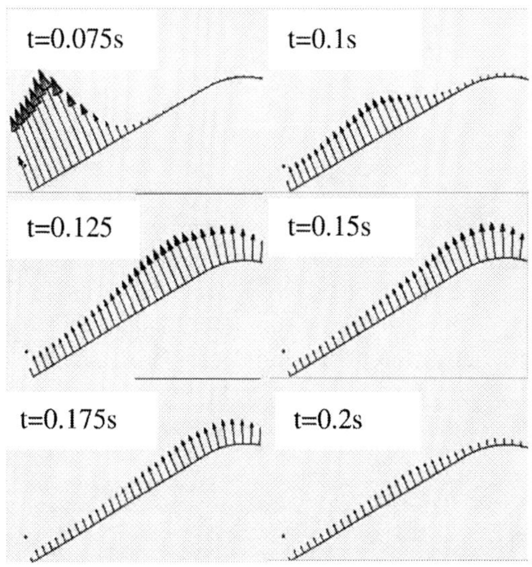

Figure 8. Pressure distribution at different times of 80mm cross section before fault terrace with $V_Z = 0.6$m/s

5.3 Motion response and pressure

Figure 7 shows the vertical velocity variation law under different sinking speeds. It can be seen that the larger the initial sinking speed is, the greater the impact load of water is, and the faster the vertical velocity decreases. Figure 7 shows that the smaller the initial sinking speed is, the earlier the plane hits the water again. Due to the impact force of water, the aircraft produces upward speed. Under the impact load and aerodynamic lift, the aircraft leaves the water surface and glides close to the water surface. Then under the action of aerodynamic lift and gravity, the aircraft strikes the water surface again and receives upward impact load, but it is not as large as the first impact load.

Figure 8 shows that when amphibious aircraft is connected to the water, there is a large pressure near the keel. As the aircraft lands on the water surface, the peak pressure transfers to bilge until the acceleration begins to decrease. When the acceleration is maximum, the peak pressure of bilge is 2-3 times of that of keel. Comparing Figure 6 and Figure 8, it can be seen that the resultant force of the pressure distribution is consistent with the time corresponding to the peak acceleration.

Figure 9 shows the curve of horizontal velocity changing with time. The larger the initial sinking velocity is, the greater the wetting depth is, the greater the water resistance is, and the faster the horizontal velocity decreases. Figure 10 shows the curve of attitude angle changing with time. After the plane hits the water surface, the attitude angle gradually increases.

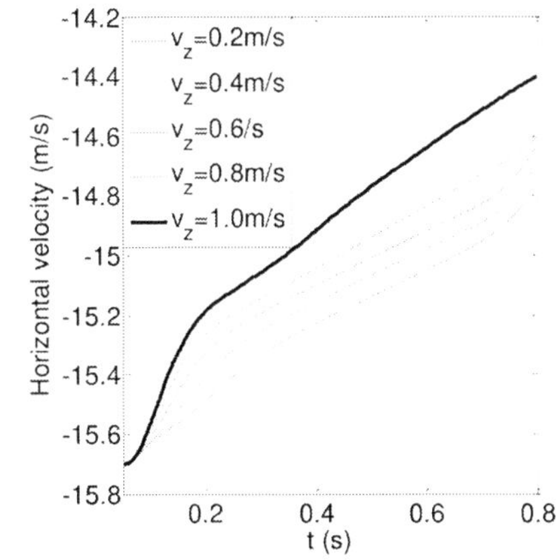

Figure 9. Time history curve of horizontal speed

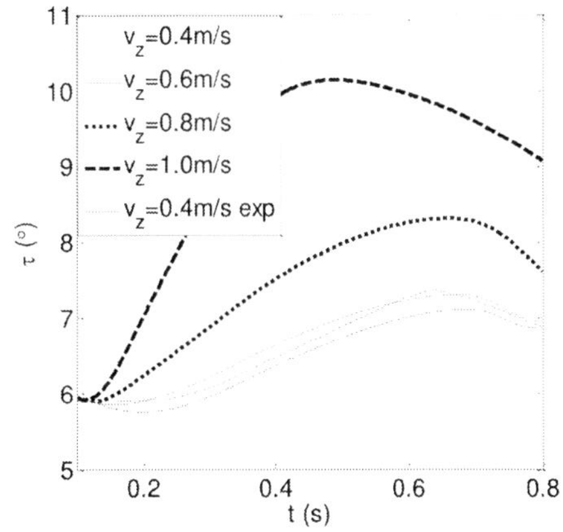

Figure10. Time history curve of attitude

6. Conclusions

Using the overset grid technology, the process of amphibious aircraft landing on the water surface at different sinking speeds is simulated, and the results are compared with the model test results, and the following conclusions are obtained:

1) The numerical simulation results are in good agreement with the test results, which shows the accuracy of the numerical simulation results;

2) When the amphibious plane lands on the water surface, the larger the initial sinking speed is, the greater the load it is subjected to, and the more obvious bounce phenomenon occurs.

3) In the process of landing, the transverse distribution of ship bottom impact pressure gradually increases from keel to bilge, but bilge is not completely immersed in the water.

References

[1] HUANG Ling-cai, YONG Ming-pei. Key technologies and industrial application prospects of amphibian aircraft [J]. Acta Aeronautica ET Astronautica Sinica , 2019,40(1):18-34 (in Chinese)

[2] MA Zeng-hui, LIU Li-sheng, YUE Zhen, et al. Numerical investigation on seakeeping performance of amphibious aircraft landing on waves [J]. Journal of Computational Mechanics, 2018, 35(3):380-386. (in Chinese)

[3] SUN Pei-cheng, GAO Xiao-peng, DONG Zu-shun, Simulation analysis of amphibious aircraft water landing [J]. Ship & ocean engineering, 2014, 43(06):171-174. (in Chinese)

[4] ZENG Yi, YANG Shi-fu, LUO Lin-yin. The structural response of amphibian landing onto wave [J]. Journal of mechanical design, 2015, 32(09):96-100. (in Chinese)

[5] Yao Xiao-hu, Huang Yu-tai, Ou Zhi-cheng, et al ,CEL algorithm-based analysis of dynamic characteristics of amphibious aircraft landing on water [J]. Journal of South China University of Technology (Natural Science Edition), 2015, 43(06):110-115. (in Chinese)

[6] QU Qiu-lin, HU Ming-xuan, GUO Hao, et al. Application of Global Moving Mesh in aircraft ditching simulation[J].Aeronautical Science & Technology, 2015,26(11):1-9. (in Chinese)

[7] ZHANG Sheng, YAN Jia- yi, ZHU Shu- hua. Numerical Simulation of Light Aircraft Ditching Attitude Based on SPH Method [J]. Aeronautical Computing Technique, 2014, 44(04):76-79. (in Chinese)

[8] CHENG Ping, WAN De-cheng. Analysis of wind turbine blade-tower interaction using overset grid method [J]. Chinese journal of Hydrodynamics (part a), 2017, 32(01):32-39. (in Chinese)

[9] XU Wei-guang, ZHAO Fa-ming, HE Shu-long. Research on calculation method for ship motion based on dynamic overset approach [J]. Journal of Ship Mechanics, 2016, 20(07):824-832. (in Chinese)

IOP Conf. Series: Materials Science and Engineering **751** (2020) 012033 doi:10.1088/1757-899X/751/1/012033

Interior Inspection for Equipment and Furnishings System of Civil Aircraft

Zhou Jianbin*, Zhang Xinyuan and Wang Xiaozheng

No. 5188 Jinke Road, Pudong District, Shanghai China

*zhoujianbin@comac.cc

Abstract. Equipment and Furnishings System of civil aircraft has many sub-systems, including Pilot Seat, Observer Seat, Cabin Attendant Seat, Galley, Lavatory, Passenger Seat, Overhead Bin, Cockpit Lining, Cabin Lining, Cargo Lining, Emergency Equipment, Passenger Service Unit and Placards/Markings. Inspection (MOC7), as one of means of compliance for certification, is usually used for the certification of the regulations of FAR25.562, 785, 815, etc. The sub-systems in Equipment and Furnishings System are very different at every aspect. They have different functions, working principles and system architectures. These differences make the inspection of Equipment and Furnishings System complicated. A lot of details need attention. This article described the inspection requirement and Pass/Failure criteria for Equipment and Furnishings System.

1. Introduction

There are 10 Means of Compliance (MOC) for certification, as shown in Table 1:

Table 1. Means of Compliance

MOC									
0	1	2	3	4	5	6	7	8	9
Compliance Statement	Design Review	Analysis / Calculation	Safety Assessment	Laboratory Test	Ground Test	Flight Test	Compliance Inspection	Simulation	Equipment Qualification

MOC7, Compliance Inspection is widely used in the certification for Equipment and Furnishings System of civil aircraft. This article, based on the experience of civil aircraft design and certification and combining the requirement in AC25-17A [1], made a summary of MOC7 work for Equipment and Furnishings System.

2. Regulations

According to TSO-C127b, a passenger seat has to pass Dynamic Test and Static Test [2]. FAR/CS/CCAR 25.561, 562 and 785 have the similar requirement. The difference is that TSO-C127b's concerning is only the performance of the seat itself, but FAR/CS/CCAR concern the interface strength between the seat and the track as well.

Dynamic Test includes 16g Dynamic test, 14g Dynamic test, Head-path test, Front-row HIC test and Row-to-row HIC test.

Related regulations for Equipment and Furnishings SystemMOC7 are: FAR25.562, 785, 787, 789, 791, 803, 807, 809, 811, 812, 813, 815, 853 and 1411. They are generally related to Evacuation Exit, Aisle, Passageway, Placards and Emergency Equipment which have influence on aircraft occupants' emergency evacuation [3]. The execution methods for MOC7 for these regulations usually are Measurement, Visual Inspection and Operation Inspection.

3. Inspection Requirement and Pass/Failure Criteria

3.1. FAR25.562

a) All seats (including flight crew seats, cabin attendant seats and passenger seats) shall have the approval of TSO-C127a or TSO-C127b. If required, the TSO seats need to obtain another kind of approval from the authority, in the form of VDA, LODA, etc.;

b) TSO tag shall have the proper seat type (e.g., type A) marking, seat cushion part numbers, restraint system part numbers, and forward or aft facing. The part numbers of seat cushion are often to be inspected;

c) The testing and installation of seats shall consider all limitations, including: the range of occupants, front-row setback and row-to-row pitch. Seat testing shall consider 5th to 95th human body. For the front-row seat, the front-row setback usually is 42 inches (for Economy Class Passenger Seat) or 45 inches (for Business Class Passenger Seat), but it can be adjusted according to the test result of Head Path Dynamic Test [4]. The row-to-row pitch is determined by passenger row-to-row HIC test;

d) Each occupant in the aircraft shall have a seat place;

e) Seat tracks part numbers on the airplane shall be as the same as the seat tracks used in the dynamic seat testing. If not, it must be proved that the real track used in the aircraft are not weaker than the track used in dynamic test. Besides, it's acceptable to do a supplemental test for the real track using the 6 component loads measured in dynamic test [5] to the show the interface strength between seat and track is strong enough. This test is too complicated, so it's usually suggested to use real track in seat dynamic test;

f) The permanent deformations of the seats shall not exceed the limits required in AC25.562-1B [4];

g) Injury criteria shall be met for each seat place, including: HIC, Femur load, Lumbar load and Belt load.

3.2. FAR25.785

a) All seats shall have the approval of TSO-C127a or TSO-C127b;

b) There shall have no potentially injurious objects within striking radius of the head. The striking radius is a arc scope of 35 inches radius [1], as shown in Figure 1;

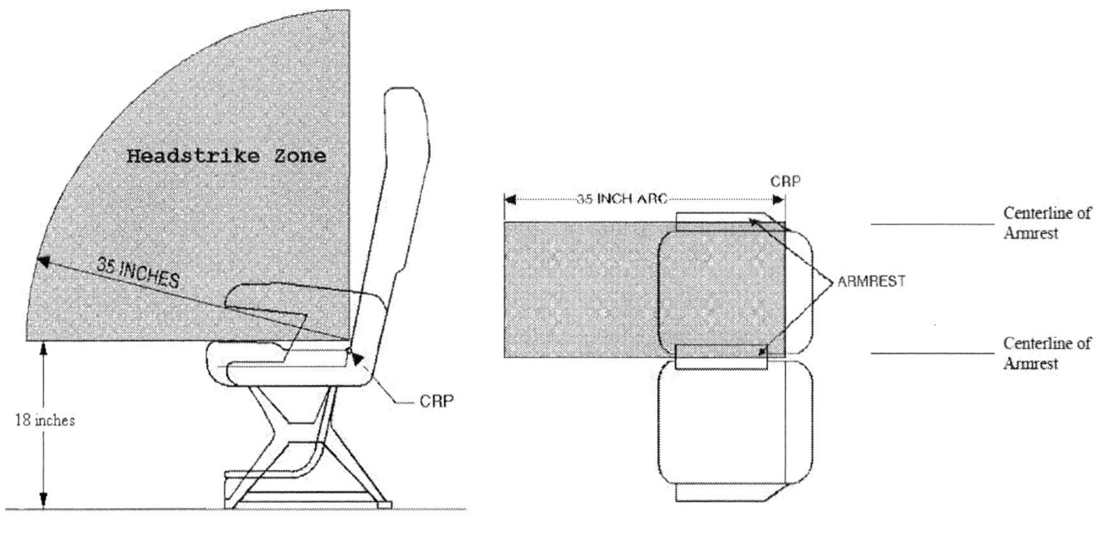

Figure 1. The definition of "35 inches"

c) Armrests when folding up shall not be beyond the seat back;

d) Footrests, legrests, or headrests shall have no potential injurious features;

e) If the footrest or legrest deploy into required crossaisles or passageways, they shall have a mechanical lock-out in the stowed position;

f) If the footrest or legrest is electrically operated, they shall have a manual override to facilitate stowage in the event of loss of electrical power. Otherwise, they may have negative influence on emergency evacuation;

g) Seat belt shall have TSO approval and there's no tendency for the seat belt shackle to become tangled or hung up on seat structure;

h) Cabin attendant seat shall have shoulder harnesses and lap belts. Usual attendant seat belt is 4-point constraint system and can comply with this requirement;

i) Handholds shall be provided to help occupants to steady themselves. Usually the slot in Overhead Bin is used as the handholds. If no slot in Overhead Bin, the seatback of passenger seat shall comply with following requirement [6]:

1) Row-to-row pitch shall be no more than 65 inches;

2) The seatback shall not break over with a 111N load applied on the top of the seatback.

j) All projecting objects, which may be contacted in flight, shall be padded. The door structure needs more attention. Usually, Doors have big hinges projecting into the cabin. It's hard to make the HIC value not to exceed 1000 as required by FAR25.562 if let the ATD hit the door structure, according to engineering experience. The easiest way is to cancel the seat place or adjust the seat dimension to make the passenger's headpath keep away from the door structure. At the meantime, sometimes we may have to move the seat track to ensure the seat width.

3.3. FAR25.787

a) Each compartment shall have a weight limit placard or it's dedicated for emergency equipment only or no stowage;

b) Compartments are completely enclosed;

c) Double latches shall be provided where necessary;

d) There shall have provisions to account for wear and tear in service. To satisfy this requirement, it's necessary to multiply a factor (1.33 or 1.15) when calculating static load;

e) Means of latching shall be positive with a positive indication when latched or unlatched.

3.4. FAR25.789

a) All objects shall be constrained at each direction. We also need to do static test or analysis to prove that the constraints are solid enough when bearing the emergency landing load and aircraft flight/ground load;

b) For convenience, the aft baggage bar of passenger seat is not necessary.

3.5. FAR25.791

a) A passenger information sign shall be visible from each flight attendant and passenger seat;

b) If there are seats that translate or swivel, a sign shall be visible from each seat position.

3.6. FAR25.803

a) There shall have no tripping hazards present in the aisle, crossaisles or passageways;

b) There shall have no other impediments (projecting objects) to rapid evacuation.

3.7. FAR25.807

a) All clear exit openings shall equal or exceed the minimum required dimensions, including any protrusions from linings, hinges etc.;

b) A flight attendant seat shall be positioned adjacent to each Type A exit.

3.8. FAR25.811

a) All of the required signs (locator, bulkhead, marking) shall be present and visible to persons in the main aisle. And the evaluation shall include consideration of a range of occupants (5th percentile female to 95th percentile male);

b) The next exit sign shall be visible from each point in the aisle;

c) All exit signs shall be positioned such that they lead persons to exits and not into galleys or other "dead ends";

d) Curtains or other features, e.g. video monitors, shall not interfere with exit sign visibility;

e) Exit operating instructions shall be clear.

3.9. FAR25.812

The interior modification shall not cause influence on emergency lighting, e.g. creating new shade for emergency lighting or making new shadows.

3.10. FAR25.813

a) All passageways shall be unobstructed from the aisle to the exit opening and have assist space. In aircraft design, it's common to find Galley and cabin attendant seat intrude passageways, so it's necessary to consider this problem when making cabin LOPA;

b) A assist space of 12 inches × 20 inches shall be provided near required exit;

c) Assist space shall have assist handhold;

d) there shall be an unobstructed projected opening of overwing exits for the width of a seat, including the seatback in any position. That's why the seat in front of exit usually is designed to only have a fixed seatback.

3.11. FAR25.815

a) When measuring aisle width, seatback recline, break over and rotation of seat shall be considered, especially at divided zones;

b) When measuring aisle width, the rubstrips on seat, Galley and so on shall be considered;

c) Curtain tiebacks shall be readily movable, where they project into the required aisle;

d) Movable armrests that protrude into the required aisle return to the normal position when released; Are latches discreet (not easily discovered by passenger) when used to control a movable armrest that protrude into the required aisle.

3.12. FAR25.853

a) Waste compartments are completely enclosed;

a) Lavatory outside shall have ashtrays;

b) All wiring shall be protected from abrasion and crushing;

c) Seats are fire-blocked. Cockpit flight crew seats only need to comply with FAR25.853(a), but Cabin seats have to comply with FAR25.853(a) and 853(c). If F/C and B/C passenger seat and E/C passenger use large nonmetal panels, the seats shall comply with FAAR25.853(d). SAE ARP6199 provides clear instructions for evaluating if the seats need to do Smoke Density and Heat Release test required by FAAR25.853(d).

3.13. FAR25.1411

a) It shall be easy to fetch emergency equipment. Usually a 50th participant will be used to fetch emergency equipment to show the compliance;

b) Emergency equipment stowage locations shall be conspicuously and conveniently marked.

c) Placards as close to eye level height as practicable.

d) Additional arrows shall be needed to locate the specific stowage location;

e) Curtains shall not block access to, or markings of, emergency equipment;

f) Emergency Equipment shall be protected from damage in its stowage location. So belts or brackets are usually used to fix emergency equipment;

g) There shall have sufficient type and quantity of required items;

h) Life vests shall be easily removable by a seated, belted, and untrained person, at all locations;

i) There shall have a placard for all seats, including the forward rows, indicating the location of the vest. For front-row, the related placards are stick to the bulk in front. For other rows, the related placards are printed or stick to the backrest table for forward seats. Some aircraft use flotation cushion to replace life vest. If yes, there shall have a placard of "cushion can be used as flotation".

3.14. Others

3.14.1. Galley. a) There shall have no compartment doors that could interfere with exit opening. They can be design to be spring loaded closed or they can be easily pushed out of the evacuation path in emergency evacuation;

b) For the folding cart ramps that could be left down for takeoff and landing they shall not pose a tripping hazard;

c) All waste compartment doors self shall be closing or marked to be closed when not in use. If waste containers are required to meet fire containment requirements, the waste compartment shall be placarded to require the waste container;

d) Fixed items shall be installed for inspection, such as ovens, coffee makers;

e) All wiring shall be protected from abrasion, especially from rotable items;

f) The "close for taxi, takeoff, and landing" placards shall be conspicuous, even when compartment doors are open;

g) The load limit placards shall be conspicuous when the compartment doors are open.

3.14.2. Lavatory. a) If the lavatory door open into the aisle, it shall have a mechanism to make it back to stowed position. If not, there shall have a placard instructing to close and latch when not in use. To avoid that the lavatory door impacts emergency evacuation, a folding door is the best choice;

b) Oxygen drops shall be compatible with both standing and seated occupants;

c) There shall have no potential stowage areas that could lead to a fire hazard. If yes, they shall have "NO STOWAGE" placards;

d) There shall have an ordinance (i.e., No Smoking, Return To Seat) sign;

e) There shall be a means to unlock the lavatory door from the outside, without the use of tools;

f) Waste compartments shall be designed with wear and tear in mind.

3.14.3. Monitor and Overhead Bin. a) Aisle mounted monitors shall be at least 73-inches off the floor, or retractable and so placarded?

b) All sharp corners shall all been eliminated from the monitor shroud;

c) The monitors shall not obscure any required exit sign;

d) There shall have a manual means to retract monitors that are normally powered;

e) In-arm monitors shall be easily break away if contacted by a passenger during turbulence;

f) Possible head contact surfaces shall be padded;

g) Monitors shall be located under sidewall stowage bins retractable;

h) In-arm monitors in front rows, at cross aisles or at exits shall be stowed, or become unstowed, such that they interfere with exit passageways or other egress routes;

i) In-arm video monitors shall break away easily without breaking off or, if they do break, are there any sharp or hazardous protrusions? Is the monitor capable of being re-stowed for taxi, takeoff and landing;

j) Required placarding for stowage shall be visible to the seated occupant;

k) The in-arm cavity shall be "completely open or completely closed" to address the collection of flammable materials.

IOP Conf. Series: Materials Science and Engineering **751** (2020) 012033 doi:10.1088/1757-899X/751/1/012033

4. Conclusion

Based on the experience of civil aircraft design and certification, this article summarized the Compliance Inspection requirement, pass/failure criteria and related design proposals for Equipment and Furnishings System. Some inspections need 5th-95th person to participate, so it's suggested to find 3 testing personnel: 5th, 50th and 95th for the inspection activity, but it's only necessary to concern their body heights. Above all, the content of inspection needs to be discussed and confirmed with the authority and described in Certification Plan which is used as the guide for all certification activities.

Reference

[1] AC25-17A Transport Airplane Cabin Interiors Crashworthiness handbook, 05/18/2009.
[2] TSO-C127b Rotorcraft, Transport Airplane, And Small Airplane Seating Systems, 06/06/2014.
[3] 14 CFR Part 25 - AIRWORTHINESS STANDARDS: TRANSPORT CATEGORY AIRPLANES (FAR25). 11/18/2003.
[4] AC25.562-1B Dynamic Evaluation of Seat Restraint Systems and Occupant Protection on Transport Airplanes. 01/20/2006.
[5] SAE AS8049B Performance Standard for Seats in Civil Rotorcraft, Transport Aircraft, and General Aviation Aircraft, 01/2005.
[6] SAE ARP5526C Aircraft Seat Design Guidance and Clarifications. 05/2011.

IOP Conf. Series: Materials Science and Engineering **751** (2020) 012034 doi:10.1088/1757-899X/751/1/012034

Integrated Requirements Management of Civil Aircraft

Xinyu Zhang[a] and Xinai Zhang[b]

No 5188, Jinke Road, Pudong District, Shanghai, China

[a]zhangxinyu@comac.cc; [b]zhangxinai@comac.cc

Abstract. Civil Aircraft is a typical complex product system, which has high integration and complex management. In the initial stage of the project, identifying market and customer needs and defining high quality requirements are the key to develop products. Meanwhile, civil aircraft needs to meet the airworthiness regulations to ensure the safety of the aircraft. From the perspective of commercial operation and airworthiness compliance, it is necessary to use requirements management method to develop civil aircraft. For civil aircraft, the requirements have the characteristics of hierarchy, variety and traceability. In this paper, an integrated requirements management method and platform are proposed based on the characteristics of civil aircraft requirements. And an application of certain type aircraft is given.

1. Introduction

The Standish Group has been produced their CHAOS report in project success and failure since 1994, which shows that the reasons for project failure are 21.8% related to requirements and the project success factors are 44% related to requirements[1]. For civil aircraft development, the requirements are the basis and keys throughout the whole lift cycle, defining what the stakeholders(users, customers, suppliers, developers)need and what the systems must do[2]. There are thousands of requirements in aircraft development. How to ensure that stakeholders' requirements are effectively delivered to the right person and how to control the change of requirements with suppliers?

The requirements management can help us to manage the requirements, support the development of a safe and certifiable civil aircraft in limited time, and minimise the development costs. Requirements management is a part of design assurance system which can also help us to show the satisfaction of regulations to the airworthiness and reduce the risk and cost of certification[3]. It can also help capture requirements of all stakeholders correctly and completely and control the change of requirements throughout the whole life cycle.

The research of requirements management starts from the field of software. Yvonne Bijan's paper gives the current state of the methodology for developing complex systems requirements, including the methods of facilitation, the methods of prioritizing needs, the methods of transforming fathered data into requirements and other methods[4]. In Annabella Loconsole's paper[5], a requirements management measure method is introduced to help control changing software requirements and estimate the cost of changing requirements. She mentioned that requirements development is a learning process rather than a gathering process because we can not specify exactly what customer wants at the beginning of a project. In Guo bozhi's paper[6], the application of requirements management in civil aircraft development is introduced. Also, there are some research about the requirements management tool[7-8]. But for the civil aircraft, a typical complex product system, it is necessary to connect requirements management with other technologies and management to avoid becoming an information island. Only when all activities are connected, can we effectively implement and play the most role of requirements management.

In this paper, we analyze that the requirements management of civil aircraft have the characteristics of hierarchy, variety and traceability. The requirement information architecture, types of aircraft and traceability network shall be built. And the requirements management process is defined and the relationship between requirements management and other activities are discussed. Based on these, the integrated requirements management platform are introduced to help manage the requirements and improve quality.

2. Integrated Requirements Management

2.1. Requirements Characteristics

2.1.1. Hierarchy. For aircraft, there are different layers of requirements including stakeholder needs, aircraft level requirements, system level requirements, sub-system requirements and so on, as shown in Figure 1. The high level requirements will be allocated to the low level requirements. From the top level to the low level, the number of the requirements are rising exponentially. In the early stage of project development, the requirement information architecture of the project shall be determined during the requirements management planning, including defining the scope of management and the decomposition[9].

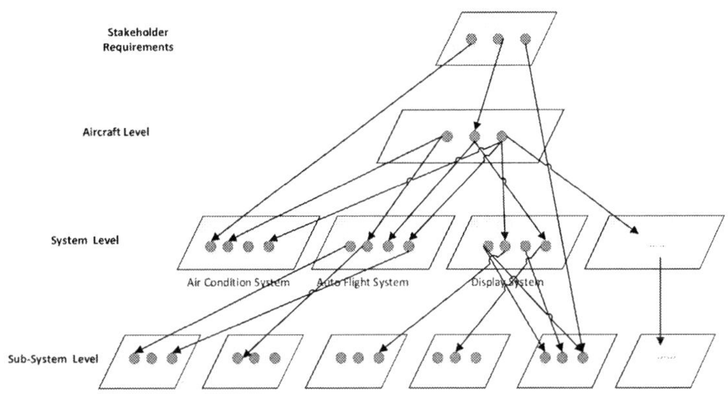

Figure 1. Requirement Information Architecture

2.1.2. Variety. As a management concept and solution centered on stakeholder requirements, the types of requirements cover the whole life cycle of design, manufacturing, service, test, and recycling. The following is included but not limited for the types of requirements:

- Safety requirements
- Functional requirements
- Customer requirements
- Operational requirements
- Performance requirements
- Physical and installation
- Maintainability requirements
- Interface requirements
- Airworthiness requirements
- Derived requirements
- Manufacture requirements
- Support requirements
- Test Requirements

IOP Conf. Series: Materials Science and Engineering **751** (2020) 012034 doi:10.1088/1757-899X/751/1/012034

2.1.3. Traceability. In the development of civil aircraft, the requirements can trace to the high-level and the low level requirements. And they can also link to the related functions, solutions and V&V matrix. The requirements can be the key to link all other data in the development of aircraft. All these data are a network from which it is easy to know the relationship of each requirement. The requirements are always changing, but the traceability relationship can help us to do the change impact analysis. And it is very important to link all of the data in whole life cycle.

2.2. Requirements Management Process

For aircraft product development, requirements are the design basis of each stage of aircraft design, which affect the design, implementation and verification of products. The definition process of requirements is the process of key factors and indicators, as shown in Figure 2, including serialization, diversity, function, performance, safety, reliability, supportability, comfort, cost, cycle, production capacity and supply chain, etc[10]. In order to implement requirements management in aircraft program, firstly the requirements management plan shall be defined. In requirements management plan, the requirement information architecture shall be defined to cover all needs of stakeholders, including customers, airworthiness, support, manufacture, suppliers and etc.

Based on fully understanding the stakeholders needs, the requirements are expressed through a concise and common engineering language. The purpose of requirements identification is to identify all requirements, their levels and characteristics in the whole lift cycle of the project. The requirements writing shall meet specific rules[9].

Validation of requirements is the process of ensuring that the specified requirements are sufficiently correct and complete so that the product will meet the needs of customers, users, suppliers, maintainers and certification authorities, as well as aircraft, system and item developers (e.g. flight crews, as users, may have a need for a certain system behavior for thrust control and a level of performance for that behavior. Certification authorities may have a need for a constraint on undesired operating behaviors)[2].

Requirements monitoring is to monitor and manage the quality and status for all levels of requirements. It helps the requirement manager to know the status and find problems so that the engineers can correct them in time.

Changes to requirements need to be thoroughly evaluate to determine the impact on architecture, design, interface, high level requirements or low level requirements. All changes must be reviewed and approved to maintain traceability and ensure that the impact of changes is fully evaluated.

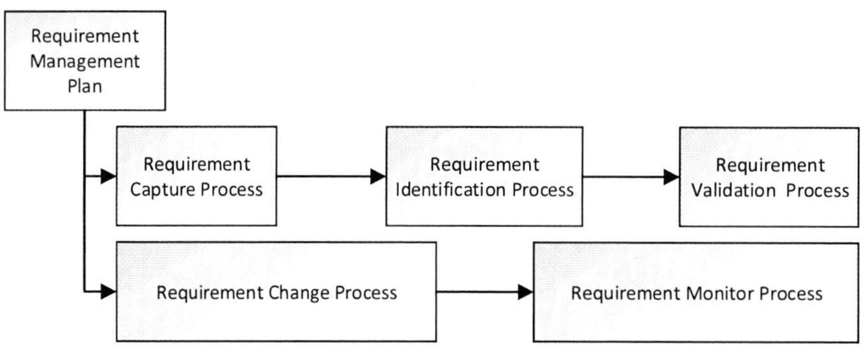

Figure 2. Requirements Management Process

3. The Relationship between Requirements Management and Others

3.1. The Relationship with Configuration Management

After the requirements are baselined, it will be controlled by the change control process. If no change control, the requirements received by the lower level will be out of data. And maybe the receivers got different versions of requirement documents which would cause massive problems. Requirement change control is a part of configuration change control. Requirements management should obey the rules of configuration management.

3.2. The Relationship with Other System Engineering Activity

The specific generation process of requirements is as follows: first, identify and capture the stakeholder requirements correctly, once the stakeholder requirements are confirmed, it is necessary to establish a requirement baseline; after the initial determination of the requirement baseline, carry out functional analysis and establish a functional architecture. The functional architecture is established by iterating in the process of functional decomposition and allocation. Then the requirement analysis shall define how to allocate the requirements. The relationship between requirements management and other system engineering activity, like function analysis and requirement analysis is as shown in Figure 3. The requirements management activity is throughout whole life cycle.

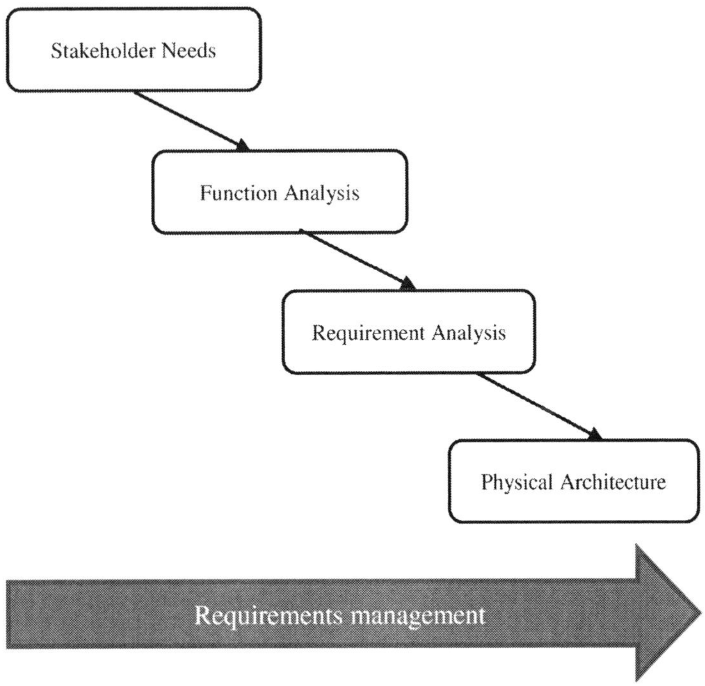

Figure 3. The Relationship with Other System Engineering Activity

3.3. The Relationship with Project Management

For the development of civil aircraft, the main manufacturer supplier mode is used. In the early stage, some suppliers like engine suppler are involved in the requirement specification. And between the developer and suppliers it is a balance between what you need and what you offer. In the critical design phase, if the wrong or incomplete requirements are passed to the suppliers, it would cause the cost rising to change requirements in contract. For the program, it will affect the deadline and risk. So the requirements management will affect the risk management and plan management.

4. Intelligent Integrated Requirements Management Platform and Application

The requirements management platform is the foundation of the parallel collaborative development mode of civil aircraft, which is mainly to manage a series of requirements data effectively and provide decision support for the development process through data analysis. The platform can support stakeholders, main manufacturers and suppliers to provide a unified requirements source, and can provide support to show the design assurance process to airworthiness. In the early stage of product design, customer interaction is realized in the form of communication, website forum, market investigation, etc. Through information acquisition, description, processing and analysis, product functions, performance and customized structure information of customer requirements are obtained[11]. The design, manufacturing, market, project and customer service are combined together to define the top-level requirements and objectives of the product under the support of integrated requirements management platform, taking the customer requirement information as a guide. The top level requirements and objectives will be decomposed layer by layer. In this paper, an integrated requirements management platform is proposed which includes template, attribute and rules database. And the coordination, decomposition tracking, change, monitoring of requirements can be realized based on the integrated requirements management platform which is able to connect with other platforms, such as PDM platform, ERP platform, etc. In order to improve the efficiency and quality of requirements, the intelligent function, like semantic analysis and knowledge database should be included in the platform.

4.1. Template, Attribute and Rules Database

Each requirement has a variety of attributes, including requirement ontology, requirement ID, requirement source, requirement type, requirement allocation, assumption, etc. These attributes are necessary for the complete and accurate description of requirements. In the process of using database to manage different attribute contents of requirements. In the process of using platform to manage requirements, different attributes must be defined to manage different kinds of requirements. There are many kinds of files and attributes, different template must be defined for different files to meet the purpose of recording different levels of requirements and different types of data.

In the attribute database, defining the attributes of all kinds of templates, realize the functions of statistics, modification, modification report of all kinds of template attributes, will greatly improve the efficiency of defining module templates in the requirements management platform. And at the same time, the visualized attribute database can effectively reduce the errors that may be caused by the original working mode.

In order to help designer improve the quality of requirements, and help requirements manager improve management efficiency, the rule database is used to automatically check the link errors, format errors and attribute errors in the integrated management platform. Compared with manually searching for the errors, the rule checking database can automatically check the errors and the error information is clearly displayed to the users, which improves the efficiency of requirements preparation, facilitates the development of requirements validation and verification, and improves the quality of requirements.

4.2. Semantic Analysis

Requirements are generally described in unstructured natural language. However, there are some disadvantages: the nonstandard expression of requirements which may be ambiguous and inconsistent; the relationship between requirements is done by the system engineers manually which takes long worktime and huge workload. These will not only lead to quality problems of requirements, but also seriously affect the design quality of complex system development, which will lead to costly design changes and project delays. The intelligent semantic analysis can help to improve the language expression quality of requirements and can help do the linking analysis.

As a large number of products, enterprises, industries and other data will be involved in the development process of civil aircraft, and there are many departments and person to update and use the requirements data and related data. It is possible to establish a cloud data canter[12] to improve the management efficiency of data, strengthen the collaborative design between data, and establish a complete design assurance system.

4.3. Knowledge Database

The development of aircraft needs a lot of professional knowledge, and the compilation of good requirements needs many years design experience. The establishment of multi aircraft requirements data knowledge database is conductive to the availability of requirements data, experience inheritance, knowledge analysis, avoidance of existing problems, and proposing better requirements for aircraft design. It can help to strengthen the communication between designers and customers, and form a good communication mechanism. The existing PDM organizes data through single product structures, which can not organize a series of data generated in the process of aircraft development, and can not effectively support knowledge management activities in the whole life cycle. Requirements knowledge database with semantic component can be established which can help overcome the deficiency of knowledge expression ability of database model. From the perspective of business process, knowledge and requirement design work are integrated to gradually form aircraft design process database and knowledge database that can support model development and future sustainable development.

4.4. Application

Based on the requirements management process and the relationship with other activities of requirements management, an integrated requirements management platform is established on a certain type of aircraft, as shown in Figure 4. Requirements capture, function analysis, requirements management, configuration management, project management, etc are effectively combined. On the basis of close combination of advanced methods and management tools, data management is carried out to ensure the only data source of requirements data in the whole life cycle, which plays a certain role in promoting the development of models.

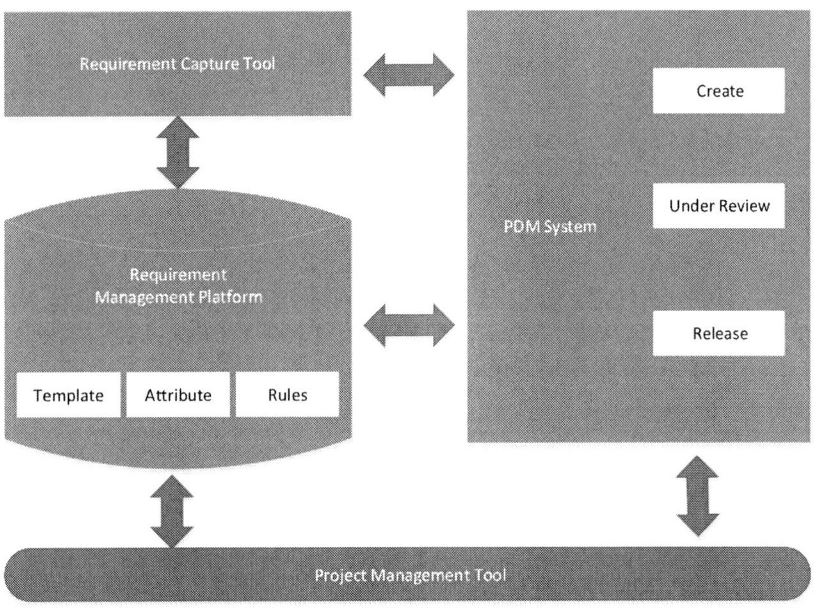

Figure 4. Integrated Requirements management Platform

5. Conclusion

In this paper, the characteristics of requirements for civil aircraft and integrated requirements management process are introduced. Meanwhile, an integrated requirements management platform is applied in civil aircraft to help manage the requirements effectively throughout the whole life.

Acknowledgments

We would like to take this opportunity to thank all of the members of the COMAC requirements management team for their incredibly inspiring work. And also a special thank you to the reviewers whose suggestions greatly improve this work.

References

[1] Jeremy Dick, and Ken Jackson, "Requirements Engineering," Springer, 2017.
[2] "SAE ARP 4754A Guidelines for Development of Civil Aircraft and Systems," Society of Automotive Engineers. December 2010.
[3] He Jing and etc. Some Considerations for Establishing Civil Aircraft Design Assurance System[J]. Advances in Aeronautical Science and Engineering. 2015,6(4)
[4] Yvonne Bijan, Junfang Yu, and Timothy Woods. Systems Requirements Engineering-State of the Methodology[J]. Systems Engineering. Vol 16. No. 3, 2013.
[5] Annabella Loconsole, and so on. Theoretical Validation and Case Study of Requirements Management Measures[R]. 1-21, 2003.
[6] Guo Bozhi, Li Haomin. The Application of Requirements Management in Civil Aircraft Development[J].2013, No4.
[5] Jerry Huller. Proposed Attribute Method: A Flexible Method of Change Control for DOORS. Copyright 2012-2013. Published and used by INCOSE with permission.
[6] Orlena Gotel and Patrick Mader. How to Select a Requirements Management Tool: Initial Steps. 17th IEEE International Requirements Engineering Conferences[C]. 365-367, 2009.
[7] Li Haomin, Zhang Xinai, Tang Chao, Zhan Chao. A Functional Requirements Development and Management Approach Applied to a Civil Aircraft Program[C]. The Proceedings of the 2018 Asia-Pacific International Symposium on Aerospace Technology (APISAT 2018).
[8] ZHOU Yong, LUO Xiaoqi, LI Rongqiang, XIONG Qingyue . Technology of lifecycle oriented aircraft product data architecture[J]. Acta Aeronautica et Astronautica Sinica,Jan.25 2016 Vol.37 No.1 324-334
[9] HE D F, ZHAO Y R, QIAN Z Y, et al. COMAC system enginering manual[M]. Shanghai: Shanghai Jiao Tong University Pres, 2017:260-276 (in Chinese).
[10] LI Jianjun, HE Weiping, YANG Haicheng, et al. Information model management study of product structure facing ob- ject [J]. Chinese Journal of Mechanical Engineering, 2001, 37 (1) : 83-87 (in Chinese) .
[11] WU G H, CHEN Z K, TIAN Y L. Architecture of Information Support Systems for Development of Large Airliners[J]. Acta Aeronautica et Astronautica Sinica, 2019, 40(1): 522736(in Chinese). Doi: 10.7527/S1000-6893. 2018. 22736.
[12] Kang Wenwen, Li haomin and so on. Cloud Knowledge Platform for Complex System Design[J]. Systems Engineering and Electronics. Vol.39, No.5, 2017.

Experimental Study on the Attitude of Civil Aircraft Ditching with High Horizontal Tail Configuration

Xu Nanbo[1,*], Wang Yuping[2], Fu Lin[3] and Zhou Qing[1]

[1]Flight Physics Department, Shanghai Aircraft Design and Research Institute
[2]Airworthiness Engineering Center, Shanghai Aircraft Design and Research Institute
[3]Requirements Engineering Department, Shanghai Aircraft Design and Research Institute

[*]xunanbo@comac.cc

Abstract. The airplane must meet the airworthiness requirements if certification with ditching provisions is requested. The probable behavior of the airplane ditching must be investigated by model tests or by comparison with airplanes of similar configuration. In this paper, aircraft with high horizontal tail configuration is considered as a research object, then the attitudes of civil aircraft in ditching is studied through a scaled model verification experiment. On this basis, this paper researches the effects of different parameters after ditching into the water, such as initial pitch angle, flap slat configuration and sinking speed. The research results in this paper can help to determine the ditching attitudes and parameters and thus develop strategies for ditching.

Keyword: Civil aircraft, Airworthiness, Ditching, Pitch attitude

1. Introduction

According to the requirements of CCAR25.801 (c) [1], this paper studies on the possible motion and states of the civil aircraft with high horizontal tail after entering into the water through the scaled model verification experiment. There have been many scaled model verification experiments on ditching[2][3]. Meanwhile, the influence factors on the attitude of the aircraft after entering into the water are studied in this paper, such as initial pitch angle, flap deflection and sinking speed. The experiment results can be used as a reference for the ditching procedure in the flight manual.

2. Experimental Equipment

The experiment of ditching was carried out in the high-speed towing tank laboratory. The experimental equipment includes high-speed trailer, delivery bracket, electromagnetic launcher, vertical motion mechanism, motion control system, camera system, gyroscope and collector installed in the model.

The aircraft model is fixed on the trailer through a bracket. During the experiment, the trailer accelerates to a preset speed and then the model is delivered after the speed is stable. The schematic diagram of the experimental equipment is shown in Figure 1.

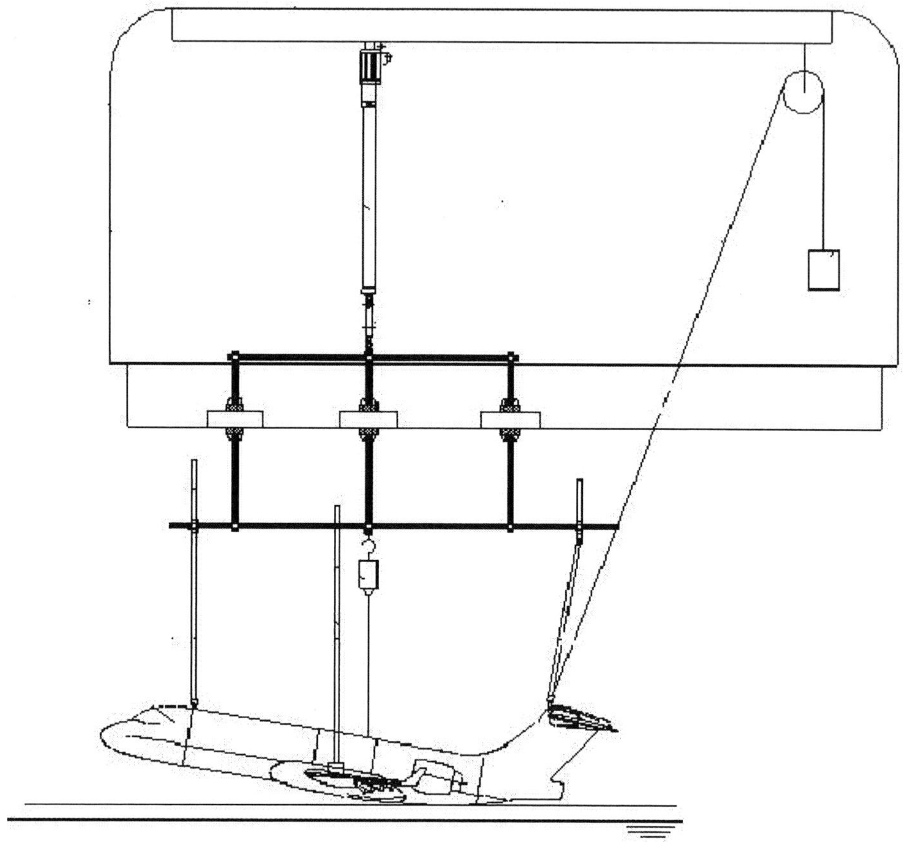

Figure 1. Diagram of Experimental Equipment

3. Experimental Model

In the process of aircraft ditching, the motion of aircraft are mainly controlled by gravity (mass force), and the influence of viscous force is secondary. Therefore, the model experiment is organized according to the Froude similarity (Fr) criterion.

Therefore, the design, manufacture and test of the experiment model meet the following similar conditions:

1) Geometric similarity:

The corresponding length of the model and the real aircraft is in the same proportion, and the corresponding angle is equal;

2) Motion similarity:

In the geometric similarity system, the velocity direction is the same and the size is in the same proportion at the corresponding instantaneous point. In the flow around the object, the streamline is similar and the acceleration distribution is similar.

3) Dynamic similarity:

In the motion similarity system, the direction of the force at the corresponding instantaneous point is the same, and the magnitude has the same proportion. If the forces are decomposed into inertial forces, mass forces, surface pressure forces and surface viscous forces, the polygonal geometry of the forces on the corresponding points of the two systems is similar.

The experiment model is composed of fuselage, wing, horizontal tail, vertical tail, nacelle and other parts, in which the flap deflection, horizontal tail and elevator deflection are adjustable.

4. Selection of Test Conditions

In order to obtain the most likely attitude when the aircraft lands into water and analyze the influence of the aircraft's attitude, the following tests are carried out :

a) The initial pitching attitude angles are set to 7°, 9°, 12°;
b) The slats are all extended and flaps are set to 15°, 25°, 41.5°;
c) The sinking speed of the aircraft model is set to 0.3m/s, 0.6m/s and 0.95m/s.

5. Attitude Analysis of ditching into water

After the ditching, the typical pitching attitude of the aircraft model is shown in Figure 2. The experiment results show that the aircraft model will produce a moment of lift up after entering into the water, which makes the aircraft lift up obviously, thus producing a large attitude angle, which is caused by a suction force on the lower surface of the aircraft tail.

However, in a NACA research memorandum [4], the author did not find this phenomenon when the Boeing 707 model ditched, and the configuration of the 707's fuselage tail is similar to this aircraft. We believe that this is due to the configuration of the horizontal tail. The horizontal tail configuration of the Boeing 707 is conventional. When the aircraft enters into the water at a certain angle of attack, the horizontal tail will also enter into the water. At this time, the horizontal tail will produce a lift which is acted by the hydrodynamic force. This lift will provide the pitch down moment, so as to offset the lift up moment caused by the suction of the lower surface of the fuselage tail. In this experiment, the aircraft model adopts T-shaped horizontal tail. So when the fuselage tail enters into the water, the horizontal tail has not entered into the water. At this time, the influence of hydrodynamic force is far more than the influence of aerodynamic force. Therefore, the suction force of the lower surface of the fuselage tail causes the aircraft to lift up. In article [5], which researches the influence of the rear fuselage's shape on the ditching, the aircraft model similar to this model (fuselage structure and T-tail) also show obvious lift-up after entering into the water.

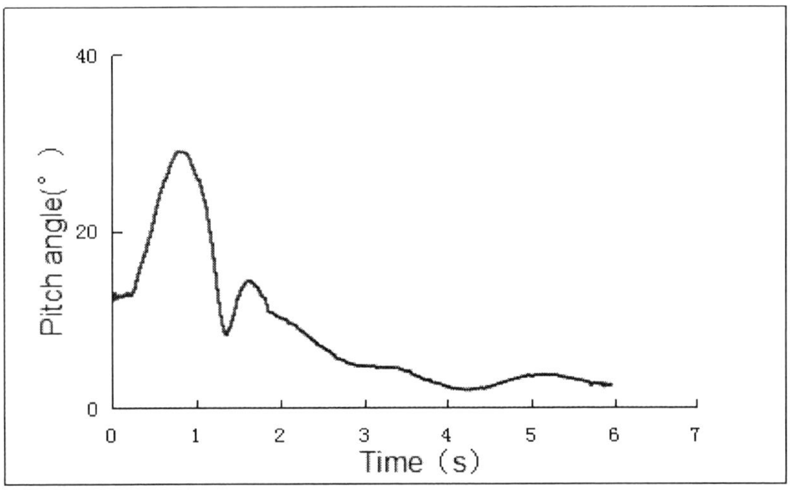

Figure 2. Typical Pitching Attitude after Entering into the Water

6. Analysis of The Influence of Different Parameters on The Attitude after Ditching

6.1. The influence of initial pitch angle on attitude after ditching

The change of aircraft attitude angle under three different initial attitude angles with 41.5°flap deflection and 0.3m/s sinking speed is shown in Figure 3. It can be seen that the aircraft model can reach to the maximum pitch angle after entering into the water at 12°pitch angle. And when the model enter into the water at 7°and 9°pitch angle, the pitch angle is similar after a little time. If considering the increase of pitch angle, the increase of pitch angle at 7°pitch angle is the largest, and the increase of pitch angle at 9°and 12°pitch angle is similar. We think that this is because the smaller the pitch angle is, the more obvious the raise of tail's lower surface is relative to the water flow, resulting in greater suction force.

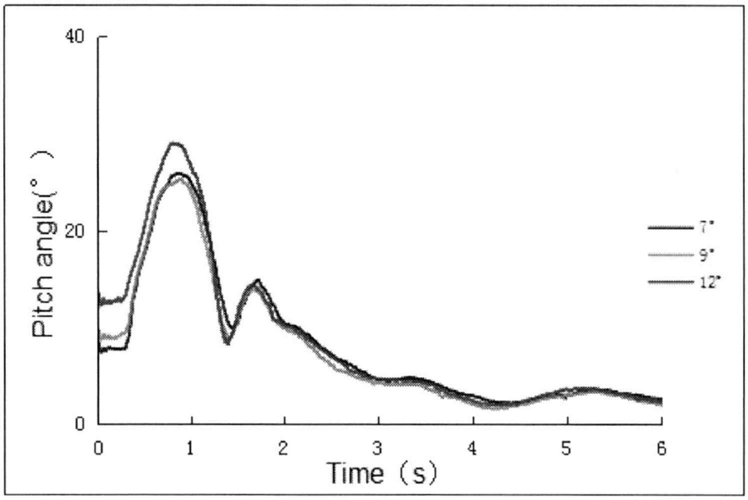

Figure 3. The Pitch Angle of Aircraft in Different Initial Attitude States

6.2. The influence of different flap deflections on the attitude after ditching

Figure 4 shows the change of aircraft pitch angle under three different flap deflections at the initial 12°pitch angle and 0.6m/s sinking speed. It can be seen that the flap deflection has no effect on the aircraft motion attitude during the whole process of nose up after the aircraft enters into the water. This is because the suction force generated by the hydrodynamic force at the tail of the fuselage is far greater than the aerodynamic force acting on the flaps. In the later stage of the whole movement, the attitude of the model in 25°and 41.5°flap deflection are somewhat different from 15°flap deflection . It is also because when the pitch angle of aircraft model becomes smaller, the flaps of 25°and 41.5°have entered into the water and are affected by the hydrodynamic force while the flaps of 15°have not been affected by the hydrodynamic force. It results in the difference of later movement attitudes.

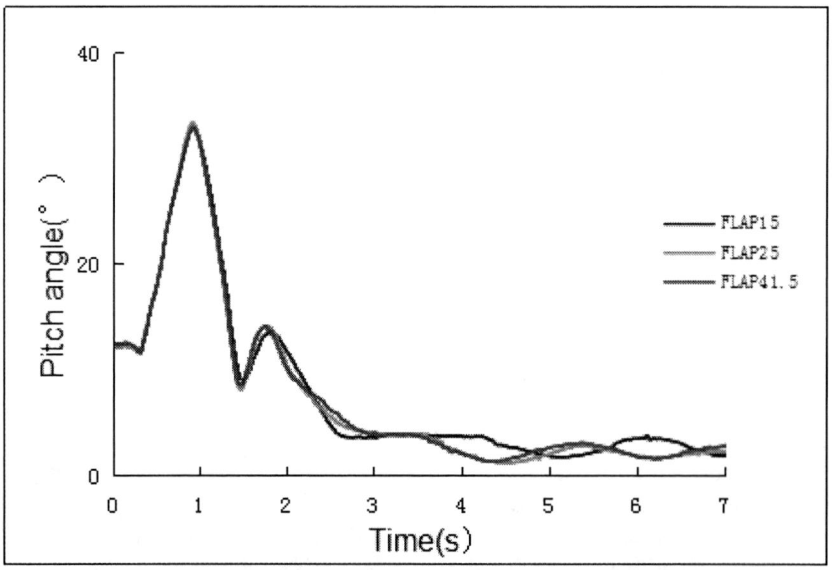

Figure 4. The Pitch Angle of Aircraft under Different Flap Deflection

6.3. Influence of different sinking speeds on attitude after ditching

The change of aircraft pitch angle under three different flap deflections at 12°initial pitch angle and 15°flap deflection is shown in Figure 5. It can be seen that the maximum pitch angle that the model can reach gradually decreases with the increase of the sinking speed. We think that under the condition of a certain horizontal forward speed, the greater the sinking speed means that the angle of attack of the model is larger relative to the water flow when entering into the water, and the suction force generated by the tail part of the fuselage is smaller.

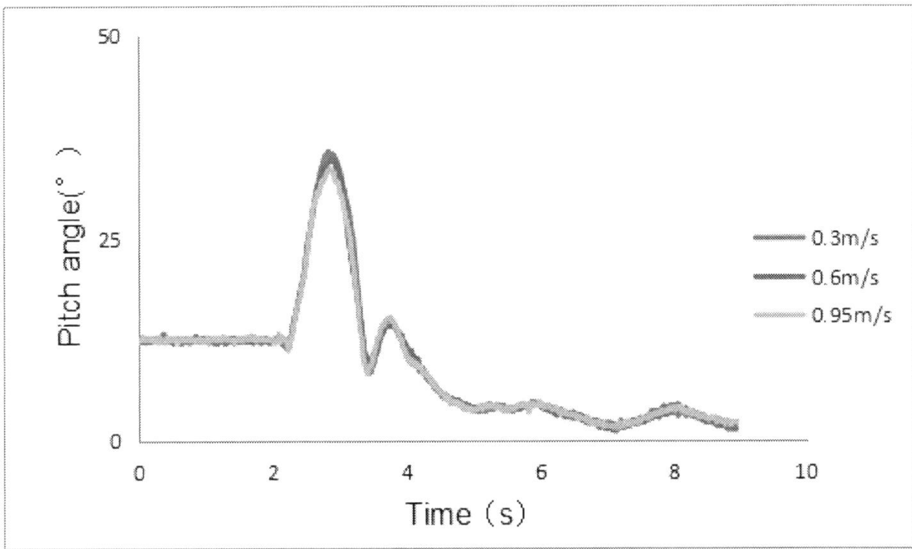

Figure 5. The Pitch Angle of Aircraft under Different Sinking Speeds

7. Conclusion

This paper analyses the attitude of a typical high horizontal tail aircraft after ditching into the water through the verification experiment with a scaled model. On this basis, this paper researches the effects of different parameters after ditching into the water, such as initial pitch angle, flap slat configuration and sinking speed. The conclusions are as follows:

a) When the aircraft model with high horizontal tail is ditched into the water, it will produce an obvious nose up moment due to the influence of suction force in the tail;

b) A smaller attitude angle when entering into the water results in a larger increment of pitch angle;

c) The configuration of flap and slat has little effect on the attitude of the aircraft model after entering into the water;

d) The smaller the sinking speed is, the larger the maximum pitch angle the model can reach after entering into the water.

References

[1] Civil Aviation Regulation of China (CCAR), Part 25-Airworthiness Standards: Transport Category Airplanes. R4.2011

[2] L.J.Fisher,G.J.Morris. Ditching tests of a 1/18 scale model of the Lockheed Constellation airplane,NACA RM L8K18,1950

[3] W.C.Thompson. Rouch water ditching investigation of a model of a jet transport with the landing gear extended and with various ditching aids. NASA TN D-101, 1959

[4] W.C.Thompson. Model ditching Investigation of the Boeing 707 jet transport,NACA RM 55K08, 1956

[5] Ellis E.Mcbride and Lloyd J.Fisher.An experimental study on the influence of the shape of the rear fuselage on the performance of ditching.*Translation of ditching*.1991: 193-223.

IOP Conf. Series: Materials Science and Engineering **751** (2020) 012036 doi:10.1088/1757-899X/751/1/012036

Comparison of Three Compensation Methods for the Touch-trigger Probe Pretravel Errors

Simi Li[1], Long Zeng[2,*], Pingfa Feng[1,2] and Dingwen Yu[1]

[1]Department of Mechanical Engineering, Tsinghua University, 100084 Beijing, P.R.China

[2]International Graduate School at Shenzhen, Tsinghua University, 518055 Shenzhen, P.R.China

[*]Corresponding author's E-mail address: zenglong@sz.tsinghua.edu.cn

Abstract. On-machine inspection (OMI) system has been used extensively for automatic setting of the workpiece and determination of kinematic error in machine tool. There is a necessity not only for knowledge of touch-trigger probing error characteristic, but also for the compensation method of probe error because probing error is an important component of system errors. The comparison of compensation accuracy of three compensation methods is presented. To compensate the touch trigger probe pre-travel errors the calibration experiments with a reference sphere are conducted. Subsequently, the calibration data are used in three compensation methods that the bilinear interpolation method, the bicubic Coons pitch interpolation method and the neural network method. Finally, the validate experiments are conducted on the OMI system, and the compensation results show that the bicubic Coons pitch interpolation method is the most accuracy compensation method.

1. Introduction

On-machine inspection (OMI) system mounts a touch-trigger probe on a CNC machine's spindle. It can inspect part quality during or after manufacturing processes without moving it, avoiding the second-clamping error [1]. It is widely applied in automatic workpiece set-up and determination of machine tool's kinematic errors. The error of touch-trigger probe usually exceeds 10 μm, larger than machine tool's geometric error. Its accuracy is usually compensated by a calibration process.

Current probing error compensation methods mainly based on two models: theoretical probe-model and data-driven based model. The formers usually develop a mechanical model of the used probe, considering various factors of calibration process, such as the measurement direction, the measurement velocity, the stylus length, gravity and so on. One classical theoretical model analyzed a touch-trigger probe's mechanical structure. It considered the elastic bending of the stylus w.r.t. triggering directions. This error is called probe pre-travel variation. This theoretical model is generalized in ref.[2], by considering the influence of friction and gravity. The Hertz deflections of the probe and the workpiece surface were considered in ref. [3]. However, Using the Lavenberg-Marquardt method[4] to identify the undetermined parameters of the mechanical model of the probe is complex. The data-driven based compensation models directly build an error map and it mainly contains two parts. The first part is the discrete error map of all pre-travel errors in calibrated directions, while the second part is a computational method (e.g., interpolation) to estimate the error of un-calibrated directions. The most widely used computational method is interpolation. Others are neural network[5], fuzzy logic[6] and Monte Carlo simulation[7]. During the inspection process, the compensation error for an arbitrary measurement directions can be easily obtained by query the error map. The interpolation method for probe error map includes two categories: the global interpolation method and the local interpolation method. The global interpolation method fits calibrated directions' pre-travels as a 3D surface that the pre-travels are increasing with polar angle decreasing and pre-travels in one latitude form a triangle as mentioned in ref.[8]. The local interpolation method, local points are selected to construct pitches, and a large number of pitches are combined to form a whole surface. For example Ref. [9] applied the bilinear interpolation method to build the probe error map. The piecewise surface from bilinear interpolation method is not accurate enough. First, the patch boundary is sharp. Second, the errors of the un-calibrated directions are lost since they are a bi-linear interpolation of known calibrated-direction errors. In this paper, the comparison of three compensation method for touch trigger probe pre-travel errors is proposed. First, the calibration method of the reference sphere by the OMI system is presented. Second, three compensation method that include the bilinear interpolation method, the bicubic Coons pitch (BCP) interpolation method and the neural network method are applied to compensate the pre-travel errors. To compare the accuracy of three compensation method, verification experiments are conducted in the measurement direction different from the calibration process, and the experiement results show that the BCP method is the most accuracy method.

2. Calibration of pre-travel errors

2.1. Definition of pre-travel error

The pre-travel error is the distance beween and ,which represents the probe characteristic (as shown in figure 1). Pre-travel error is affected by many factors, including the probe structure, the measurement direction, the measurement speed, the triggering force, length and stiffness of the probe system, time delay between probe signal and machine position sensor reading, etc. However, without changing the measuring environment, the repeatability of the probe is typically within 1.0 μm at the stylus tip. Therefore, the repeatability of compensation probe error with 3D error map is typically within 1.0 μm. Establishing 3D error map generally includes two steps: 1) calculating probe error of calibrated point (or direction), 2) calculating probe error of uncalibrated point (or direction) by the interpolation method.

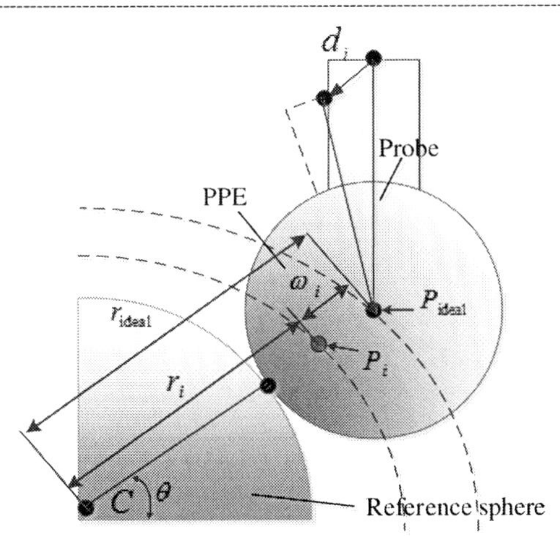

Figure 1. Probe error and pre-travel formation process

2.2. Calibration of pre-travel errors

According to the limited direction of the calibration ball, the error map can be established by modeling or numerical analysis. It is calibrated but generally distributed uniformly in the upper hemisphere of the calibration ball, as shown in figure 2.

IOP Conf. Series: Materials Science and Engineering **751** (2020) 012036 doi:10.1088/1757-899X/751/1/012036

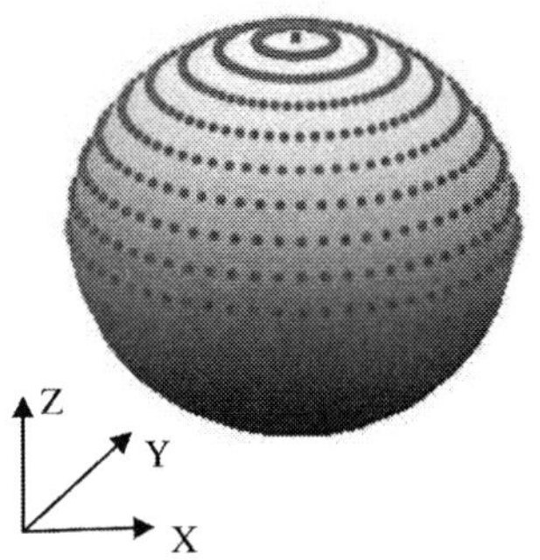

Figure 2. Probe error and pre-travel formation process

It is very important to accurately fit the centre of the sphere for calculating the pre-travel in different directions. The calibration data contain anisotropic pre-travel errors, therefore a calibration data processing strategy is purposed to calculate the accurate centre of the sphere. The pre-travel errors with different azimuth angles present triangular trend, while the pre-travel errors increase with the polar angle increase. The contact force acts on the supporting structure of the measuring rod. When the moment generated exceeds the moment generated by the spring force, the supporting structure of the measuring rod will deflect. Trigger signal is generated when the deflect angle reaches the threshold value. Therefore, the pre-travel error value is positive proportion to the arm of trigger force, and it is negative proportion to the trigger force.

As shown in figure 3, a reference sphere is mounted on the force sensor, the probe measure the reference sphere from 33 directions, the measurement points uniform distribution in the upper half of the reference sphere. The measuring points are distributed in three latitudes 30 degrees apart and the polar, while 8 points in each latitudes. The measurement results show that the deviation of points in the same latitude is much less than that of between different latitudes. And the trigger force at the polar is ten times of it at the equator. Therefore, all the calibration points except the poles are symmetrical in the XY plane. The reference sphere centre in the XY plane could be calculated by the lease square method. The location on Z axis could be calculated by assume the pre-travel at this point is zero.

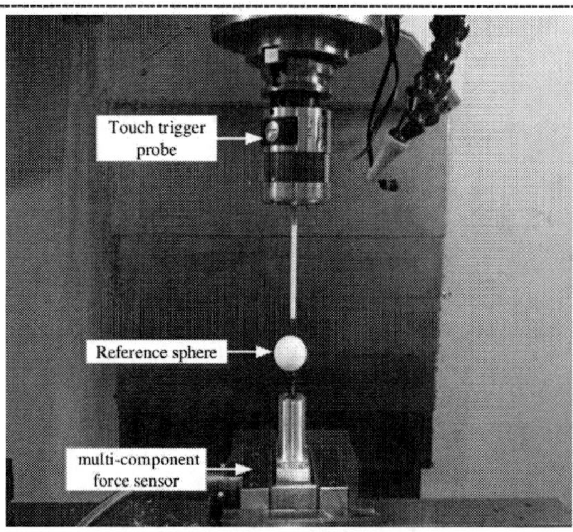

Figure 3. Measurement of trigger force in different directions

3. Compensation of pre-travel errors

The preset travel error is a vector that changes with the direction of measurement. The direction of the vector is the direction of measurement and the modulus of the vector is the pre-travel error value. All calibration data contains a distributed error map. Three compensation method are used to the same calibration data to build the continuous 3D error map. Subsequently, the residuals of error compensation of validate experiments that measuring the reference sphere from directions different from the calibration data.

First compensation method is bilinear interpolation method. As shown in figure 4, the error vector of each direction can be expressed as $Q(\theta_i,\varphi_i)$, where i and j are the indexes of the polar angle θ and azimuth angle φ of the probe, respectively. When the measuring point $Q(\theta,\varphi)$is on a surface consisting of four points $Q(\theta i,\varphi j)$, $Q(\theta i,\varphi j+1)$,$Q(\theta i+1,\varphi j)$, $Q(\theta i+1,\varphi j+1)$.The probe pre-travel error of the point $Q(\theta,\varphi)$ could be calculated by four corner points, the equation is[5]

$$\begin{aligned} Q(\theta,\varphi) &= (Q(\theta_{i+1},\varphi_j) - Q(\theta_i,\varphi_j))\theta + (Q(\theta_i,\varphi_{j+1}) - Q(\theta_i,\varphi_j))\varphi \\ &+ Q(\theta_{i+1},\varphi_{j+1}) + Q(\theta_i,\varphi_j) - Q(\theta_{i+1},\varphi_j) - Q(\theta_i,\varphi_{j+1}))\theta\varphi + Q(\theta_i,\varphi_j) \end{aligned} \quad (1)$$

Bilinear interpolation method is C^0 continuously.

Second interpolation method is bicubic Coons pitch (BCP) interpolation method. As shown in figure 5, assuming that the error vector of the uncalibrated point is $Q(\theta,\varphi)$. It can be calculated as follows:

$$Q(\theta,\varphi) = \begin{bmatrix} F_0(\theta) \\ G_0(\theta) \\ G_1(\theta) \\ F_1(\theta) \end{bmatrix}^T \begin{bmatrix} Q(\theta_i,\varphi_j) & Q^{\varphi}(\theta_i,\varphi_j) & Q^{\varphi}(\theta_i,\varphi_{j+1}) & Q(\theta_i,\varphi_{j+1}) \\ Q^{\theta}(\theta_i,\varphi_j) & Q^{\theta\varphi}(\theta_i,\varphi_j) & Q^{\theta\varphi}(\theta_i,\varphi_{j+1}) & Q^{\theta}(\theta_i,\varphi_{j+1}) \\ Q^{\theta}(\theta_{i+1},\varphi_j) & Q^{\theta\varphi}(\theta_{i+1},\varphi_j) & Q^{\theta\varphi}(\theta_{i+1},\varphi_{j+1}) & Q^{\theta}(\theta_{i+1},\varphi_{j+1}) \\ Q(\theta_{i+1},\varphi_j) & Q^{\varphi}(\theta_{i+1},\varphi_j) & Q^{\varphi}(\theta_{i+1},\varphi_{j+1}) & Q(\theta_{i+1},\varphi_{j+1}) \end{bmatrix} \begin{bmatrix} F_0(\varphi) \\ G_0(\varphi) \\ G_1(\varphi) \\ F_1(\varphi) \end{bmatrix} \quad (2)$$

where the specific calculation method can be referred to our previous research[10].

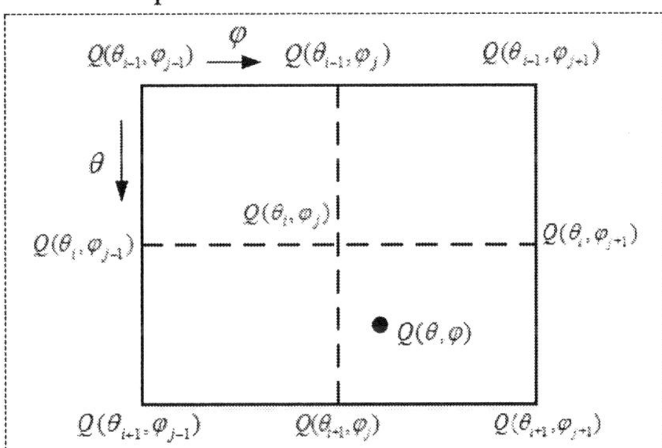

Figure 4. Mesh generation of the bicubic interpolation method

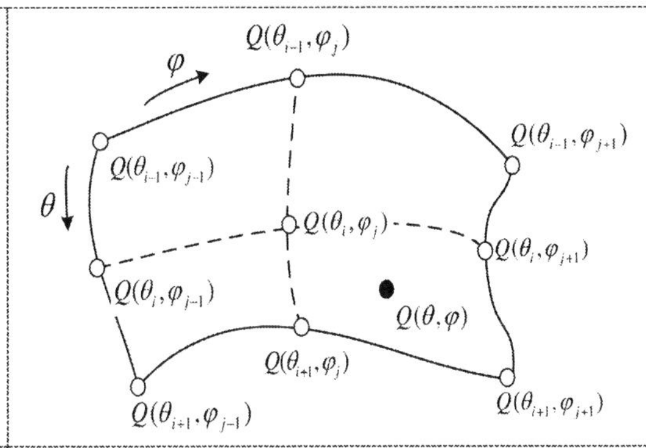

Figure 5. Mesh generation of the bilinear interpolation method

Third, neural network method is used to calculate the pre-travel errors in un-calibrated directions. A two-layer feed-forward network with sigmoid hidden neurons and linear output neurons network is built. The neural network toolbox in Matlab is used. The structure of the neural network and the fitting performance of the neural network is shown in figure 6 and figure 7, respectively.

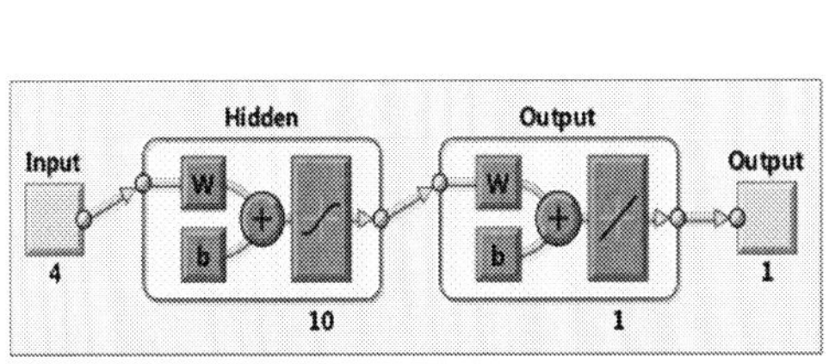

Figure 6. The neural network structure

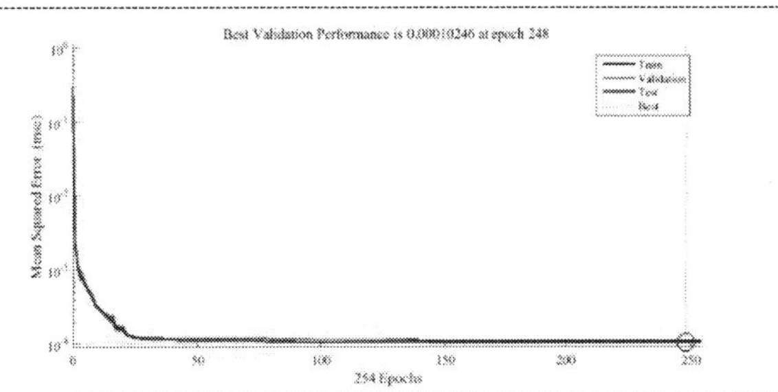

Figure 7. The fitting performance of the neural network

IOP Conf. Series: Materials Science and Engineering **751** (2020) 012036 doi:10.1088/1757-899X/751/1/012036

4. Experimental Verification and discussion

To validate the proposed method, three error maps is built to compare the accuracy of compensation. The experiments are conducted on 541 points of a reference sphere, 121 of them are used to build the error maps and the other data is used to validate the accuracy of compensation.

4.1. OMI experiment setup

A Renishaw strain gauge probe is mounted on the machine tool's spindle to calibrate a reference sphere in 541different directions (as shown in Figure 3).121 of 541 points are selected to build the compensation model in three different methods. Subsequently, the other calibration data are used to validate the accuracy of models.

4.2. Experimental Verification and discussion

To compare the accuracy of three compensation methods, 121 calibration data are used to build the error map for the probe pre-travel errors. The trend of the average residual of the three compensation methods varying with the polar angle is shown in figure 3. The average residual after compensation by BCP and neural network method is increasing with polar angle increasing and that of bilinear method is fluctuating with polar angle increasing. As shown in Table 1, the overall average absolute residual of BCP method is smallest. As shown in figure 8, the deviation of residual after all three compensation method are increasing with the polar angle increasing, which is because the uncertainty of measurement results for large polar angle points are increasing. Compare three method (as shown in figure 9), the deviation of residuals compensation after BCP method is slightly smaller than neural network method. Therefore, the accuracy of BCP method is better than other two methods.

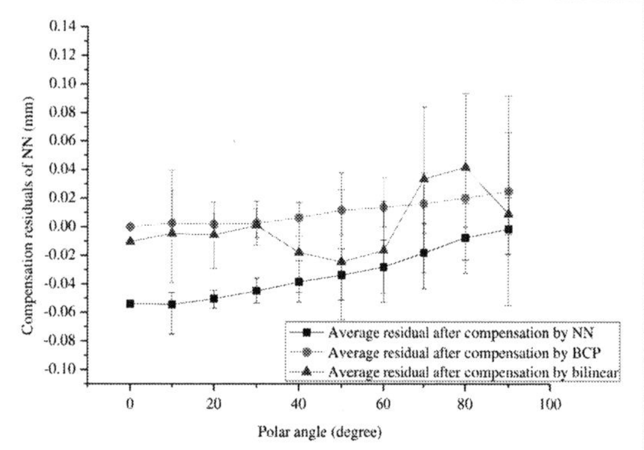

Figure 8. Compare of average residual after compensation by three methods

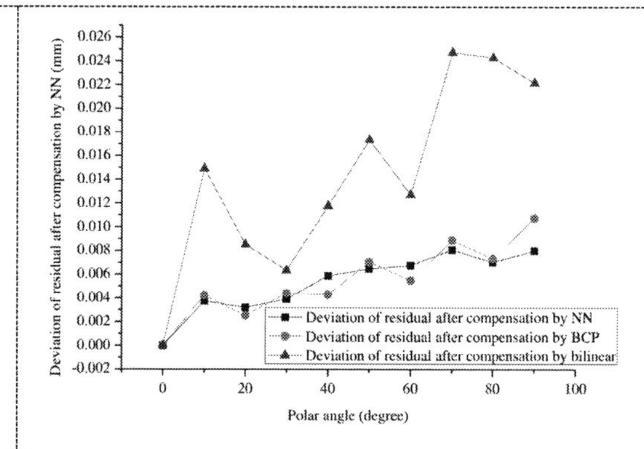

Figure 9. Compare of deviation of residual after compensation by three methods

Table 1. Formatting sections, subsections and subsubsections.

Compensation method	Neural network	BCP method	Bilinear method
Average absolute residual(μm)	33.72	11.58	19.68
Average residual deviation(μm)	19.01	10.30	26.22

5. Conclusions

This paper compares three error compensation method based on the 3D error map for an OMI system mounted with a strain gauge probe. First, the calibration experiments was conducted by measuring a reference sphere in 121 different directions. Second, three different compensation methods were applied to the same calibration data. Final, the validate experiments were conducted by measuring the same reference sphere in 541 different directions. Comparing with the experiment data the model compensation residuals are computed. No matter comparing the average absolute residual or the average residual deviation BCP method is better than another two. The reason why the uncertainty of the probe is large when the polar angle of measurement direction is large needs further study.

Acknowledgments
This paper is supported by the National Natural Science Foundation of China (Grant No. 51705281, Grant No. 61972220), Beijing Science and Technology Program (Grant No. D131100002713003), and National Defence Basic Scientific Research Program JCKY2018208B014.

References

[1] Guiassa R., Mayer J.R.R., St-Jacques P., and Engin S., Calibration of the cutting process and compensation of the compliance error by using on-machine probing. The International Journal of Advanced Manufacturing Technology, 2014. **78**(5-8): p. 1043-1051.
[2] Estler W.T., Phillips S.D., Borchardt B., Hopp T., Witzgall C., Levenson M., Eberhardt K., Mcclain M., Shen Y., and Zhang X., Error compensation for CMM touch trigger probes. Precision Engineering-Journal of the American Society for Precision Engineering, 1996. **19**(2-3): p. 85-97.
[3] Woźniak A. and Dobosz M., Metrological feasibilities of CMM touch trigger probes. Part I: 3D theoretical model of probe pretravel. Measurement, 2003. **34**(4): p. 273-286.
[4] Wozniak A., Simple method of 3d error compensation of triggering probes on coordinate measuring machine. Metrology and measurement systems, 2006. **13**: p. 289-299.
[5] Achiche S. and Wozniak A., Three-dimensional modeling of coordinate measuring machines probing accuracy and settings using fuzzy knowledge bases: Application to TP6 and TP200 triggering probes. Artificial Intelligence for Engineering Design, Analysis and Manufacturing, 2011. **26**(04): p. 425-441.
[6] Achiche S., Wozniak A., and Fan Z., 3D CMM Strain-Gauge Triggering Probe Error Characteristics Modeling Using Fuzzy Logic. Annual Meeting of the North American Fuzzy Information Processing Society, 2008: p. 194-197.
[7] Sładek J. and Gąska A., Evaluation of coordinate measurement uncertainty with use of virtual machine model based on Monte Carlo method. Measurement, 2012. **45**(6): p. 1564-1575.
[8] Mayer J.R.R., Ghazzar A., and Rossy O., 3D characterisation, modelling and compensation of the pre-travel of a kinematic touch trigger probe. Measurement, 1996. **19**: p. 83-94.
[9] Jankowski M., Woźniak A., and Byszewski M., Machine tool probes testing using a moving inner hemispherical master artefact. Precision Engineering, 2014. **38**(2): p. 421-427.
[10] Li S., Zeng L., Feng P., Li Y., Xu C., and Ma Y., Accurate compensation method for probe pre-travel errors in on-machine inspections. The International Journal of Advanced Manufacturing Technology, 2019.

Research on Risk Assessment with Uncontrollable High Thrust for Civil Airplane

Zou Candong* **and Xu Nanbo**

Flight Physics Department, Shanghai Aircraft Design and Research Institute

*zoucandong@comac.cc

Abstract. Not all single failures and probable combinations of failures that can cause UHT malfunction can be eliminated for the civil airplanes, a petition for a partial exemption from FAR25.901(c)/CCAR25.901(c) related to UHT is usually requested. The risks associated with exempting the UHT from FAR25.901(c)/CCAR25.901(c) should not be greater than those currently known and accepted for comparable existing transport category airplanes. Based on the development of a high horizontal tail airplane with 2 rear-mounted engines, this paper proposes a method for UHT malfunction risk assessment from handling qualities aspect by simulation analysis and simulator tests.

Keyword: Uncontrollable high thrust (UHT), Exemption, Simulator test.

1 Introduction

For transport category airplanes, the flight crew normally control engine thrust/power directly by means of a thrust/power/throttle lever or indirectly by means of an auto-throttle. Within traditional engine control systems, there are numerous single and anticipated combinations of failures that will result in losing these normal means of control. A subset of the resulting failure conditions may include actual thrust/power either increasing to significantly higher than commanded and/or remaining high when low thrust/power is commanded (or would have been had the failure not made that problematic, such as in the case of a stuck throttle). This subset of failure conditions is referred to herein as "uncontrollable high engine thrust/power" or "UHT" failure conditions. In other words, if the flight crew are unable to reduce excessive engine thrust/power through the normal means, that is a "UHT" failure condition.

These failures, and the hazards they pose, have long been inherent in transport airplane designs. In fact, the "fail-safe" states for engine controls have traditionally been chosen to protect high thrust capability and allow the flight crew to decide when an engine shut down is appropriate. Compliance with the relevant airworthiness regulations within parts 21, 25 and 39 have traditionally been based in part on accepting an assertion that the flight crew can recognize and safely accommodate these "uncontrollable high thrust (power)" conditions, including shutting down the affected engine via an independent fuel shutoff as required, before such a failure can pose an unacceptable hazard to the airplane. Over the last 20 years, the average rate of occurrence for the uncontrollable high thrust failure condition on turbofan-powered large transport category airplanes has remained, as reported by industry, relatively constant at about one every 2.5 million flight hours. This would indicate that to date an "Uncontrollable high thrust failure condition" has occurred hundreds of times without resulting in a single reported serious injury. Nevertheless, in recent years, several factors have raised the risks and authority concerns associated with these types of failure conditions:

1) In recent years, engineering studies and service experience, including a 1997 Saudi Arabian Airlines Boeing 737-200 accident[1], indicate that the traditional assertion that flight crews will recognize and safely accommodate loss of the normal means to control engine thrust (power) is not always valid.

2) The fact that some trends, both in design and operation, are tending to increase the risks associated with uncontrollable high thrust failures.

There are numerous regulations which may limit allowable UHT failure modes, effects and/or rates. The most limiting regulation is usually FAR25.901(c)[2]/CCAR25.901(c)[3], which states in part that: "no single failure or malfunction or probable combinations of failures will jeopardize the safe operation of the airplane". Certain aircraft manufacturers has been unable to eliminate all single failures and probable combinations of failures that can cause an engine to produce high thrust while not responding to the thrust levers. If the applicant cannot show compliance to 25.901(c), they may request a finding of equivalent safety or a petition for a partial exemption from 25.901(c) related to engine uncontrollable high thrust.

All practicable actions should be taken to minimize the adverse effects on safety associated with requesting this petition. These will include, but are not limited to, practical actions to eliminate or further reduce the risks by improving designs, procedures, training and instructions for continued airworthiness, and the risks associated with the partial exemption are no greater for the proposed type designs than those currently known and accepted for comparable existing transport category airplanes. Acceptable risk for this provision can be characterized as:

1) The airplane complies with all applicable regulations for any foreseeable uncontrollable high thrust failure conditions in flight, except possibly during approach below 400 feet.

2) The expected overall frequency of occurrence of uncontrollable high thrust failure conditions is less than once per ten million airplane operating hours.

The following of this paper will offer an example for confirming whether the aircraft is under acceptable risk level, base on the handling qualities analysis and simulator tests of a high horizontal tail airplane with 2 rear-mounted engines.

2. Risk assessment method

According to the FAA memorandum[4], close attention should be paid to the following potentially Hazardous or Catastrophic single failure scenarios when confirming whether the aircraft is under acceptable risk.

Takeoff Abort Scenario #1: An uncontrollable high thrust failure condition occurs during the takeoff roll between power set and V1; an abort is initiated; thrust on the normally operating engine retards normally while thrust on the failed engine remains high; and the resulting asymmetric thrust causes the airplane to depart the side of the active runway.

Takeoff Abort Scenario #2: Same as Takeoff Abort Scenario #1 except the resulting net excess thrust causes the airplane to overrun the end of the active runway.

Cruise Scenario #1: An uncontrollable high thrust failure condition occurs during some minimum control margin condition and the airplane becomes uncontrollable or suffers critical structural damage.

Go-Around Scenario #1: An uncontrollable high thrust failure condition occurs on approach/flare; the pilot can't hold runway center-line and initiates a go-around; however, the airplane sink-rate cannot be overcome in time and the airplane touches down off to the side of the runway.

Go-Around Scenario #2: Same as Go-Around Scenario #1 except the airplane attitude is such that the wing tip contacts the ground on (or off) the runway.

Go-Around Scenario #3: An uncontrollable high thrust failure condition occurs on final approach; the pilot attempts to compensate with significant rudder input but can not hold runway center-line and initiates a go-around; as the normally operating engine accelerates to go-around thrust the failed engine hits its over-speed trip and cuts back to minimize flow effectively resulting in loss of the failed engine; exceptional pilot skill is required to perform the rudder reversal required to maintain control; the pilot loses control of the airplane and there is insufficient altitude for recovery.

Landing Scenario #1: An uncontrollable high thrust failure condition occurs during final approach; the pilot attempts to hold runway center-line and continues landing and the airplane touches down off to the side of the runway.

Landing Scenario #2: Same as Landing Scenario #1 except wing tip contacts the ground on (or off) the runway.

Landing Scenario #3: An uncontrollable high thrust failure condition occurs during final approach or landing; thrust on the normally operating engine retards normally while thrust on the failed engine remains high; the resulting asymmetric thrust causes the airplane to depart from the active runway.

Landing Scenario #4: An uncontrollable high thrust failure condition occurs during the final approach; thrust on the normally operating engine retards normally while thrust on the failed engine remains high2, the crew compensates for the thrust asymmetry, lands long and eventually shuts down the failed engine; however, the landing performance impacts of landing long and carrying high thrust for the extra time causes the airplane to overrun the end of the active runway.

Landing Scenario #5: An uncontrollable high thrust failure condition occurs during the application of reverse thrust on landing; reverse thrust on the failed engine increases; the flight crew, concluding the lateral control problem is due to deploying the reversers, restows the reversers; the thrust on the normally operating engine retards and the reverser stows as commanded; the thrust on the failed engine remains high and the reverser is incapable of restowing due to the excessive gas path pressures; the resulting asymmetric thrust causes the airplane to depart from the side of the active runway.

The method used in risk assessment for the 11 failure scenarios from handling qualities aspect is shown in Table 1.

Table 1. The risk assessment method and condition

Scenario	Method	Criterion (should not)	Configurations and Conditions
Takeoff Abort #1	Simulator test	Depart from the side of the active runway	Takeoff configuration, after CG, low pressure altitude runway
Takeoff Abort #2	Simulator test	Overrun the end of the active runway	Takeoff configuration, forward CG, high pressure altitude runway
Cruise #1	Simulation calculation	Become uncontrollable or suffers from critical structural damage	Curse configuration, after CG, high altitude/MMO, low altitude/VMO
Go-Around #1	Simulator test	Touch down off to the side of the runway	Landing configuration, after CG, low pressure altitude runway
Go-Around #2	Simulator test	The wing tip contact the ground on (or off) the runway	Landing configuration, after CG, low pressure altitude runway
Go-Around #3	Simulator test	Lose control of the airplane and there is insufficient altitude for recovery	Landing configuration, after CG, low pressure altitude runway
Landing #1	Simulator test	Touch down off to the side of the runway	Landing configuration, after CG, low pressure altitude runway
Landing #2	Simulator test	The wing tip contact the ground on (or off) the runway	Landing configuration, after CG, low pressure altitude runway
Landing #3	Simulator test	Depart from the active runway	Landing configuration, after CG, low pressure altitude runway
Landing #4	Simulator test	Overrun the end of the active runway	Landing configuration, forward CG, high pressure altitude runway
Landing #5	Simulator test	Depart from the side of the active runway	Landing configuration, after CG, low pressure altitude runway

3. Simulation calculation method and results

Simulation models, including the airplane dynamic model and pilot model, should be built up first, and general view of pilot in loop simulation is shown in Figure 1.

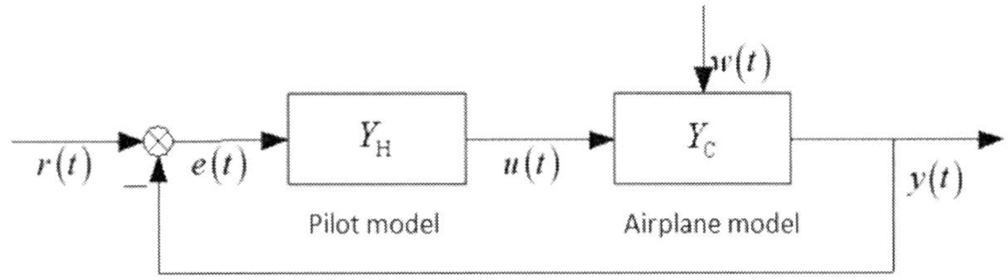

Figure 1. General view of pilot in loop simulation

The liner pilot model is adopted[5], $Y_H(s)=\frac{K}{1+t_n s}e^{-\tau s}$, where K means the gain of pilot, t_n means delay component of the pilot with the time constant of 0.1~0.2s, τ means pilot's reaction delay time that usually is 0.15s. τ is set as 1.5s conservatively considering the total pilot's reaction delay and the control system delay when simulating the airplane responses after UHT failure. The free responses after UHT failure without pilot in loop is also computed.

Thrust on the failed engine remains high as shown in Figure 2, and airplane responses after UHT failure with or without pilot in loop are shown in Figure 3. Without pilot in loop, roll angle of the airplane changes 20 degrees, yaw angle changes 13 degrees while CAS increases 10 knots and altitude descends 400 feet after 30 seconds. But roll angle remains within about 6 degrees, yaw angle changes about 4 degrees, CAS increases 5 knots and altitude descends less than 100 feet when the "pilot" attempts to maintain wing level after UHT failure using the ailerons and rudder, that indicate that the airplane is totally controllable.

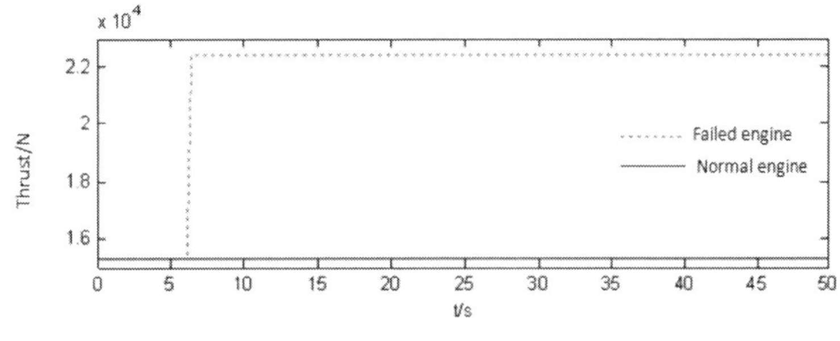

Figure 2. Thrust of the failed engine and normal engine

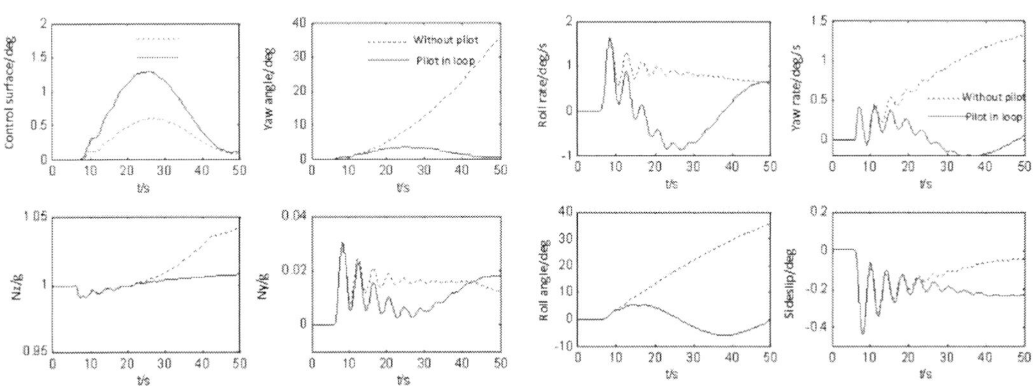

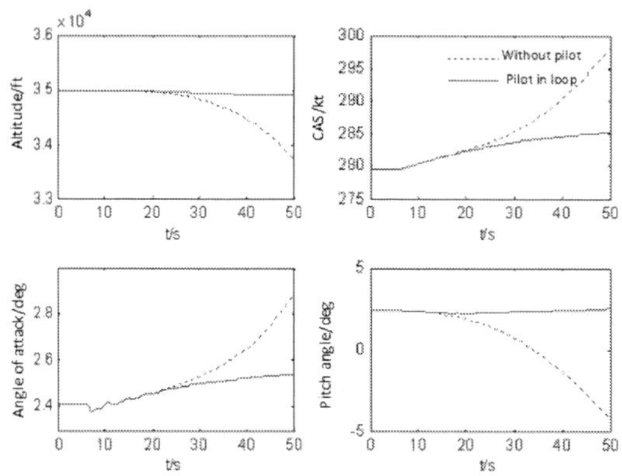

Figure 3. The airplane responses after UHT failure with or without pilot in loop

4. Simulator test method and results

Simulator tests for takeoff, go-around and landing scenarios were conducted by 2 flight crews on the flight simulator approved by the airworthiness authority and with specific UHT malfunction models. Test results consist of the airplane response and movement data and the pilot's comment on the UHT failure have impacts on handling qualities using the Cooper-Harper rating scale.

For Takeoff Abort Scenarios, tests are started with UHT failure occur at V1, and then a lower speed, until Vmcg - 30 knots. Also, 3 thrust control procedures in Takeoff Abort Scenarios are tested:

1) Not use reverse and shut down the affected engine via an independent fuel shutoff
2) Use maximum reverse, then shut down the affected engine via an independent fuel shutoff
3) Shut down the affected engine via an independent fuel shutoff, then use maximum reverse

The maximum lateral departure from the runway center-line is about 50ft and the airplane is controllable in all test points of Takeoff Abort Scenario #1 and Takeoff Abort Scenario #2. The pilots rate C-H 2~4 and recommend the third procedure. Takeoff length increase obviously in Takeoff Abort Scenario #2 , but still within the runway length of the most critical anticipated operation airport.

For Go-Around Scenario #1, tests are started with UHT failure occur at altitude of 400ft above the runway until 100ft. The maximum altitude change during go-around is less than 150ft, all controllable and without touching down off to the side of the runway, but the airplane touches down on the runway when UHT occur at altitude of 100ft. The pilots rate C-H 3~4 and suggest shutting down the affected engine after go-around to the safe altitude, then land with one engine inoperative.

For Go-Around Scenario #2, tests are started with UHT failure occur at altitude of 400ft above the runway until 100ft. The maximum altitude change during go-around is less than 170ft, lateral departure is less than 100ft, and the wing tip do not contact the ground on (or off) the runway, but the airplane touch down on the runway when UHT occur at altitude of 100ft. The pilots rate C-H 2~5 and suggest shutting down the affected engine after go-around to the safe altitude, then land with one engine inoperative.

For Go-Around Scenario #3, tests are started with UHT failure occur at altitude of 400ft above the runway until 100ft. The maximum altitude change during go-around is less than 160ft, all are controllable without wing tip contacting the ground on (or off) the runway. When UHT occurs at 100ft, the airplane succeed the go-around with touching down on the runway. The 1st pilot rate C-H 4~6 and the 2nd pilot rate C-H 7~9, that means exceptional pilot skill and more workload is required to perform the rudder reversal required to maintain control due to the failed engine thrust changes. Result of one test with UHT occurs at 100ft is shown in Figure 4.

Test configuration and condition of Landing Scenario #1, Landing Scenario #2 and Landing Scenario #3 are similar, and it is not so hard for the flight crew to hold runway center-line and continue landing. Tests are started with UHT failure occurs at altitude of 400ft above the runway until 100ft. Both flight crews land the airplane safely for all test points with lateral departure less than 170ft except that UHT occurs at 100ft (the wing tip touch down the ground). The pilots rate C-H 3~5 (9 for the test point UHT occurring at 100ft) and suggest conducting go-around when UHT appear near ground or shutting down the affected engine as soon as possible if continuing landing is necessary.

Landing Scenario #4 test are conducted at high altitude airport, and the compensation for the thrust asymmetry increase the landing length a lot but still within the runway length of the most critical anticipated operation airport. The pilots rate C-H 3~6 and suggest conducting go-around when UHT appears near ground or shutting down the affected engine as soon as possible if continuing landing is necessary, but compensation for the thrust asymmetry is not recommended.

The reverse thrust on the failed engine do not affect the taxing control a lot and the airplane is completely controllable until completely stop without departing the side of the active runway (the maximum lateral departure less than 30ft) in Landing Scenario #5. The pilots rate C-H 2~3 and suggest shutting down the affected engine as soon as possible and using the reverse thrust of the normal engine normally.

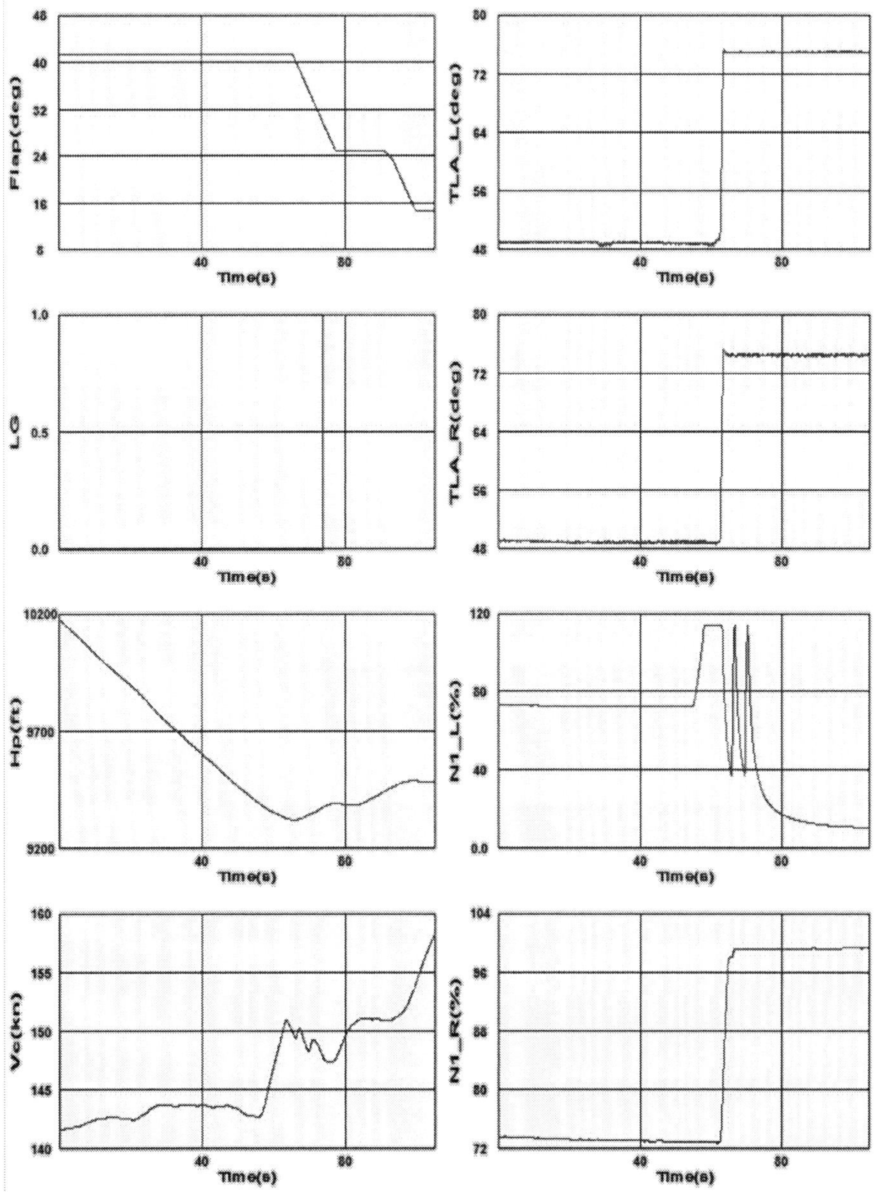

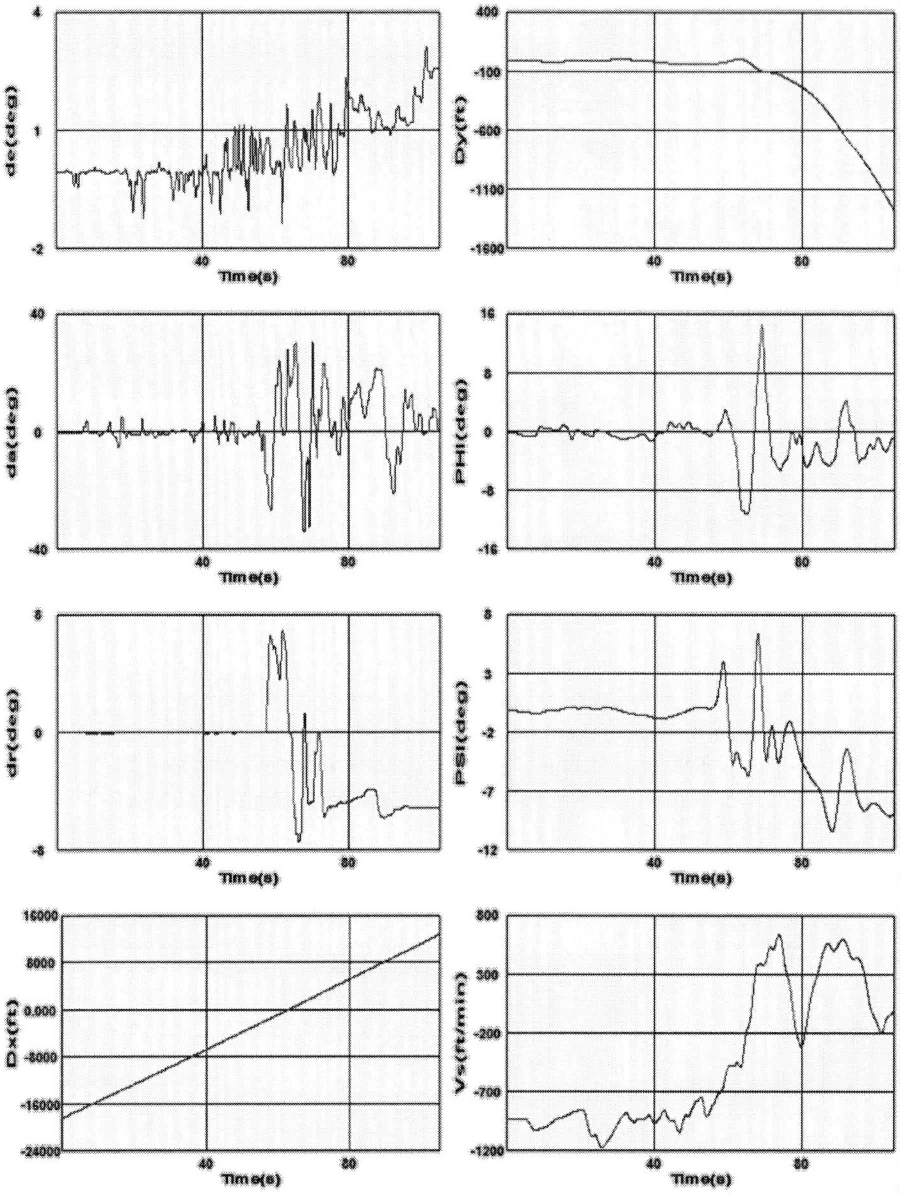

Figure 4. Test result of Go-Around Scenario #3 with UHT occur at 100ft
(Note: Flap: Flap angle, LG: Landing gear, Hp: Press altitude, TLA_L: Left throttle lever angle, TLA_R: Right throttle lever angle, N1_L: N1 speed of the left engine, N1_R: N1 speed of the right engine, de: Column angle, da: Wheel angle, dr: Pedal angle, Dx: Longitudinal position of the airplane, Dy: Lateral position of the airplane, PHI: Roll angle, PSI: Yaw angle, Vs: Vertical speed)

5. Conclusion

FAR25.901(c)/CCAR25.901(c) require in part that: "no single failure or malfunction or probable combinations of failures will jeopardize the safe operation of the airplane", but not all single failures and probable combinations of failures that can cause UHT malfunction can be eliminated, a petition for a partial exemption from 25.901(c) related to UHT is usually requested. This paper offers an example for risk evaluation for 11 potentially Hazardous or Catastrophic single failure scenarios specified in the FAA memorandum[4] by analysis and simulator tests, and the results indicating:

1) Cruise Scenario: based on the simulation calculation results, there are sufficient control capacity and altitude margin following the UHT failure at cruise condition;

2) Takeoff Abort Scenario: the maximum lateral departure from the runway center-line is about 50ft and the airplane is controllable without departing from the side of the active runway or overrunning to the end of the active runway during simulator tests.

3) Go-Around Scenario: The airplane go-around successfully for all tests points without touching down off to the side of the runway or without wing tip contacting the ground on (or off) the runway, but the airplane touch down on the runway for some test points that UHT occurs at low altitude. It is more critical for Go-Around Scenario #3 and requesting exceptional pilot skill and more workload.

4) Landing Scenario: Both flight crews land the airplane safely for all test points except one test point with UHT occurring at 100ft (the wing tip touch down the ground), and recommend conducting go-around when UHT appears near ground or shutting down the affected engine as soon as possible if continuing landing is necessary for Landing Scenario 1#, 2#, 3# and 4#. Results of Landing Scenario #5 show that the reverse thrust on the failed engine does not affect the taxing control a lot.

References

[1] NTSB, Safety Recommendation A98-70 [Z] , 1998
[2] FAA, Part 25-Airworthiness Standards: Transport Category Airplanes. R4 [S], 2009
[3] Civil Aviation Regulation of China (CCAR), Part 25-Airworthiness Standards: Transport Category Airplanes. R4 [S], 2011
[4] FAA, Policy Statement on the Thrust Control Malfunction Airworthiness Program - Uncontrollable High Engine Thrust/Power Failures [Z], 2003
[5] Qu Xiangju, Pilot-aircraft system and flight qualities [M], Beijing: Beihang University Press, 1994

IOP Conf. Series: Materials Science and Engineering **751** (2020) 012038 doi:10.1088/1757-899X/751/1/012038

Flight Path Planning Surrogate Model Based on Stacking Ensemble Learning

X Z Yang*, Z X Cui and X Y Qiu

Science and Technology on Avionics Integration Laboratory, 432 Guiping Road, Shanghai, 200233, China

*Email: yang_xizhong@outlook.com

Abstract. This paper proposes a method of establishing flight path planning surrogate model based on stacking ensemble learning, which can solve the real-time problem of complex flight mission's on-line waypoints calculation. Airborne navigation system mainly utilizes several discrete waypoints to guide flight path or flight control. These waypoints are usually calculated by a set of equations based on flight dynamics, flight kinematics and flight mission constraints, and therefore path planning for complex missions cannot guarantee real-time performance. In this paper, flight samples are generated offline by taking flight mission characteristic parameters as input and flight waypoint coordinate series as output. Then two-layer coupling model is constructed based on stacking ensemble learning. A series of base-learners are constructed to learn the quantity of waypoints or each waypoint's coordinate values respectively. At last, flight path planning surrogate model is built by combining all the base-learners, establishing the direct mapping relationship between input and output. The results show that this surrogate model can effectively calculate the aircraft flight waypoints, and meanwhile maintains ideal accuracy and real-time performance.

1. Introduction

Flight path planning is one of the main tasks of a pilot or an airborne computer in aircraft flight activities. In order to complete a specific flight mission, the corresponding flight path should be designed reasonably in advance. Aiming at different flight missions, complex mathematical analytic model of flight path planning is established via a series of simulation, optimization and iteration algorithm, and then generate a set of flight waypoints. Subsequently, the airborne navigation system conducts flight guidance and control based on these waypoints, helping to accomplish the corresponding mission process.

Path planning analytical model is generally established by the geometric optimal search theory. Meanwhile, the aircraft dynamics model, flight kinematics, mission constraints and environmental conditions are also considered. The planning space is expressed as an appropriate searching form, and then the optimal solution could be achieved via optimization algorithm or heuristic algorithm. The optimization algorithm can convert the path planning problems into mathematical programming models, and constraints such as planning time or planning area will significantly affect computational cost. The heuristic algorithm, such as genetic algorithm, particle swarm algorithm and ant colony algorithm, can simulate biological swarm intelligence, and the performance of path planning calculation still depends on the size of the problem domain. Besides, other methods such as artificial potential field method, A* method and neural network method are also in common use [1].

Path planning includes offline planning and online planning. Most of the above methods can support both offline planning and the real-time requirements of online planning when dealing with path planning problems of relatively simple flight missions. However, when the mission process gets relatively complex, the problem domain will expand rapidly, increasing the computational cost and time, even without ideal solution. For instance, in a certain flight path planning issue, it only takes several seconds to calculate 3 to 5 flight waypoints, but it will take at least ten minutes or even several hours to calculate more than 10 flight waypoints.

Aiming at solving the problem, this paper proposes a method based on artificial intelligence algorithm, which combines machine learning from flight sample data offline and invoking the training model on-line. In this way, path planning knowledge is extracted from previous offline samples to establish new online planning model, which would be used to replace the analytical model and calculate flight waypoints in real-time environment.

2. Analysis of flight path planning

2.1. Problem statement

In the research of this paper, there is a following kind of flight path planning problem. The flight mission's information which affects the flight path can be simplified and extracted as four parameters c_1, c_2, c_3 and c_4. Specifically, c_1, c_2 and c_3 are positive rational values, c_4 is a positive integer. The result of path planning is a set of waypoints' coordinate values, including longitude and latitude in earth coordinate system. The variables are defined as follows:

- $\boldsymbol{x}$ - input parameter vector of flight path planning model, $\boldsymbol{x}=[c_1,c_2,c_3,c_4]$
- $\boldsymbol{y}$ - output parameter vector of flight path planning model, which represents a set of waypoint coordinate values, $\boldsymbol{y}=\Psi(\boldsymbol{x})=[a_1,b_1,a_2,b_2,\cdots a_k,b_k]$
- k - quantity of waypoint, counted from $\boldsymbol{y}$
- a_i,b_i - longitude and latitude values of each waypoint, $i=1,2,3,\cdots,k$
- $\Psi(\cdot)$ - mathematical analytic function of flight path planning, including simulation modelling, mathematical formula and optimization iteration etc.

Since the function $\Psi(\cdot)$ could be particularly complicated, its computing process will get so slow that path planning computation will not meet real-time requirement. One solution is to build a surrogate model $\Phi(\cdot)$ to replace $\Psi(\cdot)$ and improve the calculation speed while maintaining satisfactory accuracy.

2.2 Surrogate model

The surrogate model is a statistical model similar to a black box, whose calculation results are very close to the original model, in order to obtain some better performance (for example, the calculation time is greatly reduced, the calculation costs less resources, etc.). The surrogate model can be designed by a data-driven, bottom-up approach. First, various samples' inputs are sent into the original model for calculating the outputs to set up a sample database. Then, a black-box model implying end-to-end mapping is generated by learning the samples of input-output relationship [2, 3].

Artificial neural network is applicable for learning the relationship between a kind of input-output mapping of mathematic model, algorithm or function, as well as constructing a classifying learner or a regression learner. However, there are several obstacles when using this method directly to learn the mapping relationship between flight mission input parameters and waypoint coordinate values. The main problem is that the quantity of waypoints in different flight missions is also quite different. Inconsistence of output numbers leads to the difficulty in determining the number of neural network's nodes in each layer.

In this research, the flight mission sample's input is a 4-dimention vector and the output is a $2k$-dimention vector. Because the values of k are variable in different samples, it's impossible to train all the samples in one learner. In addition, the large sample size also restricts the generalization capacity at non-sample points.

A method of ensemble learning is introduced here to solve this problem: Taking k as an intermediate eigenvalue of each sample so that all the samples are classified accordingly into different group. Samples of each group are trained separately to generate a waypoint coordinate learner at last, which containing several base-learners. When calculating waypoint coordinate, the correct base-learner is invoked on the eigenvalue of k. And k can also be calculated by a waypoint quantity learner in advance from the input data by a primary learner. The method is shown in Figure 1.

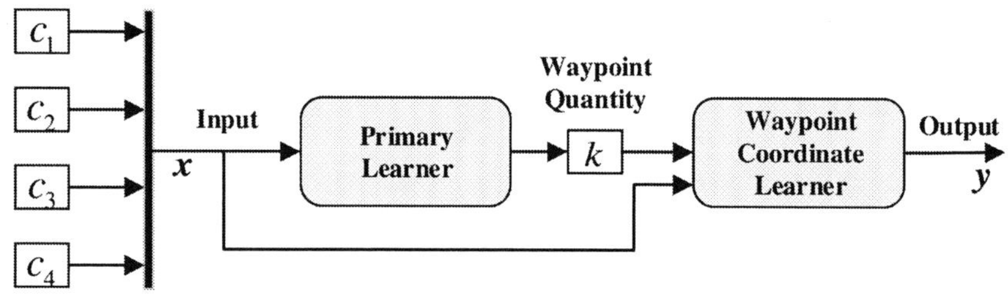

Figure 1. Waypoint calculating model by ensemble learning.

3. Stacking ensemble learning model and algorithm

Ensemble learning is a typical algorithm of machine learning, which generates multiple learners through certain rules and trains them separately, and integrates each learner with appropriate strategies, so as to obtain better performance than a single learner [4, 5, 6]. The theoretical basis of ensemble learning is probably approximately correct (PAC) theory, strong learnable and weak learnable theories. Its representative methods include Bayes optimal classifier, bagging, boosting, stacking, random forest, etc. Among them, stacking method is a technique that integrates multiple base-learners to form an integrated learner.

Stacking method classifies the sample sets through certain rules and constructs different base-learners by machine learning algorithm. Each base-learner can be trained by different algorithm types or structures. Therefore, it's compatible with the differences in structure, superparameters and training strategies while using stacking ensemble learning method for samples intelligent learning [7]. Stacking method can support heterogeneous integration, which is suitable for solving the flight path planning problem of learning inconsistent dimension waypoint samples [8]. After test and evaluation, if the ensemble learner fulfils the proper credibility level, it can be used for predicting output values from new input data.

For a set of sample data containing L groups data, each group is expressed as $\boldsymbol{P}=\{\boldsymbol{x}_i,\boldsymbol{y}_i\}_{i=1}^{L}$, where $\boldsymbol{P}$ represents the sample data set, $\boldsymbol{x}_i$ represents the input data and $\boldsymbol{y}_i$ represents the output data. $\boldsymbol{P}$ is divided into training sample data $\boldsymbol{D}=\{\boldsymbol{x}_i,\boldsymbol{y}_i\}_{i=1}^{M}$ and testing sample data $\boldsymbol{T}=\{\boldsymbol{x}_i,\boldsymbol{y}_i\}_{i=1}^{N}$, where $L=M+N$. The stacking ensemble learning algorithm is deduced as the following process.

(1) All the samples in $\boldsymbol{D}$ are classified according to a certain strategy. Assuming that the samples are divided into q cases, the sample data set of each case is:

$$\boldsymbol{D}_j=\{\boldsymbol{x}_{j,i},\boldsymbol{y}_{j,i}\}_{i=1}^{M_j}\ ,\ j=1,2,\mathrm{L}\ q \tag{1}$$

where $\boldsymbol{D}_j$ represents case j, M_j represents the quantity of samples in case j, and $\boldsymbol{D}=\sum_{j=1}^{q}\boldsymbol{D}_j$.

(2) Construct a primary learner through pre-labelled data or by unsupervised learning methods, which is used for mining eigenvalue z_i ($i=1,2,\mathrm{L}\ p$) from $\boldsymbol{D}$. The primary learner can be a single base-learner ψ or a group of base-learners $\Psi=(\psi_i)_{i=1}^{p}$, where $z_i=\psi_i(\boldsymbol{x}_{j,i})$.

(3) According to z_i, new sample data set $\boldsymbol{D}_{\psi} = \{\boldsymbol{x}'_i, \boldsymbol{y}_i\}_{i=1}^{L}$ is generated. A secondary learner containing several base-learners is designed as $\Phi = (\phi_i)_{i=1}^{P}$, where $\boldsymbol{y}_i = \phi_i(\boldsymbol{x}'_i)$ and learns all samples in $\boldsymbol{D}_{\psi}$.

(4) All the base-learners are combined and a stacking ensemble learner is established by certain strategy, as shown in Figure 2.

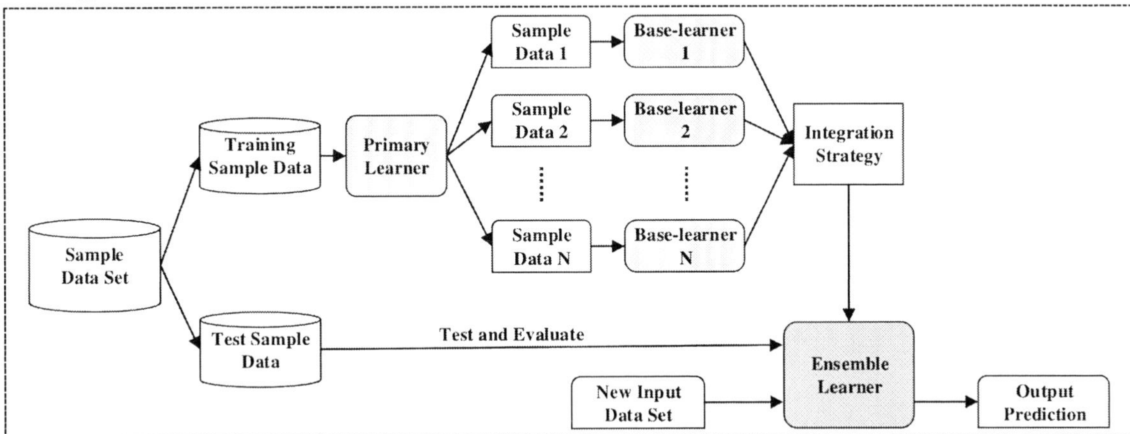

Figure 2. Basic framework of stacking ensemble learning model.

4. Flight path planning surrogate model

For aircraft's sample data sets of 'input parameters - output waypoint' sequence, divide all the sample data sets according to certain characteristics into different groups, and then each group sample can be trained by base-learners using machine learning algorithm such as neural network, support vector machine (SVM), and so on. Afterwards, base-learners can be combined orderly by stacking method, establishing the direct mapping relation from the original input to the output, which helps to obtain the waypoint coordinates from flight mission parameters intelligently. The flight path planning surrogate model is established by following steps:

- Step 1: Based on path planning analytical model $\Psi(g)$, a large number of samples are generated by setting different input parameters to obtain a sample data set $\boldsymbol{P}$, and then the training data $\boldsymbol{D}$ and test data $\boldsymbol{T}$ are extracted from $\boldsymbol{P}$.
- Step 2: For $\boldsymbol{D}$ and $\boldsymbol{T}$, take a series of data pre-processing operations, including data cleaning, data structuring, data normalization and other operations.
- Step 3: Group the data $\boldsymbol{D}$ processed in step 2 twice. Firstly, all the samples are characterized by parameter c_4, samples of the same c_4 belong to the same group, which are used for training waypoint quantity learners. Secondly, all samples are characterized by parameter k, samples of the same k belong to the same group, which are used for training waypoint coordinate learners.
- Step 4: For each group of samples in step 3, intelligent learning algorithm is used for training and then adjusting the parameters to construct each base-learner.
- Step 5: According to certain strategies, these base-learners are combined in order to form an integral surrogate model.
- Step 6: Test the integral surrogate model in step 5 with the test samples $\boldsymbol{T}$, and return to step 4 to retrain base-learners if the errors does not meet the application requirements.

5. Examples and analysis

5.1. Example of flight path planning surrogate model

IOP Conf. Series: Materials Science and Engineering **751** (2020) 012038 doi:10.1088/1757-899X/751/1/012038

Based on engineering application, a case of flight path planning is simulated and it generates more than 1.2×10^6 group of samples. In all samples, the value of parameter c_4 can be 3, 4, 5, 6 or 7, and the value of parameter k can be 4, 5, 6, 7, 8, 9, 10, 11, 12 or 13. Therefore, there are 5 base-learners in waypoint quantity learner named as Base-learner N3 to N7 and 10 base-learners in waypoint coordinate learner named as Base-learner P4 to P13. The surrogate ensemble model's structure is shown in Figure 3.

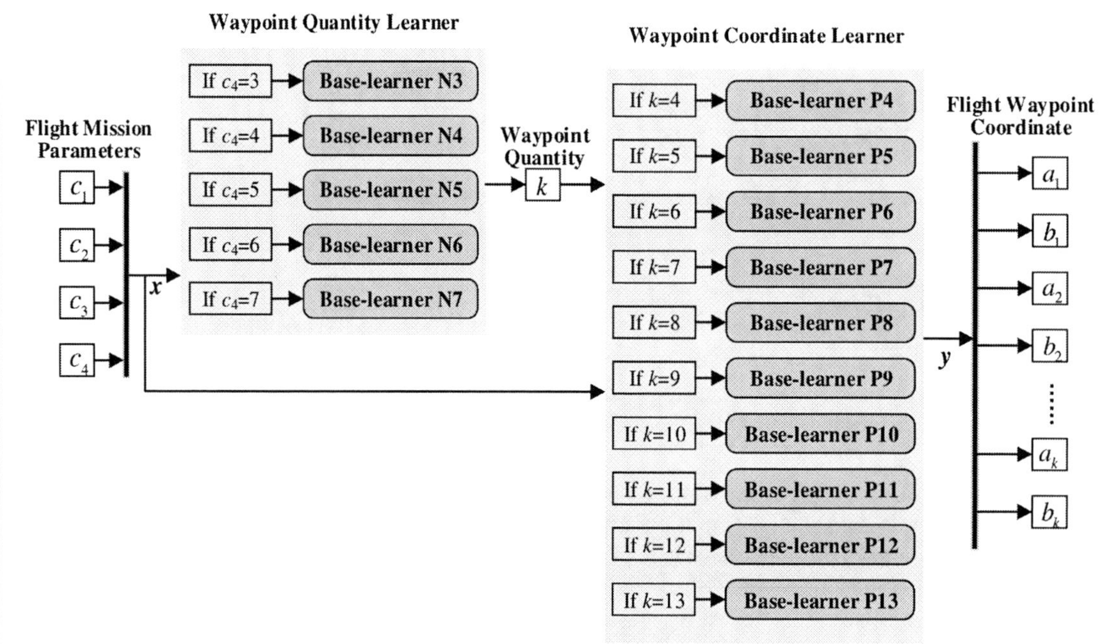

Figure 3. A case of flight path planning surrogate model's structure.

5.2. Base-learner training algorithm

In this study, RBF (radial basis function) neural network is used for training each base-learner. RBF method is a technique for high dimensional interpolation. RBF neural network has excellent function approximation ability, accurate classification ability and fast learning rate. It obtains the norm of the input sample and the hidden layer point (centre point), and then bring it into the radial basis function. The corresponding output value can be obtained after multiplying it by the weight and summing all the nodes' values. The structure of RBF neural network is shown as Figure 4, where R, S^1, S^2, IW, LW, b^1, b^2 are main calculating parameters. The neural network's calculation process is as bellows:

$$n^1 = ||\mathrm{IW} - \boldsymbol{x}||.*b^1 = \mathrm{sqrt}\left\{\mathrm{diag}\left[\mathrm{IW} - \mathrm{ones}(S^1,1)*\boldsymbol{x}^T\right]\left[\mathrm{IW} - \mathrm{ones}(S^1,1)*\boldsymbol{x}^T\right]^T\right\}.*b^1 \quad (2)$$

$$a^1 = \mathrm{radbas}(n^1) = e^{-(n^1)^2} \quad (3)$$

$$a^2 = \mathrm{purelin}(\mathrm{LW}\,a^1 + b^2) \quad (4)$$

$$\boldsymbol{y} = a^2 \quad (5)$$

Where IW, LW, b^1, b^2 will be trained after neural network's iteration. After training each base-learner's samples, flight waypoint coordinate can be calculated by formula (2) – (5) explicitly.

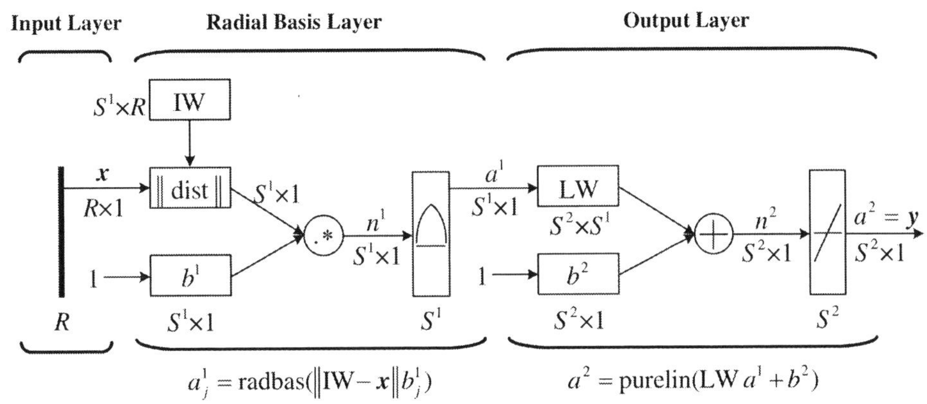

Figure 4. The structure of RBF neural network.

5.3. Results of example

According to the method in this paper, a flight path planning surrogate model was constructed, which was trained in MATLAB and developed in embedded environment to invoke the model. Each base-learner's information is shown in Table 1 and Table 2. The neural networks' training time is 158 minutes in total. The overall classification accuracy of waypoint quantity learner for k values is over 99.999%, and the overall coordinate MSR (mean squared error) is about 0.000221°(about 20.1m in distance). The single call time of this flight path planning surrogate model in embedded environment is about 0.163s. The accuracy and real-time performance both meet the application requirements.

Table 1. Each base-learner's information of waypoint quantity learner.

Values of c_4	Quantity of Sample		Training Result		
	Original Sample	After Data Pre-processing	Base-learner	Classification Accuracy (%)	Training Time (s)
3	243030	20097	N3	99.999	1021.1
4	242801	20081	N4	99.998	988.3
5	240808	20079	N5	99.999	921.1
6	242837	20081	N6	99.999	945.7
7	242641	20065	N7	99.998	963.2
Total	1214117	100400	-	99.999	4839.4

Table 2. Each base-learner's information of waypoint coordinate learner.

Values of k	Quantity of Sample		Training Result		
	Original Sample	After Data Pre-processing	Base-learner	Coordinate MSR (°)	Training Time (s)
4	350566	28990	P4	2.21×10^{-4}	1592.8
5	249459	20623	P5	1.87×10^{-4}	1032.4
6	180893	14956	P6	1.42×10^{-4}	544.5
7	226187	18706	P7	1.56×10^{-4}	797.3
8	102140	8448	P8	2.05×10^{-4}	368.1

9	35496	2934	P9	8.11×10^{-4}	141.7
10	32981	2734	P10	9.34×10^{-4}	121.9
11	31242	2582	P11	6.27×10^{-4}	65.0
12	4856	403	P12	4.35×10^{-4}	12.7
13	297	24	P13	6.23×10^{-4}	2.1
Total	1214117	100400	-	2.21×10^{-4}	4678.5

5.4. Test the surrogate model

An example was used for testing the surrogate model in Section 5.3, whose input was $\boldsymbol{x}=[120.285, 30, 6.25, 5]$. The comparison of waypoint coordinate results calculated by analytic model and surrogate model were shown in Figure 5 and Table 3. After making statistics and analysis, the calculation time of analytic model was 5132.1 seconds and the calculation time of surrogate model was about 0.12 seconds. That is to say, the cost of calculation time was reduced significantly.

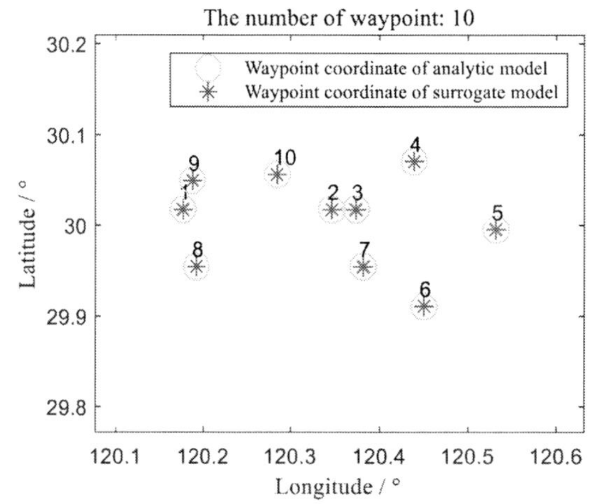

Figure 5. Comparison of waypoint coordinate results.

Table 3. Waypoint coordinate values calculated by analytic model and surrogate model.

Number of Waypoint	Longitude		Latitude	
	Analytic Model (°)	Surrogate Model (°)	Analytic Model (°)	Surrogate Model (°)
1	120.176531	120.176536	30.017327	30.017322
2	120.346733	120.346737	30.017357	30.017353
3	120.373422	120.373426	30.017342	30.017344
4	120.439254	120.439255	30.070637	30.070631
5	120.531463	120.531463	29.995221	29.995227
6	120.450346	120.450344	29.910839	29.910833
7	120.381458	120.381452	29.954492	29.954490
8	120.191981	120.191981	29.954494	29.954491
9	120.187584	120.187581	30.049527	30.049525
10	120.285005	120.285002	30.056206	30.056206

IOP Conf. Series: Materials Science and Engineering **751** (2020) 012038 doi:10.1088/1757-899X/751/1/012038

6. Conclusion

In this paper, a method based on stacking ensemble learning is introduced for establishing flight path planning surrogate model, in which several base-learners are trained for learning complex flight mission's waypoints coordinates. It solves the problem of poor real-time performance of mathematical analytic model, as well as the problem of non-uniform dimension when learning all flight samples with a single learner. The surrogate model is relatively simple and easy to operate, fast in calculation and high in accuracy. In this method, the flight waypoint coordinates can be calculated accurately in real-time environment. At last, the method is proved highly effective through lots of tests.

References

[1] Singh V and Willcox K E 2017 Methodology for path planning with dynamic data-driven flight capability estimation AIAA Journal vol 55

[2] Schwartz K N, Pfaender H, Kirby M,Mavris D N, Stouffer V L, Kumer V, Trajkov S, Hasan S and Payan A P 2016 Initial feasibility test of exploring the air traffic management design tradespace through surrogate modelling Conf. on 16th AIAA Aviation Technology,Integration, and Operations Conference (Washington, D.C.)

[3] Othman N and Kanazaki M 2015 Surrogate model of aerodynamic model toward efficient digital flight Conf. on 2014 Asia-Pacific International Symposium on Aerospace Technology vol 99 (Shanghai, China) pp 703-712

[4] Divina F, Gilson A, Gomez-Vela F, Torres M G and Torres J F 2018 Stacking ensemble learning for short-term electricity consumption forecasting energies vol 11 p 949

[5] Sikora R and AI-Laymoun O 2015 A Modified Stacking Ensemble Machine Learning Algorithm Using Genetic Algorithms (University of Texas – Arlington, USA)

[6] Yang J J, Li J Q, Shen R F, Zeng Y, He J, Bi J, Li Y, Zhang Q Y, Peng L H and Wang Q 2016 Exploiting ensemble learning for automatic cataract detection and grading Computer Methods and Programs in Biomedicine vol 124 pp 45-57

[7] Sun W and Trevor B 2018 A stacking ensemble learning framework for annual river ice breakup dates Journal of Hydrology vol 561 pp 636-650

[8] Kwak J, Park J H and Sung Y 2017 Unmanned aerial vehicle flight point classification algorithm based on symmetric big data symmetry vol 9 p 1

IOP Conf. Series: Materials Science and Engineering **751** (2020) 012039 doi:10.1088/1757-899X/751/1/012039

Research on Service-oriented UAV Route Planning Component Design

Yuanyuan Qi*, MinYang Kang, Bing Xue and Shuo Wang

Science and Technology on Avionics Integration Laboratory

432 Guiping Road, Xuhui District, Shanghai, China

*qyy_fly@163.com

Abstract. In the field of aviation equipment, with the development of airborne avionics systems towards systemized multi-platform coordination, the service-oriented architecture (SOA) has become an important trend for the development of open avionics system. Based on the idea of SOA, the service-oriented design of the UAV route planning component is proposed in this paper. Firstly, the functions of avionics are abstracted out from generally involved avionics missions with focus on the conception of avionics service. Then the function-directed servitization process of the UAV avionics resources is studied, based on which, a set of callable UAV route planning components are further constructed. The UAV route planning service component set can be flexibly deployed according to different task requirements, which solves the problem of resource-application coupling for the UAV avionics, and provides support for the intelligent and autonomous operation of UAV.

1. Introduction

In the public and civilian fields, with the complex and diversified development of system services, system resources are gradually integrated to achieve integration linkage.[1] By constructing an information system through a collection form of easily replaceable services, in the case of task changes, the configuration can be completed simply by replacing the related services, which means that the changed requirements can be responded to in a very short time. In the field of aviation equipment, from the international perspective, the US military and North Atlantic Treaty Organization are is approaching to apply a service-based architecture for their weapon systems in order to effectively adapt to the ever-changing operational environments and achicvc inter-system connectivity.[2] What's more, as UAV plays an increasingly important role in modern warfare, the avionics system that is used as an intelligent flight support technology for UAVs is becoming more and more complex. However, in traditional UAV avionics systems, the resources and applications are tightly coupled with each other, which results in the low flexibility and scalability and high maintenance cost of existing UAV avionics systems.[3] For realization of interconnection and interoperability of UAV avionics resources, and further truly reliable intelligence and autonomy of UAVs, the research on the servitization of UAV avionics system is highly desired, through which the flexible invocation of service avionics resources can be possible to meet the diversified requirements for UAV to perform various tasks. [4]

In addition, the research on the key technologies of UAV route planning is of great significance for the improvement the autonomy of UAVs.[5] However, for the UAVs in service or development, the route planning still remains in the "manual planning" or "manual planning + computer aided" mode with a low level of autonomy, in which the decision-making process under complex situations mainly depends on the prior judgment of human experience or switching to the remote control mode. The control of the UAV is mainly realized by the onboard equipment while the task management is performed on the ground control station. This separation of control and management leads to command lag and delay due to data link processing and transmission. In recent years, the drawbacks of this separation of control and management have gradually emerged, and many cases of UAV crashed due to satellite data link corruption. Therefore, it is urgent to systematically develop research on the dynamic route planning method of UAV under the conditions of full consideration of actual environmental constraints to improve the autonomy of drones.

This paper proposes a service-oriented design of the UAV route planning component library. Firstly, the functions of avionics are abstracted out from generally involved avionics missions with focus on the conception of avionics service. Then a set of callable UAV route planning components are further constructed on the basis the function-directed servitization of the UAV avionics resources. The obtained UAV route planning service component library can be flexibly deployed according to different task requirements, which is crucial for the intelligent and autonomous operation of UAV.

2. Service-oriented open avionics system architecture

Based on the service-oriented idea, the development of UAV avionics service component needs to be carried out under an open system framework. Therefore, this paper first introduces the open UAV avionics system architecture, then studies the UAV avionics task decomposition to lay the foundation for abstracting the avionics system function.

2.1. Service-oriented open avionics architecture

The core idea of the open avionics architecture is to adopt a layered concept and establish a generalized and standardized interface according to the open system interface standard, which ensures that the different layers of the system are independent of each other.[6] Therefore in this architecture, the modifications and extensions within certain hierarchy will not affect other levels of the same system, nor affect the corresponding levels of other systems, besides, module replacement of hardware and software is to be feasible, which means the systems can be upgraded and maintained with minimal redesign of existing systems in the future.

The open UAV avionics architecture can be divided into an application layer, a service layer, and a resource layer, as shown in Figure 1. The application layer encapsulates the complete system function, and provides the function of the logic control system to the upper layer user; the service layer provides data management, system configuration and other services, as well as the direct access to the hardware components; the resource layer provides a direct interface between different resources for service layer access.[7] Broadly speaking, the application layer, the service layer, and the resource layer all provide services in the form of software components.

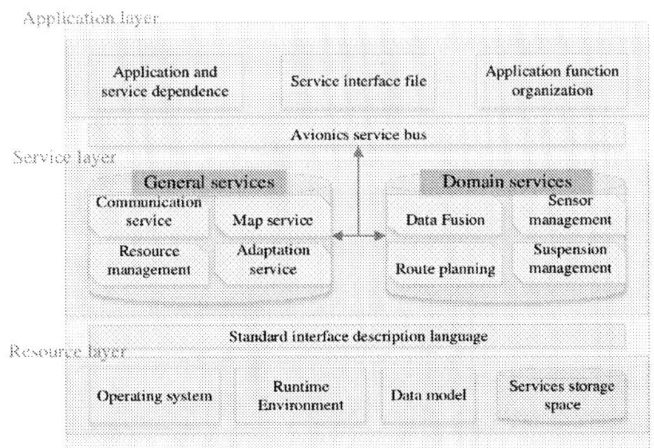

Figure 1. Service-oriented open avionics system architecture for UAV

Then an open UAV avionics service framework is constructed by applying the above open system architecture to the integration of UAV avionics resources. This framework provides basic functional components for the task application layer, as well as a resource-invoking interface. This method can improve the versatility of resources in the design stage, avoid the tight coupling between the previous avionics resources and the aircraft platform, and further realize the flexible management and use of the avionics resources.

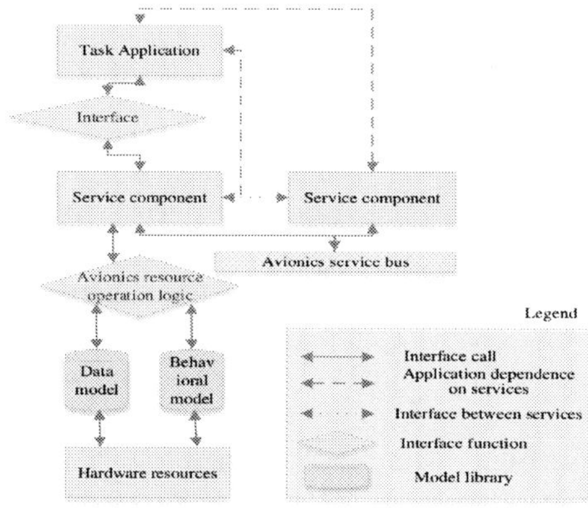

Figure 2. The open UAV avionics service framework

The open UAV avionics service framework is as shown in the figure 2, which describes the interaction and interface relationships among the UAV avionics application software, service components and hardware resources, and how the service components are organized, interacted and communicated.

2.2. Task decomposition for UAV avionics system

In fact, the UAV avionics system tasks are determined by operational requirements, therefore, in order to abstract the functions for the service layer of UAV avionics design, it is necessary to decompose the UAV avionics system tasks hierarchically.

The task decomposition needs to follow the following principles:

(1) The hierarchy is clear. From the perspective of system, UAV tasks can be divided into cooperation level tasks, platform level tasks and system level tasks. Therefore, UAV task decomposition needs to establish a hierarchical task structure, and can also break down complex tasks into task trees.

(2) Relatively independent. Different levels of tasks and different tasks at the same level should not be crossed.

(3) The system is complete. Task decomposition should be implemented according to the conception of system integration, and the overall task is regarded as the integration of specific tasks.

In this section, as shows in the Figure 3, a decomposed task tree of the UAV avionics system was created based on the above underlying principle. Firstly, the UAV tasks, namely the root of the tree, was classified into universal avionics tasks and domain avionics tasks according to mission requirements. The universal avionics tasks which support the UAV flight include communication tasks, navigation tasks, flight control tasks, etc.; The domain avionics tasks which support the UAV operation include fire control tasks, mission planning tasks, reconnaissance surveillance tasks, and electronic warfare tasks. Then the universal avionics tasks and domain avionics tasks were refined into several child nodes according to their function requirements respectively. For example, the former can be divided into route planning, sensor load planning, link use planning, task allocation, etc. After that, according to the function of the module algorithm, the granularity of the 2nd generation nodes were again refined to form 3rd generation meta-task nodes. For example, the route planning tasks were divided into search route planning, cruise route planning, tracking route planning, etc. Finally, the service layer functional requirements could be obtained through the meta-task analysis.

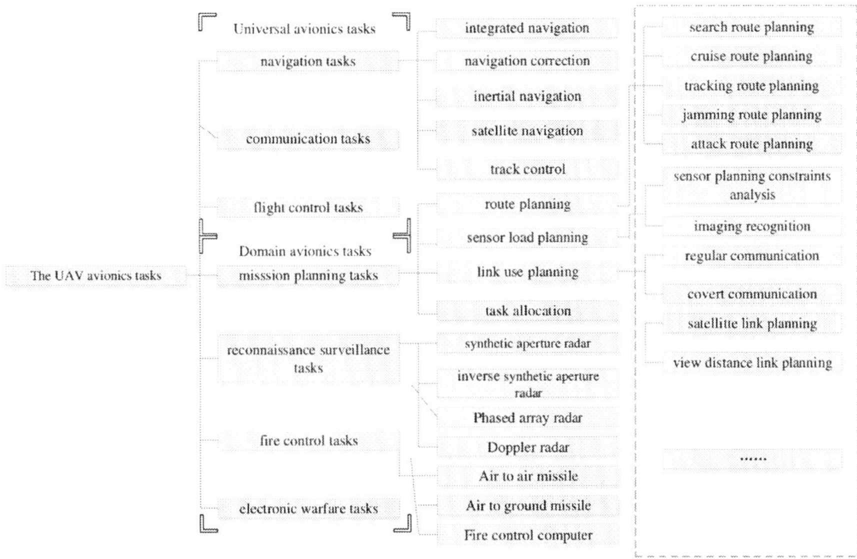

Figure 3. Decomposed task tree for UAV avionics system

3. The servitization of UAV avionics system

3.1. The process of the servitization for UAV avionics system

In the face of increasingly complex tasks, the UAV avionics system should possess dynamic, distributed, and heterogeneous features, which are difficult to port on heterogeneous platforms because of task-specific interfaces. In this section, the servitization model method of avionics system is put forward to shield the realization process and mechanism of various functions in its internal, and the avionics mission requirements abstraction into a service description is encapsulated by a unified structure, so as to achieve the purpose of effective classification and combination of avionics service, and to achieve the goal of resource on-demand and dynamic expansion in avionics system.

The service component mainly undertaking the task demands were divided into two categories: basic service components and domain service components, the domain service components are mainly related to the specific application domain and the basic service component mainly provides common services such as health management, data loading, and log management for the system. The service components design mainly includes three parts: service component interface design, component set design and management scheme. The service component interface design forms a common interface data file for the platform through abstracting and unifying the data formats of different components.

In process of service component set design, in the first place, it is necessary to decouple the organization structure of each software in the system, and abstract the functions and resources of each system. After that, the diversified functions in the system should be further abstracted into service components in the service layer. The service component set can be regarded as a "container", and different services communicate interactively according to the needs of the application to form a larger service. In order to make the design of the service component library more logical, versatile and flexible, layered design pattern was used in this paper to design the service component library. In this process, it mainly completes:

(1) Modular functional processing: isolating the relationship between the specific service components for a specific application through modular processing, so that the cross influence between each component is minimal to keep a high stability of the system.

(2) Standardization processing: after modular processing, it is necessary to establish a unified standard to realize multi-module simulation integration.

By taking the above requirements into consideration, an function-directed servitization process of the UAV avionics resources (Figure 4) was designed in this section. The integrated servitization process takes the task application as input, and refines the specific system task requirement information through task decomposition, and then the functional description of the service component can be obtained according to the corresponding task-function mapping rules. Finally the service component set was obtained through service combination.

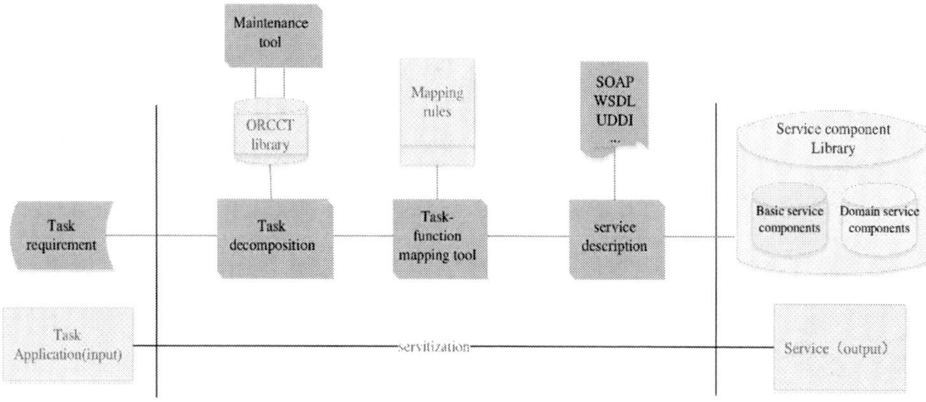

Figure 4. Process of the servitization for the UAV avionics application

3.2. Componentization design of UAV route planning service

For the existed UAV route planning software development model was unable to cope with the continuous development of combat methods and the increasingly complex battlefield environment, in this section, the componentization of the UAV route planning service was studied on the basis of the analysis of the current situation of route planning software design, which can provide ideas for the further development of the route planning component set. Figure 5 shows the design block diagram of the route planning service component set, which mainly includes three levels: the route planning function component set, the dedicated planning processing component set, and the universal data processing component set.

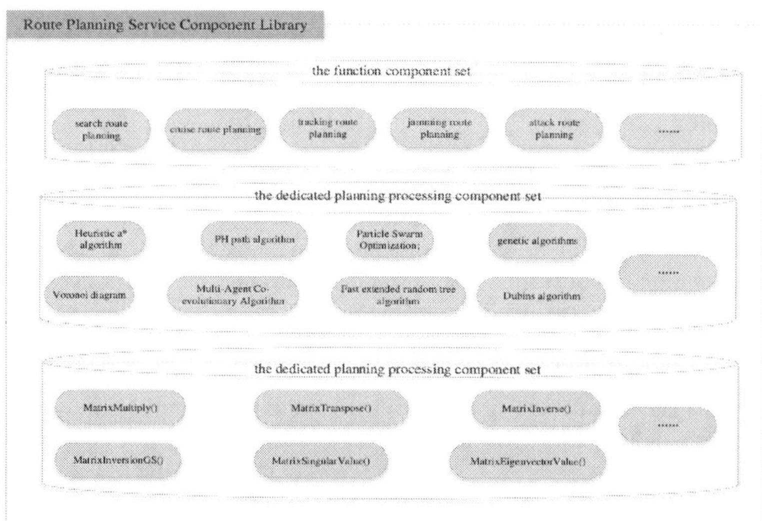

Figure 5. Route planning service component library

Route planning function component set is designed to meet the functional requirements of route planning by inductive combat applications, and corresponds to task-oriented specific route planning functions including: search route planning, tracking route planning, attack route planning, and cruise route planning, obstacle avoidance route planning, etc. The above components can be directly called by the application layer designer, which is convenient for the application layer designers to quickly perform service combination and packaging to form an application according to the combat task requirements.

Dedicated planning processing component set involving specific route planning algorithm implementation module plays a decomposition for each route planning service component. It mainly includes: heuristic A* algorithm, PH path algorithm, fast extended random number algorithm, co-evolution algorithm, genetic algorithm, Voronoi diagram and so on. [8] These algorithm components here can be called by the service layer developers, helping developers to focus on the evolution of technology, enabling incremental development for services. Service layer developers can form various processing components according to specific application requirements. However, dedicated processing components are shielded from application layer developers because there is no need for application layer developers to understand how design application-oriented processing components. In addition, due to the complexity of the route planning algorithm itself (some algorithms can be considered as a subset of other algorithms taking different advantages and disadvantages in performance, and cannot be used exclusively), the corresponding formed components own different granularity and logical relationships, and can only support calling from service developers who are familiar with the route planning profession.

Universal data processing component set mainly contains the mathematical functions invoked by the route planning processing algorithm, which is the basis for the implementation of the route planning algorithm and can provide support for the upper layer other services (such as data fusion services, sensor management services, etc.).

4. Conclusions

In this paper, a service-oriented design of the UAV route planning component library was realized though task decomposition, function abstracting and the process of the servitization for the UAV avionics system, which provides an effective approach for solving the problem of coupling the avionics resources and applications of UAV. This research could pave the way for intellectualization and autonomization of UAV.

Acknowledgment

We acknowledge the financial supports of the Aeronautical Science Foundation of China (No.20175501007)

References

[1] Tan Yongming, Su Bin, et al. Research on Service-Oriented Architecture System[J]. Computer Technology and Development, 2007, 17(3): 132-134.
[2] Gao Xiang, Li Chen, et al. Research on Open System Standards of Complex Avionics Architecture[J]. Avionics Technology, 2015(2):26-31.
[3] FACE standard and its application in UAV avionics software [C]. The 7th China Aviation Society Youth Science and Technology Forum (Volume 1). 2016.
[4] Brohaugh T . A common open avionics architecture[C]. Position Location and Navigation Symposium, 1994. IEEE. IEEE, 1994.
[5] Zhihong Ouyang, Chunrui Yu, Feng Ding. Task Oriented Methods of UAV Route Planning[C]// Control Conference. IEEE, 2017.
[6] Wojtowicz K, Sobieraj,WiesÅ,aw, Brzozowski B. Simulation Method of UAV Avionics System Designing[J]. Solid State Phenomena, 2013, 198:238-242.
[7] Majithia S , Walker D W , Gray W A . A framework for automated service composition in service-oriented architecture[C]. Semantic Web: Research & Applications, First European Semantic Web Symposium, Esws, Heraklion, Crete, Greece, May. DBLP, 2004.
[8] Jiayang Wu, Jin Yi, Liang Gao, et al. Cooperative path planning of multiple UAVs based on PH curves and harmony search algorithm[C]. Computer Supported Cooperative Work in Design (CSCWD), 2017 IEEE 21st International Conference on. IEEE, 2017.

IOP Conf. Series: Materials Science and Engineering **751** (2020) 012040 doi:10.1088/1757-899X/751/1/012040

Research on the Construction Technology of Knowledge Graph in Aviation

Xue Wang*, Xizhong Yang, Jiapan Fu and Xuyi Qiu

Science and Technology on Avionics Integration Laboratory, 432 Guiping Road, Xuhui District, Shanghai, China

*E-mail: snow_wx@163.com

Abstract. In order to alleviate the overload of airspace information under the network environment and integrate the scattered and useful airspace information, the technical architecture of knowledge graph in the aviation field is proposed. This paper describes the technical architecture of aviation knowledge graph construction, including knowledge extraction, knowledge fusion, knowledge processing, knowledge update and other key technologies and applications. Finally, an instance of knowledge graph for air combat is constructed. The knowledge graph discovers the hidden relationship between task and maneuver through resource sharing and information mining, which can be used for mission optimization and maneuver prediction to assist air combat decision-making.

1. Introduction

In the network era, avionics system not only contains the massive data of land, sea, air, and space, but also has a variety of heterogeneous data, including early warning, reconnaissance, intelligence, confrontation, etc. With the explosive growth of information, the requirements for information processing are constantly improving. How to quickly and accurately integrate, relate a large number of scattered, isolated intelligence, transform information advantages into knowledge advantage, then transform into decision-making advantage, becomes the key to fight for air power, and also the problem that knowledge graph needs to solve.

It is necessary to construct a professional aviation knowledge graph to alleviate the problem of information overload, improve efficiency and provide more useful information for commanders. Starting from the architecture, this paper expounds the construction process and application examples of aviation knowledge graph.

2. Technical Architecture of Knowledge Graph

The technical architecture of knowledge graph contains the construction and application of knowledge graph, where the construction process including knowledge extraction, knowledge fusion, knowledge processing [1], knowledge update, etc., as shown in Figure 1. The construction of knowledge graph is based on the combination of natural language processing, text mining, machine learning, artificial methods and so on, which can be applied to complex relationship query, human-computer intelligent interaction and other directions in the aviation field.

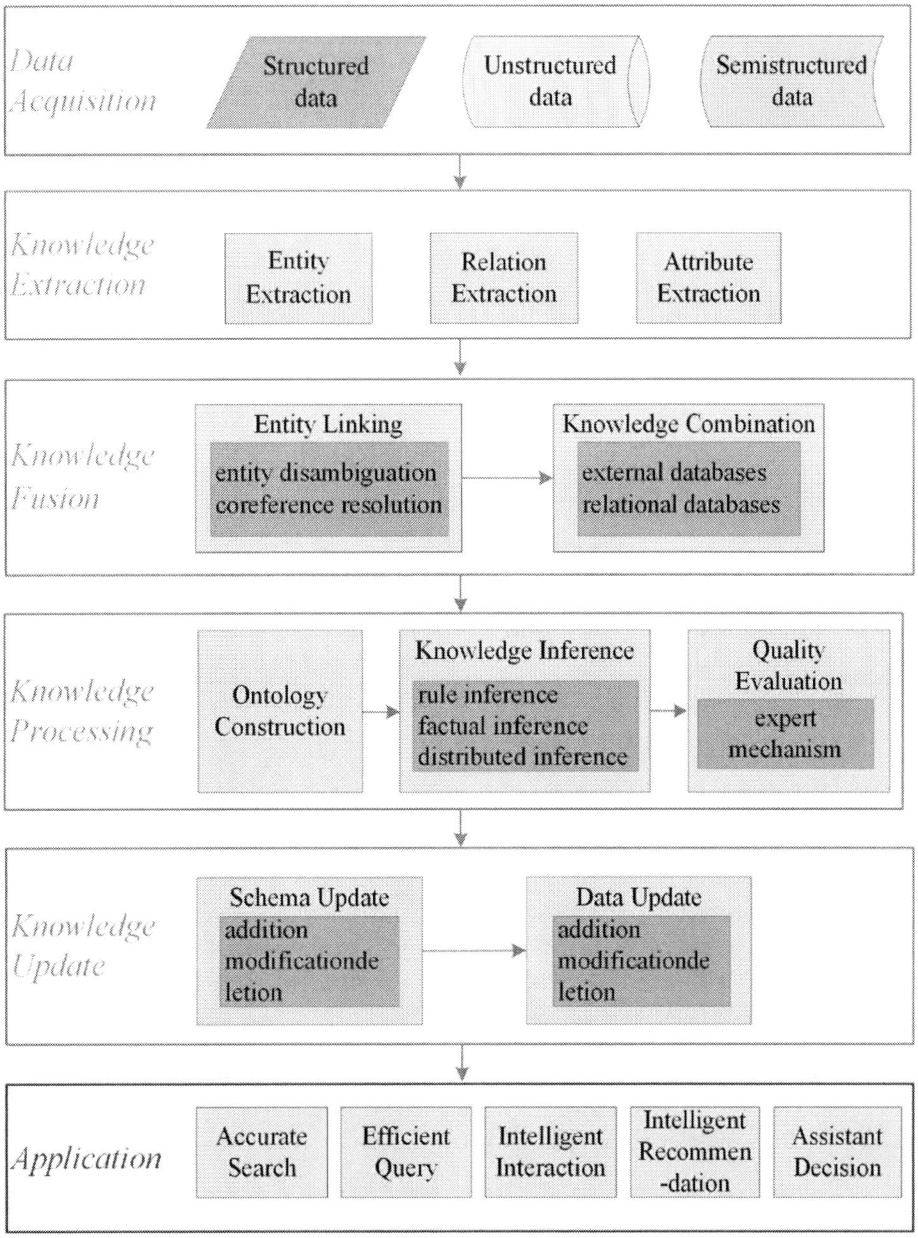

Figure 1. The technical architecture of knowledge graph.

2.1. Data Acquisition

The original data sources of knowledge graphs include structured data such as relational database, semi-structured data stored in document and web formats, and unstructured data stored in pictures, audio and video formats, etc. Therefore, making full use of these data and processing them correctly can be a good basis for the construction of knowledge graphs.

2.2. Knowledge Extraction

Knowledge extraction is an important step in the construction of knowledge graphs, it automatically extracts entity names, attributes, attribute values and relationships among entities from heterogeneous data to build domain ontology. The extraction methods include rule-based extraction [2], method based on machine learning [3] and open domain-oriented extraction method [4], etc.

Entity extraction is mainly for unstructured data such as image and text [5]. For image data, text information can be recognized by OCR technology, so that the image can be converted into text. The recognition process includes five main steps: image input, image pre-processing, feature extraction, classification and post-processing.

Relation extraction is mainly for unstructured data such as text. The associated relationship are extracted between entities from relevant corpus to solve the problem of semantic connection between discrete entities, also form a network of knowledge structure. In order to get rid of the limitation of traditional artificial method, machine learning method is introduced into the relation extraction. The common models of relation extraction [3,6]include convolution neural network, capsule neural network and other deep learning models, such as relation extraction model based on CNN, relation extraction model based on RNN and relation extraction model based on PCNN.

The purpose of attribute extraction is to collect the attribute information of specific entity from different sources, and realize the complete description of entity attribute. The Methods based on rules, heuristic algorithm, and the algorithms based on machine learning can be used for attribute extraction.

2.3. Knowledge Fusion

The purpose of knowledge fusion [7] is to eliminate the ambiguity of concepts, remove redundancy and misconceptions so as to ensure the quality of knowledge. Knowledge fusion includes two parts: entity linking and knowledge combination. Entity linking is for semi-structured data and unstructured data, and the knowledge combination is aimed at structured data.

2.3.1. Entity Linking. Entity linking make the extracted entity corresponds to the homologous entity in the knowledge base. The general process is:

1) Obtain the entity appellation through entity extraction;

2) Ensure correctness and uniqueness of the entity with entity disambiguation and coreference resolution;

3) Link the entity appellation to the corresponding entity in the knowledge base.

Entity disambiguation is used to solve ambiguity caused by the same-named appellation pointing to different entities. Clustering method is mainly used in entity disambiguation. There are four common models of clustering disambiguation: ①The model of Space vector; ②The semantic model; ③The model of Social network; ④The model of encyclopedic knowledge.

Coreference resolution is used to solve ambiguity caused by multiple appellations correspond to the same entity in the real world. Machine learning method is introduced to improve the accuracy of the algorithm, such as decision trees algorithm, Hobbs algorithm and centering theory.

2.3.2. Knowledge Combination. There are two main types of knowledge combination: ① The combination of external databases such as DBpedia [8], YAGO [9-10], Wikipedia [10], Freebase [11] and so on. This step mainly deals with two levels of problems: one is fusion in the data layer, including entity's reference, attribute, relationship and category, etc. The main purpose is to avoid relationship conflicts in the instances which can result in redundancy; another one is to integrate the new ontology into the existing ontology database through fusion in the pattern layer. ② The combination of relational databases. An important source of high-quality knowledge is the domain's own relational database. In order to integrate these structured historical data into the knowledge graph, resource description framework (RDF) [12], can be used as the data model. The essence is to convert the data of relational database into the triple data of RDF.

2.4. Knowledge Processing

Knowledge processing includes three aspects: ontology construction, knowledge inference and quality evaluation.

2.4.1. Ontology Construction. Ontology is a formal description of domain concepts. It contains the basic concepts and the relationship between concepts in the domain that can be interpreted by machines, and a general dictionary is defined for experts or machines in a specific domain to share the information. The process of ontology construction is as follows:

Step 1: Determine the domain and scope of ontology;
Step 2: Consider reusing the existing ontology;
Step 3: Enumerate important items in the ontology;
Step 4: Define the class and its hierarchy;
Step 5: Define the properties of the class, which called slot;
Step 6: Define constraints on slots;
Step 7: Create an instance.

2.4.2. Knowledge Inference. Knowledge inference [13] can mine hidden knowledge on the basis of existing knowledge base, obtain new knowledge or conclusions through computer inference, so as to meet the semantic requirements, expand and enrich the knowledge network. Its specific tasks can be divided into satisfiability, classification and materialization. Knowledge inference includes rule inference, factual inference, and distributed inference, etc.

*2.4.3. Quality Evaluation.*It is used to evaluate the credibility of new data and ensure the quality of knowledge. In view of the specialization of domain knowledge, it is necessary to introduce domain expert guidance.

2.5. Knowledge Update.

The update of knowledge graph mainly includes the update of schema layer and data layer.

The update of schema layer refers to the update of elements in ontology, including the addition, modification, deletion of concepts, the update of concept's attribute and hyponymy. It usually requires manual intervention to verify the correctness and rationality of the updated knowledge through the test method based on feedback iteration. The update of data layer refers to the update of entity elements, including the addition, modification, deletion of entities, as well as the update of relationship and attribute values, which is realized automatically through the methods of fusion, similarity matching, association mapping and so on.

2.6. Applications

- Accurate Search

In the network environment, information or situation is changing rapidly, and massive information is interconnected with complex multi-dimensional structure. By establishing such complex relationship, knowledge graph can not only search complex multi-source objects such as text, image, video, code, etc., but also realize the accurate search [14] for large data through sentence-level search and other methods, which can be used for precise classification, semantic understanding, accurate search for tactical intent, etc.

- Efficient Query

Users can quickly and directly acquire relevant knowledge by compiling information requirements into query statements in a specific format. For example, the "automatic conversion method of query statement based on Rule Template matching" can be used.

- Intelligent Human-Computer Interaction

The system can respond to pilot's questions based on the information provided by knowledge graph, thus forming an intelligent question answering system [15], in which dialogue interaction replaces text-searching as the main interaction mode.

- Intelligent Recommendation and Assistant Decision

With the help of knowledge graph, the situation information such as the flight status of target and battlefield environment is perceived, the deep relationship is inferred by mining the implicit relationship, and the intelligent recommendation is given by combining current situation information in the environment, so as to realize the function of assistant decision.

3. Aviation knowledge graph

When complete and accurate environmental information is available in advance, the optimal flight mission can be planned at one time. The actual situation is that it is difficult to ensure that the acquired environmental information does not change; on the other hand, due to the uncertainty of the mission, it is often necessary to temporarily change the flight mission. Knowledge graph can help to make better use of the coordination between domain experts and computer systems, and realize the optimal mission planning through resource sharing and information mining.

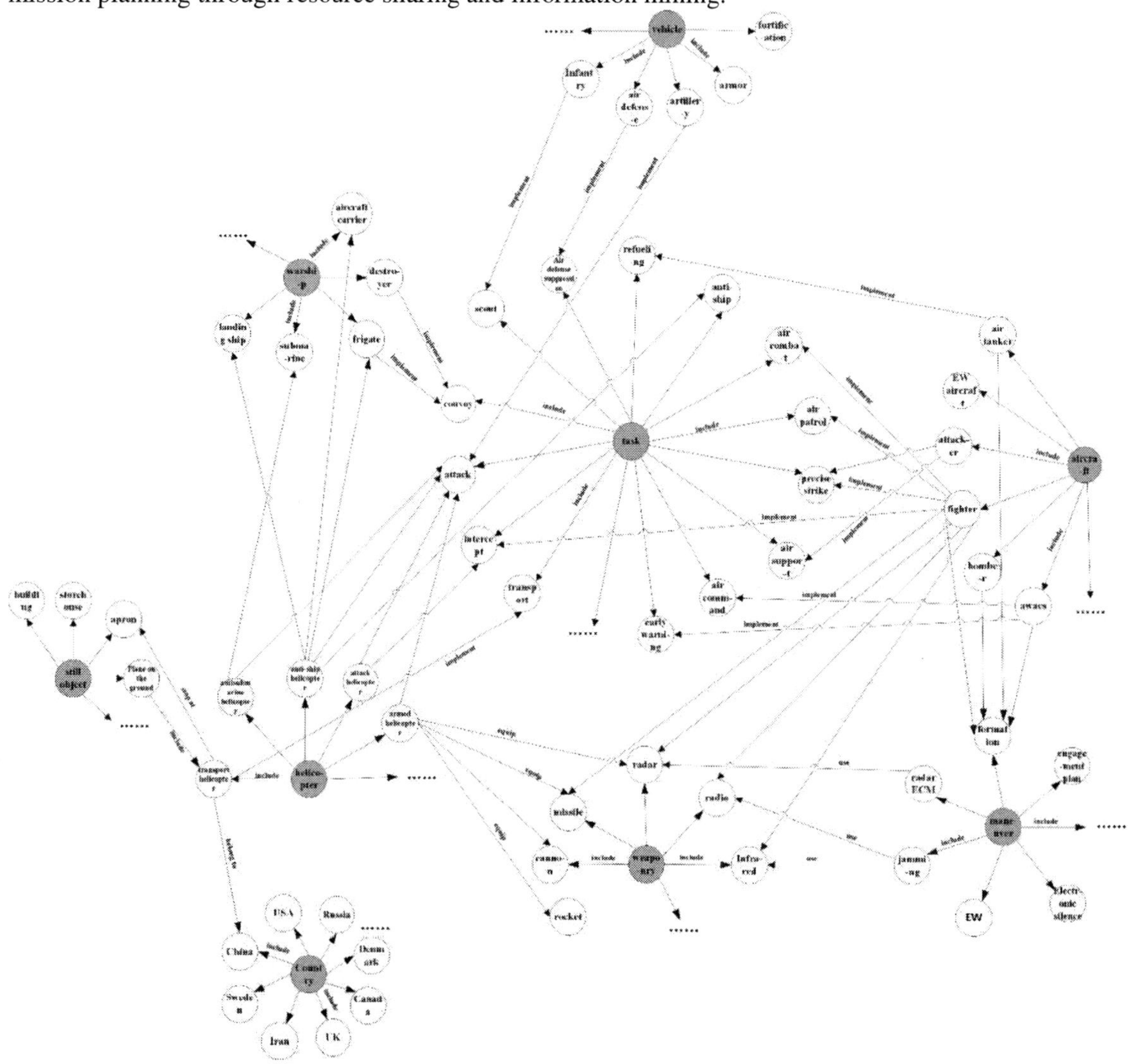

Figure 2. An instance of knowledge graph.

Figure 2 is an instance of the knowledge graph in aviation, for the air combat decision-making, which directly shows the cross-linking relationship among the types of aircraft, task, maneuver, equipment and weapons, as well as related vehicles, helicopters, warships, other objects and their attributes for air combat. With the help of knowledge graph, the hidden relationship between task and maneuver can be mined for maneuver prediction and assistant decision-making.

4. Conclusion

Information explosion brings a large amount of data overload in aviation, which greatly reduces the utilization rate of effective information and affects the commander to make correct decisions. It is vital to construct the aviation knowledge graph. This paper expounds the technical architecture of knowledge graph, and constructs an instance of knowledge graph for air combat, which can support flight planning, maneuver prediction, assistant decision and other application scenarios.

References

[1] Nadeau D, Sekine S. A survey of named entity recognition and classification[J]. Lingvisticae Investigationes, 2007, 30(1) pp 3-26.
[2] Reiss F, Raghavan S and Krishnamurthy, et al. An Algebraic Approach to Rule-Based Information Extraction[C]. IEEE International Conference on Data Engineering, 2008, pp 1-10.
[3] Zeng D, Liu K, Lai S, et al. Relation Classification via Convolutional Deep Neural Network[C]// COLING. 2014: pp 2335-2344.
[4] A. Fader, S. Soderland and O. Etzioni. Identifying relations for open information extraction[C]. Proceedings of Conference on Empirical Methods in Natural Language Processing. EMNLP, 2011, pp1535-1545.
[5] L.F. RAU. Extracting company names from text[C]. Proc of the 7th IEEE conference on Artificial Intelligence Applications. NJ: IEEE,1991, pp 29-32.
[6] Bordes A, Glorot X, Weston J, et al. A semantic matching energy function for learning with multi-relational data[J]. Machine Learning, 2014, 94(2) pp 233-259.
[7] Dong X, Gabrilovich E, Heitz G, et al. Knowledge vault: A web-scale approach to probabilistic knowledge fusion[C]. Proceedings of the 20th ACM SIGKDD international conference on Knowledge discovery and data mining. ACM, 2014, pp 601-610.
[8] S. Auer, C. Bizer, G. Kobilarov, J. Lehmann, and Z. Ives. DBpedia: A nucleus for a web of open data [C]. Processings of the 6th International Semantic Web Conference, Busan: springer, 2007, pp 722-735.
[9] F. M. Suchanek, G. Kasneci, and G. Weikum. YAGO: A core of semantic knowledge unifying wordnet and Wikipedia. Processings of the 16th International Conference on World Wide Web, Alberta: ACM, 2007, pp 697-706.
[10] J Hoffart, F. M. Suchanek, K.Berberich, YAGO2: A spatially and temporally enhanced knowledge base from Wikipedia, 2013, International Joint Conference on Artificial Intelligence. 2007:pp 1962-
[11] K. Bollacker, C. Evans, et al. Freebase: a collaboratively created graph database for structuring human knowledge. Proceedings of the 2008 ACM SIGMOD International Conference on Management of Data, SIGMOD 2008, pp 1247-1250.
[12] C. Unger, L. Buhmann and J. Lehmann, et al. Template-based question answering over RDF data. ACM Proc. www, 2012, pp 639-648.
[13] R. Socher, D. Chen, C. D. Manning. Reasoning with neural tensor networks for knowledge base completion.[C]. International Conference on Neural Information Processing Systems. 2013, pp 926-934.
[14] Jiang K, Qian K, et al. Military Information Retrieval Technology Architecture Based on Knowledge Graph. Command Information System and Technology. 2016,7(1), pp 48-52.
[15] O. Kolomiyets, M.-F. Moens, A survey on question answering technology from an information retrieval perspective. Elsevier Trans. Inf. Sci. 181(24), 2011, pp 5412-5434.

The Influence of Angle Parameters on Pulsation Characteristics of Elastomer Rotor Pump

Jiadi Lian[1,2,a], Yan Bai[3,b], Zefei Zhu[1,2,c], Jing Xu[2,d] and Jingxuan Ma[2,e]
[1]Department of Mechanical Engineering, Zhejiang Sci-Tech University Hangzhou ,31 0018 , China.
[2]Department of Mechanical Engineering, Hangzhou Dianzi University, Hangzhou, 310 018, China.
[3]Hangzhou Steam turbine company limited,Turbine Research Institute ,Hangzhou, 310 014, China.

E-mail: [a] lianjiadi@163.com; [b] baiy@htc.cn; [c] zzf.3691@163.com; [d]xujing@hdu.edu.cn; [e]331087613@qq.com,

Abstract. In view of the poor sealing performance of elastomer rotor profile and the defect of particle sticking in the tip of rotor, the pulsation characteristics of rotor cavity under 5°, 8°, 11°, 15° and 18° conditions are studied by transient calculation of the whole flow field of rotor pump based on CFX immersion solid technology and RNG $k-\xi$ turbulence model. When the wrap angle increases, the instantaneous mass pulsation rate decreases. When the wrap angle increases, the cavitation possibility at the inlet meshing point increases, and the pressure at the outlet meshing point increases. In order to improve the periodic impact of liquid on the elastomer and prevent the adhesion surface from falling off, the elastomer rotor should adopt the structure of small wrap angle under the condition of ensuring the width of sealing at the top of rotor.
Keywords: elastomeric rotor pump, profile, pulsation characteristic, immersed solid method

1. The introduction

Elastomer rotor pump is a new kind of non-contact displacement pump, using elastomeric rotor[1-4], made up of two same shape rotors, synchronous gears, pump body and sealing and so on, with the rotation of the rotors to draw the fluid from inlet to deliver to the outset. Its main performance characteristics are reflected in solid-liquid, gas-liquid, gas-solid-liquid transportation and other aspects. It has the advantages of strong passability, high self-absorption, high pressure, low noise and high efficiency, and can be operated in a wide range of rotational speed[5-7], It is widely used in agricultural irrigation, flood control and drainage, vacuum discharge of high-speed railway bullet trains and other industries. Now, domestic and foreign researches on metal rotor pump mainly focus on optimization algorithm[8-9], profile design[10-11] and method research[12-13].Tong[14]developed the method of designing the rotor profile under giving flow function, a dimensionless quantity is used to define the dimensionless flow, and the rotor profile was designed according to the characteristic parameters. Yang[15]derived formulas for profile, volume coefficient and theoretical flow of non-contact rotor pumps, the equation of rotor profile with cycloidal blade was deduced. Zhang[16] et al. deduced the rotor profile equation of cycloid profile, and proposed the parameter optimization design method and demonstration of an example for the rotor profile. Vogelsang et al[17-19]tested and analyzed the flow rate and force of the two-blade and three-blade helical rotor model pump, and elaborated the performance of hiflo rotor, which can significantly reduce pressure flow pulsation. Wang[20-21]used dynamic grid technology to simulate the working process of claw vacuum pump, and obtained the internal pressure, temperature and velocity distribution, indicating that the flow field and force of the new type of circular arc rotor are better than the existing curved claw rotor. Liu [22]deduces the leakage flow of roots blower, and conducts numerical simulation on the two-dimensional model, and analyzes the exhaust flow curve and the velocity and pressure distribution characteristics in the mixing zone, indicating that there is an obvious eddy current movement in the exhaust process.

At present, rotor pump research is mainly aimed at the rigid rotor pump, the existing literature on the elastomer rotor pump research is less. The present profile of elastomeric rotor is usually the same as that of the traditional rotor pump. The traditional rotor profile is continuous and smooth. That particle scratch easily occurs at the top of rotor for traditional rotor profile has important influence on rotor working life. Meanwhile, small seal face width for traditional rotor profile makes leakage happen easily, which has important influence on work efficiency. Therefore, a novel scraping rotor profile is proposed in this paper. The transient flow field of the novel scraping rotor pump was investigated by CFX immersed solid method and RNG $k-\xi$ turbulent model. The work in this paper may provides some parameters for further perfecting the design theory of elastomeric rotor pump, and would contribute to improve rotor mechanical properties and sealing performance.

2. Requirements for elastomeric rotor profile

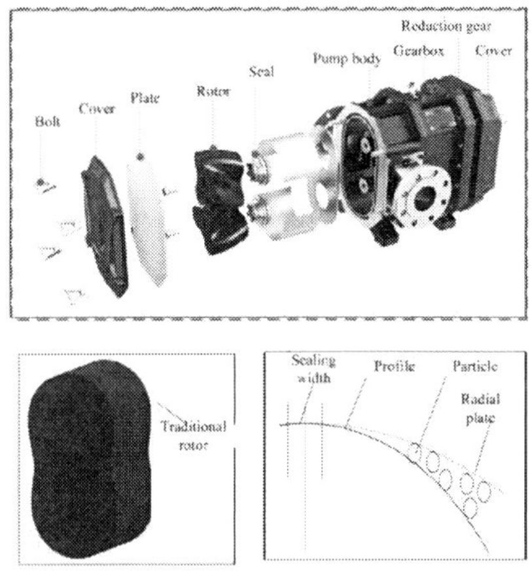

Figure 1. The structure of the elastomeric cam rotor pump.

The profile directly affects the self-priming performance and particles passing ability of the pump. Therefore, the elastomeric rotor profile needs to meet three requirements of good sealing performance, strong passing ability, and whether the rotor can effectively prevent particles from sticking and skidding. As shown in figure 1, the traditional profile cannot meet the above three requirements at the same time. Based on the above requirements, a novel scraping profile with wide head is proposed in this paper. Usually, the pulsation can be reduced by using a certain distortion of the rotor. However, the root rubber is easy to fracture for twisted impeller. It's suggested that the linear structure should be adopted in the engineering application.

3. Numerical simulation

3.1 Model simulation area

ANSYS CFX has developed a unique moving mesh method (immersed solid method) for rotor pump whose rotor movement is special. This method doesn't need any grid deformation and reconstruction. By applying the momentum source terms, the immersed entity is considered as the momentum source

IOP Conf. Series: Materials Science and Engineering **751** (2020) 012041 doi:10.1088/1757-899X/751/1/012041

terms for flow equations. The size of the momentum source terms can be adjusted by the momentum source scaling factor, whose default value is 10.0 and it can be applied in most cases. The established immersion entity needs to be partially or completely immersed in the fluid domain, without crossing any flow field boundary or colliding with other solids. Usually, the surface grid quality of the immersed entity must be very good.

In this paper, the rated parameters of rotor pump are as follows: rated speed 600r/min, rated theoretical flow rate 50m3/h, rotor center distance 136mm, rotor length 75mm, the clearance of rotor and rotor, rotor and pump cavity, rotor and liner 0.05mm. In order to make the simulation convergence and results more stable, the inlet and outlet section of the pump are properly extended. As shown in figure 2, The whole model is composed of the inlet and outlet section, the rotor solid domain and the fluid domain of the pump cavity.

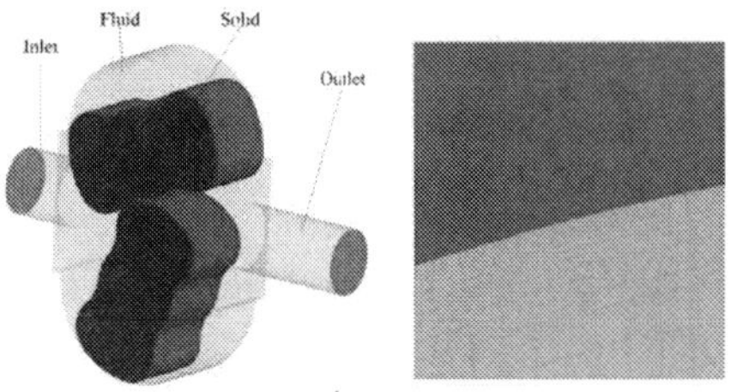

(a) Simulation model (b) mesh magnification of the solid surface

Figure 2. Assembly and grid diagram of the pump

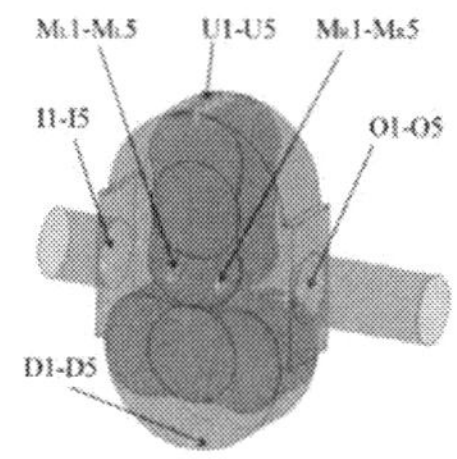

Figure 3. Measurement points

3.2 Boundary conditions

The pressure and velocity are coupled by SIMPLE algorithm, and the pressure in the flow field is absolute pressure for the reference pressure is set to 0. In order to make the calculated flow field closer to the real situation, the total pressure is used for the inlet and outlet boundary conditions. According to the actual working conditions of the elastomer rotor pump, the total pressure of the inlet is set to 40 kPa, and the total pressure of the outlet is set to 200 kpa. The rotor speed is set to 600 r/ min, and the time step is set to 5* 10^{-5} s. The residual mean square(RMS) of all the control volumes in the calculation domain is chose as the convergence criterion, and the RMS convergence accuracy is 10^{-5}.

3.3 measurement points setting

The pressure pulsations in different positions of elastomer cam rotor pump are analyzed. As shown in figure 3, Five measurement points (I1-I5) are set at the inlet of the pump cavity, and five measurement points (O1-O5) are set at the outlet of the pump cavity. 10 measurement points (ML1-ML5, MR1-MR5) are set at the meshing position. Five measurement points (U1-U5) are set in the upper rotor chamber, and five measurement points (D1-D5) are set in the lower rotor chamber.

4. Results and Discussions

Under the conditions of the same center distance and the same radius R of the pump cavity, the wrap angle of the wide circular arc and the radius of the scraping arc are the key parameters that directly affect the performance of the rotor pump. The wrap angle affects the sealing width, and the radius of the scraping arc affects the effect of mud scraping. Under the conditions of that the center distance r_p is 136mm, and the radius R is 90.5mm, figure 4 shows the rotor shape when the wrap angle θ of the wide circular arc is 5degrees, 8 degrees,11 degrees, 15 degrees and 18degrees. Under the conditions of that the center distance r_p is 136mm, and the radius R and the wrap angle θ are respectively 90.5mm and 18 degrees.

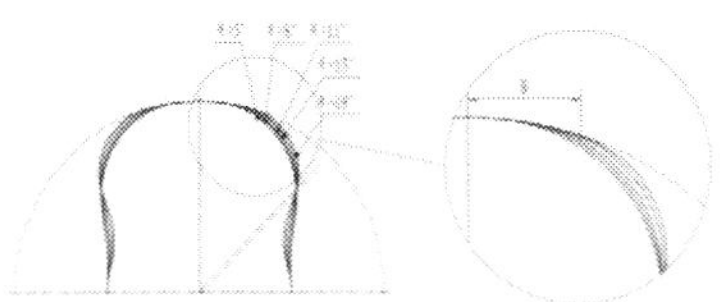

Figure 4. Change of wrap angle

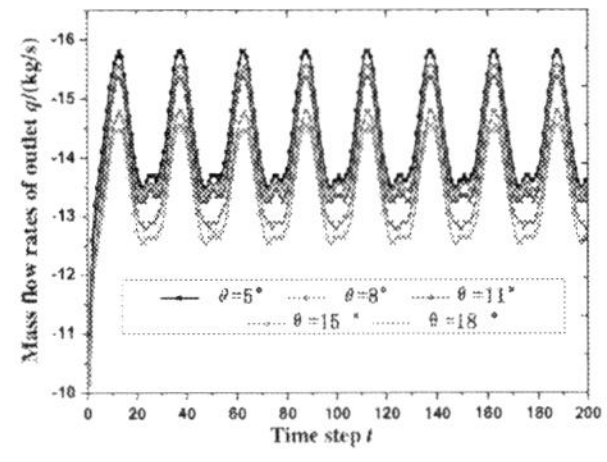

Figure 5. The pulsation curves of instantaneous mass flow rate of the outlet with different wrap angles

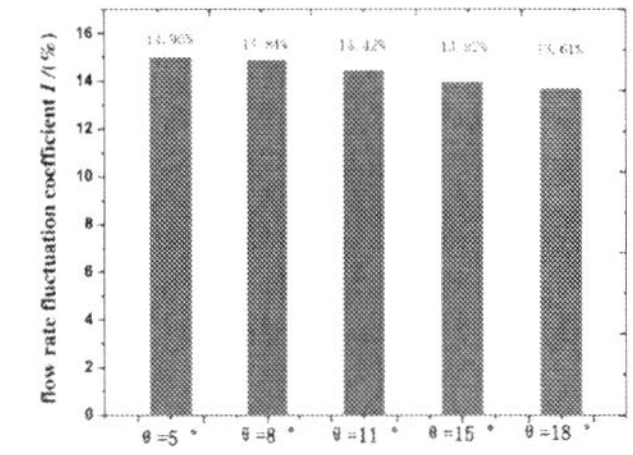

Figure 6. The pulsation coefficient of instantaneous mass flow rate of the outlet with different wrap angles

4.1. Pulsation of instantaneous mass flow rate

The pulsation of instantaneous mass flow rate affects pump outlet vibration and noise. The pulsation coefficient is defined as follows:

$$\eta = \frac{q_{\max} - q_{\min}}{q_{ave}} \tag{1}$$

As shown in the formula, qmax is the maximum mass flow rate, and qmin is the minimum mass flow rate. qave is the mean mass flow rate.

Figure 5 shows the pulsation curves and the coefficient diagram of the instantaneous mass flow rate of the outlet when the wrap angle θ is 5°,8°,11°,15°,18°. As shown in figure 6, the pulsation period of mass flow is the same at different wrap angles, and the pulsation curve is basically sinusoidal. The larger the wrap angle is, the smaller the amplitude is. As shown in figure 6, the pulsation of the instantaneous mass flow rate decreases with the increase of wrap angle, and the coefficient is respectively 14.96%, 14.84%, 14.42%, 13.92%, 13.61%. The width of the sealing surface between rotor and pump body increases with the increase of wrap angle, so the gap leakage of rotor pump is effectively suppressed, and the mass flow characteristics tend to be stable.

4.2 Pressure fluctuation

4.2.1Effect of wrap angle on pressure fluctuation at inlet monitoring point and inlet meshing point .Figure 7 shows the pressure fluctuations of the inlet monitoring point I_3 and inlet meshing poin ML_3 at different angles. The results show that the pressure fluctuation curves of I_3 and ML_3 points is periodically distributed, and the fluctuation period is consistent with the number of rotor blades. This shows that the pressure fluctuation period is determined by the number of rotor blades. There is no significant difference in the amplitude of pressure fluctuation at I_3 point, and the fluctuation of ML_3 point is divided into two parts: main wave and secondary wave. The pressure amplitude at I_3 point is smaller than that of ML_3, and the fluctuation period is misplaced for 1/4 cycle. This demonstrates that the pressure fluctuation originates from the meshing point ML_3 and transmits to the inlet, and the pressure decays gradually. The pressure fluctuations of ML_3 point are significantly different with different wrap angles. The minimum pressure amplitude of the main wave Pmin is 130kpa, 122kpa, 115kpa, 108kpa, 100kpa, respectively. With the increase of wrap angle, the minimum pressure of ML_3 decreases gradually. The reason is that the larger the wrap angle is, the more obvious the waist depression of the rotor is. The area of the meshing area that needs to be filled increases (shown in figure 8). Consequently, the liquid filling at the meshing site is insufficient, which increases the possibility of cavitation in the inlet meshing area.

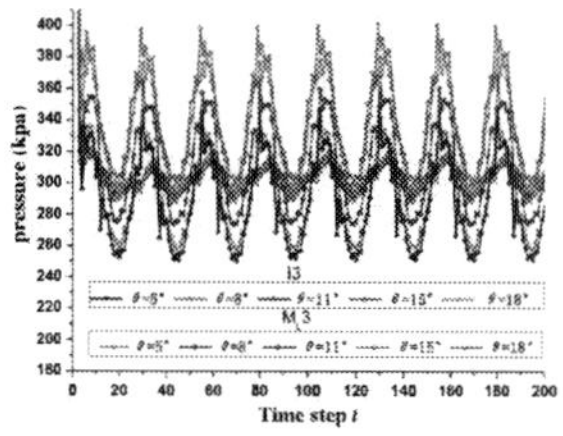

Figure 7. Pressure fluctuation at position I_3 and ML_3

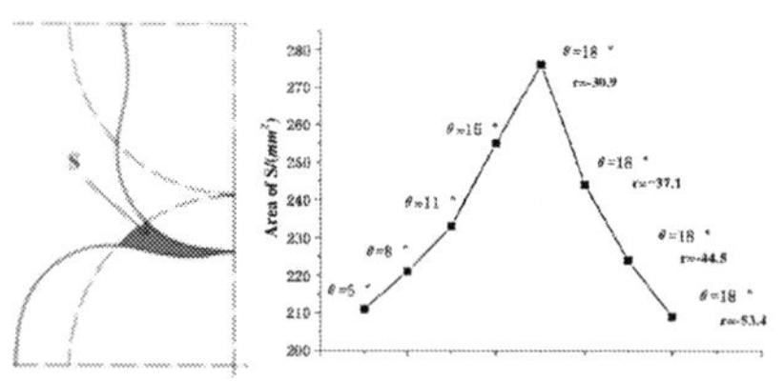

Figure 8. the area of the meshing area with different angles and radii

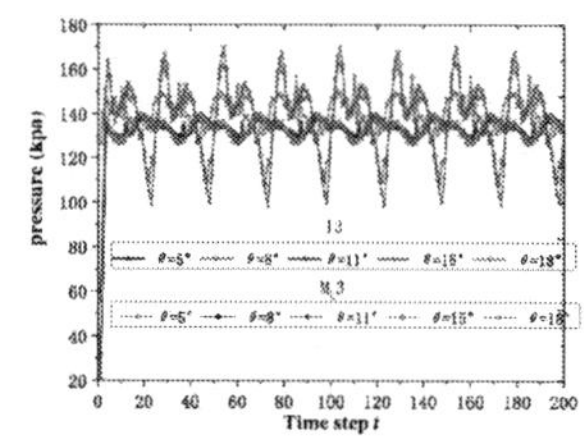

Figure 9. Pressure fluctuation at position O_3 and MR_3

4.2.2 Effect of wrap angle on pressure fluctuation at outlet monitoring point and outlet meshing point. Figure 9 shows the pressure fluctuations at outlet monitoring point O_3 and outlet meshing point MR_3. It can be seen that the pressure amplitude at O_3 point doesn't fluctuate much with different wrap angles. The pressure at MR3 point is much higher than that at O_3 point, and the fluctuation period is misplaced for 1/4 cycle. The fluctuation of O_3 point originates from the MR_3.Under the condition of different wrap angles, the maximum pressure fluctuation amplitude of MR_3 point varies greatly, which is 347kpa, 348kpa, 363kpa, 380kpa and 423kpa, respectively. This demonstrates that the larger the wrap angle is, the greater the local pressure at the outlet meshing is, and the greater the periodic pressure shock is on the rotor surface. Therefore, It is easier to cause the adhesive surface of the elastomer to fall off. Under the condition of satisfying the sealing condition at the top of the rotor, It is suggested that the small wrap angle is better.

5. Conclusions

In view of the defects of the present profile of elastomeric rotor pump, a new type of scraping rotor is proposed. The rotor profile is composed of two arcs, one eccentric scraping arc and one isometric curve of cycloid, and a wide head scraping rotor profile is obtained. The invention has the following advantages: good sealing performance, Strong passing ability, effectively preventing particles from sticking and skidding.

The wrap angle have little influence on mass flow. With the increase of wrap angle, the pulsation of instantaneous mass flow decreases.

With the increase of wrap angle, the possibility of cavitation at the inlet meshing increases, and the pressure at the outlet meshing increases. In order to improve the periodic shock of liquid on the elastomer and prevent the adhesion surface from falling off, the elastomer rotor should adopt small wrap angle under the condition of ensuring the sealing width at the top of the rotor.

References

[1] Wenner C. Running rotary lobe pumps at high speeds[J]. *World Pumps*, 2004, **2004(448)**: pp 33-35.
[2] Vogelsang Maschinenbau H. Injection housing–optimising rotary lobe pumps for liquids with

high solids content[J].*World Pumps*,2007,**2007(489)**: pp 36-38.
[3] Verhuelsdonk B. Increasing the operational lifetime of rotary lobe pumps[J].*World Pumps*,2005, **2005(468)**: pp 42-44.
[4] Eggert H. Rotary lobe pumps run like clockwork[J]. *World Pumps*, 2011, **2011(10)**: pp 34-35.
[5] Xuezhong Xu. *Design theory of rotor pump* [M]. Suzhou: suzhou university press, 2015.
[6] Xuming Du. Research and Design on Rotor Type Line of Rotor Pump[D]. Lanzhou: Lanzhou University of Technology, 2012. (in Chinese with English abstract)
[7] Y H Kang, Vu H H. A newly developed rotor profile for lobe pumps: Generation and numerical performance assessment[J]. *Journal of Mechanical Science and Technology*,2014,**28(3)**: pp 915-926.
[8] Ligang Yao, Zhonghe Ye, Haiyi Cal, et al. Geometric analysis and tooth profiling of a three-lobe helical rotor of the Roots blower [J]. *Journal of Materials Processing Technology*,2005, **(170)**: pp 259-267.
[9] Xiaojun Jian. Numerical Investigation of Positive Displacement Rotary Lobe Pump with Twisted Rotors[D]. Hangzhou: Zhejiang Sci-Tech University, 2015.
[10] Haohai Qian. Optimized model selection of rotor pump based on flow analysis [D].Hangzhou: Zhejiang Sci-Tech University, 2013.
[11] Chunhui Fang. The numerical simulation analysis and design of rotor pumps with circular curve rotor[D]. Lanzhou: Lanzhou University of Technology, 2014.
[12] Liu,Y.Y, wang,L.Q.Zhu,Z.C.Numerical study on flow characteristic of rotor pumps including caviation[J].*Proceeding of the institution of mechanical engineering* ,2015,**229(14)**: pp 2622-2638.
[13] Liu ,H.C,tong,C.H,yang,D.C.H.Trapping free rotors for high sealing lobe pumps[J].*Journal of mech design transaction of the ASME*,2005,**122(12)**: pp 536-542.
[14] S H Tong, Yang D C H. Rotor profiles synthesis for lobe pump with given flow rate function [J].*Journal of mechanical design*,2005, **127(2)**: pp 287-294.
[15] Yang Daniel C H. Tong Shih-Hsi. The specific flow rate of deviation function based lobe pump -derivation and analysis [J]. *Mechanical and machine theory* ,2002,**37(10)**: pp 1025-1042.
[16] Hongxin Zhang,Tiezhu Zhang,Jizhong Zhang,et al. Optimum design of non-contact rotor pump[J].*Transactions of the Chinese Society For Agricultural Machinery (Transactions of the CSAM)*, 2005, **36(7)**: pp 65-67. (in Chinese with English abstract)
[17] Vogelsang H, Verhulsdonk B, Turk M,et al. Pulsation problem in rotary lobe pump [J].*World pump* ,**1999(389)**: pp 1-6.
[18] Dr Meno Türk, Burkhard Verhülsdonk. Gap leakage behavior of helical vane rotary lobe pumps[J]. *World Pumps,*2006, **4**: pp 32-37.
[19] Del Campo D, Castilla R, Raush G A, et al. Numerical analysis of external gear pumps including cavitation[J]. *Journal of Fluids Engineering*,2012,134(8): 081105.
[20] Jun Wang, Kai Liu, Xitong Jiang, et al. Research on working process characteristic of claw type fluid machinery [J]. *Fluid machinery*,2012,**40(12)**: pp 39-43.
[21] Jun Wang,Kai Liu, Xitong Jiang,et al. 3D numerical simulation of the working process of claw type fluid machinery [J]. *Journal of engineering thermophysics*,2014,**35(2)**: pp 270-273.
[22] z x Liu, l H Xu. Unsteady numerical simulation and analysis of the flow fluctuation in the roots blower [J].*Journal of Aerospace Pow*,2007,**22(3)**: pp 400-406.

IOP Conf. Series: Materials Science and Engineering **751** (2020) 012042 doi:10.1088/1757-899X/751/1/012042

Experimental Study of Hybrid Deicing System

Shu Jun[1], Xu Dongguang[1*], Ye Lin[2] and Zhu dongyu[3]

[1]COMAC Shanghai Aircraft Design and Research Institute, Shanghai 201210, China
[2]Huazhong University of Science and Technology, Wuhan 430074, China
[3]AVIC Aerodynamics Research Institute, Shenyang 110034, China

[*]E-mail address: xudongguang@comac.cc

Abstract. The hybrid deicing system combines the traditional electrothermal deicing system with the electro impulse deicing system. Firstly, the prototype design of the ice wind tunnel hybrid deicing system is carried out, and the electrothermal deicing system and the electro impulse deicing system are integrated in the airfoil model. Then, through the laboratory plate deicing experiment to study the deicing effect of different heating and excitation combinations, explore the work control law of the hybrid deicing system, and finally verify the effect of the hybrid deicing system through the ice wind tunnel test. The plate panel test found that with the increase of the surface temperature of the skin, there is a minimum binding force between the ice and the skin. In the ice wind tunnel deicing experiment, the prototype of the hybrid deicing system shows a good deicing effect with less energy consumption.

1. Introduction

The traditional electrothermal deicing technology is a widely used aircraft anti-icing technology. It has high efficiency and simple structure, but it has the disadvantages of high energy consumption and easy formation of runback ice[1]. The large energy consumption of electrothermal deicing directly affects the life time of the aircraft[2]. For small, medium-sized aircraft and drones with insufficient energy, electrothermal deicing energy consumption will become unacceptable. Runback ice and ice ridge will seriously affect aircraft flight safety[3, 4].

The electro impulse deicing technology was first proposed by Soviet scientists[5]. The principle is to install a pulse coil under the metal skin and use the instantaneous discharge technique to form an electromagnetic eddy current field on the metal skin, thus generating a transient electromagnetic force, which leads to electromagnetic force. The skin quickly vibrates and deforms and the ice layer breakup and fall off, and finally removes the ice under the action of aerodynamic forces and inertial forces. The electro impulse deicing technology has the advantages of low energy consumption, simple installation and easy maintenance, and not easy to generate runback ice. The United States[6], Europe[7] and other countries[8] have conducted a lot of research but have not yet been commercialized[9]. Because the vibration of excessive frequency is easy to cause skin fatigue damage. Secondly, the strong magnetic interference generated by the electro impulse discharge is easy to cause interference to other electronic equipment of the aircraft. Currently only Russia applies the deicing technology on aircraft. The electro impulse deicing technology belongs to the mechanical deicing technology.

The hybrid deicing combines the electrothermal deicing technology with the electro impulse deicing technology, and uses the electric heating system to reduce the bonding force of the ice layer-skin surface, and then uses the electro impulse to destroy the ice layer to achieve the deicing effect, which reduces the The energy consumption of electrothermal deicing also reduces the damage of the skin by electro impulse deicing. In 1997, the United States Kamel AL-Khalil carried out experimental research in IRT, which showed that the method is suitable for de-icing of surpercooled large droplets. The electro impulse de-icing system can prevent the water generated by the heating of the leading edge heating belt from generating runback ice[10]. And the The hybrid deicing has the advantage of energy saving[11].

2. Plate deicing test

2.1. Plate deicing test model

Figure 1 shows a schematic diagram of a conventional electrothermal deicing system on a section of a wing, consisting of three pieces of heating belts. Design a hybrid deicing system based on the traditional electrothermal deicing system. As shown in Figure 2, reduce the area of the upper and lower airfoil heating belts to allow sufficient space for the impulse to operate. The two circular positions in the figure are the positions of the impulse on the surface of the skin.

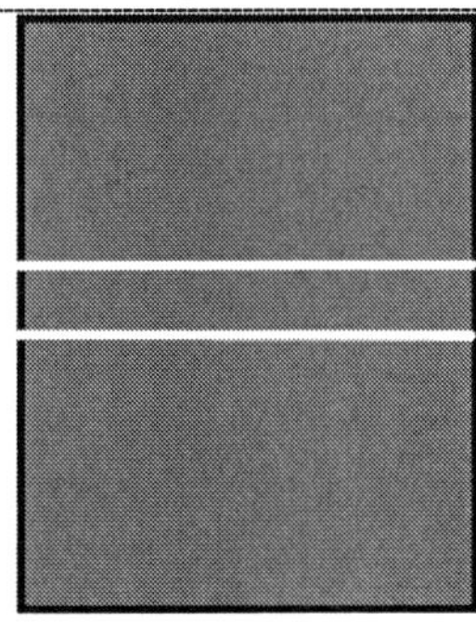

Figure 1. The schematic diagram of conventional electrothermal deicing system

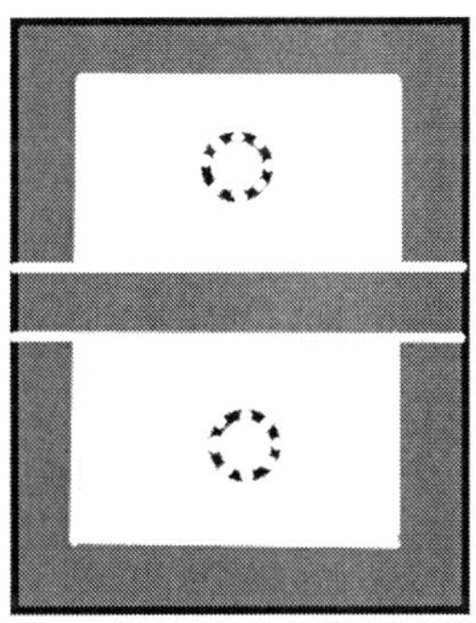

Figure 2. The schematic diagram of hybrid deicing system

The test plate is shown in the figure 3. The test plate material is made of aluminum alloy. The size of the plate is 600mm×420mm×2mm (length, width and thickness), and the back of the aluminum plate is close to 540mm×370mm×3mm (length, width and thickness).The heating film's power is 480W, and the model is equipped with an electro impulse system in the area of the unheated film in the middle of two sides, and its structural layout is as shown in the following figure.

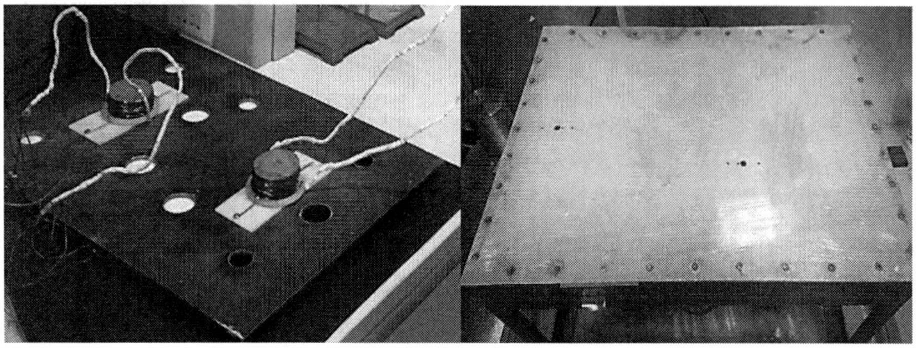

Figure 3. Plate deicing test model

2.2. Test conditions

The plate deicing test was completed in the open freezer, and the icing conditions were basically the same.The test conditions are shown in Table 1. In each test, water was sprayed layer by layer on the surface of the plate before deicing began to produce about 15 cm thick glaze ice.

Table 1. Test conditions of plate deicing tests

Temperature (°C)	Heating power (W)	Deicing time (s)	Number of Impulse
-20	480	$\approx$300	6

2.3. test results

As shown in Figure 4, the results of the electro impulse deicing test show that the ice layer has a large area crack due to the surface deformation of the plate plate, but because the ice layer and the plate surface are firmly bonded, there is no large area of ice on the plate surface falling off.

Figure 4. Result of the electro impulse deicing test

As shown in Figure 5, the result of the hybrid deicing test is that after heating for 280 s with 480 W heating power and then impulse for 6 times, most of the ice layer also has large-area cracks due to surface deformation of the plate plate, and the surface temperature rises with the electric heating plate. The adhesion between the layer and the surface of the plate is reduced, and a large area of ice is detached.

Figure 5. Result of the hybrid deicing test

3. Icing wind tunnel deicing test

3.1. Test model

Figure 6 shows the structural assembly drawing of the ice wind tunnel test model. The hybrid deicing system is arranged in the front wing section. The electro impulse system is arranged in the model fixed on the beam. Figure 7 shows the Installation position of heating film, and the heating system is arranged on the front wing surface.

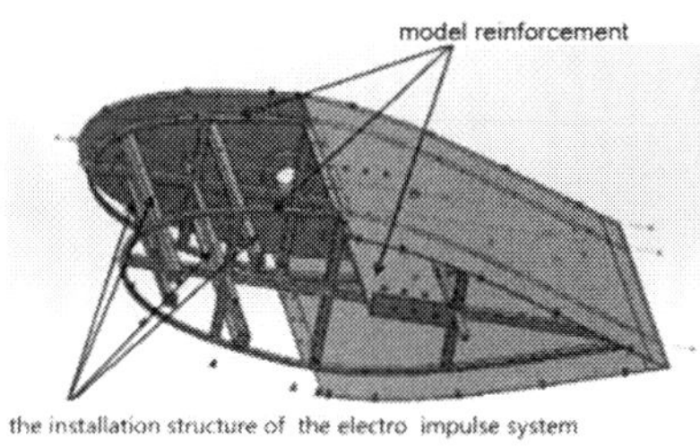

Figure 6. The structural assembly drawing of the ice wind tunnel test model

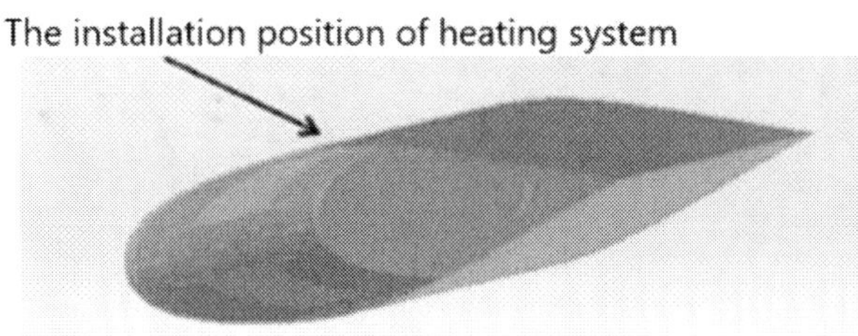

Figure 7. The installation position of heating film

3.2. Test procedure

The icing wind tunnel deicing tests was carried out in the FL-61of AVICARI. The size of the FL-61 icing test section was 0.6×0.6m.

Firstly, the hybrid deicing system is tested on the ground to confirm that the heating system and the electro impulse system are working properly, and the model is insulated. Then carry out the hybrid deicing test in the ice wind tunnel.

Based on the experience of the previous plate-plate deicing test, firstly, a certain thickness of ice is formed on the surface of the wing and then the heating system is turned on. After heating for a period of time, the electro impulse system is turned on.

3.3. Test conditions

The icing conditions of icing wind tunnel deicing tests are based on the continuous maximum icing conditions of CCAR25-R4 Appendix C. The specific icing conditions are shown in Table 2.

Table 2. Test conditions of icing wind tunnel deicing tests

Volocity (m/s)	MVD(μ m)	LWC (g/m3)	Temperature℃)
40	20	1.4	-9.4

3.4. test results

Figure 8 shows the results of the deicing experiment of the ice wind tunnel of the principle sample. It can be seen from the figure that the results of the deicing of the plate plate are basically the same. The heating system heating reduces the combination of the ice and the skin of the leading edge. After the electro impulse system act, a large area of ice layer on the front edge falls off. However, due to the large curvature of the leading edge of the airfoil, the impulse effect of the electro impulse system at different positions is different, so there is still some residual ice on the upper and lower airfoil.

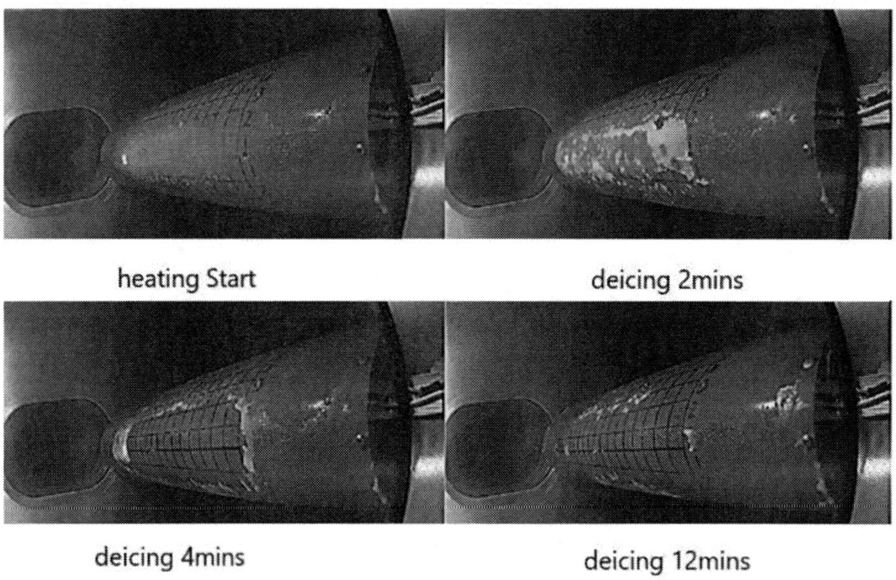

Figure 8. The results of the ice wind tunnel deicing test

4. Conclusion

The electrothermal deicing system and the electro impulse deicing system are integrated to form a hybrid deicing system, and the plate deicing test and the ice wind tunnel deicing test are carried out.

(1) The plate deicing test explores the hybrid deicing method. As the surface temperature of the plate plate rises, the bonding force between the ice layer and the plate plate decreases, and the ice layer is more likely to fall off under the action of the electro impulse system.

(2) Icing wind tunnel deicing test verifies the principle of the hybrid deicing system. For the large-area icing of the leading edge, the composite de-icing method of heating and then impulse can cause the ice layer to fall off in a large area, but Due to the large curvature of the leading edge, the impulse effect of the impulse force is uneven, and the ice in the vicinity of the icing limit is not easily shaken, and residual ice is formed.

(3) According to the results of the ice wind tunnel test, the layout and design of electrothermal system and the electro impulse system in the test model are to be optimized, so that the impulse force will be more evenly activated, and the deicing effect of the hybrid deicing system will be improved.

References

[1] CEBECIT, KAFYEKEF 2003. Aircraft Icing. AnnuRev Fluid Mech, 35: 11-21.
[2] Calay R K, Holdo A E, Mayman P 1997. Experimental simulation of runback ice. Journal of aircraft, 34(2): 206-212.
[3] Lee S, Bragg M B 2001. The effect of ridge-ice location and the role of airfoil geometry. AIAA-01-2481.
[4] Edward A W, Andy P B, Michael B B 2005. Characteristics of runback ice accretions on airfoils and ther aerodynamic effects. AIAA-2005-1065.
[5] Levin I A 1972. USSR electric impulse deicing system design. Aircraft Engineering, 7.
[6] Zumwalt G W, Friedberg R A 1986. Designing an electro impulse deicing system. AIAA-86-0545.
[7] Zumwalt G W 1988. Electro-impulse de-icing a status report. AIAA-88-0019.
[8] Reich A, Goodrich B F, Scavuzzo R, et al 1994. Survey of mechanical properties of impact ice, AIAA-94-0712.
[9] Chu M C, Scavuzzo R J 1991. Adhesive shear strength of impact ice. AIAA Journal, 29(11): 1921-1926.
[10] Al-Khalil K, Ferguson T, Phillips D, et al 1997. A hybrid anti-icing ice protection system.
[11] Fortin G, Adomou M, Perron J 2011. Experimental Study of Hybrid Anti-Icing Systems Combining Thermoelectric and Hydrophobic Coatings. SAE International Journal of Aerospace. 281-283.

Investigation on Steady State Unbalance Response of Rotor with Elastic Ring Squeeze Film Damper

Lu Zhao[a], Mingfu Liao[b] and Jihui Niu

School of Power and Energy, Northwestern Polytechnical University, Xi'an, China

[a]luzhao@mail.nwpu.edu.cn; [b]mfliao@nwpu.edu.cn

Abstract. In this paper, the method for calculating steady state response of a rotor system with elastic ring squeeze film damper(s) is studied. The finite difference method is used to calculate the oil film pressure field which excludes the influence of rotating speed and considers the action of oil orifices starting with the governing equations of oil film. The mechanical characteristic curves of ERSFD are obtained by integrating the film pressure field, based on which the steady state response of the rotor system with ERSFD(s) is calculated by the finite element method and iteration method. The calculating response is compared with the experimental data, and it is found that the calculating response is in good agreement with the actual one, with the relative error of no more than 15%, which verifies the correctness of the above algorithm, and conclusion that the elastic ring extrusion oil film damper can be regarded as a linear element when the eccentricity journal of the journal is relatively small is obtained.

1. Introduction

Complicated structure and variable working conditions restrict the performance of aero-engines. Squeeze film damper (SFD) is widely used as a kind of passive isolator in bearings of aero-engines for simple structure and excellent performance. However, the film stiffness of conventional SFDs is highly nonlinear, which may result in bistable state or uncoordinated procession and limits its application [1]. Elastic ring squeeze film damper (ERSFD) is used on Ал-31Ф engine in Russia, for it can not only improve nonlinearity of SFD but also adjust frequency of rotor system. Elastic ring is combined with squeezed film chamber in ERSFD, which divides the oil film into inner and outer parts. Each part is separated by evenly distributing bosses and is connected by radial orifices with each other. The vibration reduction mechanism of ERSFD is analysed and the experiments are carried out by M Zhou [2-4] to prove the good damping effect. The equivalent stiffness and damping of ERSFD with different eccentricity is calculated by L Cao [5-8], and conclusion that mechanical characteristics of ERSFD is more linear than SFD is got. The transformation law between structure parameters and dynamic characteristics of ERSFD is studied by J Hong [9] using finite element method.

However, investigation on dynamics characteristics of a rotor system with ERSFDs can hardly be seen. In this paper, mechanism characteristics of ERSFD is analysed using finite difference method, results of which are used to calculate steady state unbalance response of rotor system with ERSFDs. The comparison with experimental data indicates the correctness of algorithm. Thus the method established in this paper can play a guiding role for the design of ERSFD on rotor system.

2. Mechanism characteristics of ERSFD

The steady pressure field governing equation of inner and outer film in cylindrical coordinates, as equation (1) and (2) shows, can be deduced from generalized Reynolds equation using bounding conditions of oil film [2]:

$$\frac{1}{R_1^2}\frac{\partial}{\partial\theta}\left(h_1^3\frac{\partial p_1}{\partial\theta}\right)+\frac{\partial}{\partial z}\left(h_1^3\frac{\partial p_1}{\partial z}\right)=-12\mu\Omega\left(\frac{\partial h_1}{\partial\theta}-\frac{\partial r}{\partial\theta}\right)+12\mu v_d \tag{1}$$

$$\frac{1}{R_2^2}\frac{\partial}{\partial\theta}\left(h_2^3\frac{\partial p_2}{\partial\theta}\right)+\frac{\partial}{\partial z}\left(h_2^3\frac{\partial p_2}{\partial z}\right)=12\mu v_d \tag{2}$$

where R_1 and R_2 are radius of the journal and the housing, h_1 and h_2 are the film thickness, μ is the fluid viscosity, Ω is the rotating speed, r is the radial deformation of ERSFD, v_d is the net velocity through orifice, and z and θ are the axial and circumferential coordinates. In addition, L is the axial width of ERSFD and n is the number of bosses. Figure 1 is the sketch map of ERSFD components. The parameters of ERSFD are illustrated in figure 2.

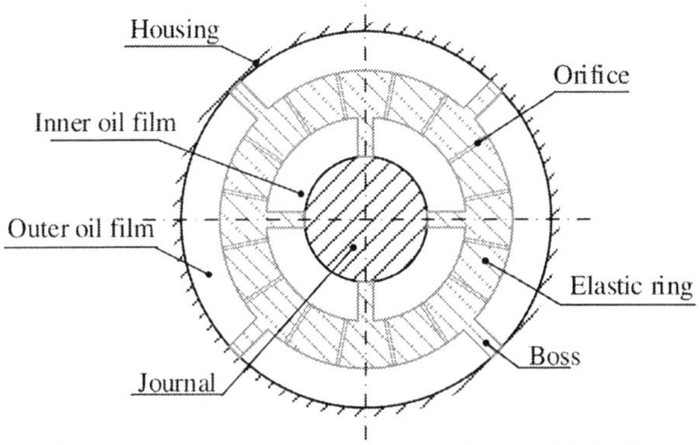

Figure 1. Structural representation of ERSFD.

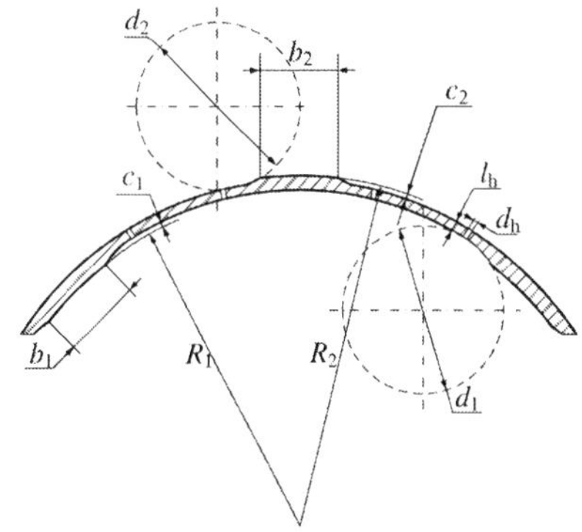

Figure 2. Structural parameters of ERSFD.

2.1. Clearance functions of ERSFD

According to equation (1) and (2), clearance functions h_1 and h_2 have effect on pressure field of ERSFD, and can be calculated from elastic deformation of ERSFD. Finite element method is accepted to solve the deformation of ERSFD with a bearing force acting on the journal. For example, a specific ERSFD with parameters in table 1 is used to explain the process of calculation.

Table 1. Structural parameters of example ERSFD.

R_1(mm)	R_2(mm)	c_1 and c_2 (mm)	l_h(mm)	L(mm)	b_1 and b_2 (mm)	d_1 and d_2 (mm)	n
65.5	67.5	0.3	1.4	16	5	10	10

First of all, three-dimensional models of ERSFD with its inner and outer bushing, just as figure 3 displays, are established.

Figure 3. Three-dimensional model of ERSFD.

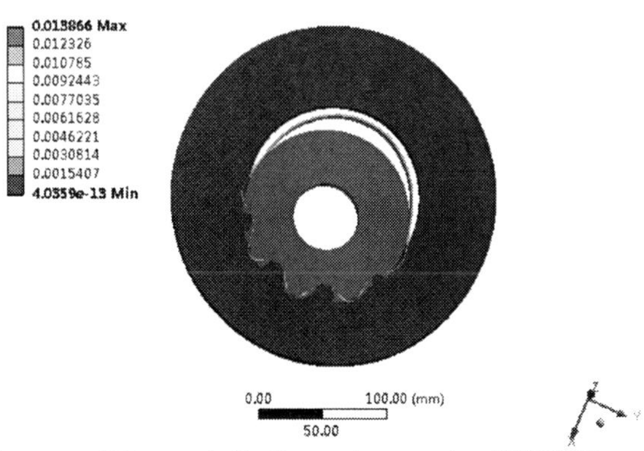

Figure 4. Deformation result of ERSFD.

After importing the material parameters into ANSYS, meshing and setting bounding conditions, problem can be solved. Here the elasticity modulus equals 2e11Pa, and the Poisson's ratio equals 0.3. Solid 186 element is used to mesh the body. The contact type between boss and lining on the direction of bearing load is bonded, while that of others is frictionless. Because boss and lining directly facing the load are bonded with each other and can hardly move relatively, but others may appear a little radial separation and circumferential slip with each other. Circumferential deformation restriction is imposed on the outer face of outer lining, and a 100N bearing load along X axis is applied to the inner face of inner lining. The deformation of ERSFD shows in figure 4, from which conclusion that the inner lining has translational motion and the eccentric side of ERSFD has gross distortion facing when the radial bearing load exists can be drawn.

The stiffness of ERSFD is calculated through equation (3), and that of example ERSFD is 7.18e6N/m.

$$K = F/s \tag{3}$$

Clearance functions can be got by exporting and disposing deformation data of ERSFD and its inner and outer lining. Relative distance between inner face of ERSFD and outer face of inner lining means the inner clearance function h_1, while that between outer face of ERSFD and inner face of outer lining means the outer one h_2. Figure 5 shows the result with 100N applied on inner lining along X axis.

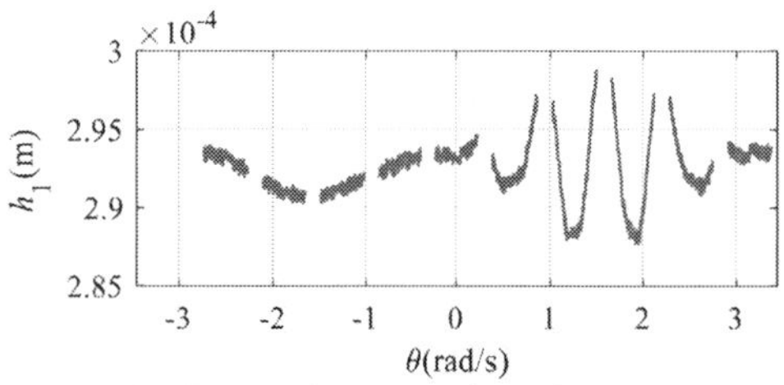

(a) Inner clearance function

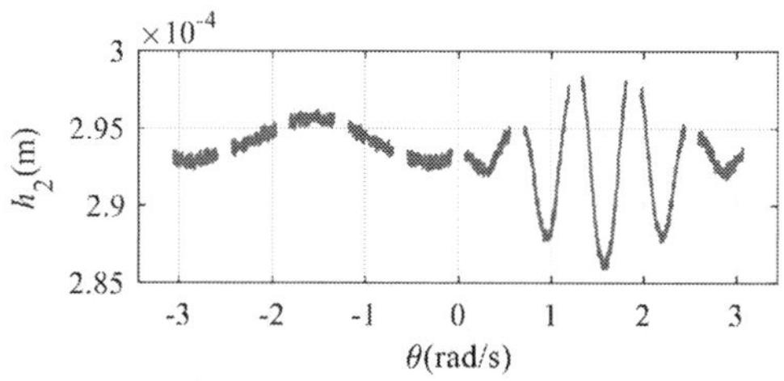

(b) Outer clearance function

Figure 5. Clearance functions of example ERSFD with 100N along X axis.

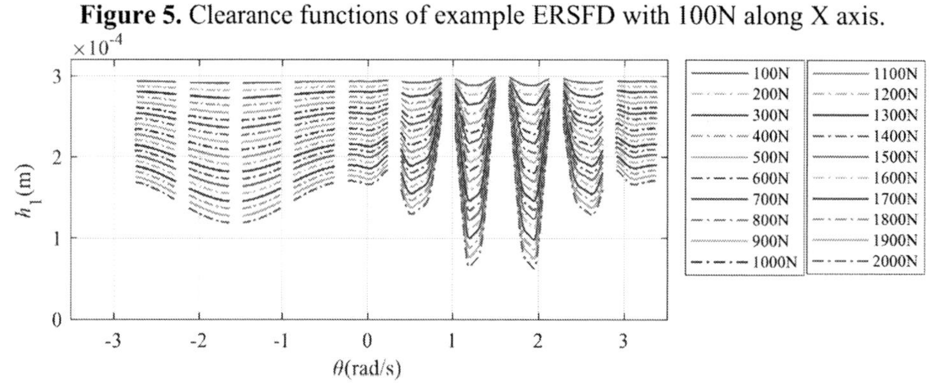

(a) Inner clearance function

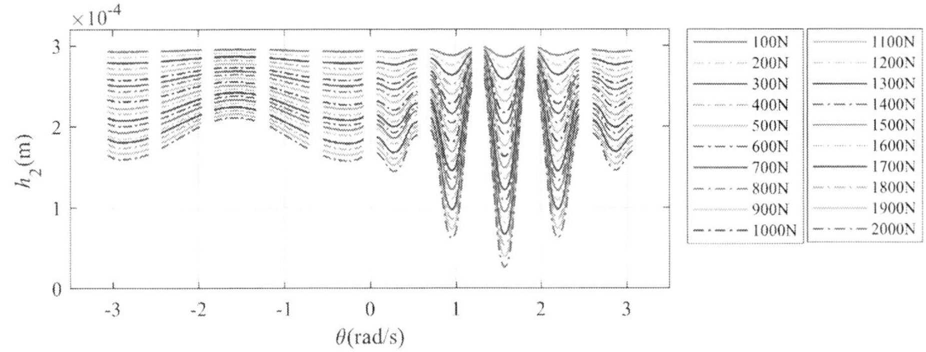

(b) Outer clearance function

Figure 6. Clearance functions of example ERSFD with different bearing loads along X axis.

Because the deformation of ERSFD and clearance functions vary with the bearing load, different clearance functions under different bearing loads on the same direction are calculated in figure 6, using which to linearly interpolate and get clearance at certain angle, and calculate the pressure field of oil film.

2.2. Pressure field of oil film

Finite difference method is accepted to get the pressure field of oil film, for it is difficult to solve equation (1) and (2) directly. Equation (5) is derived from the geometric equation of inner clearance, namely, equation (4) [2]:

$$h_1(\theta) = c_1 + r(\theta) + e\cos\theta \tag{4}$$

$$\frac{\partial h_1}{\partial \theta} = \frac{\partial r}{\partial \theta} - e\sin\theta \tag{5}$$

where e is the eccentricity of the journal. In this way, equation (1) can be written as:

$$\frac{1}{R_1^2}\frac{\partial}{\partial \theta}\left(h_1^3\frac{\partial p_1}{\partial \theta}\right) + \frac{\partial}{\partial z}\left(h_1^3\frac{\partial p_1}{\partial z}\right) = 12\mu\Omega e\sin\theta + 12\mu v_d \tag{6}$$

There is no r in equation (6), which simples the calculation of pressure field. It is obvious that the pressure of film varies with Ω. In order to eliminate the effect of Ω and improve the versatility of the result, dimensionless parameters in equation (7) are introduced.

$$\begin{cases} H = h/c \\ \lambda = 2z/l \\ P = pc^2/2\Omega\mu R^2 \\ \varepsilon = e/c \end{cases} \tag{7}$$

where ε is called the eccentricity ratio of ERSFD. Therefore, equation (6) and (2) can be written as:

$$\frac{\partial}{\partial \theta}\left(H_1^3\frac{\partial P_1}{\partial \theta}\right) + \left(\frac{2R_1}{l_1}\right)^2\frac{\partial}{\partial \lambda}\left(H_1^3\frac{\partial P_1}{\partial \lambda}\right) = 6\varepsilon\sin\theta + \frac{6v_d}{\Omega c_1} \tag{8}$$

$$\frac{\partial}{\partial \theta}\left(H_2^3\frac{\partial P_2}{\partial \theta}\right) + \left(\frac{2R_2}{l_2}\right)^2\frac{\partial}{\partial \lambda}\left(H_2^3\frac{\partial P_2}{\partial \lambda}\right) = \frac{6v_d}{\Omega c_2} \tag{9}$$

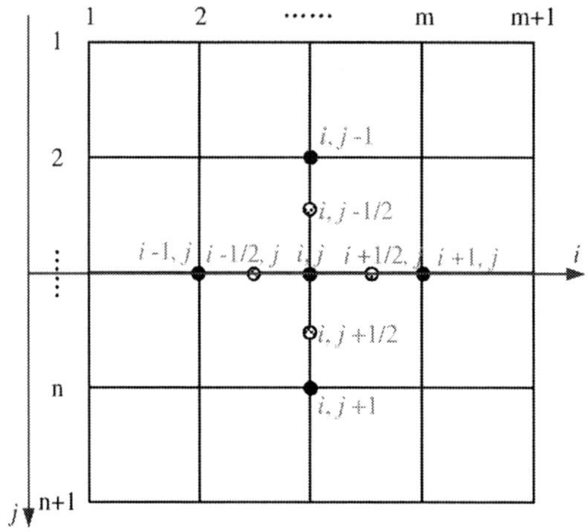

Figure 7. Meshing of oil film.

Just as figure 7 shows, inner and outer oil film can be divided into meshes and the derivative terms in equation (8) and (9) are expressed as the difference quotient form like:

$$\begin{cases} \left[\dfrac{\partial}{\partial\theta}\left(H^3\dfrac{\partial P}{\partial\theta}\right)\right]_{i,j} \approx \dfrac{H^3_{i+1/2,j}P_{i+1,j}+H^3_{i-1/2,j}P_{i-1,j}-\left(H^3_{i+1/2,j}+H^3_{i-1/2,j}\right)P_{i,j}}{(\Delta\theta)^2} \\ \left[\dfrac{\partial}{\partial\lambda}\left(H^3\dfrac{\partial P}{\partial\lambda}\right)\right]_{i,j} \approx \dfrac{H^3_{i,j+1/2}P_{i,j+1}+H^3_{i,j-1/2}P_{i,j-1}-\left(H^3_{i,j+1/2}+H^3_{i,j-1/2}\right)P_{i,j}}{(\Delta\lambda)^2} \end{cases} \tag{10}$$

where m and n are numbers of circumferential and axial nodes, $\Delta\theta=(\theta_{i+1}-\theta_i)/m$, $\Delta\lambda=2/n$.

The oil orifices are seen as capillary holes, so according to the flow equation (11) [2]:

$$Q=\frac{\pi d_h^4\Delta p_h}{128\mu l_h} \tag{11}$$

the expression equation of v_d can be got:

$$v_d=\frac{d_h^2R^2\Delta p_h\Omega}{16c^2l_h} \tag{12}$$

where d_h and l_h are the diameter and depth of orifice, and Δp_h is the pressure difference between inner and outer film at the hole. So the pressure at a certain node is expressed as equation (13) by combining equation (8) ~ (12) and adopting successive over-relaxation iteration (SOR):

$$P^{(k+1)}_{i,j}=\omega\bullet\left(A_{i,j}P^{(k)}_{i+1,j}+B_{i,j}P^{(k+1)}_{i-1,j}+C_{i,j}P^{(k)}_{i,j+1}+D_{i,j}P^{(k+1)}_{i,j-1}-F_{i,j}\right)/E_{i,j}+(1-\omega)\bullet P^{(k)}_{i,j} \tag{13}$$

where $A_{i,j}=H^3_{i+1/2,j}$, $B_{i,j}=H^3_{i-1/2,j}$, $C_{i,j}=\left[(2R\Delta\theta)/(l\Delta\lambda)\right]^2H^3_{i,j+1/2}$, $D_{i,j}=\left[(2R\Delta\theta)/(l\Delta\lambda)\right]^2H^3_{i,j-1/2}$, $E_{i,j}=A_{i,j}+B_{i,j}+C_{i,j}+D_{i,j}$, $F_{1i,j}=6\varepsilon\sin\theta-3d_h^2R^2\Delta P_h/(8l_hc^3)$, $F_{2i,j}=3d_h^2R^2\Delta P_h/(8l_hc^3)$, ω is relaxation factor. $A_{i,j}$, $B_{i,j}$, $C_{i,j}$, $D_{i,j}$ and $E_{i,j}$ are related to clearance functions, so after initializing the pressure field and getting clearance functions of ERSFD, final pressure field can be got through SOR method. The detailed process is showed in figure 8, and the final results are displayed in figure 9.

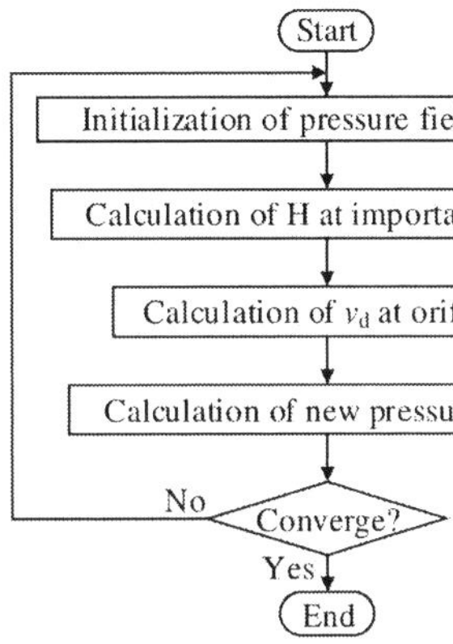

Figure 8. Calculation schedule of pressure field.

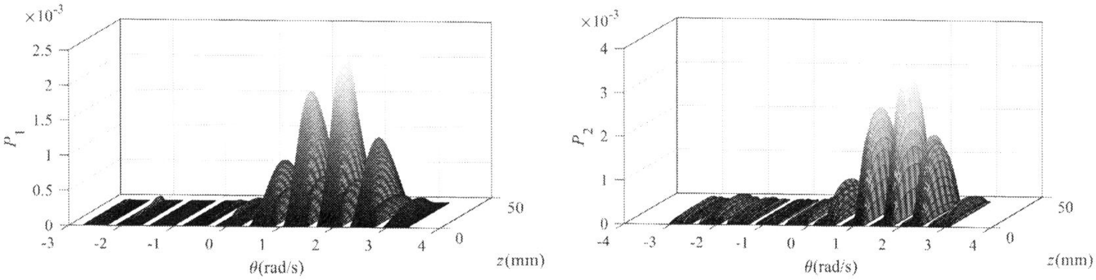

(a) Inner oil film pressure (b) Outer oil film pressure

Figure 9. Pressure field of oil film.

2.3. Equivalent stiffness and damping of ERSFD

Relative pressure fields are integrated by equation (14) to get relative radial and circumferential force.

$$\begin{bmatrix} \hat{F}_r \\ \hat{F}_t \end{bmatrix} = \sum_{1}^{m+1}\sum_{1}^{n+1} R\Delta\theta_{i,j}\Delta z_{i,j}P_{i,j}\begin{bmatrix} \cos\theta_{i,j} \\ \sin\theta_{i,j} \end{bmatrix} \tag{14}$$

Taking equation (7) in to consideration, absolute film forces can be written as:

$$\begin{bmatrix} F_{\Omega r} \\ F_{\Omega t} \end{bmatrix} = \frac{2\mu R^2}{c^2}\begin{bmatrix} \hat{F}_r \\ \hat{F}_t \end{bmatrix} \tag{15}$$

where $F_{\Omega r} = F_r/\Omega$, $F_{\Omega t} = F_t/\Omega$, F_r and F_t are absolute radial and circumferential force.

Equivalent stiffness and damping can be obtained by the following equation [10]:

$$K_n = -\frac{F_r}{e},\ C_n = -\frac{F_t}{e\Omega} \tag{16}$$

By acting different bearing loads on ERSFD, its mechanical characteristics can be got like figure 10, which shows the nonlinearity of ERSFD and is necessarily used to calculate steady state unbalance response of a rotor system with ERSFD(s). It can be seen that ERSFD is approximately a linear element when the eccentricity ratio of journal is less than 0.7.

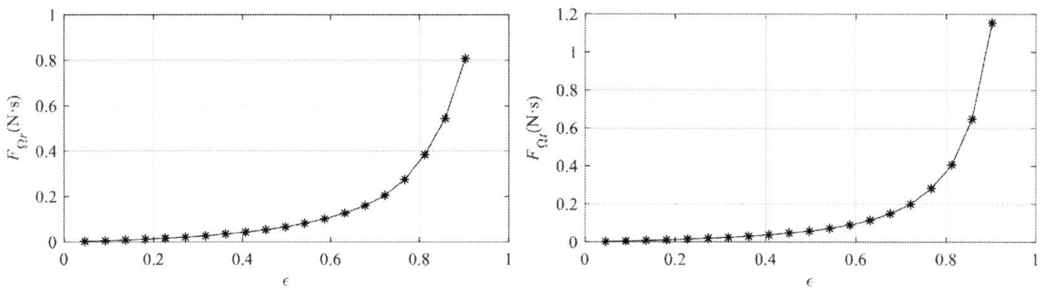

(a) Radial force (b) Circumferential force

Figure 10. Mechanical characteristics of ERSFD.

3. Dynamic characteristics of rotor system with ERSFD(s)

Finite element method is chosen to calculate dynamic characteristics using equation (17) [11]:

$$\boldsymbol{M}\ddot{\boldsymbol{q}} + (\boldsymbol{C} - \Omega\boldsymbol{G})\dot{\boldsymbol{q}} + \boldsymbol{K}\boldsymbol{q} = \boldsymbol{Q} \tag{17}$$

where $\boldsymbol{M}$ is mass matrix of the rotor system, $\boldsymbol{C}$ is damping matrix, $\boldsymbol{G}$ is gyroscopic matrix, $\boldsymbol{K}$ is stiffness matrix, $\boldsymbol{q}$ is generalized displacement vector, and $\boldsymbol{Q}$ is external force vector.

When ERSFD is installed on a rotor, it brings nonlinearity to the system as well. Nonlinear stiffness K_n is in parallel with linear stiffness of bearing K_0, and so is the damping. Therefore, equivalent stiffness and damping at bearing nodes can be written as:

$$\begin{cases} c' = C_n + C_0 \\ k' = K_n + K_0 \end{cases} \tag{18}$$

Consequently, the dynamic response of a rotor can be calculated using equation (17) if the nonlinear stiffness and damping at a certain rotating speed are known. However, just as figure 10 shows, eccentricity of journal has influence on K_n and C_n, and different eccentricity appears at different rotating speed. So the key part is to calculate eccentricity e.

Interactive method is adopted to get eccentricity and system response. When rotating speed staying at Ω_i, directly interaction method that using unbalance response with weak damping as initial value is given priority because of its simplicity and quick computation. Using q_0 to get equivalent stiffness and damping to form new matrix $\boldsymbol{K}$ and $\boldsymbol{C}$, then get new interactive response q by solving equation (17). When error between q and q_0 is less than its margin, process terminates. However, ERSFD may come into its bistable area, which makes aforesaid method dis-convergent and out of operation. Then comes to the method that calculating eccentricity of journal interactively at Ω_i firstly and using e to get unbalance response. Give e_0 a certain value bigger than 0 and smaller than 1, get matrix $\boldsymbol{K}$ and $\boldsymbol{C}$ to form equation (17), and examine whether e_0 is the root of motion equation. If not, give e_0 a new value to interact, otherwise, save the result of e or e_0 and get the response of the system. Detailed steps are showed in figure 11. By traversing all rotating speeds in solution of internal, dynamic response of the rotor with ERSFD(s) can be calculated.

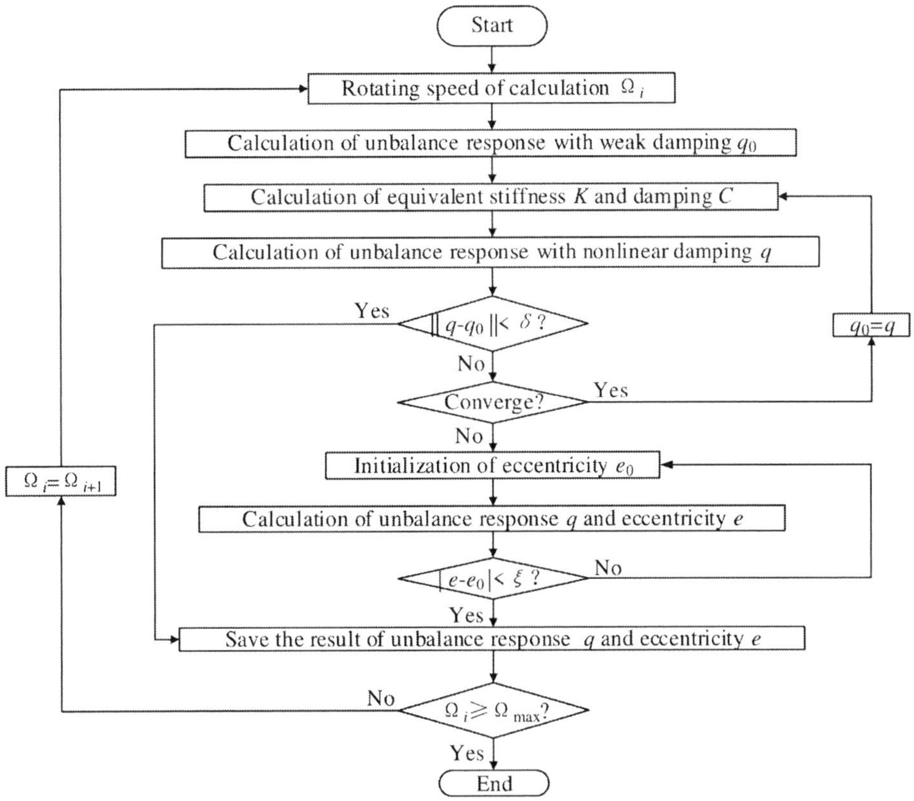

Figure 11. Calculation steps of eccentricity and system response.

4. Experiments on rotor with ERSFDs

In order to evaluate the validity of the arithmetic mentioned in chapter 2 and 3, an experiment system in figure 12 has been established, which is divided into 21 beam elements, 5 disk elements, 2 bearing elements and 1 bifurcate element. There are 2 ERSFDs at bearing 1 and bearing 2 relatively. Use finite element method to calculate dynamic characteristics of rotor system and get results in figure 13 and 14.

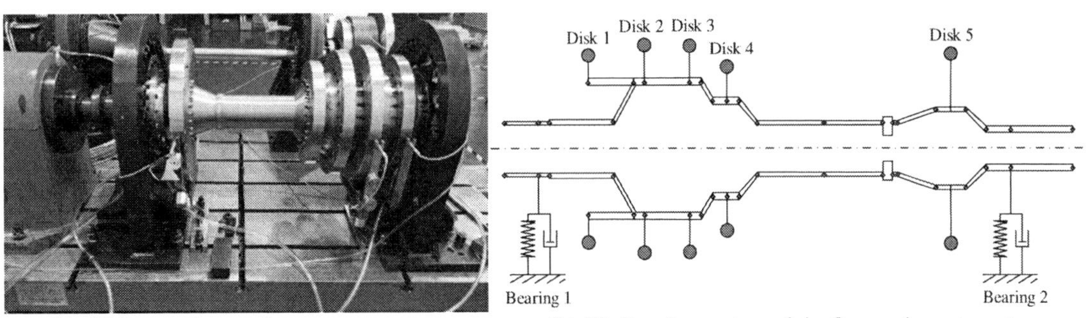

(a) Objects of experiment system (b) Finite element model of experiment system

Figure 12. Experiment system.

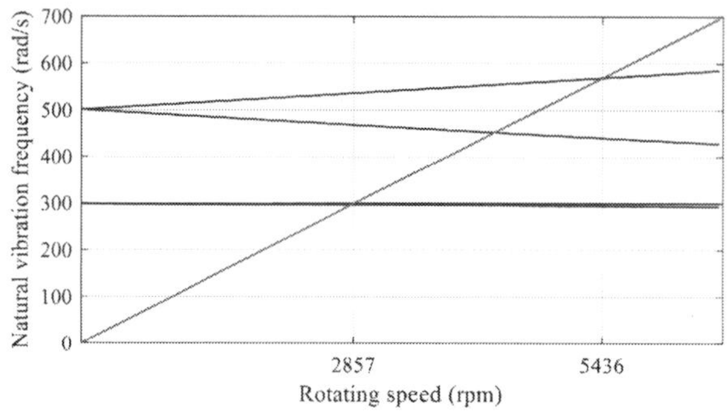

Figure 13. Campbell map of experiment system.

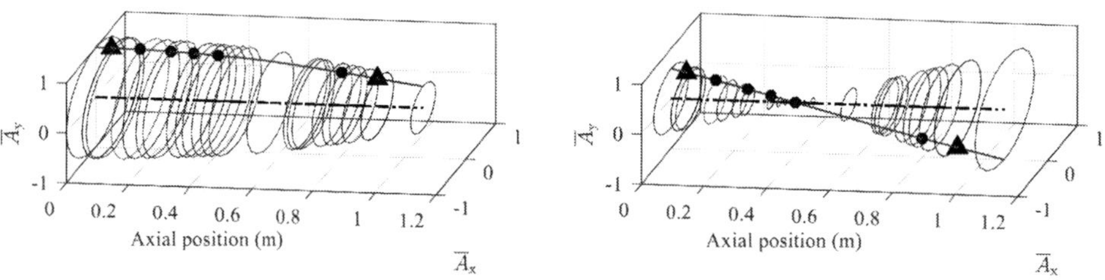

(a) The first translational mode of vibration (b) The second pitching mode of vibration

Figure 14. Mode shape of experiment system.

In figure 14, $\overline{A}_x$ and $\overline{A}_x$ are dimensionless displacement along X and Y axis. It can be saw that there is first one order critical in operating speed range 0 ~ 4000rpm. Vibration of disk 1 is larger than other disks, which provides obvious view to monitor and analyze the system, so displacement of disk 1 is chosen to verify the algorithm mentioned above. The parameters of ERSFDs used in experiment are listed in table 2, and actual components of ERSFDs are shown in figure 15.

Table 2. Structural parameters of experimental ERSFDs.

	R_1(mm)	R_2(mm)	c_1 and c_2 (mm)	l_h(mm)	L(mm)	b_1 and b_2 (mm)	d_1 and d_2 (mm)	n	K_0(N/m)
ERSFD 1	65.5	67.5	0.3	1.4	16	5	0.5	10	7.18e6
ERSFD 2	75.5	77.7	0.3	1.6	22	5	0.5	10	8.62e6

(a) ERSFDs (b) Inner lining (c) Outer lining

Figure 15. Components of ERSFDs.

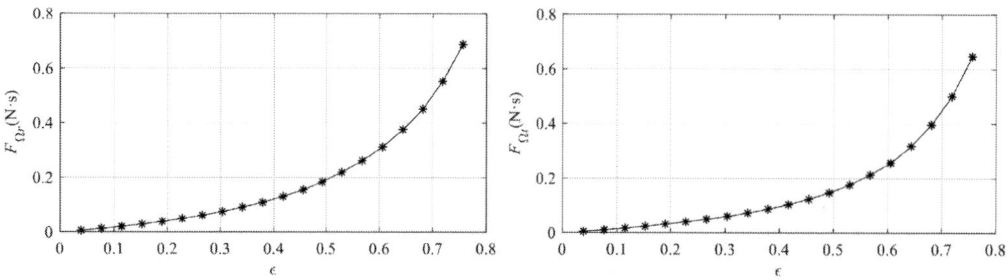

(a) Radial force (b) Circumferential force

Figure 16. Mechanical characteristics of ERSFD 2.

Mechanism characteristics of ERSFD 1 is illustrated in figure 10, and that of ERSFD 2 in figure 16.

Experiments about rotor system with ERSFDs are conducted at around 30℃. There is 77.4g·cm unbalance on disk 1, 54g·cm on disk 4 and 63g·cm on disk 5, and unbalance on disk 4 has opposite phase with the others. The results of experiment and calculation are drawn in figure 17. It can be seen from figure 17 and table 3 that calculation response agrees well with the experimental one, with the biggest relative error on damping ratio D less than 15%. The errors come from several resources. Firstly, the assembling relationships of system components when calculating are regarded to be tight, while they can be loose in practical and lead to errors. Secondly, cooperative relationship of ERSFD components will influent practical film stiffness and damping and result in errors. Thirdly, equation (1) and (2) are deduced under several assumptions, which may exists differences with factual condition.

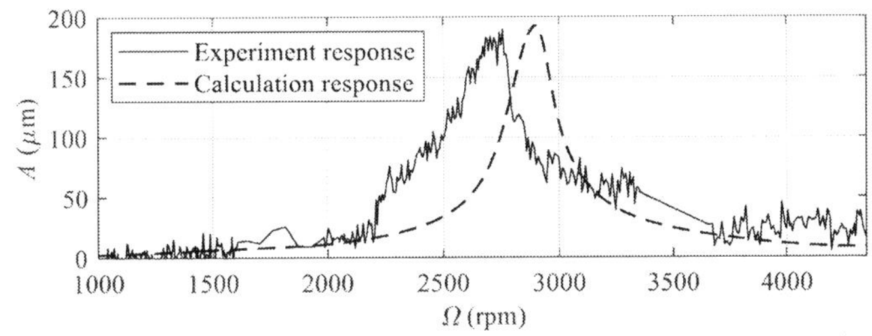

Figure 17. Comparison diagram of experiment and calculation.

Table 3. Parameters comparison of experiment and calculation.

	Peak of vibration A (μm)	Relative error of A (%)	Critical speed ω_{cr} (rpm)	Relative error of ω_{cr} (%)	Damping ratio D (%)	Relative error of D (%)
Experiment	379	1.58	2761	4.93	3.6	13.8
Calculation	384.9		2897		3.1	

Figure 18 shows the journal eccentricity ratios calculating result of two ERSFDs. It can be seen that the biggest ε of ERSFD 1 and ERSFD 2 are around 0.6 and 0.4, where they can still be regarded as linear elements. Figure 19 shows that there is not much difference between two kind of response, which means ERSFDs on the rotor system perform well and can be seen as linear elements at this operating condition.

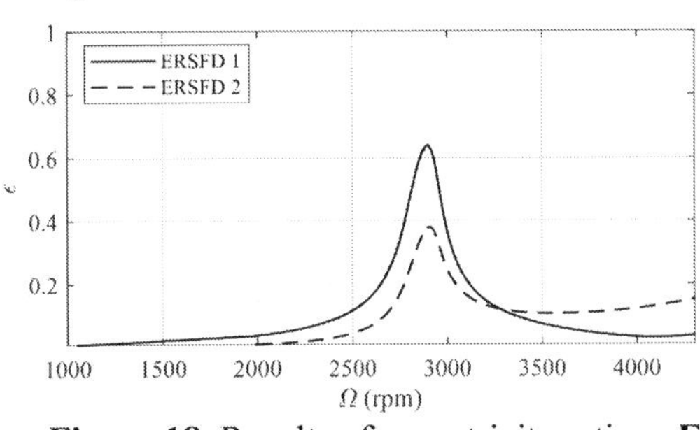

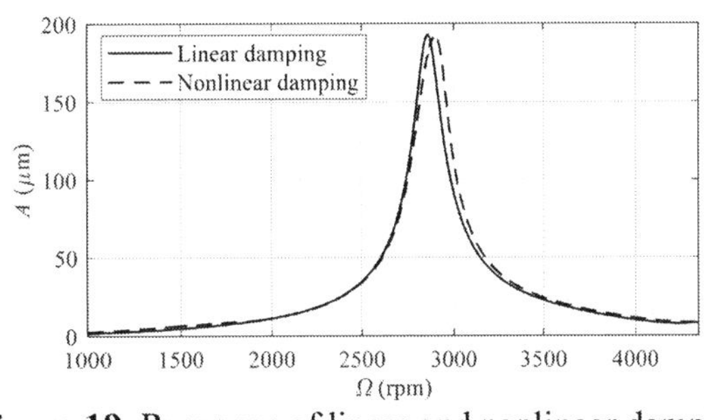

Figure 18. Results of eccentricity ratios. **Figure 19.** Response of linear and nonlinear damping.

5. Conclusion

In this paper, Theoretical research and experimental verification on steady state unbalance response of rotor system with ERSFD(s) has been carried out. Major conclusions are listed as follows:

1) A method for calculating mechanical characteristics of ERSFD with the influence of orifices taken into consideration is established using finite difference method. The variation of equivalent stiffness and damping with eccentricity ratio with the influence of rotating speed eliminated is obtained, which improves the universality of the results.

2) A method for calculating steady state unbalance response of a rotor system with ERSFD(s) is established using finite element method based on the characteristics curves of ERSFD(s). The correctness of the method is verified by comparing the calculation result and the experimental one.

3) ERSFD has good mechanical characteristics, and can be regarded as linear element with big eccentricity ratio.

References

[1] F J Liu. Failure problem analysis of squeeze film damper [J]. *Journal of Beijing University of Aeronautics and Astronautics*, 1998 **(3)**:295-300.
[2] M Zhou, Q H Li, L T Yan. Study on vibration reduction mechanism of elastic ring squeeze film damper [J]. *Journal of Aerospace Power*, 1998 **(4)**:403-407.
[3] M Zhou, Q H Li, L T Yan. Study on oil film force characteristics of elastic ring squeeze film damper [J]. *Journal of Aerospace Power*, 1998 **(3)**:17-23.
[4] M Zhou, Q H Li, L T Yan. Experimental research and application of elastic ring squeeze film damper [J]. *Journal of Aerospace Power*, 1998 **(4)**:409-412.
[5] L Cao, D P Gao, H F Jiang. Preliminary study on vibration reduction mechanism of elastic ring squeeze film damper [J]. *Journal of Vibration Engineering*, 2007 **20(6)**:584-588.
[6] L Cao, D P Gao, H F Jiang. Investigation on critical speed characteristics of elastic ring SFD-rotor system [J]. *Journal of Propulsion Technology*, 2008 **29(2)**:235-239.
[7] L Cao, D P Gao, H F Jiang. Investigation on design factor of elastic ring squeeze film damper [J]. *Gas Turbine Experiment and Research*, 2006 **19(4)**:30-34.
[8] L Cao, D P Gao, H F Jiang. Study on stiffness characteristics of elastic ring squeeze film damper [J]. *Gas Turbine Experiment and Research*, 2007 **20(1)**:36-38.
[9] J Hong, Y Deng, D Y Zhang. Dynamic design method of elastic ring squeeze film damper [J]. *Journal of Beijing University of Aeronautics and Astronautics*, 2006 **32(6)**:649-653.
[10] W Zhang, Q Ding. Elastic ring deformation and pedestal contact status analysis of elastic ring squeeze film damper [J]. *Journal of Sound and Vibration*, 2015 **346**:314-318.
[11] M F Liao. *Aeroengine Rotordynamics* [M]. Press of Northwestern Polytechnical University, 2015:209-223.

IOP Conf. Series: Materials Science and Engineering **751** (2020) 012044 doi:10.1088/1757-899X/751/1/012044

Measurement and Analysis of the Key Metrics of Information System Resilience

Wei Jinyi[1], Li Ning[1], Shi Jing[2] and Xu Luo[1]

[1]Innovation Center, North China Institute of Computing Technology, Beijing, China
[2]Center for Peace and Development Studies, Beijing, China

E-mail: 13020077476@163.com;

Abstract. In view of the current situation that the resilience of information systems cannot be accurately assessed, the key metrics that affect the resilience of information systems—the overall system capability, the system's maximum absorptive capacity, the system recovery ability and the system recovery ratio, are established. And the paper shows data calculation model of the key metrics. Finally, the information system resilience analysis method is proposed, and an aerospace simulation system is taken as an example, combined with its dual redundancy and non-redundancy mechanism to test and verify, which provides strong support for the analysis of system resilience ability.

1. Introduction

The information system is the core of the formation of a networked warfare system, and it is the focus of the enemy and the enemy to compete for the dominance of war. Therefore, the information system must be the main target of enemy reconnaissance monitoring and combating damage [1]. At present, the main attacks on information systems include physical attacks, electronic attacks, and cyber-attacks. Any of these attacks may cause damage to information system nodes, degradation, and other risks, not to mention the actual environment. The comprehensive use of attack means makes the information system invincible and faces greater threats. The US Air Force's "Suter" system is a typical example. It uses a variety of attack methods such as electronic warfare, cyber warfare and physical strikes to make the protection more difficult.

In response to the above problems, experts and scholars analyzed a series of complex characteristics such as information system robustness [2], adaptability [3] and invulnerability [4], and found that Resilience is the key attributes of system to deal with risks and damage. So, resilience has received widespread attention in the industry.

The concept of resilience [5] was first proposed by the US military in the Mission-Oriented Resilient Clouds (MRC) project [6], when resilience was the ability to continue to run cloud applications and infrastructure in the event of an attack. The proposed concept of resilience for infrastructure, and then our military proposed a concept of system-oriented resilience, which refers to a capability or quality characteristic that the system presents in response to various disturbances and changes, that is, the system predicts interference from natural or man-made events. The ability to resist, absorb, react, adapt and recover [7]. Therefore, the resilience ability indicates that the information system can be degraded and restored as soon as some nodes fail, collapse, and damage, ensuring the ability to complete critical tasks.

Information systems need resilience, and how to evaluate system resilience becomes a key issue. Testing is an effective means of quantitatively testing system capabilities. However, the current state of the art is that existing testing techniques tend to focus on functional, performance, and reliability testing, and lack of test methods and tools for system resilience.

This paper establishes the key metrics that affect the resilience of information systems, and gives the data calculation model of key metrics. Finally, the information system resilience analysis method is proposed to quantitatively measure the resilience of information systems.

2. Related works

At present, experts and scholars at home and abroad have conducted preliminary research on the resilience of information systems.

In 2014, the Systems Engineering Research Institute with the theme of Resilience Engineering System (ERS) was held in California, USA, reflecting the importance of system resilience. In the same year, Zachary A. Collier et al proposed the NCO resilience decision-making framework based on network-centric actions. However, it only compares the resilience of different systems from a conceptual perspective [8].

In 2015, resilience received extensive attention. Huy T. Tran et al. used the architecture method to evaluate the resilience of the command and control network, but it only analyzed the structural parameters affecting the network resilience, and did not measure the overall resilience of the system [9]; Lan Yushi et al proposed resilient command the concept of information system, and its construction mechanism and engineering implementation method are studied, but it lacks the evaluation method of system resilience [10]; Zhao Hongli et al. proposes a network science-based networked allegation information system resilience analysis method. The residual network performance after the failure event is used as the basis of resilience measurement, and a new idea is proposed for the resilience measurement, but it has obvious shortcomings in resilience response, recovery and regeneration [11].

In 2016, Tao Zhigang et al. proposed a C4ISR system resilience research method based on the weighted space exploration. This method provides a direction for the study of system resilience, but lacks research on system resilience measurement [12].

In 2018, a major breakthrough was made in the study of system resilience. Ding Feng et al. proposed a quantitative evaluation method for resilience ability to the mission-oriented command information system, and established a resilience system capability evaluation metrics system and model. However, the evaluation metrics are redundant and difficult to perform digital calculation, and cannot perform a quick analysis to information systems. [13].

The key metrics and resilience analysis methods of information system resilience proposed in this paper can effectively make up for the shortcomings of the above research, which is efficient and fast, and can quantitatively analyze the resilience of resident information systems.

3. The key resilience metrics for information system

According to the method described in the US space system resilience process [14], the information system resilience process can be divided into four states: original state, degradation, recovery and steady state, as shown in Figure 1. The t0-t1 time period is an original state phase, which shows the system capability state of the information system during normal operation. In the normal working process, the information system needs to maintain a relatively stable level F(t1) in order to meet its own simulation test tasks. Of course, due to its complex structural composition, small disturbances are inevitable in the normal working process, but the disturbance is in the range that the resilience can absorb, so the whole is in a dynamic and stable state. The t1-t2 time period is a degradation phase, which shows the state of the system capability of the information system in the event of an attack. When the information system is subject to small disturbances, the system capability is still in a dynamic and stable state due to its resilience ability. When the information system is attacked more seriously, its resilience ability cannot maintain a dynamic stability for the first time, and the system capability will decline, enter the degradation phase of system capability, and drop to F(t2). The degradation rate is related to the system resilience ability and the attack strength suffered. The stronger the resilience ability, the smaller the attack intensity, the slower the rate of decline, and vice versa. The t2-t3 time period is the recovery phase, which shows that after the information system is degraded by the attack system, the resilience plays a role in the recovery of the system capability. The recovery rate in the recovery phase is closely related to the system resilience ability, but the process of change is often not well measured, so the relatively stable F(t2) and F(t3) are selected to reflect the recovery state. The t3-t4 time period is a steady state phase, which shows the capability state of the information system after the attack system is recovered. After the system is recovered, we hope that the system capability can be the same as before the attack, but the system is often impossible to implement, only re-planning and adjustment, constructing a new internal component, enabling itself to adapt to external attacks and achieving normal working status.

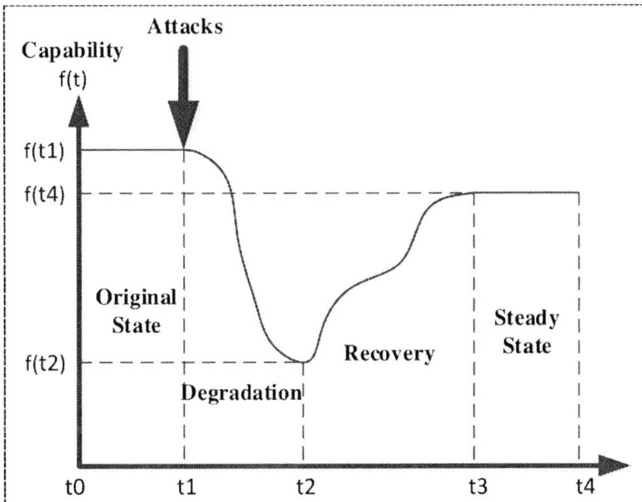

Figure 1. Information system resilient model.

By analyzing the resilience process of the information system, we chose to attack the information system at point t1 and identified four key metrics: the overall system capability, the system's maximum absorptive capacity, the system recovery ability and the system recovery ratio.

3.1. The overall system capability

The overall system capability of the resilience information system refers to the system in which the system breaks down from the original state, enters to the degradation phase, and then goes to the recovery level, and then passes through the steady state and finally reaches a relatively stable state. The overall performance, recorded as P, can be used as the first assessment metric of the resilience of information systems.

Based on the above analysis, the cumulative normalized value of the system resilience process integral is used to represent P, namely:

$$P = F(Original\ State,\ All) = \frac{F(All)}{F(t_1)} = \frac{\int_{t_0}^{t_4} f(t)/(t_4-t_0)}{\int_{t_0}^{t_1} f(t)/(t_1-t_0)} \quad (1)$$

Among them, Original State is the first phase of system capability, all is the whole process of system capability, and P is the ratio of F(All) to F(t1). It should be noted that in Figure 1, we can accurately see the values of f(t1), f(t2), f(t3), and f(t4), but in the actual situation, the points of the original state phase, the degradation phase, the recovery phase, and the steady state phase are difficult to determine, and the system capacity of the degradation phase and the recovery phase varies greatly. Therefore, we select the cumulative normalized value of f(t) for a period of time t0-t4 as the value of

IOP Conf. Series: Materials Science and Engineering **751** (2020) 012044 doi:10.1088/1757-899X/751/1/012044

F(All). Similarly, the cumulative normalized value of f(t) for a period of time t0-t1 is selected as the value of F(t1), and the ratio of the two represents the overall system capability P of the evaluation metric.

3.2. *The system's maximum absorptive capacity*

In the process of networked operations, the resilience information system will break the original state due to the attack and enter the degradation phase. In this process, the system capability drops sharply, and even itself may be affected. The less the degree of decline in the system degradation, the stronger the system's absorption capacity and the stronger the system resilience; the weaker the opposite. In addition, the system degradation rate is also an important parameter of the system's absorption capacity. The longer the system decline time, the stronger the system resilience ability; the weaker the opposite. Eliminating human maintenance and other interferences on the system, the degree of degradation and the rate of decline can be used to reflect the resilience of the information system.

Based on the above analysis, the system's maximum absorptive capacity is represented by the degree of degradation of the system capability in the degradation phase and the negative index of the rate of decline, and is used as the second evaluation metric of the resilience capability of the information system, which is denoted as D, and its relationship is:

$$D = F(Original\ State, Degradation) = e^{-d_1 \cdot v_1} = e^{-\frac{f(t_1)-f(t_2)}{f(t_1)} \cdot \frac{t_2-t_1}{t_3-t_1}} = e^{\frac{f(t_2)-f(t_1)}{f(t_1)} \cdot \frac{t_2-t_1}{t_3-t_1}} \quad (2)$$

Among them, Original State is the first phase of system capability, Degradation is the second phase of system capability, d1 is the degree of system degradation, v1 is the system's rate of decline, and D is the negative index of d1v1. The ratio of the difference between f(t1) and f(t2) to f(t1) represents d1, and v1 is represented by the ratio of the time interval of t1-t2 to the time interval of t1-t3. In addition, it is also necessary to select the cumulative normalized value of f(t) for a period of time t0-t1 as the value of f(t1), and finally use the degradation degree of the system capability of the degradation phase and the negative index of the rate of decline as an evaluation metric.

3.3. *The system recovery ability*

In the process of networked operations, the resilience information system is attacked to break the original state and enter the degradation phase, and then enter the recovery phase. The system recovery ability is closely related to the system resilience. The more the degree of ascension is when the system is recovered, the stronger the system recovery ability and the stronger the system resilience; the weaker the opposite. Similarly, the system recovery rate is also an important parameter of the system recovery ability. The shorter the system rise time, the stronger the system recovery ability; the weaker the opposite. In addition, human maintenance guarantees have also had an important impact on the recovery of system capability. Therefore, it is necessary to remove the influence of these external factors on the system itself, and use the degree of recovery and the rate of increase to reflect the resilience of the information system.

Based on the above analysis, the system recovery ability is represented by the recovery degree of the system capability in the recovery phase and the negative index of the rising rate, and is regarded as the third evaluation metric of the information system resilience ability, which is denoted as R, and its relationship is:

$$R = F(Original\ State, Recovery) = e^{-d_2 \cdot v_2} = e^{-\frac{f(t_4)-f(t_2)}{f(t_1)} \cdot \frac{t_3-t_2}{t_3-t_1}} = e^{\frac{f(t_2)-f(t_4)}{f(t_1)} \cdot \frac{t_3-t_2}{t_3-t_1}} \quad (3)$$

Among them, Original State is the first phase of system capability, Recovery is the third phase of system capability, d2 is the recovery degree of the system, v2 is the rising rate of the system, and R is the negative index of d2v2. The ratio of the difference between f(t4) and f(t2) to f(t1) represents d2, and v2 is represented by the ratio of the time interval of t2-t3 to the time interval of t1-t3. In addition, it is also necessary to select the cumulative normalized value of f(t) for a period of time t0-t1 as the value of f(t1), and select the cumulative normalized value of f(t) for a period of time t3-t4 as the value of f(t4). Finally, the recovery degree of the system capacity in the recovery phase and the negative index of the rising rate are taken as an evaluation metric.

3.4. The system recovery ratio

In the process of networked warfare, the resilience information system will break the original state due to the attack, enter the degradation phase, and then go to the recovery level, and then go through the steady state phase, and finally reach a relatively stable state. In this process, the system will be damaged, so the first and fourth, two relatively stable states are different. The stronger the resilience of the information system, the smaller the damage it receives, and the stronger the system capability of the stable state after the steady state; the weaker the opposite. Similarly, human maintenance support plays an important role in the system capability to stabilize the information system. In the information system resilience test, the impact of human subjective activities should be removed as much as possible. Therefore, the information system reflects the resilience of the information system in two different relative stable states before and after the attack.

Based on the above analysis, the ratio of the system capability of the steady state phase to the system capability of the original state phase, which is the system recovery ratio, is another evaluation metric of the information system resilience ability, and is expressed as:

$$\varepsilon = F(Original\ State, Steady\ State) = \frac{F(t_4)}{F(t_1)} = \frac{\int_{t_3}^{t_4} f(t)/(t_4 - t_3)}{\int_{t_0}^{t_1} f(t)/(t_1 - t_0)} \quad (4)$$

Among them, the Original State is the first phase of the system capability, and the Steady State is the fourth phase of the system capability, which is the ratio of F(t4) to F(t1). It should be noted that in Figure 1, we can accurately see the values of f(t4) and f(t1), but in the actual situation, the points of the original state phase and the steady state phase are difficult to determine, so we choose the cumulative normalized value of f(t) for a period of time t3-t4 is taken as the value of f(t4), and the cumulative normalized value of f(t) for a period of time t0-t1 is also selected as the value of f(t1). The ratio of the two is used as an evaluation metric.

3.5. The key metrics calculation

Information system resilience consists of four states: original state, degradation, recovery and steady state. Therefore, the four metrics of system resilience ability are combined to reflect the resilience of information systems. Set the resilience ability of the information system to S, remember that the overall system capability of the system is P, the system's maximum absorptive capacity of the system is D, the system recovery ability is R, and the system recovery ratio is ε, then $S = P \cdot D \cdot R \cdot \varepsilon$. The specific steps are:

- First, calculate the cumulative normalized value F(All) of the t0-t4 phase, the cumulative normalized value F(t1) of the t0-t1 phase, and the cumulative normalized value F(t4) of the t3-t4 phase;
- Second, calculate the degree of degradation d1, the rate of decline v1, the degree of recovery d2, and the rate of increase v2;
- Next, calculate the overall system capability of the system P, the system's maximum absorptive capacity of the system D, the system recovery ability R, and the system recovery ratio ε;
- Finally, calculate the resilience ability of the t0-t4 state information system $S = P \cdot D \cdot R \cdot \varepsilon$.

4. Testing and analysis of information system resilience

Take an aerospace simulation system as the resilience test object, as shown in Figure 2. According to the model established in the paper, the chaotic method [15] is used to simulate electromagnetic interference, network attack, CPU overload, malicious code attack, power failure, physical damage and other faults, resulting in damage to the simulation system, which cannot work properly. After the system enters the state of degradation, recovery and steady state, the data is recorded, and the resilience ability of the simulation system is analyzed by using the key metrics of resilience.

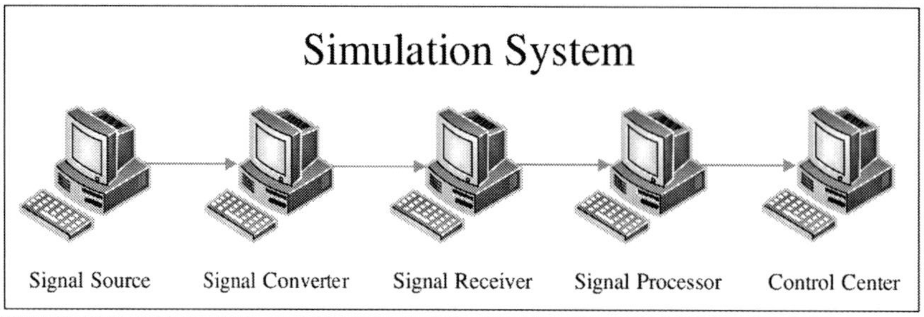

Figure 2. Information system resilience test object.

The simulation system consists of a signal source, a signal converter, a signal receiver, a signal processor, and a control center. For the composition of the system, the testing process can be designed as: design, battle simulation, attack simulation, intelligence analysis, and resilience assessment. The process is shown in Figure 3.

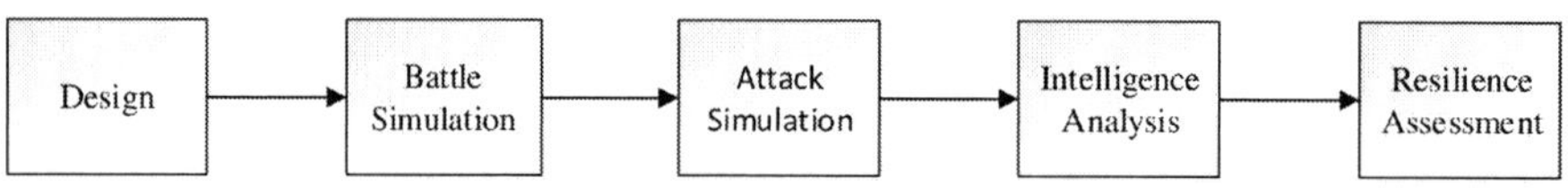

Figure 3. Information system resilience testing process.

For the simulation system, the actual combat intelligence short message is sent and received, and the malicious code is used to attack it. Select different attack strengths, and determine the overall system capability, the system's maximum absorptive capacity, the system recovery ability and the system recovery ratio, so as to test and evaluate the system.

Set the attack strength of 10, 20, 30, 40, 50 and 60 Mbit/s of malicious code, and continuously send short messages at 600 pieces per second, which yields 6 curves of the system short message processed ability as shown in Figure 4. It can be seen from Figure 4 that the greater the attack strength suffered by the system, the greater the decline in the processed ability of the short message. When the attack strength reaches 40 Mbit/s, the short message processed ability curve basically coincides, which is mainly caused by the dual redundancy of the system service [16]. At this time, the reference value of the 50 and 60 Mbit/s attack strength curves has decreased. Therefore, we select the 40Mbit/s attack intensity curve data for the analysis and calculation of the key metrics of resilience ability.

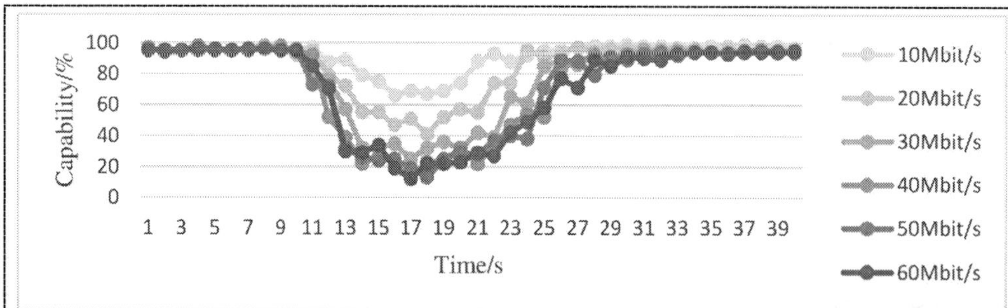

Figure 4. Dual redundant system's short message processed ability change curves under different attack strengths.

The system was modified to test under the condition of non-redundancy [17]. Similarly, in the above basic environment, the 6 curves of the short message processed ability of the system shown in Figure 5 was obtained. It can be seen from the figure that when the system is attacked at 60 Mbit/s, its short message processed ability is once lost and cannot be recovered for a while; when the system is at 50 Mbit/s attack strength, its short message processed ability is slowly maintain a stable level after a while. Therefore, we select the 50Mbit/s attack intensity curve data for the analysis and calculation of the key metrics of resilience ability.

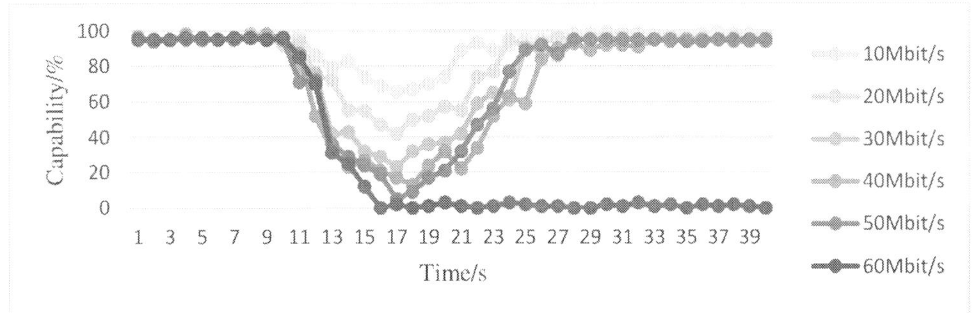

Figure 5. Non-redundant system's short message processed ability change curves under different attack strengths.

Under the dual redundancy condition of 40 Mbit/s, F(All), F(t1), F(t4), d1, v1, d2, and v2 are calculated by curve fitting, and then $\begin{cases} P = 0.73 \\ D = 0.76 \\ R = 0.55 \\ \varepsilon = 0.99 \end{cases}$ is obtained, and its system resilience ability is: $S = P \cdot D \cdot R \cdot \varepsilon \approx 0.30$.

Under the non-redundancy condition of 50 Mbit/s, F(All), F(t1), F(t4), d1, v1, d2, and v2 are calculated by curve fitting, and then $\begin{cases} P = 0.68 \\ D = 0.69 \\ R = 0.56 \\ \varepsilon = 0.99 \end{cases}$ is obtained, and its system resilience ability is: $S = P \cdot D \cdot R \cdot \varepsilon \approx 0.26$.

Since the calculation of data is based on the premise of minimizing the impact of redundancy on the system, the system recovery ability R and the system recovery ratio ε of the dual redundant and single redundant systems are basically equal; in addition, the dual redundant system is greater than non-redundant system int the overall system capability P and the system's maximum absorptive capacity D, so dual redundant system has better resilience than non-redundant systems.

The analysis results can intuitively determine the resilience of the information system. At the same time, the system's dual redundancy mechanism greatly enhances its system resilience, which also proves the practicability of our information system's dual redundancy and even multiple redundancy mechanisms.

5. Conclusion

This paper analyzes the system resilience processing, proposes the key metrics and resilience analysis methods of information system, and uses the actual system as an example to verify the effectiveness of the key metrics of information system resilience and the practicability of the system's dual redundancy mechanism.

Acknowledgments

Funded by the "13th Five-Year" Pre-research Project of the Ministry of Equipment Development (9140A15050208DZ25). Wei Jinyi is the corresponding author.

References

[1] Cui Qiong, Li Jianhua. Elastic Measurement Method of Networked Command Information System[J]. Military Planning and Systems Engineering, 2016, 30(4): 18-24
[2] Zhang Jieyong, Yi Wei, Wang Wei. Dynamic Robustness Measurement Method for C4ISR System Structure Considering Cascading Failure[J]. Systems Engineering and Electronics, 2016, 38(9): 2072-2079
[3] WANG Jie, ZHANG Lin-lin, ZHAO Wei et al. Adaptation Measurement Method of Aspect Oriented Software Architecture[J]. Microelectronics & Computer, 2015(6): 41-45
[4] Yi Yi, Wang Wei, Mao Shaojie et al. Optimization Method of Network C4ISR System Structure Destruction Ability[J]. Command Information Systems & Technology, 2015, 6(3): 9-15
[5] HOLLING C S. Resilience and stability of ecological systems [J]. Annual Review of Ecology and Systematics, 1973, 4(4): 1-23
[6] DARPA-BAA 11-55. Mission-oriented Resilient (MRC) [R]. Information Innovation Office, 2011: 12
[7] GOERGER S R, MADNI A M, ESLINGER O J. Engineered resilient systems: a DoD perspective [J]. Procedia Computer Science, 2014, 28: 865-872
[8] ZACHARY A COLLIER, LGOR LINKOW. Decision making for resilience within the context of network centric operations [C] // 18thICCRTS, 2014
[9] TRAN H T, DOMERCANT J C, MAVRIS D N. A SoS approach for assessing the resilience of reconfigurable command and control networks [R]. American Institute of Aeronautics and Astronautics SciTech, Kissimmee, Florida, 2015: 1-14
[10] Lan Yushi, Zhou Guangxia, Wang Wei et al. Research on the Construction Mechanism and Implementation of Resilience Command Information System[J]. Journal of Command and Control, 2015, 1(3): 284-286
[11] Zhao Hongli, Yang Haitao, Fu Wei. Research on Elastic Analysis Method of Networked Allegation Information System[J]. Journal of Command and Control, 2015, 1(1): 14-19
[12] Tao Zhigang, Xu Hao, Yi Wei et al. Research on the resilience of C4ISR system based on the analysis of trade-off space [C] // Proceedings of the 4th China Command and Control Conference. Beijing: Publishing House of Electronics Industry, 2016: 110-113
[13] Ding Feng, Zhou Fang, Cheng Wendi et al. Quantitative Evaluation of Resilience of Command Information System for Task Support[J]. Journal of Command and Control, 2018, 9(6): 13-18
[14] BURCH R. Measures of resilience for space systems [EB/OL]. (2013-08-14) [2017-02-20].http://www.ndia.org/Divisions/Divisions/Space/Documents/Space_forum-Boeing-NDIA-Resilience-Forum-081413-RevA.pdf
[15] PRINCIPLES OF CHAOS ENGINEERING. [2018-05]. http://principlesofchaos.org/
[16] Ding Jin, Xu Guoqiang. Design and Implementation of Automatic Switching Test Module for Dual Redundant Network Interfaces[J]. Ship Electronic Countermeasure, 2017, 40(3)
[17] Yu Meijie. Research on MMC-STATCOM control system with fault tolerance; [D]. Harbin University of Science and Technology, 2018

IOP Conf. Series: Materials Science and Engineering **751** (2020) 012045 doi:10.1088/1757-899X/751/1/012045

Modal Analysis of Variable Structure of Gate Rotor of CP-type Single-screw Structure

Xiaofeng Song*, Wei Wang, Yuting Wu and Chongfang Ma

Key Laboratory of Enhanced Heat Transfer and Energy Conservation, Ministry of Education and Key Laboratory of Heat Transfer and Energy Conversion, Beijing municipality, College of Environmental and Energy Engineering, Beijing University of Technology, Beijing 100124, China

*Email: sxfgoing@163.com

Abstract. Single screw technology, due to its superiority of structure, has been widely used in many fields of gas compression, refrigeration, power and so on at present. However, specialized research on the dynamics of single screw structure has not been systematic or in-depth yet. As for single screw compressors and expanders, the dynamics of the gate rotor are the key factors influencing its performance, and the variable-structure modal analysis of the gate rotor is an important task to study its dynamic characteristics. In this paper, the modal parameters of the CP-type single-screw structure gate rotor are compared by modal simulation and modal experiment, and the accuracy of the finite element method is verified. And mainly carried out the simulation modal analysis of the gate rotor variable structure. The results show that the change of structural parameters is an important factor affecting the natural frequency of the gate rotor mode. The variable structure modal analysis can be used to provide reference for the structural optimization of the gate rotor.

1. Introduction

At present, under the background of severe energy and environmental problems in China, the research and application of various energy-saving technologies have attracted extensive attention of the government and society [1]. In the field of medium-low temperature waste heat utilization, small and medium-sized Organic Rankine Cycle (ORC) is considered as one of the most potential technologies [2], and a lot of relevant research work has been carried out. Kang et al.[3] used R245fa working medium. The evaporating temperature and condensing temperature were 80 °C and 30 °C respectively. The expander was a radial turbine with an efficiency of 78.7%, an expansion ratio of 2.65 and a system cycle thermal efficiency of 5.22%. Li et al. [4] used R123 working medium, of which the evaporation temperature and condensation temperature were 90.7 °C and 39.4 °C, respectively. The expander was a radial turbine, the efficiency was 58.53%, the expansion ratio was 4.19, and the system cycle thermal efficiency was 6.15%. Lei Biao et al.[5]used R123 working medium, and the expander was a single screw type. The evaporation temperature and condensation temperature were 126.1 °C and 67.9 °C, respectively, the efficiency was 56%, the expansion ratio was 8.5, and the net efficiency of the system was 7.98% (excluding the consumption of auxiliary machines). Through the above comparison, it can be seen that the small and medium capacity expander has the problems of small expansion ratio and low internal efficiency at the same time. Based on the existing research methods and achievements of compressor, some scholars have carried out theoretical analysis and experimental research on various irreversible factors of expander, including leakage, heat transfer, mechanical friction, etc. However, the thermal performance can only reflect the actual performance of the expander, and can not provide a direct technical basis for improving the performance and optimizing the structure of the expander. In order to further improve the performance of the expander, the systematic study of the expander dynamics needs to be carried out.

Because of the simple structure of the single screw, the basic structure is that the screw and the gate rotor form a spatial meshing relationship, and the movement form of the meshing pair is similar to sliding, and the soft and hard meshing of the metal rotor and the organic material gate rotor is usually used, and it is the best type of vibration, noise and seal in all kinds of positive displacement power machinery. Therefore, since F. B. Zimmern proposed the original structure of single screw compressor in 1965 [6], the single screw technology has developed rapidly, and has been widely used in refrigeration compressor, process gas compressor, extruder and other fields. Beijing University of technology is the first unit to apply the single screw technology to the field of expander in China. Through relevant research, it has been proved that the single screw expander is one of the most potential types of small and medium capacity expander. How to further improve the performance of expander is the focus of technical research.

Single screw has the advantages of simple structure, and the key parts are only screw and gate rotor. However, this leads to a more complex structure of key parts, which is difficult to process, especially the gate rotor. In order to ensure a good meshing relationship, the gate rotor structure and materials are required to have good mechanical properties (certain elasticity, strength and good wear resistance), accurate processing and precise spatial positioning, and the determination of the gate rotor structure parameters is mainly based on the spatial geometric relationship, and the mechanical aspects are more dependent on engineering experience, which makes the adjustment and optimization of single screw structure lack of theoretical basis. It is an important part of dynamic analysis to obtain the modal natural frequency of the gate rotor structure through the variable parameter modal analysis, which can provide the basic data for the vibration characteristic analysis and dynamic optimization design of the variable structure system. At present, the research on the variable parameter mode of gate rotor is not systematic, which can not provide technical guidance for the application of new materials, structural parameter adjustment and optimization of gate rotor. In this paper, the variable structure simulation modal analysis of the gate rotor is carried out. Through the simulation analysis, the reasonable structural parameters of the gate rotor are explored to provide the basic data of dynamics and technical support for the optimization of the gate rotor.

2. Finite element analysis verification

As an important part of single screw structure, gate rotor plays a key role in the performance of single screw expander. The single screw structure is composed of one screw and several gate rotor pieces. It is generally divided into four types: CP, PC, PP and CC. Because the latter three types are complex, the single screw structure used in this paper is CP type, which is a conventional structure. According to different work requirements, the gate rotor pieces with a certain number of teeth are processed.

In order to verify the accuracy of the finite element method and provide technical support for the modal analysis of the variable structure of the gate rotor, the modal analysis method of the combination of simulation and experiment is used to verify the accuracy of the finite element results. The modal analysis module of ANSYS Workbench software is used for the finite element simulation, and the vibration test points of DASP V11 provided by Beijing Oriental Institute of vibration and noise are used for the experiment. Analysis instrument to carry out the experimental test.

2.1 Modal analysis theory

Modal analysis is the basic content of dynamic analysis. It is a dynamic analysis method used to analyze modal characteristics such as structural modal parameters and modal shapes [7]. In practice, the system where the main structure is located is a continuous medium system, whose modal parameters can be calculated by discrete equation. Assuming that the gate rotor system is divided into a N dimensional degree of freedom system, and the vibration displacement of each position of the gate rotor is u, its basic motion equation [8] can be expressed as:

$$[\boldsymbol{M}]\{\ddot{u}\}+[\boldsymbol{C}]\{\dot{u}\}+[\boldsymbol{K}]\{u\}=\{F(t)\} \tag{1}$$

In the formula, $F(t)$ is N dimensional excitation force vector; u , $\dot{u}$, $\ddot{u}$ are respectively N dimensional displacement, velocity and acceleration; ***M, C*** and ***K*** are the mass, stiffness and damping matrices of the main structure, usually n-order real symmetric matrix.

In the modal analysis of a continuum, the change of external load does not affect the modal characteristics of the continuum itself, and the damping of the continuum has little effect on the modal analysis. Therefore, the vibration equation of the continuum can be simplified as:

$$[\boldsymbol{M}]\{\ddot{u}\}+[\boldsymbol{K}]\{u\}=\{0\} \tag{2}$$

When simple harmonic vibration occurs, equally to $u=Usin(\omega t)$,the equation is:

$$([\boldsymbol{K}]-\omega_r^2[\boldsymbol{M}])\{\varphi_r\}=0 \tag{3}$$

For the modal analysis of a main structure, the r order natural frequency ω_r and the corresponding modal shape $\{\varphi\}_r$ can be obtained from the above matrix equation.

According to the calculated modal shape, the transfer function between two positions in the continuum can be obtained, that is, the frequency response function between the excitation point and the response point in the experimental mode, which can be expressed by the following formula:

$$H_{ab}(\omega)=\sum_{r=1}^{N}\frac{\varphi_{ra}\varphi_{rb}}{m_r[(\omega_r^2-\omega^2)]+j2\xi_r\omega_r\omega} \tag{4}$$

Where $H_{ab}(\omega)$ is the transfer function between the excitation point and the response point in the system, N is the total order of the measured mode, and φ_{ra} and φ_{rb} are the r order modes of the a point and the b point respectively; and m_r, ξ_r and ω_r are the r order mode mass, damping ratio and natural frequency respectively.

2.2 Finite element modal analysis

At present, in order to obtain a better meshing performance of CP structure, the structure of 6 screw slots and 11 teeth are usually used. Table 1 shows the structural parameters of the standard CP type gate rotor currently used for working conditions. The diameter of the gate rotor is determined by a certain working condition, the tooth length is determined by the depth of the gate rotor engaged in the screw, the tooth width is determined by the engagement depth, the constraint of the screw rib and experience, while the thickness of the gate rotor is completely determined by experience, and the diameter of the inner hole is determined by the diameter of the gate rotor shaft.

Table 1. Structure parameter of gate rotor.

Diametet/mm	Tooth length/mm	Tooth width/mm	Thickness/mm	Inner hole diameter /mm
117	24.9	17.23	6	26

The three-dimensional solid model of the gate rotor is drawn by the three-dimensional modeling software Pro / E, and the generated model is imported into the finite element software. First, the actual material properties of the gate rotor are defined. In this paper, the gate rotor is formed by the injection molding of Peek (Poly Ether Ether Ketone) which is a new self-lubricating material, and its material properties are shown in table 2.

Table 2. Material properties of gate rotor.

Diameter/mm	Material	Density/(g/cm^3)	Modulus of Elasticity/GPa	Poisson ratio
117	PEEK	1.4	10.87	0.44

Secondly, the finite element analysis and solution of the gate rotor model are carried out. Combined with the modal truncation theory, the first six modal frequencies are mainly studied to obtain the first six modal natural frequencies of the gate rotor as shown in table 3.

Table 3. Finite element modal frequency of gate rotor.

Order number	Frequency/Hz
1	966.42
2	966.44
3	1775.0
4	1775.7
5	1863.7
6	2278.0

2.3 Experimental modal analysis

In order to verify the results of finite element modal simulation analysis, experimental modal analysis is used to test and analyze the actual gate rotor, as shown in figure 1 is the actual structure diagram of the CP type gate rotor and screw.

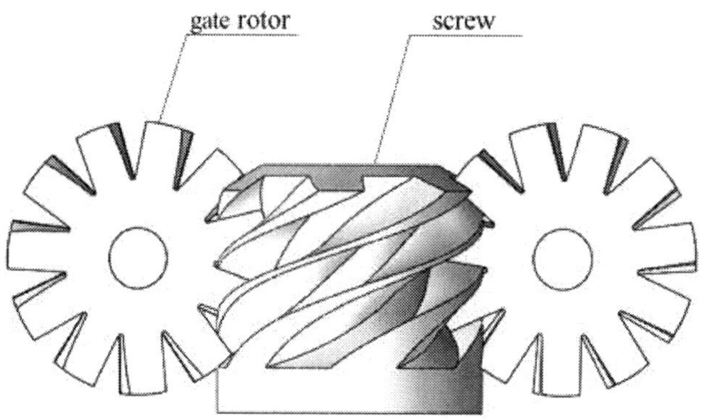

Figure 1. Actual structure diagram of CP type gate rotor and screw.

The experimental modal analysis method is to acquire the input and output signals of the system through the experiment, and obtain the modal frequency of the gate rotor through the parameter identification. That is to say, it can be seen from the reciprocity assumption that when two or more sensors are fixed at the same position, the moving hammer is used to measure the excitation force at each point of the gate rotor, and the complete modal analysis results of the gate rotor can be obtained by each sensor. Therefore, in this experiment, the moving knock point excitation is used, and the response signal is collected by two sound pressure sensors, and the more input and more output (MIMO) modal analysis experiment is carried out. The flow chart of this experiment is shown in figure 2.

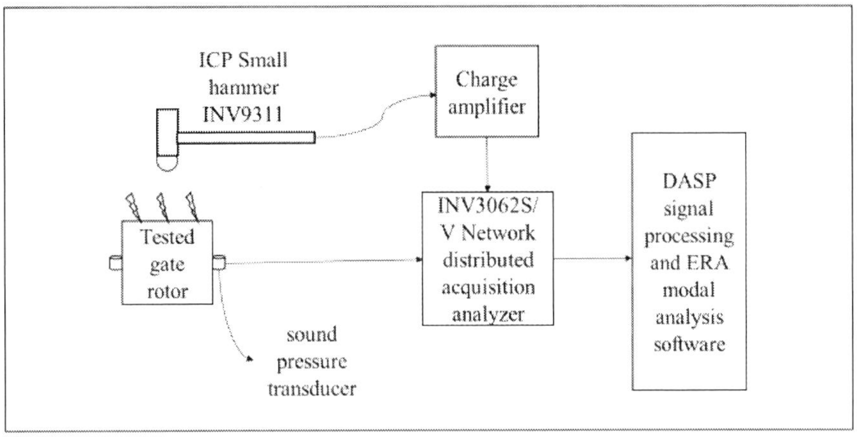

Figure 2. Experimental flow diagram.

In order to effectively compare with the results of the finite element analysis, the first six modal natural frequencies are also extracted. Table 4 shows the first six experimental modal natural frequencies of the gate rotor.

Table 4. Natural frequency measured by gate rotor experiment.

Order number	Frequency/Hz
1	981.80
2	983.90
3	1822.2
4	1823.5
5	1925.4
6	2396.1

2.4 Comparative analysis and discussion

It can be seen from the comparison between the above simulation and experimental frequencies that the natural frequencies of simulation and experimental modes have high consistency, with the maximum error of 4.7% and the minimum error of 1.5%, meeting the error range. The reason for the error may not be the same as the simulation due to the boundary conditions and other constraints in the experiment, and may be due to the rubber band suspension and gate rotor materials (special engineering plastics) injection molding process, etc. This shows that the finite element software ANSYS Workbench can effectively simulate the modal characteristics of the actual gate rotor, which provides an effective analysis method for the later variable structure modal analysis of the gate rotor.

3. Analysis of the influencing factors of structural parameters

Through the optimization analysis of the structural parameters of the gate rotor, the volume of the screw groove of the single screw expander can be increased, and a higher content product ratio can be obtained, so as to improve the efficiency of the expander. The existing literature research has proved that the main structural parameters affecting the gate rotor include: the thickness, diameter, tooth width, unequal diameter ratio. In this section, the variables of these four structural parameters are analyzed to optimize the structure of the gate rotor.

3.1 Effect of thickness

At present, the structural parameters of the thickness of the gate rotor used in the single screw expander are completely determined by experience, without any data reference. Although increasing the tooth thickness can increase the mechanical strength of the gate rotor in the meshing process, the corresponding cost will also increase, and the friction between the gate rotor and the screw groove will also increase, and the gate rotor is too thin and easy to break. Therefore, when the diameter of the gate rotor is defined as a fixed value of 350mm and the fundamental frequency is 50Hz, figure 3 shows the change curve of the natural frequency of the gate rotor mode with the increase of the thickness.

It can be seen from figure 3 that the tooth thickness is in direct proportion to the frequency. When the diameter of the gate rotor remains unchanged, increasing the thickness of the gate rotor will help to improve the natural frequency of the gate rotor mode. When the thickness of the gate rotor is 2 ~ 3mm, the natural frequency of the gate rotor mode is close to the fundamental frequency of 50Hz.

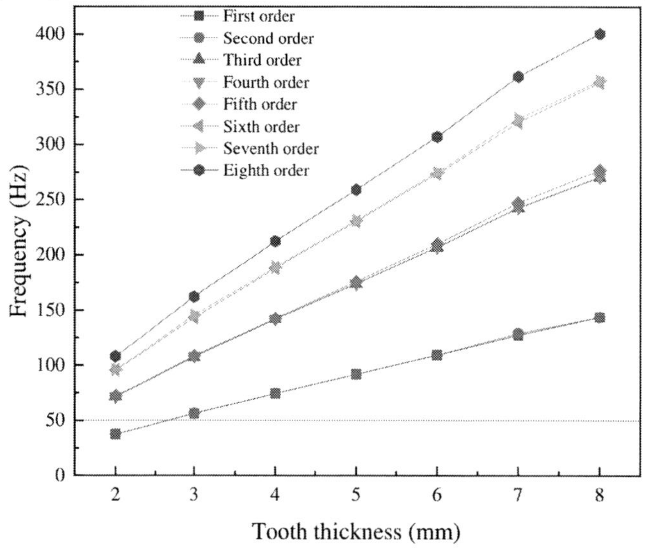

Figure 3. The change curve of the natural frequency of the mode with the tooth thickness of the gate rotor.

3.2 Effect of diameter

The diameter of the gate rotor has a great influence on the volume of the screw groove. When the thickness of the gate rotor is fixed, increasing the diameter of the gate rotor can effectively improve the volume of the screw groove, reduce the weight of the screw, reduce the centrifugal force of the screw, and reduce the overall volume. However, the disadvantage is that when running in a large displacement machine, due to the increase of the diameter of the gate rotor, the length of the gate rotor teeth is too long, which leads to the excessive force on the gear and is easy to be broken. Therefore, when the thickness of gate rotor is defined as fixed value of 3mm and the fundamental frequency is 50Hz, the change curve of natural frequency of the gate rotor mode with the increase of diameter is shown in figure 4.

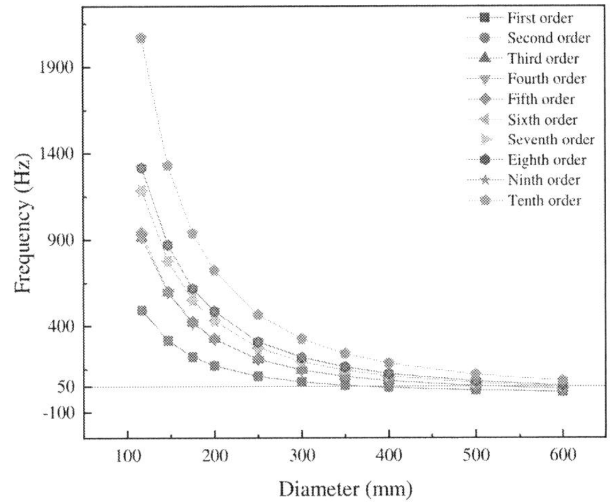

Figure 4. The change curve of the natural frequency of the mode with the diameter of the gate rotor.

It can be seen from figure 4 that the diameter is inversely proportional to the frequency. When the thickness of the gate rotor tooth remains unchanged, the natural frequency of the mode decreases in a parabola trend with the increase of the diameter of the gate rotor. When the diameter increases to 350mm, the natural frequency of the gate rotor mode is close to the fundamental frequency region.

3.3 Effect of width of gear

The width of gate rotor is generally equal width, which is mainly determined by center distance, diameter and screw rib thickness. Increasing the tooth width of the gate rotor can increase the volume of the screw groove of the single screw expander, but the thickness of the screw rib is correspondingly reduced. When the thickness of the screw rib is too small, the tooth top of the screw groove becomes sharp and the strength is not enough, and the stress concentration at the root of the gate rotor increases because the fillet radius is too small, which is easy to produce cracks and cause damage. Therefore, the width of the gate rotor cannot be increased infinitely. When the thickness of the screw rib is required, the increase of the width of the gate rotor will increase the length of the screw and the length of the whole casing, thus increasing the cost. When the thickness of gate rotor is defined as fixed value of 3mm and the fundamental frequency is 50Hz, the change curve of natural frequency of the gate rotor mode with the increase of tooth width is shown in figure 5.

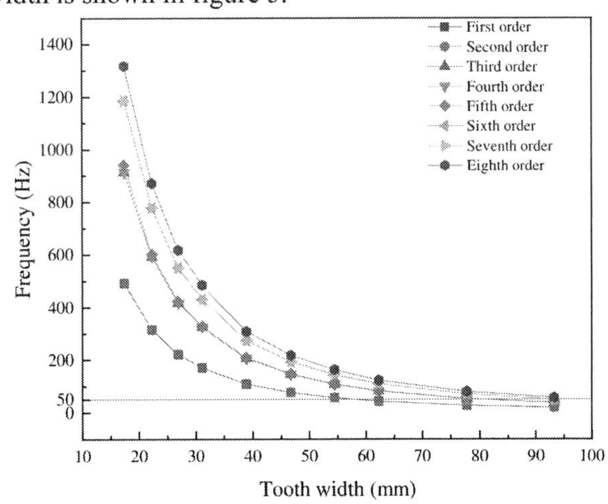

Figure 5. The change curve of the natural frequency of the mode with the tooth width of the gate rotor.

It can be seen from figure 5 that the influence of tooth width on the modal natural frequency of the gate rotor is the same as the diameter. When the tooth width is increased to 55mm, the modal natural frequency is close to the fundamental frequency of 50Hz.

3.4 Effect of unequal diameter ratio

The existing CP structure is equal diameter structure, which is, the diameter of screw and gate rotor is equal, and the unequal diameter structure is put forward on this basis, aiming to improve the efficiency by changing the unequal diameter of screw and gate rotor. Unequal diameter structure means that there is a certain proportional relationship between the diameter of the gate rotor and screw, that is, unequal diameter ratio. Previous studies have shown that increasing the unequal diameter ratio can greatly increase the volume of the screw groove. When the screw diameter and the thickness of the screw rib are fixed, the increase of the unequal diameter ratio will reduce the width of the gate rotor teeth, so that the gate rotor is easy to break. To analyze the unequal diameter ratio, it is necessary to define the screw diameter as 117mm and the thickness of the gate rotor tooth as 2mm. Figure 6 shows the change curve of the natural frequency of the gate rotor mode with the increase of the unequal radius rate.

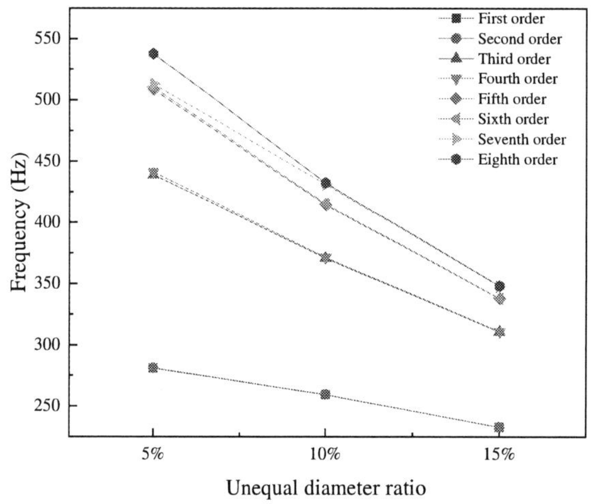

Figure 6. The change curve of the natural frequency of the mode with the unequal diameter ratio of the gate rotor.

It can be seen from figure 6 that the unequal diameter ratio is inversely proportional to the frequency. When the screw diameter and the thickness of the gate rotor remain unchanged, the increase of the unequal diameter ratio reduces the modal natural frequency. However, from the perspective of the decreasing trend, the current analysis of the unequal diameter ratio will not be close to the fundamental frequency of 50Hz.

3.5 Analysis and discussion

(1) From the above figure, it can be seen that there is a phenomenon of multilevel modal overlap. This is because the number of gate rotor teeth is 11, which is generally defined as an approximate center symmetric structure. Therefore, it can be seen from the figure that there is a phenomenon of multilevel multiple order roots, that is, the two-order frequencies are approximately equal.

(2) From the appeal figure, it can be seen that increasing the thickness of the gate rotor helps to improve the modal natural frequency while keeping certain structural parameters unchanged, while increasing the diameter, tooth width and unequal diameter ratio of the gate rotor, the modal natural frequency decreases gradually.

(3) In the process of optimizing the structural parameters of the gate rotor, all areas within the fundamental frequency 50Hz should be avoided, so as to avoid causing resonance of the gate rotor and damaging the normal operation of the machine.

4. Conclusion

In this paper, the influence of structural parameters on the natural frequency of the gate rotor is analyzed. These structural parameters mainly include: tooth thickness, diameter, tooth width and unequal diameter ratio. Through the simulation analysis, the following conclusions are obtained:

(1) According to the modal data of different structures, the structural parameters have a great influence on the natural frequency of the gate rotor. The general trend is that the natural frequency increases with the increase of the thickness of the gate rotor teeth, and decreases with the increase of the diameter, tooth width and unequal diameter ratio of the gate rotor.

(2) Minimizing vibration is a good method to design the structure of the gate rotor successfully), and it is also helpful to reduce the cost and improve the stability. However, in order to optimize the structure of the gate rotor and minimize the vibration, it is necessary to avoid the area within 50Hz of the fundamental frequency, and also to consider the separation from harmonic vibrations of other components to prevent resonance.

(3) Through the variable structure modal analysis of the gate rotor, it provides the basic data for the study of the mechanical characteristics of the gate rotor, and provides the reference for the design, verification and optimization of the subsequent gate rotor structure from the perspective of dynamics, as well as the experience accumulation for the variable structure modal analysis of similar products.

Reference

[1] Gu W, Wen Y W, Cao G Y, et al 2007 The latest research findings concerning low-temperature heat energy-based power generation and its development trend[J]. *Journal of Engineering for Thermal Energy and Power*, 2007(2):115-119.
[2] Quoilin S, Broek M V D, Declaye, Sébastien, et al 2013 Techno-economic survey of Organic Rankine Cycle (ORC) systems[J]. *Renewable and Sustainable Energy Reviews*, **22**(Complete):168-186.
[3] Kang S H 2012 Design and experimental study of ORC (organic Rankine cycle) and radial turbine using R245fa working fluid[J]. *Energy*, **41**(1):514-524.
[4] Li M, Wang J, He W, et al 2013 Construction and preliminary test of a low-temperature regenerative Organic Rankine Cycle (ORC) using R123[J]. *Renewable Energy*, 2013, **57**:216-222.
[5] Development and experimental study on a single screw expander integrated into an Organic Rankine Cycle[J]. *Energy*, 2016, **116**:43-52.
[6] B. Zimmern 1965 Worm rotary compressors with liquid joints. *US 3180565* [P].
[7] Fu Z F, Hua H X 2000 Modal Analysis Theory and Application[M]. *Shanghai Jiao Tong University Press*,
[8] Liu X J, Zhang S X 2016 Engineering vibration and testing technology[M]. *Beijing: Machinery Industry Press*.

IOP Conf. Series: Materials Science and Engineering **751** (2020) 012046 doi:10.1088/1757-899X/751/1/012046

Model-based Balancing Method of Rotors using Differential Evolution Algorithm

Yun Zhang[1,*], Meng Li[1] and Yang Hu[2]

[1] School of Mechano-Electronic Engineering, Xidian University, Xi'an 710071, China
[2] Chang Guang Satellite Technology Co.,Ltd, Changchun, 130000, China;

[*] yunzhang@xidian.edu.cn

Abstract. As a major source of undesired vibration, unbalance is a ubiquitous phenomenon for many types of rotor system. To suppress the unbalance vibration and avoid the complicated and dangerous process of taking trial weights during the balancing process, a model-based balancing method of rotors is proposed in this paper. The rotor model is built to analyse the dynamic characteristics of the unbalance rotor, an unbalance identification method is presented and the optimization objective function of the unbalance vectors is given, then the differential evolution algorithm is employed to acquire the optimization solution. Finally, the accuracy of the rotor model is verified by modal tests, and the performance of the proposed method is evaluated in balancing experiment, results reflect its effectiveness.

1. Introduction

Due to materials anisotropic, installation eccentricity, wear and other factors, the unbalance problem is unavoidable. Since even a slight unbalance can cause severe vibration of the rotor under high-speed operation condition, which leads to great impact on the quality of the rotating machinery [1]. Therefore, how to suppress the unbalance vibration becomes a key problem of rotor system.

Extensive work has been carried out on the identification of unbalance over the past several decades[2], and the trial weights is needed in the traditional methods. In order to improve the efficiency, the dynamics models are introduced into the balancing procedure. Based on this view point, Jalan et al.[3] described a model-based fault diagnosis of a rotor–bearing system for misalignment and unbalance under steady-state. Deepthikumar et al.[4] applied the polynomial curve for eccentricity distribution with finite element modeling to identify the distributed unbalance. Khulief et al.[5] developed a low-speed unbalance identification method relies on knowledge of the modal characteristics of the rotor. Tiwari et al.[6] proposed an identification method of bearing dynamic parameters and unbalance states using an unified finite element model. Wang et al.[7] proposed a novel unbalance identification strategy by the measurement point vector method, which based on the numerical rotor model. Moreover, in view of the convenience and flexibility of implementation, scholars also focus on the balancing methods based on intelligent algorithm. The genetic algorithm[8], the sequential quadratic programming[9], artificial neural network[10], least angle regression technique[11] and fuzzy logic[12] are employed in the optimization process of the balancing method

However, it should be noted that the accuracy of the model-based methods are heavily dependent on the model accuracy, and identification results only based on intelligent algorithms is valid only in current condition, when the rotor working state changes, the results may not be effective. Hence, scholars research on the methods which integrate the dynamics models and the intelligent algorithms to improve the efficiency and accuracy of the identification results. Zhang et al.[13] identified four types of faults including imbalance based on rotor dynamics and fuzzy support vector machine. Saldarriaga et al.[14] introduced genetic algorithms into the unbalance identification process, in which the response is obtain by a finite element model. Morais et al.[15] identified the unbalance distribution in linear and non-linear conditions through pseudorandom optimization method and the finite element model. Oke et al.[16] applied evolutionary algorithms to suppress the unbalance vibration of a rotor model based on simulation analysis.

Although remarkable balancing effects are obtained by above approaches, there are still some issues that can be worth studying. For example, only the amplitude is used for constructing the optimization objective function, and do not make full use of the rotor vibration information in some approaches, which may result in the identified unbalance vector being an equivalent solution rather than a true solution. In additional, there are some methods that have only been analyzed by simulation and have not been verified by actual experiments. Therefore, the contribution presented in this paper is an alternative balancing method for rotating machinery, aiming at making full use of the rotor space vibration information to construct the objective function and obtain the optimal solution, the model accuracy and the effectiveness of the proposed method are verified by experiment.

2. Unbalance identification method

The dynamics model of the rotor-bearing system can be divided into two parts, the finite element model of the rotor and the journal bearing. Assumed that the FE model may have N nodes, the radial displacement and rotation along the radial coordinates of a single element in the generalized coordinates can be expressed as $\delta_n = \{y_n + iz_n, \theta_{zn} - i\theta_{yn}\}^T$, $(n = 1,2,\mathrm{L}\ ,N)$, where y_n and z_n are the displacement of Y and Z axis, θ_{yn} and θ_{zn} are the rotation angle of Y and Z axis, respectively. The dynamic differential equation of the rotor is shown as

$$\boldsymbol{M}\ddot{\delta} + \boldsymbol{C}\dot{\delta} + \boldsymbol{K}\delta = \boldsymbol{F}^u + \boldsymbol{F}^b \tag{1}$$

where $\boldsymbol{M}$ is the mass matrix, $\boldsymbol{K}$ is the stiffness matrix, $\boldsymbol{C}$ is the damping and gyroscopic matrix. $\boldsymbol{F}^u$ represents the unbalance force. $\boldsymbol{F}^b$ stands for the oil film force which defines as $\boldsymbol{F}_n^b = \{F_y^b + iF_z^b, 0\}^T$, where F_y^b and F_z^b is the component of oil film force in different directions, which can obtained as

$$\begin{Bmatrix} F_y^b \\ F_z^b \end{Bmatrix} = \sigma A \begin{bmatrix} 3y & 3z \\ -2\cos\alpha & -2\sin\alpha \\ -\sin\alpha & \cos\alpha \end{bmatrix}^T \begin{Bmatrix} \dfrac{2}{1-y^2-z^2} + \dfrac{2(z\cos\alpha - y\sin\alpha)}{(1-y^2-z^2)^{3/2}}[\dfrac{\pi}{2} + \tan^{-1}\dfrac{z\cos\alpha - y\sin\alpha}{(1-y^2-z^2)^{1/2}}] \\ y\cos\alpha + z\sin\alpha / [1-(y\cos\alpha + z\sin\alpha)^2] \\ \dfrac{2}{(1-y^2-z^2)^{1/2}}[\dfrac{\pi}{2} + \tan^{-1}\dfrac{z\cos\alpha - y\sin\alpha}{(1-y^2-z^2)^{1/2}}] \end{Bmatrix} \tag{2}$$

where $A = [(y-2\dot{z})^2 + (z+2\dot{y})]^{1/2} / (1-y^2-z^2)$, α is the initial dynamic boundary angle, σ is the Summerfield correction coefficient, $\sigma = \kappa_b \omega R L^3 / (2c_b)^2$, and c_b is the radial clearance, κ_b is the viscosity of the oil film, ω is the rotating speed.

If the simulated unbalance response curve matches the measured curve accurately, the $\boldsymbol{F}$ obtained by Equation(1) can be considered as equal to that acquired by experiment. That means, the real $\boldsymbol{F}$ can be identified by the comparison analysis of the response curves obtained by simulation and experiment, respectively. Since the oil film force $\boldsymbol{F}^b$ can be calculated, the unbalance force $\boldsymbol{F}^u$ can be identified by $\boldsymbol{F} - \boldsymbol{F}^b$. Based on this viewpoint, the optimization variables can be set as the amplitude U and phase Π of the unbalance vector, and the sub objective functions can be built as

$$\begin{cases} f_1 = \left\| \left(v_{ep(ij)}^{\max} - v_{ep(ij)}^{\min}\right) - \left(v_{su(ij)}^{\max} - v_{su(ij)}^{\min}\right) \right\|_2, f_2 = \left\| \left(v_{ep(ij)}^{\max} - v_{ep(ij)}^{\min}\right) - \left(v_{su(ij)}^{\max} - v_{su(ij)}^{\min}\right) \right\|_\infty \\ f_3 = \left\| \chi_{ep(i)} - \chi_{su(i)} \right\|_2, f_4 = \left\| \chi_{ep(i)} - \chi_{su(i)} \right\|_\infty \end{cases} \tag{3}$$

where v is the unbalance vibration response at rotating frequency, the subscripts ep and su represent experimental and simulated data, respectively, the subscript i represents measuring plane, and the subscript j represents the measuring direction, j=1 represents Y direction and j=2 represents Z direction. χ represents the phase of the unbalance vibration response, and χ_{ep} can be obtained by vibration response and the signal of phase datum mark[17].

Consequently, the general objective function of the unbalance vector can be given as

$$\begin{cases} minimize \quad F_{obj} = \sum_{i=1}^{4}(\lambda_i F_i) \\ subject\ to \quad \sum_{i=1}^{4}\lambda_i = 1, \ U_p \subseteq (0, U_{\max}), \ \Pi_p \subseteq (-\pi, \pi) \qquad p = 1, 2, \ldots, P \end{cases} \tag{4}$$

where λ_i is the weight coefficient, $U_{\max}$ represents the upper limit of the correction mass, F_i is the decreasing semi-trapezoidal fuzzy distribution function, if $\eta_i^{\min} < f_i < \eta_i^{\max}$, F_i=$(f_i - \eta_i^{\min})/(\eta_i^{\max} - \eta_i^{\min})$, and $\eta_i^{\max}$ and $\eta_i^{\min}$ is the maximum and minimum value in the feasible region of f_i.

3. Procedure of the balancing method

DE algorithm[18, 19] is a random parallel direct search algorithm, which simulates the biological evolution process based on a stochastic model. this algorithm has strong global convergence ability and robustness. During the operation of DE algorithm, the population size remains constant, and the main steps are basically the same as other evolutionary algorithms, including mutation, crossover and selection. The mutation individual $v_i^{g+1} = x_{best}^g + H(x_{r1}^g - x_{r2}^g)$, where x_{r1}^g, x_{r2}^g ($r_1, r_2 \in [1, 2, \mathrm{L}, NP]$) are two different individuals, and x_{best}^g is the best individuals at current generation. H is the self-adaptive scaling factor which can be set as $2^{\lambda} H_0$, where $\lambda = e^{1-G/(G+1-g)}$, and H_0 is the initial factor, and G and g represents the maximum and current evolutional generation, respectively, $g = 1, 2, \mathrm{L}, G$. In the crossover procedure, if $j_{rand} / D \le CR$ or $j = j_{rand}$, the trial individual u_{ji}^{g+1} can be set as v_{ji}^{g+1}, otherwise, u_{ji}^{g+1} can be set as x_{ji}^g, where j represents the space dimension, $j = 1, 2, \mathrm{L}, D$, $j_{rand} \in [1, 2, \mathrm{L}, D]$, CR is the crossover probability, $CR \in [0, 1]$. Then, the x_i^{g+1} would be selected from the trial individual u_i^{g+1} and the parent individuals x_i^g based on the comparison result between $f(u_i^{g+1})$ and $f(x_i^g)$.

The DE algorithm is employed to acquire the optimum parameters of the U and Π, and the flow chart of the unbalance optimization solution process is shown as Figure 1.

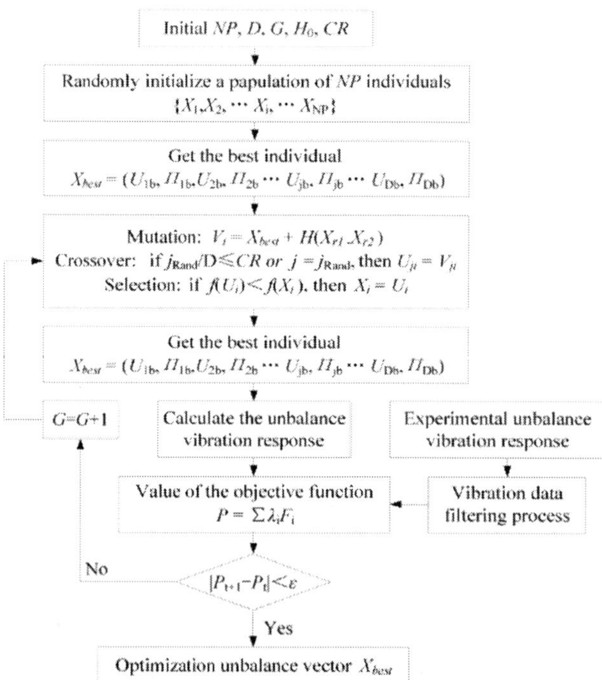

Figure 1. Flow chart of unbalance optimization solution

4. Verification and discussion

The experimental system shown in Figure 2 consists of rotor test platform, data acquiring and processing device, photoelectric and displacement sensors, force hammer, and motor controller. The photoelectric and displacement sensors are fixed on the radial axis, and the output analog signals generated by sensors are transformed into digital signals by data acquisition device.

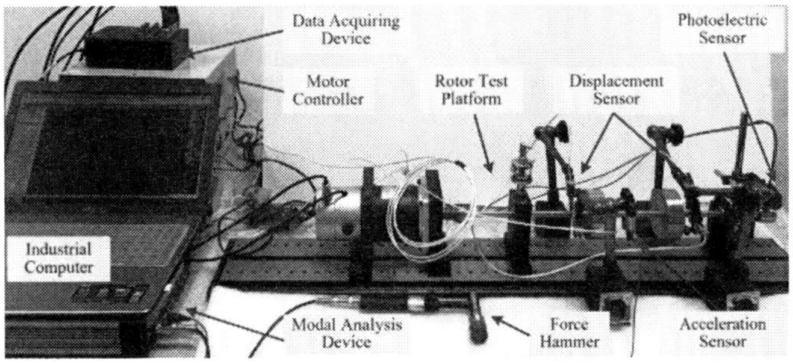

Figure 2. Experimental rotor test system

The finite element model of the rotor system is established based on 50 elements and 50 nodes with 5-DOF,. The mass disk units are located at the 21th and 33th nodes, the measuring plane are located at the 18th and 36th nodes, and the journal bearing units are located at 9th and 45th node, respectively. The modal test is carried out on the experimental platform to analyze the accuracy of the model, and the measured FRFs of the rotor is shown in Figure 3.

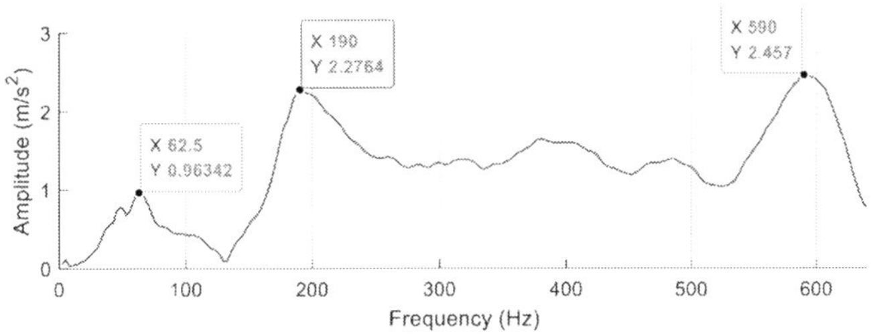

Figure 3. The measured FRFs of the rotor

According to the excitation and response curve, the structural damping ratio can be identified as 0.033, the first three order nature frequencies can be calculated as 62.5Hz, 190Hz and 590Hz in modal experiment, respectively. Meanwhile, that solved by Equation(1) are 62.3Hz, 189Hz and 565Hz, respectively. It is obviously that the error between simulation and experiment of the first-order nature frequency are 0.32%, 0.53% and 4.24%, respectively, which reflects the accuracy of the model.

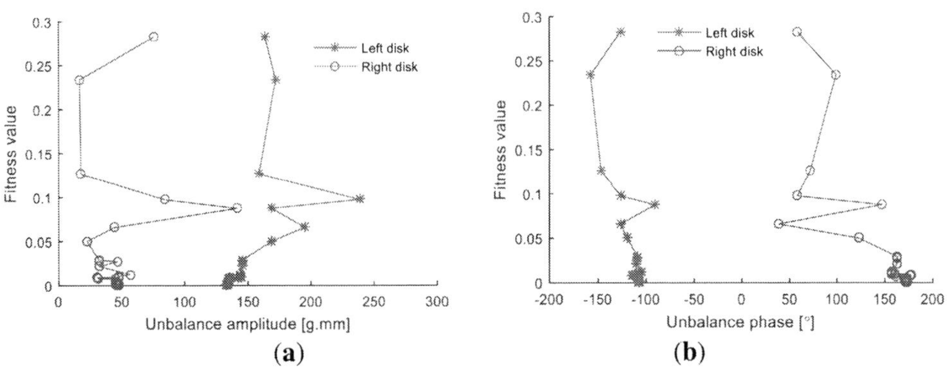

Figure 4. Evolution process for the unbalance vector: (**a**) Amplitude; (**b**) Phase

The balancing experiments are performed. The balancing speed is set to 2400 r/min, the measuring plane are on 18 and 36 nodes, and the sampling frequency is set to 10 *k*Hz. The initialization parameters in Figure 1 are set as: NP=30, D=4, H_0=0.5, CR=0.3. The weight coefficients are set as: $\lambda_1=\lambda_2=0.15$, $\lambda_3=\lambda_4=0.35$. The evolution curve of the unbalance vector with the objective function value is shown in Figure 4. It can be seen that the amplitude of the unbalance vector on both disks are converged after iterations, and the convergence of the unbalance amplitude and phase on two disks can be observed in Figure 4. Then, the optimal solution of the unbalance vector can be obtained. In here, the identified unbalance vector at the 21th and 33th nodes that represent the mass disks are 133.3g.mm ∠-107.7°and 46.60g.mm∠172.9°, respectively.

The rotor startup process are acquired and indicated as Figure 5(a), in which the green solid and blue solid lines represent the vibration response before and after the balancing procedure, respectively. For safety consideration, the experimental speed did not reach the first critical speed before balancing, after adding the counterweight, the rotor system can pass the first order critical smoothly, and the amplitude of the vibration is only about 43.5μm when it reaches the first critical speed. It can be seen that the proposed method can effectively identify the unbalance, and the residual vibration below and beyond the first critical speed is very small after the balancing procedure, which proves the effectiveness of the method. Moreover, the first critical speed in this experiment is about 3714r/min, the error between the test result and Equation(1) is 0.62%, this indicates a good agreement between theoretical analysis and experimental test.

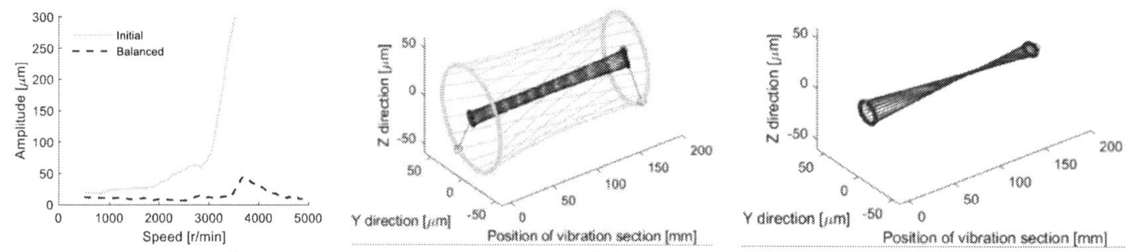

(a) (b) (c)

Figure 5. The vibration information before and after the balancing procedure: (**a**) full speed range (**b**) at 2400r/min; (**c**) at 4800r/min

The 3D holospectrum [20] of vibration at 2400r/min and 4800r/min on the measuring plane are shown as Figure 5(b)-(c). The axis orbit at both sides are reduced significantly at 2400 r/min, and after passing the critical speed, the residual vibration of the rotor at 4800 rpm is also very small. Overall, the Figure 5 show that the proposed dynamic balancing procedure is effective, which means that the unbalance mass and phase solved by the dynamics model and the DE algorithm are accurate.

5. Conclusions

In this paper, a model-based unbalance identification method of rotors has been proposed to tackle the unbalance problem of the rotor system. The rotor model with hydrodynamic bearings is built, the optimization function of unbalance are constructed, and then, the DE algorithm employed to identify the optimum results. The experiment results show that the unbalance vectors can be identified by collecting vibration data only at operating speed without trial weights, and the vibration amplitudes of the rotor around the first order critical speed are reduced greatly after adding the counterweights.

Acknowledgments

This research was funded by the Key Research and Development Program in Shaanxi Province of China (Grant No. 2017GY-051), the National Key Research and Development Program of China (Grant No. 2016YFB1102502), and the National Natural Science Foundation of China (Grant No. 51305324).

References

[1] Zhang, Y., Mei, X., Shao, M. & Xu, M. 2013. An improved holospectrum-based balancing method for rotor systems with anisotropic stiffness. *Proceedings of the Institution of Mechanical Engineers, Part C: Journal of Mechanical Engineering Science,* **227,** 246-260.

[2] Foiles, W. C.Allaire, P. E. & Gunter, E. J. 1998. Review of Rotor balancing. *SHOCK AND VIBRATION,* **5,** 325-336.

[3] Jalan, A. K. & Mohanty, A. R. 2009. Model based fault diagnosis of a rotor-bearing system for misalignment and unbalance under steady-state condition. *JOURNAL OF SOUND AND VIBRATION,* **327,** 604-622.

[4] Deepthikumar, M. B.Sekhar, A. S. & Srikanthan, M. R. 2013. Modal balancing of flexible rotors with bow and distributed unbalance. *Journal of Sound and Vibration,* **332,** 6216-6233.

[5] Khulief, Y. A.Oke, W. & Mohiuddin, M. A. 2014. Modally Tuned Influence Coefficients for Low-Speed Balancing of Flexible Rotors. *Journal of Vibration and Acoustics,* **136,** 024501.

[6] Tiwari, R. & Chougale, A. 2014. Identification of bearing dynamic parameters and unbalance states in a flexible rotor system fully levitated on active magnetic bearings. *Mechatronics,* **24,** 274-286.

[7] Wang, A., Cheng, X., Meng, G., Xia, Y., Wo, L. & Wang, Z. 2017. Dynamic analysis and numerical experiments for balancing of the continuous single-disc and single-span rotor-bearing system. *Mechanical Systems and Signal Processing,* **86,** 151-176.

[8] Xu, B.Qu, L. & Sun, R. 2000. The optimization technique-based balancing of flexible rotors without test runs. *Journal of sound and vibration,* **238,** 877-892.

[9] Wang, X. 2007. SQP algorithms in balancing rotating machinery. *Mechanical Systems and Signal Processing,* **21,** 1469-1478.

[10] Walker, R. B., Vayanat, R., Perinpanayagam, S. & Jennions, I. K. 2014. Unbalance localization through machine nonlinearities using an artificial neural network approach. *Mechanism and Machine Theory,* **75,** 54-66.

[11] Chatzisavvas, I. & Dohnal, F. 2015. Unbalance identification using the least angle regression technique. *Mechanical Systems and Signal Processing,* **50–51,** 706-717.

[12] Carvalho, V. N., Rende, B. F. R., Silva, A. D. G., Ap Cavalini, J. A. & Steffen, J. V. 2018. Robust Balancing Approach for Rotating Machines Based on Fuzzy Logic. *Journal of Vibration and Acoustics,* **140,** 051018-051018-9.

[13] Zhang, J., Ma, W., Lin, J., Ma, L. & Jia, X. 2015. Fault diagnosis approach for rotating machinery based on dynamic model and computational intelligence. *Measurement,* **59,** 73-87.

[14] Saldarriaga, M. V., Steffen, V., Hagopian, J. D. & Mahfoud, J. 2011. On the balancing of flexible rotating machines by using an inverse problem approach. *JOURNAL OF VIBRATION AND CONTROL,* **17,** 1021-1033.

[15] Morais, T. S., Der Hagopian, J., Steffen Jr., V. & Mahfoud, J. 2014. Optimization of unbalance distribution in rotating machinery with localized non linearity. *Mechanism and Machine Theory,* **72,** 60-70.

[16] Oke, W. A.Abido, M. A. & Asafa, T. B. 2015. Balancing of flexible rotors based on evolutionary algorithms. *Mechanics \& Industry,* **16,** 406.

[17] Zhang, Y., Qiao, W., Zou, D. & Shao, M. 2016. Study on the Relationship between the Imbalance Status and its Vibration Response in Balancing Method of Rotors.) *The 23rd International Congress on Sound and Vibration (ICSV23).* Athens, Greece.

[18] Qin, A. K.Huang, V. L. & Suganthan, P. N. 2009. Differential Evolution Algorithm With Strategy Adaptation for Global Numerical Optimization. *IEEE Transactions on Evolutionary Computation,* **13,** 398-417.

[19] Storn, R. & Price, K. 1997. Differential Evolution – A Simple and Efficient Heuristic for global Optimization over Continuous Spaces. *Journal of Global Optimization,* **11,** 341-359.

[20] Qu, L. S., Liu, X., Peyronne, G. & Chen, Y. D. 1989. The holospectrum: A new method for rotor surveillance and diagnosis. *Mechanical systems and signal processing,* **3,** 255–267.

IOP Conf. Series: Materials Science and Engineering **751** (2020) 012047 doi:10.1088/1757-899X/751/1/012047

Tiltrotors Position Tracking Controller Design Using Deep Reinforcement Learning

Yujia Huo[1,2,3,*], Yiping Li[2,3] and Xisheng Feng[2,3]

[1]State Key Laboratory of Robotics, Shenyang Institute of Automation, Chinese Academy of Sciences, Shenyang, 110016, China

[2]Institutes for Robotics and Intelligent Manufacturing, Chinese Academy of Sciences, Shenyang 110169, China

[3]University of Chinese Academy of Sciences, Beijing, 100049, China

[*]E-mail: huoyujia@sia.cn

Abstract. In this paper, a quad-tiltrotors air-water trans-domain robot is introduced. The nonlinear dynamic behaviours with uncertainties require a robust controller for multi-tasks. For this robot, controllers are designed using deep reinforcement learning method solving position and attitude control when operating as a UAV in the air. A ROS combining Gazebo simulation platform is designed to train the robot. The simulation results show the tiltrotors robot gets capabilities of spots tracking as a quad-rotors, and trajectory tracking as both the quad-rotors and tiltrotors.

Keywords – deep reinforcement learning, tiltrotors robot, trans-domain robot, position tracking

1. INTRODUCTION

Unmanned vehicles have rapidly extended their fields of application for decades, including unmanned aerial vehicles(UAVs) and unmanned underwater vehicles(UUVs). These two kinds of unmanned vehicles each advances differently. UAVs in the air have a fast speed of motivation when compared with UUVs operated underwater but have weak stealthiness of both observable performance and acoustic performance because of the high rate of rotors rotation. To improve environmental adaptabilities of UAVs and UUVs to expand their application scenarios and to merge the benefits, scientists and engineers on each side have focused on developing aquatic–aerial amphibious vehicles with a trans-domain capability operating in the air and underwater[1]. The trans-domain conception will significantly increase the range of sensing, communication performance, and mission scenarios available to the vehicles.

These trans-domain unmanned vehicles or trans-domain robots should at least need the capabilities of flight in the air as fixed-wing airplanes or multirotors, and the skills of diving underwater like UUVs do. Also, some trans-domain robots have the capability of surfing on the water surface as normal operation mode or just as a transition mode when taking off or landing. Moreover, the robots' cross-domain transition between air and water should be repeatedly and bidirectional. Thus, seaplanes and submarine-launched UAVs aren't considered to be with the trans-domain conception.

Due to the uncertainties of environmental conditions from ocean current and wind, as well as the uncertainties of robots' model with nonlinearity, suitable motion controllers for tasks of attitude steady control or position tracking are required for autonomous operation. Some institutions have achievements in autonomous control in the two kinds of fluid media. Maria et al. presented an octocopter able to operate in the air and underwater with a PD controller[2]. W. Weisler et al. demonstrated a fixed-wing cross-domain robot operating in both domains[3].

Unlike the trans-domain robots mentioned, in this paper, we present a tiltrotors trans-domain robot with four ducted propeller thrusts and wings foldable into the vehicle body. This robot has three operation modes, the underwater mode, the quadrotors mode with constant tilt thrust angles, and the tiltrotors mode with variable tilt thrust angles and wings providing the lift force in the air. The classical PID method was used for a tiltrotor UAVs during the VTOL and transition from vertical flight to horizontal flight[4]. MPC controller was designed with an unmanned quad-tiltrotors[5].

For our trans-domain robot with tiltrotor shape and more complex kinetics, classical methods hard to get excellent performance, neither functional or costly with time. Within the growth of artificial intelligence, reinforcement learning(RL), particular deep reinforcement learning(DRL) when combined with deep neural networks, has achieved several exciting works such as playing games, and this type of method interacting with the environment and learning from experience also performs well in robotic fields. Hwangbo from Ethz used reinforcement learning to control a quadrotors' position and attitude steady from a hand throw[6]. In the underwater area, deep reinforcement learning has been applied for depth control[7] and position tracking[8].

In this paper, the trans-domain quad-tiltrotors is only considered simulated in the air condition operating in quadrotors mode and quad-tiltrotor mode. In Section 2, the kinetic model will be presented. DRL method controller explanation will be given in Section 3. In Section 4, simulation of both two flight modes and the transitions between modes are shown, and consequently, some conclusions will be presented in Section 5.

2. KINETICS OF TRANS-DOMAIN QUAD-TILTROTORS ROBOT

The reinforcement learning method learns from interacting with the environment repeatedly and gets an optimal control policy. However, great times of training the robot physically in the real world cost a lot. It is necessary to simulate the train tasks before deployment. In this section, the kinetics model of the trans-domain quad-tiltrotors robot is presented, and this model will be used for designing the controller and RL simulation training.

Trans-domain quad-tiltrotors robot has four ducted propeller thrusts(Figure 1) with variable angels $\delta_{i=1,2,3,4}$ and $\delta_i \in [0,\pi/2]$. When the angles of propeller thrusts are 0, the robot behaves like a simple quad-rotors UAV, and the propeller thrusts provide the lift(Figure 2).

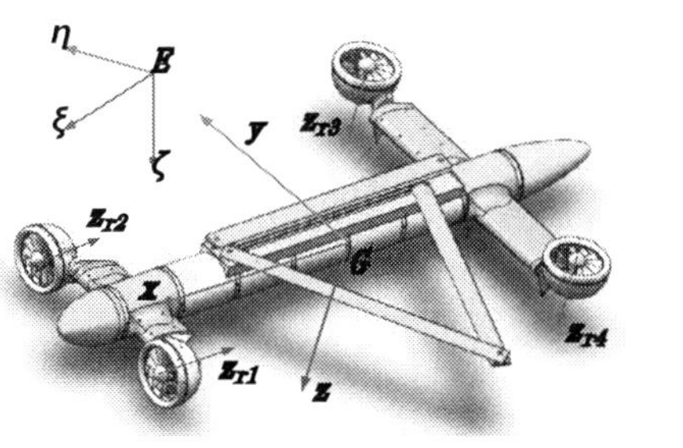

Figure 1 The Trans-domain Quad-Tiltrotors robot

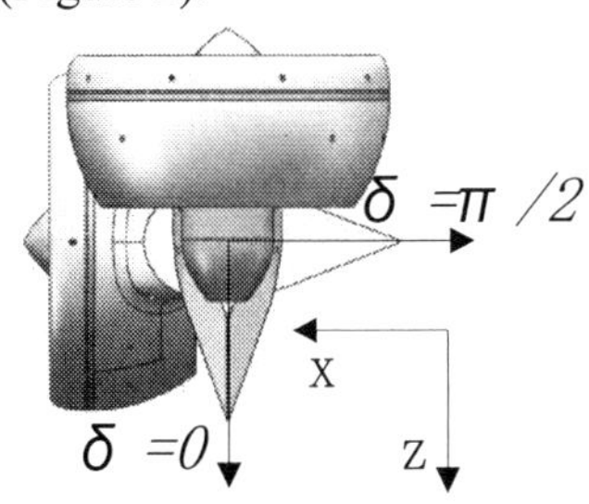

Figure 2 Tiltrotor propeller angle

In Figure 1, G is the center mass of the robot, x points to the bow, y points to the starboard, and z is determined by the right-hand rule. The body coordinate is fixed relative to the fuselage and rotates as the aircraft rotates. The attitude angles ψ , θ , ϕ represent yaw, pitch and roll, respectively[5]. The robot mainly effects from the force of four rotors $T_{i=1,2,3,4}$, gravity mg , wing force F_W . The wing force F_W can be decoupled into three orthogonal forces, the lift of wings F_{lift}, side force of wings $F_{sideforce}$ and air drag of wings F_{drag}. And in the same way, moments of the robot are gotten.

2.1 Force and moment provided by the tilt propeller thrusts

The force provided by thrusts in the Body-Fixed coordinates Frame can be explained in Equation(1) [9], [10]. where T_i presents the thrust force of the rotor i, and $T_i = C_r\omega_i^2$ with the rotor thrust coefficient C_r. For the moments $L_{x,f}$, $L_{y,f}$ and $L_{z,f}$ show the rotor position vector component of the rotor in front of the robot along with the axes x y, z respectively. And $L_{x,b}$, $L_{y,b}$, $L_{z,b}$ in the same way for the rotors behind the robot. Moreover, the rotating propellers with gyroscopic effects can be described as $M_{G,i} = C_q\omega_i^2$.

$$\begin{cases} F_{T,x} = T_1\sin\delta_1 + T_2\sin\delta_2 + T_3\sin\delta_3 + T_4\sin\delta_4 \\ F_{T,y} = 0 \\ F_{T,z} = T_1\cos\delta_1 + T_2\cos\delta_2 + T_3\cos\delta_3 + T_4\cos\delta_4 \\ M_{T,x} = [T_1\cos\delta_1 L_{x,f} + T_2\cos\delta_2 L_{x,f} + T_3\cos\delta_3 L_{x,b} + T_4\cos\delta_4 L_{x,b}] + [\sin\delta_1 M_{q,1} - \sin\delta_2 M_{q,2} + \sin\delta_3 M_{q,3} - \sin\delta_4 M_{q,4}] \\ M_{T,y} = T_1 L_{y,f} + T_2 L_{y,f} + T_3 L_{y,b} + T_4 L_{y,b} \\ M_{T,z} = [T_1\sin\delta_1 L_{z,f} + T_2\sin\delta_2 L_{z,f} + T_3\sin\delta_3 L_{z,b} + T_4\sin\delta_4 L_{z,b}] + [\cos\delta_1 M_{q,1} - \cos\delta_2 M_{q,2} + \cos\delta_3 M_{q,3} - \cos\delta_4 M_{q,4}] \end{cases} \tag{1}$$

2.2 Force and moment provided by the fixed-wings

The force provided by the fixed-wings can be decoupled into three orthogonal forces, the lift of wings F_{lift}, side force of wings $F_{sideforce}$ and air drag of wings F_{drag}. The force and moment effect of the fixed wings can be rewritten as:

$$F_w = \begin{bmatrix} F_{w,x} \\ F_{w,y} \\ F_{w,z} \end{bmatrix} = \begin{bmatrix} C_D QS \\ C_Y QS \\ C_L QS \end{bmatrix}, \quad M_W = \begin{bmatrix} M_{w,x} \\ M_{w,y} \\ M_{w,z} \end{bmatrix} = \begin{bmatrix} C_l QSl \\ C_m QSl \\ C_n QSl \end{bmatrix} \tag{2}$$

Dynamic pressure $Q = \rho V^2 / 2$, where ρ presents the air density, $V = \sqrt{v_x^2 + v_z^2}$ represents the vector sum of x axis and y axis velocity. In addition, let S be the area of the wing. C_D, C_Y, C_L are the drag coefficient, side force coefficient and the lift coefficient, and C_l, C_m, C_n means the moment coefficient respectively[11].

2.3 Force and moment effect on the robot

Thus, in the World Frame, the force and moment effect on robot can be rewritten as follows:

$$\begin{cases} F_x = F_{w,x} + F_r\sin\theta \\ F_y = F_{w,y} + F_r\cos\theta\sin\phi \\ F_z = F_{w,z} - F_r\cos\theta\cos\phi + mg \end{cases}, \begin{cases} M_x = (M_r + M_w)\cos\theta \\ M_y = (M_r + M_w)\sin\theta\sin\phi \\ M_z = (M_r + M_w)\sin\theta\cos\phi \end{cases} \tag{3}$$

where m is the mass of the robot and $g = 9.8m/s^2$.

3. ROBOT POSITION TRACKING CONTROLLER DESIGN

The trans-domain quad-tiltrotors robot has multiple operation conditions and needs controller suitable for these operation conditions with different media and operation modes. But it is hard to design only one controller to fix the whole control problem. So, an acceptable way is to create departed controllers for each condition. And in this way, the learned control policy may get void of being ruined when other condition's control policies are under training.

For our recent robot, three sub-controllers are designed for three modes, the quad-rotors mode, the quad-tiltrotors mode with fixed wings and the underwater mode(Figure 3). The others mean it could have more operation mode in the future, for example, now the robot takes off/lands as a helicopter-like quad-rotors, but may have the capability of surfing taking-off on the surface.

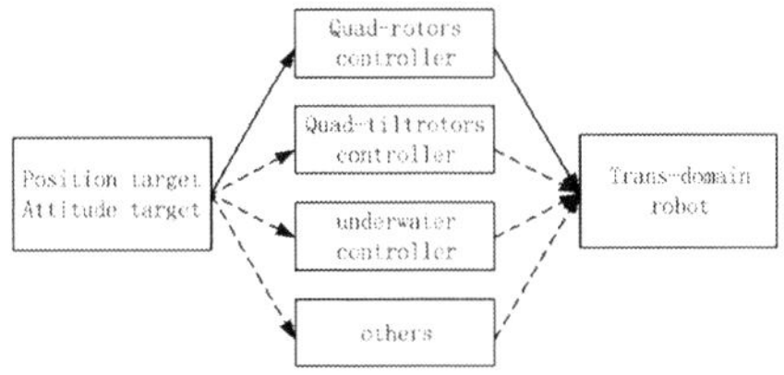

Figure 3 Controllers of the trans-domain robot

3.1 Deep reinforcement learning

To solve the uncertainties of the robotics kinetics modeling and the environmental disturbs, DRL method controllers are designed. The reinforcement learning method can learn autonomously by exploring the environment space and evaluating the action policy that has been made. The processes of making action policies can be described as a Markov Decision Process(MDP), within a tuple: $\{S, A, P, R\}$, where S is the state space, A the action space, P the state transition probability, and R the reward. Therefore, the goal of MDP is to find an action policy to maximize the long-term cumulative cost value[12]. The policy gradient is an approach to solve reinforcement learning problems. It maximizes the expected total reward by repeatedly estimating the gradient $g := \nabla_\theta \mathrm{E}[\sum_{t=0}^{\infty} r_t]$. Using the advantage function $A^\pi(s_t, a_t)$ as the rewards, the gradient can be described in Equation (4). With the state-value V^π and the action-value Q^π, the advantage function $A^\pi(s_t, a_t) = Q^\pi(s_t, a_t) - V^\pi(s_t)$.

$$g = \mathrm{E}[\sum_{t=0}^{\infty} A^\pi(s_t, a_t) \nabla_\theta \log \pi_\theta(a_t \mid s_t)] \tag{4}$$

3.2 Proximal policy optimization reinforcement learning(PPO)

The PPO[13] method is sort of an improved algorithm from the TRPO(Trust Region Policy Optimization)[14]. When considering TRPO, the problem can be described as follows:

$$\text{maximize} \frac{1}{N} \sum_{t=1}^{N} [\frac{\pi(a_t \mid s_t)}{\pi_{old}(a_t \mid s_t)} A_{\pi_{old}}(s_t, a_t)] \tag{5}$$

And the problem improved by PPO with Adaptive KL Penalty can be described as follows:

$$\text{maximize} \frac{1}{N} \sum_{t=1}^{N} [\frac{\pi(a_t \mid s_t)}{\pi_{old}(a_t \mid s_t)} A_{\pi_{old}}(s_t, a_t) \text{-} \beta \mathrm{KL}[\pi_{\theta_{old}}(\cdot|\mathrm{s}_t), \pi_\theta(\cdot|\mathrm{s}_t)]] \tag{6}$$

where β controls the weight of the penalty. It penalizes the objective if the new policy is different from the old policy. we can dynamically adjust β with d and a threshold d_{targ}, and d can be described as:

$$d = \frac{1}{N} \sum_{t=1}^{N} [\mathrm{KL}(\pi_{\theta_{old}}(\cdot|\mathrm{s}_t), \pi_\theta(\cdot|\mathrm{s}_t))] \tag{7}$$

d is the KL-divergence between the old and the new policy. If it is higher than a target value, we shrink β. Similarly, if it falls below another target value, we expand the trust region:

$$\begin{aligned} &if: d < d_{\text{targ}} / \delta, \beta \leftarrow \beta / \alpha \\ &if: d > d_{\text{targ}} \times \delta, \beta \leftarrow \beta \times \alpha \end{aligned} \tag{8}$$

Algorithm1

for $i \in \{1, \cdots, N\}$ **do**

run policy π_θ for T timesteps, get feedback $\{s_t, a_t, r_t\}$ and $A_t = \sum_{t'>t} \gamma^{t'-t} r_{t'} - V(s_t)$

$\pi_{old} \leftarrow \pi_\theta$

for $j \in \{1, \cdots, M\}$ **do**

$J(\theta) = \sum_{t=1}^{T} \frac{\pi_\theta(s_t \mid s_t)}{\pi_{old}(a_t \mid s_t)} A_t - \beta \text{KL}[\pi_{old} \mid \pi_\theta]$ and update θ by gradient method with $J(\theta)$

end for

for $j \in \{1, \cdots, B\}$ **do**

$L(\phi) = -\sum_{t=1}^{T} (\sum_{t'>t} \gamma^{t'-t} r_{t'} - V(s_t))^2$ and update ϕ by gradient method with $L(\phi)$

end for

if $d < d_{\text{targ}} / \delta, \beta \leftarrow \beta / \alpha$

if $d > d_{\text{targ}} \times \delta, \beta \leftarrow \beta \times \alpha$

end for

3.3 Reinforcement learning for point tracking control

For the point tracking tasks, the desired position of the point can be described as a three-dimension vector p_{target}, where $p_{\text{target}} = [x_{\text{target}}, y_{\text{target}}, z_{\text{target}}]^T$. The point-tracking problem is to make the robot move to a desired point and remain with a steady state of position, attitude, linear speed and angular speed. As only the quad-rotors has the hover capability with zero linear velocities, the point tracking tasks are only for quad-rotors operation mode. The actions are the rotation rate of four propeller thrusts. A quadratic reward function at each step is designed:

$$r(t) = -c_1 \left\| p_t - p_{target} \right\|_2^2 - c_2 \left\| \Omega_t \right\|_2^2 - c_3 \left\| V_t \right\|_2^2 - c_4 \left\| \omega_t \right\|_2^2 - c_5 \left\| a_t^i \right\|_2^2 \tag{9}$$

where, p_t present the robot's current position vector $[x_t, y_t, z_t]^2$, and L2 norm means as follows:

$$\| p_t - p_{\text{target}} \|_2^2 = (x_t - x_{\text{target}})^2 + (y_t - y_{\text{target}})^2 + (z_t - z_{\text{target}})^2 \tag{10}$$

So in the same way with the L2 norm and the vector, Ω_t means the current attitude angle vector $[\phi_t, \theta_t, \psi_t]^T$, V_t the linear velocity vector $[u_t, v_t, w_t]^T$, ω_t the angular velocity $[p_t, q_t, r_t]^T$, and the a_t^i means the current action of thrust i at the time of t. c_1, c_2, c_3, c_4, c_5 weights for each vector. The goal is of the RL algorithm is to find an optimal action policy, that under this policy, the robot can move to a desired point autonomously and stay steady at that position.

3.4 Reinforcement learning for quad-rotors trajectory tracking control

For the quad-rotors trajectory tracking tasks, $l = [[x_1, y_1, z_1]^T, [x_2, y_2, z_2]^T, \cdots, [x_i, y_i, z_i]^T, \cdots]$, the actions are the rotation rate of four propeller thrusts. The reward function is designed as follows:

$$\begin{aligned} r(t) = \; & -c_1 \| \psi_t^{heading} - \psi_{t,\text{target}}^{heading} \|_2^2 - c_2 \| \psi_t^V - \psi_{t,\text{target}}^V \|_2^2 - c_3 \| z_t - z_{t,\text{target}} \|_2^2 \\ & -c_4 \| \theta_t \|_2^2 - c_5 \| \phi_t \|_2^2 - c_6 \| \omega_t \|_2^2 - c_7 \| a_t^i \|_2^2 \end{aligned} \tag{11}$$

In this task, the velocity of the robot is not considered, and $\psi^{heading}$ is the attitude ψ, yaw of the robot. ψ^V is the angle of the velocity in the horizontal plane:

$$\psi_t^V = \arctan \frac{v_t}{u_t} \tag{12}$$

And proposing

$$\psi_{t,\text{target}^V} = \arctan\frac{y_{t+1} - y_t}{x_{t+1} - x_t} \tag{13}$$

In the reward function, z_t is the hover-hight/depth of the robot. While θ_t, ϕ_t are the pitch and roll of the robot, respectively.

3.5 Reinforcement learning for quad-tiltrotors trajectory tracking control

For the quad-tiltrotors trajectory tracking task, the propeller angles vary from 0 to $\pi / 2$, and the action is the tuple. And the reward function is designed as follows:

$$r(t) = -c_1\|\ \psi_t - \psi_{t,\text{target}}\|_2^2 - c_2\|\ z_t - z_{t,\text{target}}\|_2^2 - c_3\|\ \theta_t\|_2^2 - c_4\|\ \phi_t\|_2^2 - c_5\|\ \omega_t\|_2^2 - c_6\|\ a_t^i\|_2^2 \tag{14}$$

Where, a_t^i is the element of the tuple of actions. When comparing the reward functions (11) and (14), we suppose that $\psi_t^{heading} \approx \psi_t^V$ with a little drift angle.

4. ROBOT POSITION TRACKING SIMULATION

It is often costly when the robot interacts with the environment repeatedly using the reinforcement learning method. Thus, simulations are necessary before physical deploying the controller to the robot. In this section, ROS, combined with Gazebo, is designed to implement a platform for controllers' training of the trans-domain quad-tiltrotors with the method of deep reinforcement learning, and the kinetics parameters are listed in table1. And the simulation of reinforcement learning is code in Python2.7 with ROS and Python3.5 with Tensorflow1.14 on Ubuntu 16.04 system.

Table 1 kinetics parameters

parameter	unit	value	parameter	unit	value
m	kg	33	$L_{z,f}$	m	0.063
$L_{x,f}$	m	0.717	$L_{z,b}$	m	0.109
$L_{x,b}$	m	0.472	S	m^2	0.47
$L_{y,f}$	m	0.265	l	m	0.7
$L_{y,b}$	m	0.474	ρ	$kg \cdot m^{-3}$	1.293

4.1 Quad-rotors position control simulation

In this part, the robot is operated as a quad-rotors with fixed propeller angles. Usually, the fixed propeller angles are set as zeros. Thus, the robot's action simplified to the rotation rate tuple four thrusts. The quad-rotors control tasks can be mainly separate as one position tracking and trajectory tracking. While the robot tracks the goals, attitudes are also under control. For the one-point tracking task, the robot takes off from $[0,0,0]$ and achieves the desired point, we set the desired goal at $[1,1,1]$.

Figure 4-8 show the point tracking task, the robot arrives at the goal [1,1,1] in 2 seconds and remains. Meanwhile, the robot gets steady linear velocities, angular velocities and attitudes. For trajectory tracking task, as mentioned in Sec. 3.4, the continuous trajectory is described as several waypoints. In the horizontal plane, the line of sight method is adopted to design the reward function. However, in the vertical plane, due to the quad-rotors dynamics model, the attack angle always exists. The control in the vertical plane uses the height error rather than the angle error like in the horizontal plane.

We design a trajectory described as lemniscate of Huygens in Equation (15), the robot takes off and begins to track this trajectory.

$$\begin{cases} x = 5\cos(\varphi) \\ y = 5\sin(2\varphi) \\ z = 15 - 5\cos(\varphi) \end{cases} \tag{15}$$

IOP Conf. Series: Materials Science and Engineering **751** (2020) 012047 doi:10.1088/1757-899X/751/1/012047

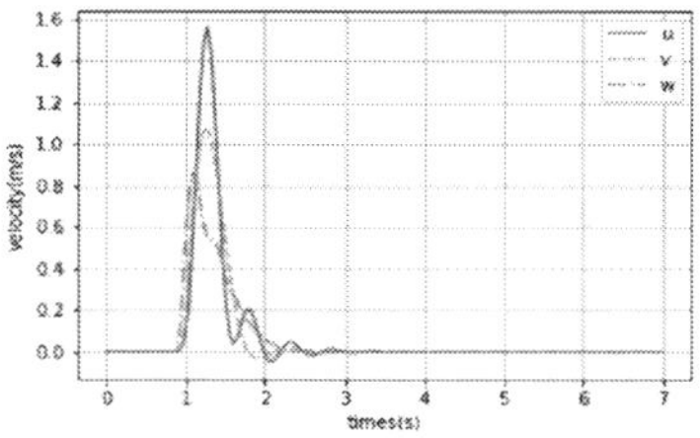

Figure 4 Linear velocity

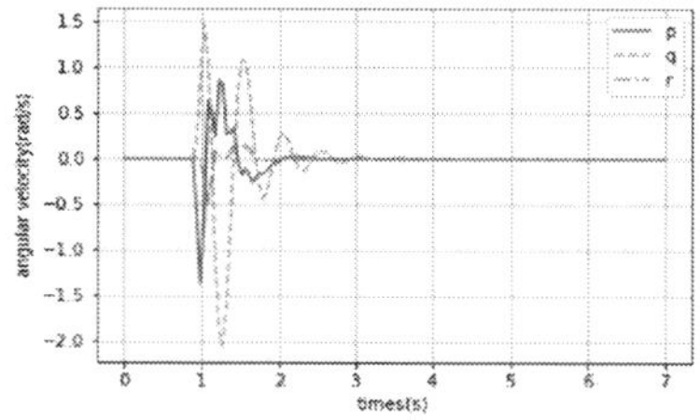

Figure 5 Angular velocity

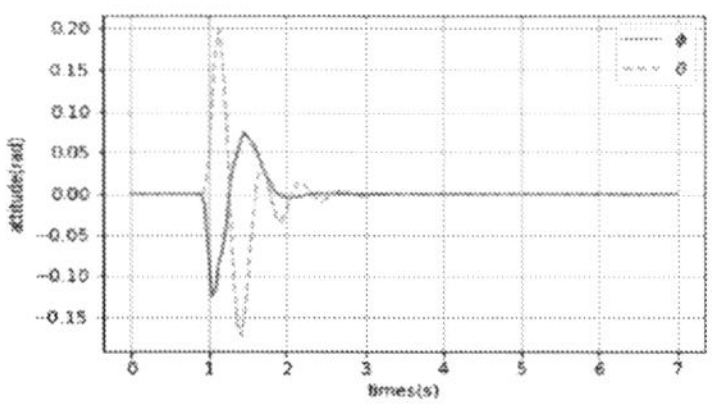

Figure 6 Attitude

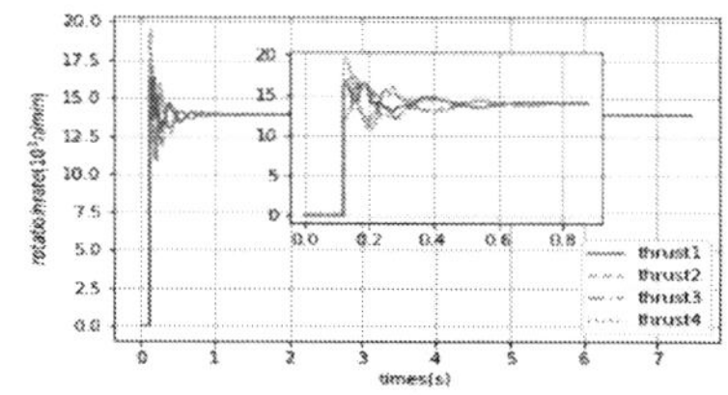

Figure 7 Thrust rotation rate

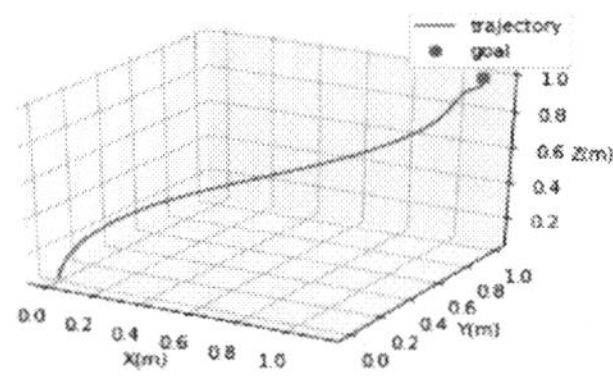

Figure 8 Position tracking [0,0,0]→[1,1,1]

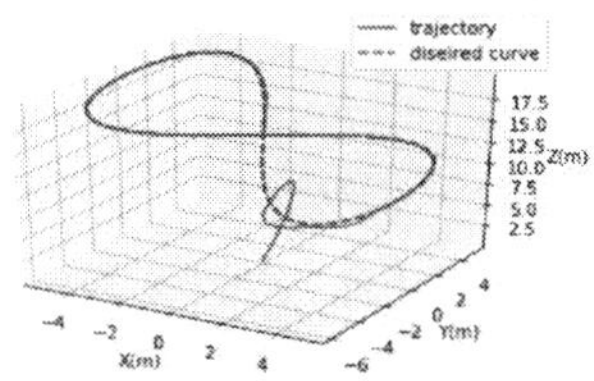

Figure 9 Trajectory tracking

Figuer 9-13 shows the task that the robot is operated as a quad-rotors and tracks a trajectory. The robot takes off from land at [0,0,0], and begins to track the trajectory at the same time. In Figure 10, the robot slows down to avoid falling due to tough attitude angles when the robot turns rapidly.

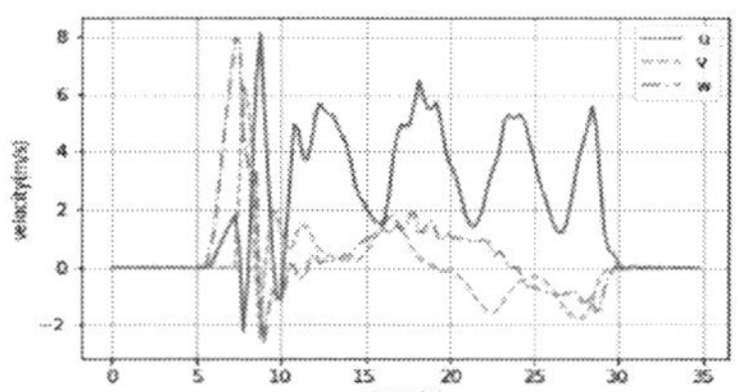

Figure 10 Linear velocity

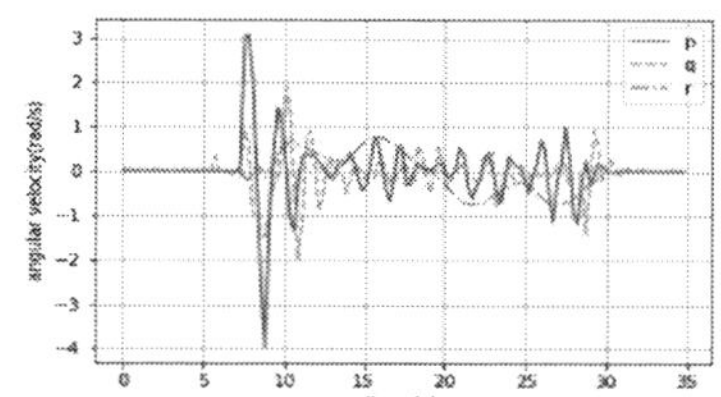

Figure 11 Angular velocity

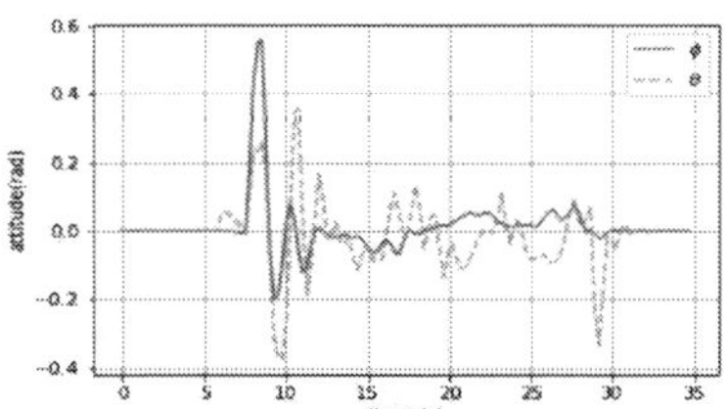

Figure 12 Attitude

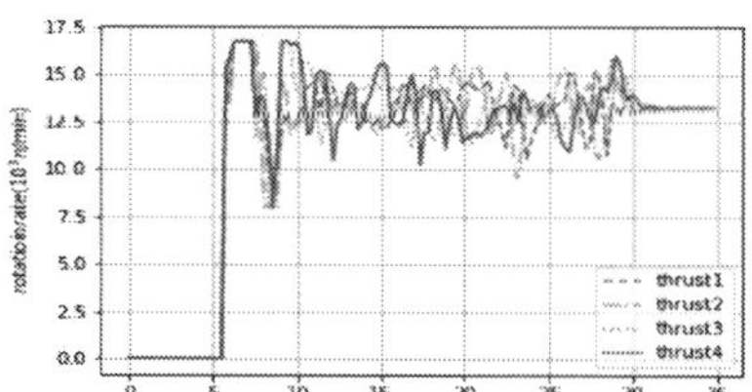

Figure 13 Thrust rotation rate

4.2 Tiltrotors position control simulation

In this part, the robot is operated as a quad-tiltrotors with variable propeller angles. Some traditional tiltrotors, attitude control is mainly provided by the elevators and rudders like fixed-wing planes.

Unlike these tiltrotors, our tiltrotors controls the attitude by changing the propeller angles and rotation rates at the same time. The control action of the tiltrotors is a tuple $[a_1, a_2, a_3, a_4, \delta_1, \delta_2, \delta_3, \delta_4]$.

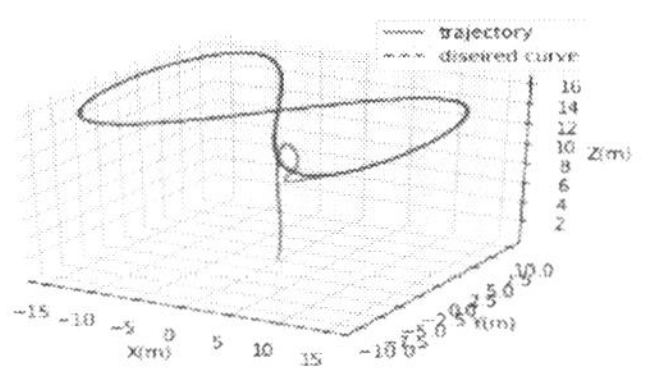

Figure. 14 Tiltrotors trajectory tracking

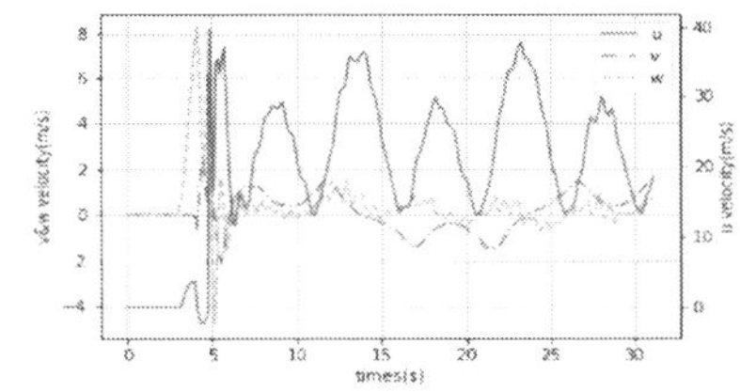

Figure. 15 Tiltrotors linear velocity

For this task, we design a similar trajectory lemniscate of Huygens in Equation(16).

$$\begin{cases} x = 15\cos(\varphi) \\ y = 10\sin(2\varphi) \\ z = 15 - \cos(\varphi)/10 \end{cases} \tag{16}$$

Figure 14-19 depict the response of the robot in the tiltrotors mode operation. The robot takes off from $[0,0,0]$, and begins to track the trajectory by controlling the variable propeller angles and the thrust rotation rates. The robot gets higher speed and lowers attitude angle when compared with the quad-rotors mode. As the fixed-wing provides lift force, the thrusts perform efficiently with lower rotation rates. Due to the roll angle gets tough when turning, less lift force in the vertical plane is not enough. To avoid falling down, propeller angles are controlled to provide the differential angle and consequently provides force compensating the lost lift force in Figure 19.

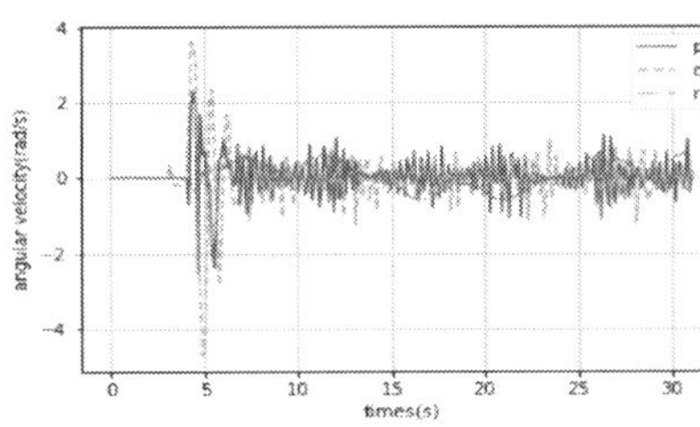

Figure 16 Tiltrotors angular velocity

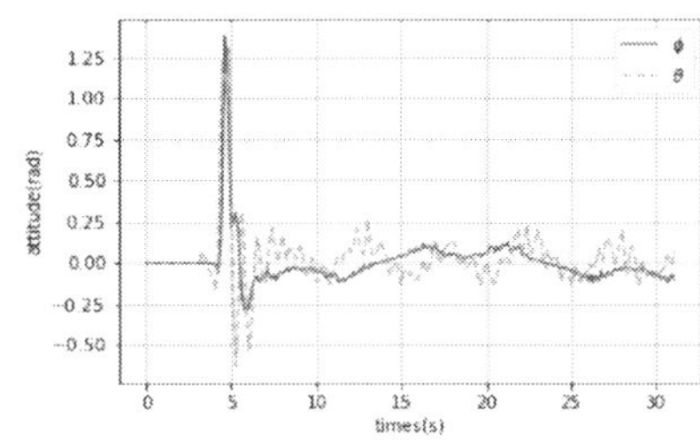

Figure 17 Tiltrotors attitude

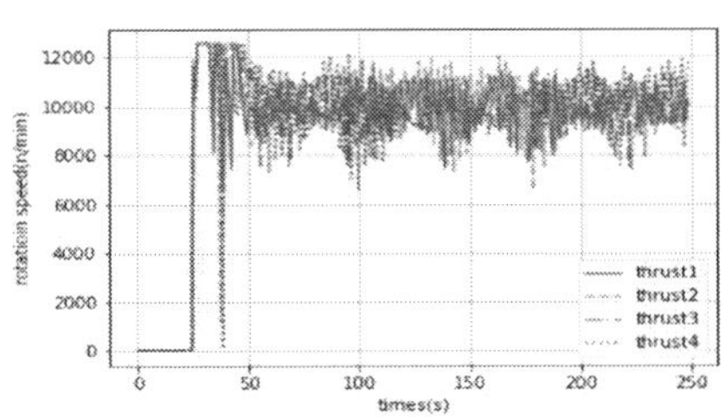

Figure 18 Tiltrotors thrust rotation rate

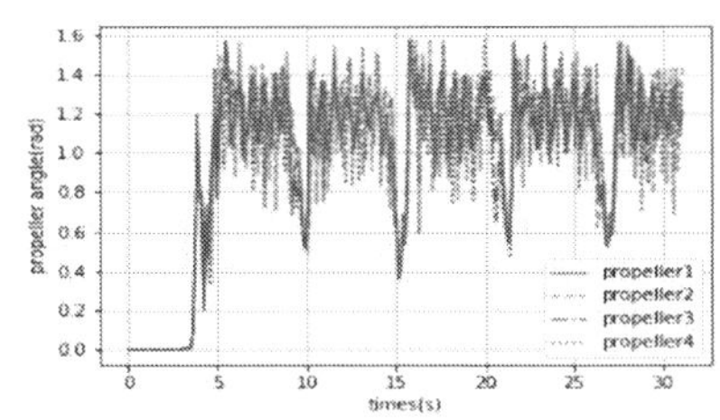

Figure 19 Tiltrotors propeller angle

5. CONCLUSION

The nonlinear model dynamics of a quad–tiltrotors UAV type were extracted, and a ROS combined with Gazebo simulation platform of the multi-mode flight envelope was developed. A control algorithm has been learned using reinforcement learning and fitted value iteration using the nonlinear dynamic model of the quad-tiltrotors. The effectiveness of the proposed controller in tackling the complexity of the conversion task, were demonstrated via simulation studies.

IOP Conf. Series: Materials Science and Engineering **751** (2020) 012047 doi:10.1088/1757-899X/751/1/012047

References

[1] X. Yang, T. Wang, J. Liang, G. Yao, and M. Liu, "Survey on the novel hybrid aquatic–aerial amphibious aircraft: Aquatic unmanned aerial vehicle (AquaUAV)," *Prog. Aerosp. Sci.*, vol. 74, pp. 131–151, Apr. 2015.

[2] M. M. Maia, P. Soni, and F. J. Diez-Garias, "Demonstration of an Aerial and Submersible Vehicle Capable of Flight and Underwater Navigation with Seamless Air-Water Transition," p. 9.

[3] W. Weisler, W. Stewart, M. B. Anderson, K. J. Peters, A. Gopalarathnam, and M. Bryant, "Testing and Characterization of a Fixed Wing Cross-Domain Unmanned Vehicle Operating in Aerial and Underwater Environments," *IEEE J. Ocean. Eng.*, vol. 43, no. 4, pp. 969–982, Oct. 2018.

[4] Z. Kong and Q. Lu, "Mathematical Modeling and Modal Switching Control of a Novel Tiltrotor UAV," *J. Robot.*, vol. 2018, pp. 1–12, Aug. 2018.

[5] C. Papachristos, K. Alexis, and A. Tzes, "Hybrid model predictive flight mode conversion control of unmanned Quad-TiltRotors," in *2013 European Control Conference (ECC)*, Zurich, 2013, pp. 1793–1798.

[6] J. Hwangbo, I. Sa, R. Siegwart, and M. Hutter, "Control of a Quadrotor with Reinforcement Learning," *IEEE Robot. Autom. Lett.*, vol. 2, no. 4, pp. 2096–2103, Oct. 2017.

[7] H. Wu, S. Song, K. You, and C. Wu, "Depth Control of Model-Free AUVs via Reinforcement Learning," *IEEE Trans. Syst. Man Cybern. Syst.*, pp. 1–12, 2018.

[8] I. Carlucho, M. D. Paula, S. Wang, B. V. Menna, Y. R. Petillot, and G. G. Acosta, "AUV Position Tracking Control Using End-to-End Deep Reinforcement Learning," in *OCEANS 2018 MTS/IEEE Charleston*, 2018, pp. 1–8.

[9] M. Miller and J. Narkiewicz, "TILTROTOR MODELLING FOR SIMULATION IN VARIOUS FLIGHT CONDITIONS," p. 26.

[10] K. M. Kleinhesselink, "STABILITY AND CONTROL MODELING OF TILTROTOR AIRCRAFT," p. 255.

[11] J. D. Anderson, "Fundamentals of Aerodynamics," p. 1131, 2009.

[12] R. S. Sutton and A. G. Barto, *Reinforcement Learning: An Introduction*. 1988.

[13] J. Schulman, F. Wolski, P. Dhariwal, A. Radford, and O. Klimov, "Proximal Policy Optimization Algorithms," *ArXiv170706347 Cs*, Jul. 2017.

[14] J. Schulman, S. Levine, P. Moritz, M. I. Jordan, and P. Abbeel, "Trust Region Policy Optimization," *ArXiv150205477 Cs*, Feb. 2015.

IOP Conf. Series: Materials Science and Engineering **751** (2020) 012048 doi:10.1088/1757-899X/751/1/012048

Analysis of the Influence of On-Board Temperature and Pressure Control System on Inert Gas Generating Performance of Hollow Fiber Membrane

ZHANG Jing-yang[1,*], LUO Bi-tai[1], HU Wen-chao[2] and SHEN Ping[1]

[1] College of Astronautics, Nanjing University of Aeronautics and Astronautics, Nanjing 210016, China

[2] Aviation Key Laboratory of Science and Technology on Aero Electromechanical System Integration, Nanjing Engineering Institute of Aircraft Systems, Aviation Industry Corporation of China, Nanjing 211106. China

[*]E-mail address: zjy@nuaa.edu.cn

Abstract. Based on the typical on-board membrane separation system for fuel inerting, the mathematical models of temperature and pressure control subsystem and hollow fiber membrane were built with the coupled iterative numerical solution methods. While the temperature and pressure were controlled, the design area, nitrogen production effect of hollow fiber membranes were analyzed and compared with the situation of no temperature and pressure control in the full flight envelope by numerical simulation. Moreover, the root causes for those changes were found out. It shows that the method used in temperature and pressure control can increase the bleed air pressure and enhance the stability of temperature and pressure effectively. Compared with the situation of no temperature and pressure control, the temperature and pressure control method are able to decrease the demanded area of membrane by 42.3%~51.7%. And there is no need to adjust the membrane area for condition of low nitrogen production efficiency. But it also decreases the highest nitrogen production efficiency by 9.4% and increase the minimum oxygen concentration by 0.65%. The rate of flow has the most significant effect on nitrogen production. These phenomena are caused by the effects of the rate of flow, pressure, temperature and membrane area.

1. Introduction

According to the data of air crash in 50 years, the viability of the aircraft is affected by the safety of fuel tank greatly. Since the 1980s, the United States had begun research on aircraft fuel tank inerting systems [1] and made great achievement [2]. With the development of new type in China[3], the theoretical and experimental research on aircraft fuel tank inerting systems had been carried out without hesitation [4]. Among all kind of aircraft fuel tank inerting technology, onboard inert gas generating technology received extensive attentions because of its high reliability, long working life, low repair costs and few needs for support equipment on ground. This technology asks for bleeding air from an aircraft engine or an environmental control system and obtains nitrogen-enriched gas through an onboard air separation device, which is used for fuel tank washing [5]. After that, the oxygen concentration is able to keep a lower level than the oxygen demand for the possibility of combustion in the upper space of the aircraft fuel tank [6]. At present, many military aircrafts and civil aircrafts, such as C5A, C17, KC135, F22, B737, B747 and A320 [7], have applied this technology.

The air separation device is the most critical part of the onboard fuel inerting system. In 1950, Weller was the first to establish a mathematical model of the cross-flow binary gas separation process based on the working mechanism of the fiber membrane for gas separation [7]. After that, Zhu Yulin and Jiang Guoliang established a mathematical model of the process of separating nitrogen and hydrogen with hollow fiber membranes and performed some experiments to confirm the theory [8]. Based on their research, Xu Renxian analyzed the pressure drop in the hollow fiber membrane with the gas separation model and contributed to verified the effectiveness of the mathematical model of the gas separation [9]. At the same time, the US Air Force conducted a large number of experiments on the inert gas generating performance of hollow fiber membrane in the 1980s. The data showed that the active temperature control can help to improve the nitrogen production efficiency by mostly about 15% and decrease the oxygen volume fraction of the product gas by mostly about 4%, and the active pressure control can help to improve the nitrogen production efficiency by mostly about 20% and decrease the oxygen volume fraction of the product gas by mostly about 6% [10]. Liu Xiaofang simulated the effects of ambient pressure, ambient temperature, altitude and gas temperature on the performance of inert gas generating system based on the inert gas generating system of some type of aircraft, which showed that factors such as pressure, temperature and mass flow determine the nitrogen production of the air separation device. George also found that the mass flow and gas temperature have a significant impact on nitrogen production by the experiments in 41 different working conditions [11]. Thomas made some research on the feasibility of onboard fuel inerting system applied to commercial aircraft. The result showed that it is closely related to the bleed air parameters[12]. Therefore, the active gas temperature control is an effective way to improve the nitrogen production effect of the onboard membrane separation system for fuel inerting [13]-[14]. However, there is still short of quantitative research on the combined effects of gas temperature control and gas pressure control for membrane separation system. The future design of onboard membrane separation system for fuel inerting needs to know well about the influence of flight envelope on effectiveness of nitrogen production and its internal cause.

This paper took the typical membrane separation system for fuel inerting as the research object. And it established the model of temperature and pressure control subsystem coupled with hollow fiber membrane which is used to compare the change of hollow fiber membrane design area, nitrogen production efficiency and oxygen volume fraction in the whole flight envelope with or without temperature and pressure control and analyze their internal factors. This paper will be able to provide technical reference for the design of onboard membrane separation system for fuel inerting.

2. Model building

2.1. Global structure

The onboard membrane separation system for fuel inerting designed in this paper has a temperature and pressure control subsystem with turbine and compressor. And the specific scenario is the following: this system bleeds air from the engine or the environmental control system and deliveries the air to the air separation device after temperature and pressure control in order to obtain inert gas, and finally deliveries inert gas to the fuel tank for washing. The system structure is shown in figure 1. This device bleeds air from the engine or the environmental control system. After that, the gas is divided into two part. One part of the gas is used to drive the turbine in order to run the compressor to adjust pressure, and the other one is divided into two part again. One part of the divided gas flows through the heat exchanger while the other one not. Then, these gases are mixed and delivery to the compressor. This device can adjust the gas flow delivery to the turbine by the valves V1 and V2 in order to control the pressure of gas products. And it can adjust the gas flow delivery to the radiator by the valve V4 in order to control the temperature of gas products. After temperature and pressure control, the final outlet gas is the gas product.

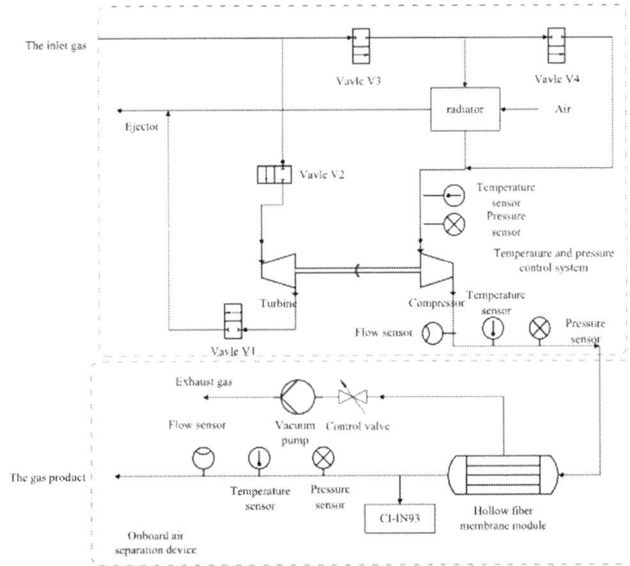

Figure 1. The diagram of system structure.

2.2. *Mathematical Model*

2.2.1. Mathematical model and parameters of temperature and pressure control subsystem. Turbine: The characteristic curve of turbine is shown in Figure 2. The relevant parameters can be obtained by this curve. For saving space, the process won't be present here.

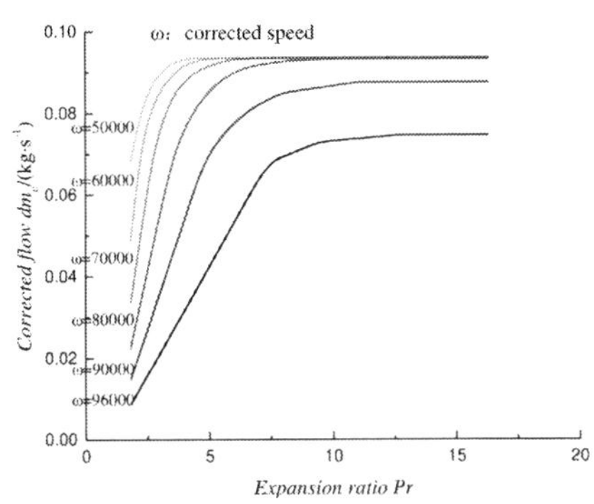

Figure 2. The expansion ratio and relative mass flow characteristic curve of turbine.

Compressor: The characteristic curve of the compressor is shown in Figure 3. The relevant parameters can be obtained by this curve. For saving space, the process won't be present here.

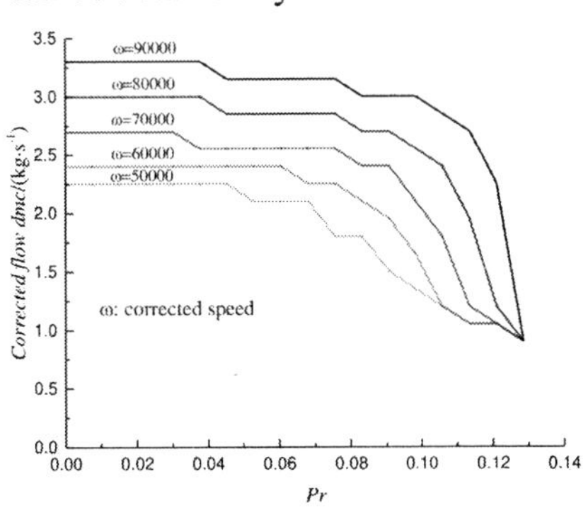

Figure 3. The pressure ratio and corrected flow characteristic curve of turbine.

Heat exchanger: The heat exchanger adopts a counter-flow plate type. This one is similar with the heat exchanger in paper [15]. The space between the parallel plate is 30mm, and the heat exchange area is 20cm^2. The Nu number of heat transfer is calculated by the Dittus-Boelter formula, and drag coefficient is obtained by the Blasius formula:

$$Nu = 0.023 \cdot Re^{0.8} \cdot Pr^{0.3} \tag{1}$$

Where Re is the Renault number($10^4 \leq Re < 1.2\times10^5$), Pr is the pressure ratio($0.6 < Pr < 120$).

$$f = 0.079Re^{-0.25} \tag{2}$$

Where Re is the Renault number($4000 \leq Re < 10^5$).

2.2.2. Mathematical model and parameters of hollow fiber membrane. Hollow fiber membrane can be divided into parallel flow, counter flow and cross flow according to the direction of intake and exhaust gas [15]. And this paper adopts the parallel flow type, which means the directions of intake and exhaust gas are the same. And the inner diameter of the hollow fiber membrane is 0.45 mm. The outer diameter of the hollow fiber membrane is 0.55 mm, and the length of the hollow fiber membrane is 0.98 m. The structure of the hollow fiber membrane is shown in figure 4. And it's possible to adjust the membrane area by increasing or decreasing the number of hollow fiber membrane silk.

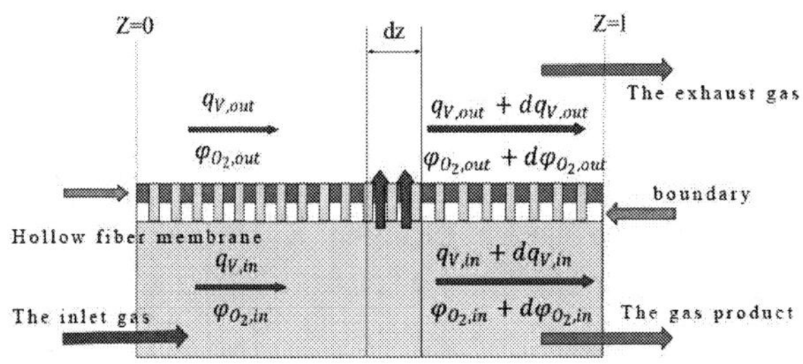

Figure 4. Parallel flow type of the hollow fiber membrane.

According to conservation of mass, permeation law and the pressure drop of equation, their equations can be obtained. The first equation is about flow mass conservation:

$$q_{V,in} + dq_{V,in} + q_{V,out} + dq_{V,out} = q_{V,in} + q_{V,out} \tag{3}$$

Where $q_{V,in}$ is the inlet flow, $q_{V,out}$ is the outlet flow.

The second equation is about conservation of oxygen mass:

$$\left(q_{V,in} + dq_{V,in}\right)\left(\varphi_{O_2,in} + d\varphi_{O_2,in}\right) - q_{V,out}\varphi_{O_2,out} = -\left(q_{V,out} + dq_{V,out}\right)\left(\varphi_{O_2,out} + d\varphi_{O_2,out}\right) + q_{V,in}\varphi_{O_2,in} \tag{4}$$

Where $\varphi_{O_2,in}$ is the oxygen content of inlet flow, $\varphi_{O_2,out}$ is the oxygen content of outlet flow.

The third equation is permeability equation of oxygen:

$$\frac{J_1\left(p_f\varphi_{O_2,in} - p_p\varphi_{O_2,out}\right)Adz}{L} = -d\left(q_{V,in}\varphi_{O_2,in}\right) \tag{5}$$

Where J_1 is the penetration rate of oxygen, A is the area of hollow fiber membrane, L is the length of the hollow fiber membrane, p_f is the pressure within the membrane, p_p is the pressure of the membrane outside.

The fourth equation is permeability equation of nitrogen:

$$\frac{J_2\left[p_f\left(1-\varphi_{O_2,in}\right) - p_p\left(1-\varphi_{O_2,out}\right)\right]Adz}{L} = -d\left[q_{V,in}\left(1-\varphi_{O_2,in}\right)\right] \tag{6}$$

Where J_2 is the penetration rate of nitrogen.

The fifth equation is about pressure drop:

$$\frac{dp_f}{dz}=-\frac{128RTq_{V,in}}{N\pi di^4 p_f} \tag{7}$$

Where N is the number of hollow fiber membrane silk, T is the temperature of membrane, di is the inner diameter of membrane silk, is the deal gas constant

The change of penetration rate of nitrogen and oxygen with temperature conforms to the pattern of Arrhenius [16]. So, the penetration rate can be obtained by formula (8) and formula (9).

$$J_1=1.81\times10^{-7}\times e^{-\frac{19.3}{RT}} \tag{8}$$

$$J_2=8.76\times10^{-7}\times e^{-\frac{27.6}{RT}} \tag{9}$$

In the boundary condition, the initial flow within membrane is equal to the bleed air flow. And initial oxygen content within membrane silk is 21%. The initial flow outside the membrane is 0. And the initial oxygen content outside the membrane can be obtained by formula (10).

$$\frac{\varphi_{O_2,out}\big|_{z=0}}{1-\varphi_{O_2,out}\big|_{z=0}}=\frac{J_1\left(p_f\,\varphi_{O_2,in}\big|_{z=0}-p_p\,\varphi_{O_2,out}\big|_{z=0}\right)}{J_2\left(p_f\,\varphi_{N_2,in}\big|_{z=0}-p_p\,\varphi_{N_2,out}\big|_{z=0}\right)} \tag{10}$$

Where $\varphi_{N_2,in}$ is the nitrogen content of inlet flow, $\varphi_{N_2,out}$ is the nitrogen content of outlet flow.

2.3. Calculation flow

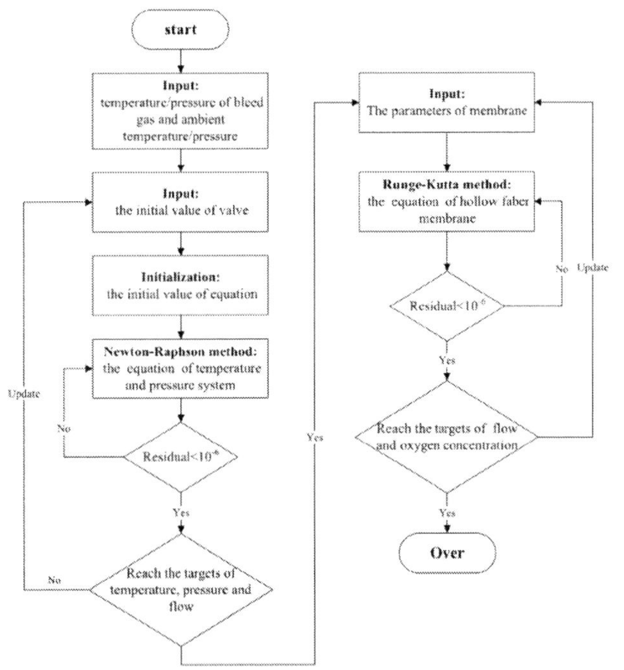

Figure 5. The Calculation flow.

The calculation flow is shown in Figure 5. The inlet temperature, inlet pressure, ambient temperature and ambient pressure are known conditions. When the temperature and pressure control subsystem is working, the mathematical model of the hollow fiber membrane can be solved by the classical Runge-Kutta method. And the model of the temperature and pressure control subsystem can be solved by Newton's iteration method. If the residuals of these equations are less than 10^{-6}, the efficiency of nitrogen production and oxygen content of gas product can be obtained by formula (11) and formula (12).

$$\eta = \frac{q_{V,in}\big|_{z=l}}{q_{V,in}\big|_{z=0}} \times 100\% \tag{11}$$

$$\varphi = \varphi_{O_2,in}\big|_{z=l} \times 100\% \tag{12}$$

Where $q_{V,in}\big|_{z=0}$ is the initial flow within membrane, $q_{V,in}\big|_{z=l}$ is the outlet flow within membrane (m^3/s), $\varphi_{O_2,in}\big|_{z=l}$ is the oxygen content of the outlet flow within membrane.

2.4. Model validation

This part is used to verify the reliability of the model. The structural parameters of the device are the same with the parameters in the paper [17]. The inlet flow of the product gas is 13kg/s, and the temperature of the membrane is 298K. Comparing the results of calculating with the data in the paper, the maximum error is less than 10%. The data is shown in Figure 6. Therefore, this model is able to reflect the change of efficiency of nitrogen production and oxygen content of gas product very well.

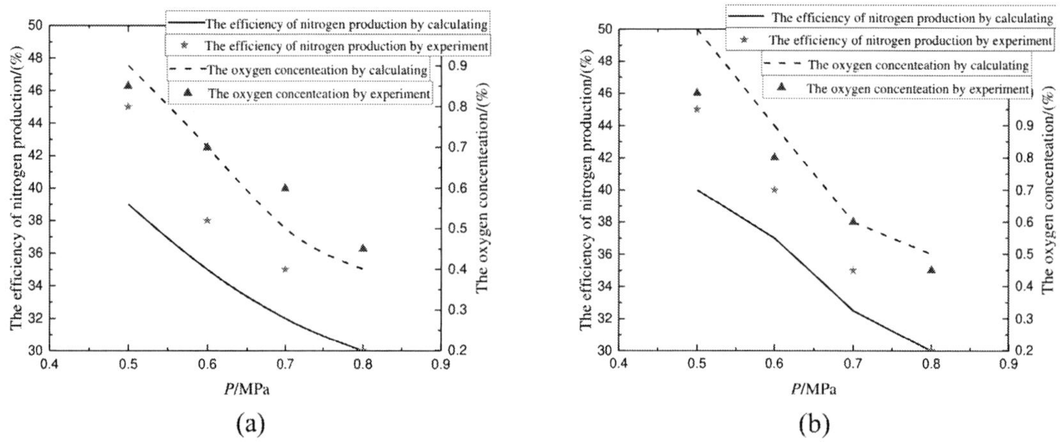

Figure 6. Comparison between calculating results

3. Simulation condition

Based on the flight envelope of large transport airplane and its flight parameters, the flight path is designed. And this flight path can be divided into static stage of the ground, takeoff roll stage, ascent stage, cruise stage and dive stage, which is shown in Figure 7.

The flight state parameters at every stage and the parameters of gas product are shown in Table 1. According to the requirements of fuel tank inerting in each flight phase, this system designs flow control device [18] in order to achieve the different working modes of flow and oxygen content of gas product. There are two kinds of working mode in this paper [19]. One of them is small flow mode with low oxygen concentration, and the other one is large flow mode with high concentration oxygen. No stage except dive stage adopts large flow mode with high oxygen concentration. Large flow mode with high oxygen concentration requires that oxygen concentration is less than 9%, and small flow mode with low oxygen concentration requires that oxygen concentration is less than 5%. In the flight of aircraft, the working mode of flow is changing. And different types of aircraft have different requirements of flow. In this paper, the required flow is changing between $0.0064m^3/s$~$0.172m^3/s$. And this paper sets that the flow in small flow mode with low oxygen concentration is no less than 0.118 m^3/s, while the flow in large flow mode with high oxygen concentration is no less than 0.200 m^3/s. In the static stage of the ground, the flow is limited because of static condition of aircraft. In this stage, the flow is no more than 0.0901 m^3/s.

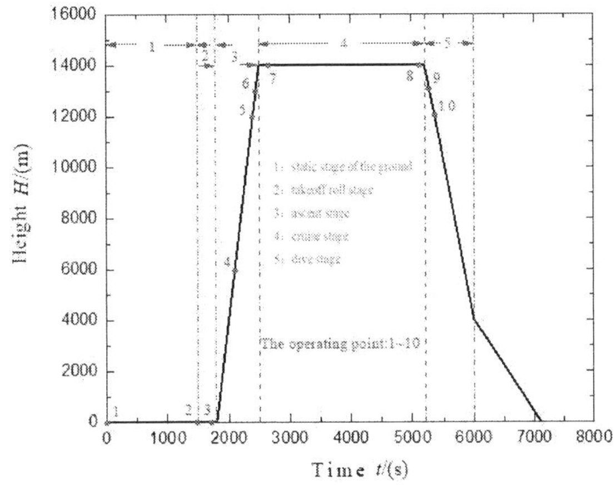

Figure 7. Flight path

Table 1. Selected point parameter table

Number	t/s	the required flow/$m^3 \cdot s^{-1}$	the required oxygen concentration
1	0	<0.0901	<5%
2	1500	⩾0.0901	<5%
3	1800	⩾0.118	<5%
4	2100	⩾0.118	<5%
5	2440	⩾0.118	<5%
6	2640	⩾0.118	<5%
7	2700	⩾0.118	<5%
8	5150	⩾0.118	<5%
9	5250	⩾0.200	<9%
10	5350	⩾0.200	<9%

4. Simulation result

4.1. Contrastive analysis of temperature and pressure

Figure 8 is a graph showing changes in ambient temperature, temperature of bleed gas and temperature after controlled. The change of the temperature of bleed gas is similar to ambient temperature. And these changes fluctuate greatly, while the temperature after controlled changes smoothly. The maximum temperature of bleed gas is 493K, which exists in the takeoff roll stage. The minimum temperature of bleed gas is 413K, which exists in the ascent stage. The maximum ambient temperature is 299.6K, which exists in the static stage of the ground. The minimum ambient temperature is 216K, which exists in the cruise stage. The temperature control objectives in this system is 365±5K. The temperature of air flow can be controlled to 365K with closed-loop control of temperature. This temperature is suitable for the working of hollow fiber membrane.

Figure 9 is a graph showing changes in ambient pressure, pressure of bleed gas and pressure after controlled. The change of the pressure of bleed gas and ambient pressure fluctuate greatly, especially in the takeoff roll stage and ascent stage. The maximum pressure of bleed gas is 4.5bar, which exists in the takeoff roll stage. The minimum pressure of bleed gas is 2.4bar, which exists in the dive stage. The ambient pressure changes with altitude. The most unstable situation of pressure after controlled exists in adopting open-loop control [20]. The maximum pressure after controlled is 5.4bar, which exists in the takeoff roll stage and cruise stage. The minimum pressure after controlled is 4.2bar, which exists in the static stage of the ground. The pressure of air flow can be controlled to 5.6bar with closed-loop control of pressure [21]. But the pressure after controlled is higher than pressure of bleed gas, no matter adopting closed-loop control or open-loop control of pressure.

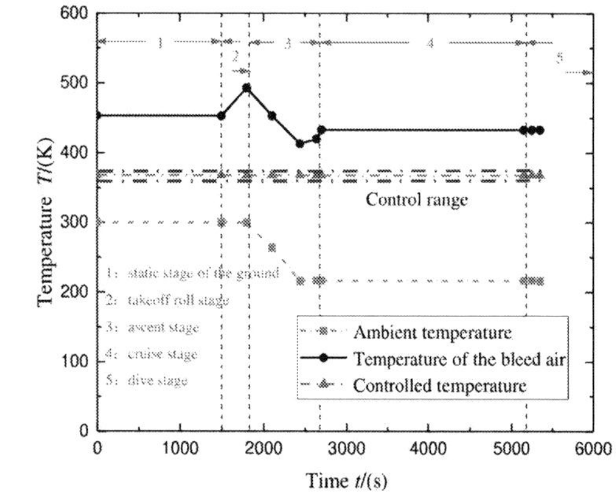

Figure 8. The characteristic curve of temperature

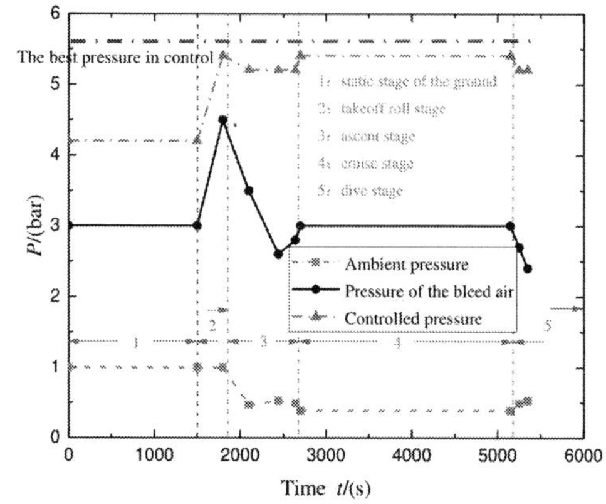

Figure 9. The characteristic curve of pressure

4.2. Contrastive analysis of the area of hollow fiber membrane

Based on the simulation results, the minimum required fiber membrane area in the whole flight envelope is calculated. The calculation results are shown in Figure 10. And the maximum value will be taken as the actual area of the hollow fiber membrane. If the temperature and pressure were not controlled, the maximum required area of the hollow fiber membrane is 1.49 m^2, which exists in the static stage of the ground. And the minimum required area of the hollow fiber membrane is 0.78 m^2, which exists in the ascent stage. The required area of the hollow fiber membrane fluctuates with the flight stage greatly. If the temperature and pressure were controlled, the maximum required area of the hollow fiber membrane is 0.72 m^2 in the static stage of the ground with open-loop control. And the minimum required area of the hollow fiber membrane is 0.41 m^2 in this situation, which exists in the cruise stage. Compared with the situation without temperature and pressure control, the required area decreases by 42.3%~51.7%. And the change of required area of hollow fiber membrane with altitude is smooth.

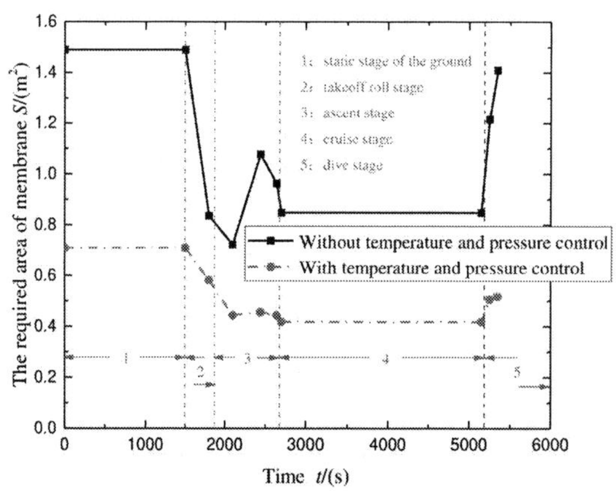

Figure 10. The chart of the required area of hollow fiber membrane

4.3. Contrastive analysis of the effects of hollow fiber membrane on nitrogen production

The nitrogen production efficiency of the hollow fiber membrane and the oxygen concentration of the gas product are analyzed with or without temperature and pressure control. Since the requirements in every stage have to be met, the area of hollow fiber membrane is set as the maximum required area in the whole flight envelope. When temperature and pressure are not controlled, the temperature and pressure of the membrane inlet are the temperature and pressure of the bleed air. When temperature and pressure are controlled, the temperature and pressure of the membrane inlet are the temperature and pressure controlled. The results are shown in Figure 11 and Figure 12. When temperature and pressure are not controlled, the nitrogen production efficiency of the hollow fiber membrane will change between 7.71% and 68%. And the oxygen concentration of the gas product will change between 0.1% and 8.5% in this situation. They fluctuate with altitude greatly, and the maximum values all appear in the dive stage. Although the results all meet the requirements in Table 2, the nitrogen production efficiency in the operating point 3 and 4 is lower than expectation. When temperature and pressure are controlled, the nitrogen production efficiency of the hollow fiber membrane will change between 31.44% and 58.6%. And the oxygen concentration of the gas product will change between 0.75% and 5.7% in this situation. The maximum values also appear in the dive stage. Though the minimum oxygen concentration of the gas product is larger while the nitrogen production efficiency of the hollow fiber membrane is smaller, the fluctuation becomes smoother than the situation of temperature and pressure are not controlled. And the nitrogen production efficiency will not be lower than expectation after temperature and pressure control.

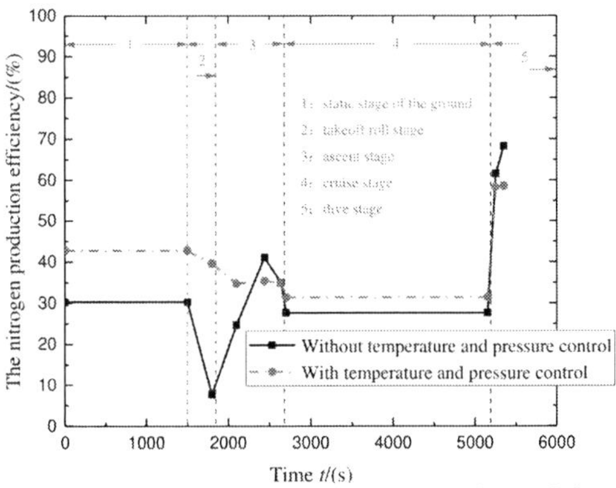

Figure 11. The oxygen concentration of the gas product in the whole flight envelope

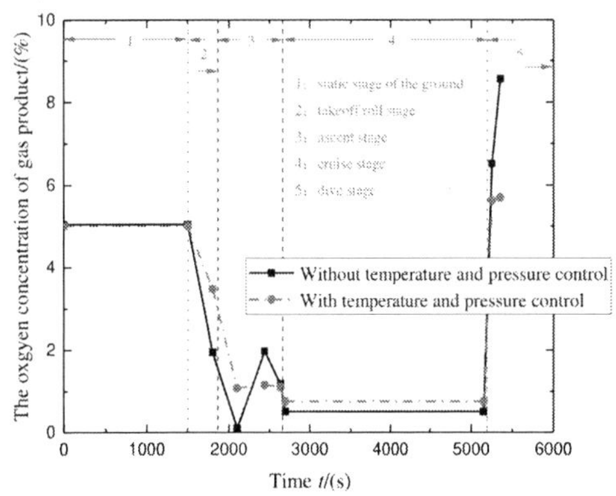

Figure 12. Full flight envelope line rich nitrogen gas volume fraction change chart

4.4. Analysis of causes

The changes of the nitrogen production effects with flow, area of hollow fiber membrane, pressure ratio of membrane can be obtained by the model of hollow fiber membrane, which are shown in Figure 13, Figure 14, and Figure 15. With the increase of flow, the nitrogen production efficiency of the hollow fiber membrane and the oxygen concentration of the gas product keep increasing. But they will reduce with the increase of pressure ratio. And they will also reduce with the increase of the area of hollow fiber membrane. The temperature of membrane effects them not that much[22].

According to these results, it's possible to analyze the reason why the temperature and pressure control system plays such a special rule in nitrogen production. When the temperature and pressure are controlled, the required area of hollow fiber membrane is much less than the required area of membrane without temperature and pressure control. And its fluctuation is smoother. The reason why the area of membrane will reduce can be explained by the changes of temperature and pressure. With temperature and pressure control, temperature and pressure are obviously stable and the pressure in the hollow fibre module is significantly increased. Then the oxygen concentration of the gas product will reduce. Therefore, the required area of membrane is less than before with the same requirement of oxygen concentration of the gas product. The nitrogen production efficiency of the hollow fiber membrane and the oxygen concentration of the gas product with temperature and pressure control fluctuate much less than the results without temperature and pressure control, and the situation that nitrogen production efficiency becomes less than expectation won't happen. This can be explained by follows: the temperature is much more stable and the pressure is higher with temperature and pressure control than the situation without temperature and pressure control, which means it's still possible to get the high nitrogen production efficiency and the low oxygen concentration of the gas product after decreasing the required area, but the maximum nitrogen production efficiency is less than before while the oxygen concentration of the gas product becomes higher. And the nitrogen production efficiency and the oxygen concentration of the gas product are obviously increased in the dive stage, no matter the temperature and pressure are controlled or not.

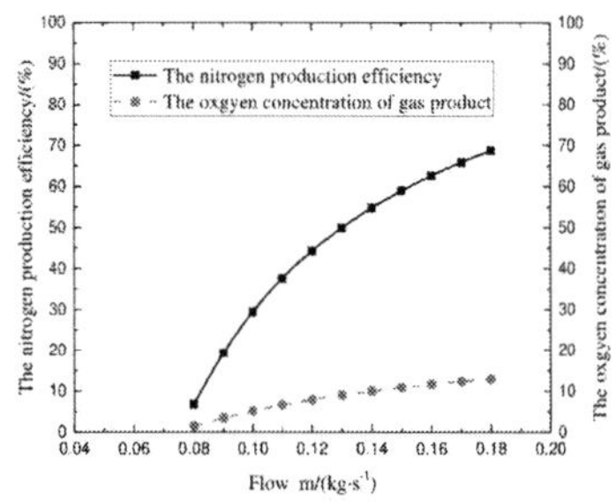

Figure 13. The diagram of nitrogen production effect with flow rate change

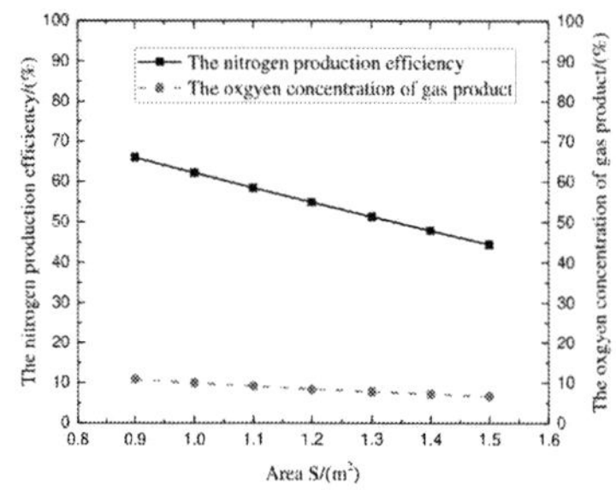

Figure 14. The diagram of nitrogen production effect with area change

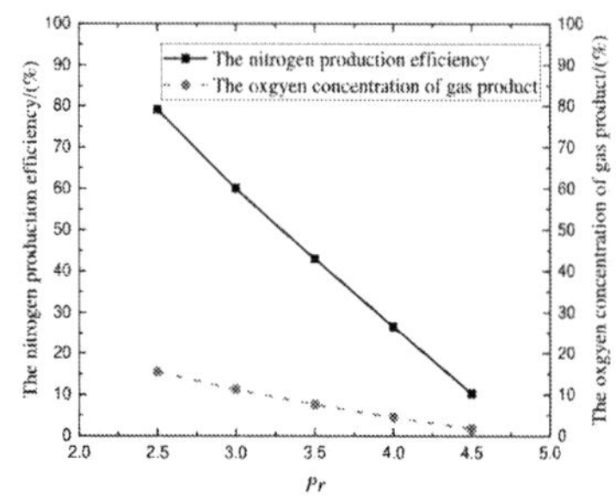

Figure 15. The diagram of nitrogen production effect with area change

5. Conclusion

In this paper, the typical onboard membrane separation type of fuel tank inerting system is taken as the research object, and its model of temperature and pressure control subsystem and hollow fiber membrane are established with the coupled iterative numerical method. Through contrastive analysis of the effects of hollow fiber membrane on nitrogen production in the whole flight envelope, the required area of membrane, nitrogen production effect and their internal factors are analyzed. The main conclusions are as follows:

1) When adopting open-loop temperature and pressure control, it's possible to control the temperature from 413K~493K to 365K and increase the pressure of the bleed air from 2.4bar~4.5bar to 4.2bar~5.4bar. And the pressure is able to controlled to 5.6bar stably with closed-loop temperature and pressure control.

2) Compared with the case of no temperature and pressure control, the required area of membrane will decrease by 42.3%~51.7% after temperature and pressure control. And it will change with the flight state.

3) In the dive stage, due to the obvious increase in flow, the nitrogen production efficiency and oxygen concentration of gas product will be increased greatly with or without temperature control.

4) These phenomena are caused by the positive and negative effects of flow, pressure, temperature and area of the membrane.

Acknowledgement

The authors gratefully acknowledge the financial supports for this project from the Aviation Foundation of China (grant No: 20162852035)

References

[1] Johnson R L & Gilleman J B. (1993). Aircraft Fuel Tank Inerting System. *Airesearch manufacturing company*, (5),154~176

[2] Wang Mingming. (2010). Comparative Analysis of Fuel Washing and Scrubbing in Aircraft Fuel Tank, *Journal of Nanjing University of Aeronautics& Astronautics*, 42(5), 615-619.

[3] Gao Yongting. (2005). Study on nitrogen inerting characteristic of military aircraft fuel tank. *Journal of Shenyang Institute of Aeronautical Engineering*, 22(5):12-14.

[4] Wei Shuzhuang 2014 *Design and simulation of fuel tank Inerting system for aircrafts*(Nanjing:

Nanjing University of Aeronautics and Astronautics)

[5] Lu Ji 2012 *Theoretical study of onboard air separation unit and inerting system*(Nanjing: Nanjing University of Aeronautics and Astronautics)

[6] Xue Yong 2010 *Study on separation performance of onboard hollow fiber membrane and flush inerting for fuel tank of civil aircraft* (Nanjing: Nanjing University of Aeronautics and Astronautics)

[7] Weller S & Steiner W A.(1950). Engineering Aspects of Separation of Gases-Fractional Permeation through Membranes. *Journal of Machine Learning Research*, 11(18):3571-3594.

[8] Zhu Baoling & Jiang Guoliang. (1987). Calculation method of separation performance of hollow fiber membrane N_2-H_2 separation. *Journal of Chemical industry and Engineering*, 38(3):281-292.

[9] Xu Renxian & Zhu Baoling. (1993). Study on the law of internal pressure drop in hollow fiber gas separation filaments. *Membrane Science and Technology*, (2):22-28.

[10] Liu Xiaofang. (2010) Separation characteristics in onboard membrane air separation module. *Journal of Nanjing University of Aeronautics& Astronautics*, 42(2):250-255.

[11] Airesearch manufacturing company. (1974). *Catalytic reactor for Inerting of aircraft fuel tanks.. June,1974*

[12] Thomas L R, Daniel F L & Delbert B B. (2001) Onboard Inert Gas Generation System/Onboard Oxygen Gas Generation System (OBIGGS/OBOGS) Study, *Aircraft System Requirements* Part I(5)

[13] Lei Yansheng. (2011) Research of aircraft fuel tank explosion prevention technique basing on FAA airworthiness requirement. *Civil Aircraft Design and Research*, (3):23-27.

[14] Howell T J. (1987) *Fighter aircraft OBIGGS (on-board inert gas generation system) study*(US: Aero Propulsion Laboratory Fire Protection Branch)

[15] He Guangwen. (2015) *Application of the study on configuration optimization of plate heat exchanger and its heat transfer performances* (Guangzhou: South China University of Technology)

[16] Ivory J, Feng X & Kovacik G. (2002) Development of hollow fiber membrane systems for nitrogen generation from combustion exhaust gas: Part II: Full-scale module test and membrane stability. *Journal of Membrane Science*, 202(1):195-204.

[17] Xue Yong, Liu Weihua & Guo Xiufeng. (2011) Experimental study on separation performance of hollow fiber membrane for onboard inerting generating system. *Journal of Xi'an jiaotong University*, 45(3):107-111

[18] Thorman J M, Rhim H & Hwang S T. (1975). Gas separation by diffusion through silicone rubber capillaries. *Chemical Engineering Science*, 30(7):751-754.

[19] Liu Yulin. (2015). *Engineering calculations and numerical simulation of fuel inerting flow field in a military aircraft fuel tank* (Nanjing: Nanjing University of Sciense & Technology)

[20] Knight T C & Ritter J E. (1984) *Conf. The AH-64A nitrogen inerting system.* (SanDiego: aircraft Design System and Operations Meeting.)

[21] Burnell S & Leigh T. *Aircraft fuel tank inerting system* (America)

[22] Cai Yan. (2015) Mathematical modeling analysis of hollow fiber membrane onboard inert gas generation system. *Journal of aerospace Power*, 30(9):2100-2107.

IOP Conf. Series: Materials Science and Engineering **751** (2020) 012049 doi:10.1088/1757-899X/751/1/012049

Vibration signal Processing of Cutting Gearbox Based on Haar-Wavelet Denoise

X F Zhang, F W Ma, N N Wang* and Z F Guo

Taiyuan Research Institute, China Coal Technology & Engineering Group, Taiyuan 030000, China

*E-mail: wangnn_123@126.com

Abstract. Cutting gearbox with complex structure is difficult to obtain mechanical parameters and conduct fault diagnosis, aiming at this problem, a novel handling method is proposed after studying vibration theory of gearbox. Firstly, traditional calculation method and modal analysis are combined to acquire mechanical parameters of gearbox, especially inherent frequencies of decelerator housing and gear transmissions which are difficult to calculate. Based on Haar-Wavelet, vibration test signal of cutting gearbox is de-noised. And Fast Fourier Transform (FFT) is conducted for de-noised signal, thus operation status of gearbox can be analyzed in time domain and frequency domain. Taking a mine cutting gearbox as an example, the method proposed is used to detect parts damage in maintenance process. It is found that the tooth surface of second-stage gear transmission arise even wear, and this is consistent with actual situation. Therefore the practicability and effectiveness of this method are verified, and a new solution is provided for other gearboxes with complex structure.

1. Introduction

As key component to bear load and transmit power, gearbox is widely used in production. With the development and application of computer aided design, researchers can ensure performance of gearboxes in the design stage for multi-stage transmission gearboxes with complex structure. However, in next sessions such as components production, gearbox assembly, and application in equipment would generate irreversible degradation for gearbox performance, this may lead to real performance is inconsistent with the design performance of gearbox. Some studies have shown that the intensity of vibration and noise generated by mechanical equipment under dynamic conditions as well as its main frequency are closely related to the type, degree, location, and reason of failure of gearbox [1]. Therefore, how to obtain mechanical characteristic parameters of the multi-stage gearbox and diagnose its operation state using vibration signals obtained from vibration test are key problems to be solved urgently in the process of acceptance inspection and maintenance.

In the aspects of obtaining mechanical parameters and dealing with vibration signals, many researchers have conducted studies. In order to get mechanical characteristic parameters of gearbox, especially inherent frequencies of every gear transmission and the gearbox housing, Wang Feng [2] and Li Yafeng [3] established the finite element models of herringbone gear box and gear box of draught fan respectively in ANSYS software, and they obtained inherent frequencies and corresponding vibration modes of the system by conducting modal analysis. Qian Lulu [4] et al. got the inherent frequency of single-stage gear transmission system by establishing the dynamic model, and they compared the results with those obtained by establishing finite element model. Weis Peter [5] et al. used Workbench software to carry out modal analysis for gearbox housing, and the inherent frequency of gearbox housing with load was obtained. He Zeyin [6] et al. developed a coupled gear-rotor-bearing-housing system, and the mode of vibration as well as modal frequency is obtained.

In addition, Puchalski Andrzej [7] tested vibration acceleration signals to diagnose internal combustion engine faults online and offline, respectively. Afia Adel et. al [8] performing autogram to handle signal, and Fourier transform is applied to extract the fault signature. Sun Ruobin [9] et al. conducted dynamics simulation to obtain the prior information of a chipped planetary gear set, and utilize the information in the time domain analysis. Fang Yuan [10] et al. analyzed vibration signal of electric vehicle power system by means of spectrum analysis, which provided experimental support for the optimal design of power system. Dong Lei [11] et al. applied local mean decomposition (LMD) to decompose vibration signals of the hoist spindle system. When conducting bearing fault diagnosis, Goncalves, Mario J. M. [12] et al. used various methods to deal with vibration signals of motor bearings, and results were compared. Formberger Max [13] and others collected angular displacement of axle in gearbox and diagnosis it.

A combined method to handle vibration test signals is proposed in this paper. Vibration mechanism of the cutting gearbox is researched, and frequency components of vibration signal as well as signal characteristics corresponding to gearbox fault are analyzed. Then taking a mine cutting gearbox with complex structure as an example, inherent frequencies of gearbox housing and every gear transmission are obtained. In addition, other mechanical parameters of cutting gearbox are calculated by traditional formulas. Finally, vibration test of the gearbox is carried out in maintenance stage. Based on Haar-Wavelet de-noised method, time-domain vibration signal is handled. And then the method of Fast Fourier Transform is conducted to the signal, thus performance of cutting gearbox and damage of the original parts can be diagnosis from both time-domain and frequency-domain perspectives.

2. Vibration mechanism of cutting gearbox

Vibration signal of mine gearbox is complex, especially for planetary gearboxes, they will excite highly modulated vibration [14]. Aiming at multi-stage cutting gearbox with planetary gear transmission, its vibration is usually affected by varieties of factors and it will also be influenced by cross-modulation. If the cross-modulation component in test signal is neglected, the vibration signal can be shown in equation (1):

$$\mathrm{Y}(t)=G(t)+\sum X_K(t)D_E(t)+\sum X_{\mathrm{GA}}(t)D_G(t)+\sum X_{\mathrm{X}}(t)D_G(t)+\sum X_{\mathrm{B}}(t)D_B(t)+n(t) \tag{1}$$

In the equation, $G(t)$ is vibration component produced by rotating frequency of axes; $\sum X_K(t)D_E(t)$ is modulation signal result from meshing frequencies of every gear transmission; $\sum X_{\mathrm{GA}}(t)D_G(t)$ is modulation signal result from inherent frequencies of every gear transmission; $\sum X_{\mathrm{X}}(t)D_G(t)$ is modulation signal result from inherent frequency of gearbox housing; $\sum X_{\mathrm{B}}(t)D_B(t)$ is signal component produced by abnormal vibration of rolling bearing; $n(t)$ is other interference signal; $X(t)$ is carrier signal; $D(t)$ is modulation signal.

If there is no fault and tooth profile error is slight in gearbox, the trend of vibration signal would be relatively stable, and there is no obvious shock phenomenon. If gear transmissions in gearbox exists serious faults such as broken gear teeth and serious teeth profile error, it will lead to violent vibration. And it not only may appear modulation phenomenon result from gear meshing frequency, but also may cause resonance phenomena. Hence, it can be found that there is side-band near inherent frequencies of gear transmissions with fault in frequency spectrum. When more serious faults occur in cutting gearbox, it may generate large energy and cause resonance phenomena of gearbox housing. And it can be found that there is side-band near inherent frequency of gearbox housing in frequency spectrum. Furthermore, if rolling bearings are damaged and produce abnormal vibration, the phenomenon of modulation will appear and side-band would occur near the inherent frequency of rolling bearing with fault, and modulation signal are passing frequencies.

IOP Conf. Series: Materials Science and Engineering **751** (2020) 012049 doi:10.1088/1757-899X/751/1/012049

3. Obtaining mechanical parameters of cutting gearbox

Before handling vibration test signal, mechanical parameters of cutting gearbox should be obtained. Based on these parameters, cutting gearbox can be diagnosed combined with frequency spectrum. Traditional formulas and modal analysis are combined in this paper. On the one hand, formulas are used to calculate rotation frequencies of axles, gear meshing frequencies and others, on the other hand, modal analysis can be conducted to calculate inherent frequencies of every gear transmission and gearbox housing. The following will introduce how to obtain mechanical parameters of cutting gearbox in detail.

3.1. Obtaining mechanical parameters through formulas

Assuming that the speed of an axle in gearbox is n (r/min), the number of gear teeth on this axle is z, and rotation frequency of this axle can be calculated by equation (2):

$$f_n = n/60 \tag{2}$$

The meshing frequency of the gear can be calculated as follows:

$$f_z = z \cdot f_n = z \cdot n/60 \tag{3}$$

In addition, passing frequencies of rolling bearings need to be calculated, and specific formulas are as follows:

Passing frequency of inner ring:

$$f_i = \frac{1}{2} m f_n \left(1 + \frac{d_0}{D}\cos\alpha\right) \tag{4}$$

Passing frequency of outer ring:

$$f_O = \frac{1}{2} m f_n \left(1 - \frac{d_0}{D}\cos\alpha\right) \tag{5}$$

Passing frequency of bearing ball:

$$f_g = \frac{1}{2} m f_n \left[1 + \left(\frac{d_0}{D}\cos\alpha\right)^2\right] \cdot \frac{D}{d_0} \tag{6}$$

Passing frequency of bearing retainer:

$$f_b = \frac{1}{2} m f_n \left[1 + \left(\frac{d_0}{D}\cos\alpha\right)^2\right] \tag{7}$$

In above formulas (4) to (7), m is the number of bearing ball, d_0 is the diameter of bearing ball, D is the diameter of bearing, and α is the contact angle.

3.2. Obtaining inherent frequencies by conducting modal analysis

3-D models are built for gearbox housing and every gear transmission, which are difficult to get inherent frequencies. And then these models are imported to Workbench software to conduct modal analyses in this study. Because stiffness of structure will be affected by stress generated by load, in order to make the simulation result more realistic, pre-stressed modal analyses are conducted to gearbox housing and each gear transmission to obtain their inherent frequencies. The specific analysis process and results are described in detail as follows.

It is necessary to perform static analysis to obtain stress distribution firstly when mechanical structure needs to conduct pre-stressed modal analysis. The solid model of gearbox housing built in

Solidworks is imported into Workbench software, and material parameters are set. Meshing model can be built using free mesh method as shown in figure 1.

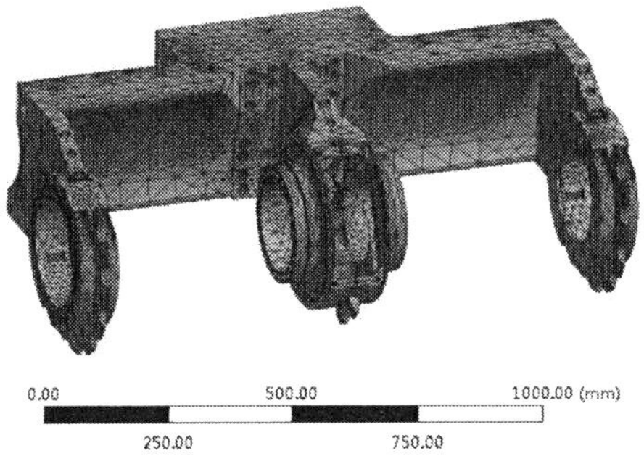

Figure 1. Meshing model of decelerator housing.

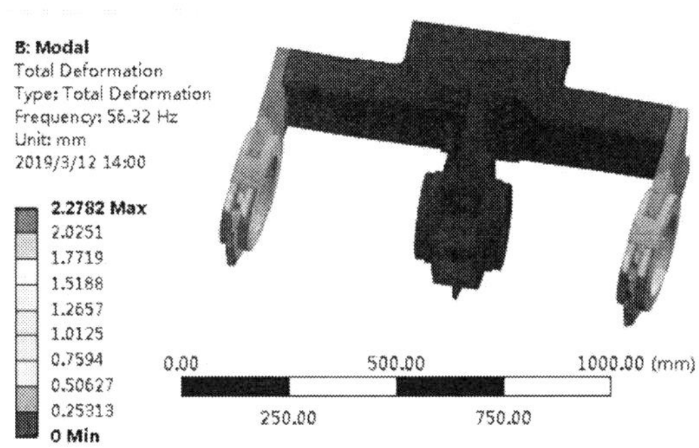

Figure 2. Vibration diagram of first mode.

According to work condition when conducting vibration test to gearbox housing, boundary condition is set for meshing model, and then static analysis is performed. After static analysis is completed, modal analysis is performed to obtain inherent frequencies and corresponding vibration diagram. Vibration diagram of first-mode is shown as figure 2, and inherent frequencies of first six modes are shown in table 1.

Table 1. Inherent frequency of decelerator housing.

Mode	Frequency/Hz
1	56.32
2	56.48
3	75.64
4	95.00
5	95.14
6	137.84

In addition, modal analyses are performed to every gear transmission refer to above process in turn. And their inherent frequencies are shown in table 2.

Table 2. Inherent frequency of gear drive.

Mode	Frequency/Hz				
	Frist stage	Second stage	Third stage	Fourth stage	Fifth stage
1	166.44	89.57	41.91	52.49	39.82
2	208.77	102.38	51.99	82.61	52.73
3	528.10	587.36	648.24	469.27	405.26
4	805.70	814.53	752.34	641.48	721.34
5	1099.45	981.58	972.51	883.47	903.24
6	2961.1	1685.3	1288.4	1095.6	1249.1

Therefore, cutting gearbox can diagnosed via de-noised signal from vibration test based on these mechanical parameters calculated from formulas and obtained from modal analysis.

4. Dealing with vibration signal of cutting gearbox

Conducting vibration tests to mining gearboxes, it is necessary to arrange measuring points at input, location corresponding to bearings, output of gearbox. Xu Liang [15] et al. monitored vibration condition of wind turbine gearbox by acceleration sensors. And acceleration sensors are used in this paper too, thus vibration signals of each measuring point can be got varies with time. What's more, time-domain vibration signal will be de-noised based on Haar-wavelet [16], and then frequency-domain signal can be obtained by performing Fast Fourier Transform (FFT) to handling signal.

There are totally 7 measuring points which are arranged on the cutting gearbox in this study, and they are described as figure 3. And acceleration vibration signals are tested with full load.

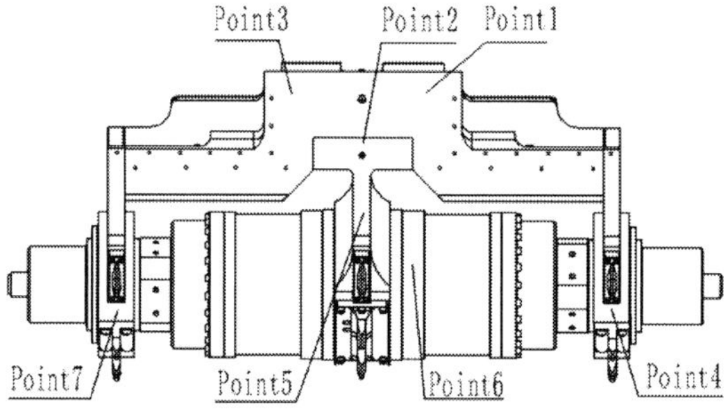

Figure 3. Arrangement of measuring points.

Time-domain vibration signals of each measuring point are analyzed, and the effective value, maximum value, and minimum value of each vibration signal are counted as table 3.

Table 3. Statistics of time domain data.

Measuring point	Effective value(m/s^2)	Maximum value (m/s^2)	Minimum value (m/s^2)
1	1.3	5.6	-5.4
2	1.1	4.3	-4.2
3	1.4	5.5	-5.4
4	0.87	3.7	-3.7
5	1.5	6.6	-7.7
6	2.1	8.5	-8.6
7	2.5	11.1	-10.6

It can be found that the effective value and the amplitude value of measuring point 7 are larger than other measuring points. That means, vibration energy of measuring point 7 is largest, and time-domain vibration signal of point 7 can be described as waveform in figure 4.

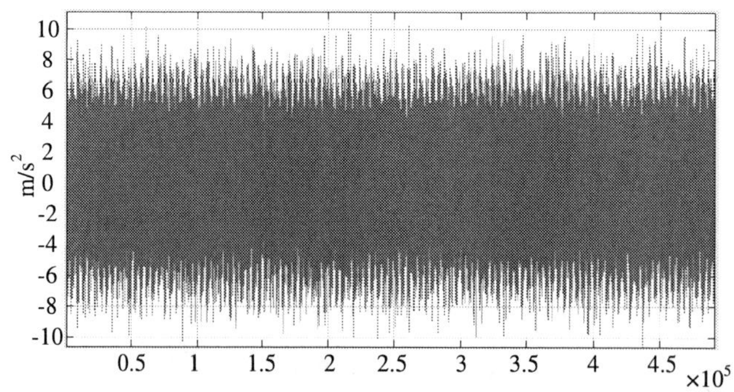

Figure 4. Vibration waveform.

Usually, vibration signals obtained by data collector often contain interference signals, so it is important to filter out noise signal before handling vibration signals. Noise reduction handling can be realized as decomposing and reconstructing signals, and time-domain signal shown in figure 5 is de-noised based on Haar-Wavelet. It can be seen from figure 5 that the signal is decomposed into 5 parts.

Appropriate threshold coefficients are chosen, and de-noised time-domain signal can be reconstruct from decomposed signal as shown in figure 5. Intercept a part of signal, it is shown as black part in figure 6.

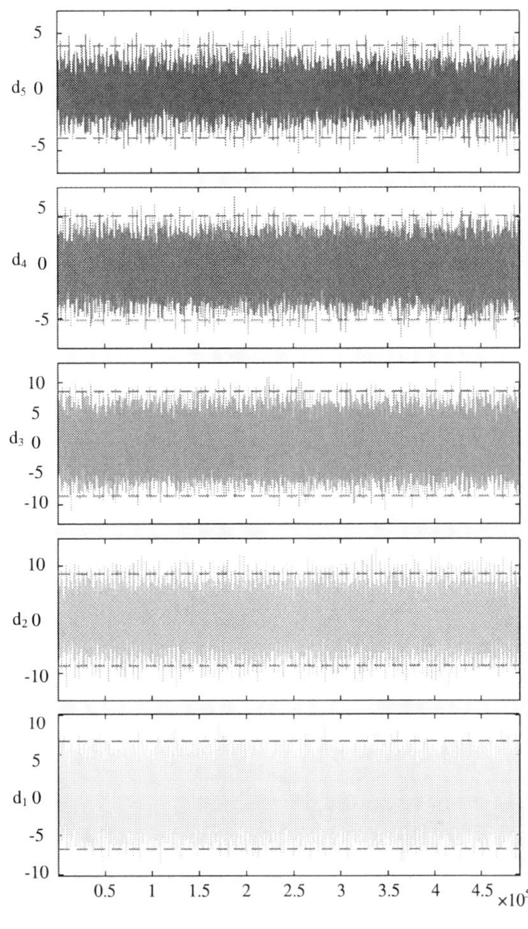

Figure 5. Decomposition of time domain signal.

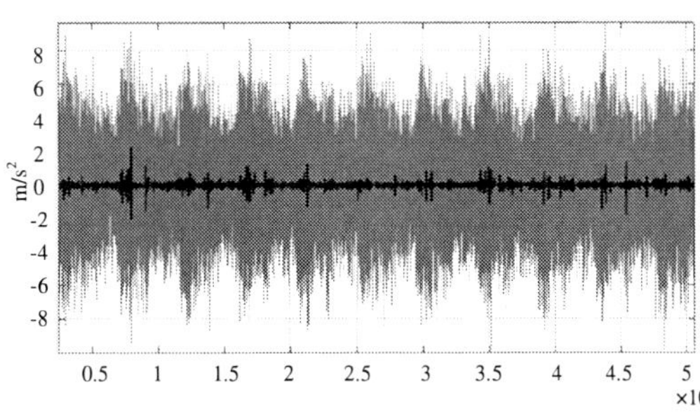

Figure 6. Comparison of original signal and de-noised signal.

It can be seen from figure 6 that slight periodic vibration is happened in the cutting gearbox. But there is no phenomenon of regular or irregular impact. Therefore, it can be judged basically that there is no serious fault such as severe shaft bending and broken teeth. And excitation energy is not large enough to cause resonance of gear transmissions and gearbox housing.

FFT is conducted to de-noised signal, and the diagram of frequency spectrum can be shown as figure 7. It can be found from spectrogram that main carrier frequencies of vibration signal are meshing frequency and its multiple frequency of each gear transmission. Inherent frequencies of gearbox housing and each gear transmissions are not main carrier frequency. And this shows that there are no serious faults such as broken teeth and severe shaft bending.

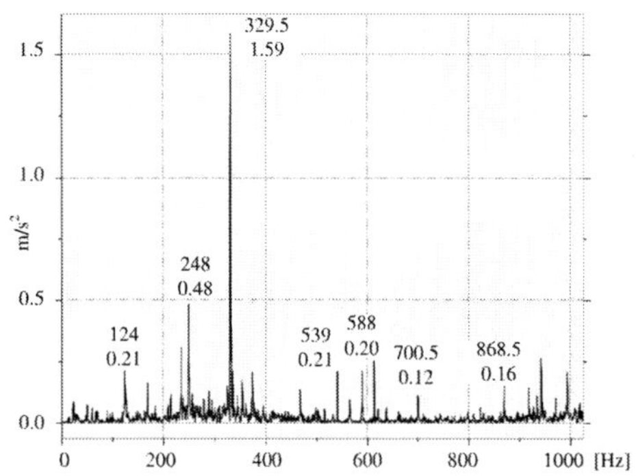

Figure 7. Spectrogram of vibration signal.

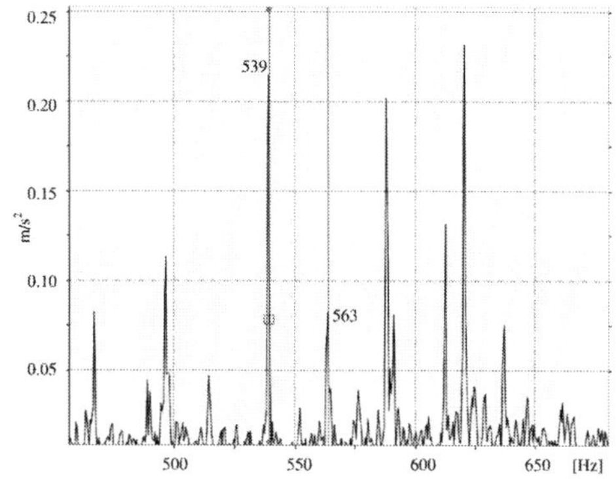

Figure 8. Refining spectrum.

Refining spectrum can be obtained by enlarging spectrogram of vibration signal near carrier frequencies, which is easy to observing side frequency. And it is helpful to research modulation phenomenon. Observing spectrogram near main carrier frequencies one by one, it is found that there is side frequency near the carrier frequency of 539 Hz, and its refining spectrum is shown in figure 8.

It can be seen from figure 8 that modulation frequency is 24Hz, and it is rotation frequency of first-stage axle and input axle of second-stage. Moreover, 539Hz is consistent with meshing frequency of second-stage gear transmission. This phenomenon shows that gear transmission of second-stage exist fault.

Therefore, the cutting gearbox is disassembled, and the gear pair of second-stage is checked, it is found that teeth surface of the gear pair is uniformly worn. And actual gear of second-stage transmission is shown as figure 9.

Figure 9. Gear of second-stage transmission.

5. Conclusions

The following conclusions can be drawn:

(1) A novel method combining traditional formula and modal analysis to obtain mechanical parameters is proposed, and it provides a new idea for mining gearboxs with complex structure which are difficult to obtain mechanical parameters, especially inherent frequencies;

(2) Time-domain vibration signal is de-noised based on Haar-Wavelet, and Fast Fourier Transform is conducted to transform time-domain signal into frequency-domain signal. Thus vibration test can be researched from these two aspects, and failure of cutting gearbox can be diagnosed.

(3) Taking a cutting gearbox with complex structure as an example in this paper, vibration signal handling method proposed in this paper is performed when it is maintenance. And it is found that teeth surface of gear pair on second-stage transmission has uniform wear, which is consistent with actual situation and the effectiveness of the method is verified.

Acknowledgments

This work was supported by Special Project of Science and Technology Innovation of Tiandi Science & Technology Co., LTD (2018-TD-MS048).

References

[1] Liao B Y 1995 *Basis of mechanical fault diagnosis* (Beijing: Metallurgical Industry Press)

[2] Wang F, Fang Z D 2013 The herringbone gear box modal analysis and experimental study *Applied Mechanics and Materials* **387** 328-331

[3] Li Y F, Xu Y X 2013 Operational modal analysis of wind turbine speed-increase gearbox *Telkomnika-Indonesian Journal of Electrical Engineering* **11** 6699-6705

[4] Qian L L, Tang J Y 2016 Dynamic modeling of a one-stage gear system by finite element method and the dynamic analysis in high speed *Journal of mechanical engineering* **52** 155-161

[5] Weis P, Kučera L, Pecháč P and Močilan M 2017 Modal analysis of gearbox housing with applied load *Procedia Engineering* **192** 953-958

[6] He Z Y, Tang W Y, Lin T J and Sun S Z 2018 Modal analysis and an experimental study into a marine gearbox featuring *Transactions of famena* **42** 43-52

[7] Puchalski A 2015 A technique for the vibration signal analysis in vehicle diagnostics *Mechanical systems and signal processing* 173-180

[8] Afia A, Rahmoune C and Benazzouz D 2018 Gear fault diagnosis using autogram analysis *Advances in mechanical engineering* **10**

[9] Sun R B, Yang Z B, Luo W, Qiao B J and Chen X F 2019 Weighted sparse representation based on failure dynamics simulation for planetary gearbox fault diagnosis *Measurement science and technology* **30**

[10] Fang Y, Zhang T, Yu P, Chen F F and Guo R 2015 Experimental study on vibration and noise of electric vehicle power assembly *Journal of Vibration, Measurement & Diagnosis* **35** 218-224

[11] Dong L, Shi R M and Zeng Z Q 2016 Diagnosis of hoist spindle system based on complex network clustering *Journal of Vibration, Measurement & Diagnosis* **36** 688-693

[12] Goncalves M J M, Creppe R C, Marques E G and Cruz S M A 2015 Diagnosis of bearing faults in induction motors by vibration signals-Comparison of multiple signal processing approaches *IEEE International Symposium on Industrial Electronics* 488-493

[13] Fromberger M, Weinverger U, Sc M, Kohn B, Utakapan T, Otto M and Stahl K 2017 Evaluating signal processing methods for use in gearbox condition monitoring *24th International Congress on Sound and Vibration*

[14] Ploeger D F, Zech P and Rinderknecht S 2019 Vibration signature analysis of commodity planetary gearboxes *Mechanical systems and signal processing* **119** 255-265

[15] Xu L, Zhu C C, Liu H J, Chen G and Long W 2019 Dynamic characteristics and experimental study on a wind turbine gearbox *Journal of mechanical science and technology* **33** 393-402

[16] Zheng Z Z, Shen P and Yang X H 2001 *Wavelet transform and its MATLAB tool application* (Beijing: Seismological Press)

IOP Conf. Series: Materials Science and Engineering **751** (2020) 012050 doi:10.1088/1757-899X/751/1/012050

Dynamic Optimization Design of Transverse Axis Cutting Reducer

Minghu Yin[1, 2], Hao Liu[1,*], Yahui Cui[1], Feifei Sun[3] and Daixiang Zhang[4]

[1] School of Mechanical and Precision Instrument Engineering, Xi'an University of Technology, Xi'an, China
[2] Yankuang Group Postdoctoral Research Station, Jining, China
[3] Fangyuan Group Co., Ltd. Yantai, China
[4] Yanzhou Coal Industry Co., Ltd. Equipment Management Center, Jining, China

[*]Corresponding e-mail: liuhao@stu.xaut.edu.cn

Abstract. Focused on the gear transmission system of a certain type of transverse axis cutting reducer, this paper firstly establishes the ADAMS virtual prototype model for calculating the dynamic load coefficient of the system, and the influence of gear modulus and tooth width factor on the dynamic load coefficient of the system is determined by multiple linear regression. Then the GA-BP neural network model for calculating the dynamic load coefficient of the system is constructed based on Genetic Algorithm (GA) and BP neural network, which makes the decoupling of the system dynamics analysis and optimization design process be realized. Finally, based on the above analysis, a multi-objective optimization design with the goal of minimizing the system volume and dynamic load coefficient is carried out. The calculation results show that the gear transmission system of the transverse axis cutting reducer is optimized efficiently, and its volume is reduced by 18.8%, and the dynamic load coefficient is decreased by 11.5%. This paper starts with the system dynamics analysis, makes the GA-BP neural network model as the bridge, and finally establishes the technical method for dynamic analysis optimum design of the transverse axis cutting reducer. The analytical technique and method can provide reference for the optimization design analysis of complex gear transmission system.

1. Introduction

Boom-type roadheader is the main machinery for tunneling in coal mine roadway. According to its cutting head arrangement, it can be divided into longitudinal axis type and transverse axis type. Compared with the longitudinal axis type, the unstable moment produced by the transverse axis roadheader when cutting coal is opposite to the stable moment formed by the machine gravity, which has the advantages of enhancing the cutting efficiency and stability. Thus, the development of transverse axis roadheader with high performance and high power has become the mainstream direction of today's mining industry[1].

In the boom-type roadheader, the cutting reducer is the most important power transmission factor, its working performance determines the actual capacity and efficiency of the roadheader[2], the design and manufacture of high power and high reliability cutting reducer is one of the important research topics in this field. Jiao Xinxin et al. verified the transmission ratio of EBZ-135 boom-type roadheader cutting reducer based on virtual prototype technology, and carried out quality optimization and fatigue life analysis of some key parts[3]. Yin min et al. studied the vibration of a cutting reducer by analysis, they found the conclusion that the noise of cutting reducer is mainly caused by the dynamic meshing force of gears, and adding staggered stiffener can reduce the vibration[4]. LV Xiaodan et al. completed the structural optimization design of cutting reducer with the goal of simplifying structure, reducing volume and gear module[5]. Wu Shiwen et al. took the minimum volume as the objective, and used MATLAB optimization toolbox to optimize the volume of cutting reducer[6]. Zhang Qiang et al. used the multi-objective fuzzy reliability optimization strategy of genetic algorithm to optimize the motion parameters of the cutting head, and it worked[7]. Wang Fuqiang et al. tested the functon parameters of the EBH-315 road-header and provided a data reference for its performance study[8]. Zhu Shengqi et al. improved the pr-oduction process of the planetary carrier of the longitudinal cutting reducer to enhance its production efficiency and performance[9].

It can be found that the researches on the design and analysis of the cutting reducer are mainly focused on the longitudinal axis type, but less on the transverse axis type with more complex structure. Moreover, as a power transmission device mainly composed of gear transmission, under the action of internal and external excitations such as cutting resistance and dynamic tooth meshing force, the cutting reducer generates strong vibration and noise inevitably[10,11]. Therefore, the dynamic analysis and optimization design of the gear system of cutting reducer based on the dynamic analysis have great significance for improving the working performance of the cutting reducer and the reliability of the whole roadheader. However, the dynamic analysis of the gear system is carried out with solving the nonlinear equations or virtual simulation generally, it makes the coupling of nonlinear equations solution or virtual simulation and optimization when considering the dynamic performance on optimization process. This is also the reason that hinders the relevant optimization work.

Therefore, this paper integrates ADAMS virtual prototype, multiple linear regression, BP neural network, Genetic Algorithm (GA), multi-objective optimization design and other theories and techniques to develop characteristic analysis and optimization design of the gear transmission system for the transverse axis cutting reducer. Its starting from system dynamics analysis, using GA-BP neural network as a bridge and finally optimizing the system structure, which provides a reference for the optimization design analysis of complex gear transmission system.

2. Optimization design variable decision based on dynamics simulation and multiple linear regression

For the optimization design problem, the selection of optimized design variables has a direct impact on its efficiency and effect. In order to determine the design variables for the dynamic performance optimization of the transverse axis cutting reducer gear transmission system. Firstly, based on ADAMS, the system dynamics simulation model is established, and the dynamic load of the system in work process is calculated. Then, the multivariate linear regression is applied to analyse the influence degree of each gear structural parameter on the dynamic performance of the gear system. At the same time, the optimized design variables are determined.

In this paper, the gear transmission system of a transverse axis cutting reducer produced by Yankuang Group is taken as the research object which is shown in Figure 1. Its input power is 20kW, the input speed is 1470rpm, and the output speed is 21rpm.

The virtual prototype model of the gear transmission system is established in ADAMS. In order to improve the simulation efficiency, the model is simplified as follows:

- The planetary gears on the left and right sides are symmetrical, so only one side is retained;
- Ignore the parts such as gaskets, retaining rings, keyways, chamfers, etc.
- The supporting effect of bearings are simplified as spring-damper elements operating at the journal.

And, the impact model is applied to define the impact contact force of each gear pair in the system, and its function expression is as follows:

$$F_c = \mathrm{MAX}\{0, K(q_0 - l)^e - C \times (dl/dt) \times STEP(l, q_0 - d, 1, q_0, 0)\} \quad (1)$$

Where K is the contact stiffness coefficient; q_0 is the initial distance between two contact objects;

l is the actual distance between the two contact objects; e is the rigid force index; C is the damping rate; dl/dt is the relative speed between the two objects; d is the penetration depth when the damping rate reaches the maximum. In this paper, the contact of gear meshing pair is the metal-metal contact, thus, the rigid force index is set to 1.5, the damping ratio is 0.01%, and the penetration depth is 0.1 mm. In addition, the Coulomb friction model is applied to express the friction between the gear pairs, the dynamic friction coefficient is 0.08, and the static friction coefficient is 0.1.

After the above processing and setting, the virtual prototype model is shown in Figure 2.

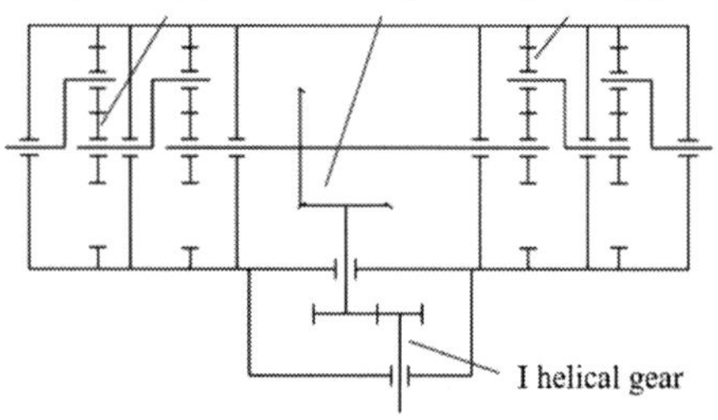

Figure 1. Structure of transverse axis cutting reducer

Figure 2. Cutting reducer gear transmission system virtual prototype model

Based on the above virtual prototype model, the dynamic simulation analysis is carried out, and the dynamic meshing force, also called dynamic load, of each gear pair in the working process are obtained, which is an important symbol to evaluate its dynamic characteristics[12]. Thus, in this paper, the dynamic load is made as one of the objectives of the following dynamic performance optimization of the cutting reducer. The dynamic load is expressed with the dynamic load coefficient in this paper, and the expression for the dynamic load coefficient is:

$$K_D = G/G_D \tag{2}$$

Where G is the dynamic load of the gear pair obtained by simulation; G_D is the nominal load of the gear pair calculated based on the relevant gear transmission design theory.

In order to further simplify the subsequent analysis and the optimization design process, the integrated maximum dynamic load coefficient of the system (the average of the maximum dynamic load coefficients of gear pairs in the system) is used to characterize the comprehensive dynamic performance of the system. Its expression is:

$$K_D^{\max} = [K_{D_{\rm I}}^{\max} + K_{D_{\rm II}}^{\max} + K_{D_{\rm III}1}^{\max} + K_{D_{\rm IV}1}^{\max} + K_{D_{\rm III}2}^{\max} + K_{D_{\rm IV}2}^{\max}]/6 \tag{3}$$

Where $K_D^{\max}$ is the comprehensive maximum dynamic load coefficient of the system; I, II, III and IV represent the first, second, third and fourth gear stage; $K_{D_{\rm I}}^{\max}$ 、 $K_{D_{\rm II}}^{\max}$ is the maximum dynamic load coefficient of the first, second gear stage; $K_{D_{\rm III}1}^{\max}$ 、 $K_{D_{\rm IV}1}^{\max}$ is the maximum dynamic load coefficient of the third and fourth planetary gear transmission in external meshing; $K_{D_{\rm III}2}^{\max}$ 、 $K_{D_{\rm IV}2}^{\max}$ is the maximum dynamic load coefficient of the third and fourth planetary gear transmission in internal meshing.

Related researches illustrate that the influence of gear modulus on the dynamic performance of the gear transmission system is significant[13,14], so this paper determines the gear modulus of each stage m_i (i=I, II, III, IV) as one of the optimized design variables. Furthermore, the sensitivity analysis between some other structural parameters (containing the tooth width factor Φ, the pressure angle α and the helix angle β_I of the first helical gear stage) and the dynamic performance (the integrated maximum dynamic load coefficient) of the gear system is carried out to determine the rest optimization design variable with the multiple linear regression, which is a statistical analysis method that uses the regression analysis in mathematical statistics to determine the quantitative relationship between multiple variables[15]. Its mathematical model is:

$$y = \lambda_0 + \lambda_1 x_1 + \lambda_2 x_2 + \lambda_3 x_3 \tag{4}$$

Where the y is the object variable, which is corresponding to comprehensive maximum dynamic load coefficient in this paper; and the design variables are x_1、x_2 and x_3, which are corresponding to Φ, α, and β_I respectively; $\lambda_0, \lambda_1, \lambda_2, \lambda_3$ are the regression coefficients.

To obtain the samples used in the multiple linear regression analysis, in a certain value range of tooth width factor, pressure angle and helix angle of the first helical gear stage (where Φ takes 0.7~1.1, α takes 14.5°~22.5°, β_I takes 9.064°~11.564°), The orthogonal experimental design method is used to plan the combination of the three parameters reasonably, and the system comprehensive maximum dynamic load coefficients under these parameter combinations are calculated based on the dynamics simulation model constructed above.

Through the multiple linear regression analysis of the samples, equation is obtained as follows:

$$y = 5.0915 + 7.06x_1 - 0.228x_2 - 0.4140x_3 \tag{5}$$

The correlation coefficient of the regression equation is R=0.9998, which indicates that the regression effect of the equation is significant. And the regression coefficient of the tooth width factor is much larger than the latter two. Therefore, in this paper, the final optimized design variables are the gear drive modulus of each stage m_i (i=I, II, III, IV) and the tooth width factor Φ.

3. GA-BP neural network for calculating dynamic performance

The gear transmission system of transverse axis cutting reducer is complicated, and its dynamic simulation takes a long time. If make the system dynamics simulation embed in the optimization design process directly, the computational efficiency will be greatly reduced. That's why the dynamic performance is ignored or simplified in the optimization design of the complex gear systems. Therefore, based on the genetic algorithm and BP neural network technology, this paper constructs the GA-BP neural network to calculate the dynamic performance (the integrated maximum dynamic load coefficient) of the gear system. The GA-BP neural network can provide the integrated maximum dynamic load coefficient under different structural parameters for the optimization design accurately and instantaneously, which relieves the coupling between dynamic simulation and optimization design.

BP neural network is widely used in the nonlinear fitting and pattern recognition. But it has the disadvantages of slow convergence and easy to fall into local minimum. So, this paper introduces Genetic Algorithm to optimize the weights and thresholds of BP neural network, which makes the BP neural network trains operate near the global minimum, then increases its convergence speed and avoids from falling into local minimum.

Similar to the multivariate linear regression, the GA-BP neural network also needs some samples. Thus, the orthogonal experimental design method and dynamic simulation models established in this paper are also used to obtain the samples (while the value ranges of the parameters are as following: m_I is from7 to 16, m_{II} is from 20 to 28, m_{III} is from 7 to 16, m_{IV} is from 10 to 18, and Φ is from 0.7 to 1.1). Then, with these simples, an eligible GA-BP neural network for calculation of dynamic performance is obtained.

Figure 3 shows the comparison between the results of the verification samples calculated with the simulation models and the GA-BP neural network. It can be seen that the relative error between the two results is not more than 2%, which indicates that the GA-BP neural network for calculating the dynamic performance of the gear system has great generalization ability, it can provide the integrated maximum dynamic load coefficient under different structural parameters for the optimization design accurately and instantaneously.

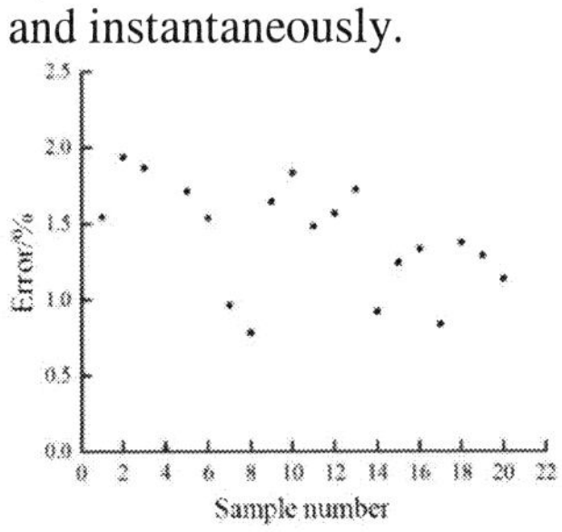

Figure 3. Verification sample error

4. Structural parameter optimization design

In view of the above work, with the multiple objects of decrease the volume and dynamic load for the gear system of transverse axis cutting reducer, a multi-objective optimization design model is established in this paper based on the linear weighted sum method.

With the linear weighted sum method, the design variables are written as the followed vector form:

$$x=[x(1),x(2),x(3),x(4),x(5)]^{T}=[m_{I}\ m_{II}\ m_{III}\ m_{IV}\ \Phi]^{T} \tag{6}$$

And the total volume of the gear system can be expressed as V:

$$\begin{aligned}V=[&517\times\pi(x(1)\times17)^{2}\times x(1)x(5)+41\times\pi(x(1)\times41)^{2}\times x(1)x(5)+17\times\pi(x(2)\times17)^{2}\times x(2)x(5)+\\&28\times\pi(x(2)\times28)^{2}\times x(2)x(5)+79\times\pi(x(3)\times79)^{2}\times x(3)x(5)+58\times\pi(x(4)\times58)^{2}\times x(4)x(5)]/4\end{aligned} \tag{7}$$

According to the established GA-BP neural network, the integrated maximum dynamic load coefficient of the gear system can be written as follows:

$$K_{D}^{\max}=\text{postmnmx}(\text{sim}(net\_tp\_0,\text{tramnmx}(x,\min p,\max p)),\min t,\max t) \tag{8}$$

Based on the linear weighted sum method, the objective function is written as follows:

$$F=\omega_{1}V+\omega_{2}K_{D} \tag{9}$$

Where ω_1 is the weight of the objective variable of the total volume; ω_2 is the weight of the objective variables of the integrated maximum dynamic load coefficient. In this paper, take $\omega_1=0.4$ and $\omega_2=0.6$.

In this paper, the constraints of optimal design are as follows:

$$\begin{cases}7\le m_{III}\le16\quad 20\le m_{II}\le28\quad 7\le m_{III}\le16\quad 10\le m_{IV}\le18\quad 0.7\le\Phi\le1.1\\ \dfrac{2KTY_{Fa}Y_{Sa}}{bm^{2}z}-[\sigma_{F}]\le0\qquad Z_{E}Z_{H}\left(\sqrt{\dfrac{2KT}{bm^{2}z^{2}}\cdot\dfrac{i\pm1}{i}}\right)^{-1}-[\sigma_{H}]\le0\end{cases} \tag{10}$$

Where K is the load coefficient; T is the torque ; Y_{Fa} is the tooth shape coefficient; Y_{Sa} is the stress coefficient when the load acts on the tooth top; $[\sigma_F]$ is the allowable bending fatigue strength of the tooth root; Z_E is the elastic influence coefficient; Z_H is the area coefficient; i is the transmission ratio; $[\sigma_H]$ is the allowable contact fatigue strength of the tooth surface.

Based on the above-mentioned multi-objective optimization design model, this paper uses MATLAB to realize the dynamic performance optimization for the gear transmission system of the transverse axis cutting reducer (its calculation process is shown in Figure 4).

Table 1 and Figure 5 give the comparison of structural parameters, volume and integrated maximum dynamic load coefficient before and after optimization. So, after the optimization, the volume and dynamic load of the gear system are also decrease obviously. Totally, the volume is reduced by 18.8%, and the integrated maximum dynamic load coefficient is decreased by 11.5%. The results prove that the optimization design carried out in this paper is effective, which can be promoted to solve the dynamic optimization design for complex gear systems.

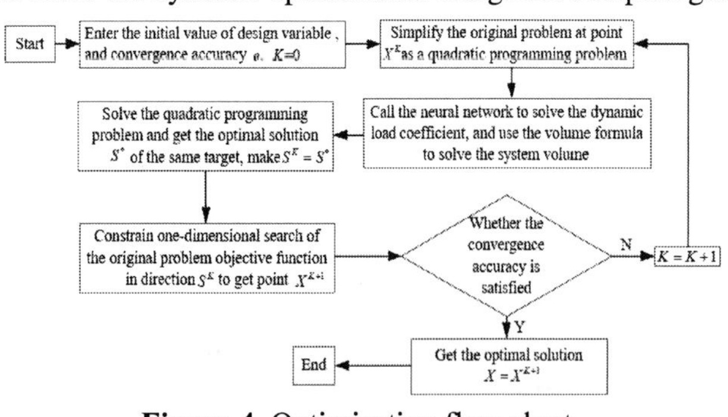

Figure 4. Optimization flow chart

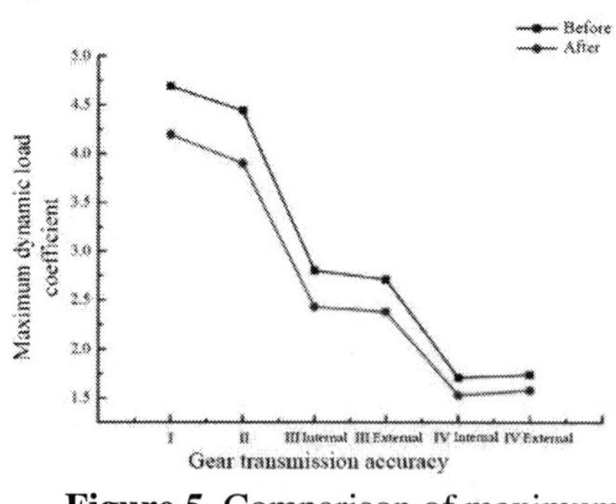

Figure 5. Comparison of maximum dynamic load coefficients at various stages before and after optimization

Table 1. Comparison before and after optimization

	m_{I} /mm	m_{II} /mm	m_{III} /mm	m_{IV} /mm	Φ	V/mm^3	$K_D^{\max}$
Before	10	26	10	14	1	1.2366×10^9	3.015
After	7.9653	21.1691	7.6568	10.6676	0.7742	1.0037×10^9	2.669
Round/Reduction ratio	8	22	8	11	0.8	18.8%	11.5%

5. Conclusions

With the combinations of the ADAMS virtual prototype, multiple linear regression, BP neural network, genetic algorithm, and multi-objective optimization design, the dynamic optimization design for the gear system of a transverse axis cutting reducer is carried out in this paper. The technical line of the dynamic optimization design proposed in this paper starts from the dynamic simulation of the gear system, uses the GA-BP neural network as a bridge to relieve the coupling between dynamic simulation and optimization design, and finally realizes the lightweight design and enhances the stability of the gear system. The technical line and model established in this paper can provide a reference for the dynamic optimization design of some other complex gear systems.

Acknowledgments

This work was supported by China Postdoctoral Science Foundation (Grant No. 2019M653879XB) and Natural science foundation of Shaanxi province (Grant No. 2019JQ-307).

References

[1] Liu JW and Liu QH Comparative analysis of Boom-type roadheader 2012 *Manage. J.* **17** 388
[2] Ma P Technical development of EBZ series horizontal axis roadheader 2006 *Coa. Technol.* **10**
[3] Jiao XX Research on EBZ-135 cantilever type roadheader cutting speed reducer based on virtual prototype technology 2011 *Taiyuan Univers. of Technol.*
[4] Yin M Vibration reduction and noise reduction analysis and improved design of reducer for cutting machine 2015 *Wuhan Univers. Technolo.*
[5] Lv XD Structural optimization design of EBZ135 roadheader planetary reducer 2010 *Taiyuan Univers. Technol.*
[6] Wu SW Research on Optimization of Planetary Reducer in Cutting Section of EBZ-135 Roadheader 2015 *Shando. Coa. Technol.* **2** 110-2
[7] Zhang Q Mao J and Tian DF Multi-objective fuzzy reliability optimization of cutting head of roadheader based on genetic algorithm 2008 *J. China Coa. Soci.* **33** 1435-37
[8] Wang FQ 2016 Dynamic test analysis of EBH315 roadheader cutting speed reducer *Coa. Technol.***35** 268-71
[9] ZhuSQ Improvement of the planetary frame of the roadheader reducer 2018 *Inner Mongolia Coa. Econo.* **09** 119
[10] Kang MR and Kahraman A An experimental and theoretical study of the dynamic behaviour of double-helical gear sets 2015 *J. Soun. Vib.* **350** 11-29
[11] Yin M, Chen G, Su H. Theoretical and experimental studies on dynamics of double-helical gear system supported by journal bearings 2016 *Ad. Mech. Engi.* **5** 1-13
[12] Cheng YL Xiao ZM and Wang X Dynamic Characteristics Analysis and ADAMS Simulation of Spur Gear 2016 *J. Mech. Stren.* **04** 667-74
[13] Wang GL Pan XD and Li YF One-time mechanical design concept and practice 2014 *J. Mech. Engi.* **01** 152-6
[14] Sondkar P and Kahraman A dynamic modle of a double-helical planetary gear set 2013 *Mech. Mach. Theor.* **70** 157-174
[15] Wang HW and Meng Hao Predictive modeling method for multiple linear regression 2007 *J. Beij. Aerospa. Univers.* **04** 500-4

IOP Conf. Series: Materials Science and Engineering **751** (2020) 012051 doi:10.1088/1757-899X/751/1/012051

Application of Kalman Filter in Satellite Propellant Remaining Prediction Technology

Wenjuan Yin[1,*], Zhiwei Cao[1] and Mengjie Wang[1]

[1]Beijing institute of Control Engineering, 100094 Beijing, China

[*]yinwenjuan1221@126.com

Abstract. The prediction of the propellant remaining is related to the use efficiency of the propellant, it is of great significance for the estimation of the remaining life of the spacecraft. The gauging uncertainty of the commonly used propellant remaining prediction method is high, and the single gauging method can not meet the higher requirements of long life satellites on-orbit management. In order to further improve the accuracy of the spacecraft propellant remaining prediction, combined with the characteristics of the existing propellant remaining estimation method, a real-time estimation method of propellant remaining is proposed. The algorithm combines two gauging methods to further improve the reliability of propellant consumption estimation. The practical application shows that the algorithm can effectively reduce the disadvantages of the BK method and the PVT method, and provide a more reasonable estimation of the real-time propellant remaining.

1. Introduction

With the development of spacecraft technology, especially the application of long-life satellites, higher and higher requirements are put forward for the on-orbit management level during the life of spacecraft. The prediction of the propellant remaining is closely related to the working life of the spacecraft, and the exhaustion of the propellant means the end of the life of the spacecraft. The accurate estimation of the propellant remaining can improve the use efficiency of the propellant and prolong the life of the spacecraft, thereby obtaining huge economic benefits.

The PVT method and the BK method are the earliest applied satellite liquid propellant remaining prediction technologies, and their applications are very common [1~5]. The gauging techniques of these two methods are simple, and there is no special requirement for the hardware of the satellite propulsion system. However, the short-time accuracy of the BK method is high, but there is a cumulative error, and the PVT method is suitable for long-term estimation, but the accuracy is low. Obviously, no matter which calculation method is used, there is a lack of long-term high measurement accuracy.

Based on the analysis of the measurement characteristics of PVT method and BK method, a real-time estimation algorithm based on kalman filter is proposed. The method is based on the existing gauging method and does not require redundant measurement data, and has strong practicability. The on-orbit data is used to verify the method, and the calculation results show that the method effectively reduces the disadvantages of PVT and BK methods, and improves the estimation accuracy and stability of propellant consumption.

2. Propellant remaining gauging method

2.1. Bookkeeping method
The BK method is based on the telemetry data including temperature, pressure and orbital velocity of the spacecraft propulsion system, combined with the ground test data of the engine and the thrusters, and calculates the propellant consumption of the engine and each thruster at each specific working sequence. The consumption is cumulatively added to obtain the total propellant consumption for a period of time.

2.2. PVT method
Based on the pressure and temperature telemetry data of the pressurizing gas in the propellant tanks and helium tanks, the PVT method calculates the gas volume in the propellant tanks using the gas equation of state, and then calculates the liquid volume in the tanks from the total volume of the tanks and the propellant density. The PVT method relies on the initial propellant mass in the tanks and cannot obtain the propellant remaining in each tank of the parallel tanks. If an accurate initial propellant mass can be obtained, the PVT method can achieve relatively high accuracy combined with high precision pressure sensor.

3. Calculation model

3.1. Kalman filter
Kalman filter is an algorithm that uses the linear system state equation to estimate the state of the system through the input and output observation data of the system [6]. The Kalman filter algorithm takes the minimum mean square error as the optimal estimation principle and uses the recursive method to solve the linear filter problem of discrete systems. The core content is three recursive formulas, which mainly describe two processes, namely prediction and correction, which are calculated by continuous "predictive-correction" recursive method, based on the previous time estimate and the most recent observation data. Estimate the current value of the signal to achieve the purpose of filter.

Consider a discrete-time linear dynamic system as follows:

$$\begin{cases} X_k = AX_{k-1} + BU_{k-1} + W_{k-1} \\ Z_k = HX_k + V_k \end{cases} \tag{1}$$

The Kalman filter equations are respectively state update equations and observation correction equations.

The state update equation is as follows:

State further prediction: $\hat{X}_{k/k-1} = A\hat{X}_{k-1/k-1} + BU_{k-1}$ (2)

One-step prediction of the covariance matrix: $P_{k/k-1} = AP_{k-1/k-1}A^{\mathrm{T}} + Q$ (3)

The measurement correction equation is as follows:

Gain matrix: $K_k = P_{k/k-1}H^{\mathrm{T}}\left(HP_{k/k-1}H^{\mathrm{T}} + R\right)^{-1}$ (4)

Covariance update: $P_k = \left(I - K_k H\right)P_{k/k-1}$ (5)

Status update: $\hat{X}_k = \hat{X}_{k/k-1} + K_k\left(Z_k - H\hat{X}_{k/k-1}\right) = \left(I - K_k H\right)\hat{X}_{k/k-1} + K_k Z_k$ (6)

3.2. Calculation model
According to the characteristics of the propellant remaining gauging methods, the BK method is used to construct the measurement equation and the PVT method is used to construct the observation equation. For the BK and PVT methods, it is assumed that the oxidant consumption and the fuel

consumption calculation are completely independent.

Defined at the time t_k, the system state x_k is the propellant consumption and the state model can be obtained.

$$x_k = x_{k-1} + u_{k-1} + w_{k-1} \tag{7}$$

Where, $x = [m_o \quad m_f]^T$, m_o is the oxidant consumption and m_f is the fuel consumption.

u_{k-1} is the control signal, which is the propellant consumption mass flow measured by the BK method within the sampling period.

$$u = [u_o \quad u_f]^T,\ u_o(k) = \int_{t_{k-1}}^{t_k} \omega_o \times dt,\ u_f(k) = \int_{t_{k-1}}^{t_k} \omega_f \times dt \tag{8}$$

$w_k = [w_o \quad w_f]^T$, w_o, w_f are the process noise of the propellant consumption calculation model, which is assumed to be white noise with a mean of zero.

The covariance matrix of w_k is Q_k, which is the measurement error variance of BK method within the sampling period.

$$Q_k = \mathrm{diag}\left[q_o T^2, \quad q_f T^2\right] \tag{9}$$

T is the sampling period, and q_o and q_f the correlation noise intensity of the BK method for the oxidant and the fuel consumption calculation.

The observation equation is:

$$z_k = x_k + v_k \tag{10}$$

The observed value z_k is obtained by the PVT method, and $z = [z_o \quad z_f]^T$, z_o and z_f are the propellant consumption calculated by the PVT method, respectively.

v_k represents propellant consumption error (observation noise), assuming it is also white noise with a mean of zero, and the covariance matrix is R_k, which is the measurement error variance of the PVT method.

$$R_k = \mathrm{diag}[r_o, \quad r_f] \tag{11}$$

r_o and r_f are the error variance of the oxidant and fuel consumption measurements by the PVT method, respectively.

4. Results analysis

Set the propellant consumption measurement interval time $T = 1s$, and the three orbital firing time is 4280s, 4921s, 3976s, respectively. During the ignition period, the calculation error of the BK method within one sampling period is small, and the theoretical error of PVT method is large, and the helium state during the orbital transfer is in an unbalanced state, further increasing its measuring error. During the non-ignition period, since the BK method no longer accumulates propellant consumption, the corresponding calculation noise is zero, and for the PVT method, the helium in the tanks will gradually return to a relatively stable state, the calculation error is close to theoretical error. Therefore, different process noise and measurement noise intensity need to be set separately for different stages.

During the ignition period: $Q_k = \mathrm{diag}\left[0.0004^2, \quad 0.00027^2\right], R_k = \mathrm{diag}\left[8.4^2, \quad 4.8^2\right]$

During the non ignition period: $Q_k = \mathrm{diag}[0, \quad 0], R_k = \mathrm{diag}\left[11.67^2, \quad 6.67^2\right]$

System initial state, $x_0 = [0.23 \quad 0.137]^T$

Initial state variance : $P_0 = \mathrm{diag}[0.001, \quad 0.0006]$

The kalman filter is used to calculate the propellant consumption during the orbital transfer, and the kalman filter result is compared with the calculation results of the BK method and the PVT method. The propellant consumption calculation results of the three algorithms are shown in Figure 1-Figure 6 and table 1.

It can be seen from F Figure 1-Figure 6 that since the accuracy of the BK method within the sampling period is much higher than that of the PVT, the value of kalman filter is closer to the BK method. However, kalman filter can not completely eliminate the cumulative error of BK method. Therefore, with the accumulation of time, the measurement error of BK method becomes larger, and the filtered data gradually approaches the predicted value of PVT method.

Affected by the steady state of helium, the PVT measurement during the ignition period is quite different from that of the non-ignition period. According to the common weighted fusion algorithm, the fusion result variation is consistent with the PVT method. Although the calculation value of the PVT method is relatively stable, the fusion results will be ideal at a moment after the engine ignition, but the error of the real-time fusion result during the ignition period will be too large. For kalman filter, there is no such problem. Since the error of the observation value is much larger than the model error, the change of the prediction value has little effect on the filter result.

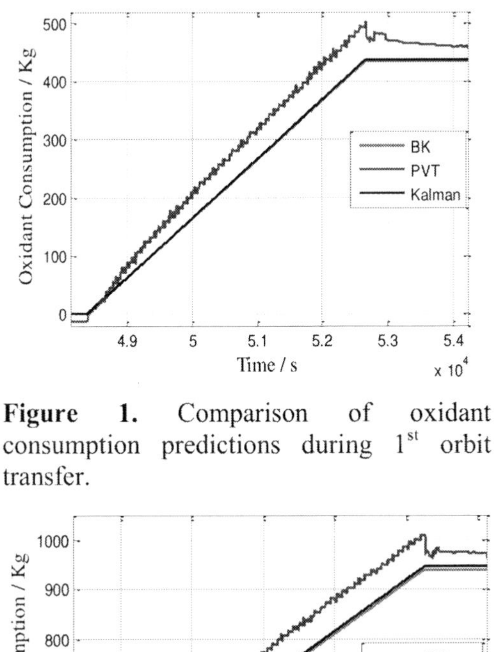

Figure 1. Comparison of oxidant consumption predictions during 1st orbit transfer.

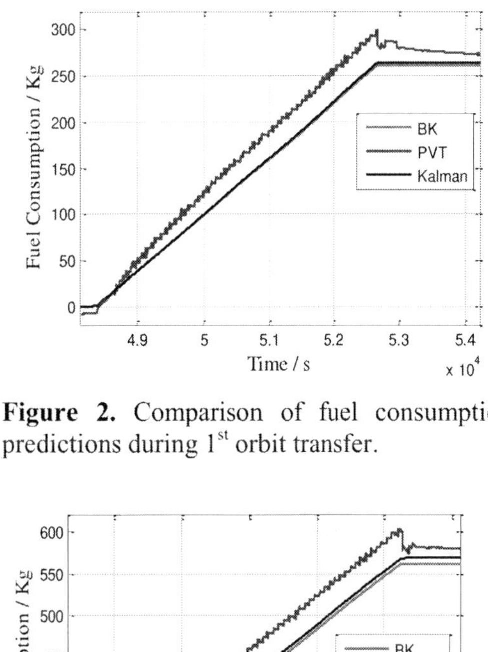

Figure 2. Comparison of fuel consumption predictions during 1st orbit transfer.

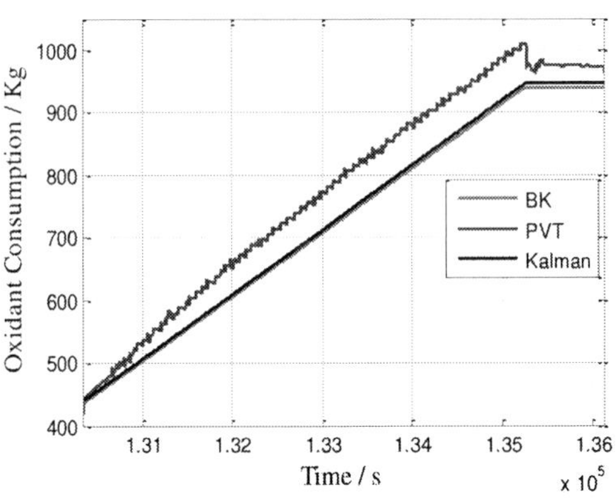

Figure 3. Comparison of oxidant consumption predictions during 2st orbit transfer.

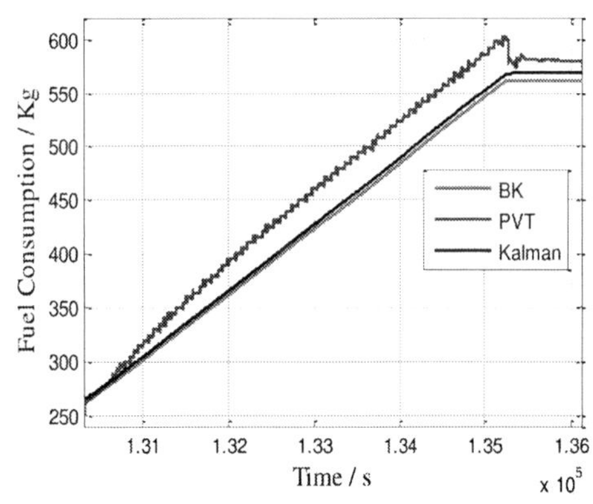

Figure 4. Comparison of fuel consumption predictions during 2st orbit transfer.

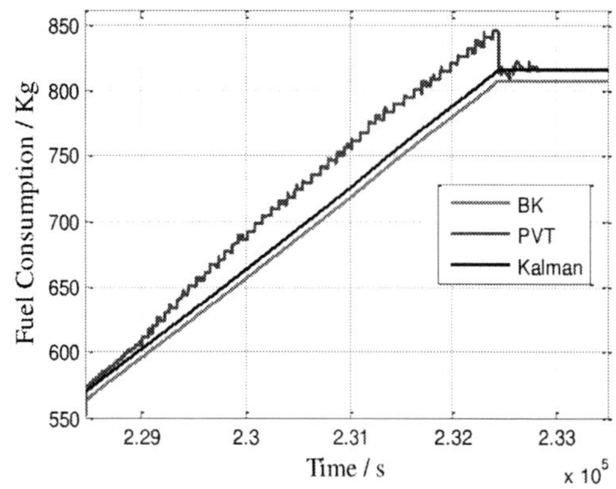

Figure 5. Comparison of oxidant consumption predictions during 3st orbit transfer.

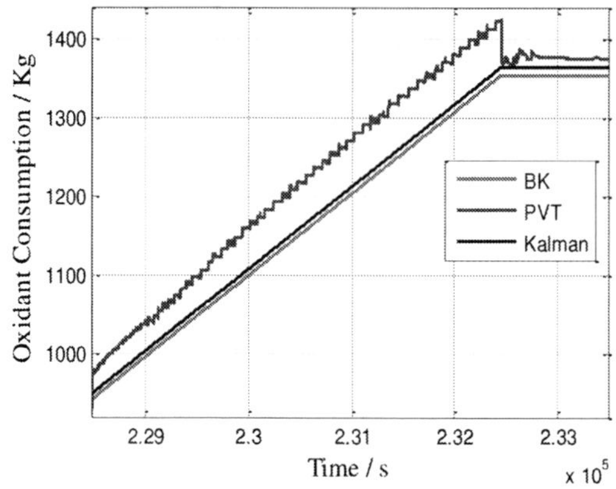

Figure 6. Comparison of fuel consumption predictions during 3st orbit transfer.

Table 1. Comparison of prediction results with theoretical values (units/kg)

	$O_{Theoretical}$	O_{BK}	O_{PVT}	O_{Kalman}	$F_{Theoretical}$	F_{BK}	F_{PVT}	F_{Kalman}
1	439.723	434.464	456.191	437.925	262.783	260.247	271.404	263.545
2	951.097	940.196	960.569	947.466	567.105	562.099	572.309	568.565
3	1369.558	1353.732	1374.849	1363.951	815.362	807.228	815.745	815.655

Figure 1 to Figure 6 all show that the PVT value decreases drastically after the end of the orbit change, but the change of the kalman filter value is very small.

Table 1 shows the comparison between the calculated values of three algorithms and the true values of the propellant consumption. It can be noted that the kalman filter value is closer to the true value than the BK method and the PVT method as a whole. Therefore, the kalman filter algorithm can be used as an effective state estimation method for propellant consumption prediction. This method is not limited to the fusion of the BK method and the PVT method, and can also be used for other gauging methods such as the ultrasonic flow meter method and the PVT method.

5. Conclusion

Based on the characteristics of satellite propellant consumption prediction algorithms, this paper proposes a real-time prediction algorithm based on kalman filter. The algorithm combines two gauging methods to further improve the reliability of propellant consumption estimation. The simulation results show that the proposed method reduces the cumulative error effect of the BK method to some extent, and is not affected by the stability fluctuation of the PVT method, which improves the estimation accuracy and stability of the propellant consumption. The method is based on the existing gauging method, does not require redundant measurement data, and is also suitable for fusion of other propellant consumption estimation methods, and has strong practicability.

References

[1] Matthijssen R, Van P P. State-of-the art gauging components for improved propellant management on 3-axis stabilized spacecraft[C]. AIAA 2006-4714
[2] Yendler B. Review of propellant gauging methods[C]. AIAA 2006-939
[3] Tso-Ping Y. Bipropellant propulsion system performance and propellant remaining

prediction[C]. AIAA 89-2512
[4] Chang Y K, Yong M B, Chae Y S, et a1. Error analyses for spacecraft propellant gauging techniques[C]. AIAA 94-2792
[5] Lal A, Raghunandan B N. Uncertainty analysis of propellant gauging system for spacecraft[J]. Journal of Spacecraft and Rockets, Vol. 42, No. 5, September-October 2005, pp 943-946
[6] Kalman R E. A new approach to linear filtering and prediction problem[J]. Journal ofBasic Engineering Transactions, 1960, 82: 35-45.

IOP Conf. Series: Materials Science and Engineering **751** (2020) 012052 doi:10.1088/1757-899X/751/1/012052

Simulation Analysis of Influence of the Blade Elasticity on Propeller Performance

Yuhang TAN, Weicai PENG and Lijiao SHEN

China Ship Development and Design Center, National Key Laboratory on Ship Vibration & Noise, Wuhan 430064, China

tanyuhang@163.com

Abstract. The blade elasticity can influence hydrodynamic and structural performance of the propeller. The elasticity of the propeller is considered in the hydrodynamic simulation based on the CFD/FEM fluid-structure interaction method. And the hydrodynamic results of the elastic are compared with the rigid. Which show that the accuracy of elastic hydrodynamic prediction improved significantly compared with the rigid. In addition, the hydrodynamic prediction considering the elasticity also shows the transient fluid flow considering the influence of blade vibration. The study indicates considering blade elasticity can significantly improve the accuracy of propeller hydrodynamic prediction, and the prediction of fluid pulsation and propeller response can be used to analyse the flow noise and structural vibration.

1. Introduction

The study of propeller performance is extremely important as the mainstream thruster in the ship. How to accurately predict the propeller in the flow field has very important engineering meaning. The influence factors of ship such as propeller hydrodynamic performance, cavitation, structure vibration and flow noise and so on are studied based on the accurate calculation of propeller hydrodynamics.

The early hydrodynamic prediction is mainly based on the theoretical method of lifting line, lifting surface and panel method. And then CFD technology is used to predict the hydrodynamic performance. However, the calculation model often regards the propeller as a rigid structure without considering the deformation. With the improvement of prediction requirements and the development of composite propeller research, the influence of propeller elasticity on hydrodynamic performance and structural performance has to be taken into account.

Considering the propeller elasticity, the fluid-solid coupling effect should be analyzed in hydrodynamic prediction. In 1996, Lin et al. [1] considered the effect of fluid-solid coupling and firstly used VLM/FEM fluid-solid coupling method to analyze the propeller hydrodynamic performance. Then Young et al. [2] analyzed the composite propeller based on BEM/FEM fluid-solid coupling method, considering the influence of fluid-solid coupling and non-uniform wake. With the development of CFD hydrodynamic prediction, HE et al. [3] analyzed the hydroelastic behavior of composite propellers in non-uniform flow based on CFD/FEM fluid-solid coupling analysis method. In recent years, the CFD/FEM fluid-solid coupling method has become the mainstream to analyze fluid-solid coupling problems.

Considering the influence of elastic effect on hydrodynamic and propeller's performance, the CFD/FEM fluid-solid coupling method is adopted based on ANSYS-CFX modules. The influence of propeller elasticity on hydrodynamic prediction and its own response characteristics are analyzed through the simulation.

2. Calculation model

The calculation model is established based on the CFD/FEM method. The two-way fluid-structure coupling calculation of elastic propeller is realized by computational fluid mechanics software CFX and structural calculation module ANSYS Mechanical in Workbench platform. The fluid-structure coupling calculation model is composed of the elastic propeller model and the external flow field model. The two-way fluid-structure coupling calculation method is used to realize the transfer of fluid pressure and structural deformation data in the coupling process.

2.1. Propeller model

A marine MAU four-blade propeller is used as the model. Design parameters of the propeller are given in Table 1. 3-D model of the propeller is shown in Figure 1(a).

The propeller model is meshed by mesh module of the ANSYS Mechanical. And make the blade surface mesh denser. The propeller mesh is shown in Figure 1(b). The material of the propeller is copper alloy. The density is 8300kg/m^3, Yang's modulus is 110GPa, and the Poisson's ratio is 0.34.

Table 1. Design parameters of the propeller

Propeller Type	MAU
Diameter	4.78m
Number of Blades	4
Advanced Speed	15.48kn
Efficiency	0.545
Pitch/Diameter Ratio	0.682
A_E/A_0	0.544
Hub/Diameter Ratio	0.18
Direction of Rotation	Right

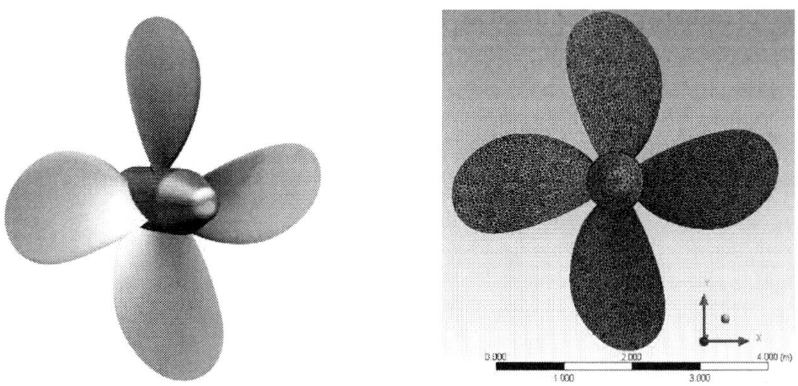

(a) 3-D model (b) 3-D mesh

Figure 1. 3-D model and mesh of the propeller

2.2. Flow field model

The whole flow field is divided into two cylindrical computing fields, which are a inner rotating domain surround by a outer static one. The diameter of the outer is 5D (D is propeller diameter) and long 7.5D. The upstream from the inlet to the propeller is 2.5D, and the downstream from the propeller to the outlet is 5D. The diameter of inner rotating domain is 1.2D, which is 0.63D (about 3m) long.

Mesh generation adopts the form of combination of structured grids and unstructured grids by using ICEM software, which is shown in Figure 2. And fluid-solid coupling interface of the flow domain adopts boundary layer mesh generation method.

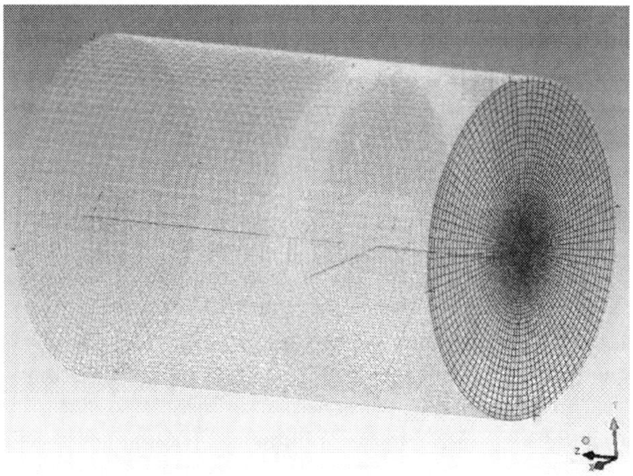

Figure 2. 3-D Mesh of the fluid model

In the flow field, the Z axis is the flow direction, the flow field inlet is speed entrance and the outlet is pressure exit where pressure is zero. The outside boundary of outer domain is set as Free Slip Wall and the rotate speed of inner domain is 155r/min. Those two domains are connected by using General Connection Interface model.

The simulation of flow field model is based on CFD method, which uses incompressible unsteady RANS equation as the governing equation of the flow field. The turbulence model adopts SST k- ω turbulence model, which has good performance on wall bounded boundary layer flows [4].

2.3. Fluid-solid coupling model

The coupling simulation is solved by using the separation method, based on the CFX-ANSYS coupling module in ANSYS Workbench. CFX module is used for flow field simulation, and Transient Structural module for propeller's transient calculation. Two modules transmit the data on the Fluid-Structure Interface (FSI) to achieve the two-way unsteady fluid-structure coupling calculation.

In the constraint of the structure part, the head face of propeller shaft is set as the fixed constraint, and the blade as the FSI. In the flow field, the blade contact surface in the rotating domain is set as the coupling surface for data transmission, and the shaft as the No Slip Wall. In the time step setting of flow field calculation, the time step is set to 0.002s, the total time of coupling calculation is 2s, and the time setting of structure transient calculation is the same as the flow field.

In the coupling simulation, Transient Rotor Stator is adopted for the interface between the rotating field and the outer field. The advance coefficient J is set at the condition of 0.3, which means the inflow velocity is set at 3.7m/s. The hydrodynamic results of the steady calculation under this condition are taken as the initial value of the coupled calculation of the flow field. The Second Order Implicit Euler method is used to solve the transient problem. The solving process monitors propeller's thrust and torque.

3. Hydrodynamic validation

First, CFX simulation software is used to verify the reliability of steady hydrodynamic simulation model of rigid. The turbulence model and the boundary conditions are the same as the coupling calculation Settings. The Frozen Rotor is selected for the interface between the rotating field and the outer field, and the propeller boundary surface in the rotating field is set as No Slip Wall.

By changing the speed of the inlet, steady hydrodynamic results of the propeller advance coefficient from 0.1 to 0.7 are calculated in CFX, including the thrust coefficient K_t, the torque coefficient K_q and the open-water efficiency η.

The simulation results are verified by using the experimental regression formula of open-water performance of AU series propellers in Shanghai Jiao Tong University [5]. The thrust coefficient and torque coefficient regression formulas are as follows:

$$K_T = \sum_{i=0}^{n_1}\sum_{j=0}^{n_2}\sum_{k=0}^{n_3} A_{ijk}(P/D)^i (J)^j (A_E/A_o)^k \tag{1}$$

$$10K_Q = \sum_{i=0}^{n_1}\sum_{j=0}^{n_2}\sum_{k=0}^{n_3} B_{ijk}(P/D)^i (J)^j (A_E/A_o)^k \tag{2}$$

where A_{ijk} and B_{ijk} are the regression coefficients of thrust coefficient and torque coefficient respectively.

Through the experimental regression formula, curves of the thrust coefficient, torque coefficient and open water efficiency of the propeller are calculated. The open-water performance comparison of CFX results and experimental curves are shown in Figure 3. It is shown that the results of the CFX calculation are slightly less than the experimental results, but the trend is consistent. The error of the thrust coefficient is within 5%, and the error of the torque coefficient is within 10%. Which indicates that the calculation results of the CFX are reliable and can be used for the hydrodynamic simulation of the propeller.

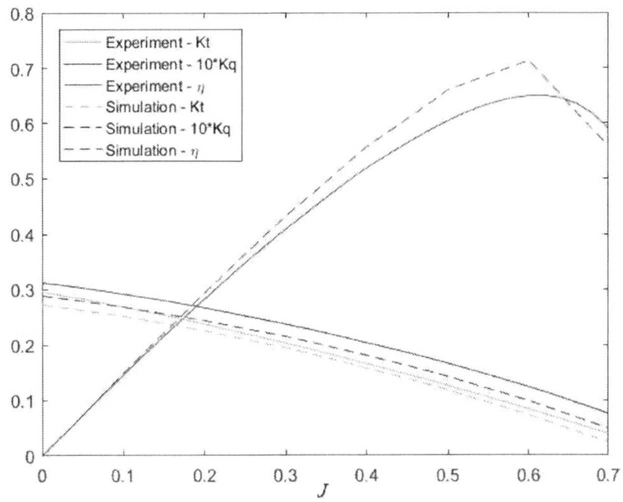

Figure 3. Open-water performance comparison between CFX and experiment

4. Simulation analysis

4.1. Coupling hydrodynamic analysis

After the coupling calculation is completed, the monitoring data results are shown in Figure 4. The coupling calculation reaches stability after 0.2 s. The absolute average values of thrust and torque after stabilization are 7.275×10^5 N and 4.110×10^5 N·m respectively. Corresponding to the test values, the comparison results of thrust coefficient K_t and torque coefficient K_q of steady and coupling calculations are shown in Table 2. The coupling calculation results are 7.19% and 14.74% higher than the steady hydrodynamic results respectively, and 2.81% and 4.09% higher than the errors respectively. Compared with the rigid hydrodynamic calculation, the thrust of the coupling results decreases from 4.09% to 2.81% relative to experiment results, and the torque from 9.28% to 4.09%. Which indicate that the coupling hydrodynamic prediction is more accurate than the steady hydrodynamic prediction, and the prediction accuracy is obviously improved.

Table 2. Comparison of the open-water performance of simulations and test on J=0.3

J=0.3	K_t	Error (%)	10*K_q	Error (%)
Elasticity	0.2088	2.81	0.2468	4.09
Rigidity	0.1948	-4.09	0.2151	-9.28
experiment	0.2031		0.2371	

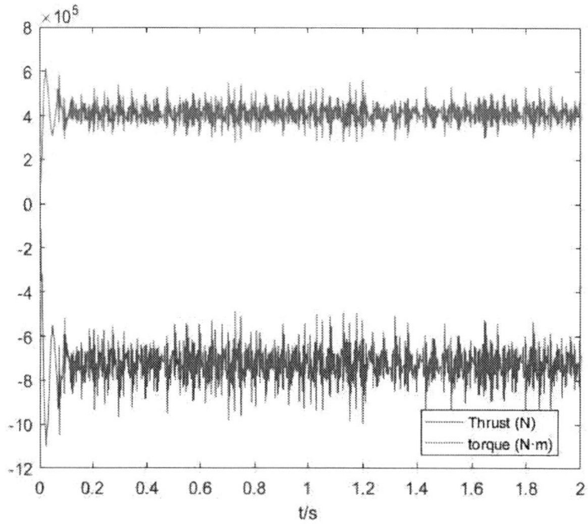

Figure 4. Monitor data of the force, and torque

The contour of elastic blade's pressure at 2s is shown in Figure 5, and the rigid hydrodynamics in Figure 6 for comparison. Compared with the rigid hydrodynamics, the pressure distribution of the coupling simulation remains unchanged. The local pressure on the guide of the blade thrust surface is the largest. And the pressure near the tip of the suction surface is the smallest, which results in negative pressure. The extreme pressure is shown in Table 3. The maximum pressure of the elastic blade decreases by 3.5% relative to the rigid blade, but the absolute value of the minimum negative pressure increases by 3.1%.

In addition, the transient results of coupling contain the flow field pulsating information, which can be used to analyse flow noise and stern vibration induced by flow and propeller vibration.

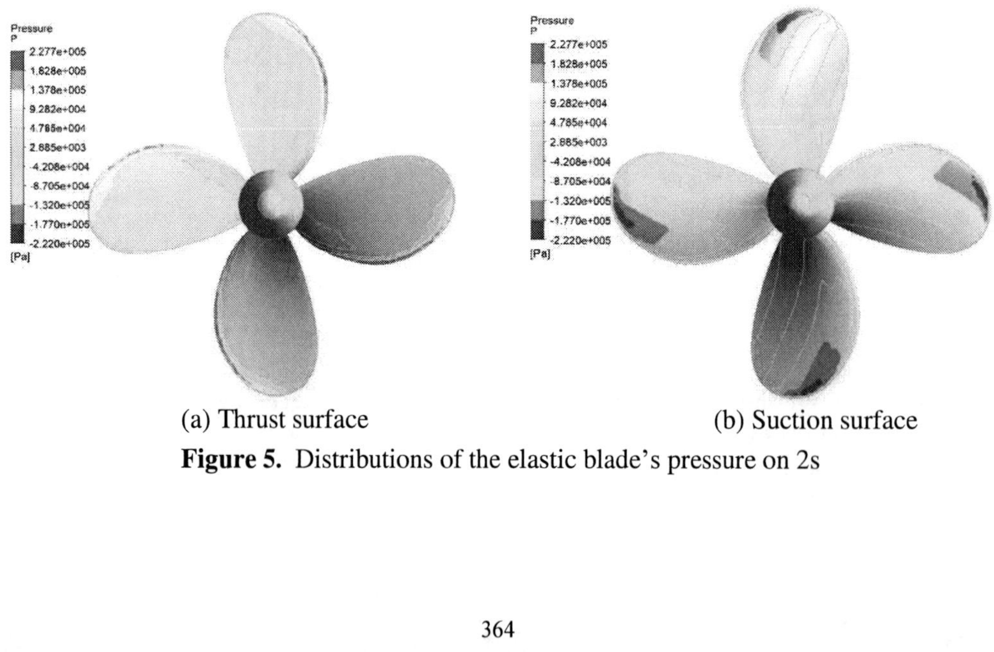

(a) Thrust surface (b) Suction surface

Figure 5. Distributions of the elastic blade's pressure on 2s

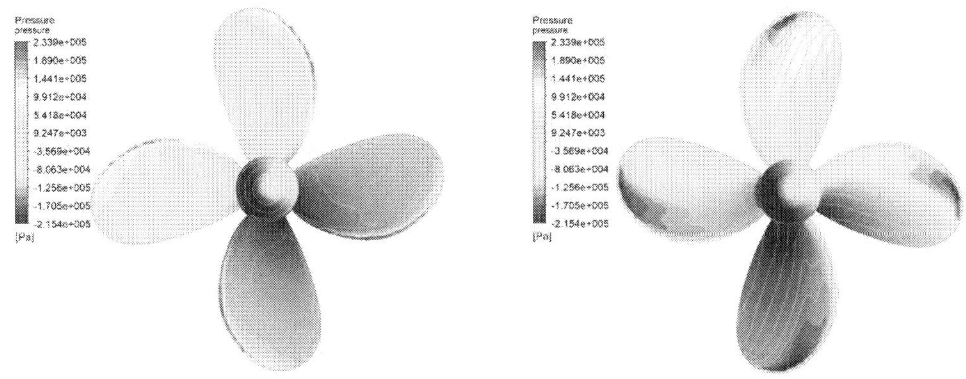

(a) Thrust surface (b) Suction surface

Figure 6. Distributions of the rigid blade's pressure on 2s

Table 3. Comparison of the blade's extreme pressure

	MAX (Pa)	*MIN* (Pa)
Elasticity	2.2773×10^5	-2.2220×10^5
Rigidity	2.3393×10^5	-2.1544×10^5

4.2. Propeller response analyse

Selecting the results of structural calculation at 2s, the distributions of Von Mises equivalent stress on blade surface are shown in Figure 7. It can be seen from the stress distributions that the maximum stress occurs in the middle of the blade root and gradually decreases along the radial direction of the blade surface. But the distributions of the stress on thrust and suction of the blade surface are different. At the same radius, the stress distributions of the thrust and the suction are opposite.

As the blade tip deformed the most, the displacement-time curve at the blade tip is shown in Figure 8. The average value of the tip displacement is 18.8mm, 0.4% of the diameter, and the amplitude is about 0.1mm, 5.3% of the maximum deformation. Which can be seen that considering the propeller elasticity, in addition to improving the accuracy of hydrodynamic prediction, the predicted structural vibration response cannot be ignored.

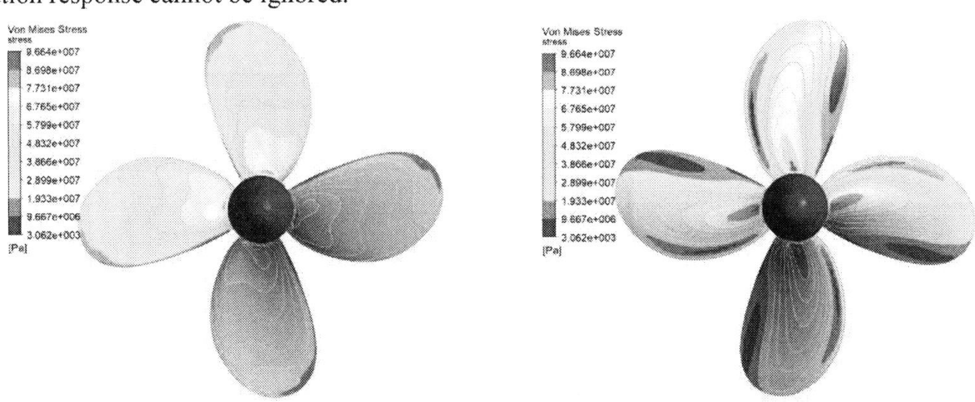

(a) Thrust surface (b) Suction surface

Figure 7. Distributions of the blade's Von Mises stress on 2s

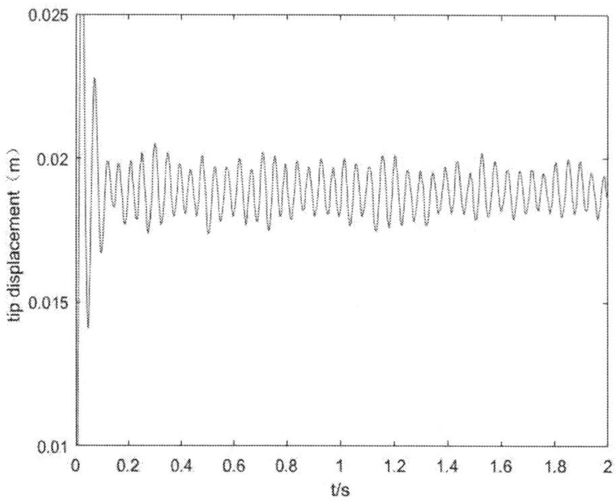

Figure 8. Displacement graph of the tip

5. Conclusion

In this paper, we consider the elasticity of the propeller to simulate the hydrodynamic performance and response characteristics of the MAU four-blade copper alloy propeller by using CFD/FEM fluid-solid coupling method based on CFX and Transient Mechanical in Workbench platform. The main conclusions are as follows:

1) The two-way fluid-solid coupling hydrodynamic results with considering structural elasticity are closer to the test data compared with the rigid hydrodynamic results, which means the prediction accuracy increases obviously.

2)After considering the propeller elasticity, the pressure distribution of the coupling simulation remains unchanged. But the maximum predicted pressure of the blade decreases, and the absolute value of minimum negative pressure increases.

3) Considering the structural elasticity, the transient hydrodynamic simulation results can show the flow field pulsation information which considers the influence of propeller response compared with rigid transient hydrodynamics. Which can predict structural response and flow noise more accurately.

The study also provides a research approach for the prediction of propeller flow noise and structural response.

Acknowledgments

The authors are grateful for supports from the National Natural Science Foundation of China (No.51609226)

References

[1] Lin H J and Lin J J 1996 Nonlinear hydroelastic behavior of propellers using a finite-element method and lifting surface theory *J. Mar. Sci. Technol.* **1(2)** 114-24

[2] Yong Y L 2007 Hydroelastic behavior of flexible composite propellers in wake inflow *16th Int. Conf. on Composite Materials (Kyoto)*

[3] He X D, Hong Y and Wang R G 2012 Hydroelastic optimisation of a composite marine propeller in a non-uniform wake *Ocean Eng.* **39** 14-23

[4] Li D Q 2006 Validation of RANS predictions of open water performance of a highly skewed propeller with experiments *J. Hydrodyn. Ser.* B **18(3)** 520-28

[5] China State Shipbuilding Corporation 1998 *Practical Manual for Ship Design (General Catalogue)* (Beijing: National Defence Industrial Press) pp 316-17

Electrochemical Kinetic Property Determination of MIECs by Comprehensive Multi-physical Coupling Model

Jianhua Wang[1], Salman Khalid[1], Fei He[1,*] and Wanfan Wu[1,2]

[1] CAS Key Laboratory of Mechanical Behavior and Design of Materials, Department of Thermal Science and Energy Engineering, University of Science and Technology of China, Jinzhai Road 96, Hefei 230027, PR China
[2] Datang east China electric power test and research institute, Chuangxin Street 666, Hefei 230088, PR China

[*]**Email:** hefeihe@ustc.edu.cn

Abstract. Surface exchange and bulk diffusion coefficients are important electrochemical kinetic properties of mixed ionic-electronic conductors (MIECs), which are usually obtained through electrical conductivity relaxation (ECR) experiments. However, the extensively employed method is built upon the ideal diffusion described by Fick's second law with linear absorbing boundary, which is almost impossible to achieve in actual experiments. To simulate the actual experiment system and improve the estiamtion accuracy, a comprehensive mathematical model is established to describe the coupled multi-physical processes involved, such as convection-diffusion in the surrounding atmosphere, surface exchange, bulk diffusion of MIECs, and heat transfer. A comparison between different models is carried out, and the results show that the new model established is more accurate and compatible with more complex conductivity curves.

1. Introduction

Mixed ionic and electronic conductors (MIECs) have occupied an important place as the materials for the cathode of solid oxide fuel cells (SOFCs) [1], oxygen sensors [1] and gas separation membranes [3]. Electrical conductivity relaxation (ECR) is one of the most widely used techniques to determine surface exchange coefficient and bulk diffusivity for MIECs, which has attracted more and more attention for its simplified experiments and reduced costs [4],[5].

Oxygen diffuses into or out of MIECs due to the imbalance of oxygen partial pressure, resulting in the variation of oxygen ions concentration in bulk material. The conductivity of the sample varies with the oxygen concentration and can be measured directly by using ECR technique. Therefore the information on surface exchange coefficient k and bulk diffusivity D is usually extracted by analyzing the relaxation behavior of conductivity. In actual ECR measurements, a sample fixes in a sealed container with a flowing gas mixture. The oxygen partial pressure of the gas mixture can be changed. When the sample is in equilibrium with the surrounding atmosphere, the oxygen partial pressure of the gas mixture is subject to a step change to record the transient response of the bulk material conductivity. Many physical mechanisms are involved in the ECR, and thereby, many factors may influence sample conductivity value. As the kinetic parameters k and D are determined by fitting the measured conductivity curve with corresponding theoretical predictions under identical operating conditions, modeling in combination with inverse analysis covering all the influence factors is of significant importance to obtain accurate estimations.

In literatures, the Fick's second law with linear absorbing boundary are usually used to describe the bulk diffusion and surface exchange processes in materials [5],[7]. This model focuses on the diffusion process within the bulk material, without regard to the influences from the surrounding atmosphere, hence only applicable when the surrounding atmosphere is stagnant, evenly distributed and its response time switched to another steady state is instantaneous or negligible. There are a large number of investigations on kinetic parameters predictions from an analytic solution of such model [8], as well as error analysis and optimization of the estimated results [7]. However, accurate estimations are hard to be determined and controlled because the ideal state is almost impossible to achieve in actual experiments. Some ECR curves obtained in the experiments deviate from the standard curve derived by the Fick's second law [9],[10],[11].In this situation, the Fick's second law is invalid. The influences of gas velocity and operating temperature are still unknown. Therefore it would be of great significance to develop a comprehensive model that is close to actual experiments and able to be compatible with more complex conductivity curves.

In this work, a new and comprehensive model is established to describe the coupled multi-physical processes involved in the real ECR experiment system, and a inverse algorithm with the direct numerical approach coherently embedded is developed to determine the bulk diffusivity D and surface exchange coefficient k for MIECs. A comparison between different models is carried out, and the results show the new model established is more accurate and compatible with more complex conductivity curves.

2. Synthesizing modeling and inverse algorithm

2.1. Coupled modeling of multi physical-chemical processes in the material and the surrounding atmosphere

The schematic illustration of the core experimental section is shown in figure 1. A MIEC cuboid bulk sample lies in the center of a cylindrical tube with the two ends sealed. A gas mixture with two components is supplied into the tube, flowing over the sample surface and providing the sample with the surrounding atmosphere, with certain oxygen partial pressure distribution. Two couples of silver wires (yellow and green wires) are attached to either end of the sample through silver paste to supply current and measure voltage respectively. The whole tube is heated by a test furnace, which can offer uniform heating and temperature control. At first, the sample is in equilibrium with the surrounding atmosphere, then the gas flow rate ratio of the two components is changed, and a new oxygen partial pressure distribution in the surrounding atmosphere forms. As a result, the equilibrium state at the gas/solid interface breaks, and oxygen is either released from or incorporated into the crystal lattice through surface exchange and bulk diffusion. The relaxation time of the sample to the new equilibrium depends on the response time of the surrounding atmosphere approaching a steady state. Therefore the relaxation process is governed by the flow of the gas mixture, oxygen exchange at the gas/solid interface, oxygen diffusion in the surrounding atmosphere, and bulk material.

Suppose the velocity of oxygen is identical with that of the gas mixture in the tube. The Navier-Stokes equation for compressible fluid is applied to describe the velocity and stress distributions of the oxygen in the surrounding atmosphere.

Momentum conservation equation:

$$\rho\frac{\partial \mathbf{u}}{\partial t}+\rho(\mathbf{u}\cdot\nabla)\mathbf{u}=\nabla\cdot\left[-p\mathbf{I}+\mu(\nabla\mathbf{u}+(\nabla\mathbf{u})^T)-\frac{2}{3}\mu(\nabla\cdot\mathbf{u})\mathbf{I}\right]+\mathbf{F} \tag{1}$$

Mass fraction conservation equation of oxygen in the gas mixture:

$$\frac{\partial c_g}{\partial t}+\nabla\cdot\left(-D_{OA}\nabla c_g\right)+\mathbf{u}\cdot\nabla c_g=0 \tag{2}$$

D_{OA} represent the diffusivity of oxygen in the surrounding atmosphere, which depends on the temperature and pressure of gas mixture:

$$D_{OA} \sim p^{-1} T^{3/2}$$

The diffusion process of oxygen within the bulk material is described as:

$$\frac{\partial c}{\partial t} + \nabla \cdot (-D\nabla c) = 0 \quad (4)$$

$$-D\frac{\partial c}{\partial n}\bigg|_{\Gamma} = k(c_2 - c) \quad (5)$$

Linear reaction kinetic is assumed to describe the surface exchange between the gaseous and solid phases. A test furnace offers uniform heating and temperature control to the tube. In this situation, radiation heat transfer and heat-transfer are neglected due to diffusion in the system so that we can write the heat conservation equation as:

$$\frac{\partial}{\partial t}(\rho C_p T) + \nabla \cdot (\rho C_p \mathbf{u} T) = \nabla \cdot (k_T \nabla T) + Q \quad (6)$$

Where Q is the ohmic loss depending on the sample conductivity.

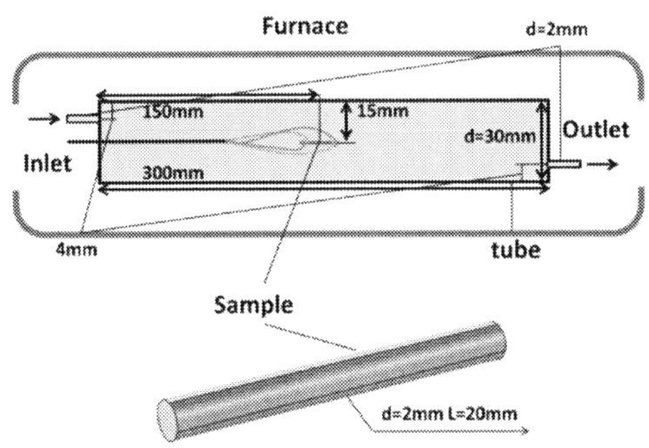

Figure 1 Schematic illustration of the physical model.

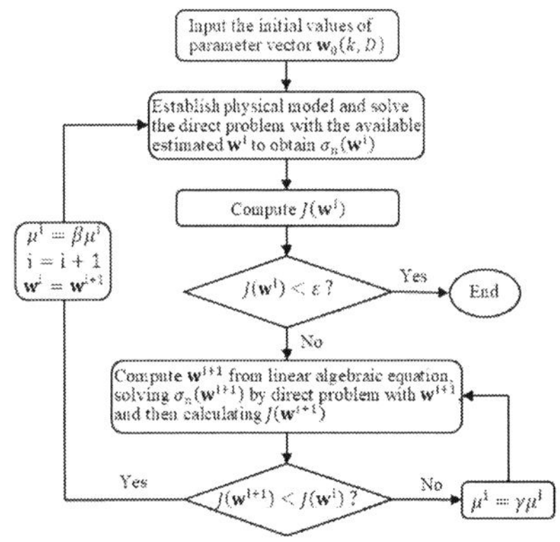

Figure 2 Flowchart of the numerical simulation process, where $\boldsymbol{\gamma} < \mathbf{1}, \boldsymbol{\beta} > \mathbf{1}$.

Suppose that the initial oxygen concentration is c_1 everywhere in the tube, and then gas mixture with oxygen concentration c_2 / oxygen partial pressure $p_{O,2}$ is supplied into the tube by changing the volume flow rate ratio of the two components. The boundary conditions are listed in Table 1.

Table 1 All tube surface boundary conditions.

Boundary	Inlet	Outlet	Other boundaries
Laminar flow	Normal velocity U	Pressure and non-viscous stress $p_0 = 0$	No-slip
Mass transfer	Oxygen partial pressure$p_{O,2}$ /Oxygen concentration c_2	$-\mathbf{n} \cdot D_{OA} \nabla c_g = 0$	Insulation

The Fick's second law with linear absorbing boundary, which has been extensively employed in exsting literatures, is summarized in Table 2.

Table 2 Fick's second law with linear absorbing boundary.

Fick's second law with linear absorbing boundary	
Diffusion equation	$\frac{\partial c}{\partial t} = D\nabla^2 c$
Boundary condition	$-D\frac{\partial c}{\partial n}\Big\|_{\Gamma} = k(c_2 - c)$
Initial condition	$c(t, x, y, z)\|_{t=0} = c_1$

The conductivity of MIEC is in proportion to the charge carrier density, which are either oxygen ion or electron-hole here. We could obtain the oxygen ion concentration field *c(t,x,y,z)* by solving the coupled Equations (1)-(6). If the apparent conductivity is normalized as:

$$\sigma_n(t) = \frac{\sigma(t)-\sigma_1}{\sigma_2-\sigma_1} \tag{7}$$

Where $\sigma(t), \sigma_1$ and σ_2, are the apparent conductivities of the sample at time t, initial and final equilibrium states, respectively. Then the oxygen ions concentration c_O can be linked to the normalized conductivity [7]:

$$\sigma_n(t) = \frac{\sigma(t)-\sigma_1}{\sigma_2-\sigma_1} = \frac{c_O(t)-c_{O,1}}{c_{O,2}-c_{O,1}} \tag{8}$$

2.2. *Inverse algorithm*

As illustrated above, we can get the apparent conductivity relaxation history of a sample directly in ECR experiments. Then the determination of the k and D values is turned into a parameter estimation/inverse problem, i.e., finding the closest k and D so that the calculated conductivity is matchable with the experimental data under identical operating conditions. The matching error between experimental measurement and model predictions is defined as:

$$J(\mathbf{w}^{\mathrm{T}}) = \sum_{i=1}^{\mathrm{Nt}}[\sigma_n(t_i, \mathbf{w}^{\mathrm{T}}) - Y_{meas}(t_i)]^2 \tag{9}$$

where $\mathbf{w} = [k, D]^{\mathrm{T}}$ is the estimated kinetic parameter vector; the discrete time t_i, $i = 1, 2 \cdots$, Nt,represent the sampling points; $\sigma_n(t_i, \mathbf{w}^{\mathrm{T}})$and $Y_{meas}(t_i)$ are normalized conductivites predicted by the model and measured in the experiment at the sampling time t_i. To minimize the matching error, the derivative of $J(\mathbf{w}^{\mathrm{T}})$ with respect to vector $\mathbf{w}$ is set to be zero:

$$\mathbf{X}^{\mathrm{T}}(\mathbf{w})[\boldsymbol{\sigma}_n^{\mathrm{T}}(\mathbf{w}^{\mathrm{T}}) - \mathbf{Y}_{meas}^{\mathrm{T}}] = 0 \tag{10}$$

where $\boldsymbol{\sigma}_n = [\sigma_n(t_1, \mathbf{w}), \sigma_n(t_2, \mathbf{w}), \cdots, \sigma_n(t_{Nt}, \mathbf{w})]$, $\mathbf{Y}_{meas} = [Y_{meas}(t_1), Y_{meas}(t_2), \cdots, Y_{meas}(t_{Nt})]$, and $\mathbf{X}(\mathbf{w})$ is the sensitivity matrix defined as,

$$\mathbf{X}(\mathbf{w}) = \left[\frac{\partial \boldsymbol{\sigma}^{\mathrm{T}}(\mathbf{w})}{\partial \mathbf{w}}\right]^{\mathrm{T}} = \begin{bmatrix} \frac{\partial \sigma_n(t_1)}{\partial k}, & \frac{\partial \sigma_n(t_1)}{\partial D} \\ \frac{\partial \sigma_n(t_2)}{\partial k}, & \frac{\partial \sigma_n(t_2)}{\partial D} \\ \vdots & \vdots \\ \frac{\partial \sigma_n(t_{Nt})}{\partial k}, & \frac{\partial \sigma_n(t_{Nt})}{\partial D} \end{bmatrix} \tag{11}$$

A group of nonlinear algebraic equations can be obtained by expanding Eqn. (10). However, it is hard to solve for kinetic parameter vector $\mathbf{w}$ from Eqn. (10) analytically, instead, the Levenberg-Marquardt iterative method [12] is resorted. The kinetic parameter vector $\mathbf{w}$ is expanded in a Taylor series, only the first-order terms remain and thus an iterative formula of Eqn. (10) is obtained:

$$\left[(\mathbf{X}^{(s)})^{\mathrm{T}}\mathbf{X}^{(s)} + \mu^{(s)}\boldsymbol{\Omega}^{(s)}\right]\mathbf{w}^{(s+1)} = \left[(\mathbf{X}^{(s)})^{\mathrm{T}}\mathbf{X}^{(s)} + \mu^{(s)}\boldsymbol{\Omega}^{(s)}\right]\mathbf{w}^{(s)} - (\mathbf{X}^{(s)})^{\mathrm{T}}[\boldsymbol{\sigma}_n^{\mathrm{T}}(\mathbf{w}^{\mathrm{T}}) - \mathbf{Y}_{meas}^{\mathrm{T}}] \tag{12}$$

where s is the number of iterative steps, μ is the damping factor, and:

$$\boldsymbol{\Omega} = diag(\mathbf{X}^{\mathrm{T}}\mathbf{X}) \tag{13}$$

One can obtain the k and D values through the iteration of eq. (12). However, the estimation of k and D is a typical ill-posed problem which has no unique solution but a set of solutions depending on the measurement noise and expected confidence levels. The measurement noise of the experimental conductivity $\mathbf{Y}_{meas}$ is assumed to satisfy normal distribution with standard deviation S_Y, then one can calculate the standard deviation of estimated parameters S_w through the following expression:

$$S_{w_j^*} = S_Y\sqrt{[\mathbf{X}^{\mathrm{T}}\mathbf{X}]_{j,j}^{-1}}, j = 1,2 \tag{14}$$

A 95% confidence level is generally considered in engineering applications. Accordingly, the uncertainty of the estimated k and D can be known as:

$$w_j^* - 1.96S_{w_j^*} \le w_j \le w_j^* + 1.96S_{w_j^*}, j = 1,2 \quad (15)$$

2.3. Numerical solution and computational algorithm

The commercial software COMSOL Multiphysics 5.3 is employed as the direct numerical approach to solve the mathematical model described by equations (1)-(6). A structured mesh with refined grids near gas/solid interfaces is adopted after the grid independence is validated. Then the whole direct numerical calculation of the model is embedded into the iterative loop of the inverse algorithm, which is implemented via MATLAB in combination with COMSOL. Figure 2 illustrates the flow chart of the inverse algorithm, and parameter ε is the prescribed tolerance of the matching error while parameters β, γ are employed to adjust the damping factor of the iteration procedure.

3. Results and discussion

Through above analysis, there are obvious differences between the model established and the Fick's second law employed in the existing method. To understand the difference, we determine kinetic parameters for a set of normalized conductivity data respectively with these two models and the related inverse algorithm. The set of normalized conductivity data to be fitted was assigned assumed measurement noise in advance. First, we carried out direct numerical simulations using the comprehensive mode and assumed values of k and D. Then the generated conductivity relaxation responses were added with normally-distributed random noise, of which the mean was 0 and the variance was 5% of the conductivity difference between the initial and final equilibrium states. The initial guesses of k and D values are chosen in a wide range to avoid local minimums of matching error. A set of estimated k and D values with the corresponding matching errors are finally obtained. Figure 3 shows the conductivity curves estimated by the two different models when the smallest matching error is reached.

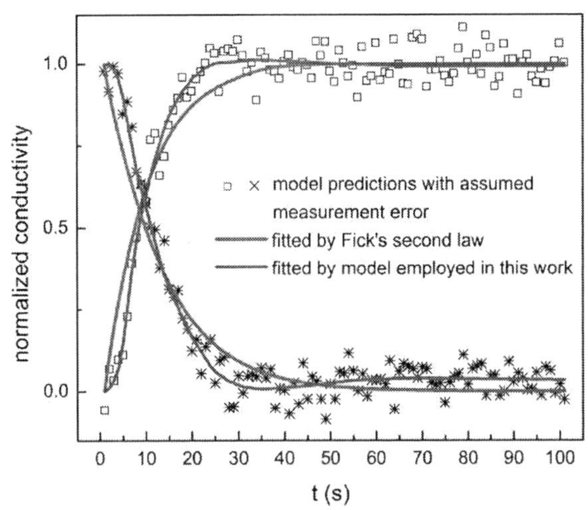

Figure 3 Estimated results comparison between different models.

Clearly, the set of preset normalized conductivity data somewhat deviates from perfect monotone increasing or decreasing trend. In this situation, the fitted conductivity curves by coupled multi-physical processes model are more consistent than those by Fick's second law, which indicates the model employed in this work is more compatible with complex curves. The corresponding estimated values of k and D under certain operating parameters are then summarized in table III. It can be found that the estimated values of k by the two models are both very close to the true value, and the estimation by new multi-physical coupling model is a little more accurate. Although both the estimated values of D significantly deviate from the true value, the deviation of the estimation by new multi-physical model are much smaller and can be controlled within an order of magnitude. Since the two models are embedded into the same inverse algorithm to determine surface exchange coefficient k and bulk diffusivity D, it can be concluded that the comprehensive model employed in this work is more reliable and the corresponding estimated results obtained are more accurate.

Table 3 Comparison of Fick's law and the newly developed model estimated result.

Relaxation process	Oxidation	Reduction
Operating parameters	Inlet velocity:1200ml/min Oxygen partial pressure: 0.021atm→ 0.062atm Temperature: 800K	Inlet velocity:1200ml/min Oxygen partial pressure: 0.062atm→ 0.021atm Temperature: 800K
Preset values	$k = 1 \times 10^{-4}\ m/s$ $D = 1 \times 10^{-6} m^2/s$	$k = 1 \times 10^{-6}\ m/s$ $D = 1 \times 10^{-10} m^2/s$
Estimations by model in this work	$k = 8.06 \times 10^{-5}\ m/s$ $D = 1.14 \times 10^{-5} m^2/s$	$k = 9.99 \times 10^{-7}\ m/s$ $D = 7.07 \times 10^{-10} m^2/s$
Estimation by Fick's second law	$k = 5.52 \times 10^{-5}\ m/s$ $D = 2.10 \times 10^{-4} m^2/s$	$k = 3.92 \times 10^{-5}\ m/s$ $D = 1.76 \times 10^{-6} m^2/s$

4. Conclusions

A generic inverse analysis method is developed to determine the bulk diffusivity D and surface exchange coefficient k for MIECs from ECR measurement data. A comprehensive model is established and employed to describe the coupled multi-physical processes during the actual experimental procedure that is different from ideal diffusion described by Fick's second law with linear absorbing boundary. A comparison study between different models is carried out, and the results show the new model established is more accurate and able to be compatible with more complex conductivity curves; in turn, the estimated results fitted by the new model are more reliable.

Acknowledgment

National Natural Science Foundation of China supported this work (Contract no. 51806206) and the Fundamental Research Funds for the Central Universities (WK2090130022).

References

[1] He F, Jiang Y, Ren C, Dong G, Gan Y, Lee M, Green D-R, Xue X*, Generalized electrical conductivity relaxation approach to determine electrochemical kinetic properties for MIECs. *Solid State Ionics*, 297 (2016) 82-92.

[2] Tuller L-H, Defect engineering: design tools for solid-state electrochemical devices, *Electrochimica Acta,* 48(20-22) (2003) 2879-2887.

[3] Lee T, Cho D, Kwona H, Yoo H, Isothermal transport properties and majority-type defects of BaCo0.70Fe0.22Nb0.08O3−δ, *Phys. Chem. Chem. Phys.,* 17 (2015) 2598-2607.

[4] Lane J-A, Kilner J-A, Measuring oxygen diffusion and oxygen surface exchange by conductivity relaxation, *Solid State Ionics,* 136-137 (2000) 997-1001.

[5] He F, Jin X, Tian T, Ding H, Green D-R, Xue X*, Determination of electrochemical kinetic property for mixed ionic electronic conductors from electrical conductivity relaxation measurements. *Journal of The Electrochemical Society,* 162(9) (2015) F951-F958.

[6] Chen D, Shao Z, Surface exchange and bulk diffusion properties of Ba0.5Sr0.5Co0.8Fe0.2O3-δ mixed conductor, *Int. J. Hydrogen Energy,* 36(11) (2011)6948-6956.

[7] Li Y, Gerdes K, Horita T, Liu X, Surface exchange and bulk diffusivity of LSCF as SOFC cathode: electrical conductivity relaxation and isotope exchange characterizations, *J. Electrochem. Soc.,* 160(4)(2013) F343-F350.

[8] Gopal C-B, Haile S-M, An electrical conductivity relaxation study of oxygen transport in samarium doped ceria, *J. Mater. Chem. A,* 2 (2014) 2405-2417.

[9] Zheng K, Klimkowicz A, Świerczek K, Malik A, Ariga Y, Tominaga T, Takasaki A, Chemical diffusion and surface exchange in selected Ln–Ba–Sr–Co–Fe perovskite-type oxides, *Journal of Alloys and Compounds,* 645(1)(2015) S357-S360.

[10] Choi S-M, Lee J, Hong J, Yoon K-J, Son J-W, Kim B-K, Lee H-W, Lee J-H, Degradation of hydration kinetics of proton-conducting Ba(Zr0.84Y0.15Cu0.01)O3−δ during conductivity-relaxation experiment, *Journal of Power Sources,* 332 (2016) 299-304.

[11] Zhu H, Ricote S, Coors W-G, Chatzichristodoulou C and Kee J-R, Equilibrium and transient conductivity for gadolium-doped ceria under large perturbations: II. Modeling, *Solid State Ionics,* 268 (2014) 198-207.

[12] Beck J-V, Arnold K-J, Parameter Estimation in Engineering and Science, (1st edition, John Wiley, New York, 1977)

IOP Conf. Series: Materials Science and Engineering **751** (2020) 012054 doi:10.1088/1757-899X/751/1/012054

Aircraft Inflight Icing Detection Based on Statistical Decision Theory

D Ding[1,2,*], W Q Qian[1] and Q Wang[1]

[1] Computational Aerodynamics Institute, China Aerodynamics Research and Development Centre, Mianyang, 621000, China

[2] Research supported by CARDC Fundamental and Frontier Technology Research Fund

[*]E-mail: dingdi1981@cardc.cn

Abstract. Aircraft inflight icing detection is very important for aviation flight safety. The ice alarm signal given by detectors can help the pilot to actuate the ice protection system. The application of a Neyman Pearson (NP) based statistical decision approach in icing detection is discussed in this paper. First, the flight dynamics for icing aircraft is modelled, the dynamic behaviour of an airbus influenced by ice accreting on the leading edge of wings is analysed. Then the residuals which are directly affected by the aerodynamic parameters are presented and their generation way is given. The Generalized Likelihood Ratio Test (GLRT) method is utilized in detecting the changes of residuals inflight, its performance is assessed by a flight scenario with wing ice. The results show that the fault of wing icing can be detected by every residual under different window sizes. By comparing the results of the residuals, it can be concluded that the small window size will degrade the detection performance with the longer alarming time delay and the lower probability of detection; appropriately increasing the window size can reduce the alarming time delay; diagnosing by residual of normal overload is more reliable than by the other two residuals.

1. Introduction

Aircraft in-flight icing is one of the most severe threats to flight safety, and sometimes it even leads to fatal accidents and casualties. Based on statistic data given by the American Safety Advisor, from 1990 to 2000, there are 12% of all the flight accidents which resulted from adverse weather conditions that occurred in icing and 92% of the ice induced accidents that took place in in-flight icing [1]. Thus, the icing detection and protection technologies become more and more important in the aviation industry. The ice protection system usually needs the ice alarm signal to actuate the de-ice or anti-ice devices. There are two main means to detect the alarm signal: the direct detection using icing detectors or sensors [2-3], and the indirect detection using mathematical algorithms. Although many aircrafts have utilized the direct detection technology successfully, there still is a strong demand to develop detection algorithms as an auxiliary means for the insufficient or incapable usage of the direct detection method.

Since the concept of Smart Icing System [4] presented by NASA icing research group, the studies on indirect detection methods have been developed rapidly. There are three main types of algorithms for aircraft in-flight icing detection: the parameter identification method, the data-based modelling method, and the observer-based detection method. The parameter identification method estimates the flight status and the changing aerodynamic parameters jointly by filtering algorithms, such as Extended

Kalman Filter (EKF) [5-6], H_∞ filter [6-8] and Multiple Model Adaptive Estimator (MMAE) [9-10]. The parameterized ice accretion process can be estimated by these filters or estimators and can be used for giving icing alarm or for reconfiguring the flight control [11-13] to ensure the aircraft stability in icing conditions. The data-based modelling method models the correlations between flight status and the detailed icing information, such as ice-influenced parameters, icing severity factor, position and time. Caliskan [13] presented a Neural Network (NN) method for identifying icing parameters from the measured state data, Schuchard [14] also developed an artificial NN method to detect the presence of icing and classify its severity, Dong [15-16] utilized a probabilistic NN to decide the icing location. For icing detection, data-based modelling methods can obtain as many icing details as parameter identification methods, but the problems of the two approaches, like the over-fitting and generalization problems of data-based modelling methods and the parameter identifiability and accuracy problems of identification methods, limit the usages of the two methods in some circumstances.

On the other hand, the observer-based detection method is more effective in giving an icing alarm. There are two applications of this method: the model-based fault diagnosis technique and the statistical diagnosis technique. The model-based fault diagnosis technique models the aircraft failure (including actuator [17], sensor [18], structure and icing [19-21]) and then detects this failure by estimators. The statistical diagnosis technique uses hypothesis testing theory to detect the changes in the observed flight data. This diagnosis approach does not require a priori statistical characteristics of the faults and has low computational cost. The innovation approach [12,22-23] and the Neyman Pearson based statistical change detection approach [24] are the commonly used statistical diagnosis techniques for aircraft icing detection, and the former one directly uses the hypothesis testing on the innovation sequence to determine the changes of the aircraft dynamics, its performance relies on the predetermined confidence coefficient. However, the NP based diagnosis technique has a predictable performance under the NP theory, and is proved to be the optimal test subject to a constant probability of false alarm.

This paper focuses on the issue of giving an icing alarm in aircraft flight, the NP based diagnosis technique is chosen because of its detection efficiency and predictable performance. The diagnosis residuals of overloads and angular accelerations are discussed, and their generation way is considered in this paper. A Generalized Likelihood Ratio Test method is utilized to detect the changes of residuals caused by the variation of any aerodynamic parameter. Then an airbus inflight wing icing scenario is assumed and used to assess the performance of the icing detector. Different window sizes are considered for diagnosing the residual, the icing alarming time and the probability of detection are concerned about for the inflight detection performance assessment.

2. Aircraft icing dynamics

2.1. Icing research airbus

An airbus model is specifically presented for the aircraft icing research by the research group. The prototype has the similar aerodynamic and flight dynamic characteristics as Boeing 737 and Airbus A320. Figure 1 shows the configuration structure of the airbus, and Table 1 lists its physical parameters.

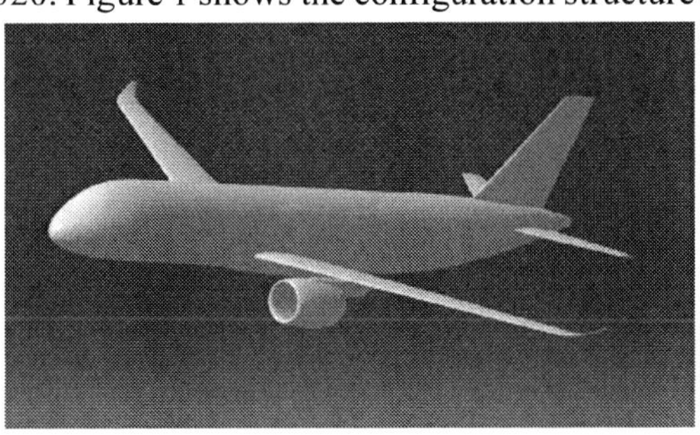

Figure 1. Icing research airbus 3D structure.

Figure 2. Severe wing icing.

The aerodynamic characteristics of the airbus are calculated by computational fluid dynamics (CFD) tool. RANS and multi-block structured grid are used to obtain the longitudinal aerodynamic data under clean and severe wing icing configuration (Figure 2). Then a polynomial equation (1) is used to fit the longitudinal characteristics, where C_D, C_L, and C_M represent drag, lift and pitch moment coefficients respectively, α is the angle of attack (AOA), q is the pitch rate, δ_e is the elevator angle. The aircraft's stability and control derivatives in two configurations are compared in Table 2. Ignoring the small influence of wing icing on C_{D0}, C_{L0}, and C_{M0} in this table, the other longitudinal derivatives are influenced by ice accretion. The fitted results indicate wing icing increases drag coefficients, and decreases lift and pitch moment coefficients and control surface efficiency.

$$\begin{aligned} C_D &= C_{D0} + C_{D\alpha}\alpha \\ C_L &= C_{L0} + C_{L\alpha}\alpha + C_{L\delta e}\delta_e \\ C_M &= C_{M0} + C_{M\alpha}\alpha + C_{M\dot{\alpha}}\dot{\alpha} + C_{Mq}q + C_{M\delta e}\delta_e \end{aligned} \quad (1)$$

Table 1. Parameters of icing research airbus.

Aircraft Parameters	Values
Mass m	72000 kg
Wing reference area S	124 m^2
Wing span l	34.1 m
Mean aerodynamic chord b_A	4.15 m
Moment of inertia I_{xx}	1658755 kgxm^2
Moment of inertia I_{yy}	2392630 kgxm^2
Moment of inertia I_{zz}	3846326 kgxm^2

Table 2. Longitudinal derivatives of icing research airbus in clean and severe iced configurations.

Longitudinal Derivatives	Clean	Severe Iced
C_{D0}	0.0277	0.0277
$C_{D\alpha}$	0.1069	0.7701
$C_{D\delta e}$	0	0.0149
C_{L0}	0.1415	0.1415
$C_{L\alpha}$	6.2326	3.0637
$C_{L\delta e}$	0.448	0.2848
C_{M0}	0.0353	0.0353
$C_{M\alpha}$	-1.7873	-1.2762
$C_{M\dot{\alpha}}$	-13.7477	-13.0491
C_{Mq}	-41.513	-39.4878
$C_{M\delta e}$	-1.9035	-1.3476

2.2. Aircraft icing flight dynamics

The longitudinal flight dynamics considering ice accretion process for aircraft is discussed. The motion and rotation equations (2) are established in the velocity and the body coordinate respectively.

$$
\begin{aligned}
\dot{V} &= -\frac{q_\infty S}{m} C_D + \frac{P_x}{m} + g_x \\
\dot{\alpha} &= -\frac{q_\infty S}{mV\cos\beta} C_L + \frac{P_z}{mV\cos\beta} + \frac{g_z}{V\cos\beta} - \\
&\quad \frac{1}{\cos\beta}\left(p\cos\alpha\sin\beta - q\cos\beta + r\sin\alpha\sin\beta\right) \\
\dot{q} &= \frac{q_\infty S b_A}{I_{yy}} C_M + \frac{M_y}{I_{yy}} + \frac{I_{zz} - I_{xx}}{I_{yy}} pr - \frac{I_{xz}}{I_{yy}}\left(p^2 - r^2\right) \\
\dot{\theta} &= q\cos\phi - r\sin\phi \\
\dot{h} &= V\cos\alpha\cos\beta\sin\theta - V\sin\beta\cos\theta\sin\phi - \\
&\quad V\sin\alpha\cos\beta\cos\theta\cos\phi
\end{aligned}
\tag{2}
$$

where V is the aircraft velocity, β is the angle of sideslip, p, q, and r represents roll, pitch and yaw rates respectively, θ is pitch angle, ϕ is roll angle, h is aircraft height, q_∞ is dynamic pressure, S is wing reference area, b_A is the mean aerodynamic chord, m is the aircraft mass, P_x and P_z represent engine thrusts, M_y is the thrust moment, g_x and g_z are the gravitational acceleration components.

The changes of drag, lift and pitch moment coefficients during ice accretion process are needed to be modelled for analysing the icing influence on aircraft flight dynamics. Bragg [4] of Illinois University presented a mathematical model to describe the icing influence on aerodynamic derivatives. It has already been utilized in the development of Ice Management System (IMS) and aircraft icing online detection [7,9,16]. This model is not an accurate icing influence model, but it can indicate the variation trend of the aerodynamic derivatives during ice accretion. The ice influence on aerodynamic derivatives can be expressed by:

$$C_*^{\text{iced}} = C_*^{\text{clean}}\left(1 + K_{C_*}\eta_{\text{ice}}\right) \tag{3}$$

C_* represents arbitrary aerodynamic derivative, the superscript "iced" indicates the iced derivative, and "clean" indicates the clean derivative. K_{C_*} represents the coefficient slope which depends on the modified parameter. η_{ice} is the icing severity factor, with $\eta_{\text{ice}}=0$ denoting a clean configuration, and $\eta_{\text{ice}}=1$ denoting wing iced configuration. The curve of η_{ice} represents the ice influence on aerodynamic derivative. Melody [7] gave a continuous accretion model of ice over time. The ice accretion rate is considered as a function of both atmospheric conditions and the amount of ice already accreted. The differential equation is given by:

$$\dot{\eta}_{\text{ice}} = N_1\left(1 + N_2\eta_{\text{ice}}\right)d_\eta \tag{4}$$

$$d_\eta(t) = \frac{1}{2}\left[1 - \cos\left(\frac{2\pi t}{T_{\text{cld}}}\right)\right] \tag{5}$$

$$
\begin{aligned}
N_1 &= \frac{2}{N_2 T_{\text{cld}}}\ln\left[1 + N_2\eta_{\text{ice}}\left(T_{\text{cld}}\right)\right] \\
N_2 &= \frac{\eta_{\text{ice}}\left(T_{\text{cld}}\right) - 2\eta_{\text{ice}}\left(T_{\text{cld}}/2\right)}{\eta_{\text{ice}}^2\left(T_{\text{cld}}/2\right)}
\end{aligned}
\tag{6}
$$

T_{cld} is the ice accretion time. When T_{cld}, $\eta_{\text{ice}}(T_{\text{cld}}/2)$ and $\eta_{\text{ice}}(T_{\text{cld}})$ are given, the curve of arbitrary aerodynamic derivative varying over time can be described by equations (3)~(6). The parameters of $\eta_{\text{ice}}(T_{\text{cld}}/2)$ and $\eta_{\text{ice}}(T_{\text{cld}})$ denote the different ice accretion rates.

2.3. Analysis of icing influence on flight dynamics

A steady-level cruise scenario with ice accreting on the leading edge of wings is studied to analyse the ice influence on dynamic behaviour. Assuming the aircraft maintains a steady-level cruise on the trimmed status, the engine thrusts are used to keep the aircraft stable in the absence of a control system.

The flight status and the ice accretion parameters of this scenario are shown in Table 3. The process noises are considered and assumed to be the independent vertical and horizontal acceleration perturbations. White Gaussian noise with the intensity of $[0.1\text{m/s}, 0.1/V_0]^T$ is used for both V and α, and the other flight status is assumed to have zero process noise.

The aircraft icing flight dynamics model is used to analyse the dynamic behaviour influenced by ice accretion. Figure 3 depicts the simulated flight states of velocity, AOA, pitch rate and height. The results indicate wing ice would raise the aircraft nose, reduce the speed, increase the height and cause the periodic vibration of the aircraft flight states.

Table 3. Flight status and ice accretion parameters of simulation scenario.

Flight States	Values
Height H	5000 m
Velocity V	0.3 Ma
Trimmed elevator angle d_e	1.06 deg
Angle of attack α	0 deg
Pitch angle θ	0 deg
Pitch rate q	0 deg/s
Ice occurring time t_{ice}	200 s
Ice accretion time T_{cld}	200 s
Ice accretion rate $\eta_{\text{ice}}(T_{\text{cld}})$	1.0
Ice accretion rate $\eta_{\text{ice}}(T_{\text{cld}}/2)$	0.7

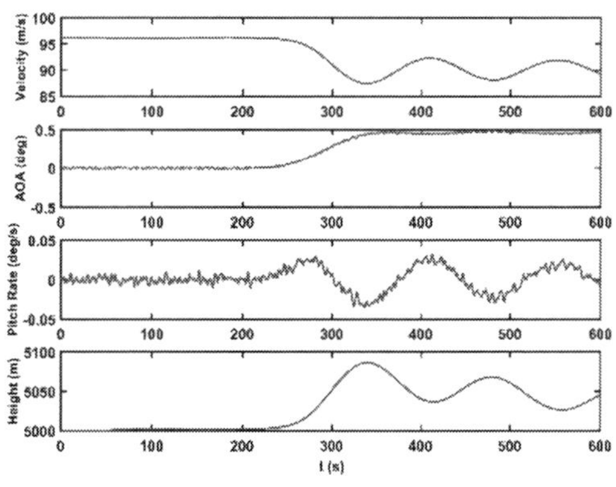

Figure 3. Flight states influenced by wing ice.

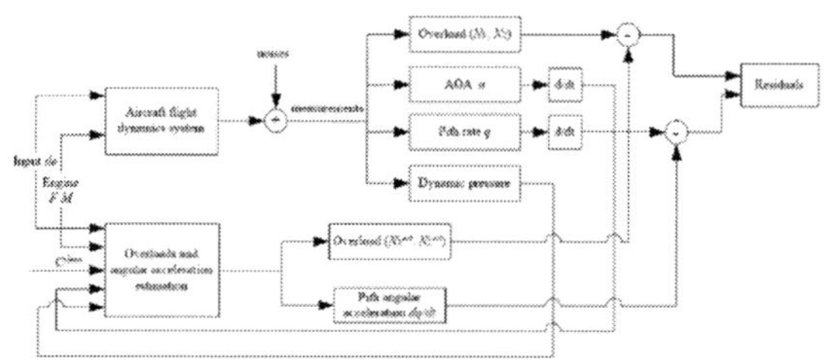

Figure 4. A scheme for generating diagnosis residuals.

3. Aircraft icing statistical diagnosis approach

3.1. Generalized likelihood ratio test

For the realistic problem, the probability density function (PDF) of a signal will be partly unknown due to some unknown parameters. The GLRT method estimates the unknown parameters by Maximum Likelihood estimators (MLEs) to solve these problems [25]. The detection problem can be mathematically expressed as:

$$\begin{aligned} \mathcal{H}_0 &: x[n] = w[n], \quad n = 0,1,\ldots,N-1 \\ \mathcal{H}_1 &: x[n] = A + w[n], \quad n = 0,1,\ldots,N-1 \end{aligned} \tag{7}$$

Where x denotes the testing signal, A is unknown with $-\infty<A<\infty$ and $w[n]$ is white Gaussian noise with unknown variance σ^2. N is the window size. The $\mathcal{H}_0$ hypothesis describes the case where the unknown parameter A equals zero, whereas the alternative hypothesis $\mathcal{H}_1$ describes the case where the unknown parameter has an offset from zero. Meanwhile, the signal contains noise with unknown statistical characteristics in both hypotheses.

The GLRT can be used to distinguish between the two hypotheses. Based on the likelihood ratio between the probability of the two hypotheses, the test statistic for the problem in equaiton (7) decides $\mathcal{H}_1$ in a given window size of data if

$$L_G(x) = \frac{p\left(x;\hat{A},\hat{\sigma}_1^2,\mathcal{H}_1\right)}{p\left(x;\hat{\sigma}_0^2,\mathcal{H}_0\right)} > \gamma \tag{8}$$

Where $\left(\hat{A},\hat{\sigma}_1^2\right)$ is the MLE of the parameters $\left(A,\sigma^2\right)$ under $\mathcal{H}_1$ and $\hat{\sigma}_0^2$ is the MLE of the parameter σ^2 under $\mathcal{H}_0$. γ denotes the threshold. With the MLEs the following test statistic can be derived and its asymptotic PDF is

$$T(x) = N\ln\left(1+\frac{\bar{x}}{\hat{\sigma}_1^2}\right) : \begin{cases} \chi_1^2 & \text{under } \mathcal{H}_0 \\ \chi_1^2(\lambda) & \text{under } \mathcal{H}_1 \end{cases} \tag{9}$$

Where the MLE of the unknown parameters can be written as

$$\begin{aligned} \hat{A} &= \bar{x} \\ \hat{\sigma}_1^2 &= (1/N)\sum_{n=0}^{N-1}\left(x[n]-\bar{x}\right)^2 \end{aligned} \tag{10}$$

The noncentrality parameter $\lambda = NA^2/\sigma^2$. Then the threshold can be determined according to the NP theorem, that is for a given $P_{FA}=\alpha$, the threshold γ is found from equation (11) and maximizes the probability of detection PD under $\mathcal{H}_1$.

$$P_{FA} = \int_{\{x:L_G(x)>\gamma\}} p\left(x;\mathcal{H}_0\right)dx \tag{11}$$

Where $p(\cdot)$ is the probability distribution function of a given test statistic.

3.2. Residual analysis

Ice accreting on the leading edge of wings changes the aircraft structure and causes the variation of the aerodynamic parameters. Although the ice accretion causes the obviously periodic vibration of the aircraft flight states, it doesn't mean the flight states are suitable for the statistical diagnosis. Because the state variables are the integral or multiple integral results of the changing aerodynamic parameters, which means the bias existing in state signals is not only related to the variation of the aerodynamic parameters, but also related to the noises and accumulative errors. Thus, the signal directly related to the changing aerodynamic parameters is needed for statistical diagnosis, such as the axial and normal overloads presented by Sorensen [24]. The overloads and angular accelerations are directly affected by aerodynamic parameters, their changes indicate the changing of forces and moments. If the engine forces and moments are all known, the overloads and angular accelerations are the perfect observing objects for icing detection. Although the icing statistical diagnosis based on overloads has already been proved to be efficient and effective, the feasibility of diagnosing by angular accelerations has not been discussed yet.

Figure 4 gives a scheme of generating diagnosis residuals for realizing the in-flight icing detection. In this scheme, the inputs, engine forces/moments and measurements of aircraft dynamic system are used to estimate the overloads and angular accelerations based on the clean aerodynamic characteristics, then the estimated values are compared with the measured values to generate the diagnosis residuals.

The estimation of overloads and angular accelerations is directly based on the dynamic equations, and for the problem in this paper, it can be calculated by equation (12).

$$
\begin{aligned}
N_x^{est} &= \frac{\left[-q_\infty S\left(C_D^{clean}\cos\alpha - C_L^{clean}\sin\alpha\right)+P_x\right]}{mg_0} \\
N_z^{est} &= \frac{\left[-q_\infty S\left(C_D^{clean}\sin\alpha + C_L^{clean}\cos\alpha\right)+P_z\right]}{mg_0} \\
\dot{q}^{est} &= \frac{q_\infty S b_A C_M^{clean} + M_y}{I_{yy}}
\end{aligned}
\qquad (12)
$$

Where g_0 is the gravitational constant. If the ice accretion causes the aerodynamic coefficients displaying unexpected changes, a bias will be introduced into the residuals, and then the statistical diagnosis method can be used to detect the changes in the residual signals for giving an icing alarm.

The simulation scenario in Table 3 is utilized to analyse the residuals. Here the measurement noises are considered additionally. The white Gaussian noises are assumed, and the standard deviations are listed in Table 4 by referring to the measurement noises of A340 [12]. Figure 5 gives the three residuals of this simulation scenario. Except for the residual of N_z, the ice influence can be barely seen from the other two residuals under the random noises. Then the normal distribution parameters under clean and iced intervals are estimated to illustrate the ice influence, the mean μ and standard deviation σ are given in the figure respectively for the clean intervals of [0, 200]s and iced intervals of [400, 600]s. The results show the residuals have obvious offsets of mean values after the ice accretion, but the standard deviations are almost unaffected.

Table 4. Standard deviations of measurement noises.

Standard Deviations	**Values**
Overload N_x	0.01
Overload N_z	0.01
Dynamic pressure q_∞	1e-3 kg/(m·s^2)
Angle of attack α	0.0056 deg
Pitch rate q	0.0069 deg/s

3.3. Inflight icing detection performance assessment

Due to the lack of prior information on the noises, the inflight icing detection is a typical statistical decision problem with unknown parameters. The GLRT method is suitable for giving an inflight icing alarm. The assessment of GLRT performance in inflight icing detection is focused on giving a timely alarm. The three generated residuals are used to monitor the icing situation. The residual data segments of a given window size are successively taken out for statistical diagnosis. When the test statistic exceeds the threshold, the icing situation is decided.

The simulation residuals in Figure 5 are used to assess the performance of GLRT method. The trade-off between a high probability of detection P_D and a low probability of false alarm P_{FA} is needed to be concerned for GLRT diagnosis. In this paper a small P_{FA} of 1e-6 is used, then the right-tail probability of χ_1^2 distribution under $\mathcal{H}_0$ is used to calculate the threshold. For this problem, the threshold is constant and equals 23.93. The influence of window size is considered. Long time intervals are not good for alarming timely. Here the window size range of [1, 40]s is chosen and the sampling frequency is 100Hz. The results of alarming time and probability of detection P_D under different window sizes are shown in Figure 6, in this figure the detection results of three residuals are compared. The results indicate the small window size will degrade the detection performance with the longer alarming time delay and the lower probability of detection. Detecting by the residual of N_x has the longer time delay than by the other two residuals. Diagnosing by the residual of N_z has higher probability of detection P_D than by the other two residuals. Increasing the window size reduces the alarming time to the range of [270, 290]s by the residuals of N_z and dq/dt. The results also show that diagnosing by residual N_z is more reliable than by the other two residuals, it is probably because the aircraft wing ice causes the more obvious normal offset.

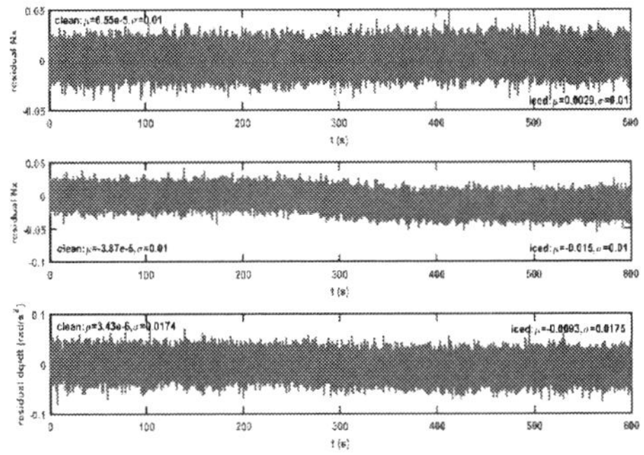

Figure 5. Diagnosis residuals analysis.

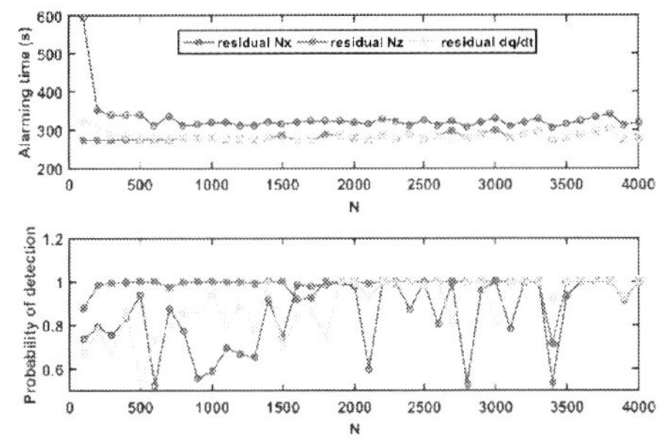

Figure 6. Icing alarming time and probability of detection under different window sizes.

4. Conclusion

This paper addresses the issue of detecting ice when it occurs in the aircraft flight. A NP based statistical decision approach is presented to give a quick alarm of icing. The dynamic behaviour under icing is modelled and analysed for the icing detector design. Then the axial and normal overloads and pitch angular acceleration which are directly affected by the aerodynamic parameters are considered to be the diagnosis residuals, and their generation way is discussed. The GLRT method is used in the inflight icing detection, its performance is assessed by a flight scenario of an airbus with ice accreting on the leading edge of wings. The results show the three residuals are all effective in icing detection. Due to the normal offset is more obvious, diagnosing by residual N_z is more reliable than by the other two residuals.

The research in this paper is a preliminary work of introducing the statistical decision approach into inflight icing detection, the primary purpose of this paper is to verify the feasibility of this statistical decision approach. However, considering the complicated environment factors in real flight, there still have many studies to be done before applying the approach in real inflight icing detection.

Acknowledgments

We are very grateful to the research group of Professor Jie Li of Northwestern Polytechnical University to provide the aerodynamic characteristics of the airbus for this project.

References

[1] Cao Y H, Wu Z L, Su Y and Xu Z D 2015 Aircraft flight characteristics in icing conditions *Progress in Aerospace Sciences* **74** pp 62–80
[2] Goraj Z 2004 *Proc. the 24th Congress of International Council of the Aeronautical Sciences* (Yokohama, Japan) Paper ICAS 2004-7.5.1 (I.L.)
[3] Jarvinen P 2007 *Proc. the 45th AIAA Aerospace Sciences Meeting and Exhibit* (Reno, NV) Paper AIAA-2007-696.
[4] Bragg M B, Basar T and Perkins W R 2002 *Proc. the 40th AIAA Aerospace Sciences Meeting and Exhibit* (Reno, NV) Paper AIAA-2002-0813
[5] Wenz A and Johansen T A 2016 *Proc. 2016 IEEE Conf. on Control Applications* (Buenos Aires) pp 230-236
[6] Melody J W, Basar T and Perkins W R 2000 Parameter identification for inflight detection and characterization of aircraft icing *Control Engineering Practice* **8(9)** pp 985-1001
[7] Melody J W, Hillbrand T and Basar T 2001 H∞ parameter identification for inflight detection of aircraft icing: The time-varying case *Control Engineering Practice* **9(12)** pp 1327-1335

[8] Melody J W 2004 *Inflight characterization of aircraft icing Ph.D. dissertation* (Illinois Urbana: Dept. Elect. Eng., Univ. of Illinois at Urbana-Champaign)
[9] Cristofaro A, Johansen T A and Aguiar A P 2015 *Proc. European Control Conf.* (Linz) pp 1651-1656
[10] Cristofaro A, Johansen T A and Aguiar A P 2017 Icing detection and identification for unmanned aerial vehicles using adaptive nested multiple models *International Journal of Adaptive Control and Signal Processing* **31(11)** pp 1584-1607
[11] Ying S B, Ge T and Ai J L 2013 H∞ parameter identification and H2 feedback control synthesizing for inflight aircraft icing *J. Shanghai Jiaotong Univ (Sci)* **18(3)** pp 317-325
[12] Caliskan F, Aykan R and Hajiyev C 2008 Aircraft Icing Detection, Identification, and Reconfigurable Control Based on Kalman Filtering and Neural Networks *Journal of Aerospace Engineering* **21(2)** pp 51–60
[13] Caliskan F 2007 Neural network based icing identification and fault tolerant control of a 340 aircraft *International Journal of Electrical, Computer, and Systems Engineering* **1(2)** pp 98–103
[14] Schuchard E A, Melody J W and Basar T 2000 *Proc. 38th Aerospace Sciences Meeting and Exhibit* (Reno, NV) Paper AIAA-2000-0361
[15] Dong Y Q and Ai J L 2013 Research on inflight parameter identification and icing location detection of the aircraft *Aerospace Science and Technology* **29(1)** pp 305-312
[16] Dong Y Q and Ai J L 2014 Inflight parameter identification and icing location detection of the aircraft: The time-varying case *Journal of Control Science and Engineering* **2014** pp 1-11
[17] Shen Q, Jiang B and Cocquempot V 2013 Adaptive fault-tolerant backstepping control against actuator gain faults and its applications to an aircraft longitudinal motion dynamics *International Journal of Robust and Nonlinear Control* **23(15)** pp 1753-1779
[18] Shen Q, Jiang B and Shi P 2014 Adaptive fault diagnosis for T-S fuzzy systems with sensor faults and system performance analysis *IEEE Trans. On Fuzzy Systems* **22(2)** pp 274-285
[19] Miller R H and Ribbens W B 1999 *Proc. 37th Aerospace Sciences Meeting and Exhibit* (Reno, NV) Paper AIAA-99-0637
[20] Tousi M M and Khorasani K 2009 *Proc. IEEE SysCon 2009* (Vancouver) pp 302–307
[21] Tousi M M and Khorasani K 2011 *Proc. IEEE Int. SysCon 2011* (Montreal) pp 428–434
[22] Caliskan F and Hajiyev C 2013 A review of in-flight detection and identification of aircraft icing and reconfigurable control *Progress in Aerospace Sciences* **60** pp 12–34
[23] Mehra R K and Peschon J 1971 An innovations approach to fault detection and diagnosis in dynamic systems *Automatica* **7(5)** pp 637-640
[24] Sorensen K L, Blanke M and Johansen T A 2015 Diagnosis of wing icing through lift and drag coefficient change detection for small unmanned aircraft *IFAC-PapersOnLine* **48(21)** pp 541-546
[25] Kay S M 1998 *Fundamentals of Statistical Signal Processing: Detection Theory.* (New Jersey, USA: Prentice Hall PTR) pp 341–349

IOP Conf. Series: Materials Science and Engineering **751** (2020) 012055 doi:10.1088/1757-899X/751/1/012055

Feasibility and Uncertainty Analysis of Constitutive Parameters Identification for Composite Materials Using Displacement Field Data

Zhaoyan Liu[a], Liu Liu[b*] and Tiren He[c]

[a] Beijing Institute of Spacecraft Environment Engineering, Beijing, 100094, China
[b] School of Aerospace Engineering, Beijing Institute of Technology, Beijing, China
[c] Beijing Key Laboratory of Commercial Aircraft Structures and Composite Materials, BASTRI, COMAC

*Email: liuliu@bit.edu.cn

Abstract. Due to the normalized average strain error is remarkably higher than the displacement error, a new cost-effective method based on the Levenberg-Marquardt method, which is a gradient-based optimization method, was presented in this paper. The unknown constitutive properties are determined through minimization of the square difference between the reconstructed DIC-measured and the FEM-calculated displacements. The feasibility analysis of this method is presented by synthetic displacement data. The covariance matrix of the extracted material constitutive parameters has been given explicitly. It has been proved that this method can significantly reduce the uncertainty of the extracted material constitutive parameter induced by the approximation error due to reconstruction algorithm and the measurement noise. The material constants show strong robustness to the measurement noise. And the optimization procedure is not sensitive to the formulation and parameters in the initial approximation of the constitutive model. A sensitive matrix was derivation and evaluation iterative in this method to improve the convergence of the optimization procedure.

1. Introduction

It is a typical inverse problem to obtain the constitutive relationship of composite materials by mechanical experiments. In solving this inverse problem, a lot of related research has been done and lots of important results have been achieved [1].

Grédiac and Pierron [2] proposed the virtual field method (VFM), and first carried out a series of theoretical and experimental studies. The most significant advantage of this method is that the explicit relationship between measured data and unknown material parameters can be established, which avoids a large number of numerical iterations and correction processes and greatly improves the computational efficiency[3]. However, the VEM method depends strongly on the initial assumptions of the constitutive parameters. If the initial assumptions deviate greatly, the correct recognition results may not be obtained.

The finite element model updating technique identifies multiple parameters in the constitutive relationship iteratively by minimizing the objective function, which is established by the variance between experimental results and numerical results. Molimard[4] presents a first step towards a global approach of mechanical identification for composite materials by open-hole tensile test using

Levenberg-Marquardt(L-M method) algorithm. Makeev[5] applies the non-standard short beam shear test combined with the full-field deformation data to identify the mechanical properties of composite firstly. Comparing with the V-notched shear test, the results validate the superiority and reliability of SBS test in obtaining the non-linear in-plane shear constitutive relation of materials. He [6] presents an uncertainly analysis about the measurement noise and reconstruction parameters on extracted material properties. It is evidenced that there resist an appropriate ROI and the number of images, from which reliable material parameters can be identified.

Various types of constitutive models has been used in the identification process, such as Johnson-Cook model[7-9], power-law model[10], Zerilli-Armstrong model[11]. In previous study, Ramberg-Osgood expression has been used to characterize nonlinear shear relationship[5,6].

Based on the above research results, this paper mainly studies the feasibility analysis of identification constitutive parameters by constructing the objective function with measurement and numerical displacement data. The robustness of the identification method is carried out from the aspects of the composition of the objective function, the identification method and the initial value of the identification process, respectively.

2. Experiment

In a short beam shear test, which following the ASTM D2344 standard [12], a customized specimen is loaded under three point bending at room temperature (figure.1 (a)). The specimen is 6.4mm thick and 6.4 mm width, the diameter of loading nose is 50.8mm. The support diameter is 3.2mm. The specimen length is 44.5mm and the support length is 30.5mm. The specimen were loaded at 1.27mm/min crosshead displacement rate in the transverse (1-3 plane) material direction. The material direction is shown in figure.1 (b).

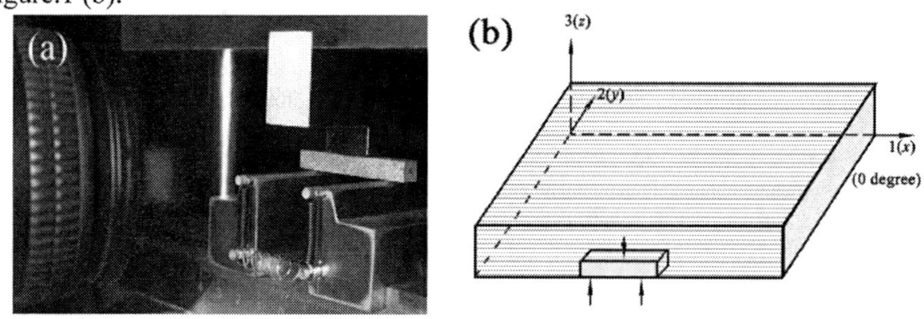

Figure 1. (a) SBS test setup combined with DIC measurement system; (b)Direction of composite.

In order to quantify the magnification of random errors, a loading free test is performed, and then the reconstructed strain field is obtained by full-field finite element approximation reconstruction method[6,13]. The ratio of the average normalized errors of displacement data and strain data under various reconstruction parameters with three different random error levels is shown in figure 2. It is illustrated that the error level of strain field is much higher than that of displacement field. Therefore, it can be inferred that using displacement data instead of strain data to identify constitutive parameter can significantly reduce the impact of errors on identification results.

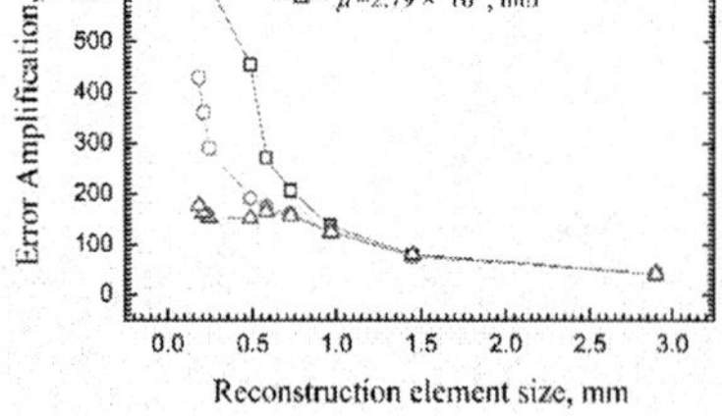

Figure 2. Error amplification scale of normalized average error.

3. Synthetic displacement data of SBS test

A 1/4 three-dimensional finite element model with symmetry boundary condition along width direction and length direction is established. The loading nose and supports are assumed as analytical rigid body. There are totally 58140 elements in the model. The element type is C3D8I. The model and the location of ROI is shown in figure 3. ROI is a 2mm width rectangular in the middle of loading nose and support. There are totally 3072 nodes on the surface of ROI. The maximum load is P=421N after 10 continuous loading steps. The constitutive parameters are provided by material suppliers. The random error with mean value of 0 and standard deviation of $\mu=2.79\times10^{-5}$mm are superimposed on the displacement data to generate the synthetic data.

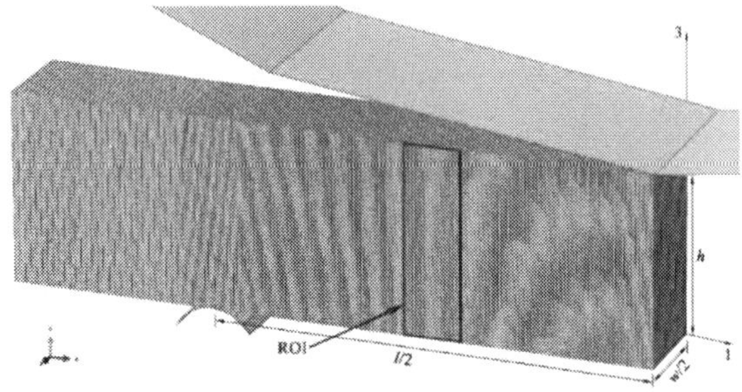

Figure 3. Finite element model and location of ROI.

4. Finite element model updating

The flowchart in figure 4 shows the identification process. The objective function is formed by the variance of synthetic displacement and numerical displacement. The convergence criteria to stop the iterative procedure is the relative value of identified parameters updates is less than 0.5% or the variation of the objective function value is less than 1%. The process of calculation and formula derivation can be referred to previous study[6,14].

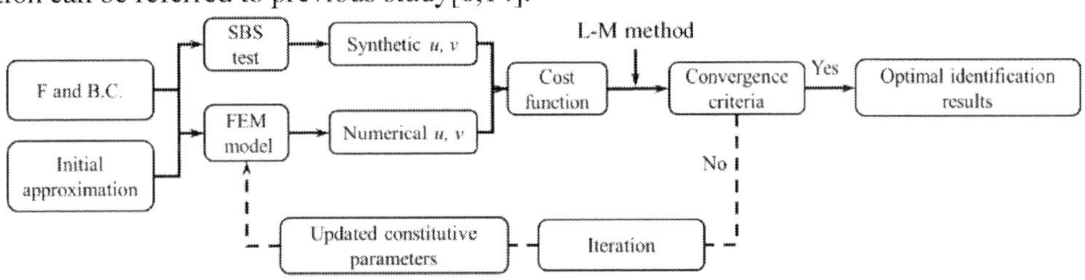

Figure 4. Schematic flowchart for the material constant identification procedure

5. Results

The parameters of the identification process are shown in table 1.

Table 1. True values, initial values, identification results, standard deviations and relative deviations

	E_T(GPa)	E_C(GPa)	G_{13}(MPa)	υ_{13}
Truth value	187.93	142.35	4368.79	0.36
Initial approximation	239.11	102.24	5240.62	0.27
Convergence results	187.6916	142.2857	4370.518	0.359
Standard deviation	0.104	0.0283	0.532	0.00068
COV	0.13%	0.05%	0.04%	0.28%

Figure 5(a) shows the variation of the objective function value with iterations. The value decreases after one iteration and convergence after 36 iterations. Figure 5(b) shows iteration results for the Young's module in tension and compression direction, the shear module and Poisson's ratio with the iteration number. The Young's module in tension and compression direction and the shear module convergence after 12 iterations.

6. Uncertainly analysis of the extracted constitutive parameters

6.1. Influence of the cost function

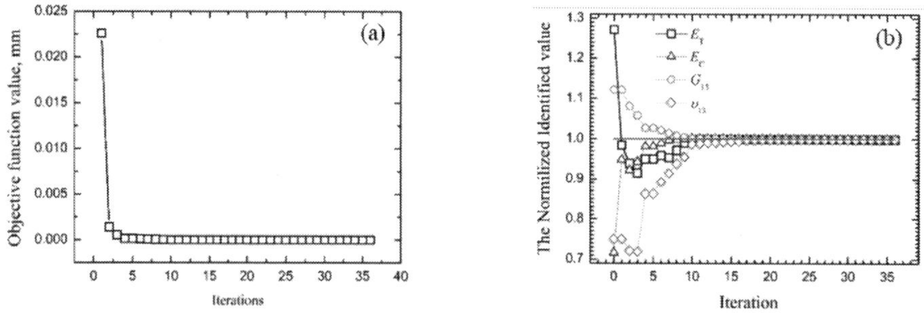

Figure 5. Variation of values with the iteration numbers (a) Objective function value; (b) Normalized identified results.

The identification results of displacement data and strain data are compared in table 2. It can be seen that the standard deviation of the identification results based on reconstructed displacement data is much smaller than that of reconstructed strain data.

Table 2. Comparison of the identified results from two different objective function

		E_T(GPa)	E_C(GPa)	G_{13}(MPa)	υ_{13}
	Real parameters	187.93	142.35	4368.79	0.36
displacement	mean	187.69	142.29	4371.24	0.36
	Standard deviation	0.00104	0.000283	0.000532	0.0068
	Relative deviation	0.13%	0.04%	0.06%	0.28%
strain	mean	197.41	142.13	4273.01	0.34
	Standard deviation	0.1026	0.0632	0.0897	0.082
	Relative deviation	5.04%	0.15%	2.19%	5.56%

6.2. Influence of the identification method

In table 3, the CPU time of L-M method, which is 36 iterations and 370minutes, are compared with those of simplex search method without sensitivity matrix which is 44 iterations and 1800 minutes. The L-M method with sensitivity matrix saves 4/5 of the computation time since the introduction of sensitivity matrix greatly improves the optimization efficiency.

Although the iterative process costs 370 minutes, it has advantages than the analytical method such as the accuracy aspect[15,16]. Besides, it is required to update the properties in the identification process[17].

Table 3. Comparison of identified results and CPU time of different algorithm

		E_T(GPa)	E_C(GPa)	G_{13}(MPa)	υ_{13}	Iterations	CPU time (minutes)
	Real parameters	187.93	142.35	4368.79	0.36		
Simple search method	Convergence results	187.61	142.27	4371.07	0.362	44	1800
	Relative deviation	0.17%	0.06%	0.05%	0.67%		
L-M method	Convergence results	187.69	142.29	4370.24	0.359	36	370
	Relative deviation	0.12%	0.04%	0.03%	0.27%		

6.3. Robustness to the initial approximation and random error

In order to analyze the sensitivity of the method to initial approximation, various initial parameters in ±30% range of the true values are used as the initial values to identify the constitutive parameters. The convergence results, standard deviation, relative deviation are shown in figure 6. It is shown that the identification results show strong robustness to the initial approximation. The relative deviation of Poisson's ratio is slightly larger, but also less than 10%.

Figure 7 shows the mean and standard deviation from synthetic displacement data with 1 to 10 times of Gauss distribution random errors. It can be seen from figure 7 that the mean values of E_T, E_C and G_{13} are less sensitive to the random error. Even if the random error is increased to 10 times, the mean values of the constitutive parameters obtained by inversion are basically consistent with the true values of the constitutive parameters, and the standard deviation is less than 0.5%.

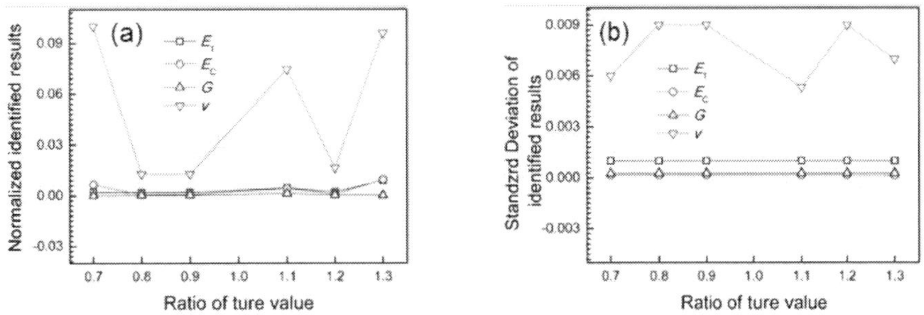

Figure 6. Parameters of various initial values (a) Normalized mean value (b) Standard deviation

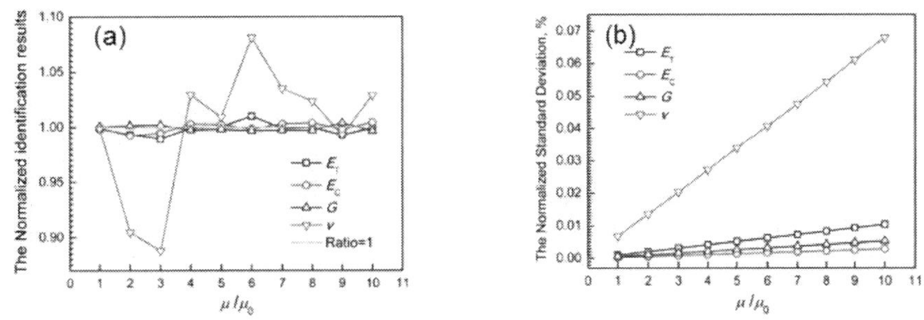

Figure 7. Identified results with various random error. (a) Mean value; (b) Standard deviation.

On the contrary, the mean value of identification results of Young's modulus, which identified by strain field data, is changing with the level of random noise due to the lower positive strain level involved in identification of Young's modulus. When the noise level increases to three times, it is impossible to identify the theoretical true value while the theoretical true value of the parameters can still be obtained when the noise level reaches 10 times. This shows that the identification with displacement as the objective function has strong robustness to random errors in the data, especially for the identification of constitutive parameters under low strain conditions.

On the other side, the standard deviation of identification results increases linearly with the increase of the level of random error. As for the Poisson's ratio, which is more difficult to invert, it can be seen from the figure 7 that the normalized mean of Poisson's ratio identification results fluctuates obviously, and the standard deviation from the real value is close to 10%, which is much larger than the standard deviation of the normalized mean of Young's modulus and shear modulus. This is because Poisson's ratio is determined by the ratio of displacement along the thickness direction to displacement along the fiber direction. Compared with other constitutive parameters, the identification uncertainty is higher, which results in a larger standard deviation between the inversion Poisson's ratio and the real value.

7. Conclusion

In this paper, the constitutive parameters of carbon fiber reinforcement composite (IM7/8552) unidirectional composite laminates are identified by using finite element model updating method combined with L-M method with synthetic data of random error of μ=2.79×10^{-5}mm superimposed in the 2 mm width area between loading nose and support point. Different from the previous identification process, the objective function is constructed by using the variance between the numerical displacement data of the finite element model and the reconstructed displacement data. In the identification process, the non-linear L-M optimization method is adopted, and the explicit expression of sensitivity matrix in the optimization process is given by numerical difference. Four elastic constants including material Poisson's ratio are identified by this method.

The study shows that this method has strong robustness to the random error and the initial approximation, can reduces the influence of errors on the identification process significantly comparing with the identification method which used the strain field data and the explicit expression of sensitivity matrix can greatly speeds up the convergence speed of the identification process. Since it is not necessary to obtain strain field data by differential calculation from displacement field data, directly constructing objective function from displacement data can save calculation cost and improve efficiency.

This identification method will be improved by induced user subroutine UMAT to FEM model established by ABAQUS to identify shear nonlinear stress-strain relationship in the future. As the identification method shows strong robustness to the random error and the initial approximation, it can solve the problem of constitutive parameters identification in low strain or high noise level experiments, such as ceramic matrix composites.

Reference

[1] Sutton M A, Wolters W J, et al 1983 *Determination of Displacement Using an Improved Digital Correlation Method.* The New International Economic Order and UNCTAD IV: Bangladesh Institute of Development Studies, 133-39.

[2] Pierron F, Grédiac M 2012 *The Virtual Fields Method.* New York: Springer.

[3] Pierron F, Avril S and Tran V T 2010 Extension of the virtual fields method to elasto-plastic material identification with cyclic loads and kinematic hardening *J. International Journal of Solids & Structures*, 47(22): 2993-3010.

[4] Molimard J, Riche R L and Vautrin A, et al 2005 Identification of the four orthotropic plate stiffnesses using a single open-hole tensile test *J. Experimental Mechanics*, 45(5): 404-11.

[5] Carpentier, A P 2013 Advanced materials characterization based on full field deformation measurements D. PhD thesis. University of Texas at Arlington.

[6] He T, Liu L and Makeev A 2017 Uncertainty Analysis in Composite Material Properties Characterization Using Digital Image Correlation and Finite Element Model Updating *J. Composite Structures*, 184: 337-51.

[7] Jinqiang N, Vinh N and Yong H 2018 Inverse determination of Johnson–Cook model constants of ultra-fine-grained titanium based on chip formation model and iterative gradient search *J. Advanced Manufacturing Technology,* 99:1131-40.

[8] Ning J, Liang S Y 2018 Model-driven determination of Johnson-Cook material constants using temperature and force measurements *J. Advanced Manufacturing Technology,* 97:1053-60.

[9] Ning J, Liang S Y 2019 Inverse identification of Johnson-Cook material constants based on modified chip formation model and iterative gradient search using temperature and force measurements *J. Advanced Manufacturing Technology,* 102:2865-76.

[10] Qi C, Feng F and Xiao-Song W U 2015 Numerical Research on Flow Characteristics of Kerosene Gel in Different Kinds of Pipeline Based on Power-law Constitutive Model *J. Acta Armamentarii*, 36:242-49.

[11] He A, Xie G and Zhang H 2014 A modified Zerilli–Armstrong constitutive model to predict hot deformation behavior of 20CrMo alloy steel *J. Materials & Design*, 56:122-27.

[12] ASTM Standard D 2344/D 2344M, Standard test method for short-beam strength of polymer

IOP Conf. Series: Materials Science and Engineering **751** (2020) 012055 doi:10.1088/1757-899X/751/1/012055

matrix composite materials and their laminates S. West Conshohocken, PA: ASTM International, 2006.
[13] Bornert M, Brémand F and Doumalin P, et al 2009 Assessment of digital image correlation measurement errors: Methodology and results *J. Experimental Mechanics*, 49(3): 353-370.
[14] Gras R, Leclerc H and Hild F 2015 Identification of a set of macroscopic elastic parameters in a 3D woven composite: Uncertainty analysis and regularization J. International Journal of Solids & Structures, 55: 2-16.
[15] He Y, Makeev A and Shonkwiler B 2012 Characterization of nonlinear shear properties for composite materials using digital image correlation and finite element analysis *J. Composites Science and Technology*, 73:64-71.
[16] Makeev A, He Y and Carpentier P 2012. A method for measurement of multiple constitutive properties for composite materials *J. Composites Part A: Applied Science and Manufacturing*, 43(12):2199-2210.
[17] He Y, Makeev A 2014 Nonlinear shear behavior and interlaminar shear strength of unidirectional polymer matrix composites: A numerical study *J. Solids & Structures*, 51(6):1263-1273.

IOP Conf. Series: Materials Science and Engineering **751** (2020) 012056 doi:10.1088/1757-899X/751/1/012056

Characterization and Uncertainty Analysis of the Interlaminar Inelastic Properties of Unidirectional Fibre-reinforced Composites

Tiren He[a,*] , Liu Liu[b*] and Jifeng Xu[a]

[a]Beijing Key Laboratory of Commercial Aircraft Structures and Composite Materials, BASTRI, COMAC
[b] School of Aerospace Engineering, Beijing Institute of Technology, Beijing, China

[*]Email: liuliu@bit.edu.cn

Abstract. The compression failure mode has been achieved through changing the SBS (Short Beam Shear) experiment parameters, and the nonlinear constitutive relationship along thickness direction for the typical specimens loaded in 1-3 principal material plane has been computed iterative by FEMU with the reconstruction strain field and the numerical simulated stress data of FEM model. The uncertainly of the constitutive parameters is induced by the measurement system noise in the DIC technique and the approximation error in the displacements and strains smoothing algorithm. The covariance matrix of the extracted material constitutive parameters has been given explicitly. Three material constitutive parameters were identified from a customized short-shear experiment simultaneously using an estimated optimal reconstruction mesh size as an illustration. Sensitivity of measurement noise and reconstruction parameter on extracted material properties has been investigated. The effects of region of interest (ROI) and DIC image number on uncertainties of extracted material properties have been addressed. It is suggested that there exist an appropriate ROI and the number of images, from which reliable material parameters can be identified, but much more data used in identification process always lead to smaller standard deviation and COV. It is observed that the material constants used to characterize the interlaminar stress-strain behavior show strong robustness to the measurement noise. Another key finding is that the reconstruction parameter in the global finite-element based approximation approach is critical for reliable material properties identification. Its value has to stay close to optimum for guaranteeing reliable identification of material properties.

1. Introduction

The mechanical properties of composite laminates under interlaminar loads have an important influence on the overall mechanical properties and failure modes of composite laminates[1,2]. Lateral failure often occurs first in the loading process and may lead to further damage of other layers of laminates, such as interlaminar delamination failure. With the increasing complexity of composite components under actual working conditions, the analysis of strain distribution and deformation mechanism of composite materials under interlaminar loads is the basis for further study of mechanical properties of composite materials[3-6]. In the short beam shear test[7], the material under the loading nose has a large displacement and strain gradient, and the deformation is more complex[8-11]. Therefore, in the study of identifying the mechanical constitutive parameters of materials by combining the short beam shear test with digital image correlation technology, the ROI far from the loading nose and the supports is mainly used. According to the previous study[12], the reliable constitutive parameters of materials can be obtained by a local region, so the complex high gradient strain region captured by digital image correlation directly under the loading nose is adopted.

In this paper, the constitutive parameters of carbon fiber reinforced resin matrix (IM7/8552) unidirectional composite laminates representing stress-strain relationship along thickness direction is identified by short beam shear test and finite element model updating method. Just like the previous study, the displacement data and strain data under the loading nose are reconstructed by a full-field finite element approximation method and the constitutive parameters describing the non-linear stress-strain relationship along the thickness direction of the composite material are identified by minimizing the variance between the reconstructed strain field data and the numerical computation strain field data of FEM model. The effects of the random error level and the reconstructed parameters on the identified constitutive parameters are discussed.

2. Experiment

The test parameters and equipment are mostly as the same as that in the previous study [12] expect the loading nose diameter, which is 50.8 mm to ensure the compression-shear combined failure mode. The test equipment setup and the material direction are shown in figure 1. The deformation images is captured by the VIC-2D setup in every 5 seconds and a total number of 48 images are acquired for material constitutive constant extraction consequently.

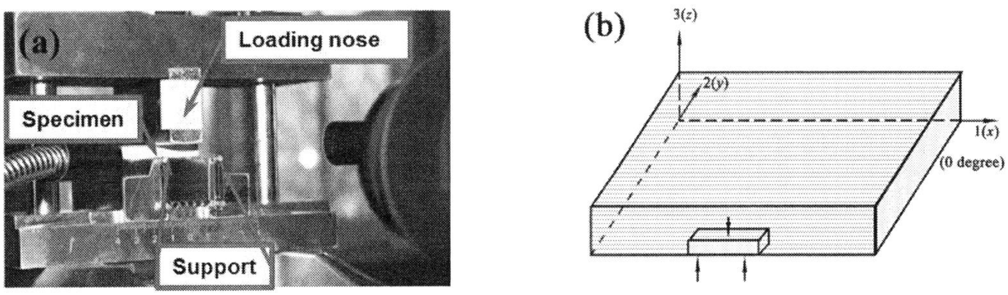

Figure 1 (a) Short beam shear (SBS) experiment setup [12]. (b) Specimen configuration of the SBS test.

The location of the ROI area on the specimen surface is showed in figure 2(a). ROI-2 supplies the stress-strain data for material parameters characterization. Considering edge effect of the reconstruction process, deformation data in a larger area ROI-1 is reconstructed to make sure the explicit of the data in ROI-2. Figure 2(b) shows the shear-compression failure mode of specimen.

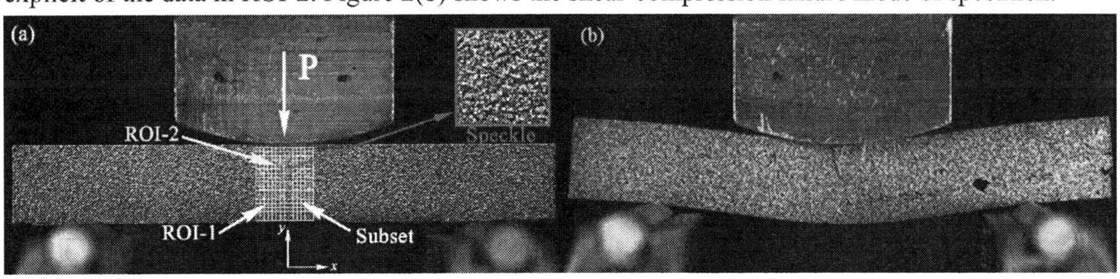

Figure 2. (a) The ROI on the specimen surface; (b) The shear-compression failure mode in SBS test

3. Finite element model updating

The finite element model updating method is applied to identify the interlaminar compressive nonlinear stress-strain relation. The initial approximation of the constitutive parameters are supplied by the material manufacturer. The convergence criteria of the identification process is that the relative value of the parameters updates is less than 0.5% and the variance ratio of the objective function value is less than 1%. The flowchart in figure 3 illustrated the identification process. The objective function and the identification process are as similar as the previous study[11,12].

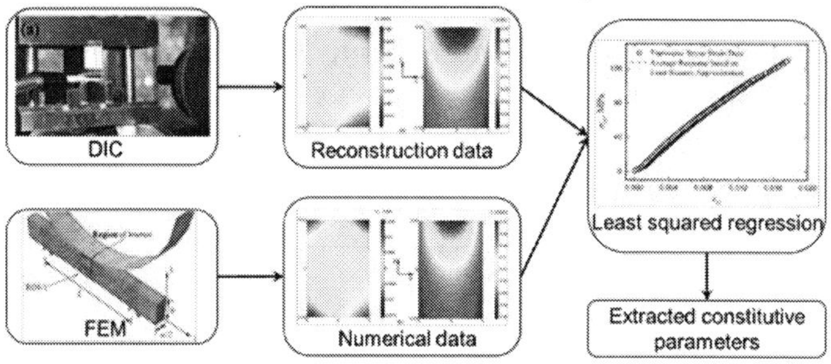

Figure 3. Flowchart of the identification process

4. Finite element sub model of the local high-scale area

4.1. Global model of finite element method

As shown in figure 4(a), a half 3D numerical finite element model with 42000 elements and 48081 nodes is established by ABAQUS. The model involves geometric nonlinearly, material nonlinearly and contact interaction. The symmetric boundary condition is defined at middle width (z=0). The FEM model is meshed with first-order continuum rectangular shape element (C3D8I) and consists of 48 steps to match the number of the images which captured by DIC in SBS test. In order to simulate the material nonlinearly, user subroutine UMAT is applied to the FEM model. The nonlinearly shear properties and interlaminar constitutive relationship are illustrated by Ramberg-Osgood expression.

4.2. Finite element sub model of the specimen

Considering the high level of the stress state and the large gratitude of the strain data under the loading nose of a specimen in SBS test, in this study, a sub model is established to improve the precision of stress calculation and enhance the efficiency of iterative calculation. Only the stress-strain data in a narrow band area (ROI-2), which is shown in figure 4(b), right under the loading nose is considered in the identification process to avoid the uncertain influence of shear stress on the interlaminar stress-strain relationship. The driving variable of the sub model is the displacement of global model. The boundary condition and the loading process of the sub model is the same as the global model[11,12].

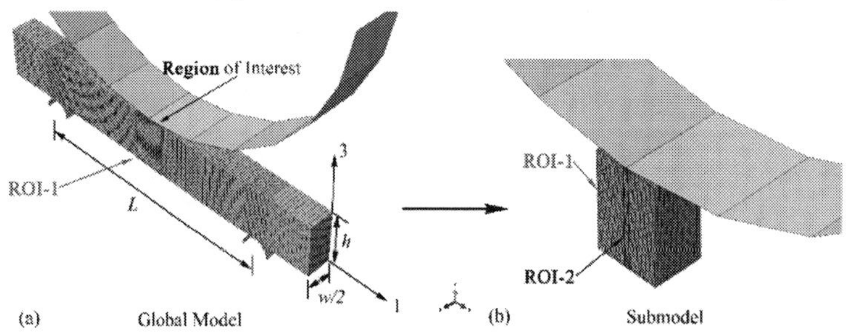

Figure 4. Global modal and sub model of the specimen in SBS test by finite element method

5. Uncertainly analysis

5.1. Displacement measurement and reconstruction

The reconstruction process of full-field displacement and strain field data is followed by the previous study[12]. The standard deviation of the Gaussian white noise is $\mu=5\times10^{-5}$ mm, which is determined by the average confidence margin given by VIC-2D software refer to the analysis between $\eta_{\exp}$ and η, the analysis process is as the same as the previous study[12].

5.2. Results

A least-square regression is used to extract the stress-strain relationship through the FEM-calculated stress and the DIC-reconstructed strain data of all images in the loading process. The convergence expectation, standard deviation and the coefficient variation (COV) of each identified material parameter are listed in table 1. Figure 5 shows the typical nonlinear converged results.

Table 1. Exception and standard deviation of identified material parameter

	$1/E_{33}$(1/MPa)	$1/K_{33}$(1/MPa)	n_{33}
Exception	1/8603.36	1/786.15	3.59
Standard Deviation(SD)	1/(8603.36×84.75)	1/(786.15×68.97)	0.016
COV	1.18%	1.45%	0.45%

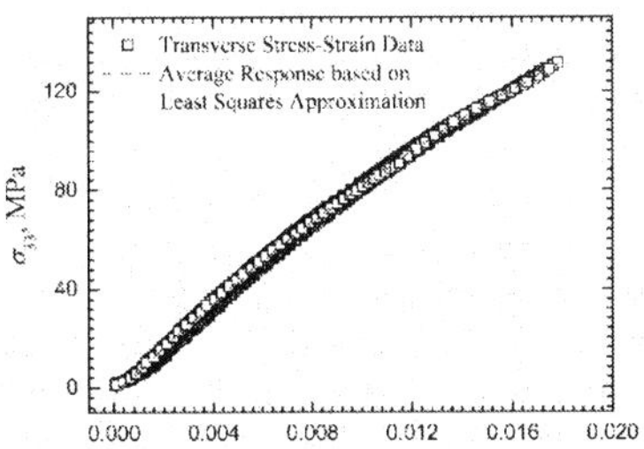

Figure 5. Converged result for material constitutive relationship derived through least-square regression coupling of reconstruction strain and FEM stress

Figure 6 shows the displacement field obtained from DIC and FEM model with the converged identification results respectively.

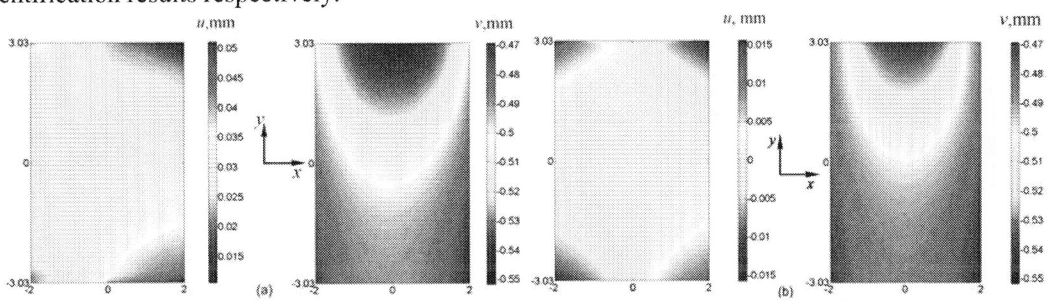

Figure 6. (a) Displacement data obtained from DIC; (b) Displacement data obtained from FEM

The identification process achieved the convergence criteria after 5 iterations, the stress-strain relationship and the objective function value of each iteration are indicated in figure 7(a) and figure 7(b), respectively. The A curve in figure 7(a) shows a linear estimation of the constitutive relationship and it changes to three parameters Ramberg-Osgood relationship after one iteration through nonlinear least-square regression between FEM-calculated stress and reconstruction strain data. It is clearly indicated in figure 7(b) that the identification procedure achieved the convergence criteria after 5 iterations, the objective function value downward rapidly after one iteration and converges gradually.

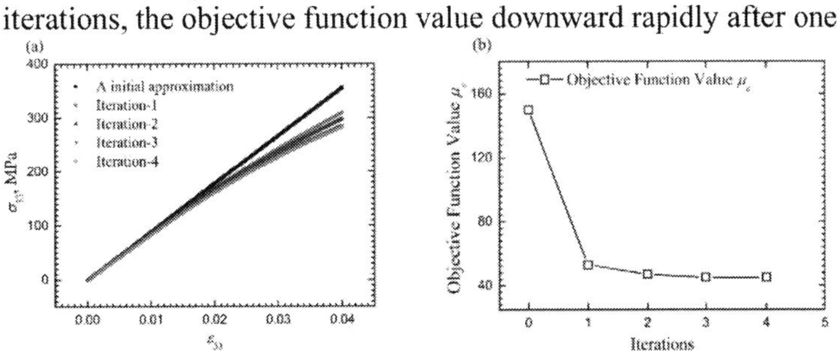

Figure 7. (a) Iteration results for the interlaminar nonlinear constitutive relationship; (b) variance of the nominalized objective function value with iterations

5.3. Uncertainly of the initial approximation

The converged nonlinear interlaminar stress-strain curves obtained for a unidirectional SBS specimen from linear initial approximation (curve A) and over nonlinear initial approximation (curve B) of the material parameters are illustrated in figure 8. The agreement between the convergence results illustrated that the identification procedure is not sensitive to the initial approximation.

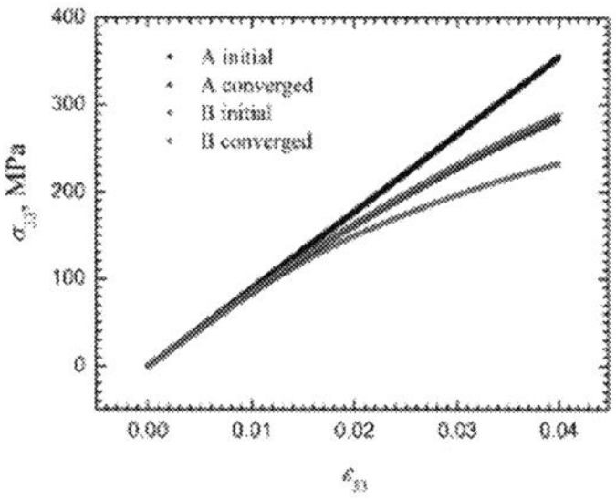

Figure 8. Converged stress-strain curves obtained from two different initial approximation A and B. A uses E_{33}= 8000MPa with an assumption of linear shear behavior in the all three material planes; B uses E_{33}=*8000*MPa, K_{33}=100MPa, n_{33}=2.5 with an assumption of over-nonlinear behavior in interlaminar direction

5.4. Sensitivity to the measurement error and the reconstruction error

The uncertainly of identification results to the random error and reconstruction element mesh are conducted as the same as the previous study[12].

The influence of the displacement measurement error on the identified constitutive parameter are listed in figure 9. The synthetic data is combined with the FEM numerical displacement data and a Gaussian white noise with a standard deviation of μ=5×10^{-5} mm, μ=1.5×10^{-4} mm, μ=2.5×10^{-4} mm, μ=3.5×10^{-4} mm, μ=5×10^{-4} mm, respectively. The reconstruction mesh size remains h=0.18mm in the reconstruction process. It is clearly illustrated that all the three material parameters show strong robustness with the displacement measurement error and keeps very agreement with the nominal value with the μ < 10 μ_0.

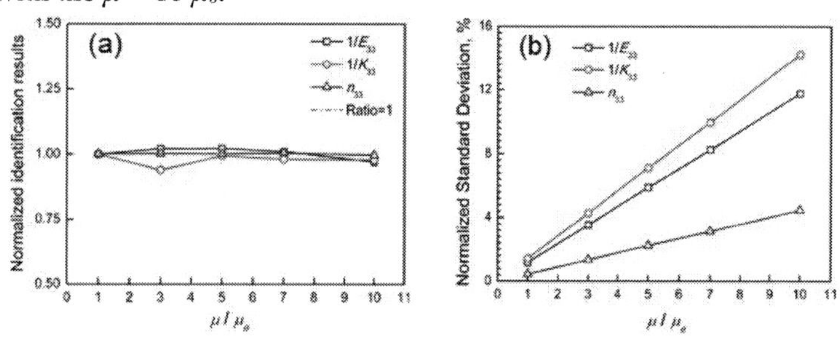

Figure 9. Variations of the normalized identified material constant and the standard deviation with the standard deviation of the measurement error. (a) Mean value; (b) Standard deviation.

Besides, the effects of the reconstruction mesh size on the identified material parameter has been investigated and presented in figure 10. The synthetic data is combined with the FEM numerical displacement data and a Gaussian white noise with a standard deviation of μ_0=5×10^{-5} mm. The results indicated that the identified material parameter are sensitive to reconstruction mesh size. The standard deviation of the identified material parameter decreases as the enlargement of the reconstruction mesh size. It has been pointed out that no parameter can be extracted reliably if an inappropriate mesh size is used for reconstruction. Thus an estimation of an appropriate reconstruction parameter, based on minimization of average strain reconstruction error using synthetic data or other complementary prior information is necessary for improving reliability of material constant identification procedures.

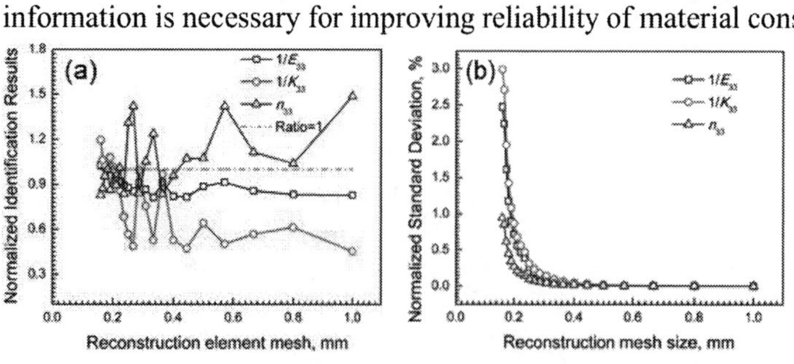

Figure 10. Variations of the normalized identified material constant and its standard deviation with the reconstruction mesh size. (a) Mean value; (b) Standard deviation.

6. Concluding remarks

The constitutive parameters of carbon fiber epoxy resin based unidirectional composite laminates along the thickness direction are obtained by the non-linear least squares regression through the reconstructed strain field data and the numerical stress calculated by finite element model. The Ramberg-Osgood expression is used to assume the constitutive relation along the thickness direction directly below the loading nose. Finally, the mean, standard deviation and variation coefficients of the constitutive parameters are calculated. The mean of identification results is insensitive to random error, and the nonlinear coefficient n_{33} has the best robustness to random error. The standard deviation of identification results increases linearly with the level of random error. The mean of recognition results is very sensitive to the reconstruction parameters. Only when the optimal reconstruction parameters are obtained, the accurate identification results can be obtained. In the process of identification, the simulation data should be obtained from the initial constitutive values and then the approximate optimal size of reconstructed elements should be estimated. This step is indispensable in the process of identifying constitutive parameters by using finite element model updating method and digital image correlation technology.

Acknowledgments
This work is sponsored by the Project funded by China Postdoctoral Science Foundation under Grant NO. (2019M651629). Such support is greatfully acknowledged.

References

[1] Zhou Y, Hosur M, Jeelani S, et al 2012 Fabrication and characterization of carbon fiber reinforced clay/epoxy composite *J. Journal of Materials Science*, 47(12): 5002-12.
[2] Avril S, Pierron F, Pannier Y, et al 2008 Stress reconstruction and constitutive parameter identification in plane-stress elasto-plastic problems using surface measurements of deformation fields *J. Experimental Mechanics*, 48(4): 403-19.
[3] Grédiac M, Toussaint E and Pierron F 2002 Special virtual fields for the direct determination of material parameters with the virtual fields method. 2—Application to in-plane properties *J. International Journal of Solids & Structures*, 39(10): 2707-30.
[4] Grédiac M, Pierron F and Avril S, et al 2010 The virtual fields method for extracting constitutive parameters from full - field measurements: A review *J. Strain*, 42(4): 233-53.
[5] Gibson R F 2010 A review of recent research on mechanics of multifunctional composite materials and structures *J. Composite Structures*, 92(12): 2793-2810.
[6] Zhang Y X, Yang C H 2009 Recent developments in finite element analysis for laminated composite plates *J. Composite Structures*, 88(1): 147-57.
[7] ASTM Standard D 2344/D 2344M, Standard test method for short-beam strength of polymer matrix composite materials and their laminates S. West Conshohocken, PA: ASTM International, 2006.
[8] Carpentier, A P 2013 Advanced materials characterization based on full field deformation measurements D. PhD thesis. University of Texas at Arlington.
[9] He Y, Makeev A and Shonkwiler B 2012 Characterization of nonlinear shear properties for composite materials using digital image correlation and finite element analysis *J. Composites Science & Technology*, 73: 64-71.
[10] He Y, Makeev A 2014 Nonlinear shear behavior and interlaminar shear strength of unidirectional polymer matrix composites: A numerical study *J. International Journal of Solids & Structures*, 51(6): 1263-73.
[11] He T, Liu L, Makeev A, et al 2016 Characterization of stress–strain behavior of composites using digital image correlation and finite element analysis *J. Composite Structures*, 140: 84-93.
[12] He T, Liu L and Makeev A 2017 Uncertainty Analysis in Composite Material Properties Characterization Using Digital Image Correlation and Finite Element Model Updating *J. Composite Structures*, 184: 337-51.

IOP Conf. Series: Materials Science and Engineering **751** (2020) 012057 doi:10.1088/1757-899X/751/1/012057

Stress Analysis and Strength Test of 2500 Fracturing Truck Frame

Lei Li[a], Bin Zhang[b] and Chunming Fan[c]

Baoji Oilfield Machinery Co., Ltd., National Engineering Research Center of Oil and Gas Drilling Equipment, Baoji, China

[a] bslil@cnpc.com.cn, [b] zb_sc@cnpc.com.cn, [c] bsfcm@ cnpc.com.cn

Abstract. For the fracturing trucks, which is an important member of oil and gas fracturing, the stress and strength of fracturing truck's frame have a very significant influence on the continuous fracturing operation. This paper takes the frame of 2500 fracturing truck as the research object, analyses its stress condition, and conducts the static strength test and dynamic strength test of the whole frame of fracturing truck, respectively tests the strain of 15 measuring points of the frame under the fully loaded static condition and the right front wheel lifting condition, and finds out the distribution rule of the stress and strain of the frame. The results show that the maximum stress position of the frame under migration condition is at the root of the rear support of the tank, the total stress is 146.76 MPa, the minimum safety factor reaches 2.35, and the strength of measuring point meets the requirements.

1. Introduction

The fracturing truck is the core equipment of the fracturing operation. Its role is to inject high-pressure fracturing fluid with large displacement into oil and gas wells, so as to fracturing oil and gas. The on-site fracturing puts a high standard on the technical performance of the fracturing truck. It is required that the fracturing truck must have large displacement, high pressure, strong abrasion resistance and corrosion resistance [1]. The frame is the main load-carrying component of the fracturing truck. During the fracturing operation, the fracturing pump will generate great vibration to the frame.

In 2005, Zhao Han and Wei Ying applied ADAMS to simulate the dynamics of a passenger car air suspension and then related mechanical analysis of the spring suspension bracket in ANSYS [2]. In 2009, Zhu Sihong and Xiao Zhijin of Nanjing Agricultural University analyzed multi-body dynamics combined with finite element analysis for a certain type of vehicle frame [3]. In 2011, Wang Jun-qiao and others calculated and studied the frame of 2500 fracturing truck. According to the driving and fracturing conditions of fracturing truck, transient dynamic finite element calculation was carried out on the frame to obtain the dynamic stress distribution of the frame under the two working conditions [4]. In 2013, Sang Meng solved the fatigue load spectrum of 2500 fracturing truck under fracturing and migration conditions based on dynamic simulation [5].

Whether it is a random load on the road surface or the simple harmonic excitation load of the fracturing pump, it is possible to cause fatigue damage to the frame. Therefore, it is necessary to calculate the fatigue life and safety factor of the frame under various working conditions to ensure the safety of the fracturing truck's service. This paper analyzes the strength of various types of working conditions of the 2500 fracturing frame during the fracturing process and conducts strength tests.

IOP Conf. Series: Materials Science and Engineering **751** (2020) 012057 doi:10.1088/1757-899X/751/1/012057

2. Finite element static analysis of 2500 fracturing truck frame

2.1. Stress analysis of 2500 fracturing truck frame

In order to analyze the force of the fracturing frame efficiently and accurately, it is necessary to first understand the structure of the truck, analyze the load of the frame, analyze the force of the frame and simplify it, and establish a mechanical model of the frame.

2.1.1. Structural Analysis of 2500 Fracturing Truck. The structure of the 2500 fracturing truck is shown in Figure 1 and the secondary frame and mainframe are combined by connecting plates. The lower part is connected to the axle and wheels by the leaf spring suspension system. The upper part bears the weight of various mounted equipment, which mainly includes the cab, the engine of the truck, the torque converter, the gearbox, and the fracturing pump.

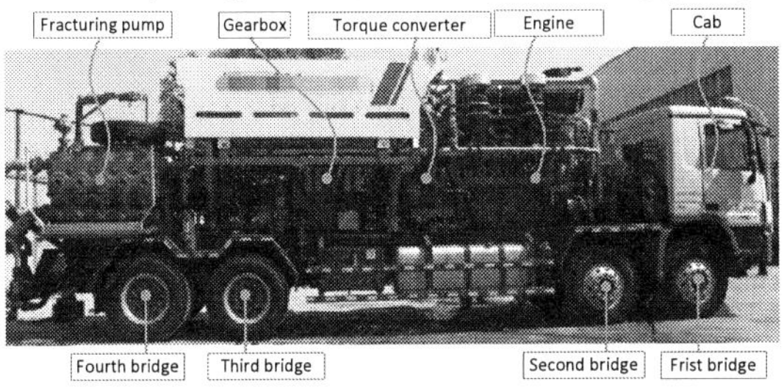

Figure 1 Structure of 2500 fracturing truck

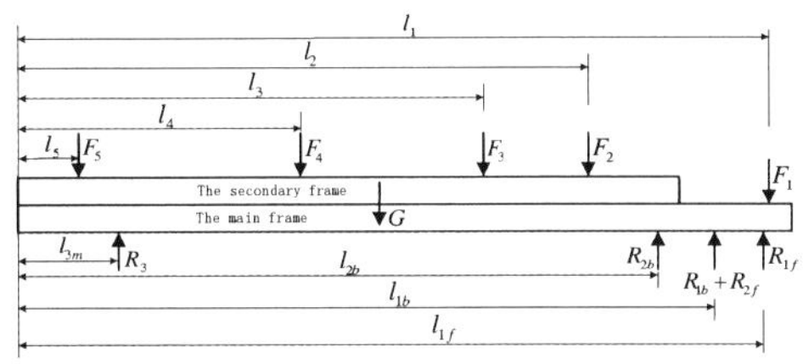

Figure 2 Mechanical model of 2500 fracturing truck

2.1.2. Stress analysis on the frame of model 2500 fracturing truck. Before the statics analysis of the frame, the stress condition of the frame is firstly analyzed and the mechanics model of the frame is established, which will be used as the basis to apply load and boundary conditions in the finite element model of the frame. In order to simplify the frame mechanics model, the following assumptions are made:

(1) The frame is a flat structure and the longitudinal beam is treated as a simply supported beam.

(2) The left and right sides members of the frame each bear half of the frame load.

(3) The load of the top-loading equipment passes through the center of the frame longitudinal section.

(4) Due to the small size of the top-loading equipment, the load of the top-loading equipment can be treated as concentrated quality, and the point of action is the position of the center of gravity of each equipment.

(5) The main and auxiliary frames are considered as rigid connections, that is the deflection changes synchronously.

The force analysis of the mainframe and secondary frame is shown in Figure 2. According to the system balance conditions, there are:

$$\begin{cases} F_1+F_2+F_3+F_4+F_5+G=R_{1f}+R_{1b}+R_{2f}+R_{2b}+R_3 \\ F_5(l_{3m}-l_5)+R_{2b}(l_{2b}-l_{3m})+R_{2f}(l_{2f}-l_{3m})+R_{1b}(l_{1b}-l_{3m})+R_{1f}(l_{1f}-l_{3m})= \\ F_1(l_1-l_{3m})+F_2(l_2-l_{3m})+F_3(l_3-l_{3m})+F_4(l_4-l_{3m})+G(l_G-l_{3m}) \\ F_5(l_{2b}-l_5)+F_4(l_{2b}-l_4)+G(l_{2b}-l_G)+F_3(l_{2b}-l_3)+F_2(l_{2b}-l_2)+ \\ R_{1b}(l_{1b}-l_{2b})+R_{2f}(l_{1b}-l_{2b})+R_{1f}(l_{1f}-l_{2b})=F_1(l_1-l_{2b})+R_3(l_{2b}-l_{3m}) \\ F_5(l_{1b}-l_5)+F_4(l_{1b}-l_4)+G(l_{1b}-l_G)+F_3(l_{1b}-l_3)+F_2(l_{1b}-l_2)+ \\ R_{1f}(l_{1f}-l_{1b})=F_1(l_1-l_{1b})+R_3(l_{1b}-l_{3m})+R_{2b}(l_{1b}-l_{2b}) \end{cases} \quad (1)$$

Since the tires of the first two bridges are installed at the center of the corresponding leaf spring, there are:

IOP Conf. Series: Materials Science and Engineering **751** (2020) 012057 doi:10.1088/1757-899X/751/1/012057

$$\begin{cases} R_{1f} = R_{1b} \\ R_{2f} = R_{2b} \end{cases} \tag{2}$$

2.2. Finite element analysis of 2500 fracturing truck frame

Firstly, the model of the fracturing frame is simplified. The body model is built in SolidWorks and then imported into ANSYS Workbench. For the full load static and the right front wheel elevation 120 mm, the finite element statics are respectively carried out. The stress and strain distribution of the frame is obtained.

2.2.1. Modeling of the three-dimensional frame. The fracturing truck has a large vehicle and a complicated structure. The chassis is mainly connected by the mainframe and the secondary frame through the connecting plate. Under the conditions of precision, the following assumptions and simplifications are made on the frame model:

(1) Processing the frame rails as simply supported beams;

(2) This paper mainly considers the overall stress and deformation distribution of the frame, ignores the influence of welding seam, and considers that the force transfer between the mainframe and the secondary frame is average [3];

(3) This paper mainly analyzes the static frame stress, temporarily do not consider the role of the suspension system;

(4) Parts with small mass and low load are removed, such as bolts;

(5) To divide the ideal mesh, the small rounded corners in the structure are turned into right angles.

According to the above simplified principle, the solid model of each component of the frame is respectively constructed in three-dimensional mechanical design software SolidWorks. The model of 2500 fracturing truck frame is shown in Figure 3.

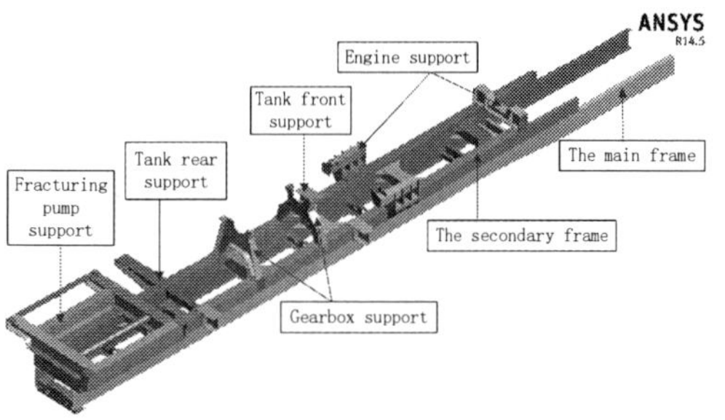

Figure 3 Surface model of 2500 Fracturing Truck frame

Figure 4 Meshing for 2500 Fracturing Truck frame

2.2.2. Definition of fame material. The frame of 2500 fracturing truck is made of carbon manganese steel 16MnL, which is a kind of low alloy and high strength structural. Its comprehensive mechanical properties are outstanding.

2.2.3. Application of load and boundary conditions

(1) Full load static condition boundary condition

(2) Full load static condition boundary condition

(3) The right front wheel raises the boundary condition of 120 mm working condition

2.2.4. Frame meshing. The frame structure of fracturing truck is complex, and the two grid generation methods, automatic grid and sweeping grid are combined to improve the dividing efficiency. A total of 122972 grid nodes and 121725 grid cells were obtained in Figure 4.

2.2.5. Results of frame static analysis

(1) Static analysis results of the frame under full load static condition

The static analysis of the 2500 fracturing frame under full load static conditions is shown in Figure 5 and Figure 6. It can be seen that, except for the stress concentration due to the simplification of the boundary conditions, the overall stress level of the fame of the fracturing truck under this working condition is low; there are three main bearing areas on the frame: the engine bearing area on the platform, the front bearing area on the water tank and the rear bearing area on the water tank. The stress range is about 10~30 MPa. The strain distribution rule of the frame is the same as the stress, and the strain range of the main bearing area is 1.0×10^{-5}~5.5×10^{-5} mm/mm. In the next chapter, the measuring points of frame strength test will be mainly selected in these three areas.

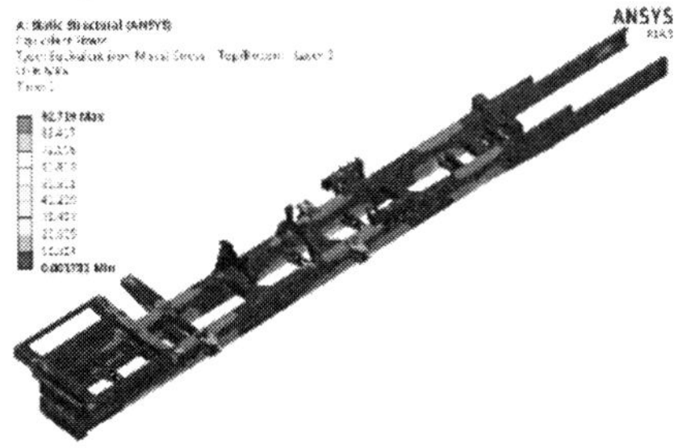

Figure 5 The stress of the full-loaded frame

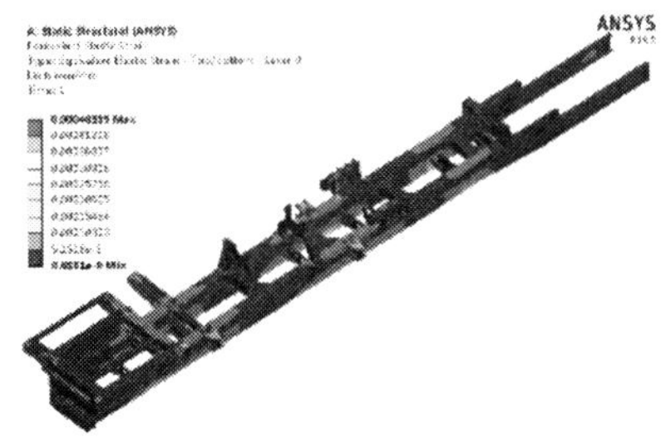

Figure 6 Strain of the full-loaded frame

(2) The results of statics analysis of right front wheel lift frame

The static analysis results of 2500 fracturing truck frame under the condition of 120 mm elevation of the right front wheel are shown in Figure 7 and Figure 8. It can be seen that the overall stress level of the frame of the fracturing truck under this condition is still not high, but a certain increase compared with the full-load static condition. The stress range of the main bearing of the frame is about 30~60 MPa. In addition, the strain range of the main bearing area of the frame is 7.0×10^{-5}~6.0×10^{-4} mm/mm.

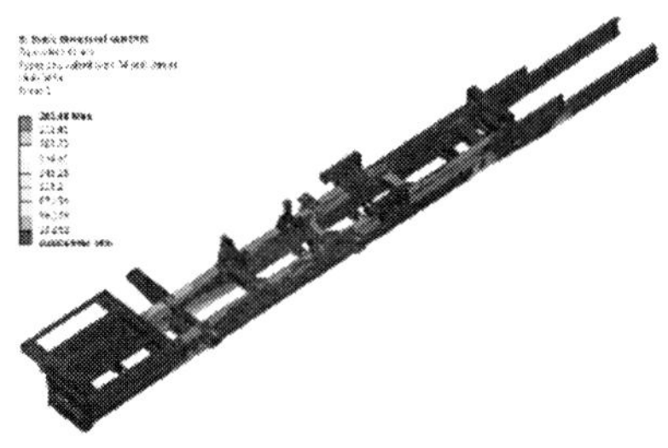

Figure 7 The stress of the frame under bending and torsion mode

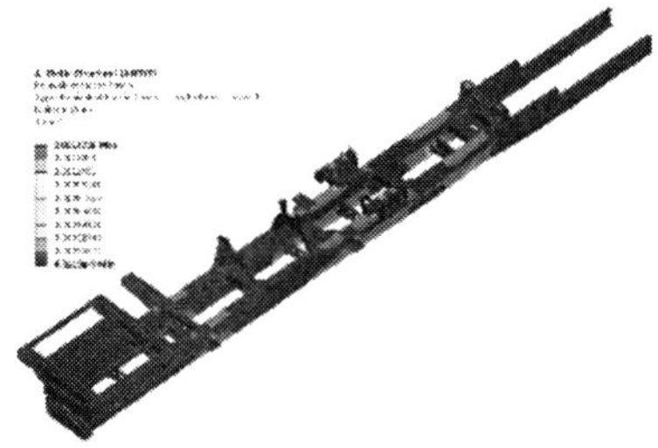

Figure 8 Strain of the frame under bending and torsion mode

3. Strength test of 2500 fracturing truck frame

Static strength test and dynamic strength test were carried out on the 2500 fracturing frame by resistance strain measurement.

3.1. 2500 fracturing truck frame strength test program

3.1.1. Frame strength test program

IOP Conf. Series: Materials Science and Engineering **751** (2020) 012057 doi:10.1088/1757-899X/751/1/012057

Figure 9 Static strength test scheme of 2500 Fracturing Truck frame

Figure 10 Dynamic strength test scheme of 2500 Fracturing Truck frame

(1) Frame static strength test plan: the static strength test of the frame refers to a test of the left side of the frame when the right front wheel of the fracturing truck is raised by 120 mm and the other wheels are in contact with the horizontal ground. Due to the limited conditions at the test site, the wooden wheels are used to raise the wheel, as shown in Figure 9.

(2) The frame dynamic strength test program means that the fracturing truck travels on the selected off-road surface at a speed of about 20 to 30 km/h to test the frame strain, as shown in Figure 10.

3.1.2. Strain gauge arrangement and strength test equipment. First of all, according to the statics analysis of the fracturing of the owner's secondary frame, the areas with large surface stress on the upper surface of the sub-frame and under the mainframe are selected as the preliminary position of strain gauge arrangement. Then, according to the operability of pasting strain gauge on-site, the final arrangement scheme of strain gauge is determined. This test adopts the resistance strain measurement method to test the strength of the fracturing truck frame. According to the relationship between stress and strain, the stress value of the measuring point of the truck frame is finally determined.

3.2. Frame strength test results of 2500 fracturing truck

3.2.1. Frame static strength test results. The purpose of the static strength test is to verify the correctness of the finite element model of the frame. The strain information collected in the test can only reflect the strain change of the loading equipment (initial strain). The statics calculation results for comparison can only be the strain increment after subtracting the initial strain form the total strain after raising the right front wheel of the frame. The calculation results of the strain points on the left side of the fracturing truck frame are shown in Figure 11. when the frame is only subjected to a single load of the top-loading equipment, the overall strain is very small; when the right front wheel is raised, the strain increases to different types. In general, the closer to the position of the right front wheel, the greater the strain increase. This is in line with the actual loading situation; the maximum strain increase and the maximum total strain occur at the measuring point 7, 0.000514 and 0.000529, respectively, because the measuring point 7 is closest to the right front wheel and also close to the heavy-duty platform engine.

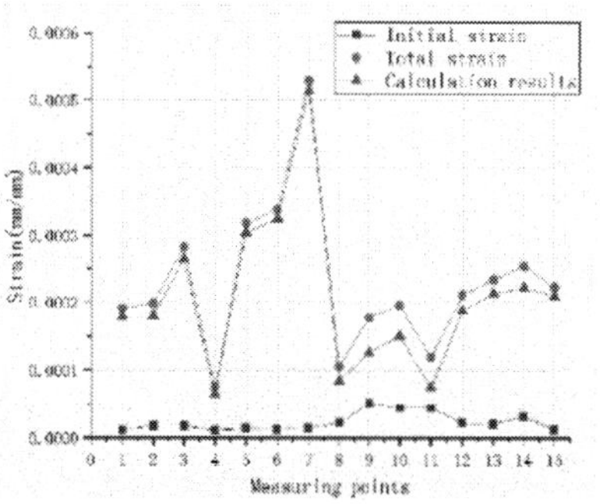

Figure 11 Strain calculation of the left side of the frame with its right front wheel raised up

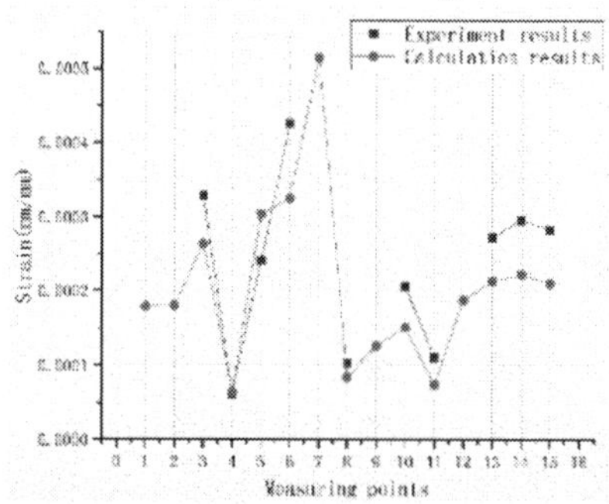

Figure 12 Comparison of strain result of the frame's left side with its right front wheel raised up

The comparison between the calculation results of the strain points on the left side of the fracturing truck frame and test results is shown in Figure 12. although there is a certain error between the effective test results and the corresponding calculation results, the overall trend of the two is consistent, that is, the closer to the position of the right front wheel, the greater the strain increase.

3.2.2. Frame dynamic strength test results. When the fracturing truck runs on the off-road at a speed of 30km/h, strain test results of each measuring point on the left side of the frame are shown in Fig.13(a)~(j). It can be seen that among all the effective points, the stress change at the measuring point is the largest, and the peak value reaches 87.54MPa. The possible reason is that the measuring point 3 is located at the root of the cantilever tank support and adjacent to the fracturing pump, which bears a large load in the process of migration. However, both point 4 and point 6 are far away from the main bearing area, and the range of stress variation is the smallest, about -8~8 MPa.

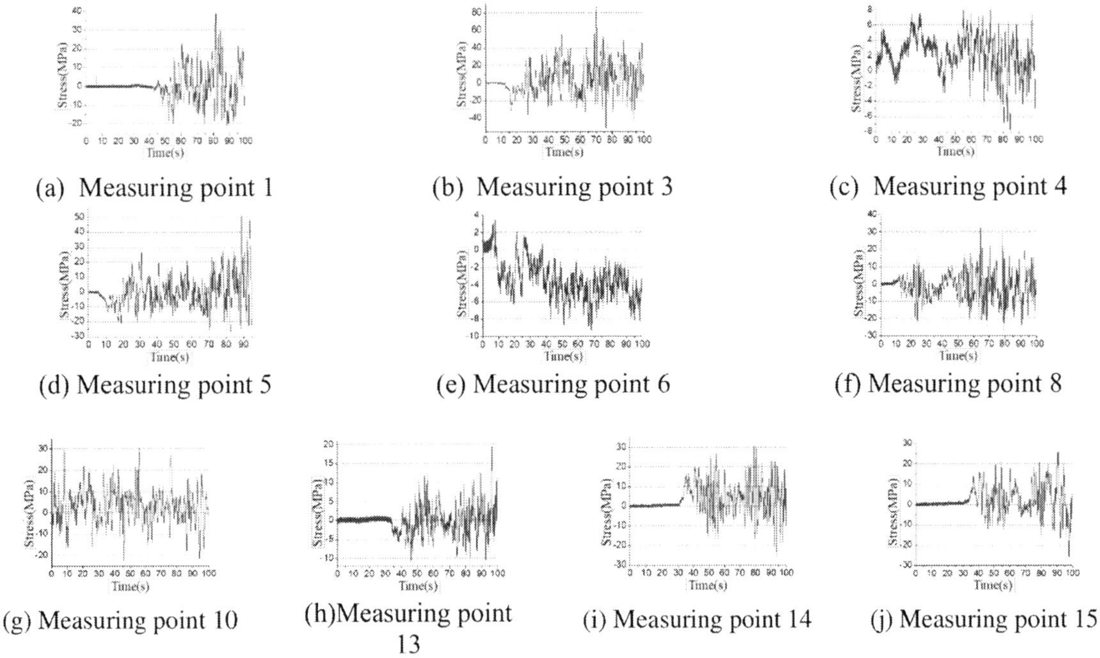

(a) Measuring point 1 (b) Measuring point 3 (c) Measuring point 4

(d) Measuring point 5 (e) Measuring point 6 (f) Measuring point 8

(g) Measuring point 10 (h)Measuring point 13 (i) Measuring point 14 (j) Measuring point 15

Figure 13 Dynamic strength test result of the frame's left side: $v = 30km/h$

It can be seen that when the fracturing truck runs at the test section at a speed of 30km/h, the overall stress level on the left side of the frame is relatively low, about 23.73~146.76 MPa; among them, the total stress at the two-point of measuring point 3 and measuring point 5 is larger. What they have in common is that they are all at the root of the tank support (cantilever structure), so care should be taken to improve the support structure and minimize the length of the boom. The total stress of measuring point 3 is the largest, which is 146.76 MPa and does not exceed the yield limit of the frame material (345MPa), and the safety factor reaches 2.35. Then the measuring point 3 satisfies the strength requirement under this working condition, so other measuring points also meet the requirements.

4. Conclusion

Based on the analysis of the overall structure of the 2500 fracturing truck, this paper extracts the simplified mechanical model of the main and subframes and provides a reference for the selection of the strength test points of the fracturing frame. The static strength test plan of the fracturing frame (120mm right front wheel lift) and the dynamic strength test plane (off-road transport) are designed. Then the strength test of the 2500 fracturing frame was completed, and the strain strength test results of the frame are compared with the strain calculation results, and the error is about 4.69%~33.33% (the average error is only 22.01%). Finally, the maximum stress position of the frame under the moving condition is measured point 3 (the root of the rear support of the water tank), the total stress is 146.76 MPa, and the minimum safety factor is 2.35. The strength of each measuring point meets the requirements.

Acknowledgments
The researchers gratefully acknowledge the financial support of Shaanxi Post-doctoral Research Funding Project "Research on Key Technologies of Intelligent Control System for Electric Drive Fracturing Device" (Project No.: 2018BSHQYXMZZ04) and Major Engineering and Technology Field Test Projects of CNPC "Integration and field test of 7000hp electric drive fracturing skid and on-line monitoring system for diesel drive fracturing unit" (Project No.: 2019F-30)

References

[1] Zhao Xu-ping, Kong Dan, Chang Liang. Research on the development of 2500 ultra-high-pressure shale gas pressure cracking vehicle[J]. Petroleum Planning and Design,2017,28(03):12-14+43+52.
[2] Zhao Han,Qian Demeng,Wei Ying. Kinematics Simulation Analysis and Optimization Design of Automotive Air Suspension[J]. China Mechanical Engineering, 2005,16(4):361-364.
[3] Xiao Zhi-jin, Zhu Si-hong. Fatigue Life Prediction Method for Light Truck Body Frame Based on Virtual Prototype Technology [J]. Mechanical Design, 2010,27(1):59-63.
[4] Wang Jun-qiao. Analysis and Prediction of Fatigue Life of 2500 Fracturing Truck Frame[J]. Petroleum Machinery, 2011(39): 20-22.
[5] Sang Meng. Research on the bearing capacity of high-power fracturing truck frame: (Master's thesis). Dalian: Dalian University of Technology, 2013.

Copula-based Estimation Method of the Expected Value Function for Robust Design Optimization

Xinyao Li*, Weihong Zhang and Liangli He

School of Mechanical Engineering, Northwestern Polytechnical University, Xi'an, Shaanxi, 710072, China

*lxy-09103@163.com

Abstract. The expected value function (EVF), which defined as function of mean value with respect to the design parameters, can simplify the solving process of the RDO with nested-double loop. The traditional method for estimating the expected value function requires a nested procedure and the computational cost depends on the product of the optimal steps and the numbers of the random variable samples. In this paper, a novel method based on copula is proposed to estimate the EVF. In the proposed method, the EVF is derived as the copula function related to the design parameters and the model output, which can be estimated with a single loop samples and the computational cost of the copula-based method does not depend on the total number of design parameters or the number of the optimal steps. Additionally, a vine-copula is introduced to the EVF so that the estimation for the EVF by two-dimensional kernel density estimation (KDE) method can be extended to the multi-dimensional design parameters problems. The present example with two cases demonstrate the effectiveness of the proposed copula-based method.

1. Introduction

Robust design optimization (RDO) [1-3], as one important branch of the optimization design, aims to find the optimal solutions that are relatively insensitive to input uncertainty. A typical RDO problem can be formulated as follows:

$$\begin{aligned} &\text{find} && \boldsymbol{\theta} \\ &\text{minimizing} && \{\mathrm{E}(\varphi(\boldsymbol{\theta},\boldsymbol{x})),\ \sigma(\varphi(\boldsymbol{\theta},\boldsymbol{x}))\} \\ &\text{subject to} && \mathrm{E}(g_i(\boldsymbol{\theta},\boldsymbol{x}))+\beta_i\sigma(g_i(\boldsymbol{\theta},\boldsymbol{x}))\le 0 \ \ (i=1,2,...,k) \\ & && \sigma(h_j(\boldsymbol{\theta},\boldsymbol{x}))\le \bar{\sigma}_j \ \ (j=1,2,...,l) \\ & && \underline{\boldsymbol{\theta}}\le \boldsymbol{\theta}\le \bar{\boldsymbol{\theta}} \end{aligned} \tag{1}$$

where $\boldsymbol{\theta}\in\mathbf{R}^{m\times 1}$ denotes the vector of design variables as well as the design parameters of the random variables $\boldsymbol{x}$, $\underline{\boldsymbol{\theta}}$ and $\bar{\boldsymbol{\theta}}$ are the lower and upper bounds of the design variables, respectively. φ and g_i denote the objective function and the constraint function in the deterministic design optimization problem. $\mathrm{E}(\cdot)$ and $\sigma(\cdot)$ are the expected value and the standard deviation, respectively. The notation h_j is the structural performance to which constraints on standard deviations of function values are applied. The scalar $\beta_i>0$ is a prescribed feasibility index for the ith original constraint and the $\bar{\sigma}_j$ denotes the upper limit for the standard deviation of the performance. It can be seen that a RDO problem is a nested-loop problem, where the inner is a robustness analysis problem and the outer is an ordinary constrained optimization problem.

For estimating the expected value regarded as one of robustness measures, various methods have been reported in the literature. These methods can be divided into three major classes [4]: (i) Taylor series expansion methods [5]; (ii) sampling-based methods [6] and (iii) point estimate methods [7]. However, all of these methods for RDO are double-loop, i.e., candidate optimal solutions are first proposed in the outer loop which is followed by the evaluation of robustness analysis in the inner loop. The process tends to pose prohibitive computational burden when complex models are involved. Thus, there is a strong need to estimate the expected value to simplify the solving process of the RDO.

The expected value function (EVF), defined as function of mean value with respect to the m-dimensional design parameter vector $\boldsymbol{\theta}=(\theta_1,\theta_2,...,\theta_m)$, is concerned in this paper. Theoretically speaking, once obtaining the EVF, the standard deviation function can be acquired simultaneously. Then the traditional RDO with nested-loop is decoupled into a deterministic and ordinary optimization problem which can be solved by the existing optimization techniques. The direct solution for estimating the EVF is Monte Carlo simulation (MCS). However, the computational cost is expensive for complex models calling for the finite element analysis. To reduce the computational burden and to solve the multi-dimensional design parameters problems, the copula-based method is proposed.

The proposed method in this work uses the idea of approximating the conditional probability density function (PDF) replaced by the copula density function. As has been known to many researcher, copula is a very useful tool for understanding the relationship between random variables [8-10]. Moreover, after deriving the EVF as copula form, a vine-copula can be introduced to process multi-dimensional design parameters, which breaks the limitation of the 2-dimensional PDF estimation by the widely used kernel density estimation (KDE) method [11,12]. A vine-copula function disassembles the multidimensional random variables' joint distribution construction problem independently into the estimations of the marginal probability distributions and a dependence structure among marginal distributions, thus makes the construction of a joint probability distribution easier and more flexible [13].

The paper is outlined as follows. Section 2 reviews the definitions of the original conditional expected value. Section 3 establishes the relationship of the copula and the EVF, introduces the vine- copula to process the multi-dimensional design parameters. Section 4 proposes the copula-based method for estimating the EVF. Section 5 employs a typical examples to demonstrate the effectiveness of the proposed method. Section 6 briefly draws the conclusions. Sections 7 discusses some issues.

2. Formulation of the EVF

According to the definition of the conditional expected value, considering a model with the output Y, the EVF $\mu(\boldsymbol{\theta})$ can be written as

$$\mu(\boldsymbol{\theta}) = \mathrm{E}(Y|\boldsymbol{\theta}) = \int_{-\infty}^{\infty} yf(y|\boldsymbol{\theta})dy \tag{2}$$

where $y|\boldsymbol{\theta}$ represents the output of the model considering both input random variables $\boldsymbol{X}$ and design parameters $\boldsymbol{\theta}$ of $\boldsymbol{X}$. $f(y|\boldsymbol{\theta})$ denotes the conditional joint PDF of $\boldsymbol{\theta}$, and it can be estimated by

$$f(y|\boldsymbol{\theta}) = f_{Y,\theta}(y,\boldsymbol{\theta})/f_{\theta}(\boldsymbol{\theta}) \tag{3}$$

in which based on Bayes' theory, the design parameters $\boldsymbol{\theta}$ are regarded as random variables, and the joint PDF $f_{\theta}(\boldsymbol{\theta})$ of $\boldsymbol{\theta}$ is assigned subjectively. In practical application, the distribution of $f_{\theta}(\boldsymbol{\theta})$ is usually chosen as simple forms as possible, such as uniform distribution, Gaussian distribution and so on. $f_{Y,\theta}(y,\boldsymbol{\theta})$ denotes the joint PDF of the output y and design parameters $\boldsymbol{\theta}$.

Based on equation (2), the EVF $\mu(\boldsymbol{\theta})$ can be obtained if the conditional joint PDF $f(y|\boldsymbol{\theta})$ is estimated. Based on equation (3), the conditional joint PDF $f(y|\boldsymbol{\theta})$ is evaluated according to the joint PDF $f_{\theta}(\boldsymbol{\theta})$ and $f_{Y,\theta}(y,\boldsymbol{\theta})$. Because the joint PDF $f_{\theta}(\boldsymbol{\theta})$ has been prescribed beforehand, $f_{\theta}(\boldsymbol{\theta})$ is obtained without computational cost. Well then, the joint PDF $f_{Y,\theta}(y,\boldsymbol{\theta})$ is required. Although several articles have demonstrated that the widely used KDE method [11, 12] is an appropriate technique in estimating the joint PDF, one should be noted that the KDE method is very inaccurate when the dimensionality of the random variables is greater than two. Therefore, to accurately estimate the EVF $\mu(\boldsymbol{\theta})$ with multi-dimensional design parameters, the copula-based method is proposed in the next section.

3. The copula-based method for estimating the EVF

3.1. The relationships of copula and the EVF

It is well known that copula is a function that links the multivariate joint cumulative distribution function (CDF) with its one-dimensional marginal CDFs. By Sklar theory [8, 9], if two marginal CDFs $F_{\theta_1}(\theta_1)$ and $F_{\theta_2}(\theta_2)$ are continuous, and $F_{\theta_1\theta_2}(\theta_1,\theta_2)$ is the joint CDF of $\boldsymbol{\theta}_1$ and $\boldsymbol{\theta}_2$, then there is only one unique copula $\mathrm{C}\left(F_{\theta_1}(\theta_1),F_{\theta_2}(\theta_2)\right)$ to combine $F_{\theta_1\theta_2}(\theta_1,\theta_2)$ with marginal $F_{\theta_1}(\theta_1)$ and $F_{\theta_2}(\theta_2)$. It is defined as:

$$F_{\theta_1\theta_2}(\theta_1,\theta_2)=\mathrm{C}\left(F_{\theta_1}(\theta_1),F_{\theta_2}(\theta_2)\right) \tag{4}$$

where if $F_{\theta_1}(\theta_1)$ and $F_{\theta_2}(\theta_2)$ are continuous, then C is uniquely determined. $F_{\theta_1\theta_2}(\theta_1,\theta_2)$ is a joint CDF with marginal CDFs $F_{\theta_1}(\theta_1)$ and $F_{\theta_2}(\theta_2)$.

Let $u_i=F_{\theta_i}(\theta_i)$ and $v=F_Y(y)$, $F_{\theta_i}(\theta_i)$ and $F_Y(y)$ are the CDF of the ith design variable and the CDF of the model output respectively, then the copula function $C(u_i,v)$ of θ_i and Y is a joint CDF with uniform marginal distribution in $[0, 1]^2$.

The copula density $c(u_i,v)$ is given as:

$$c(u_i,v)=\partial^2 C(u_i,v)/\partial u_i\partial v \tag{5}$$

By the relationship between the joint two dimensional CDF and PDF, we have

$$f_{Y,\theta_i}(y,\theta_i)=\frac{\partial^2 F_{Y,\theta_i}(y,\theta_i)}{\partial y\partial\theta_i}=\frac{\partial^2 C(u_i,v)}{\partial u_i\partial v}\frac{\partial v}{\partial y}\frac{\partial u_i}{\partial\theta_i}=c(u_i,v)f_Y(y)f_{\theta_i}(\theta_i) \tag{6}$$

Substituting equations (3) and (6) into equation (2), for single design parameter θ_i, $\mu(\theta_i)$ yields:

$$\mu(\theta_i)=\int_{-\infty}^{\infty}y\frac{f_{Y,\theta}(y,\theta_i)}{f_{\theta_i}(\theta_i)}dy=\int_{-\infty}^{\infty}y\frac{f_{Y,\theta_i}(y,\theta_i)}{f_{\theta_i}(\theta_i)\cdot f_Y(y)}f_Y(y)dy=E_V\{y\cdot c(u_i,v)\}=E_V\{F_Y^{-1}(v)\cdot c(u_i,v)\} \tag{7}$$

where according to the property of the CDF, $f_V(v)=1$ and $y=F_Y^{-1}(v)$ hold. And $E_V(\cdot)$ denotes the expected value for v.

The current relationship of the copula and the EVF, mainly aims toward two-dimensional correlation problems with single design parameter θ_i and the model output y. For the multi-dimensional design parameters $\boldsymbol{\theta}=(\theta_1,\theta_2,...,\theta_m)$, a more flexible mathematical tool, i.e., vine-copula function, has been developed in the uncertainty analysis domain [14-16]. It is developed to process multi-dimensional design parameters for estimating the EVF in the next section.

3.2. Vine-copula processing multi-dimensional design parameters

The core idea of a vine-copula is to perform decomposition on the joint PDF of multi-dimensional random variables and convert the joint PDF to several two-dimensional copula functions related to the original variables and their conditional variables for processing.

Based on Sklar's theorem [8, 9], for the m-dimensional random vector $\boldsymbol{\theta}=\{\theta_1,\theta_2,...,\theta_m\}$, its CDF $F_{1,2,...,m}(\theta_1,\theta_2,\mathrm{L}\ \theta_m)$ for any marginal CDF $F_1(\theta_1),F_2(\theta_2),...,F_m(\theta_m)$ can be expressed as

$$F_{1,2,\ldots,m}(\theta_1,\theta_2,\mathrm{L}\ \theta_m)=C(F_1(\theta_1),F_2(\theta_2),\ldots,F_m(\theta_m)) \quad (8)$$

in which C is the copula function of the corresponding CDF.

For the m-dimensional design parameter vector $\boldsymbol{\theta}$, its joint PDF $f(\theta_1,\theta_2,\ldots,\theta_m)$ can be decomposed as outlined below:

$$f(\theta_1,\theta_2,\ldots,\theta_m)=f_1(\theta_1)f_{2|1}(\theta_2|\theta_1)f_{3|1,2}(\theta_3|\theta_1,\theta_2)\ldots f_{k|1,2,\ldots,k-1}(\theta_k|\theta_1,\theta_2,\ldots,\theta_{k-1})\ldots f_{m|1,2,\ldots,m-1}(\theta_m|\theta_1,\theta_2,\ldots,\theta_{m-1}) \quad (9)$$

where $f_{k|1,2,\ldots,k-1}(\theta_k|\theta_1,\theta_2,\ldots,\theta_{k-1}),k=1,2,\ldots,m$ is the conditional PDF. In a case with four variables, its PDF $f_{1,2,3,4}(\theta_1,\theta_2,\theta_3,\theta_4)$ is decomposed by D-vine [14] model as follows

$$\begin{aligned}f_{1,2,\ldots,4}(\theta_1,\theta_2,\ldots,\theta_4)&=f_1(\theta_1)\cdot f_2(\theta_2)\cdot f_3(\theta_3)\cdot f_4(\theta_4)\\&\cdot c_{1,2}(F_1(\theta_1),F_2(\theta_2))\cdot c_{2,3}(F_2(\theta_2),F_3(\theta_3))\cdot c_{3,4}(F_3(\theta_3),F_4(\theta_4))\\&\cdot c_{1,3|2}(F_{1|2}(\theta_1|\theta_2),F_{3|2}(\theta_3|\theta_2))\cdot c_{2,4|3}(F_{2|3}(\theta_2|\theta_3),F_{4|3}(\theta_4|\theta_3))\\&\cdot c_{1,4|2,3}(F_{1|2,3}(\theta_1|\theta_2,\theta_3),F_{4|2,3}(\theta_4|\theta_2,\theta_3))\end{aligned} \quad (10)$$

where

$$F_{1|2}(\theta_1|\theta_2)=\frac{\partial C_{1,2}(F_1(\theta_1),F_2(\theta_2))}{\partial F_2(\theta_2)},\quad F_{1|2,3}(\theta_1|\theta_2,\theta_3)=\frac{\partial C_{1,3|2}(F_{1|2}(\theta_1|\theta_2),F_{3|2}(\theta_3|\theta_2))}{\partial F_{3|2}(\theta_3|\theta_2)} \quad (11)$$

Let $u_{i|j}=F_{i|j}(\theta_i|\theta_j)$ and $u_{i|j,k}=F_{i|j,k}(\theta_i|\theta_j,\theta_k)$, thereof, for the three-dimensional design parameters $\boldsymbol{\theta}=\{\theta_1,\theta_2,\theta_3\}$ and one output y of the model, the joint PDF $f_{\theta,Y}(\boldsymbol{\theta},y)$ is deduced as

$$f_{\theta,Y}(\boldsymbol{\theta},y)=f_1(\theta_1)\cdot f_2(\theta_2)\cdot f_3(\theta_3)\cdot f_Y(y)\cdot M \quad (12)$$

where

$$M=c_{1,2}(u_1,u_2)\cdot c_{2,3}(u_2,u_3)\cdot c_{3,Y}(u_3,v)\cdot c_{1,3|2}(u_{1|2},u_{3|2})\cdot c_{2,Y|3}(u_{2|3},u_{Y|3})\cdot c_{1,Y|2,3}(u_{1|2,3},u_{Y|2,3}) \quad (13)$$

$$u_{Y|3}=\frac{\partial C_{Y,3}(u_3,v)}{\partial u_3},\quad u_{Y|2,3}=\frac{\partial C_{Y,3|2}(u_{Y|2},u_{3|2})}{\partial u_{3|2}} \quad (14)$$

Thus, $\mu(\boldsymbol{\theta})$ for the three-dimensional design parameters based on equation (2) is derived as:

$$\mu(\boldsymbol{\theta})=\int_{-\infty}^{\infty}y\frac{f_{Y,\theta}(y,\boldsymbol{\theta})}{f_\theta(\boldsymbol{\theta})}dy=\int_{-\infty}^{\infty}y\cdot M\cdot f_Y(y)dy=\int_0^1 y\cdot M dv=E_V(F_Y^{-1}(v)\cdot M) \quad (15)$$

The above analysis indicates that through a vine-copula, a multi-dimensional random variables joint distribution is converted to a product of the marginal distributions and multiple two-dimensional copula functions. Thereof, a vine-copula can eventually be processed to estimate the EVF $\mu(\boldsymbol{\theta})$ for the multi-dimensional design parameters.

4. The copula-based solution for estimating the EVF

According to the equations (7) and (15), it can be found that the bivariate copula density function is pivot for estimating $\mu(\boldsymbol{\theta})$. For estimating these bivariate copula density functions, generally, there are two categories for this purpose: the parametric estimation and the nonparametric estimation. Comparing with the former, because it does not need to assume any specific family for copula function, the nonparametric estimation is more general and flexible [17]. Many nonparametric methods can be employed to estimate the bivariate copula density. In this paper, the widely used two dimensional KDE method [11, 12] is employed to evaluate the bivariate copula density functions.

In summary, the following steps of the copula-based method are implemented for estimating the EVF $\mu(\boldsymbol{\theta})$.

Step.1 Prescribe a prior PDF $f_\theta(\boldsymbol{\theta})$ for the design parameters $\boldsymbol{\theta} = (\theta_1, \theta_2, ..., \theta_m)$, where the distribution type is assumed as independent and uniform distribution in this paper. And generate N samples of the design variables $\boldsymbol{\theta}$ by $f_\theta(\boldsymbol{\theta})$.

Step.2 Generate one sample of the random variables X for each sample of the design parameters by the conditional PDF $f_X(x|\boldsymbol{\theta})$. Then, calculate the output of the model $y = \varphi(\boldsymbol{x}|\boldsymbol{\theta})$.

Step.3 Obtain the CDF samples of $\boldsymbol{u}_j = F_\theta(\theta^{(j)}) = \{u_{j1}, u_{j2}, ..., u_{jm}\}$ $(j = 1, 2, ..., N)$ by $f_\theta(\boldsymbol{\theta})$ and $v = F_Y(y)$ by the one-dimensional KDE method.

Step.4 Calculate each conditional CDF in equation (13).

According to the equation (14), two kinds of conditional CDFs, i.e., $u_{i|j} = F_{i|j}(\theta_i|\theta_j)$ and $u_{i|j,k} = F_{i|j,k}(\theta_i|\theta_j, \theta_k)$, are required. Based on the strategy of estimating the conditional CDFs proposed in Ref.[18], these two kinds of conditional CDFs can be derived as

$$F_{\Theta_j|\Theta_i}(\theta_j|\theta_i) = P(\Theta_j \le \theta_j|\theta_i) = \frac{f_{\Theta_i}(\theta_i|\Theta_j \le \theta_j)}{f_{\Theta_i}(\theta_i)} \cdot F_{\Theta_j}(\theta_j) = c(u_i|U_j \le u_j) \cdot u_j \tag{16}$$

where $c(u_i|U_j \le u_j)$ is one dimensional conditional copula density on the event $U_j \le u_j$.

$$F_{\Theta_j|\Theta_i}(\theta_i|\theta_j, \theta_k) = P(\Theta_i \le \theta_i|\theta_j, \theta_k) = \frac{f_{\Theta_j,\Theta_k}(\theta_j, \theta_k|\Theta_i \le \theta_i)}{f_{\Theta_j}(\theta_j) \cdot f_{\Theta_k}(\theta_k)} \cdot F_{\Theta_i}(\theta_i) = c(u_j, u_k|U_i \le u_i) \cdot u_i \tag{17}$$

where the two dimensional conditional copula density function $c(u_j, u_k|U_i \le u_i)$ is estimated by the two-dimensional KDE method.

Step.5 Estimate the bivariate copula density functions in equation (13) by the two-dimensional KDE method.

Step.6 Estimate the EVF $\mu(\boldsymbol{\theta})$ based on equation (7) or equation (15), or one that can be derived by the vine- copula for other multi-dimensional design parameters.

5. Illustrative example: an aero engine turbine disk

In this section, comparing with the direct MCS and the copula-based method proposed in this paper for estimating EVF, an aero engine turbine disk is analyzed to demonstrate the accuracy and efficiency of the proposed method based on copula. The results estimated by the direct MCS are regarded as the reference values. For the two methods, the low-discrepancy sequence sampling method for generating samples of input variables is used, which may accelerate the convergence with the increase of sample size.

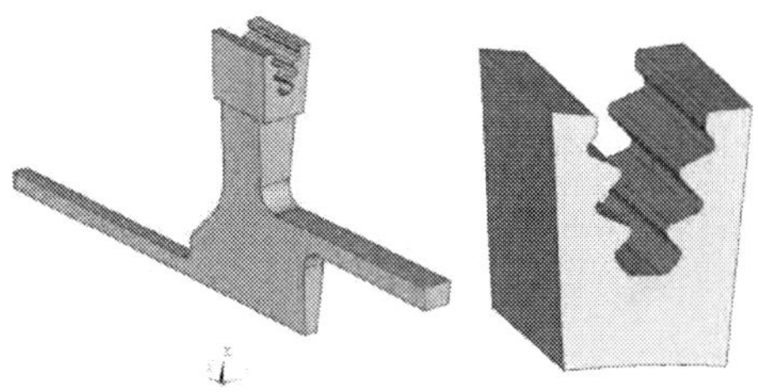

Figure 1. An aero engine turbine disk.

Turbine disk [19], which plays a big role in supporting, connecting, transmitting torque and power, and meanwhile, undergoes the complex loads, is a very important part in engine turbine. Because of the complex structural shape, the stress concentration easily takes place in some parts during the working process, such as tongue, groove bottom and pinhole, etc. After working for a period of time, some crack may appear. As shown in Figure 1, it shows the crack of an aero engine turbine disk.

The load applied on an aero engine turbine disk is $F = C\omega^2 / 2\pi + 2\rho\omega^2 J$, in which ρ, C, ω and J are mass density, coefficient, a rotation angular velocity and moment of inertia, respectively. And $\omega = 2\pi n$, n is rotational frequency. The performance function of the structure is

$$g(\sigma_s, \rho, C, A, J, n) = \sigma_s A - F \tag{18}$$

where σ_s is the ultimate strength, A is the sectional area. And the distribution types and parameters of input variables are listed in Table 1.

Table 1. Distribution parameters of the example.

Variables	σ_s	ρ	C	A	J	n
Distribution	normal	lognormal	lognormal	normal	normal	normal
Mean	1.1×10^9	8240	5.67	6.2×10^{-3}	1.22×10^{-4}	200
C.o.v	0.1	0.2	0.2	0.1	0.1	0.1

Suppose the mean values of σ_s, ρ and C are the design parameters, i.e., $\theta_1 = \mu_{\sigma_s}$ and $\theta_1 \in [0.9\times10^9, 1.3\times10^9]$, $\theta_2 = \mu_\rho$ and $\theta_2 \in [6240, 10240]$, $\theta_3 = \mu_C$ and $\theta_3 \in [3.67, 7.67]$. For this problem with three design parameters $\boldsymbol{\theta} = [\theta_1, \theta_2, \theta_3]$, the EVF $\mu(\boldsymbol{\theta})$ is evaluated by equation (15). The MCS and the copula-based method are employed to evaluate $\mu(\boldsymbol{\theta})$. The corresponding results on the edges of the design parameters are plotted in Figure 2. As shown in Figure 2, the results of the copula-based method with 100063 samples are consistent with those of the MCS over the whole region.

In order to further check the accuracy of the proposed method, the mean relative error (MRE) of the estimates is employed. Figure 3 shows the MREs of the proposed method vs the sample size. It can be found that the proposed method is convergent gradually with the sample size increasing.

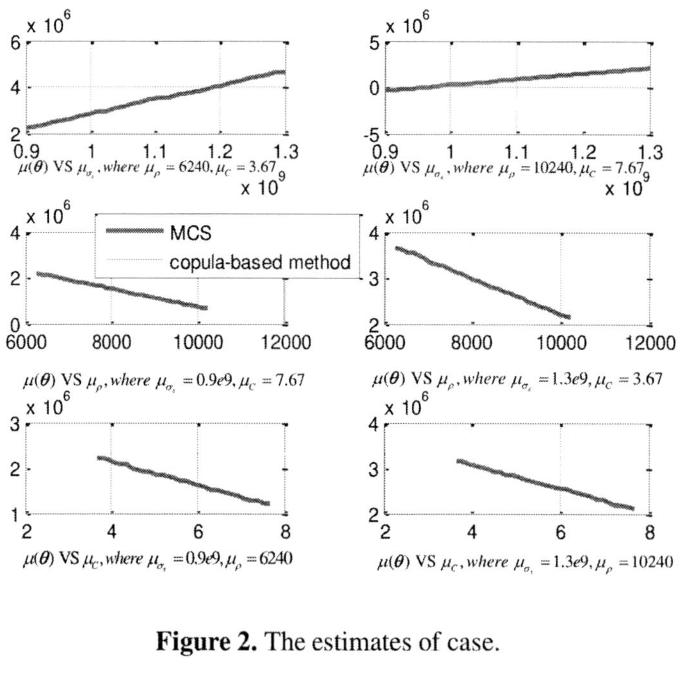

Figure 2. The estimates of case.

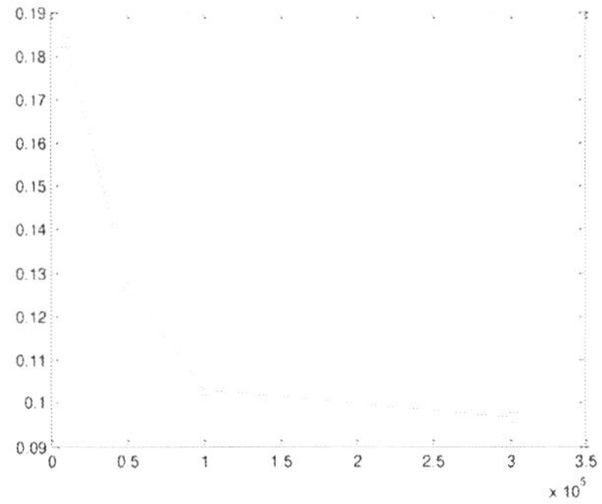

Figure 3. MREs vs sample size.

6. Conclusion

EVF, defined as a function of mean value with respect to the design parameter, is one key to decoupling RDO problem into an ordinary deterministic optimization problem. In this paper, based on copula theorem, we propose a novel EVF expression. In the novel form of EVF, the joint PDF is replaced by the copula density function, then the EVF is derived as the copula function related to the design parameters and the model output. Additionally, a vine-copula is developed to disassemble the multi-dimensional joint PDF into some bivariate copula density functions which can be estimated by the widely used two-dimensional KDE method. Thus, based on vine-copula, the EVF estimated by the KDE method is no longer limited to one single-dimensional design parameter. In contrast to the traditional MCS concerns the double-nested-loop simulation, the novel method only concerns single-loop simulation, even the computational cost of the copula-based method does not depend on the total number of design parameters or the number of the optimal steps.

The example with two cases illustrates that the EVF estimated by the copula-based method has good agreement with one obtained by MCS, however, the proposed method requires less model evaluations than MCS.

7. Discussion

In theory, once the EVF is estimated, the standard deviation function can be acquired simultaneously. However, the standard deviation function is more difficult to be obtained than the EVF. Since the variance as the 2-order moment is tougher to be computed accurately than the expected value as the first-order moment. In order to estimate the standard deviation function more accurate, more sample size may be needed to describe the tail features of distribution function.

References

[1] Beyer HG and Sendhoff B 2009 Robust optimization - a comprehensive survey *Comput Methods Appl Mech Eng* **96(33)** 3190-3218
[2] Doltsinis I and Kang Z 2004 Robust design of structures using optimization methods *Comput Methods Appl Mech Eng* **193(23)** 2221-2237
[3] Papadrakakis M, Lagaros N and Plevris V 2005 Design optimization of steel structures considering uncertainties *Eng Struct* **27(9)** 1408-1418
[4] Kais Z, Mark M, Sankaran M and Lawrence G 2011 Robustness-based design optimization under data uncertainty *Struct Multidisc Optim* **44** 183-197
[5] Lee KH and Park GJ 2001 Robust optimization considering tolerances of design variables. *Comput Struct* **79** 77-86
[6] Cheng H and Sandu A 2009 Efficient uncertainty quantification with the polynomial chaos method for stiff systems *Math Comput Simul* **79** 3278-3295
[7] Lee I, Choi KK, Du L and Gorsich D 2008 Dimension reduction method for reliability-based robust design optimization *Comput Struct* **86** 1550-1562

IOP Conf. Series: Materials Science and Engineering **751** (2020) 012058 doi:10.1088/1757-899X/751/1/012058

[8] Schweizer B and Wolff EF 1981 On nonparametric measures of dependence for random variables *Annals of Statistics* **9** 879-885

[9] Wolff EF1980 N-dimensional measures of dependence *Stochastica* **4** 175-188

[10] Wei PF, Lu ZZ and Song JW 2014 Moment-independent sensitivity analysis using copula *Risk Analysis* **34(2)** 210-222

[11] Botev ZI, Grotowski JF and Kroese DP 2010 Kernel density estimation via diffusion. *Annals of Statistics* **38** 2916-2957

[12] Resenblatt M 1956 Remarks on some nonparametric estimates of a density function *Annals of Mathematical Statistic* **27(6)** 832-837

[13] Jiang C, Zhang W and Han X 2015 A vine-copula-based reliability analysis method for structures with multidimensional correlation *J. of Mech. Design.* **137** 1-13

[14] Frederik M and Scherer M 2014 Pair Copula Constructions, Simulating Copulas: Stochastic Models *Samp Algo & Appl* 185-230

[15] Bedford T and Cooke RM 2001 Probability density decomposition for conditional dependent random variables modeled by vines *Ann. Math. Artic. Intell.* **32** 245-268

[16] Bedford T and Cooke RM 2002 Vines-a new graphical model for dependent random variables *Ann. Stat.* **30(4)** 1031-1068

[17] Scaille O, Fermanian and Jean D 2002 Nonparametric estimation of Copulas for time series *FAME Research paper* **57**

[18] Li XY, Zhang WH and He LL 2019 A novel nonparametric estimation for conditional copula functions based on Bayes theorem *under review*

[19] Yun WY, Lu ZZ and Jiang X 2018 A modified importance sampling method for structural reliability and its global reliability sensitivity analysis *Struct Multidisc Optim* **57(4)** 1625-1641.

IOP Conf. Series: Materials Science and Engineering **751** (2020) 012059 doi:10.1088/1757-899X/751/1/012059

Analysis of Geometric Characteristics of Cycloidal Transmissio

Xiangmao Huang[a] and Jing Zhang[b]

Shuanghuan Driveline Co.,Ltd,Zhejiang,China

E-mail: [a]xiangmaohuang@gearsnet.com, [b]zhangjing@gearsnet.com

Abstract. In order to improve the transmission performance of RV reducer,the equation of cycloid tooth profile is deduced according to the transmission principle of cycloid.The relationship between the curvature radius of cycloid tooth profile and the rotation angle is deduced. The rotation angle corresponding to the maximum curvature radius is solved and the value of the radius of needle tooth under the circumstance of avoiding top cutting is given. Pressure angle is an index to evaluate transmission performance. This paper deduces the relationship between pressure angle and rotation angle, simulates the influence of eccentricity, radius of pin teeth and radius of distribution circle of pin teeth on pressure angle. The simulation show that the position with pressure angle of 90 degrees appears at the top and root of cycloidal gear, which is not conducive to the transmission performance of cycloidal gear, the position with the best transmission performance appears in the middle of cycloidal gear.The change speed of pressure angle near the root of tooth is faster than that near the top of tooth. The results show that increasing the eccentricity properly or reducing the radius of the center circle of the pin teeth can make the minimum curvature close to the root of tooth, which is beneficial to improving the load-carrying capacity of cycloidal gear.

1. Introduction

RV reducer is the key component of industrial joint robots.It is used in robot joints because of its high load-carrying capacity and impact resistance.RV reducer includes two stages of deceleration.The first stage is involute spur gear planetary transmission and the second stage is cycloidal transmission. Cycloid pinwheel transmission is driven by the meshing of cycloid gear and a number of pin teeth.Friction heat will be generated in the transmission process and then thermal deformation will occur, which will affect the overall transmission accuracy.Therefore,it is necessary to analyze the geometric meshing of cycloidal transmission for improving the overall performance of RV reducer.

Scholars have done a lot of research on cycloidal pinwheel transmission such as RV reducer.Among them, YANG et al. has studied the relationship between processing error, design parameters and transmission performance in cycloidal transmission and deduced the calculation formula of transmission backlash and torque fluctuation[1].He Weidong et al. established the mechanical model of RV reducer, and put forward the mathematical model of cycloid gear optimization according to the clearance distribution of each tooth meshing[2-3].In reference[4-6],the error analysis and experimental study of RV reducer are systematically carried out by using the equivalent model of geometry and mass spring. The main factors affecting the transmission accuracy of RV reducer are revealed and the conclusion beneficial to improving the transmission accuracy of the system is drawn.In this paper, the geometric characteristics of cycloidal transmission are analyzed to provide guidance for the modification of cycloidal wheel.

2. Cycloid Gear Profile Equation

The principle of cycloidal transmission is shown in Figure.1.The track of the meshing point of the cycloidal pinwheel is a part of the circular arc of the pin teeth in the space coordinate system.

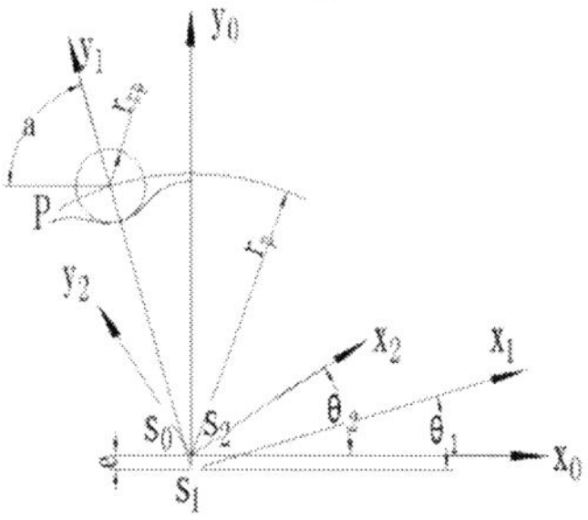

Figure 1. Cycloid-pin drive coordinate system

Let s_0 be the reference coordinate system, s_1 is a pinwheel coordinate system, s_2 is a cycloid coordinate system.he origin of the cycloid coordinate system coincides with the origin of the reference coordinate system.The distance between the origin of the pinwheel coordinate system and the origin of the cycloid coordinate system is e. r_{rp} is needle roller radius, r_p is the distribution circle radius of pin teeth. The position of point p on the pin tooth in the coordinate system of the pin wheel can be obtaine.

$$r_1(\alpha)=\begin{bmatrix}-r_{rp}\sin\alpha\\ r_p+r_{rp}\cos\alpha\\ 1\end{bmatrix} \tag{1}$$

Let θ_1 be the rotation angle of the needle wheel around the reference coordinate system. θ_2 is the rotation angle of the cycloid wheel around the reference coordinate system. z_p is the number of pinwheel teeth. i^H is cycloidal pinwheel transmission ratio.The equation can be obtain.

$$i^H=z_p/z_c=\theta_2/\theta_1 \tag{2}$$

The p-point trajectory on the pinwheel profile is transformed into the cycloid coordinate system by coordinates, the conjugate cycloid profile with the pinwheel can be obtained.

$$r_2(\theta_1,\alpha)=M_{21}(\theta_1)\cdot r_1(\alpha) \tag{3}$$

In the equation (3),coordinate transformation matrix $M_{21}(\emptyset_1)$

$$M_{21}(\emptyset_1)=\begin{bmatrix}\cos(\theta_1-\theta_2) & -\sin(\theta_1-\theta_2) & -e\sin(\theta_2)\\ \sin(\theta_1-\theta_2) & \cos(\theta_1-\theta_2) & -e\cos(\theta_2)\\ 0 & 0 & 1\end{bmatrix} \tag{4}$$

By substituting the relevant parameters and formulas (1) (3) (4) into (2),the tooth profile equation of cycloid wheel in reference coordinate system s_0 can be obtain.

$$\begin{cases}x_0=-(r_p-r_{rp}S)\sin(1-i^H)\theta_1-(e-k_1r_{rp}S)\sin(i^H\theta_1)\\ y_0=(r_p-r_{rp}S)\cos(1-i^H)\theta_1-(e-k_1r_{rp}S)\cos(i^H\theta_1)\end{cases} \tag{5}$$

In the equation (5), $S=(1+k_1^2-2k_1\cos\theta)^{-1/2}$, $k_1=ez_p/r_p$.

3. Curvature radius of cycloid wheel

According to the basic definition of curvature radius of curve,the curvature radius of cycloid gear tooth profile at any point can be calculated as follows:

$$\rho_c = \frac{(\dot{x}^2+\dot{y}^2)^{3/2}}{\dot{x}\ddot{y}-\ddot{x}\dot{y}} \tag{6}$$

The radius of curvature of cycloid gear profile can be obtain by deriving formula (5) with respect to θ_1 and substituting formula (6).

$$\rho = \frac{\left(1+k_1^2-2k_1\cos\theta_1\right)^{\frac{3}{2}} r_p}{k_1\left(1+z_p\right)\cos\theta_1-\left(1+z_p k_1^2\right)} + r_{rp} \tag{7}$$

Let $d\rho/d\theta_1=0$,The rotation angle corresponding to the maximum curvature radius of cycloidal gear profile can be obtain as follows:

$$\begin{cases} \theta_{11} = \arccos\frac{z_p(2k_1^2-1)-k_1^2+2}{k_1(z_p+1)} \\ \theta_{12} = 2\pi - \arccos\frac{z_p(2k_1^2-1)-k_1^2+2}{k_1(z_p+1)} \end{cases} \tag{8}$$

By substituting formula (8) into formula (7),the maximum curvature radius of the cycloid gear profile at the outer convex end is obtained as follows.

$$\rho_{max} = r_p\sqrt{\frac{27\left(1-k_1^2\right)(z_p-1)}{(z_p+1)^3}} \tag{9}$$

In the design of cycloid gear profile,it is necessary to avoid the phenomenon of tooth tip sharpening and tooth top cutting.For the concave profile,no matter what the radius of the pin tooth is,there will be no undercutting phenomenon.For outer convex profile,in order to avoid top cutting,the radius of needle teeth must be satisfied.

$$r_{rp} < r_p\sqrt{\frac{27\left(1-k_1^2\right)(z_p-1)}{(z_p+1)^3}} \tag{10}$$

From the above relations,it can be concluded that reducing the eccentricity or increasing the radius of the distribution circle of needle teeth can avoid the top cutting phenomenon and improve the transmission performance.The curvature radius of the cycloid wheel varies with the rotation angle as shown in Figure. 2 by simulating the parameters of the cycloid wheel in Table 1.

Table 1. The parameters of the cycloid wheel

Parameters	Value
Eccentricity e/mm	6
needle roller radius r_{rp}/mm	12
distribution circle radius of pin teeth r_p/mm	90
the number of pinwheel teeth	7

Because of the infinite curvature radius of the cycloid gear profile,in order to describe the change rule conveniently,the relationship between curvature and rotation angle is taken as shown in Figure 3.

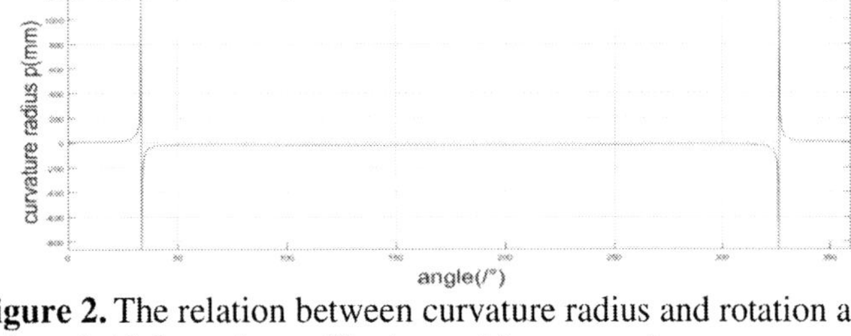

Figure 2. The relation between curvature radius and rotation angle

The curvature of cycloidal tooth profile is positive near the root, concave,negative near the top,convex. From figures 2 and 3,it can be seen that there are inflection points in the middle of the

IOP Conf. Series: Materials Science and Engineering **751** (2020) 012059 doi:10.1088/1757-899X/751/1/012059

tooth side. The curvature of cycloid tooth profile changes quickly near the tooth root, and the curvature change near the tooth top is smaller than that near the tooth root.In the convex tooth profile section,the minimum curvature is not at the tooth top,but near the tooth top.

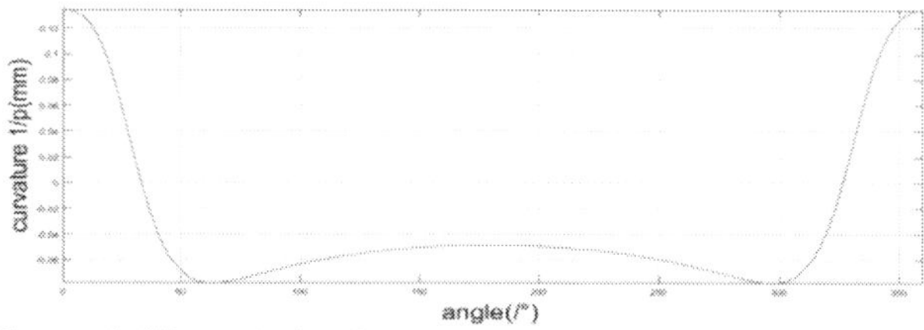

Figure 3. The relation between curvature and rotation angle

4. Pressure Angle of Cycloidal Tooth Profile

Pressure angle is an important index for judging transmission performance. From figure 4,In order to describe the transmission characteristics between cycloid gear and pin teeth,the angle between the relative speed of rotation of cycloid profile relative to geometric center and the force direction of cycloid profile at meshing point is defined as the pressure angle between cycloid profile and pin teeth.

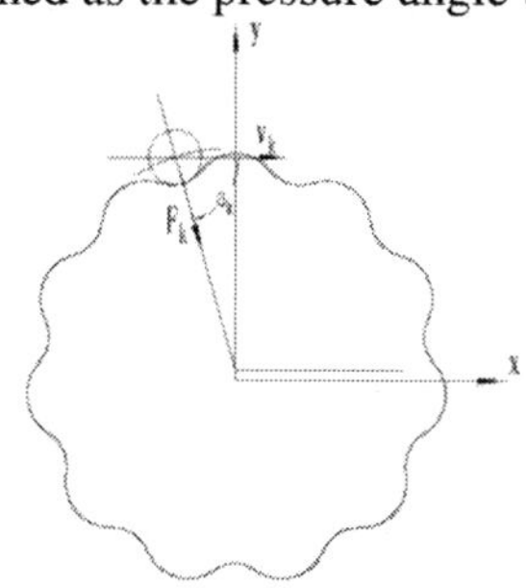

Figure 4. Pressure Angle of Cycloidal Tooth Profile

By definition,the expression of pressure angle can be describe as follows:

$$\alpha_k = \arccos\left(\frac{F_K \cdot v_k}{|F_K||v_k|}\right) \quad k = 1,2,3 \ldots z_c \tag{11}$$

In the equation (11), F_K is the force acting on the cycloid tooth profile at the meshing point K,The direction is the same as the normal vector n_K of the cycloid profile at point K.v_k is the velocity of the meshing point K relative to the center of cycloid wheel rotation.The expression of pressure angle can also be written as follows:

$$\alpha_k = \arccos\left(n_k \frac{v_k}{|v_k|}\right) \quad k = 1,2,3 \ldots z_c \tag{12}$$

Bring the relevant values into equation (12),the equation can be obtain as follows:

$$\alpha_k = \arccos\frac{-\sin\theta_k(1+\lambda^2-2\lambda\cos\theta_k)^{-\frac{1}{2}}}{(1-2r_{rp}(\lambda-\cos\theta_k)(1+\lambda^2-2\lambda\cos\theta_k)^{-\frac{1}{2}}+(\frac{r_{rp}}{r_p})^2)^{-\frac{1}{2}}} \quad k = 1,2,3 \ldots z_c \tag{13}$$

In the equation (13),$\lambda = \frac{r_p}{ez_p}$, $\theta_k = \theta + \frac{k-1}{z_p}2\pi$

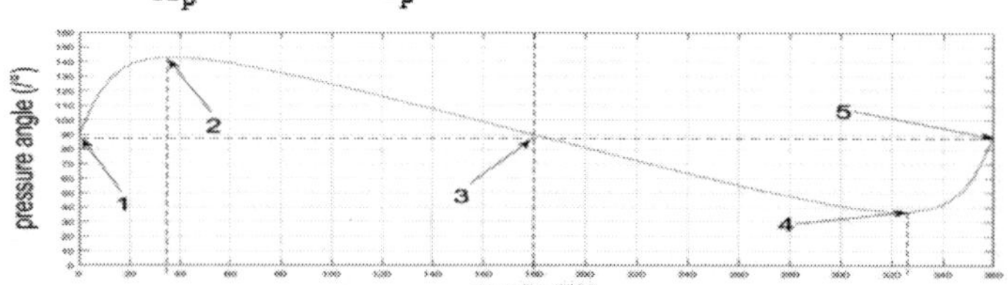

Figure 5. The relationship between pressure angle of single tooth and rotation angle

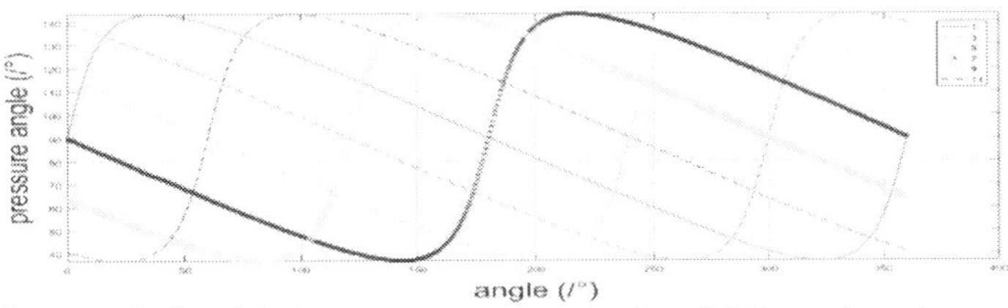

Figure 6. The relationship between pressure angle of full tooth and rotation angle

Figure 5 shows that the pressure angle equals 90 degrees when the rotation angle equals 0 degrees,180 degrees and 360 degrees respectively. The three positions are at the root 1, the top 3 and the root 5 respectively, which are dead points. When the rotation angle is between 0 and 180 the pin tooth 1 has no transmission effect on the cycloid wheel. When the rotation angle is between 180 and 360 the pin tooth 1 has transmission effect on the cycloid wheel. When the rotation angle is equal to $2\pi - \arccos\frac{z_p(2k_1^2-1)-k_1^2+2}{k_1(z_p+1)}$, the position 4 shown in Figure 5 has the smallest pressure angle, the best transmission performance and the smallest curvature. With this point as the demarcation point, along the root direction, the pressure angle increases rapidly, while along the top direction, the change rate is opposite. From the simulation results in Figure 6, the pressure angle of each pin tooth and cycloid gear profile has the same trend, but the phase angle exists in each curve. Of all the needle teeth, half of the pressure angle of the needle teeth is less than 90 degrees, only half of the needle teeth act as transmission.

5. Effect of Geometric Parameters on Pressure Angle

From the simulation results in figure 7 and figure 8,When the rotation angle is between 0 and 180 degrees,the pressure angle increases with the increase of eccentricity and decreases with the increase of the radius of the center circle of the pin teeth.When the rotation angle is between 180 and 360 degrees,the pressure angle is opposite to the eccentricity and the radius of the center circle of the pin teeth.Increasing the eccentricity or reducing the radius of the center circle of the pin teeth is beneficial to improving the transmission characteristics of cycloidal transmission.In addition, it can be seen from the simulation results that increasing the eccentricity or reducing the radius of the center circle of the pin teeth appropriately makes the rotation angle corresponding to the minimum curvature close to the root of the tooth,which is conducive to improving the load-carrying capacity of cycloidal gear.As can be seen from Figure 9,the radius of the pin tooth does not affect the pressure angle.

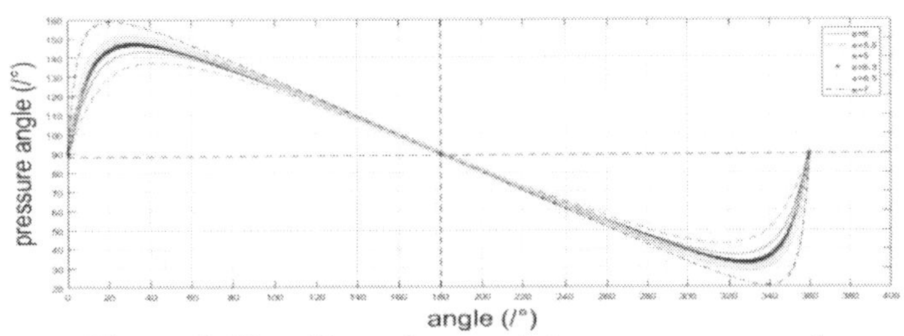

Figure 7. The effect of eccentricity on pressure angle

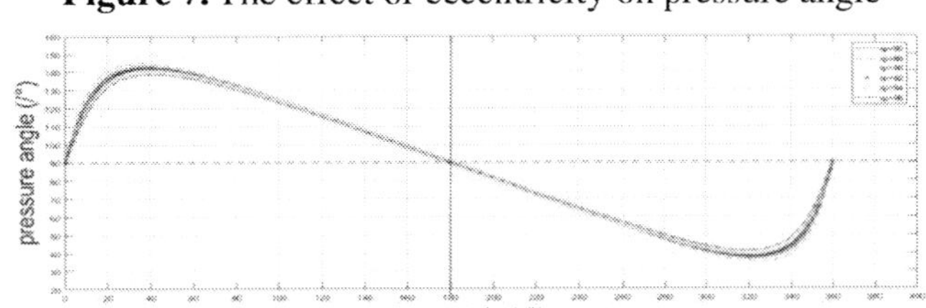

Figure 8. The effect of the radius of the distribution circle of the pin teeth on the pressure angle.

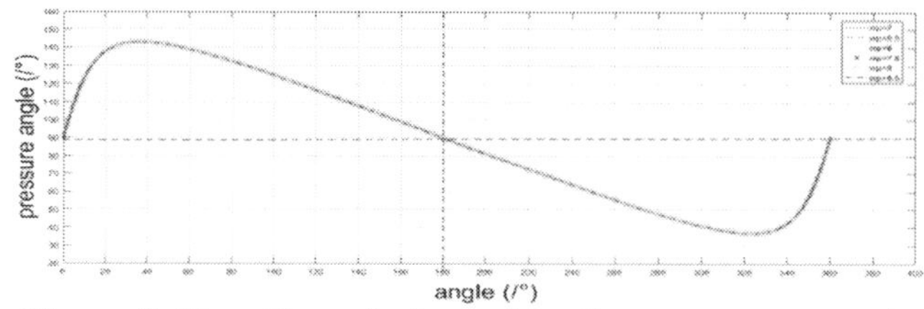

Figure 9. The effect of pin tooth radius on pressure angle

According to the principle of cycloidal transmission,there are relative sliding and rolling between cycloidal gear and pin teeth.The friction heat at the meshing point between cycloidal gear teeth and pin teeth is mainly caused by the relative sliding between them.Where the pressure angle is large is where the relative sliding occurs.From the above simulation results,it is known that the pressure angle of the top and the root of the tooth is the largest,the transmission performance is the worst,and the heat is the largest.This sub-region is not conducive to transmission.Therefore,the tooth top and root can be modified.

6. Conclusion

In view of the characteristics of cycloidal transmission,the equation of cycloidal tooth profile is established.The curvature radius of cycloidal tooth profile and the expression of pressure angle of cycloidal transmission are derived.The influence of geometric parameters on pressure angle is studied.From the above analysis,the following conclusions can be drawn:

1.Appropriate increase of eccentricity or reduction of the radius of the center circle of the pin teeth makes the rotation angle corresponding to the minimum curvature close to the root of the teeth, which is beneficial to improving the load-carrying capacity of cycloidal gear.

2.The dead-point position of cycloidal transmission occurs at the top and root of cycloidal gear,and the transmission performance is the worst. Therefore, the top and root of cycloidal gear can be modified.

3.The position with the best transmission performance appears in the middle of the cycloid profile, not in the top and root of the teeth.

Acknowledgments

This work was supported by the National High Technology Research and Development Program of China (Grant No.2015AA043002).

References

[1] YANG D C H, BLANCHE J G C. Design and application guidelines for cycloid drives with machining tolerances[J].Mechanism and Machine Theory,1990,25(5):487-501.

[2] HE Weidong, LI Lixing, XU Yongxian, et al. Analysis on the forces and efficiency of RV reducer transmission for high precision[J]. Chinese journal of mechanical engineering,1996, 32(4):104-110.

[3] HE Weidong,LI Lixing,LI Xin.New optimized tooth-profile of cycloidal gear of high precision RV reducer on robot[J]. Journal of Mechanical Engineering,2000(3):51-55.

[4] Hidaka T,Wang H,Ishida T,et al. Rotational transmission error of K-H-V planetary gears with cycloid gear(1st report, analytical method of the rotational transmission error)[J].JSME International Journal:Series C.1994,60(570):645-653(in Japanese).

[5] Ishida T,Wang H,Hidaka T,et al.Rotational transmission error of K-H-V planetary gears with cycloid gear (2nd report,effects of manufacturing and assembly errors on rotational transmission error)[J]. JSME International Journal: Series C,1994,60(578):278-285(in Japanese).

[6] Wang H,Ishida T,Hidaka T,et al.Rotational transmission error of K-H-V planetary gears with cycloid gear (3rd,mutual effects of errors of the elements on the rotational transmission error)[J]. JSME International Journal:Series C,1994,60(578):286-293(in Japanese).

IOP Conf. Series: Materials Science and Engineering **751** (2020) 012060 doi:10.1088/1757-899X/751/1/012060

Static and Dynamic Analysis of a 2T3R Five-degree-of-freedom Parallel Mechanism

Bofan Lu*, Guohua Cui and Qiang Guo

Shanghai University Of Engineering Science

*E-mail: 544328781@qq.com

Abstract. taking a 2T3R five-degree-of-freedom 4-SPRR-SPR parallel mechanism as the research object, a three-dimensional model and a finite element model of the parallel mechanism are established. Through the static analysis of the mechanism, the weak link of the whole mechanism is analyzed. It is then modally analyzed to analyze the natural frequency and mode shape of the mechanism. On the basis of the harmonic response analysis, the vibration resistance performance and the maximum influence frequency of the mechanism under external force are obtained.

1. Introduction

The five-degree-of-freedom parallel robot is a more important type of parallel mechanism with less degrees of freedom. It is more flexible than the three- or four-degree-of-freedom parallel robot. It is characterized by simple structure and easy control compared with the six-degree-of-freedom parallel robot. There are good application prospects in polishing, etc. [1-4], but there are relatively few studies on 4-SPRR-SPR five-degree-of-freedom parallel mechanism at home and abroad[5-6].

In this paper, a 2T3R five-degree-of-freedom 4-SPRR-SPR parallel mechanism is used for static analysis and modal analysis for harmonic response analysis. Analyze the weak points of the mechanism's force and anti-vibration, and provide theoretical support for building the prototype. The 3D model of the 4SPRR-SPR parallel mechanism is shown in Figure 1. The mechanism consists of a rotating shaft platform, a fixed platform frame and five branches connected to the fixed platform frame. The first branch is a -SPR branch, which consists of a rotating pair (R) and a moving pair (P). A ball pair (s). The 2, 3, 4, and 5 branches are -SPRR branches, which are composed of two rotating pairs (R), one moving pair (P), and one ball pair (s)[7].

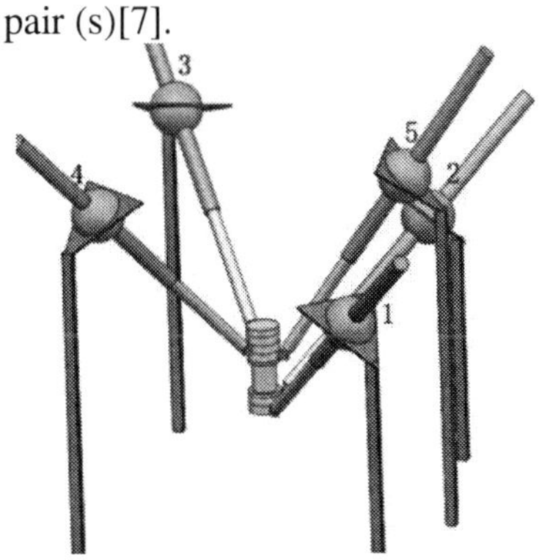

Figure 1. 4-SPRR-SPR Three-dimensional model of parallel mechanism

IOP Conf. Series: Materials Science and Engineering **751** (2020) 012060 doi:10.1088/1757-899X/751/1/012060

2. Static Analysis of 4-SPRR-SPR Parallel Mechanism

2.1. Building a finite element model

By considering the application environment of the 4-SPRR-SPR parallel mechanism, since the 4-SPRR-SPR parallel mechanism is a complex space geometry mechanism[8], the 3D model is first modeled in SolidWorks, and then the built complex 3D model is saved in xt format. And imported into the ansys, set the material to 45 steel, the density is 7890, the modulus of elasticity is 209 Gpa, and the Poisson's ratio is 0.269. The meshing is automatically divided by the system, and the finite element model of the 4-SPRR-SPR parallel mechanism is shown in Figure 2.

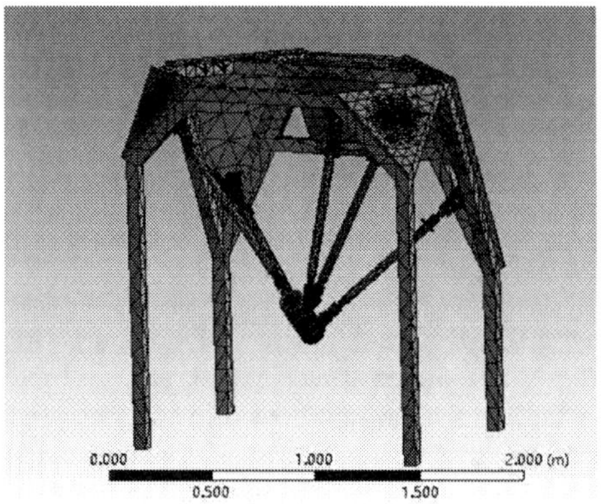

Figure 2. 4-SPRR-SPR Parallel mechanism finite element model

2.2. Static analysis

In ansys, simulate the loading of the organization in the actual working environment. Apply a force of 500 N to the moving platform of the mechanism and then fix it at the four low angles of the frame. Then the stress strain diagram of the 4-SPRR-SPR parallel mechanism can be obtained as shown in Figure 3.

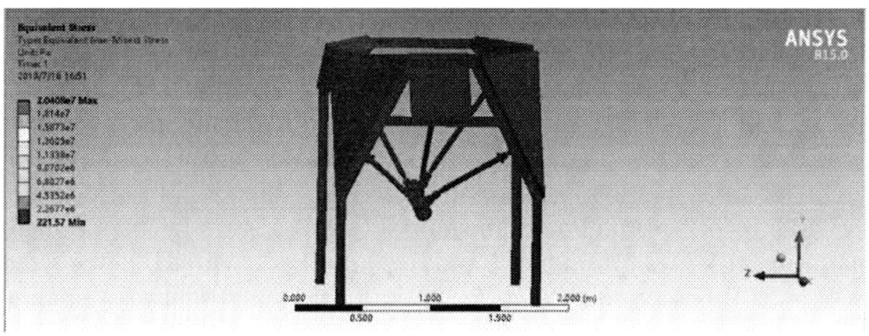

(a) Parallel mechanism stress map

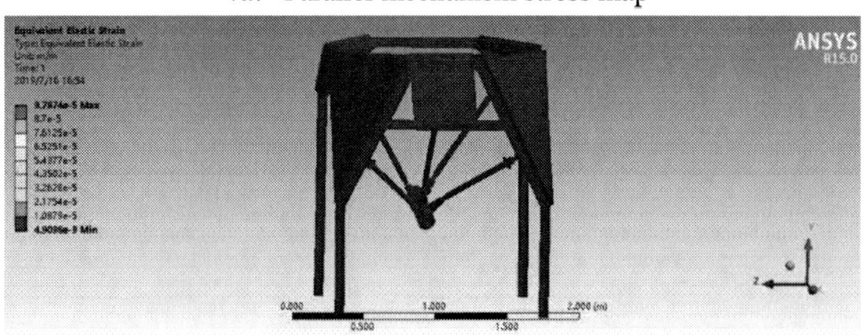

(b) Parallel mechanism strain diagram

Figure 3. 4-SPRR-SPR Parallel mechanism stress strain diagram

IOP Conf. Series: Materials Science and Engineering **751** (2020) 012060 doi:10.1088/1757-899X/751/1/012060

It can be seen from the stress diagram of the parallel mechanism that the overall force of the mechanism is relatively small, and the stress from the frame to each branch is relatively small. Only the stress on the moving platform is larger, and the maximum stress appears at the branch 3. But they are all within reasonable limits of the design. It can be seen from the strain diagram of the parallel mechanism in Figure. 3(b) that the overall strain of the mechanism is very small. It mainly occurs on each branch and moving platform, and the maximum strain occurs at the end of the branch 3 and the moving platform. The maximum strain is 0.0009mm. Fully in line with the actual application, the 4-SPRR-SPR parallel mechanism meets the static stiffness design requirements.

3. Modal analysis of 4-SPRR-SPR parallel mechanism

3.1. Modal analysis results and conclusions

Using the Modal module in ANSYS, the static platform is fixed according to the actual working condition of the 4-SPRR-SPR parallel mechanism, and then the undamped linear structure is freely vibrated. Only the payload on the zero constraint is considered in the module. Other loads will be can be ignored[9].

The nonlinear modal analysis solver analyzes and solves the 4-SPRR-SPR parallel mechanism, and obtains the first 6 natural frequencies of the mechanism (as shown in Table 1 and its corresponding mode (as shown in Figure 4).

Table 1. 4-SPRR-SPR Parallel mechanism natural frequency

Order	Frequency /Hz
1	17.367
2	17.917
3	27.418
4	65.439
5	79.048
6	87.078

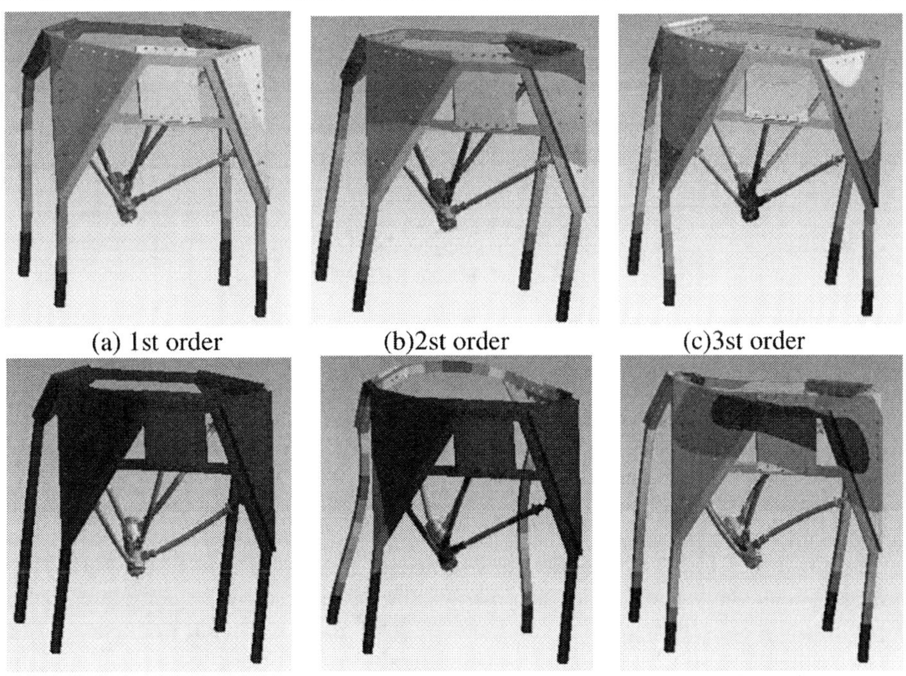

(a) 1st order (b)2st order (c)3st order

(d)4st order (e)5st order (f)6st order

Figure 4. 4-SPRR-SPR The 6 modes of the parallel mechanism

The first-order modal pattern is shown in Figure 4a. The vibration mode is mainly reflected in the parallel mechanism swinging left and right along the ZX plane. The displacement mainly occurs on the thrust ball bearing at the tail of the moving platform and the 4th and 5th branches, wherein the ball joint of the 4th branch is connected with the frame. The displacement is the largest.

The second-order modal pattern is shown in Figure 4b. The vibration mode is mainly reflected in the parallel mechanism swinging back and forth along the ZX plane. The displacement mainly occurs on the thrust ball bearing at the tail of the moving platform and the 4th, 5th and 3rd branches, wherein the displacement is maximum at the 5th branch.

The third-order modal pattern is shown in Figure 4c. The vibration mode is mainly reflected in the fact that the parallel mechanism is twisted along the Y axis, and the displacement mainly occurs on the parallel mechanism frame and the moving platform, wherein the displacement at the ball sub-fixing plate of the frame is the largest.

The fourth-order modal pattern is shown in Figure 4d. The vibration mode is mainly reflected in the parallel mechanism swinging up and down along the x-axis. The displacement mainly occurs on the moving platform and each branch, and the displacement of the fifth branch is the largest.

The fifth-order modal pattern is shown in Figure 4e. The vibration mode is mainly reflected in the fact that the parallel mechanism is twisted along the YZ plane, and the displacement mainly occurs on the mechanism frame and each branch, wherein the displacement of the fifth branch is the largest.

The 6th-order modal pattern is shown in Figure 4f. The vibration mode is mainly reflected in the left and right swing of the parallel mechanism along the YZ plane. The displacement mainly occurs on the moving platform and the 4th and 5th branches, wherein the displacement of the 5th branch is the largest.

In summary, the 1st and 2nd order modes show that the moving platform swings left and right along the ZX plane. There is little difference between the natural frequencies of the two. It shows that the stiffness of the moving platform along the ZX plane is similar. However, the natural frequencies of the 3rd to 6th order are much larger than the 1st to 2nd order. It is indicated that the motion stiffness of 3 to 6 is much larger than the motion stiffness of 1 to 2 steps. It can be seen from the vibration patterns of the various stages that the stiffness of the five branches and the moving platform and the frame joint is relatively weak, especially at the fifth branch (as shown in Table 2. However, the parallel mechanism runs at a lower speed during the operation, which indicates that the mechanism meets the application standards in terms of rigidity and stability.

Table 2. 4-SPRR-SPR Maximum displacement of natural frequency of each step of parallel mechanism

Order	Maximum displacement	Displacement/m
1	Ball pair of the 4th branch	0.047709
2	5th chain	0.042294
3	Ball sub-fixing plate	0.049533
4	5th chain	0.19242
5	5th chain	0.11305
6	5th chain	0.12537

4. Harmonic response analysis of parallel mechanism

Calculate the response of the linear structure at different frequencies and obtain an image of the response as a function of frequency. Under the premise of modal analysis of the parallel mechanism, the harmonic response analysis of the mechanism will be carried out by the complete method[10]. For the dynamic characteristics of the parallel mechanism, the high-order modal frequency is attenuated by the damping effect, so the influence of the low-order modal frequency or the fundamental frequency is mainly considered. The harmonic response analysis only considers the response of the mechanism in the range of 0 to 90 Hz. Figure 5 to Figure 7 show the displacement response curves of the 4-SPRR-SPR parallel mechanism moving platform along X, Y, and Z.

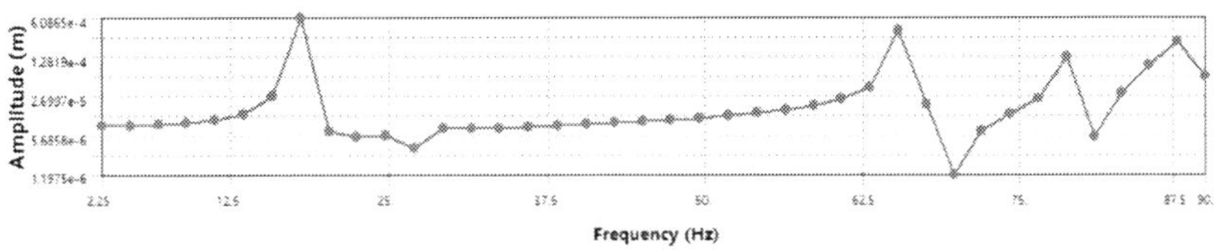

Figure 5. Displacement response curve of moving platform along X axis

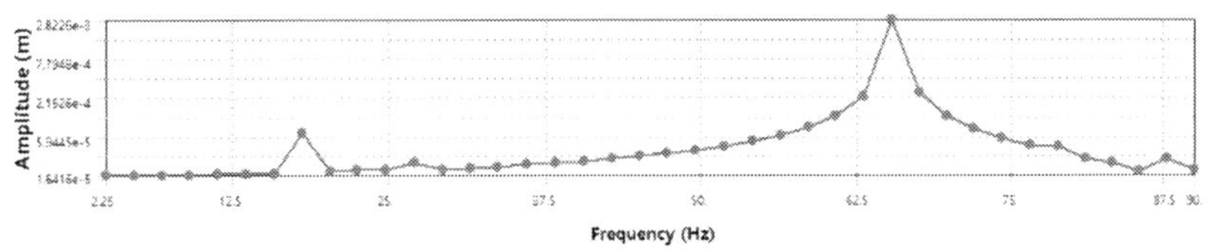

Figure 6. Displacement response curve of moving platform along Y axis

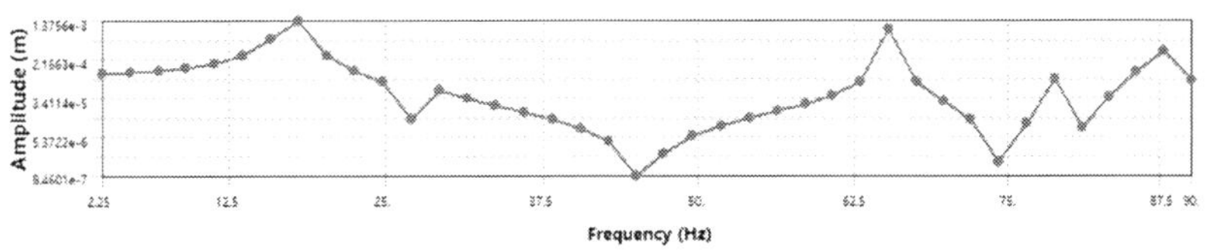

Figure 7. Displacement response curve of moving platform along Z axis

From the displacement response curves of X, Y, Z, it can be seen that in the vicinity of the natural frequency of the 1st and 2nd orders, the moving platform is easy to influence along the X and Z directions, and near the natural frequencies of the 3rd and 4th order, the moving platform is along the X, Y, The Z-direction has the most obvious effect, indicating that the 17Hz and 65Hz frequencies are most susceptible to damage to the 4-SPRR-SPR parallel mechanism. Therefore, in order to make the organization work properly, the frequency should be avoided as much as possible. Other natural frequencies have less resonance for the mechanism and do not affect the normal operation of the mechanism.

5. Conclusion

The static analysis of a 5 degree-of-freedom parallel mechanism is carried out by the finite element method. The maximum deformation of the mechanism when the mechanism is under force can be obtained, and the theoretical basis for the structural design and improvement of the mechanism is provided. Through the modal analysis of the 4-SPRR-SPR parallel mechanism, the first 6 natural frequencies and mode shapes of the mechanism can be obtained, and the natural frequency of the parallel mechanism meets the normal operation requirements. On this basis, the harmonic response analysis of the whole mechanism is carried out, and the weak links in the design of the mechanism are obtained, and the two frequencies that the mechanism should avoid in the work are pointed out. It provides a theoretical basis for the control of the 4-SPRR-SPR parallel mechanism.

References

[1] Gao F , Li W , Zhao X , et al. New kinematic structures for 2-, 3-, 4-, and 5-DOF parallel manipulator designs[J]. Mechanism & Machine Theory, 2002, 37(11):1395-1411.
[2] Piccin O , Bayle B , Maurin B , et al. Kinematic modeling of a 5-DOF parallel mechanism for semi-spherical workspace[J]. Mechanism & Machine Theory, 2009, 44(8):1485-1496.
[3] Gao F , Peng B , Li W , et al. Design of a novel 5-DOF parallel kinematic machine tool based on workspace[J]. Robotica, 2005, 23(1):35-43.

IOP Conf. Series: Materials Science and Engineering **751** (2020) 012060 doi:10.1088/1757-899X/751/1/012060

[4] Cui Guohua,Cui Kangkang,Wu Haizhen,et al.Reliability analysis of the pressing force of pre-stressed steel cylinder concrete pipe port grinding robot[J].Journal of Engineering Design,2018,25(06):647-654.
[5] Yu Yu. Development and application prospects of horizontal well production logging technology [J]. Mechanical Design, 2014(15) : 282-283.
[6] Gao Sheng, Sun Wen, Ni Wei, Chang Yulian. The current situation of oil well tractor design and some understanding [J]. Mechanical Design, 2014, 31 (02) : 1-8.
[7] Feng Zhiyou, Wang Meng Dynamic Analysis of a New 4-DOF Parallel Mechanism Based on ANSYS[J].Machinery Design,2016(3):88-91.
[8] Guo Qiang, Cui Guohua, Zhang Fan, et al. Kinematics and workspace analysis of 4-SPRR-SPR five-degree-of-freedom parallel mechanism[J],2019(1):34-40.
[9] Tian Meizi, He Fang, Wang Diandong. Modal analysis of palletizing robot based on finite element method [J]. Machine Tool & Hydraulics, 2015, 43(9) : 80-83.
[10] Li Hua, Zhang Dabin, Luo Yu, et al. Modal analysis of rotating roller cutter system for small mountain roasting tobacco rod puller [J]. Modern Machinery, 2015 (6): 46-49.

IOP Conf. Series: Materials Science and Engineering **751** (2020) 012061 doi:10.1088/1757-899X/751/1/012061

Numerical Simulation Method for Wave Surface Landing of Seaplane

Sun Feng[1, 2], Wang Mingzhen[1, 2], Zhang Jiaxu[1, 2] and Hu Qi[1, 2]

[1]Key Aviation Scientific and Technological Laboratory of High-Speed Hydrodynamic, Hubei, Jingmen, 448035, China
[2]China Special Vehicle Research Institute, Hubei,Jingmen, 448035, China

*E-mail: sunfeng686@126.com

Abstract: A seaplane is an aircraft that takes off and land on the water. The water surface is divided into a static surface and a wave surface. The numerical simulation technology of seaplane landing on wave surface is carried out. The accuracy of the simulation method of static water is verified by the comparison of simulation and experiment. The numerical wave generation is realized based on the pseudo-physical method, and the wave surface landing simulation is realized by combining the numerical simulation method with the numerical wave-making method. It is found that the pseudo-physical wave-making method can realize numerical wave-making and the precision is adjustable; then the water-water simulation of the wave surface is realized.

1. Introduction

Amphibious aircraft are aircrafts that can take off and land on both water and land. Their uses include transportation, water search and rescue, forest fire fighting, tourism and so on. They can also be used for maritime reconnaissance, patrol and anti-submarine missions. The amphibious aircraft will be impacted by the water surface during the surface landing, and the water impact load is one of the main load conditions of the amphibious aircraft. The initial analysis of the amphibious aircraft landing problem relies on engineering estimation methods and model tests. In recent years, with the development of computer hardware and numerical simulation software, numerical simulation has made great progress in solving large-scale engineering problems. There are more and more numerical simulation studies on water impact.

StreckWall et al [1] established the hydrostatic surface model of the aircraft in two programs Ditch and Comet respectively and obtained the change of the movement posture during landing on water. Nathalie [2] used the FE, SPH and FE-SPH hybrid methods to establish the water-entry model of an aircraft and studied the damage of the aircraft during the ditching and the influence of structure energy absorption on dynamic response. Zeng Yi et al [3]. carried out numerical simulation calculations for the response of amphibious aircraft landing on wave-water and analyzed the load characteristics. Ma Zenghui et al [4] carried out numerical analysis on the seakeeping resistance of amphibious aircraft landing on the wave and analyzed the influence of wave height and wavelength on the seakeeping resistance. Yang Rong et al [5] used the fluid-solid coupling technique to analyze the overload of the center of gravity of the aircraft and the pressure response of the bottom of the hull. Qiu [6] uses CFD technology to calculate the aerodynamic and hydrodynamic forces of the amphibious aircraft during take-off by decoupling. The information is exchanged at each time step, and the resistance calculation results have better accuracy. R. Ortiz [7] used the SPH method to simulate the landing of the aircraft on the water and analyzed the structural deformation and damage under the action of the ditching load.

At present, the research on the simulation of amphibious aircraft landing on the wave is still in the method exploration stage, and it can't meet the need to solve engineering problems. The research on this problem is divided into three steps, and the simulation method of the hydrostatic surface is first verified. Then numerical simulation of the wave environment is achieved. Finally, the simulation method of aircraft landing on wave is formed by combining with the static water surface simulation method and numerical wave method to form a wave surface.

2. The verification of hydrostatic surface simulation method

The water-entry process of the single ship hull is simulated using the nonlinear finite element software LS-DYNA. The aircraft structure adopts the Lagrange element algorithm, and the fluid adopts the ALE element algorithm. The element size is 0.02m, which satisfies requirements of the calculation result convergence, the simulation method uses the penalty function method to solve the contact force at the coupling interface. Figure 1 shows a calculation model for a single hull, including a single hull model, water and air. The research focuses on the water load and motion response of the aircraft. The deformation of the structure has little effect on the water load and motion response, only a few percent. Therefore, the single hull simulation model is rigidly processed, which is beneficial to improve the calculation efficiency. It is also possible to ensure the consistency of the weight, the position of the center of gravity, and the inertia of the calculation model by the definition of the rigid body attribute.

Figure 1. FE model

Figure 2 and Figure 3 show the simulation and experimental comparison of the acceleration and pitch attitude angles at the center of gravity of the model. It can be seen from the figure that the simulation and test data have a high degree of coincidence in the time history, which verifies the effectiveness of the simulation method.

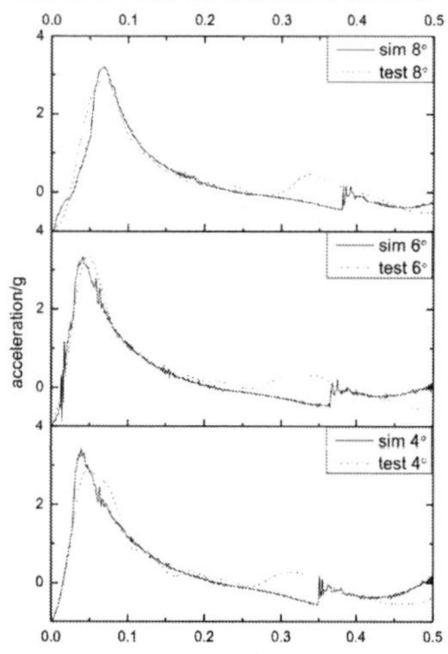

Figure 2. the acceleration at different cases

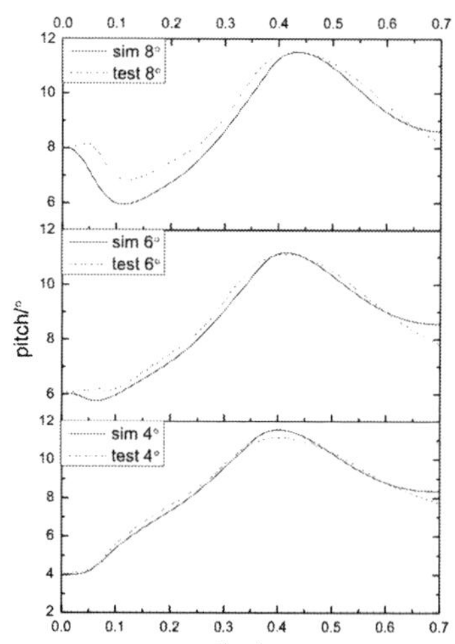

Figure 3. the pitch angle at different cases

3. Numerical wave

Pushing waves is one of many imitation physics waves. The wave-making idea is like the real experimental wave pool, that is, the forced vibration of the simulated solid boundary is used as the disturbance source, thus generating the waves required for the experiment.

H is the wave height, h is the depth of the still water, that is, the height from the original still water surface to the bottom of the water tank, S is the push plate stroke, the equation of motion of the push plate:

$$x = \frac{S}{2}\sin \omega t \tag{1}$$

Based on the assumption of the micro-wave, the relationship between the velocity potential and the speed of the push plate can be obtained:

$$\begin{aligned} \frac{\partial \phi}{\partial x} &= u = \omega \frac{S}{2}\cos \omega t \\ \frac{\partial \phi}{\partial z} &= w \end{aligned} \tag{2}$$

In the process of numerical wave making, when the fluid is treated as an incompressible fluid, the velocity potential satisfies the Laplace equation throughout the calculation region:

$$\frac{\partial^2 \phi}{\partial x^2} + \frac{\partial^2 \phi}{\partial z^2} = 0 \tag{3}$$

Using the separation variable method to solve the Laplace equation, the expression of the velocity potential function can be obtained.

$$\phi = C_0 \frac{\omega}{k_0}\cos(k_0 z)\sin(k_0 x - \omega t) - \sum_{n=1}^{\infty} C_n \frac{\omega}{k_n}\cos(k_n z)e^{-k_n x}\cos(\omega t) \tag{4}$$

The following wave equation can be obtained:

$$\eta(x,t) = C_0 S\cos(k_0 x - \omega t) + \sum_{n=1}^{\infty} C_n \sin(k_n z)e^{-k_n x}\sin(\omega t) \tag{5}$$

among them: $C_0 = \dfrac{4\sinh^2 k_0 h}{2k_0 h + \sinh 2k_0 h}$

The first term on the right side of the upper middle is the expression of the traveling wave. The latter is the standing wave, which is generated during the movement of the wave-making plate. The standing wave disappears with its own propagation, so it can be ignored. The linear wave is obtained at a distance from the wave plate. For a wave with a given wave height, the stroke of the push plate is

$$S = \frac{A}{C_0} \tag{6}$$

The Euler mesh is used in the calculation of the flow field in this study. The geometrical dimensions of the water is 60m long, 1m wide and 1.5m high. The air field is 60m long, 1m wide and 1.5m high. The upper part is the air field. The lower part is water, and the finite element model information is shown in Table 1. The target wavelength is 6.25m, the wave height is 0.6m, and the period is 2s.

Table 1. Model information

	Water domain	Air domain
Element type	8 node Euler elements	8 node Euler elements
material	*MAT_NULL	*MAT_NULL
Number of Element	90000	90000

Boundary conditions have an important influence on wave generation. For models without a wave-eliminating zone, except for the moving boundary and the upper surface of the flow field, the other boundaries must be bound in normal.

Figure 4 shows the waveforms at different times. By observing the wave generation process, the first wave is continuously consumed as the waveform propagates. This is due to the energy loss caused by the viscosity and numerical dissipation of the fluid. The waveform distortion closest to the wave-making boundary is more serious, but it will get closer and closer to the target waveform as the propagation distance increases.

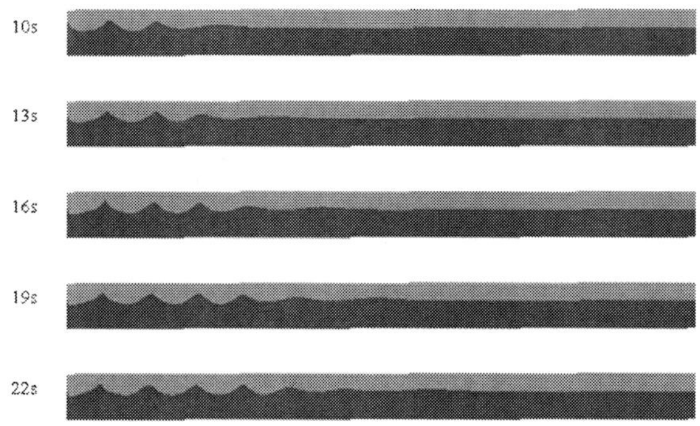

Figure 4. wave form at different times

The peaks of the four waveforms are selected to take the distance between the two adjacent peaks. The simulated wavelength is compared with the target wavelength. As shown in Table 2, The wavelength will decrease as the propagation distance increases. the overall error is small.

The waveform at different positions will be stable after 3-4 waves along the wave propagation direction. The average wave height and period at a position farther from the wave boundary will have a small decrease. The wave height and period at the four positions is like each other. The wave height and period of the waveform at 14m are compared to the target wave parameters to verify the quality of the generated wave. As shown in Table 2, the errors of the period and the wave height are Within 10%, the wave quality can be further adjusted by adjusting the parameters of the boundary motion equation. As time goes by, the error tends to increase gradually. Therefore, when the simulation of the wave surface is forced to drop, the time with better wave precision can be selected as the water-entry time.

Table 2. Wave quality

Wave length/m		error	period /s		error	Wave height /m		error
simulation	target		simulation	target		simulation	target	
6.5	6.25	4%	2.18	2	9.0%	0.56	0.6	6.7%
6.2	6.25	0.8%	2.18	2	9.0%	0.55	0.6	8.3%
6.1	6.25	2.4%	2.17	2	8.5%	0.55	0.6	8.3%

4. Simulation of aircraft landing on wave

Figure 5 shows the simulation of aircraft landing on wave simulated by the hydrostatic surface simulation method and the numerical wave-making technique. First, the model is stationary, the motion boundary conditions are set, and the wave is started. When the wave reaches the designated position, the aircraft starts to move achieving aircraft and wave encounters. It can be seen from the figure that after the aircraft encounters the wave, the aircraft gradually begins to rise. When the aircraft encounters the second wave briefly, it is bounced out of the water. Figure 6 shows the wave surface acceleration curve. Compared with the static water surface, the acceleration peak is reduced, but the duration is longer.

Figure 5 Wave water surface watering process

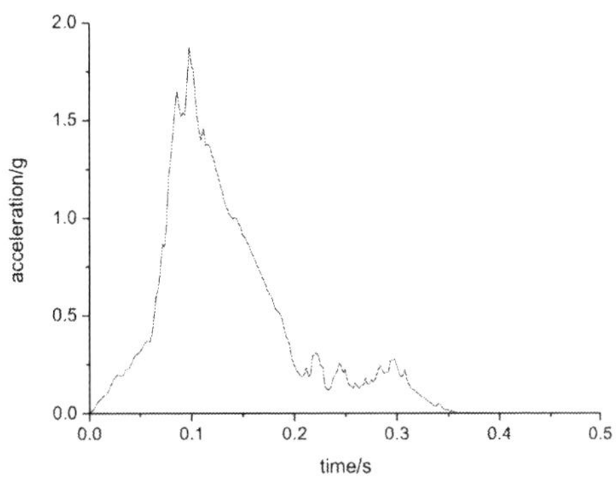

Figure 6 Wave water surface acceleration

5. Conclusion

The main research conclusions are as follows.

1) The effectiveness of the simulation method is verified by comparing the simulation results of the hydrostatic surface with the experimental results.

2) Applying the pseudo-physical wave-making theory to the numerical wave-making process, the numerical wave is realized. The wave quality is quite well;

3) Based on the static water surface simulation technology and numerical wave-making technology, the simulation of aircraft landing on waves is realized. The simulation motion process and acceleration time history curve are calculated.

References

[1] H Streckwall, O Lindenau, L Bensch. Aircraft ditching: a free surface/free motion problem [J]. Archives of Civil & Mechanical Engineering,2007,7(3):117-190.

[2] R S TNathalie. Contribution to the modeling and simulation of aircraft structures impacting on water[J]. Applied & Environmental Microbiology,2009,64(12):4774-4781.

[3] ZENG Yi,YANG Shi-fu,LUO Lin-yin. The structural response of amphibian landing onto wave[J]. JOURNAL OF MACHINE DESIGN,2015,32(9):96-100.

[4] MA Zeng-hui, LIU Li-sheng, YUE Zhen, et al.Numerical investigation on seakeeping performance

IOP Conf. Series: Materials Science and Engineering **751** (2020) 012061 doi:10.1088/1757-899X/751/1/012061

of amphibious aircraft landing on wave[J].Chinese Journal of Computational Mechanics, 2018,35(3):380-386.

[5] YANG Rong, LV Ji-hang. Dynamic load characteristics of large firefighting amphibious aircraft [J]. Journal of Shenyang Aerospace University,2016,33(3):18- 24.

[6] Liangjun Qiu. Efficient Decoupled Hydrodynamic and Aerodynamic Analysis of Amphibious Aircraft Water Takeoff Process[J]. JOURNAL OFAIRCRAFT,2013,50(5):1369-137 9.

[7] R. Ortiz, J.L. Charles, J.F. Sobry. Structural loading of a complete aircraft under realistic crash conditions: generation of a load database for passenger safety and innovative design[C].24TH INTERNATIONAL CONGRESS OF THE AERONAUTICAL SCIENCES,2012

Design and Experimental Research of Rope Missle

Haoting Wu[1], Lei Zhao[5], Pingjun Wu[2], Feng Xiao[1], Yongxiu Guo[2], Ming Jiang[3], Junlei Geng[5] and Lei Yang[3,*]

[1] Communication construction branch of State Grid Corporation of China, Beijing, China.
[2] Beijing ZhongDian Golden Bridging & Technology Development Co., Ltd, Beijing , China
[3] China Electric Power Research Institute Co., Ltd, Beijing, China
[4] Beijing Electric Power Engineering Co., Ltd., Beijing, China
[5] State Grid Beijing Electric Power Company, Beijing, China

[*]Corresponding author: 379247828@qq.com

Abstract. The UAV (unmanned aerial vehicle) unfolding guide rope is the construction process adopted in the tension stringing of overhead power transmission line. At present, due to the limit of UAV control precision, the function of threading the guide rope into the paying-off sheave groove directly cannot be realized, and we can only unfold the guide rope of UAV to the cross arm of iron tower, and then, make the operation personnel at height transfer the guide rope into the steel wire groove of paying-off sheave. Therefore, the construction efficiency of this process is relatively, and there is operation at height, with a certain safety risk. In this project, it is planned to develop a launching device carried by UAV, and this device uses the compressed carbon dioxide (CO_2) which is relatively safe as power source, which realizes the instantaneous release of compressed air through electromechanical servo control, and then generate the power which promotes the "rope missle" to fly. The tail of "rope missle" is connected with guide rope, and when it flies forward, the traction guide rope will move forward synchronously. After the "rope missle" threads into the sheave, the compressed air will be released completely, and then, the "rope missle" will transfer into the motion of a free falling body, thus further driving the falling of guide rope, so as to realize the function of "rope missle" tracts the guide rope to realize threading into the paying-off sheave.

Key words: guide rope; compressed air; UAV; launching device

1. Introduction

With the continuous development of our country's economy, and as the foundation of social production, the demand of the society for power supply is increasing day by day, and the matched power grid construction tasks are also more and more arduous. However, during the process of power grid erection construction, the power grid erection is a decisive part, which directly influences the construction progress of power grid erection and completion quality of engineering. [1-3] During the construction, to erect the power transmission line to the power transmission iron tower, workers are usually required to thread the guide rope into the paying-off sheave. On one hand, this limits the construction efficiency severely, and on the other hand, this also increases the hazards during the construction severely. [4-9]

IOP Conf. Series: Materials Science and Engineering **751** (2020) 012062 doi:10.1088/1757-899X/751/1/012062

To avoid the operation risks of power workers and enhance the working efficiency, the research reported in this paper focuses on the operations including utilizing rope missle to carry the guide rope, and then substitute the power workers to complete the erection of guide rope. Take the rope missle as the actuator of stringing operation, so as to substitute the power workers to complete the action of threading guide rope into the paying-off sheave.

2. System Structure

The rope missle system mainly refers to dedicated "rope missle" subsystem, which consists of missle body case, CO_2 air cylinder, striker, release system, control circuit and other main parts. The details of the rope missle are shown in Figure 1.

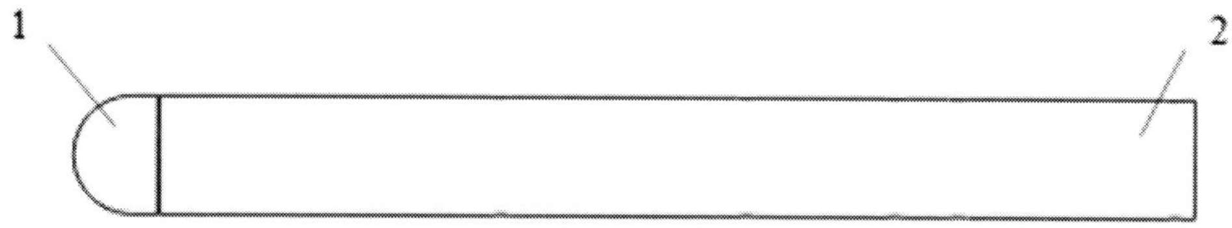

(a) Appearance Diagram of Rope Missle

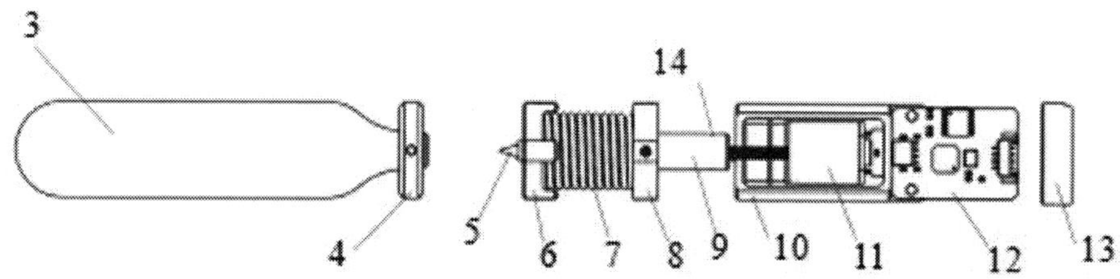

(b) Internal Structure of Rope Missle

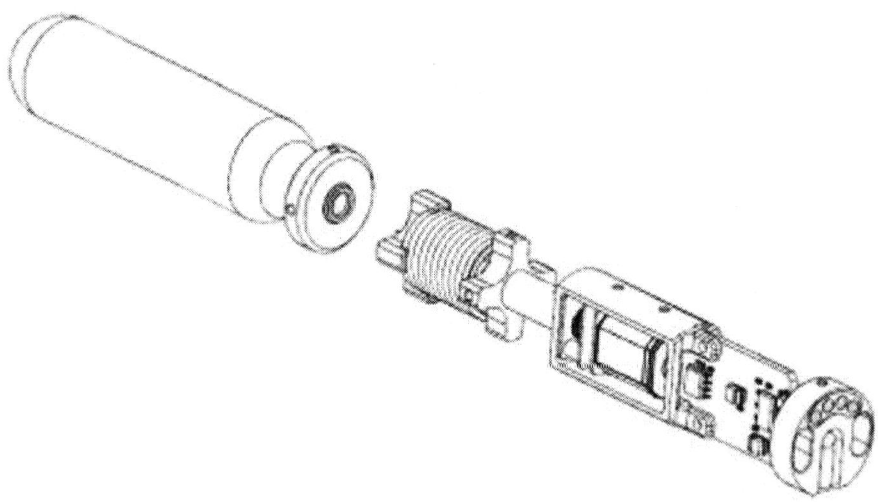

(c) Side View of Rope Missle

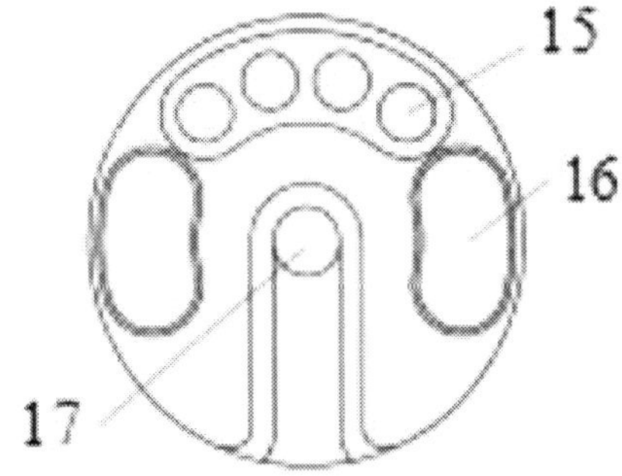

(d) Tail Structure Diagram of Rope Missle

Front cover (1), missle body (2), compressed air cylinder (3), air cylinder washer (4), striker (5), striker base (6), spring (7), spring limit base (8), locking nut (9), motor base (10), speed reduction motor (11), main control circuit board (12), tail flow electrical combination (13), jackscrew (14), electronic contact (15), holes on both sides (16), connection hole (17)

Figure 1. Details of the Rope Missle

3. Working Principle of "Rope Missle" System

The rope missle is protected with external case, and the appearance diagram of rope missle is shown in Figure 1(a). The external case includes front cover and missle body, and the thread mode is adopted for rotary joint between front cover and missle body. Screwing off the front cover, as shown in Figure 1(b) and Figure 1(c), insert the compressed air cylinder, and the rear part of air cylinder is fixed with air cylinder washer, and the air cylinder washer is fixed with screw and missle body.

The tail of striker is thread, fixed on the striker base via rotary joint, and a spring is installed between striker base and spring limit base. There is a hole in the middle of spring limit base, the screw rod at the tail of striker base threads into the hole, which is to be fixed with locking nut. The spring limit base is fixed with missle body via screw.

The tail of locking nut is connected with the output shaft of speed reduction motor, and the jackscrew is adopted for fixing. The speed reduction motor is installed in the motor base, and the motor base is fixed with the missle body via screw. The main control circuit board is installed on the motor base, connected with the tail flow electrical combination via wire arranging. The tail flow electrical combination, as shown in Figure 1(d) has four electronic contacts, with holes on both sides, used to release gas. There is a connection hole in the middle, which is used to install fixed guide rope.

In case of launching, the rope missle is connected with the control system via four contacts on the tail flow electrical combination, so as to obtain electric energy and control command. The control interface adopts RS-232 serial port connection to receive the control command issued from upper computer, and then give feedback information to the upper computer.

If the rope missle receives the effective launching commands (including unlocking of authority, command launching, etc.) issued by the upper computer, the main control circuit board will drive the speed direction motor for rotation. When the motor is rotating, the output shaft will drive the locking nut for rotation, and the tail thread of striker base will be screwed off from locking nut gradually till that it is released.

At this time, the spring will promote the striker base to move towards the air cylinder mouth rapidly, and the striker which moves at a high speed will pierce the sealing aluminium foil at the air cylinder mouth, and then the high pressure gas will be released and exhausted from the air exhaust outlet, so as to promote the rope missle move forward in the launching tube, thus realizing the function of rope missle launching.

4. Design Thought of Structure of "Rope Missle" System

To solve the problem of personnel at height threading guide rope into sheave during the stringing, the function of UAV carrying rope missle to tract the guide rope to thread into the sheave is taken into consideration, and there has been no example of this technology at home and abroad at present.

4.1. Model Selection of Power

With regard to the design of rope missle, it is necessary to consider the model selection of launching power source firstly, which shall not only launch the rope missle, but conform to relevant national rules, and the specific kinetic energy of rope missle shall be smaller than 1.8J/cm2; the management and using of gunpowder are strict, and the spring rubber band is too powerful, so these two power sources are excluded. Through referring to the power source of net gun in the public security system, finally, the compressed CO_2 gas is adopted as the power source of rope missle, and this kind of power is safe and controllable, with convenient storage. There are two modes of steel-type air cylinder of CO_2 compression: 12g and 45g; the weight and external diameter of 12g air bottle are more convenient for UAV in installation and carrying.

The internal diameter of missle body is 18.7mm, with external diameter of about 22mm, length of about 198mm, and weight of missle body of about 140g.

Through analysis and calculation of above parameters, when the launching tube is 30cm in length, the missle body can obtain the initial speed of about 10-15m/s, which basically conforms to the system design requirements.

4.2. Structure of Striker Device

The compressed air cylinder refers to steel structure, and the mouth of air cylinder has a layer of aluminium film, and only after the aluminium film is pierced can the gas be released. In this case, it is necessary to research a kind of electronic and mechanical structure to pierce the aluminium film via the numerical control device.

1. Confirm that the striker is to be adopted to pierce the aluminium film.
2. After testing, it is known that only the striker of 50N can be used to pierce the aluminium film, and through calculation with Hooke's law, we can select the spring which can generate thrust required by striker.
3. The release structure adopts thread unlocking device, and the striker and thread unlocking device adopts top-quality steel material, which not only guarantees that the thread of striker structure will have no thread slipping during the using process; in the static status, the thread device is used to stick the spring; in the working status, the spring is to be released via controlling the rotation of thread shaft.
4. The thread shaft is to be driven via motor, and after adopting the motor of n20 which is provided with reduction gearbox, the speed reduction ratio of thread shaft to motor is 1:60, and the power supply voltage of motor is 5V.

4.3. Control Circuit

The control circuit can store the secrete key of launching of "rope missle", and it can drive the striker release mechanism to complete the execution of launching command;

Considering safety, each main board of the circuit part which is driven by the striker has sole ID No. and secrete key for launching unlocking, and the tail of missle body of circuit is reserved with power supply and data interface; the launching system must very the secrete key of launching via dedicated protocol, so as to drive the striker, thus ensuring safety and reliability of missle body launching.

The control circuit can store the secrete key of launching of "rope missle", and it can drive the striker release mechanism to complete the execution of launching command;

Considering safety, each main board of the circuit part which is driven by the striker has sole ID number and secrete key for launching unlocking, and the tail of missle body of circuit is reserved with power supply and data interface; the launching system must very the secrete key of launching via dedicated protocol, so as to drive the striker, thus ensuring safety and reliability of missle body launching.

The design diagram of control circuit is shown in Figure 2.

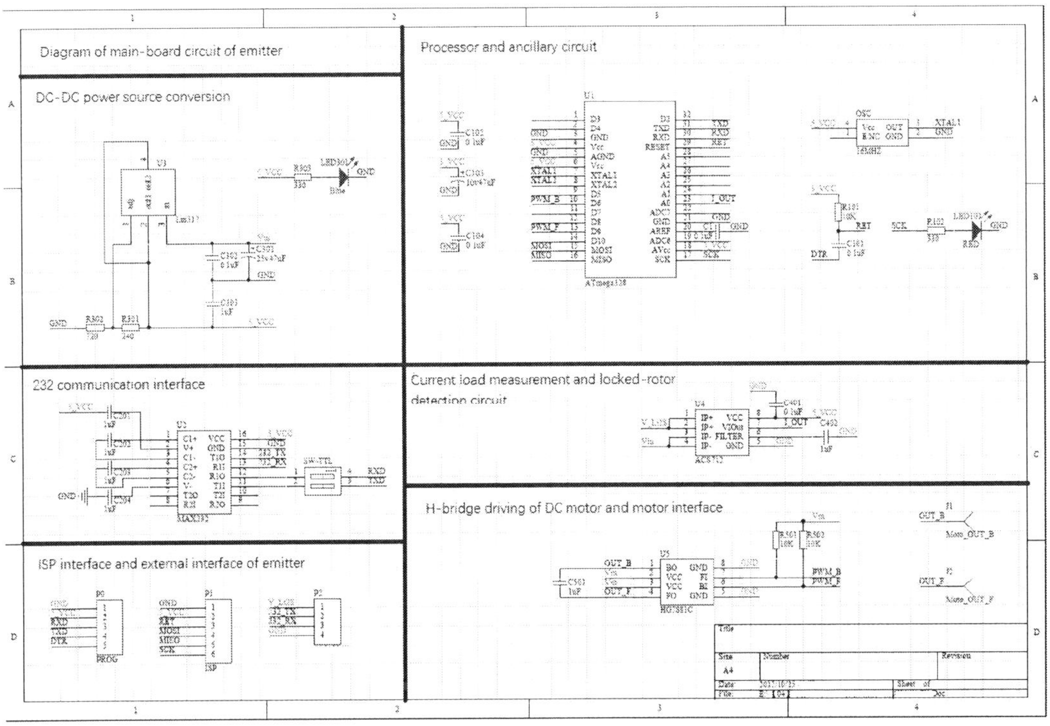

Figure 2 The design diagram of control circuit

The inner of missle body has driving circuit for micro-processor and electro-mechanical system, and the power supply and data connection of rope missle and launching system are realized via four electrical contacts at the tail of missle body. The micro-compressor is mainly used to complete the logic function control and control of secrete key of launching, which can avoid risks brought by wrong operation and unauthorized launching.

4.4. External Case of Missle Body

Through survey and testing, the external case adopts aerial composite material, featured by light weight and solid structure; the external case is used to package and fix all internal components; the surface electroplating technology of all aluminium alloys is adopted to reduce the friction with launching tube to the greatest extent.

The external case tail is designed with a washer, and the washer is designed with an open hole for gas exhaust; the electronic contact is used for power supply and data connection; the middle connection hole is used to carry the guide rope.

5. Key Points of the System

(1) Adopt compressed air as power, and adopt the electronic circuit to control mechanical servo device, so as to release the striker; the striker which moves at a high speed is adopted to pierce the sealed aluminium foil of air cylinder, so as to realize the gas release, thus completing the launching of rope missle and traction of guide rope.

(2) Adopt the thread at the tail of striker base, and connect it with the locking nut driven by speed reduction motor, so as to realize the locking and release of striker.

IOP Conf. Series: Materials Science and Engineering **751** (2020) 012062 doi:10.1088/1757-899X/751/1/012062

6. System Test

6.1. Function Testing

6.1.1 Force Required by Striker to Pierce Aluminium Film of Compressed Air Cylinder Connect the striker and the common spring, and place them into a sleeve tube. Adopt 8 pieces of thin rope to connect at the edge of striker, and connect the other end of thin rope with dynamometer. In case of testing, fix the compressed air cylinder in front of the sleeve tube mouth, and pull the dynamometer towards the back of sleeve tube. After pulling to a certain value, shear the thin rope at the dynamometer end, and then the spring will promote the striker to pierce the aluminium film.

Conduct testing repeatedly, and then test the force required by striker for piercing the aluminium film of compressed air cylinder.

6.1.2 Testing of Spring inside Rope Missle Select the spring radius according to air cylinder size; calculate the elasticity modulus and length of spring as required according to the data obtained from the test in 5.1.

6.1.3 Testing of Thread Locking Device Rotate the thread column, make the thread column separate away from the thread cap, release the spring, and during the testing, it is found that the thread has been polished gradually after the thread column and thread cap made of aluminium have been used for many times, so there is safety hazard. Later, after the thread locking device is changed into that made of steel, and after multiple times of testing, it is proved to be reliable and stable.

6.2 Performance Testing

Install an instrument for speed measurement in front of the rope missle device. The speed measurement instrument consists of two optoelectronic switches, one of which emitting infrared light, and the other receiving infrared light. When there is an object in the middle of two switches, there will be a signal, and the distance between two optoelectronic switches is 5cm.

When the rope missle is launched out of the tube, the head of shell will successively shut out the first optoelectronic switch and the second optoelectronic switch, and these two optoelectronic switches will generate two signals; through the time points recorded by two signals, the initial speed of rope missle can be worked out, and through the diameter and weight of rope missle, we can work out the specific kinetic energy of missle.

It is worked out that the specific kinetic energy of rope missle leaving the shell is smaller than 1.8J/cm2, which conforms to relevant legal rules of the State.

7. Conclusions

The using of UAV with the view of threading by carrying rope missle provides technical guarantee for automation of guide rope unfolding in the power stringing engineering, and targeting the problems including low automation degree of guide rope unfolding, labor force and time waste and that there is a certain danger, this paper analyzes the actual task demand of unfolding of guide rope, researches the mechanism design of a kind of rope missle, designs and realizes the rope missle for realizing automatic unfolding of guide rope, and verifies the method correctness under the simulation environment. This method supplements the insufficiency of existing construction method of guide rope unfolding, which has significant meaning for enhancing construction efficiency of power stringing and reducing construction cost.

Main research achievements of this paper are shown as below: according to actual demands of stringing tasks, it has been proved that when the rope missle is launched, a sufficient initial speed will be obtained in the launching tube, and after the rope missle leaves the launching tube, there will still be a certain power, which makes the rope missle fly forward, thus driving the guide rope to fly effectively, and then the paying-off operation function by threading the guide rope into the sheave is realized. It can be used to substitute the construction technology of manual operation of artificial tower climbing. Targeting demand, the traditional method of threading guide rope into sheave is improved by combining actual situation. A corresponding experimental platform is set up, and under the laboratory environment, the experimental verification is conducted to the feasibility of "rope missle" guide; in addition, the method correctness is verified, the stringing process is simulated, and the result shows that the "rope missle" meets the stringing requirements.

Acknowledge
This Research was supported by Science and Technology Research Project of State Grid (Research on the construction technology of launching guide rope through pulley based on ballistic theory. GCB17201800306).

References

[1] Zhang Jin. Research on Stringing Construction Technology in Power Transmission Line Erection [J]. Practical Electronics, 2013.
[2] Qiao Yuanyuan. Research on Comprehensive Evaluation of Stringing Construction Scheme of Power Transmission Line [D]. Xi'an: Xi'an University of Architecture and Technology, 2016.
[3] Xu Jingcan. Study on Status and Correlative Factors of Mental Health among Electric Linemen in Hunan [D]. Changsha: Central South University, 2011.
[4] Liang Xinglong. Application of Dynamic Parachute in High-voltage Overhead Transmission Line Construction [J]. China New Telecommunications, 2013 (20).
[5] Sun Bin, Yang Yuun, Wang Min. Analysis on Cost Effectiveness of Application of UAV Stringing Based on Operation Cost Method [J]. China Power Enterprise Management, 2016.
[6] Liu Guosong , Jia Jiqiang. Application and Development Direction of UAV in Power System [J]. Journal of Northeast Electric Power University, 2012, 32(01):53-56.
[7] Wei Dongni, You Fei, Liu Xiaomin, Yu Bo. Analysis on Economy of UAV Application in Power Stringing [J]. Construction Materials & Decoration, 2018(04):243.
[8] Chen Qun. Analysis on Application of Stringing Construction Technology of Power Transmission Line [J]. Telecom Word, 2014(23):69-70.
[9] Liu Chao. Discussion on Construction of Tension Paying-off and Threading Power Transmission Line [J]. Telecom World, 2014(15):40-41.

IOP Conf. Series: Materials Science and Engineering **751** (2020) 012063 doi:10.1088/1757-899X/751/1/012063

Study on the Rate of Elastic-plastic Crack Propagation of Heterogeneous Metal Welded Joints in Nuclear Power

G B Li*, H Xue ,Y Q Bi and L Zhang

School of Mechanical Engineering, Xi'an University of Science and Technology, Beilin District, Xi'an, Shaanxi, China

*Email: ligb_xkd@163.com

Abstract. In the high-temperature water environment of nuclear power, austenitic stainless steel, nickel-based alloy materials and welded joints composed thereof are the key locations for environmental cracking (EAC) problems such as stress corrosion cracking (SCC). In order to understand the crack propagation behavior and expansion rate of welded joints under the condition of abrupt and heterogeneous material properties of welded joints, this paper uses the extended finite element technique (XFEM) in the large nonlinear finite element software ABAQUS, based on the elastoplastic material model. Under the condition of welding residual stress field, the effects of different initial crack length and strength mismatch ratio on the crack growth rate of stress corrosion crack under constant load were studied. The results show that the SCC crack growth rate increases with the increase of the initial crack length. When the crack length of the initial welded joint is constant, different mismatch ratio has a certain effect on the crack growth rate. In the low ratio range(less than 1), the crack growth rate also increase if the strength mismatch ratio increase. but the crack growth rate is gradually reduce in high ratio range(more than 1). This indicates that the mismatch ratio of 1 is favorable for crack propagation, and the larger the mismatch ratio, the crack propagation is hindered.

1. Introduction

The welding process uses a lot of processes in nuclear voltage vessels and pipes, so there are many welded joints on equipment such as nuclear voltage vessels, which are the most dangerous parts [1]. Due to the very complicated internal structure of the welded joint, mechanical properties and the residual stress generated during the welding process, various defects may occur, such as cracks, inclusions, etc., the welded joint is the main area that causes structural failure which are EAC occurrence and crack propagation[2]. Quantitative prediction of EAC crack growth rate in high temperature water environment has been an important topic for Domestic and foreign scholars. It is hoped that an accurate SCC crack growth rate prediction model can be established [3-4]. Nuclear power safety end welding process of dissimilar metal welded joints can be divided into three steps: smelting, casting and heat treatment. Each step is very complicated, forming a difficult research topic, and residual stress is generated after welding, and residual stress It will cause stress corrosion cracking behavior of nuclear power materials [5]. After studying the X52 steel in depth, Contreras-Cuevas et al. used a number of different welding methods to study the behavior of SCC. The circumferential and longitudinal residual stresses were measured by the X-ray method. The measurement results show that 98% are circumferential residual stress and 74% are longitudinal residual stress. This study provides a reference for the effects of residual stress on SCC behavior [6]. Using A533, Alloy182 alloy and -

A533B low alloy, Zhao Lingyan established an inhomogeneous material model close to the actual welded joint in the finite element software, and studied the mechanical field of crack propagation crack tip under the interaction of complex working load and welding residual stress. A quantitative prediction model of crack growth rate. The results show that the residual stress has a certain influence on the mechanical field and expansion rate of cracks in metal welded joints [7]. Wang Haitao and other researchers at East China University of Science and Technology used experimental methods to perform fracture experiments on dissimilar metal materials. Their experimental results show that the crack propagation path has a certain relationship with the strength matching ratio of the welding consumables and the base metal of the welded joints. A new method for local damage assessment [8].

Through the knowledge of fracture mechanics theory, the length of the crack is a very important factor affecting the fracture toughness of the material, which will have a certain influence on the crack tip field. Many research scholars have done a lot of research on the relationship between the mechanical field and the crack length of the relevant static crack tip. And the SCC crack growth rate is calculated by the correlation formula.[9]. But there is a little research on the relationship between mechanical field and length of dynamic crack tip. Therefore, based on the elastoplastic material model, this paper approximates the nuclear welding joint to a "sandwich" material mechanics model, considering the residual stress, using the XFEM technology in commercial finite element software ABAQUS to study homogeneous materials and dissimilar metal materials. The mechanical field and expansion rate of crack propagation under different initial crack lengths are studied, and the crack propagation of dissimilar materials with different mismatch coefficients is discussed.

2. Numerical model

2.1. Geometric models and material models

EAC experiments in high-temperature water environments often use compact tensile specimens (CT) under constant load. The specimens are compact, so cracks can extend longer distances during loading. American Society for Testing and Materials Standards ASTM E399-90 [10]. In this paper, the crack tip mechanical field at the interface position of dissimilar materials will be studied. Therefore, a pre-cracked CT specimen is selected for analysis. The geometry is shown in Figure 1, a=W/2, W=50mm, B=W/2= 25mm. The simplified "sandwich" model of the welded joint is shown in Figure 2. The specimen with the crack at the material interface is selected (Figure 2-II), and the interface crack propagation characteristics are analyzed.

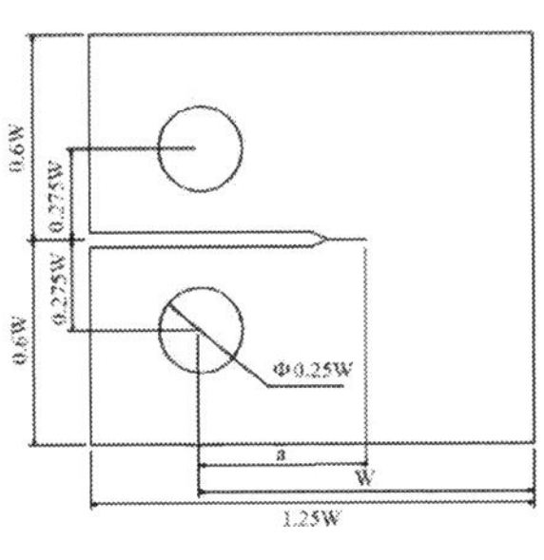

Figure 1. Geometric size.

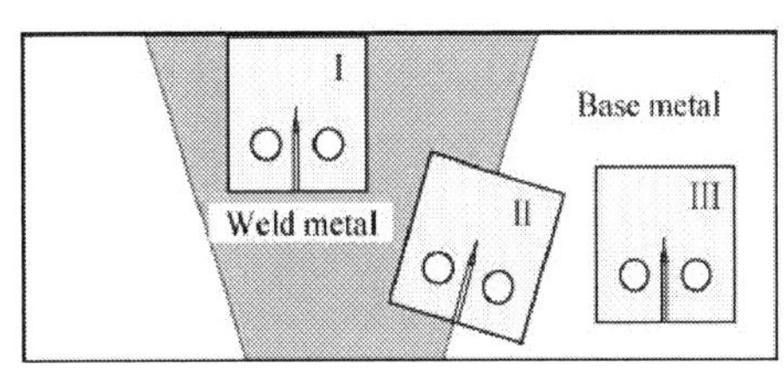

Figure 2. Sample selection position.

Different locations of nuclear power dissimilar metal welded joints have different cracking and crack propagation behaviors. The traditional EAC crack propagation rate prediction method is based on linear elastic fracture mechanics (LFEM). However, a large number of experimental evidences indicate that the plasticity of the material The EAC crack growth rate has a very important effect. According to the elastoplastic theory, the plastic properties of isotropic materials are established. We know that the deformation process of elastoplastic metal materials consists of initial elastic segments

IOP Conf. Series: Materials Science and Engineering **751** (2020) 012063 doi:10.1088/1757-899X/751/1/012063

and plastic deformation segments. In this paper, a plastic material with no obvious yielding platform is used. The material mechanics relationship is in accordance with the Ramberg-Osgood relationship.

This paper mainly analyzes the effects of two mismatched forms, high and low, on the crack propagation law. A typical high-profile welded joint, that is, the strength of the base metal is less than that of the weld metal, and vice versa. The specific material parameters are shown in Table 1, which are the parameters of the material under high temperature and high pressure.

Table 1. Material parameters under high temperature and high pressure (340 ° C).

Materials	Elasticity modulus E(MPa)	Poisson's ratio ν	Yield strength σ0(MPa)	Hardenin g exponent n	Deviation ratio α	M=σ0W/ σ0B
Base metal	206000	0.28	254	3.6	1.0	1
Welding material (high ratio)	206000	0.28	304.8	4.04	1.0	1.2
Welding material (high ratio)	206000	0.28	355.6	4.5	1.0	1.4
Welding material (low ratio)	206000	0.28	152.4	2.78	1.0	0.6
Welding material (low ratio)	206000	0.28	203.2	3.19	1.0	0.8

2.2. Finite element model

The finite element mesh of this paper is 2658 linear plane strain elements (CPE4), the number of grids is 4926, and the crack extension area mesh is refined, and the number of grids is 1652, as shown in Figure 3.

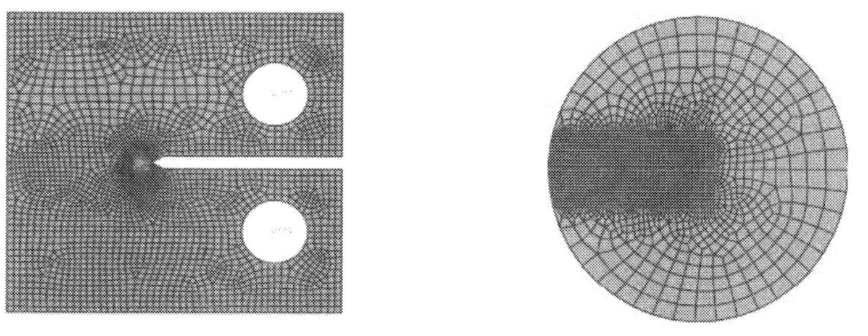

(a) CT sample grid (b) Local mesh refinement

Figure 3. Finite element model mesh.

In this paper, the XFEM technique is used to simulate the stress corrosion cracking crack propagation under the dead load condition. The amplitude curve is used to control the load application. According to the actual situation, the load is linearly increased to the maximum value through the 0.35s load within 1s of the analysis step time. All time calculations are kept constant.

Effect of Different Initial Crack Length on Crack Growth Rate

2.3. Variation of crack tip stress field of homogeneous and dissimilar materials

In this paper, the initial crack lengths are 1mm, 2mm, 3mm, and 4mm respectively. As shown in Figure 4(a→d), they are crack propagation stress field Mises with different initial crack lengths under residual stress and constant load interaction. It can be seen from the figure that the stress field on both sides of the crack tip of the extended crack is symmetrically distributed, and is the largest in the small range near the crack tip, this mean, the stress concentration zone appears in the crack tip during the expansion process, and will increase with the initial length of the crack increase. And the crack tip is still symmetrically distributed, and the stress field is enhanced. Figure 5(a→d) shows the Mises stress cloud diagram of dissimilar metal crack propagation at a different initial crack length with a mismatch ratio of 0.6. In the figure, the stress field distribution on both sides of the crack tip is asymmetrical, but the difference of the stress cloud map is not obvious. With the increase of the initial crack length, the stress field of the dissimilar metal crack tip is enhanced.

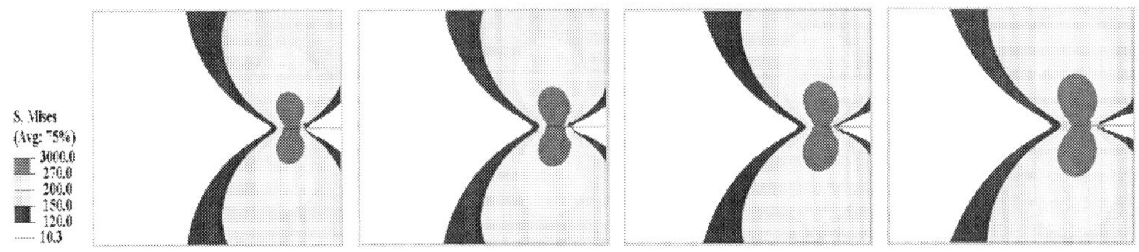

(a) a=1mm (b) a=2mm (c) a=3mm (d) a=4mm

Figure 4. Crack tip PEEQ in the process of crack propagation of dissimilar metals.

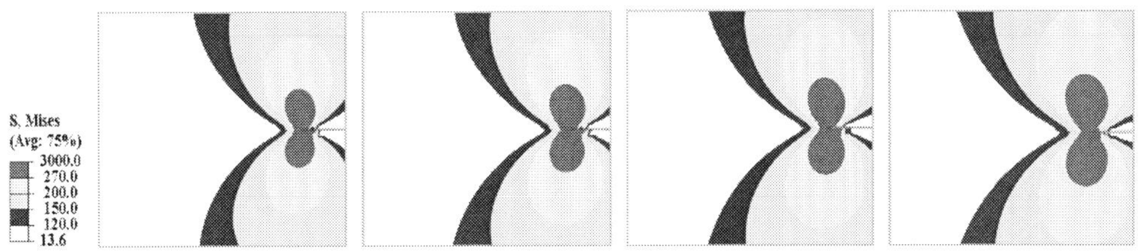

(a) a=1mm (b) a=2mm (c) a=3mm (d) a=4mm

Figure 5. Crack tip PEEQ in the process of crack propagation of dissimilar metals.

2.4. Variation of strain field in homogeneous and dissimilar materials

Figure 6(a→d) shows the crack propagation field of Homogeneous materials and dissimilar metals under different initial crack lengths. It can be seen from the figure that in the elastic-plastic material, plastic strain occurs after the crack propagates, and the plastic strain continues to remain in the crack. With the increase of the initial crack length, the strain field of the crack tip increases, the stress concentration of the crack tip is obvious, and the symmetric distribution state remains. Figure 7 (a→d) is the dissimilar metal strain field. Under the external load, due to the material mismatch ratio (The difference in strength mismatch ratio of 0.6) is clearly asymmetrically distributed on both sides of the crack tip, and we can find that the plastic strain is larger on the side where the yield strength is smaller. And because it is an elastoplastic material, its plastic strain will remain in the expanded path.

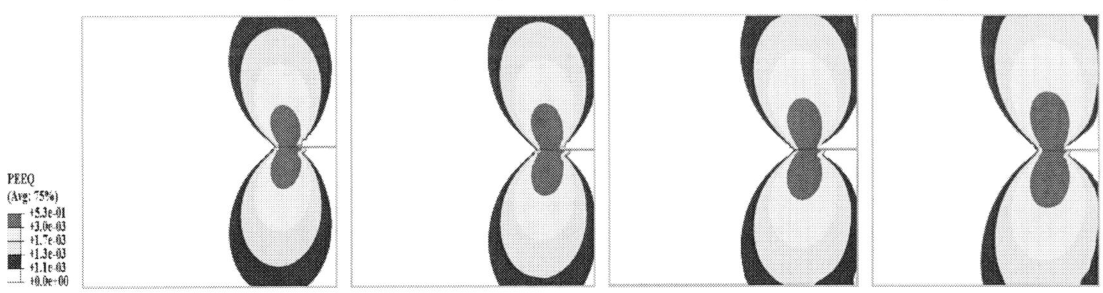

(a) a=1mm (b) a=2mm (c) a=3mm (d) a=4mm

Figure 6. Crack tip PEEQ in the process of crack propagation of dissimilar metals

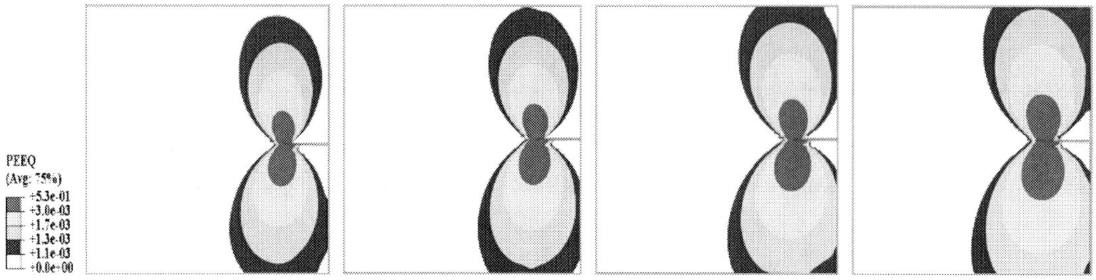

(a) a=1mm (b) a=2mm (c) a=3mm (d) a=4mm

Figure 7. Crack tip PEEQ in the process of crack propagation of dissimilar metals

2.5. Crack growth rate change

Figure. 8 is a graph showing the crack growth rate (CGR) of the homogeneous material and the dissimilar metal material as a function of the initial crack length. From the figure, we can find that when the initial length of the crack is different, the SCC crack growth rate will increase with time, and increase sharply in the later stage, which is the same as the stress-strain field cloud pattern. After cracking, the crack growth rate increases gradually; the crack growth rate of the homogeneous material is greater than the crack growth rate of the dissimilar metal, which indicates that the equivalence ratio is favorable for crack propagation of the welded joint, and as time goes by, we will find The rate of SCC crack propagation will increase faster and faster, almost exponentially increasing.

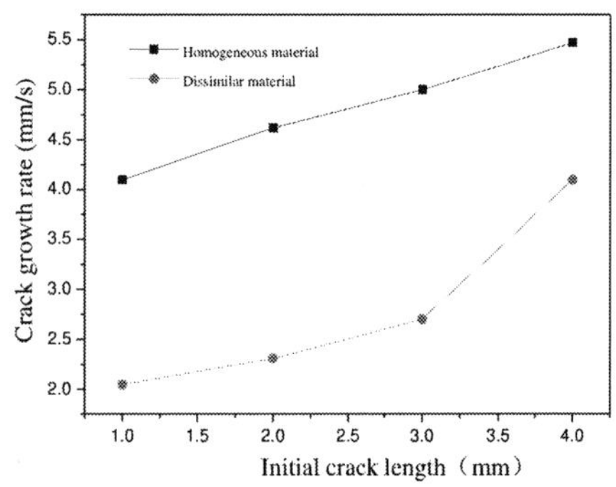

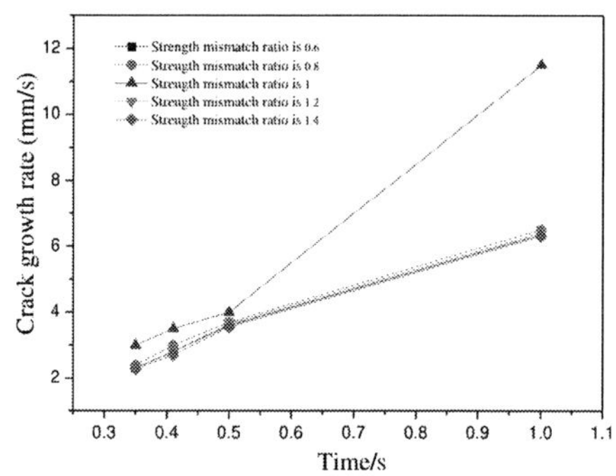

Figure 8. CGR of homogeneous and dissimilar metals **Figure 9.** CGR under different mismatch

3. Change of crack growth rate under different mismatch coefficients

Figure 9 shows the expansion rate of the extended crack at the interface of the dissimilar materials when the initial crack length is 2 mm under different mismatch ratios (0.6, 0.8, 1, 1.2, 1.4). Since the welded joints are made of different materials. There are inhomogeneities in the properties and geometry of the material, which results in different SCC crack growth rates. The crack growth rate at 0.35s, 0.41s, 0.5s, and 1s in the crack propagation history was selected. From the figure, we can see that when the intensity mismatch ratio is 1, the crack will first crack and its crack will occur. The expansion rate is the fastest, and the rate increases significantly in the late stage of crack propagation, indicating that the strength ratio is favorable for crack propagation. By comparing the crack growth rates of different intensity mismatch ratios, we found that when the mismatch ratio coefficient is high, the crack growth rate is the slowest, that is, the high composition ratio hinders the crack growth.

4. Conclusions

(1) In this paper, XFEM technology is used to study the crack growth rate of different initial crack lengths at the interface between uniform materials and dissimilar metal elastoplastic materials under constant load. It is found that the longer the initial crack length in dissimilar metal materials, the SCC crack propagation rate will be faster.

(2) By establishing different strength mismatch ratio material models, it is concluded that the different mismatch ratios of dissimilar metal materials affect the crack growth rate. When the ratio is low, the SCC crack growth rate is faster, and the high distribution time is lost. The larger the ratio, the smaller the crack growth rate and the better the crack growth rate when the mismatch ratio is 1.

Acknowledgments

This work was supported by the Structural Safety Evaluation Laboratory of School of Mechanical Engineering, Xi'an University of Science and Technology. And Gangbo Li also wishes to acknowledge the financial support the National Natural Science Foundation of China 51475362, Shaanxi Provincial Department of Education Research Project Funding Project 12JK0657

References

[1] Dai D P Numerical simulation of residual stress in austenitic stainless steel welded joints [D]

[2] Yang F Q 2014 Study on crack tip creep characteristics and environmental cracking quantitative prediction model of nuclear power structural materials. Xi'an University of Science and Technology

[3] Shoji T and Suzuki S 1995 Theoretical prediction of SCC growth behavior -- Threshold and plateau growth rate[J]. *Nace International Houston Tx*

[4] Shoji T and Lu Z 2010 Formulating stress corrosion cracking growth rates by combination of crack tip mechanics and crack tip oxidation kinetics *Corrosion Science* 52(3):0-779

[5] Ji J 2008 Austenitic stainless steel weldability analysis *Coal mine machinery* 29(7):85-86

[6] Contreras-Cuevas A and Alamilla-López 2014 The Role of Residual Stresses in Circumferential Welding Repairs of Pipelines in SCC Susceptibility[C]// Trans Tech Publications p 159-168.

[7] Zhao L Y 2014 Study on the mechanical characteristics of crack tip of nuclear power welded joints and the rate of environmental cracking crack growth *Xi'an University of Science and Technology*

[8] Huang J Y and Chiang M F 2013 Environmentally assisted cracking behavior of dissimilar metal weldments in simulated BWR coolant environments Journal of Nuclear Materials 432(1-3):189---197

[9] Wang H T and Wang G Z 2013 Local fracture behaviour of nuclear-electric 52M nickel-based alloy dissimilar metal welded joints nuclear technology 36(4):000139-144 [10] Shoji

[10] ASTM Standard E399-90. Standard test method for plane strain fracture toughness of metallic materials. Annual Book of ASTM Standards, 2002, (03.01)

IOP Conf. Series: Materials Science and Engineering **751** (2020) 012064 doi:10.1088/1757-899X/751/1/012064

Energy and Exergy Analysis of Steam Source Optimization for Boiler Soot Blowing System

Wanfan Wu[1, 2], Fangyuan Ran[1], Fei He[1*], and Jianhua Wang[1]

[1]CAS Key Laboratory of Mechanical Behavior and Design of Materials, Department of Thermal Science and Energy Engineering, University of Science and Technology of China, Jinzhai Road 96, Hefei 230027, PR China

[2] Datang east China electric power test and research institute, Chuangxin Street 666, Hefei 230088, PR China

[*]**Email:** hefeihe@ustc.edu.cn

Abstract. In order to improve the economy of boiler soot blowing system, the steam source position of boiler soot blowing system should be optimized. For a 1000MW ultra-supercritical coal-fired unit, the thermodynamic process of soot blowing steam in temperature and pressure reducing station was studied through energy analysis and exergy analysis in this paper, and it was revealed that the annual water and energy consumption of boiler soot blowing system was independent of the steam source position, the change of steam source position only affected the annual exergy loss of soot blowing system. The comparion results indicated that the annual exergy loss of soot blowing system with cryogenic reheater outlet as steam source was minimal. After adopting cryogenic reheater outlet as the steam source position of soot blowing system, it would be more economical than the original soot blowing system with platen superheater outlet as steam source. The annual exergy loss would be reduced by about 2.183×10^{10}kJ, and it was equivalent to an annual increase of power generation revenue by about 2486000CNY.

1. Introduction

Steam soot blowing is widely adopted to clean the fly ash adhering to heating surface during the operation of utility boilers due to high efficience and strong coal adaptability. The steam employed is generally taken from the outlet header of the boiler's platen superheater, which will cause large energy loss when blew into furnace in view of its high temperature and pressure. Therefore, in order to enhance the operation economy of the boilers, some power plants have already decreased the quality of soot-blowing steam and thereby the energy loss by optimizing the steam source positions of steam soot blowing [1]. However, when analyzing the economy of steam soot blowing with steam from different source positions, power generation enterprises either simply compared the steam enthalpy at different steam source positions and calculated the economic benefits by the enthalpy difference after optimizing the steam source positions [2], or simply compared the steam exergy at different steam source positions and calculated by the exergy difference after optimizing the steam source positions, but did not throughly analyze the thermodynamic process of soot-blowing steam in temperature and pressure reducing station, and thus neglected the influence of the desuperheating water there on the economy of soot-blowing steam. As a result, they are unable to measure the economic benefits of steam at the steam source comprehensively.

IOP Conf. Series: Materials Science and Engineering **751** (2020) 012064 doi:10.1088/1757-899X/751/1/012064

2. Boiler and Soot Blowing System

The boiler of a 1000MW ultra-supercritical generator is DG3060/27.46-π1 converter DC furnace manufactured by Dongfang Boiler Works and its design parameters are illustrated in Table 1. According to the requirements of soot blowers, the steam pressure and temperature should be controlled at about 2 MPa and 350 °C respectively after the steam passes through the temperature and pressure reducing station. Meanwhile, the total soot-blowing steam consumption of the boiler is about 117800 tons per year based on statistics.

Table 1 The designed steam parameters

Name	Unit	BMCR	BRL	THA	75%THA	50%THA
Outlet pressure of platen superheater	MPa	27.74	27.60	25.93	19.08	12.64
Inlet pressure of superheater	MPa	4.97	4.71	4.47	3.31	2.21
Outlet pressure of cryogenic reheater	MPa	4.88	4.63	4.40	3.27	2.19
Outlet pressure of high temperature reheater	MPa	4.77	4.53	4.31	3.22	2.17
Outlet temperature of platen superheater	°C	549	548	547	541	544
Inlet temperature of superheater	°C	344	341	343	349	356
Outlet temperature of cryogenicreheater	°C	508	508	510	514	526
Outlet temperature of high temperature superheater	°C	603	603	603	603	603

3. Optimization of steam source in soot blowing system

In order to improve the economy of the boiler, the positions of the soot-blowing steam source can be optimized by using the steam somewhere else instead of from the outlet header of the platen superheater. In this study, the platen superheater outlet, the reheater inlet, the cryogenic reheater outlet and the high temperature reheater outlet are selected as four alternative steam source positions. The parameters of steam at these positions in temperature and pressure reducing station are analyzed, and the effect of desuperheating water on the economy of soot blowing is considered comprehensively so as to obtain the most economical and reasonable soot blowing steam source position.

3.1. Energy Analysis of Soot Blowing System

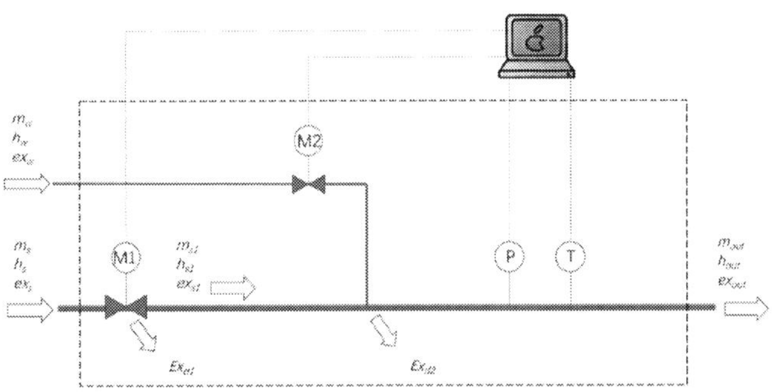

Figure 1 The thermodynamic process diagram of soot blowing steam in temperature and pressure reducing station

Table 2 The specific enthalpy of steam with different source positions

Working conditions	Steam specific enthalpy of platen superheater outlet (kJ/kg)	Steam specific enthalpy of reheater inlet (kJ/kg)	Steam specific enthalpy of cryogenic reheater outlet (kJ/kg)	Steam specific enthalpy of high temperature reheater outlet (kJ/kg)
BMCR	3303.57	3053.99	3454.46	3675.63
BRL	3301.84	3052.67	3457.22	3677.54
THA	3318.49	3064.07	3464.36	3679.29
75%THA	3380.08	3106.74	3485.65	3687.90
50%THA	3459.42	3146.71	3523.68	3696.13

Fig.1 presents the process of steam from different steam source positions changing into soot-blowing steam with specific parameters through temperature and pressure reducing station. First, steam from steam source throttles and decompresses by passing through electric pressure reducing valve M1, and then mixes with desuperheating water to reduce temperature. Next, temperature meters and pressure gauges installed in the tail pipe of temperature and pressure reducing station send back the steam temperature and pressure massages to computer. By calculating and controlling the opening of electric pressure reducing valve M1 and desuperheating water flow control valve M2, the required parameters of soot blowing steam can be attained ultimately. Under different typical working conditions, the steam pressure and temperature at the four positions, i.e., the platen superheater outlet, the reheater inlet, the cryogenic reheater outlet and the high temperature reheater outlet can be acquired from steam design parameters as shown in Table 1. When combining these parameters with IAPWS-IF97 (Industrial Formulation 1997 for the Thermodynamic Properties of Water and Steam)[3], the specific enthalpy of steam at each position can also be calculated as shown in Table 2.

Clearly, it can be seen from Table 2 that the steam specific enthalpy at the reheater inlet is the lowest under each typical working condition, based on which some power generation enterprises considered the reheater inlet as the most energy-saving soot-blowing steam source position [2]. In fact, there is mixing of desuperheating water into soot-blowing steam in temperature and pressure reducing station, so the amount of the steam out of the temperature and pressure reducing station is unequal to that from the steam source position. Therefore, the conclusion drawn by the above-mentioned enterprises is inaccurate exactly. Here, the temperature and pressure reducing station is taken as a volume-controlled open system, as surrounded by black dotted lines in Fig. 1. Eq. (1) shows the general expression of the energy equation of this system. Due to the steady flow in the system and the favorable thermal insulation, the variation of the total storage energy of the system and the heat transfer between the system and the outside can be neglected. Additionally, the system has no external work and the variations of the steam mechanical energy can be also neglected, so Eq. (1) can be simplied into Eq. (2) and then rewritten to Eq. (3). Based on the technical specifications of the steam soot blower in power plants, the temperature and pressure of the steam entering soot bowers is usually controlled at 350°C and 2 MPa by operators. Thus, the temperature and pressure of the soot-blowing steam at the outlet of temperature and pressure reducing station and the annual statistical consumption of the soot-blowing steam in power plants can be determined, then $\dot{m}_{out}$ and h_{out}in the right of Eq. (3) are stable, which means the annual consumption and energy consumption of soot-blowing steam are both independent with the position of soot-blowing steam source. Therefore, the variation of soot-blowing steam source position could not change annual water and coal consumptions caused by boiler soot blowing.

$$\dot{Q} = \frac{dE_{CV}}{d\tau} + \sum (h_{out} + \frac{1}{2}c_{out}^2 + gz_{out})\dot{m}_{out} - \sum (h_{in} + \frac{1}{2}c_{in}^2 + gz_{in})\dot{m}_{in} + \dot{W} \tag{1}$$

$$0 = h_{out}\dot{m}_{out} - (h_w\dot{m}_w + h_s\dot{m}_s) \tag{2}$$

$$\dot{m}_s h_s + \dot{m}_w h_w = \dot{m}_{out} h_{out} \tag{3}$$

3.2. Analysis of Soot Blowing System

Although the variation of soot-blowing steam source position could not change annual water and coal consumptions caused by boiler soot-blowing, the quality of the incoming steam used for producing soot-blowing steam is different due to the distinct steam parameters at each steam source position. The aim of optimizing the position of soot-blowing steam source is to use the lower quality incoming steam for producing soot-blowing steam as far as possible, by which the higher quality steam can be conserved for power generation. Compared with energy analysis, exergy analysis can acquire more information about the energy utilization of the system, evaluate the energy loss from the perspective of energy quality and then provide effective ways to reduce the energy loss [4, 5, 6, 7]. Liu et al [8] tried to study the influence of soot-blowing steam source position on the benefits of power plants by exergy analysis, but the effect of the desuperheating water in temperature and pressure reducing station was neglected, so the calculated enconomy benefit of power plant is inaccurate.

The incoming steam from the steam source has undergone two thermal processes in temperature and pressure reducing station. As illustrated in Fig.1, one is throttling through the pressure reducing valve M1 and the other is cooling by mixing with the desuperheating water. During the throttling process, the steam pressure at the outlet of the pressure reducing valve drops to 2MPa. Since the throttling can be regarded as constant enthalpy process, the temperatures of the steam from the four source positions after throttling and decompression can be obtained according to the IAPWS-IF97. As shown in Table 3, in most cases, the temperature of incoming steam from reheater will be lower than 350°C after throttling, which can not satisfy the temperature requirement of the soot-blowing steam.Therefore, the inlet of reheater is not a suitable place of soot-blowing steam source. Instead, the temperatures of the steams from the outlet of the platen superheater, the cryogenic reheater and the high temperature reheater are all more than 350°C after throttling, and thereby they could be qualified as soot blowing steam only after mixed with the desuperheated water in the temperature and pressure reducing station. The temperature and pressure of the desuperheating water are 180°C and 11MPa respectively, and the temperature and pressure of soot-blowing steam at the outlet of temperature and pressure reducing station are 350°C and 2MPa respectively.

Table 3 The temperature of steam with different source positions after throttling

Working conditions	Steam temperature of platen superheater outlet after throttling (°C)	Steam temperature of reheater inlet after throttling (°C)	Steam temperature of cryogenic reheater outlet after throttling (°C)	Steam temperature of high temperature reheater outlet after throttling (°C)
BMCR	425.19	312.91	493.82	593.29
BRL	424.40	312.33	495.07	594.14
THA	431.98	317.33	498.31	594.92
75%THA	460.03	336.19	507.95	598.75
50%THA	496.07	354.07	525.14	602.41

The following equations can be used to express the exergy balance of steam from the outlets of the platen superheater, the cryogenic reheater and the high temperature reheater when passing through temperature and pressure reducing station[4].

$$Ex_Q + Ex_W = \sum m_{out}ex_{out} - \sum m_{in}ex_{in} + \sum Ex_d \tag{4}$$

The temperature and pressure reducing station can be regarded as an adiabatic system and does not have external work, so according to Fig.1, Eq. (4) can be rewritten to Eq. (5):

$$m_s ex_s + m_w ex_w - m_{out}ex_{out} = Ex_{d1} + Ex_{d2} \tag{5}$$

Where $Ex_{d1}$and$Ex_{d2}$are the exergy losses of the throttling process and the mixing process of steam with desuperheated water respectively.

For the open flow stabilization system, the specific exergy of working fluid can be expressed as:

$$ex = (h - h_0) - T_0(s - s_0) \tag{6}$$

where *ex*, *h* and *s* are the specific exergy, the specific enthalpy and the specific entropy of working fluid respectively, subscript 0 denotes the ambient condition, and the specific enthalpy $h_0$is 84.01kJ/kg, and the specific entropy is 0.2965kJ/(kg.°C) for liquid water at ambient condition.

The specific entropies of the steam at the outlet of the platen superheater, the cryogenic reheater and the high temperature reheater can be calculated through the corresponding steam temperature and pressure and the IAPWS-IF97, and then the specific exergy of the steam at these positions can be calculated according to Eq. (6) and the data in Table 2, which are shown in Table 4. The specific exergy of desuperheating water and steam at the outlet of temperature and pressure reducing station calculated by corresponding temperature and pressure are 147.98kJ/kg and 1100.75kJ/kg, respectively. The mixing ratio of desuperheating water (the percentage of desuperheating water mass to the total mass of the steam at the outlet of temperature and pressure reducing station) can be calculated by the law of conservation of mass and Eq. (3):

$$X = \frac{\dot{m}_w}{\dot{m}_{out}} \times 100\% = \frac{h_s - h_{out}}{h_s - h_w} \times 100\% \tag{7}$$

The total exergy loss of the thermodynamic process in the temperature and pressure reducing station can be solved by Eq. (5) and Eq. (7), and is shown in Eq. (8):

$$Ex_d = Ex_{d1} + Ex_{d2} = m_{out}(ex_s - ex_{out}) - \frac{h_s - h_{out}}{h_s - h_w} m_{out}(ex_s - ex_w) \tag{8}$$

Fig. 2 shows the variations of the mixing ratio of desuperheating water and the total exergy loss of thermal process in the temperature and pressure reducing station with working condition with respect to the steams from different soot-blowing steam source positions. It is revealed that for the same soot blowing system, with the decrease of boiler load, the mixing ratio of desuperheating water increases while the annual total exergy loss caused by steam soot blowing decreases. Besides, in the three types of steam soot blowing systems, the total exergy loss of soot blowing systems using the outlet of cryogenic temperature reheater as the steam source are minimal.

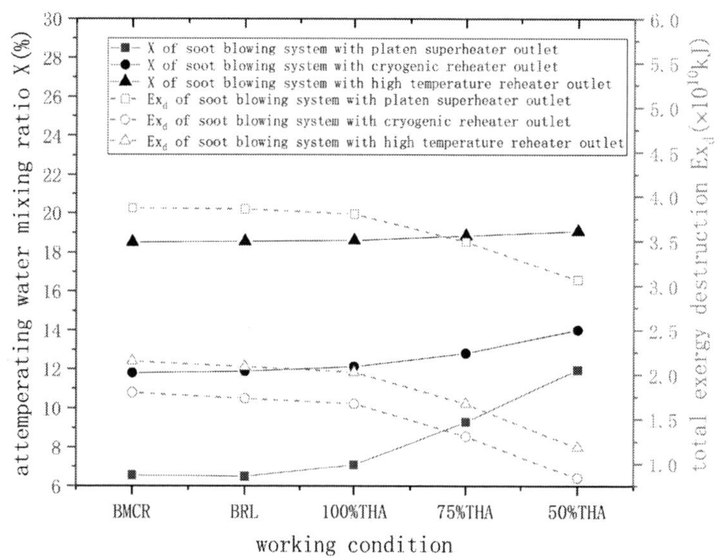

Figure 2 The variation of attemperating water mixing ratio and total exergy destruction by working conditions

Table 4 The specific exergy of steam with different source positions

Working conditions	Steam specific exergy of platen superheater outlet (kJ/kg)	Steam specific exergy of reheater inlet (kJ/kg)	Steam specific exergy of the cryogenic reheater outlet (kJ/kg)
BMCR	1518.70	1401.13	1540.98
BRL	1517.03	1395.97	1535.41
THA	1520.78	1393.79	1529.98
75%THA	1524.95	1368.09	1496.93
50%THA	1525.75	1338.84	1449.67

3.3. Economic Analysis of Soot Blowing System

According to the energy analysis, the variation of soot-blowing steam source position do not change annual water and coal consumptions caused by boiler soot blowing, and thereby the annual average material costs of the steam soot blowing systems. Moreover, according to the exergy analysis, the total exergy loss of soot blowing system with the outlet of cryogenic reheater as the steam source is the smallest among the three types of steam soot blowing system. Since the equal quality of exergy and power [6, 7], the smaller total exergy loss of the soot blowing system means that more exergy can be conserved to convert into electric energy, which brings the greater power generation benefits. Therefore, the soot blowing system with the outlet of cryogenic reheater as the steam source has the greatest power generation benefits.

The unit considered in this paper runs at 75% THA for most of the year, the data under 75% THA are used to calculate and compare the economy of different soot blowing systems. Compared with the original soot blowing system (with platen reheater outlet as steam source), the total exergy loss of soot blowing system with cryogenic reheater outlet as steam source is reduced by 2.183×10^{10} kJ, which could increase the generating capacity by $6.064\times10^{6}\mathrm{kW\cdot h}$. For instance, taking the grid purchase price $0.41\mathrm{CNY/(kW\cdot h)}$of coal-fired units in a province in 2018 as the pricing standard, the increased generating capacity is worth about 2486 000 CNY. Therefore, if the steam source position of soot-blowing system is changed from platen reheater outlet to cryogenic reheater outlet, there is an increase of annual average power generation benefits about 2486 000 CNY without changing the annual average cost of raw materials consumed by the soot blowing system.

4. Conclusions

In view of a 1000MW ultra-supercritical coal-fired unit, this work studies the thermodynamic process of soot-blowing steam in the temperature and pressure reducing station through energy analysis and exergy analysis, and conclusions can be drawn as follows:

(1) The annual raw material cost of the steam soot blowing system is independent with the soot blowing steam source position.

(2) For the same soot blowing system, with the decrease of boiler load, the mixing ratio of desuperheating water increases while the annual total exergy loss caused by steam soot blowing decreases. Besides, the total exergy loss of soot blowing systems using the outlet of cryogenic temperature reheater as the steam source are minimal in the three types of steam soot blowing systems.

(3) After changing the steam source position of soot blowing system from platen superheater outlet to cryogenic reheater outlet, there is an increase of annual average power generation benefits about 2486 000 CNY without changing the annual average cost of raw materials consumed by the soot blowing system.

Acknowledgment

National Natural Science Foundation of China supported this work (Contract no. 51806206).

References

[1] WANG Y 2016 The T91 Materials Used in the Boiler *Boiler Manufacturing* **vol 3(2)** pp 54-56

[2] ZHANG H 2015 Modification and Energy Conservation Analysis on Utility Boiler Steam Soot Blowing System *Telecom World* **vol 5(10)** pp160-161

[3] Wagner W Cooper J R Dittmann A Kijima J 2000 The IAPWS Industrial Formulation 1997 for the Thermodynamic Properties of Water and Steam *Journal of Engineering for Gas Turbines and Power* **vol 1(122)** pp 150-182

[4] Dincer I Rosen M A 2012 Exergy: energy, environment and sustainable development (Newnes)

[5] Shirazi A Aminyavari M Najafi B et al 2012 Thermal-economic-environmental analysis and multi-objective optimization of an internal-reforming solid oxide fuel cell-gas turbine hybrid system *International Journal of Hydrogen Energy* **vol 37** pp 19111-19124

[6] Arabkoohsar A Machado L Farzaneh-Gord M et al. 2015 The First and Second Law Analysis of a Grid Connected Photovoltaic Plant Equipped with a Compressed Air Energy Storage Unit *Energy* **vol 87** pp 520-539
[7] ZHANG Y 2014 Study on System Integration and Simulation of Wind Power and Advanced Adiabatic Compressed Air Energy Storage Technology (Beijing: Institute of Engineering Thermophysics Chinese Academy of Sciences)
[8] LIU Y WU Y JIANG X et al 2013 Study on Choice Problem of Utility Boiler Steam Soot Blowing Source *Huadian Technology* **vol 35(7)** pp 6-9

IOP Conf. Series: Materials Science and Engineering **751** (2020) 012065 doi:10.1088/1757-899X/751/1/012065

The Calculation of Friction Heat and Thermal Analysis of Spur Gears Based on Thermal Elastohydrodynamic Lubrication Theory

Shuyi Liu[1,*] and Chen Qi[1]

[1] China Aerodynamics Research and Development Center

*E-mail: LSYNWPU@126.com

Abstract. In order to predict the temperature distribution of gears more accurately, the influence of oil film on frictional heat generation should be taken into account in computational model. In this paper, based on the theory of thermal elastohydrodynamic lubrication, the calculation of frictional heat flux in the meshing process and the steady thermal analysis of spur gears is completed. The reliability of the model is verified by comparing the calculated results with the value of empirical formulas and experimental data. This paper provides an effective analysis method for gluing prediction, failure analysis and lubrication state analysis of aero-engine transmission gears.

1. Introduction

With the development of aero-engine toward high speed, high thrust-weight ratio and high reliability, the transmission power and working speed of gear, as an important transmission component of aero-engine, have been significantly improved. In order to ensure the stable and reliable operation of the gear, it is the most important to keep it in a full lubrication. In the sufficient lubrication, there is a layer of oil film between the two contact teeth to avoid wear caused by direct contact. In addition, the relative slip between the teeth surfaces leads to friction loss of oil film which increases the local temperature of the gear. Accurate thermal analysis before gear manufacture and operation profits reducing the possibility of gear thermal deformation and gluing failure caused by excessive local temperature.

In most current studies, the calculation of heat source in gear thermal analysis use the friction coefficient given by experience and the influence of oil film is not fully taken into account. Li Baoliang[1] got the dynamic analysis results by ADAMS and the friction coefficient changed along the meshing line from previous studies, and then calculated the frictional heat generation. The steady-state thermal analysis and transient thermal analysis of involute gears were carried out by using ANSYS. In the research of Liangheng Qiu [2], the friction coefficient was set as a fixed value to obtain frictional heat source. Fan Zhimin[3] had calculated the friction coefficients of each meshing position according to the empirical formula and obtained the temperature distribution of gear by ANSYS finite element analysis.

The lubrication problem of spur gears, as a line contact, can be solved by the increasingly developed theory of thermal elastohydrodynamic lubrication. Many studies have applied this theory to the calculation of minimum oil film thickness between gears to obtain the lubrication condition. In 1916, Martin[4] transformed gear meshing problem into rolling problem between cylinder and plane and solved Reynolds equation to get the solution. Because of its simple principle, the theory has been widely used in solving gear lubrication problems and has been constantly improved in later studies. Later, Lu

Lixin[5] used Newton-Raphson method to solve the transient elastohydrodynamic lubrication problem and obtained the pressure and thickness distribution of the oil film between two meshing gears' surfaces. Wang Youqiang[6] used the multi-grid method to accelerate the solution of the isothermal Elastohydrodynamic problems under line contacts and analysed the influence of geometric parameters of gears on the oil film thickness in the gear meshing process. There are not many studies calculate the friction coefficient of two teeth by solving the elastohydrodynamic problem. But, in fact, it can be obtained.

Remarkably, the researches on gear elastohydrodynamic lubrication mentioned above are based on the static load without considering the vibration during gear meshing. In order to reflect the actual physical situation more accurately, Barbieri[7] applied the solution of the dynamic load problem of two meshing gears to the elastohydrodynamic lubrication analysis in elliptical contact and compared the results with static load. Li[8] and Ouyang[9] carried out six degrees of freedom dynamic load analysis, considering the influence of bearing vibration, to get the load distribution along the meshing line on the teeth surfaces as the input of elastohydrodynamic lubrication calculation. Then the mixed elastohydrodynamic lubrication solution was obtained.

From the above research, it can be seen that most of the current calculation of gear frictional heat generation does not consider the influence of oil film, which will bring errors to gear thermal analysis. In this paper, the influence of oil film between meshing surfaces is taken into account when calculating friction heat generation and the difference of friction coefficient at each meshing position is considered in the thermal analysis model. Firstly, the dynamic analysis of spur gears is carried out. Then the thermal elastohydrodynamic lubrication equations are established and solved by using the multi-grid method to accelerate. The calculated frictional heat flux is loaded on the corresponding position of the tooth surface as a heat source to take the thermal analysis of the single tooth model of spur gear.

2. Dynamic load analysis of the spur gear

2.1 Kinematics analysis of the spur gear

Firstly, the kinematics analysis of spur gears is carried out to obtain the relative slip velocity between two meshing tooth surfaces. According to the geometric structure and transmission characteristics of spur gears, the geometric relationship is illustrated in figure 1.

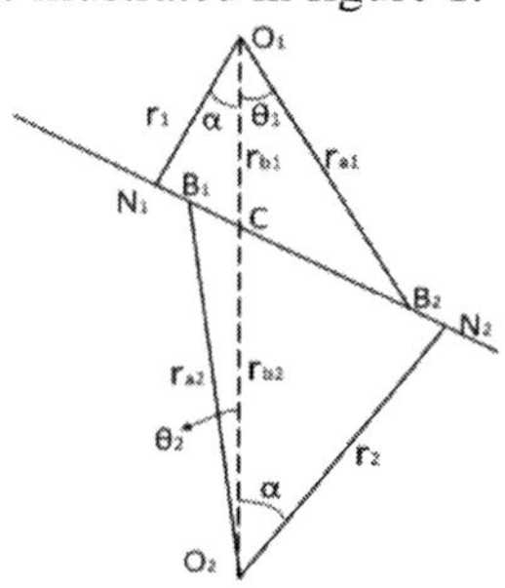

Figure 1. Geometric relationship of transmission parameters of spur gear

Where r_1 and r_2 are the pitch circle radius of the driving gear and the driven gear respectively, r_{b1} and r_{b2} are the base circle radius of the gear, r_{a1} and r_{a2} are the top circle radius of the gears, α is the pressure angle of the gear. For a pair of teeth, B_1B_2 is the actual meshing line and N_1N_2 is the theoretical meshing line, C is the meshing pitch point.

The curvature radius of the two surfaces (R_1 and R_2) is combined into a comprehensive curvature radius *R*:

$$R = \frac{R_1 R_2}{R_1 + R_2} \tag{1}$$

Where s is the distance between meshing point and pitch node. The absolute velocities of the two teeth surfaces can be calculated by the curvature radius of the tooth surface and the rotation speed of the two gears:

$$u_1 = \frac{\pi n_1 R_1}{30} \tag{2}$$

$$u_2 = \frac{\pi n_2 R_2}{30} \tag{3}$$

Where n_1 and n_2 are the speed of driving gear and driven gear respectively. The entrainment velocity (u_s) and the relative slip velocity of two meshing surfaces (u_e) can be defined as:

$$u_s = \frac{u_1 + u_2}{2} \tag{4}$$

$$u_e = u_1 - u_2 \tag{5}$$

2.2 Dynamic load analysis of spur gears

The variety of meshing stiffness and load on a tooth surface with time stimulates the gear vibration. In this paper, the single-degree-of-freedom torsional vibration model of gears, ignored the influence of vibration of bearings and transmission shafts and the tooth profile error of gears, is established by lumped mass method (figure 2).

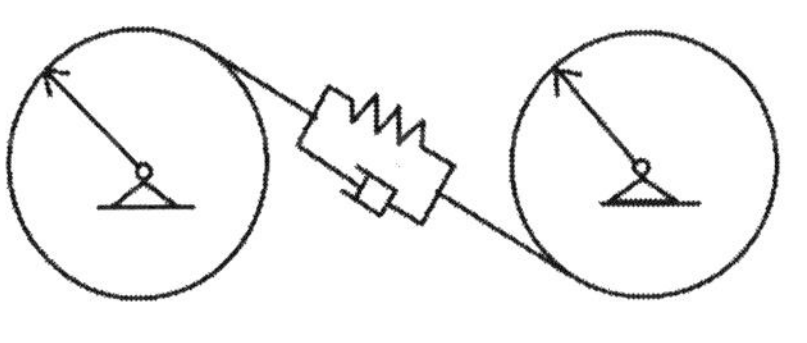

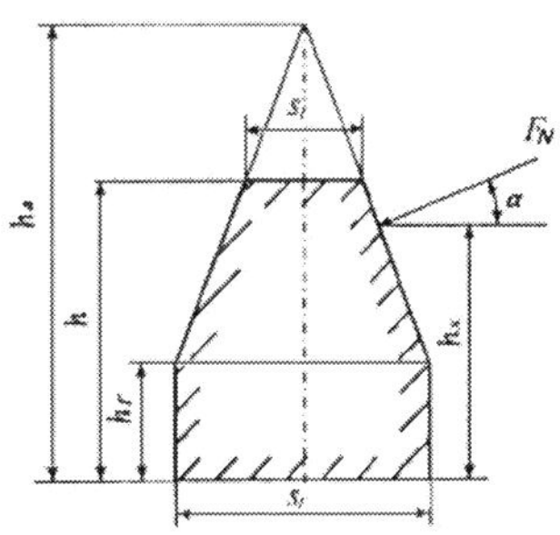

Figure 2. Single degree of freedom vibration model **Figure 3.** Simplified model of Ishikawa method

The differential equations of gear vibration are established according to the force balance:

$$J_1\ddot{\theta}_1 + c\left(r_{b1}\dot{\theta}_1 + r_{b2}\dot{\theta}_2\right)b_g + k_m(t)\left(r_{b1}\theta_1 + r_{b2}\theta_2\right)b_g = N \tag{6}$$

$$J_2\ddot{\theta}_2 + c\left(r_{b1}\dot{\theta}_1 + r_{b2}\dot{\theta}_2\right)b_g + k_m(t)\left(r_{b1}\theta_1 + r_{b2}\theta_2\right)b_g = N\frac{r_{b2}}{r_{b1}} \tag{7}$$

Where J_1 and J_2 are the inertial moment of the driving gear and the driven gear respectively, θ_1 and θ_2 are the torsion angle caused by the elastic deformation of the driving gear and the driven gear respectively, b_g is the tooth width, c is the meshing damping coefficient, k_m is time-varying meshing stiffness, N is the resistance moment of drive gear.

Let $x = r_{b1}\theta_1 + r_{b2}\theta_2$: the equation (9) and equation (10) can be simplified as follows:

$$\ddot{x} + 2\zeta\sqrt{\frac{k_m(t)}{m_e}}\dot{x} + \frac{k_m(t)}{m_e}x = \frac{N}{r_{b1}m_e} \tag{8}$$

Where m_e is induction quality of gear, ζ is relative damping coefficient, equalled 0.07[10], $k_m(t)$ is the meshing stiffness. The meshing stiffness varying with time of gears is:

$$k(t)=\frac{F_N}{\delta_\Sigma b_g} \tag{9}$$

F_N is the load on meshing surface, δ_Σ is the deformation of the tooth. Based on the theory of material mechanics and the simplified model of Ishikawa method of spur gear (figure 3), the deformation at the loading point can be that as follow:

$$\delta=\delta_{wr}+\delta_{wt}+\delta_j+\delta_q \tag{10}$$

δ_{wr} is the bending deformation of the rectangular part of the simplified model. δ_{wt} is the bending deformation of the trapezoidal part of the simplified model. δ_j denotes the amount of deformation caused by shear force on the tooth surface. Deformation due to the inclination of the simplified model is δ_q.

In the meantime, the deformation of two tooth surfaces at the meshing point due to contact can be expressed as follows:

$$\delta_c=\frac{4\left(1-\nu^2\right)F_N}{\pi E b_g} \tag{11}$$

Therefore, the sum of the deformation at the meshing point of the gear pair can be obtained as follows:

$$\delta_\Sigma=\delta_1+\delta_2+\delta_c \tag{12}$$

The change of meshing stiffness k with time can be obtained by substituting the equation (12) into the equation (9). The meshing stiffness of spur gears can be calculated by the following equation considering that it is different in the case of single and double teeth meshing.

$$k_m(t)=\begin{cases} k(t)+k(t+T_i) & \text{double teeth meshing} \\ k(t) & \text{single tooth meshing} \end{cases} \tag{13}$$

Where:

$$T_i=\frac{60}{Z_1 n_1} \tag{14}$$

In this paper, the fourth-order Runge-Kutta method is used to solve equation (8) and obtain the solution of x. Then the load on every meshing points can be obtained.

$$F_N=\begin{cases} \dfrac{k_m(t)x(t)}{2} & \text{double teeth meshing} \\ k_m(t)x(t) & \text{single tooth meshing} \end{cases} \tag{15}$$

3. Frictional heat calculation based on thermal elastohydrodynamic lubrication theory

3.1 Basic equation

The thermal elastohydrodynamic lubrication model at line contact is established to analyse the lubrication of spur gears in meshing process and obtain the frictional heat flux. In this model, the thermal effect, time-varying and non-Newtonian characteristics of lubricating oil are taken into account so that it is closer to the reality of gear meshing.

1) Film thickness equation

$$h(x)=h_c+\frac{x^2}{2R}-\frac{2}{\pi E'}\int_{s_1}^{s_2} p(s)\ln(x-s)^2\,ds \tag{16}$$

h_c is the oil film thickness in contact centre. E' is the composite modulus of elasticity. x is the direction of contact width.

2) Reynolds equation

The general time-varying Reynolds equation of line contact derived from equivalent viscosity assumption is used in this model.

$$\frac{\partial}{\partial x}\left[\left(\frac{\rho}{\eta}\right)_{ex} h^3 \frac{\partial p}{\partial x}\right] = 12u_s \frac{\partial}{\partial x}\left(h\rho_x^*\right) + 12\frac{\partial(\rho_e h)}{\partial t} \tag{17}$$

Where ρ, η, h, p, t is the density, viscosity, film thickness ,pressure of lubricating oil and time in turn. The subscript e represents the equivalent value. The pressure of oil film at the inlet and outlet is set as zero and the pressure gradient at the outlet is equal to zero, it can be expressed as follows:

$$\begin{cases} p_{in} = 0 \\ p_{out} = 0, \left(\dfrac{\partial p}{\partial x}\right)_{out} = 0 \end{cases} \tag{18}$$

3) Energy equation

Ignoring the influence of volume force and thermal radiation in oil film, the energy equation is as follows:

$$\rho c_p\left(\frac{\partial T}{\partial t} + u\frac{\partial T}{\partial x}\right) = \lambda\frac{\partial^2 T}{\partial z^2} - \frac{T}{\rho}\frac{\partial \rho}{\partial T}\left(\frac{\partial p}{\partial t} + u\frac{\partial p}{\partial x}\right) + \eta\left(\frac{\partial u}{\partial z}\right)^2 \tag{19}$$

Where c_p, λ and T is the specific heat capacity at constant pressure, heat conductivity coefficient and temperature in turn. z is the direction of film thickness. The boundary conditions are determined by the thermal interface equation.

$$\begin{cases} T_1 = \dfrac{\lambda}{\sqrt{\pi\lambda_1\rho_1 c_{p1} u_1}}\displaystyle\int_{-\infty}^{x} \left.\frac{\partial T}{\partial z}\right|_{z=0} \frac{ds}{\sqrt{x-s}} + T_0 \\ T_2 = \dfrac{\lambda}{\sqrt{\pi\lambda_2\rho_2 c_{p2} u_2}}\displaystyle\int_{-\infty}^{x} \left.\frac{\partial T}{\partial z}\right|_{z=h} \frac{ds}{\sqrt{x-s}} + T_0 \end{cases} \tag{20}$$

The subscript 1 and 2 represent the parameters of driving gear and driven gear respectively.

4) Viscosity equation

Roelands viscosity-pressure-temperature relation is used in this model：

$$\eta = \eta_0 \exp\left\{\left(\ln\eta_0 + 9.67\right)\left[\left(1 + 5.1\times10^{-9}p\right)^z \times \left(\frac{T-138}{T_0-138}\right)^{-s_0} - 1\right]\right\} \tag{21}$$

Where:

$$s_0 = \frac{\beta(T_0 - 138)}{\ln\eta_0 + 9.67} \tag{22}$$

$$z = \frac{\alpha}{5.1\times10^{-9}\left(\ln\eta_0 + 9.67\right)} \tag{23}$$

η_0 is the viscosity of lubricating oil corresponding to ambient temperature (T_0). α and β are viscous pressure coefficient and viscous temperature coefficient, respectively.

5) Density equation

Dowson-Higginson density-pressure-temperature formula is used to express the density of lubricating oil.

$$\rho = \rho_0 \left[1 + \frac{Ap}{1+Bp} + D(T - T_0) \right] \tag{24}$$

Where ρ_0 is density in ambient temperature.The three coefficients A, B and D in the equation can be set as $0.6\times10^{-9}\ \mathrm{m^2/N}$, $1.7\times10^{-9}\ \mathrm{m^2/N}$ and $-0.0007/\mathrm{K}$ respectively.

6) Load balance equation

For per unit length of the oil film, the instantaneous pressure should be balanced with the load on the gear.

$$\int_{x_0}^{x_e} p(x)dx = w \tag{25}$$

Where w is the load on the contact width. The load balance equation is used to determine the thickness of the central oil film when solving the thermoelastohydrodynamic lubrication problem.

7) Constitutive equation of non-Newtonian fluid

The Ree-Eyring rheological model represents the non-Newtonian characteristic of lubricating oil:

$$\frac{\partial u}{\partial z} = \frac{\tau_0}{\eta} \sinh\left(\frac{\tau}{\tau_0}\right) \tag{26}$$

The equivalent viscosity η^* is defined as follow and $\tau = \eta^* \frac{\partial u}{\partial z}$, so:

$$\eta^* = \eta\left(\frac{\tau}{\tau_0}\right) \Big/ \sinh\left(\frac{\tau}{\tau_0}\right) \tag{27}$$

Where τ_0 is the characteristic shear force at 10MPa.

3.2 Nondimensionalization of Equations

The equations of thermal elastohydrodynamic lubrication for line contact are dimensionless treated. The dimensionless parameters used are as follows:

$$\bar{t} = \frac{tU_0}{b},\ \bar{T} = \frac{T}{T_0},\ U_0 = \frac{E'R}{\eta_0},\ H = \frac{hR}{b^2},\ W = \frac{w}{E'R},\ U = \frac{\eta_0 u}{E'R},\ G = \alpha E',\ \bar{z} = \frac{z}{h},\ X = \frac{x}{b},\ \bar{p} = \frac{p}{p_H},$$

$$\bar{\eta} = \frac{\eta}{\eta_0},\ \bar{\rho} = \frac{\rho}{\rho_0}$$

Where b is the half contact width under w load, $b = \sqrt{\frac{8wR}{\pi E'}}$. p_H is the maximum contact stress under w load, $p_H = \frac{E'b}{4R}$.

The dimensionless equation can be obtained by introducing the above dimensionless parameters into the thermoelastohydrodynamic lubrication equation (16) - (27).

3.3 Equation discretization

The contact width is divided into n_x equal spacing parts and expressed by subscript *i*. Dimensionless film thickness direction is divided into n_z equal spacing parts expressed by subscript *j*. The time item is divided into n_t equal parts expressed by superscript m. Numerical solution

Matlab is used to solve the thermoelastohydrodynamic lubrication equations. Firstly, in the film thickness equation, the pressure value is known from the Hertzian contact theory and the central film thickness is modified according to the load balance equation, thus the corresponding oil film thickness distribution is obtained. Then the oil film thickness is taken as a known quantity into Reynolds equation to obtain a new pressure distribution. The pressure and film thickness are solved iteratively accelerated by using the five-layer multi-grid w iteration method and results are brought into the energy equation to get the temperature value by the sequential scanning method, and then the temperature value in the iteration of pressure and film thickness is updated. Finally, the convergence of pressure, film thickness and temperature are achieved.

3.4 Frictional heat calculation

The shear stress (τ) on the upper and lower surfaces of the oil film can be obtained by solving the thermal elastohydrodynamic lubrication equation. The integral of it is the friction force between the oil film and the meshing surface (F_{f1} and F_{f2}). The meshing friction coefficient of the two tooth surfaces, f , can be further calculated by using the following method:

$$F_{f1} = \int_{x_0}^{x_e} \tau\Big|_{z=0} dx \tag{28}$$

$$F_{f2} = \int_{x_0}^{x_e} \tau\Big|_{z=h} dx \tag{29}$$

$$f = \frac{|F_{f1}| + |F_{f2}|}{2w} \tag{30}$$

After the friction coefficient is obtained, the friction heat generation ($q(t)$) can be calculated and it should be allocated to the two meshing surfaces as follows:

$$q(t) = \frac{\varepsilon_h f(t) p_{ave} u_e(t)}{2b} \tag{31}$$

$$q_{1,2}(t) = \kappa_{1,2} \cdot q(t) \tag{32}$$

ε_h , the coefficient of thermal energy conversion, in this paper, equals to 0.9. p_{ave} is average contact stress, $q_1(t)$ and $q_2(t)$ the friction heat flux of two meshing tooth surfaces, respectively, κ_1 and κ_2 are heat distribution coefficients of two meshing surfaces related to thermal conductivity, density, specific heat capacity and velocity of two teeth, respectively.

$$\kappa_{1,2} = \frac{\sqrt{\lambda_{1,2}\rho_{1,2}c_{p1,2}u_{1,2}}}{\sqrt{\lambda_1\rho_1 c_{p1}u_1} + \sqrt{\lambda_2\rho_2 c_{p2}u_2}} \tag{33}$$

4. Thermal analysis model

4.1 Thermal analysis theory

This paper considers that the gear material is isotropic. This model pays more attention to the influence of frictional heat and thermal convection while ignoring thermal radiation loss and other heat sources. The steady state heat conduction and the third boundary conditions can be expressed as follows:

$$\frac{\partial}{\partial x}\left(\lambda\frac{\partial T}{\partial x}\right) + \frac{\partial}{\partial y}\left(\lambda\frac{\partial T}{\partial y}\right) + \frac{\partial}{\partial z}\left(\lambda\frac{\partial T}{\partial z}\right) = q \tag{34}$$

$$-\lambda\left(\frac{\partial T}{\partial n}\right)_w = h(T_w - T_f) \tag{35}$$

Where q is heat source, h is convective heat transfer coefficient, n is normal direction of surfaces with thermal convection. Subscripts w and f denote the gear wall and environment respectively.

4.2 Finite element model

4.2.1 Geometric model. The following assumptions are made in this paper to reasonably reduce the amount of calculation:

1. The temperature field of each tooth is the same when it reaches at steady state.
2. Because of the high speed and the short meshing period of each tooth, it can be considered that the temperature field of each tooth is the same at a time.
3. The ambient temperature is not affected by the friction heat source so it remains at 40 °C.

Based on the above assumptions, a single tooth model (figure 4 and figure 5) can be used for thermal analysis.

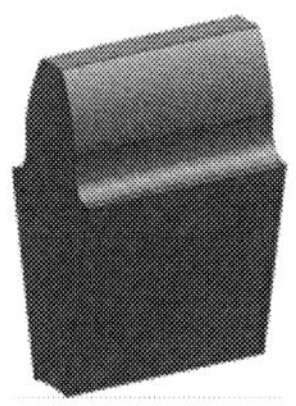

Figure 4. Single tooth model of driving gear

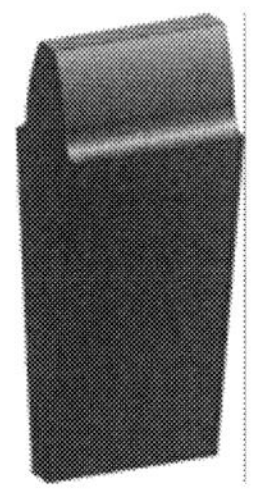

Figure 5. Single tooth model of driven gear

In this paper, the meshing tooth surface is divided into 51 rectangles along the meshing track, which is convenient to load the friction heat flux to the corresponding position.

4.2.2 Boundary condition. The ambient temperature is constant as 40 °C.

When calculating the heat transfer coefficient, referring to references[1] and [3], the tooth surface is divided into 3 parts, meshing surface, end surface and other surfaces. The empirical formulas for calculating the convective heat transfer coefficient between the meshing tooth surface and the environment are as follows:

$$h_t = 0.228\,\mathrm{Re}^{0.371}\,\mathrm{Pr}^{0.333}\,\lambda_{oil}/d \tag{36}$$

Where d is pitch diameter. Re, Pr and λ_{oil} is Reynolds number, Prandtl number and thermal conductivity of lubricating oil in turn.

In the case of jet lubrication, the convective heat transfer occurs between end surface and the mixture of oil and air when the proportion of air is large so it is regarded as the convective heat transfer to air.

$$h_s = 0.308\lambda_{\mathrm{air}}\left(m+2\right)^{0.5}\mathrm{Pr}_{air}{}^{0.5}\left(\frac{\omega}{v_{air}}\right)^{0.5} \tag{37}$$

Where Pr_{air} is the Prandtl number of air. v_{air} and λ_{air} are the kinematic viscosity and thermal conductivity of air respectively.

The convective heat transfer coefficients at the top, root and unmeshed tooth surfaces are approximately 1/3 of the end-face heat transfer coefficients:

$$h_{tt} = \frac{1}{3}h_s \tag{38}$$

The bottom of the tooth and the connecting surface with the adjacent teeth are seen as adiabatic.The frictional heat flux calculated based on thermal elastohydrodynamic lubrication theory is averaged and then load on the corresponding meshing rectangle continuously. In steady-state thermal analysis, the heat flux loaded on each meshing rectangle is as follows:

$$q_{ai} = \frac{\int_{\tau_0}^{\tau_1} q_i d\tau}{\tau_2 - \tau_0} \tag{39}$$

Where q_i is the frictional heat flux on each meshing position calculated by elastohydrodynamic lubrication theory. q_{ai} is the frictional heat flux loaded on the ith rectangle in the steady thermal analysis.

τ_0 and τ_1 are the beginning of meshing and the end of meshing at the *i*th meshing point, respectively. τ_2 is the end of the whole meshing period.

4.3 Results and analysis

The spur gear parameter[11]used in this example and the material of gear are shown in the table 1 and table 2, respectively:

Table 1. Gear geometric parameters

Tooth number, Z1/Z2	Modulus, m/mm	Pressure angle, $\alpha/°$	Tooth width, b_g/mm	Hole diameter, d_{in}/mm
16/24	4.5	20	20	25

Table 2. Gear Material Parameters

Elastic modulus, E_1, E_2/GPa	Poisson ratio, ν_1, ν_2	Density, $\rho / kg \cdot m^{-3}$	Specific heat, $c_1, c_2 / J \cdot kg^{-1} \cdot K^{-1}$
206	0.3	7850	465

The lubricating oil is gear lubricating oil SCH632.

4.3.1 Reliability Verification. In order to verify the reliability of the model established in this paper, we calculate the minimum oil film thickness under different loads when the relative slip velocities of two surfaces are 5 m/s and results are compared with those of Dowson-Higginson formula, Dowson-Higginson modified formula, Hamrock-Jacobson formula and YangPeiran-WenShizhu formula[12].

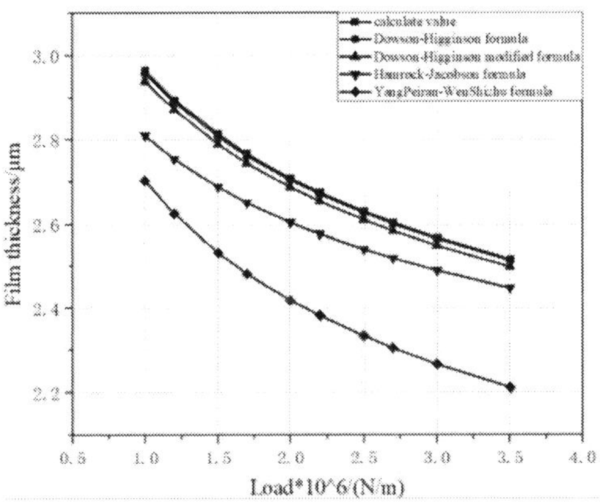

Figure 6. The minimal oil film thickness

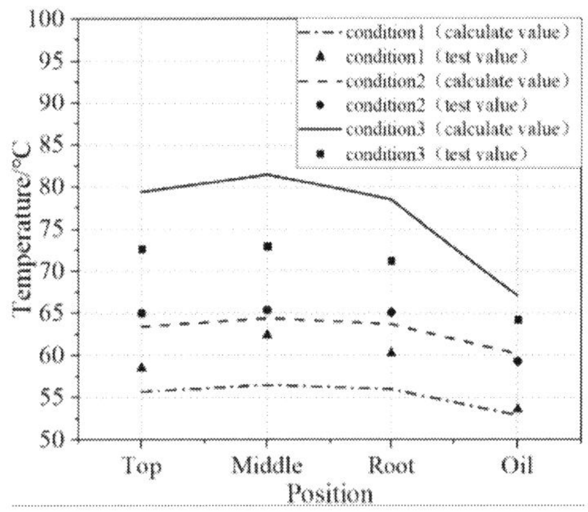

Figure 7. Temperature distribution on teeth

From figure 6, it can be seen that the results of thermal elastohydrodynamic lubrication agree well with the Dowson-Higginson formula and the Dowson-Higginson modified formula, supported by a large number of experiments and research results, verifying the reliability of the program.

On the other hand, the results of thermal analysis model are compared with the results in reference [11] and the selected experimental conditions are shown in table 3.

Table 3 Test conditions

	Speed of driving gear/rpm	Transfer power /kW
working condition 1		3.12
working condition 2	2175	8.03
working condition 3		13.84

It can be seen from figure 7 that the difference between the calculated and experimental values is acceptable in heat transfer research, which verifies the reliability of the thermal analysis model.

4.3.2 Temperature field. The steady temperature fields of the driving gear and the driven gear when the speed of the driving gear is 800rpm with 10kW load are shown in figure 8. It can be seen that under the same working conditions, the temperature of the driving gear is higher than that of the driven gear. One of reasons is that the rotating period of the driving gear is shorter, which means the time of heat loading accounts for a large proportion of the whole rotating cycle. At the same time, the larger area of the driven gear occuring convective heat transfer leads to lower temperature.

The relative slip velocity of two meshing tooth surfaces at the radius of pitch circle is zero. As a result, the temperature near the pitch circle of the driving gear is lower than that at the entry point (top) and the exit point (root) of meshing procession. On the top of the driven gear, the temperature is the highest, because the entry point is impacted and loaded heavily. In the meantime, due to the large non-meshing volume of the driven gear, more heat is transmitted to the gear wheel and the temperature at the root of the tooth is low.

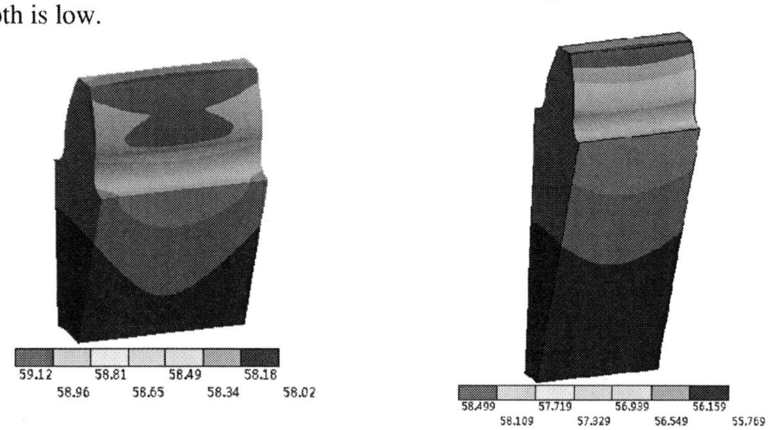

(a) driving gear (b) driven gear

Figure 8. Temperature distribution at steady state (°C)

5. Conclusion

Based on the theory of thermal elastohydrodynamic lubrication, the lubrication and heat generation in the process of gear transmission are calculated by using Matlab and steady thermal analysis of the gear is carried out. The following conclusions can be summarized from the results:

1. Compared with the empirical formula and experimental measurements, it is verified that the thermoelastohydrodynamic lubrication model and thermal analysis model in this paper are reliable.
2. At the same working conditions, the temperature of the driving gear is higher than that of the driven gear. Meanwhile, the temperature near the pitch radial of the driving gear is lower than that at the entry point and the exit point.

Reference

[1] Baoloang L, Lin B and Xuejuan T 2015 Analysis of Temperature Field of Involute Spur Gear based on Friction Influence *J. Mechanical transportation*. **4**.

[2] Liangheng Q, Yihang X 1995 A Calculation of Bulk Temperature and Thermal Deflection of Gear Tooth about Profile Modification *J. Jouranl of Shanghai Jiaotong University*. **29** 79-86.

[3] Zhimin F, Ruixue W and Wanfeng Z 2017. Simulation Analysis of The Steady Temperature Field of Double Involute Gear Drive *J. Journal of Mechanical Strength* **39** 198-203.

[4] Martin H M 1916 Lubrication of Gear Teeth *J. Engineering*. 102.

[5] Lixin L and Hehao Z 2002 Nonsteady Elasto-hydrodynamic Lubrication of Gear Transmission System under Dynamic Load *J. Lubrication and Sealing*.12-14

[6] Youqiang W, Hongqi L, Weitong J and Peiran Y 2004 EHL Solution of Involute Spur Gear with Multi-Grid Method *J. Journal of Tianjin University*. **37** 985-989.

[7] Barbieri M, Lubrecht A A and Pellicano F 2013 Behavior Of Lubricant Fluid Film In Gears Under Dynamic Conditions *J. Tribology International.* **62** 37-48.
[8] Li S, Anisetti A 2016 On the Flash Temperature of Gear Contacts Under The Tribo-Dynamic Condition *J. Tribology International.* **97** 6-13.
[9] Ouyang T, Huang H and Zhang N 2017 A Model to Predict Tribo-Dynamic Performance Of A Spur Gear Pair *J. Tribology International.*
[10] Litao W and Zhixian H 2012 Numerical Solution of Dynamic Load Of Gear Tooth *J. Journal of Jinggangshan University (Natural Science).* **33** 79-82.
[11] Jianhua X and Wei L 2014 Prediction Methods of Gear System Temperature Fields *J. Journal of University of Science and Technology Beijing.* **2** 252-259.
[12] Shizhu W, Ping 2012 *Tribological Principles. 4th Edition.* (Tsinghua university press).

IOP Conf. Series: Materials Science and Engineering **751** (2020) 012066 doi:10.1088/1757-899X/751/1/012066

Thermodynamic Modeling and Simulation Analysis of Double-tube Damper

Yijie Chen*, Xiaodong Gao, Yiqiang Wan, Xiaoling Han and Jianjian Dai

China North Vehicle Research Institute, Beijing, China

*E-mail: chenyijie1206@163.com

Abstract. At present, most of the special vehicles were also designed with a double-tube damper, which was characterized by convenient installation and low cost. However, due to the design of the oil storage cavity outside the working cylinder, the heat dissipation effect was not ideal, and the working temperature of the oil too high will accelerate the aging of the sealing member, which was also a key problem that plagues the reliability design of the damper. Aiming at the structural characteristics of the double-tube damper, the heat generation mechanism and energy conversion research of the oil were carried out, and the heat transfer coefficient and heat conduction equation of the inner and outer surfaces of the cylinder were derived. The comprehensive heat transfer expression was solved for the working medium, and the thermodynamic model of the double-tube damper was constructed. Finally, the simulation analysis of the influence of different structural parameters on the heat balance temperature were carried out, which could effectively guide the design and development of the damper.

1. Introduction

When the vehicle is driving under various road conditions, in order to reduce the influence of ground excitation on the ride comfort, the vibration damping device is usually installed between axle and frame. The current application is more widely used as a double-tube damper, mainly through the damping valve to generate damping force on the oil throttling to attenuate the body vibration, the vibration energy consumed is converted into heat energy, which causes the temperature of the cylinder to rise continuously, which has great influence on the seal. The double-tube damper is simple in process and low in cost, but it is difficult to dissipate heat. When the temperature of the cylinder reaches and exceeds the limit temperature that the seal can withstand, it is easy to cause oil leakage. Based on this situation, this paper systematically studies the thermodynamic model of the double-tube damper [1-2].

2. Thermodynamic physical model of double barrel damper

The thermodynamic physical model of the established double cylinder hydraulic damper is shown in Figure 1. The two cylinders are set together, and a working chamber of the damper is formed in the inner cylinder, and an oil storage chamber is arranged between the inner and outer cylinders for storing and compensating the oil in the working chamber. The oil storage chamber is usually filled with a certain pressure of air, according to the conventional design, the oil can be set to account for half of the volume of the oil storage chamber. A piston is arranged at the lower end of the piston rod, which supports and guides the inner cylinder, and isolates the working chamber into two upper and lower chambers. The compression and recovery valve are mounted on the piston to attenuate the vibration of the vehicle body by damping the oil throttling. The bottom valve isolates the lower chamber from the oil reservoir, and has a function of draining and replenishing the working fluid to prevent the damper from being subjected to aerodynamic distortion.

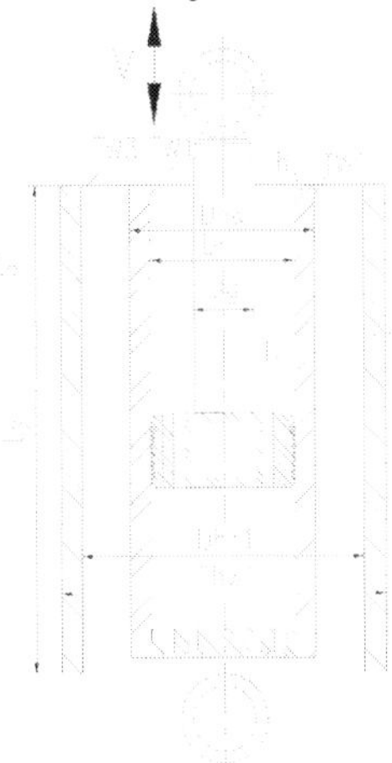

Figure 1. Shock absorber thermodynamic physical model

3. Thermodynamic Mathematical Model of Double-tube Damper

The heat transfer mode of the damper is usually dominated by heat conduction and heat convection. Heat conduction, also referred to as heat conduction, refers to the transfer of heat from the high temperature part of the object to the low temperature part, or the transfer of a high temperature object to the low temperature object in contact therewith. When the temperature of the inner and outer walls of the cylinder is different, heat conduction occurs inside the solid. The thermal differential equation of the cylindrical coordinate system is as follows [3-4]:

$$\rho c\frac{\partial t}{\partial \tau}=\frac{1}{r}\frac{\partial}{\partial r}\left(\lambda r\frac{\partial t}{\partial r}\right)+\frac{1}{r^2}\frac{\partial}{\partial \varphi}\left(\lambda\frac{\partial t}{\partial \varphi}\right)+\frac{\partial}{\partial z}\left(\lambda\frac{\partial t}{\partial z}\right)+q_v \tag{1}$$

Where ρ is the cylinder density, c is the specific heat capacity of the cylinder material, λ is the thermal conductivity of the cylinder material, q_v is the heat generation rate of the internal heat source, t is the Celsius temperature, and τ is the time of heat transfer.

In the operating temperature range of the damper, once the cylinder material is determined, its thermal conductivity can be set to a constant value, q=0 when there is no internal heat source; in addition, the temperature field of the cylinder heat conduction is one-dimensional heat conduction symmetrically distributed along the axis. So equation (1) can be simplified as follows:

$$\frac{1}{r}\frac{d}{dr}\left(r\frac{dt}{dr}\right)=0 \tag{2}$$

According to the heat conduction differential equation and Fourier's law, the heat conduction equations of the inner and outer cylinder walls are obtained [5-6].

$$Q_w=-2\pi L_y\lambda g\frac{T_{w1}-T_{w2}}{\ln\left(r_{hn}/r_{hw}\right)} \tag{3}$$

$$Q_w=-2\pi L_y\lambda g\frac{T_{w3}-T_{w4}}{\ln\left(r_{hn1}/r_{hw1}\right)} \tag{4}$$

Where Q_w is the heat transferred by the cylinder wall, r_{hn} is the inner cylinder radius, r_{hw} is the inner cylinder outer radius, r_{hn1} is the outer cylinder inner radius, and r_{hw1} is the outer cylinder outer radius.

In the oil reservoir, the upper part is the air with pressure and the lower part is the oil. Since the liquid level does not fluctuate greatly when the damper is working, the medium in the oil storage chamber mainly radiates heat to the outside in the way of heat conduction. According to formula (2) and Fourier's law, the heat conduction of air and oil can be separately derived:

$$Q_{ck} = -\pi L_y \lambda_k g \frac{T_{w2} - T_{w3}}{\ln\left(r_{hw} / r_{hn1}\right)} \tag{5}$$

$$Q_{cy} = -\pi L_y \lambda_y g \frac{T_{w2} - T_{w3}}{\ln\left(r_{hw} / r_{hn1}\right)} \tag{6}$$

Where λ_k is the thermal conductivity of air, λ_y is the thermal conductivity of the oil, Q_{ck} is the heat transferred by the air in the reservoir, and Q_{cy} is the heat transferred by the oil in the reservoir.

The simultaneous expressions (3)-(6) are combined according to the heat transfer relationship to obtain a comprehensive expression of heat conduction:

$$Q_w = \pi L_y \left(T_{w1} - T_{w4}\right)\left[\frac{\ln\left(r_{hw} / r_{hn}\right)}{2\lambda} + \frac{\ln\left(r_{hw1} / r_{hn1}\right)}{2\lambda} + \frac{\ln\left(r_{hn1} / r_{hw}\right)}{\left(\lambda_k + \lambda_y\right)}\right]^{-1} \tag{7}$$

According to the theory of heat transfer, the development of the boundary layer of air near the wall of the cylinder is not disturbed by space constraints, and it belongs to the natural convection heat transfer in large space. The expression of the heat transfer coefficient on the outer surface of the cylinder in the vertical state can be derived by the computational correlation proposed by S W Churchill [7]:

$$h_{w0} = \frac{\lambda_k}{L_y}\left\{0.825 + \frac{0.387\left(Gr \cdot \mathrm{Pr}_k\right)^{1/6}}{\left[1 + \left(0.492 / \mathrm{Pr}_k\right)^{9/16}\right]^{8/27}}\right\}^2 \tag{8}$$

Prandt number:

$$\mathrm{Pr}_k = \frac{v_k}{\lambda_k}\rho_k c_k$$

Grachev number:

$$Gr = \frac{g\beta \cdot \left(T_{w4} - T_\infty\right) \cdot L_y^{\ 3}}{v_k^{\ 2}}$$

Where c_k is the specific heat capacity of air, ρ_k is the air density, and v_k is the kinematic viscosity of air. $\beta = \frac{1}{V}\left(\frac{\partial V}{\partial T}\right)_p$ is the volume expansion coefficient, also known as the body expansion coefficient.

Heat exchange between the studio fluid and the inner cylinder is forced convection, so the forced convection heat transfer coefficient of the inner surface of the inner cylinder under the laminar flow state is derived as follows:

$$h_{wyi} = \frac{\lambda_y}{L_y} \cdot \left\{3.66 + \frac{0.0688 V D_{hn}^2 \,\mathrm{Pr}_y}{v_y L_y + 0.04 L_y \left[V D_{hn} \,\mathrm{Pr}_y \left(D_{hn} / L_y\right)\right]^{2/3}}\right\} \tag{9}$$

Where λ_y is the thermal conductivity of the oil, v_y is the kinematic viscosity of the oil, and Pr_y is the Prandtl number of the oil.

And the forced convection heat transfer coefficient of the inner surface under turbulent flow:

$$h_{wyi} = R \cdot \frac{\lambda_y}{L_y} \cdot \frac{M \cdot N \cdot \mathrm{Pr}_y}{1 + B \cdot \left(\mathrm{Pr}_y^{2/3} - 1\right)} \tag{10}$$

Where:

$$R = 1 + \left(\frac{D_{hn}}{L_y}\right)^{0.7}, \quad M = \frac{1}{8}\left(0.79 \ln\left(\frac{VD_{hn}}{v_y}\right) - 1.64\right)^{-2} \tag{11}$$

$$N = \frac{VD_{hn}}{v_y} - 1000, \quad B = 25.4\sqrt{2} \cdot \left[0.79 \ln\left(\frac{VD_{hn}}{v_y}\right) - 1.64\right]^{-1} \tag{12}$$

From the formula (7) and the Newtonian cooling formula, the heat expression of the oil to the external environment can be derived:

$$Q_w = \left(T_y - T_\infty\right) \cdot \left[\frac{\ln\left(r_{hw} / r_{hn}\right)}{2\lambda\pi L_y} + \frac{\ln\left(r_{hw1} / r_{hn1}\right)}{2\lambda\pi L_y} + \frac{\ln\left(r_{hn1} / r_{hw}\right)}{\pi L_y\left(\lambda_k + \lambda_y\right)} + \frac{1}{h_{wyi} A_n} + \frac{1}{h_{w0} A_{w1}}\right]^{-1} \tag{13}$$

Where A_{hn} is the inner surface area of the inner cylinder and A_{w1} is the outer surface area of the outer cylinder.

Taking oil as the research object and deriving according to the principle of conservation of energy:

$$m_y c_{yp} \frac{dT_y}{dt} + B_y T_y = Y + B_y T_\infty \tag{14}$$

Where:

$$B_y = \left[\frac{\ln\left(r_{hw} / r_{hn}\right)}{2\lambda\pi L_y} + \frac{\ln\left(r_{hw1} / r_{hn1}\right)}{2\lambda\pi L_y} + \frac{\ln\left(r_{hn1} / r_{hw}\right)}{\pi L_y\left(\lambda_k + \lambda_y\right)} + \frac{1}{h_{wyi} A_n} + \frac{1}{h_{w0} A_{w1}}\right]^{-1} \tag{15}$$

Damping valves are calculated by orifice and slit throttling:

$$Y = \frac{12\mu_{y0} l_f \left(A_h V\right)^2}{\pi \delta_f^{\,3} B_f} e^{\lambda_t\left(t_y - t_{y0}\right)} + \frac{\left(A_h V\right)^3}{2\left(C_d A_c\right)^2} \rho_y \tag{16}$$

Where m_y is the oil mass, c_{yp} is the constant pressure specific heat capacity of the oil, ρ_y is the oil density, A_c is the orifice area, the flow coefficient $C_d = 0.62$, δ_f is the width of the gap, B_f is the wet circumference of the gap, l_f is the length of the gap, μ_{y_0} is the initial dynamic viscosity of the oil, λ_t is the viscosity temperature index of the oil, and A_h is the effective working area of the piston.

The simultaneous (3) to (16) form a thermodynamic mathematical model of the double barrel damper.

If the outer wall of the damper is painted black, the amount of heat transfer from the heat radiation must also be considered. The expression of the amount of radiated heat transfer is directly listed below:

$$Q_{wf} = \varepsilon C_b \left[\left(\frac{T_{W4}}{100}\right)^4 - \left(\frac{T_\infty}{100}\right)^4\right] A_{w1} \tag{17}$$

Where Blackbody emissivity is $C_b = 5.67 W / m^2 \cdot K^4$, ε is the rate of emissivity for thermal radiation.

The expression of the composite heat transfer coefficient of the outer wall of the outer cylinder is derived as follows:

$$h_{wof} = \varepsilon C_b \frac{\left(T_{w4}^2 + T_\infty^2\right)\left(T_{w4} + T_\infty\right)}{(100)^4} + h_{wo} \tag{18}$$

Combined with equation (7) and Newtonian cooling formula, the comprehensive heat expression of oil released to the outside world considering heat radiation is obtained:

$$Q_w = \left(T_y - T_\infty\right) \cdot \left[\frac{\ln\left(r_{hw} / r_{hn}\right)}{2\lambda\pi L_y} + \frac{\ln\left(r_{hw1} / r_{hn1}\right)}{2\lambda\pi L_y} + \frac{\ln\left(r_{hn1} / r_{hw}\right)}{\pi L_y\left(\lambda_k + \lambda_y\right)} + \frac{1}{h_{wyi}A_n} + \frac{1}{h_{w0f}A_{w1}}\right]^{-1} \tag{19}$$

Equation (15) is also changed to the following form:

$$B_y = \left[\frac{\ln\left(r_{hw} / r_{hn}\right)}{2\lambda\pi L_y} + \frac{\ln\left(r_{hw1} / r_{hn1}\right)}{2\lambda\pi L_y} + \frac{\ln\left(r_{hn1} / r_{hw}\right)}{\pi L_y\left(\lambda_k + \lambda_y\right)} + \frac{1}{h_{wyi}A_n} + \frac{1}{h_{w0f}A_{w1}}\right]^{-1} \tag{20}$$

The above expressions constitute a thermodynamic mathematical model of a double-tube damper considering thermal radiation.

4. Thermodynamic model simulation analysis

The thermodynamic model of the damper is iteratively operated by software Matlab programming. The relevant parameters are as follows: $\delta_f = 0.1 \times 10^{-3}\text{m}$, $B_f = 0.25\text{m}$, $C_d = 0.62$, $l_f = 1 \times 10^{-3}\text{m}$, $\lambda = 23\text{W}/(\text{m} \cdot \text{k})$, $T_\infty = 300\text{K}$, $\rho_y = 890\text{kg}/\text{m}^3$, $\lambda_t = 0.02$, $\rho_0 = 1.293\text{kg}/\text{m}^3$, $T_0 = 273\text{K}$, $\mu_{y0} = 8.9 \times 10^{-3}\text{kg}/(\text{m} \cdot \text{s})$, the initial temperature of the system is equal to the ambient temperature.

Input harmonic excitation $A = 0.05\text{m}, f = 1\text{Hz}$, set the damper to exchange energy with the outside world every time it works, cycles $X = 5000$, single cycle time $t = \frac{1}{f}$.

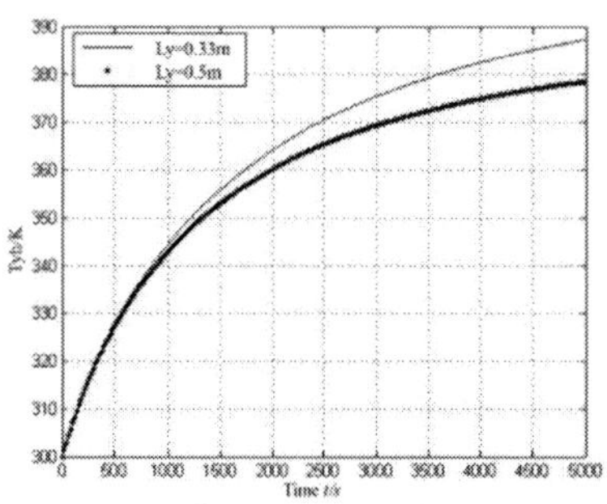

Figure 2. Different cylinder lengths

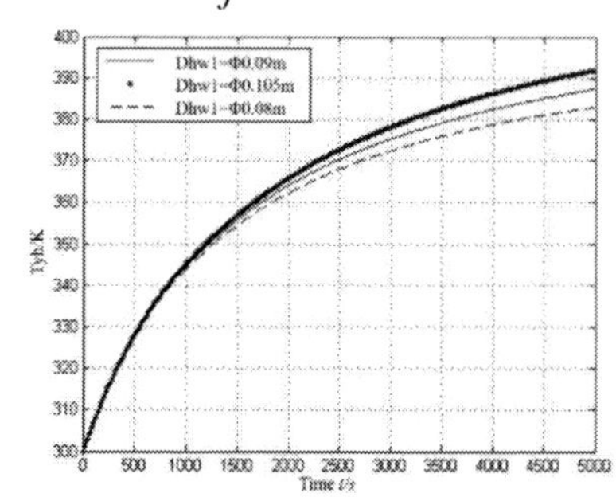

Figure 3. Different outer cylinder diameters

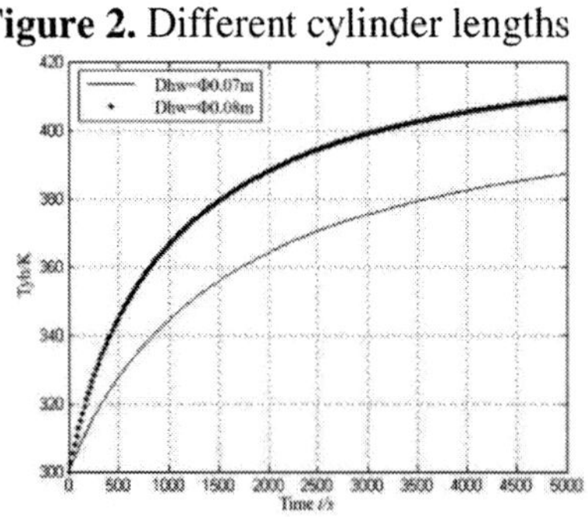

Figure 4. Different inner cylinder diameters

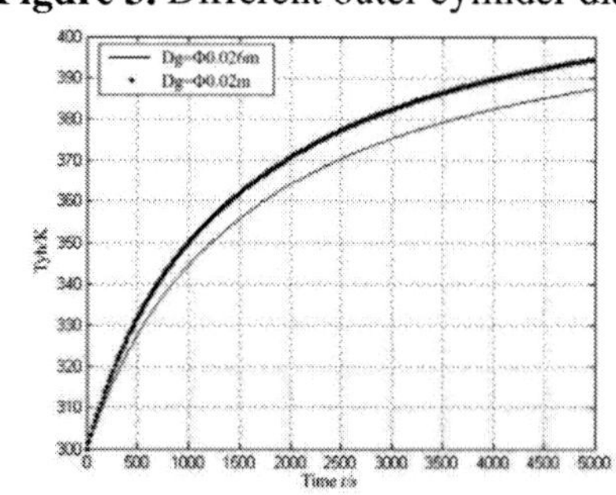

Figure 5. Different piston rod diameters

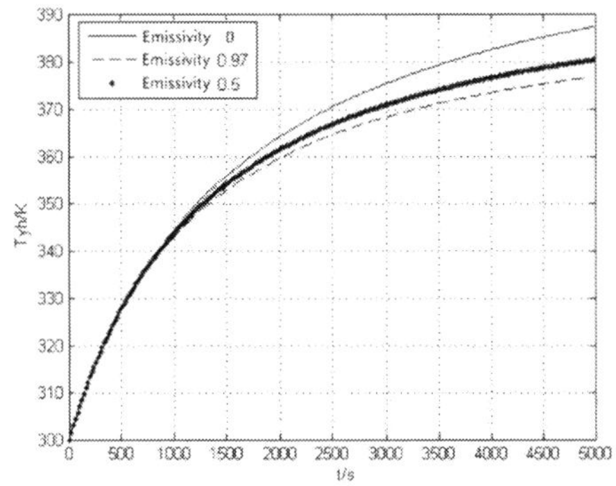

Figure 6. Thermal radiation oil temperature comparison

Through the above figiures 2-6, the following conclusions can be drawn: under the condition that the structure size of the damping valve is constant, as the axial dimension of the cylinder and the radial dimension of the piston rod increase, the temperature rise of the damper oil is gradually reduced in the same time. With the increase of the radial dimension of the inner and outer cylinders, the temperature rise of the damper oil gradually increases in the same time; when the outer surface of the damper is sprayed with dark or black paint, the surface thermal radiation emissivity increases, and the temperature rise of the damper oil gradually decreases.

5. Conclusion

(1) When the structural parameters of the damping valve are constant, increasing the axial dimension of the cylinder and the radial dimension of the piston rod, or reducing the radial dimension of the inner and outer cylinders, can reduce the temperature rise of the damper oil.

(2) In order to increase the heat dissipation of the system, dark or black paint can be sprayed on the outer surface of the damper to increase the emissivity of the heat radiation, thereby effectively slowing the temperature rise of the oil.

(3) As time goes by, the system will reach the temperature balance point. It can avoid the leakage phenomenon by adjusting the relevant parameters of the damper and controlling the balance point within the limit temperature range of the seal.

References

[1] Everaert K 2006 Heat transfer from a single tube to the flowing gas-solid suspension in a CFB riser *Heat Transfer Engineering* **27** 66-70

[2] Khan W A, Culham J R and Yovanonich M M 2006 Fluid flow and heat transfer in power-law fluids across circular cylinders *Journal of Heat Transfer* **128** 870-8

[3] Morales A L, Nieto A J and Chicharro J M 2015 An adaptive pneumatic system for the attenuation of random vibrations *Journal of Vibration and Control* **21** 907-18

[4] Konieczny L, Burdzik R and Wegrzyn T 2016 Analysis of Structural and Material Aspects of Selected Elements of a Hydropneumatic Suspension System in a Passenger car *Archives of Metallurgy&Materials* **61** 79-83

[5] Li Z L, Li X Q and Chen X D 2014 Generic vibration criteria-based dual-chamber pneumatic spring vibration isolation table design *P I Mech Eng B-J Eng.* **228** 1621-29

[6] Theron N J and Els P S 2007 Modeling of a Semi-active Hydro-pneumatic Spring-damper Unit *International Journal of Vehicle Design* **45** 501-21

[7] Chai T, Dreyer J T and Singh R 2015 Nonlinear dynamic properties of hydraulic suspension bushing with emphasis on the flow passage characteristics *P I Mech Eng D-J Aut.* **229** 1327-44

IOP Conf. Series: Materials Science and Engineering **751** (2020) 012067 doi:10.1088/1757-899X/751/1/012067

Numerical Simulation of Liquid-vapor Phase Transition Induced by a Heated Surface in a Shear Flow

Xiaopeng Shan and Deming Nie*

Institute of Fluid Mechanics, China Jiliang University, Hangzhou, China

* Corresponding author: nieinhz@cjlu.edu.cn

Abstract. The influence of the external flow on the liquid-vapor phase transition is numerically studied in this work. The bubble growth and departure from a heated surface in a shear flow is realized through a pseudo-potential based lattice Boltzmann method. The benchmark test of the pool boiling induced by a single heated plate is adopted to validate the present method. The averaged heat flux, the bubble releasing frequency and flow features are presented, along with the effects of the flow intensity on them. It is shown that the external flow significantly enhances the heat transfer of the system in two respects: increasing the averaged heat flux and the bubble releasing frequency.

1. Introduction

Two-phase flows are very common in many natural processes as well as in industrial applications. The phase transition may occur under certain conditions in numerous industrial process involving two-phase flows, which is known as boiling heat transfer. The phase change has a significant influence on the liquid-vapor flows in terms of heat transfer rate as well as flow characteristics. It is important to pay much attention to the phenomena and mechanism of phase transition between liquid and vapor, which may provide a fundamental understanding of the behavior of heat transfer in two-phase flows.

The lattice Boltzmann method (LBM), which is based on the well-known Boltzmann equations, has emerged as a powerful numerical scheme for the simulation of particle suspensions, multiphase flow, microfluidics, and turbulence due to its several remarkable advantages since it was originated. In particular, the LBM is proved to be a promising method for dealing with interfacial flows such as solid-liquid and liquid-vapor flows. So far several lattice Boltzmann models have been proposed to simulate the liquid-vapor flows, including the color-gradient model [1], the pseudo-potential model [2-4] and the free-energy model [5]. Based on the popular pseudo-potential scheme [2], Gong and Cheng [3, 4] proposed an improved lattice Boltzmann model for liquid-vapor phase change. In their model, a new form of the source term in the energy equation was derived, which was shown to improve the numerical stability. Overall, the mechanism of liquid-vapor phase change is not well-understood even though much effort [6-8] has been devoted to it, owing to highly nonlinear effects in a two-phase system.

It is known that the major body of fluid is quiescent for the pool boiling, which, however, is not the general case in the industrial applications. In comparison with the pool boiling, things may become more complex when the major body of fluid is driven to move, such as Poiseuille flow and shear flow. The external flow may have a noticeable influence on the vapor bubble departure and the behavior of heat transfer. However, literature survey shows that the study on the liquid-vapor phase change in the Poiseuille flow or shear flow is rare. Much more attention should be paid to this issue. Therefore, the objective of this work is to present a preliminary understanding of the bubble departure from a heated-surface in a shear flow. The improved LB model proposed by Gong and Cheng [3, 4] was used here.

2. Lattice Boltzmann Method

The two-phase lattice Boltzmann method [3, 4] is briefly introduced here. The single-relaxation-time lattice Boltzmann equations are expressed as,

$$f_i(\boldsymbol{x}+\boldsymbol{e}_i\Delta t, t+\Delta t) - f_i(\boldsymbol{x},t) = -\frac{1}{\tau_f}\left[f_i(\boldsymbol{x},t) - f_i^{(eq)}(\boldsymbol{x},t)\right] + \Delta f_i \tag{1}$$

where $f_i(\boldsymbol{x}, t)$ is the density distribution function corresponding to the microscopic velocity $\boldsymbol{e}_i$, Δt is the time step of the simulation, τf is the relaxation time. $f_i^{(eq)}(\boldsymbol{x}, t)$ is the equilibrium distribution function which is given by,

$$f_i^{(eq)} = w_i\rho\left[1+\frac{\boldsymbol{e}_i\cdot\boldsymbol{u}}{c_s^2}+\frac{(\boldsymbol{e}_i\cdot\boldsymbol{u})^2}{2c_s^4}-\frac{\boldsymbol{u}^2}{2c_s^2}\right] \tag{2}$$

where c_s is the speed of sound, and w_i are weights related to the lattice model. Δf_i is the discrete form of the body force $\boldsymbol{F}$, which accounts for the inter-particle interaction force $\boldsymbol{F}_{int}$, the gravitational force $\boldsymbol{F}_g$ and the interaction force between solid surface and fluid $\boldsymbol{F}_s$. The fluid density and velocity are obtained through,

$$\rho = \sum_i f_i, \rho\boldsymbol{u} = \sum_i \boldsymbol{e}_i f_i \tag{3}$$

Due to the forcing term, the real fluid velocity of fluid U is modified by,

$$\rho\boldsymbol{U} = \sum_i \boldsymbol{e}_i f_i + \frac{\Delta t}{2}\boldsymbol{F} \tag{4}$$

Similarly, the lattice Boltzmann equations are proposed [3, 4] to solve the fluid temperature T,

$$g_i(\boldsymbol{x}+\boldsymbol{e}_i\Delta t, t+\Delta t) - g_i(\boldsymbol{x},t) = -\frac{1}{\tau_g}\left[g_i(\boldsymbol{x},t) - g_i^{(eq)}(\boldsymbol{x},t)\right] + \Delta t w_i\phi \tag{5}$$

where τ_g is the relaxation time for the fluid temperature and $g_i^{(eq)}(\boldsymbol{x}, t)$ is the corresponding equilibrium distribution function,

$$g_i^{(eq)} = w_i T\left[1+\frac{\boldsymbol{e}_i\cdot\boldsymbol{U}}{c_s^2}+\frac{(\boldsymbol{e}_i\cdot\boldsymbol{U})^2}{2c_s^4}-\frac{\boldsymbol{U}^2}{2c_s^2}\right] \tag{6}$$

The source term is responsible for the phase change, determined by,

$$\phi = T\left[1-\frac{1}{\rho c_v}\left(\frac{\partial p}{\partial T}\right)_\rho\right]\nabla\cdot\boldsymbol{U} \tag{7}$$

where p is the pressure and c_v is the heat capacity. Then the temperature is obtained through,

$$T = \sum_i g_i \tag{8}$$

3. Validation

In the case of boiling, the bubble release frequency plays a key role in the process of heat transfer, which has been an everlasting topic from both numerical and theoretical points of view. According to the balance between adhesive force and buoyant force experienced by a vapor bubble, Fritz [9] obtained a formulation for the departure diameter which is related to the magnitude of gravity force (|g|),

IOP Conf. Series: Materials Science and Engineering **751** (2020) 012067 doi:10.1088/1757-899X/751/1/012067

$$D_b \sim \sqrt{\frac{\sigma}{|g|(\rho_L - \rho_G)}} \tag{9}$$

where is the surface tension, $_L$ and $_G$ is the density of liquid and vapor, respectively. Eq. (9) indicates a power-law relationship between D_b and g, .e. D_b~$|g|^{-0.5}$. Similarly, Zuber [10] developed a formulation for the release frequency of bubbles,

$$T = \frac{1}{f} \sim D_b \left[\frac{\sigma |g| (\rho_L - \rho_G)}{\rho_L^{\ 2}} \right]^{-0.25} \tag{10}$$

From (9) and (10), it is easy to reach the following relationship: T~|g|-0.75.

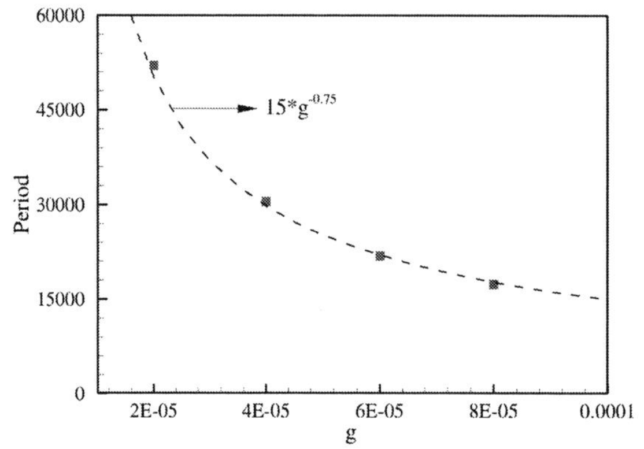

Figure 1. Bubble releasing period as a function of the gravitational force.

According to the present LBM simulations, Figure 1 shows the bubble releasing period of a single vapor bubble in the case of pool boiling. It can be clearly seen that the dependence of T on the gravity force is realized.

4. Results

This work aims to present a preliminary study on the behavior of heat transfer in a shear flow. Figure 2 shows the schematic diagram of the present problem. The liquid of density ρ_s and temperature T_s is filled with in a two-dimensional domain with dimensions LxH. A heated plate with length L_h and temperature T_w is placed in the middle of the bottom wall. The upper wall is driven to move with a constant speed U_0, which eventually results in the shear flow of the domain. The periodic boundary conditions are applied in the horizontal direction. In the simulations, the parameters are fixed as follows: L = 600, H = 400, L_h = 10, T_w=0.98T_c, T_s=0.9T_c and ρ_s = ρ_c. ρ_c and T_c denote the critical density and temperature of the liquid, respectively. It should be stated here that the parameters are all in lattice units, which is very common for the lattice Boltzmann computations.

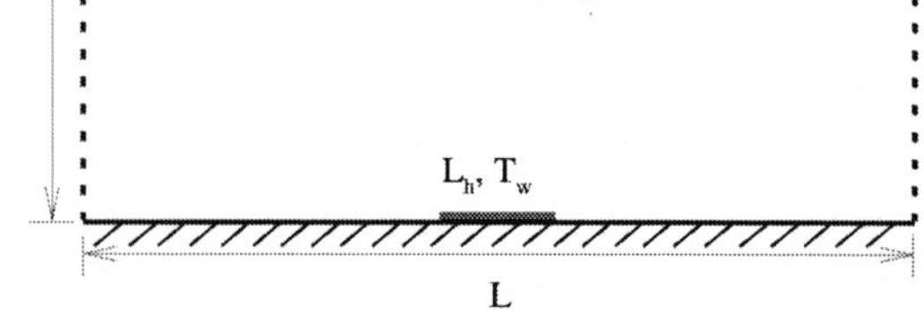

Figure 2. Schematic diagram of the present problem.

Figure 3 presents the instantaneous density contours at different times, which shows the vapor bubble growth and departure from the heated plate during one period. Three values of shearing velocity are taken into account, i.e. $U_0 = 0$ (top), 0.05 (middle) and 0.1 (bottom). Obviously, the case of U0 = 0 corresponds to the classical pool boiling. As seen in Figure 3, the symmetry in bubble shape is destroyed when there is external flow (i.e. $U_0 \neq 0$). As a result, there is a difference between the upstream and downstream contact angles of bubble, which becomes more significant at larger values of U_0. Moreover, it is seen that the existence of the external flow delays the generation of bubble neck, suggesting a different bubble releasing frequency.

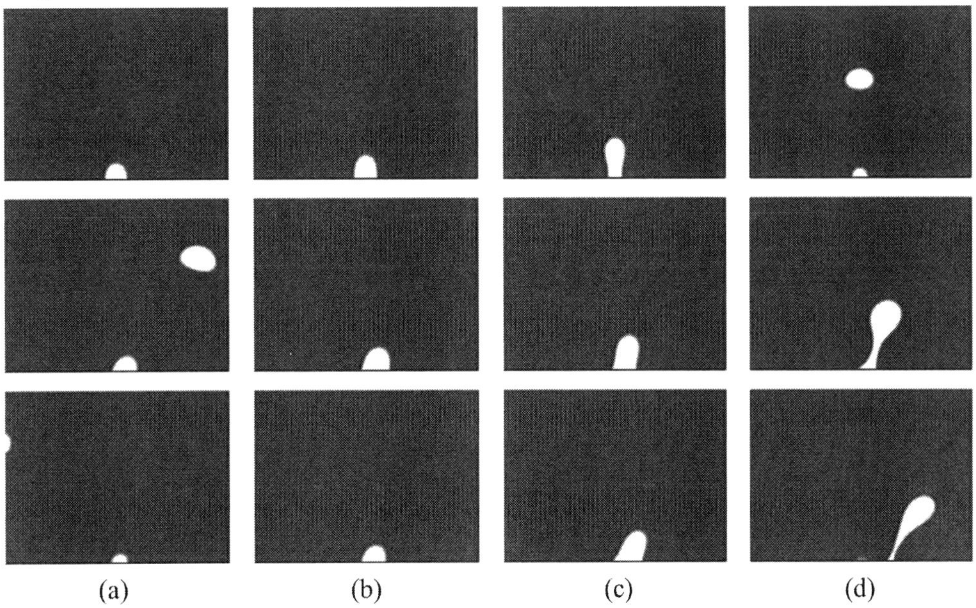

Figure 3. Instantaneous density contours showing the bubble growth and departure from a heated surface under the external flow with different intensities, i.e. $U_0 = 0$ (top), $U_0 = 0.05$ (middle) and $U_0 = 0.1$(bottom), at different times: (a) $t = 0$, (b) $t = 0.25T$, (c) $t = 0.5T$ and (d) $t = 0.75T$.

Figure 4 shows the time history of the heat flux which is averaged over the heated surface for three values of U_0. It is clearly seen that the presence of external flow significantly increases the averaged heat flux, reflecting the fact that the external flow greatly enhances the heat transfer. However, as the flow intensity (U_0) increases, the averaged heat flux does not monotonously increase. More importantly, Figure 4 also indicates that the bubble departure is accelerated when the value of U_0 is nonzero. In other word, the bubble releasing frequency is higher for $U_0 \neq 0$ than that for $U_0 = 0$. This, of course, increases the speed at which the heat is brought from the heated plate.

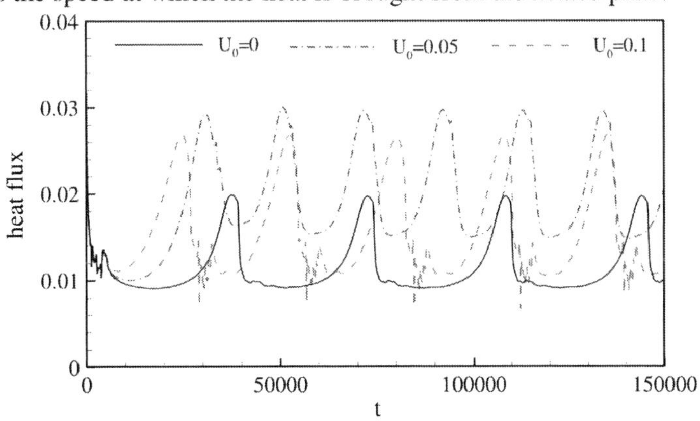

Figure 4. Time history of the averaged heat flux of the heated surface for different values of U_0.

Finally, Figure 5 summarizes the dependence of the bubble releasing period (T) on the flow intensity. It is expected that the time period for $U_0 > 0$ is smaller than that for $U_0 = 0$, which is in accord with the Figure 4. Furthermore, it is seen that the value of T decreases as U_0 increases for $U_0 \leq 0.05$. However, a sudden jump can be observed as U_0 increases from 0.05 to 0.06, as shown in Figure 5 . To be more specific, the value of T is about 21000 for $U_0 = 0.05$, which becomes almost 30600 for $U_0 = 0.06$. This sudden increase indicates a significant change in the behavior of heat transfer for the system, which is expected to owing to the change in the flow features. So far the reason is unclear. It is worth additional simulations and further investigation.

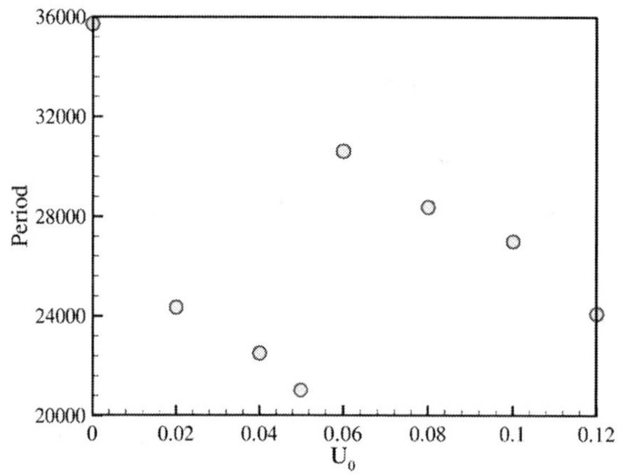

Figure 5. Bubble releasing period as a function of the value of U_0.

5. Conclusion

In this work a two-phase lattice Boltzmann method is used to numerically study the influence of the external flow on the phase transition resulted from a heated surface in a cold liquid. The method is validated by simulating the pool boiling involving a single heated plate and comparing the bubble releasing frequency with the theoretical prediction. Numerical tests indicate that the external flow noticeably enhances the heat transfer of the system in two respects, i.e. increasing the averaged heat flux of the heated plate and increasing the bubble releasing frequency. Moreover, in comparison with the case of quiescent fluid, the period is seen to be always smaller in the presence of the external flow. However, it is strange that a sudden increase is seen in the time period of bubble departure at a certain flow intensity. The reason behind this is unclear.

Acknowledgement

This work was supported by the National Key Research and Development Program of China (2017YFB0603700).

Reference

[1] A.K. Gunstensen, D.H. Rothman, S. Zaleski, G. Zanetti. Lattice Boltzmann model of immiscible fluids, Phys. Rev. A 43 (1991) 4320–4327.
[2] X. Shan, H. Chen, Lattice Boltzmann model for simulating flows with multiple phases and components, Phys. Rev. E 47 (1993) 1815–1820.
[3] S. Gong, P. Cheng, A lattice Boltzmann method for simulation of liquid-vapor phase-change heat transfer, Int. J. Heat Mass Trans. 55 (2012) 4923–4927.
[4] S. Gong, P. Cheng. Numerical investigation of droplet motion and coalescence by an improved lattice Boltzmann model for phase transitions and multiphase flows, Comput. Fluids 53 (2012) 93–104.
[5] M.R. Swift, E. Orlandini E, W.R. Osborn, J.M. Yeomans, Lattice Boltzmann simulations of

liquid-gas and binary fluid systems, Phys. Rev. E 54 (1996) 5041–5052.
[6] G. Hazi, A. Markus. On the bubble departure diameter and release frequency based on numerical simulation results, Int. J. Heat Mass Trans. 52 (2009) 1472–1480.
[7] S. Ryu, S. Ko. Direct numerical simulation of nucleate pool boiling using a two-dimensional lattice Boltzmann method. Nucl. Eng. Des. 248 (2012) 248–262.
[8] T. Sun, W. Li. Three-dimensional numerical simulation of nucleate boiling bubble by lattice Boltzmann method, Comput. Fluids 88 (2013) 400–409.
[9] W. Fritz. Maximum volume of vapour bubbles, Physik Zeitschr 36 (1935) 379–384.
[10] N. Zuber. Nucleate boiling. The region of isolated bubbles and the similarity with natural convection. Int. J Heat Mass Tran. 64 (1963) 122–132.

IOP Conf. Series: Materials Science and Engineering **751** (2020) 012068 doi:10.1088/1757-899X/751/1/012068

Molecular Dynamics Simulation of Rarefaction Effect on Shear-driven Gas Flow in Nanochannels

Ye Zhang[1], Ran Zhang[1,2,*], Wenjia Xie[1] and Hua Li[1]

[1]College of Aerospace Science and Engineering, National University of Defense Technology, Changsha 410073, China

[2]State Key Laboratory of Laser Interaction with Matter, Northwest Institute of Nuclear Technology, Xi'an 710024, China

[*]E-mail: zr07024221@126.com

Abstract. The nanoscale gas flow at normal temperature and pressure varies from the slip flow regime to the transition regime, of which the character length changes across several nanometers to several thousand nanometers. The difference in the degree of rarefaction leads to different rules of the flow in nanochannels. In order to study the influence of rare gas effect on the nanoscale gas flow, we use three-dimensional molecular dynamics method to simulate the shear flow confined by two parallel plate under different Knudsen numbers, and then obtain the distribution of flow velocity, density, normal stress and shear stress, thus calculating and analyzing the variation of the tangential momentum accommodation coefficient. Counting the gas molecules according to the velocity with positive and negative values respectively, we find that the gas flow is decomposed into two independent shear flows, which indicates strong non-equilibrium characteristics in the nanochannel.

1. Introduction

In recent years, with the rapid development of micro-nano electromechanical systems (MEMS/NEMS), nanoscale gas flow has received extensive attention and has gradually become an important research field[1-4]. Nanoscale gas flow has broad application prospects,(widely applied foreground) such as biological sample extraction, integrated circuit cooling, and active flow control[5]. However, the flow characteristics at the nanoscale are quite different from the macroscopic flows. In order to understand the nanoscale flow characteristics, it is necessary to conduct an in-depth study of its mechanism.

The degree of thinness of the gas is generally expressed by the Knudsen number Kn. In the micro-nanoscale gas flow, the mean free path of the molecule is no longer a small amount compared with the flow characteristic scale, and the continuous medium assumption is no longer valid, and the gas exhibits a significant thinning effect. At the same time, the reduction of the scale will lead to an increase in the area to volume ratio, and the physical quantity associated with the wall surface will have a more pronounced effect on the flow. The difficulties of theory and experimental approach have led to the tendency to use numerical simulation methods to study the flow of thin gases. The Direct Analog Monte Carlo (DSMC) method is a widely used numerical simulation method that can effectively simulate the flow of thin gases in a transitional flow region. However, DSMC still needs to provide momentum and energy accommodation factors when dealing with the interaction of gas and wall at the boundary. For different gases and walls, the accommodation coefficients vary widely, so this method is not suitable for studying the nanoscale flow of wall dominant factors.

Molecular Dynamics (MD) method[6], as a deterministic simulation method, from the perspective of atomic motion, can not only obtain the flow details of the flow field, but also obtain the scattering regularity of gas molecules at the wall surface. Therefore, it has unique advantages in studying the micro-nano scale flow. Based on MD simulation, many scholars have conducted extensive research on micro-nano channel flow. Cao et al[7,8] used molecular dynamics simulation method to study the gas flow in the micro-nano channel of the slip zone, and discussed the relationship between velocity slip and potential energy intensity, system temperature and wall roughness. Sun and Li[9] adopted the MD method to calculate the tangential momentum accommodation coefficient and the normal momentum accommodation coefficient in the smooth micro channel. Spijker et al[10] used the MD method to simulate the gas flow in the nanochannel, and statistically analyzed the velocity distribution before and after the collision of the gas molecules with the wall surface, and introduced the velocity correlation cloud map when analyzing the relationship between the incident and the reflection velocity. Xie and Liu[11] simulated the Poiseuille flow of the mixed gas in the nanochannel, and found that the wall force caused the gas mixture in the channel to exhibit a special phenomenon that the chemical composition and the physical structure were no longer uniform. Barisik and Beskok[12,13] introduced an intelligent wall molecular dynamics (SWMD) method. Assuming that the wall atoms are stationary, it is found that the changes in density, velocity and pressure of the gas in the near-wall region of the channel are determined only by the wall force field. Bao et al[14,15] used molecular dynamics to study the pressure-driven flow in finite-length nanochannels, and come to the conclusion that the pressure difference is independent of the temperature, the Knudsen number, the channel aspect ratio and wall properties. To et al.[16] studied the mean free path distribution of non-equilibrium gas in micro-nano channels by molecular dynamics simulation method, and found that the mean free path is strongly reduced near the wall surface and not sensitive to the flow type. Lim et al[17] simulated the flow of He, Ne, Ar, Kr and Xe gases by molecular dynamics method, studied the influence of molecular mass and adsorption effect on the tangential momentum accommodation coefficient, and found that the tangential momentum accommodation coefficient with molecular mass The changes are not monotonous.

For nanometer-scale gas flow at normal temperature and pressure, the flow varies across several nanometers to several thousand nanometers, and the flow crosses the slip flow region and the transition flow region. The degree of thinness is different, and the flow also shows different flow patterns. In order to study the effect of rare gas effect on the shear flow of nanoscale gas, the flow velocity, density, normal stress and shear stress distribution under different Knudsen conditions are investigated. The gas density of these examples remains the same , ie the mean free path of the gas molecules in all channels is the same, and the variation of the Knudsen number is achieved by changing the channel height.

2. The simulation method

In this paper, a three-dimensional non-equilibrium molecular dynamics method is used to simulate the shear flow of gaseous argon in micro-nano channels. The schematic diagram of the model is shown in Figure 1. The monoatomic gas is distributed in a nanochannel having a height of two infinitely long flat plates. The X and Y directions adopt periodic boundary conditions. The upper and lower walls move in opposite directions at a speed of 0.2 that drives the gas to make a shear flow. For monoatomic gases, their molecular mass $m_{Ar} = 6.63\times10^{-26}\,\mathrm{kg}$, molecular diameter $\sigma_{Ar} = 0.3405\,\mathrm{nm}$, and potential energy parameters $\varepsilon_{Ar} = 1.67\times10^{-27}\,\mathrm{J}$.

The gas-gas and gas-wall interactions are modeled using a truncated Lennard–Jones(L–J)6–12 potential energy function18, defined as:

$$U^{\mathrm{LJ}}\left(\mathrm{r}_{ij}\right) = 4\varepsilon_{Ar}\left\{\left[\left(\sigma_{Ar}/r_{ij}\right)^{12} - \left(\sigma_{Ar}/r_{ij}\right)^{6}\right] - \left[\left(\sigma_{Ar}/r_{c}\right)^{12} - \left(\sigma_{Ar}/r_{c}\right)^{6}\right]\right\} \tag{1}$$

Where r_{ij} is the intermolecular distance, r_c the cutoff radius which equals $3.0\sigma_{Ar}$.

3. The simulation method

3.1. The shear gas flow in the transition regime

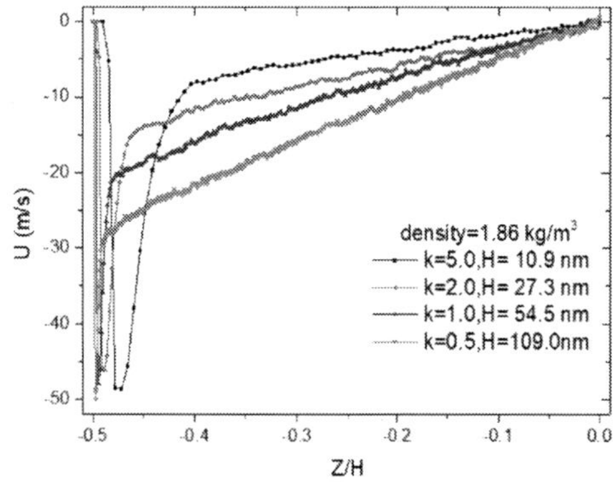

Figure 1. Dimensionless velocity distribution under different Knudsen numbers.

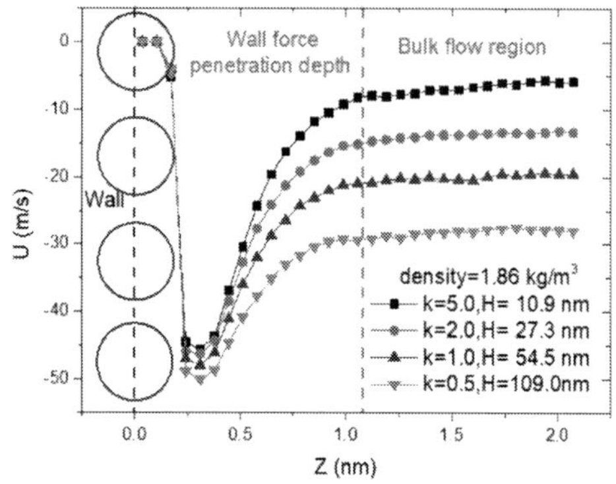

Figure 2. Density distribution of near-wall regions in each channel under different Knudsen numbers.

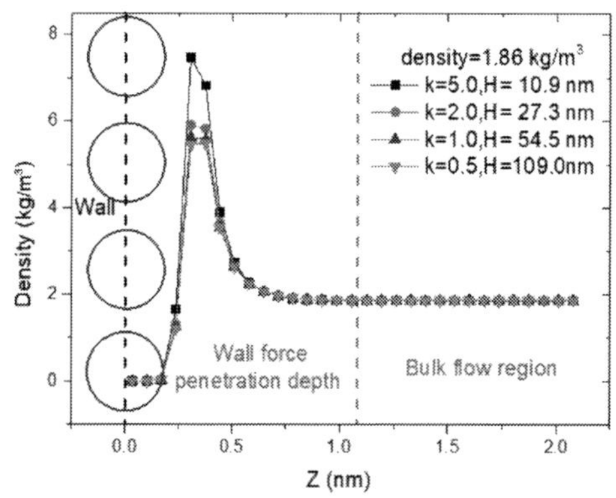

Figure 3. Dimensionless velocity distribution under different Knudsen numbers.

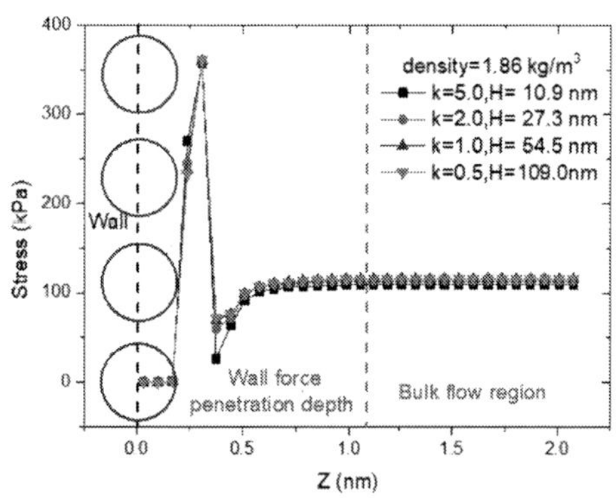

Figure 4. Normal stress distribution in the near wall area of each channel under different Knudsen numbers.

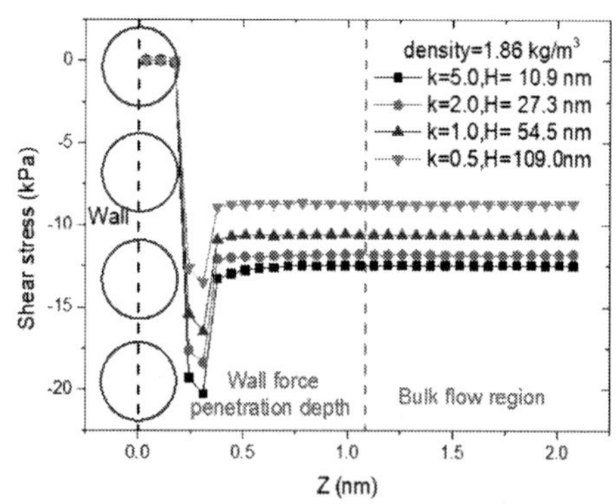

Figure 5. Shear stress distribution in the near wall area of each channel under different Knudsen numbers.

Figure.1 shows the velocity distribution under different Knudsen conditions. It can be seen from the figure that as the height of the pipe increases, the velocity shear rate in the channel decreases. The influence of the wall force is a fixed width of 1 nm. When the height of the pipe increases, the proportion of the flow field affected by the wall force is gradually reduced. For example, when the height of the channel is 109 nm, the wall force affecting area only accounts for about 2% of the entire flow field. Figure 2 shows the velocity distribution of the near-wall region under different Knudsen conditions. We find that the flow velocity in the near-wall region increases with the decrease of the Knudsen number due to the effect of the thin gas effect.

Figure 3 and Figure 4 show the density and normal stress distribution curves in the flow field under different Knudsen conditions. It can be seen from the figure that the change of channel height does not affect the density and normal stress distribution of the gas in the channel. Figure 5 is the shear stress distribution curve in the flow field under different Knudsen conditions. The shear stress in each channel shows the same distribution law, but its value decreases with the increase of Knudsen number.

3.2. The shear gas flow in the slip regime

When the height of the channel continues to increase and the average free freeness of the gas molecules is less than 0.1 compared with the pipe height, the flow gradually transitions to the slip flow field according to the division of the thin gas flow field. For the flow in the slip flow field, the influence of the wall facing flow is limited to the Knudsen layer, and the solution can be solved according to the NS equation with the additional Maxwell slip boundary condition. The velocity dimensionless velocity distribution of the slip zone is:

$$\frac{U}{U_W} = \frac{Z}{H} \frac{2}{1+2\frac{2-TMAC}{TMAC}Kn} \tag{2}$$

In this section, the channel height is 272.4nm and the gas density is 3.73 and 7.46kg/m3 respectively. The velocity distribution in the flow field is shown in Figure 6. The solid line in the figure is the theoretical solution corresponding to the Kn. The simulation results are obtained. And the theoretical solution is in good agreement with most of the flow areas. However, in the area of the Knudsen layer close to the wall, the simulated velocity curve has a certain deviation from the theoretical solution. This is due to that the first-order slip boundary conditions of the Maxwell slip boundary condition used in the theoretical solution assume the speed linear in the Knudsen layer. In fact, with the rarefied gas effect, the velocity distribution in the Knudsen layer is a typical nonlinear distribution, which is the limitation of the slip boundary condition.

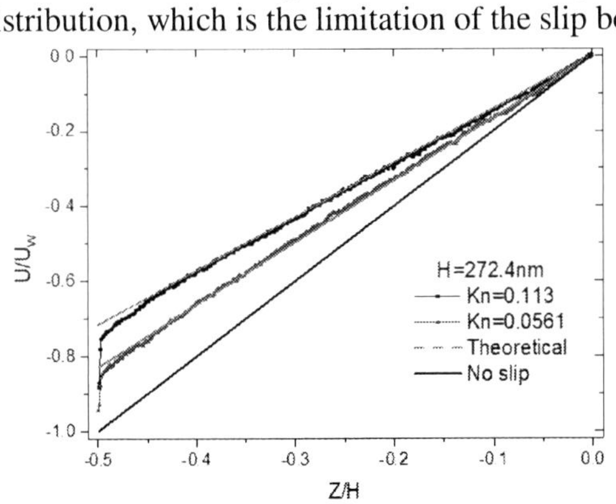

Figure 6. Comparison of gas shear flow velocity distribution and theoretical solution in the slip regime.

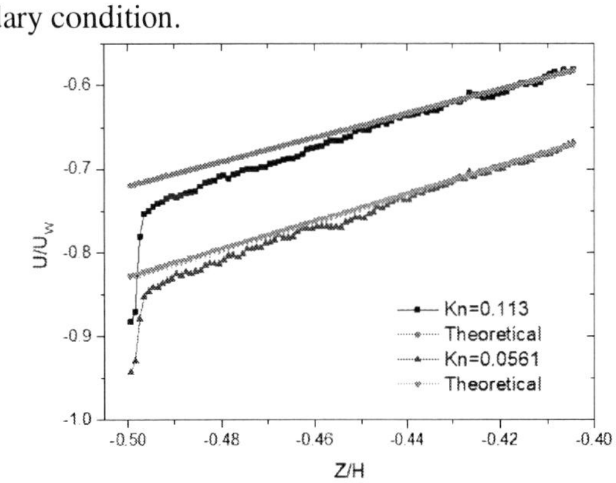

Figure 7. Differences between velocity distribution and theoretical solution in the Knudsen layer.

IOP Conf. Series: Materials Science and Engineering **751** (2020) 012068 doi:10.1088/1757-899X/751/1/012068

In the Knudsen layer, it can be seen from the Figure 7. that as Kn increases, the nonlinearity of the gas velocity becomes more and more obvious, and at the same time, the deviation from the theoretical solution becomes larger. According to the velocity profile obtained by MD simulation, the velocity curve of the central region of the flow field is fitted to obtain the dimensionless velocity shear rate of the flow under different Knudsen conditions. According to the relationship between the velocity and the Knudsen number, the corresponding Tangential momentum accommodation factor. The tangential momentum accommodation coefficient under Kn 0.113 and 0.0561 is both0.72, which is consistent with the tangential momentum accommodation coefficient of the transition flow region in the previous section, further illustrating the independence of the tangential momentum accommodation coefficient and the Knudsen number.

3.3. The analysis of the rarefaction effect

The entire flow area can be layered in the Z direction, wherein the area away from the wall surface is the wall force area, the remaining part is the main flow area, the width is , and then it is equally divided into 13 layers. Then, the velocity samples of each layer of gas molecules are subjected to classification, and the positive and negative of the velocity components of the gas molecules are classified, and the average tangential velocity of the two types of gas molecules in each layer of the flow field is obtained by statistical average. Figure.8 is a tangential velocity distribution of each layer of the flow field obtained by the above method, wherein represents the tangential average velocity of the gas molecules having a normal velocity component in each layer and represents a negative normal velocity component in each layer. The tangential average velocity of gas molecules. We found that the average tangential velocity distribution of the above two part of gas molecules is a completely antisymmetric distribution, the shear rate is exactly the same, and the two parts are added together and averaged, that is, the true velocity of each layer is obtained, which is completely consistent with the MD calculation results. Based on the above findings, it can be considered that the gas shear flow in the nanochannel is synthesized from two separate shear flows.

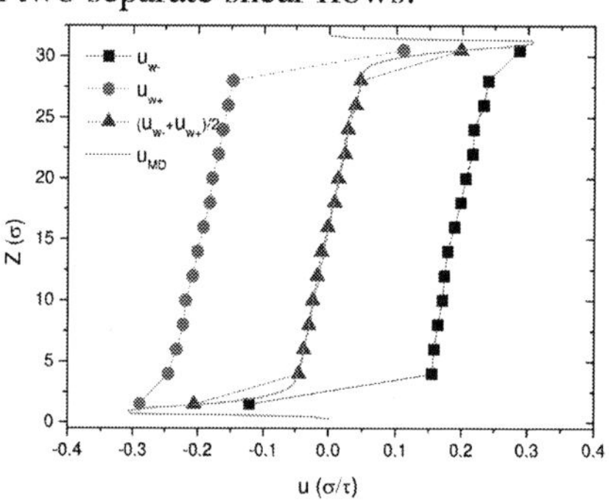

Figure 8. Average tangential velocity distribution of two types of gas molecules in each layer.

In order to explain the above phenomenon, it is necessary to proceed from the law of motion of a single gas molecule in a nanotube. We start tracking when the gas molecules are incident on the lower wall surface. After the collision with the lower wall surface, the tangential momentum is adjusted and the normal velocity direction is changed, and then the upper wall surface is moved. Since the channel height is relatively narrow, the gas molecules cannot be carried out in the channel. The collision of sufficient gas molecules, when it reaches the upper wall, still carries the tangential velocity information after collision with the lower wall surface; after that, the gas molecules collide with the upper wall surface to complete the adaptation of the tangential momentum with the upper wall surface. The speed direction changes again and the gas molecular moves to the lower wall surface; repeating the incomplete gas molecules in the channel and then reaching the lower wall again, thus forming a complete cycle. Each gas molecule in the channel follows the above-mentioned motion law. By ensemble averaging, we can obtain the gas tangential velocity distribution as shown in Figure 8.

4. Conclusion

In this paper, the three-dimensional non-equilibrium molecular dynamics simulation method is used to study the gas shear flow in the slip zone and the transition zone nanochannel. The distribution of the physical quantities of the gas in the channel is obtained. The flow characteristics of the near wall region are studied in depth and analyzed. The variation trend of the tangential momentum accommodation coefficient is found. The tangential momentum accommodation coefficients of the slip zone and the transition zone are consistent, and the conclusion that the tangential momentum accommodation coefficient is independent of the Knudsen number is obtained. The gas molecules in the channel are classified according to the normal velocity direction, and the velocity distribution of the two gas molecules in the flow field is obtained. It is found that the gas shear flow in the nanochannel can be decomposed into two independent flow laws. The discovery directly confirms the existence of a non-equilibrium effect of gas flow within the nanochannel.

References

[1] S. S. Verbridge, H. G. Craighead and J. M. Parpia, *Appl. Phys. Lett.* **92** (2008) 013112.
[2] B. Y. Cao, J. Sun, M. Chen and Z. Y. Guo, *Int. J. Mol. Sci.* **10** (2009) 4638.
[3] U. Boettcher, H. Li, R. A. Callafon and F. E. Talke, *IEEE T. Magn.* **47** (2011) 1823.
[4] H. Q. Song, M. X. Yu, W. Y. ZHU, Y. Zhang and S. X. Jiang, *Chin. Phys. Lett.* **30** (2013) 014701.
[5] W. Zhang, G. Meng and X. Wei, *Microfluid. Nanofluid.* **13** (2012) 845.
[6] D. C. Rapaport, *The Art of Molecular Dynamics Simulation* (Cambridge University Press, New York, 1995).
[7] B. Y. Cao, M. Chen and Z. Y. Guo, *Acta Phys. Sin.* **55** (2006) 5305.
[8] B. Y. Cao, *Mol. Phys.* **105** (2007) 1403.
[9] J. Sun and Z. X. Li, *Mol. Phys.* **106** (2008) 2325.
[10] P. Spijker, A. J. Markvoort, S. V. Nedea and P. A. Hilbers, *Phys. Rev. E* **81** (2010) 011203.
[11] H. Xie and C. Liu, Mod. *Phys. Lett. B* **25** (2011) 773.
[12] M. Barisik and A. Beskok, *Microfluid. Nanofluid.* **11** (2011) 611.
[13] M. Barisik and A. Beskok, *Microfluid. Nanofluid.* **13** (2012) 789.
[14] F. Bao., Y. Huang, L. Qiu and J. Lin, *Mol. Phys.* **113** (2015) 561.
[15] F. B. Bao, Y. L. Huang, Y. H. Zhang and J. Z. Lin, *Microfluid Nanofluid* **18** (2015) 1075.
[16] Q. D. To, C. Leonard and G. Lauriat, *Phys. Rev. E* **91** (2015) 023015.
[17] W. W. Lim, G. J. Suaning and D R McKenzie, *Phys. Fluids* **28** (2016) 097101.

Investigation on Application of Closed Cavity Inductively Coupled Plasma in Inlet Stealth

Han Xinmin, Zhang Wenyuan, Wei Xiaolong*, Xu Haojun and Chen Junlin

Science and Technology on Plasma Dynamics Laboratory, Air Force Engineering University, Xi'an, 710038, China

*Corresponding author's e-mail :wei18892022001@163.com

Abstract. The design of inlet into S-shaped is a stealth method with limited effect. The ICP generated in closed quartz chamber is an effective solution to improve the stealth effect. In this paper, an inductively coupled plasma (ICP) generator with the size of 30cm×30cm×2cm which could be conformal with S-shaped inlet was designed, the discharge experiment was carried out to study the discharge configuration of air ICP and the distribution of the electron density in the core area of ICP under 100Pa and different discharge power was diagnosed with microwave interference method. On this basis, the three-dimensional electromagnetic model of inlet and the plasma generator were established and the finite-difference time-domain method was used to calculate the broad-band scattering parameter of the inlet with and without the ICP generator. The variation of radar cross section area (RCS) with the azimuth angle, electromagnetic wave frequency and discharge power of ICP was studied. The result indicated that the ICP generator could effectively improve the stealth effect of S-shaped inlet in the head direction.

1. Introduction

Inlet is the strongest and most widely scattering source of the fighter in the forward area. In general single-engine aircraft the inlet can be accounted for 40% of the forward scattering, and in the double-engine aircraft the proportion could be as high as 60% or even higher [1]. Some scholars have put forward the design scheme of the S-shaped low-scattering inlet and the corresponding electromagnetic wave absorption measurement. Through the numerical calculation and the measurement of the scaling model, the optimal design of the duct parameters in inlet could effectively restrain but couldn't completely eliminate the cavity scattering, the scattering in the important direction (± 60 degrees around the head of aircraft) is still strong[2]. Therefore, it is necessary to adopt the technical means for further suppressing the scattering of the inlet cavity.

Inductively Coupled Plasma (ICP) is a discharge pattern which inputs the RF power to the non-resonant induction coil that delivers energy to the plasma through the inductive coupling style. In ICP discharge, the plasma density can reach to10^{11} ~ $10^{12}cm^{-3}$and the plasma frequency can cover the main radar wave band. The device has the advantages of simple structure, low discharge power, and the plasma parameters are easy to be controlled [3-4]. Thus ICP plasma is much potential for the plasma stealth applications. In this paper, the application of the thin-layer transmission cavity ICP in the local stealthy design of the S-shaped inlet is explored.

2. Scheme of Plasma Stealth of S-shaped Inlet

Due to the limited power provided by the aircraft's on-board conditions, it is not advisable to fully or largely cover the inlet with plasma for the stealth, but to replace the key parts locally. Figure 1 shown the application of ICP generator to achieve plasma stealth of S-shaped inlet, the overall framework of the S-shaped inlet was made of solid wood with the aluminum skin attached to the surface, and the outlet of the inlet was closed with a metal plate. In order to facilitate the processing of the ICP generator and its installation on the S-shaped pipe, the cross-section of the inlet was processed into rectangular shape. The pipeline is designed by the regular of the changing center line, which was tangentially connected with the straight one. The lower surface retains an outlet for the installation of conformal ICP generator.

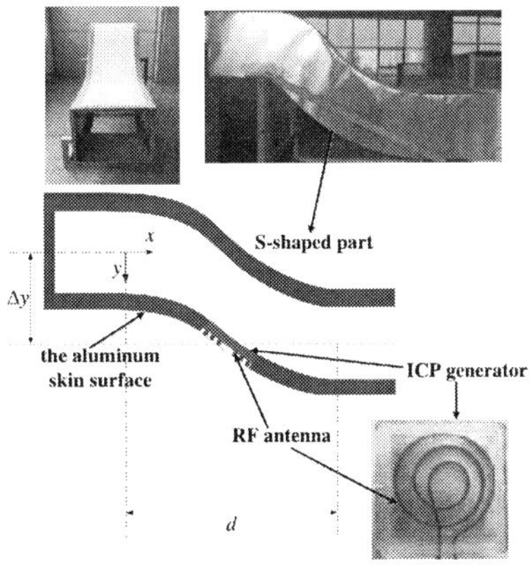

Figure 1. Details of inlet plasma stealth

The inlet parameters included: a total length of 1.5m, inlet cross-section of 60 cm × 50 cm rectangle. The change of centerline and area was the key parameter in the design of S-shaped inlet, which determined the change of the inner channel profile. In reference [5], the flow field and the total pressure recovery of the S-shaped inlet were analyzed. The results shown that the inlet with the center line that changed slowly in the front and urgent in the back would have stable flow field and high total pressure. Therefore, the change law of the center line of the road was as follow:

$$y = \Delta y \cdot \left[-3 \cdot (\frac{x}{d})^4 + 4 \cdot (\frac{x}{d})^3 \right] \quad (1)$$

Where y was the ordinate of the centerline, Δy was the longitudinal offset of the center of the inlet and outlet sections, and d was the length of the S-shaped part in the x-axis. ICP generator was installed 40cm away from the lip of the inlet which was a 0.8cm thick quartz chamber, with the coil placed under the surface of the quartz chamber.

3. Air ICP Discharge and Its Electronic Density Diagnosis

To study the effect of ICP on the electromagnetic scattering characteristics of the inlet, the specific form and parameter distribution of air ICP discharge need to be obtained. So the experiment of ICP discharge and electronic density diagnosis were carried out, as shown in Figure 2.

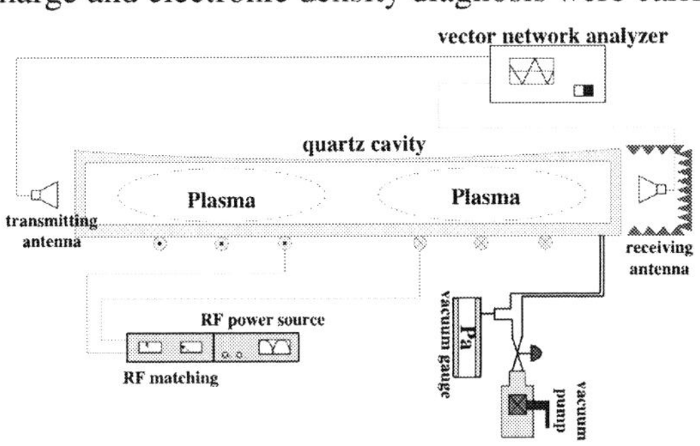

Figure 2. ICP Discharge Systems and Diagnostic Systems

IOP Conf. Series: Materials Science and Engineering **751** (2020) 012069 doi:10.1088/1757-899X/751/1/012069

ICP discharge system was mainly composed of RF power and RF matching, mosquito coils radio frequency antenna, vacuum system and the quartz chamber. The quartz cavity was made by integral welding technology. The thickness of the quartz was 0.8cm and the inner size of the cavity was 20cm × 20cm × 2cm. The upper surface of the quartz cavity had a certain curvature to conform to the key part of the S-shaped inlet. The frequency of the RF power source (MSY-1) was 13.56MHz. The matching power was adjusted by the automatic matching circuit (SP-1) to minimize the reflected power. The planar-coil RF antenna made of copper pipe with inner and outer diameter of 6mm and 8mm respectively was positioned under the bottom of quartz cavity. The diameter of RF antenna was 19.2cm and three laps around. In order to ensure the RF antenna working stably for enough time, the copper pipe with a cooling water system was provided.

Diagnosis system was mainly composed of a vector network analyzer (Anritsu-37347C) and a pair of horn antenna, the working frequency f=12GHz, the polarized mode was vertical linear polarization and the horn antenna was symmetrically fixed on the left and right sides of the quartz chamber, with 0.3m away from the cavity, the receiving antenna was placed on the back side of the semi-open absorbing black-box. In this experiment, electron density n_e was measured by the microwave transmission interference method.

Air ICP discharge process was divided into two typical stages. The first phase was E-mode discharge. At 100Pa pressure, when the discharge power increases to about 80W, the thin plasma was generated in the quartz chamber under the capacitive high voltage of the stainless steel interface and high voltage between the coils. The electron density was low and almost uniformly filled the entire cavity, not suitable for aircraft stealth [6].

As shown in Figure 3, when the RF power increased to 300W, the brightness increased in a hopping mode, and the ICP exhibited a stable ring shape. In the presence of negative ions in air ICP, an electronegative core region with large electron density and even distribution was formed in the central region of the discharge, surrounded by an electropositive edge halo around its periphery [7]. As shown in Figure 3, with the increasing of power, the width and brightness of the ICP ring region both increased significantly.

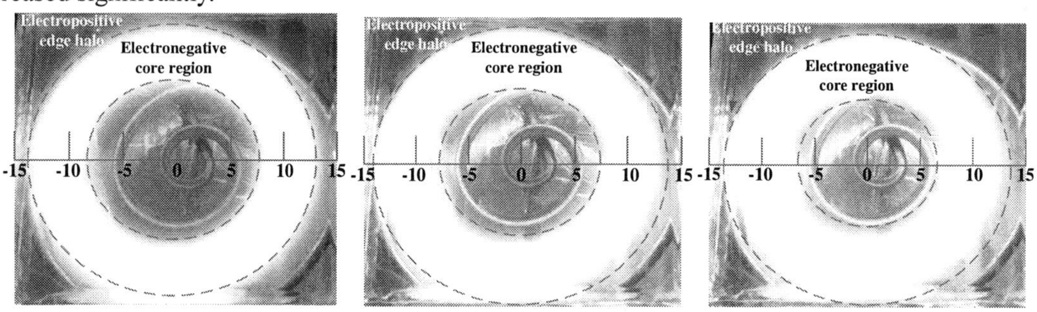

(a) 300W (b) 400W (c) 500W

Figure 3. The shape and structure of plasma changed versus and input power

The vector network analyzer was used to record the phase shift of the microwave through the plasma [8]. The electron density in the electronegative core region was high and uniform, while the electron density in the edge halo decay fast, the phase shift $\Delta\phi$ measured by the vector analyzer was mainly caused by the electron density in core region, the phase shift caused by the edge halo was negligible, so the electron density n_e in the electronegative core region could be expressed as [9]:

$$n_e = \frac{2\varepsilon_0 m_e c\omega}{e^2 l}\Delta\phi \tag{2}$$

where ε_0 was the dielectric constant in vacuum, m_e was the electron mass, c was the velocity of light in vacuum, ω is the angular frequency of the microwave, e is the individual electron charge and l is the width of the plasma ring.

Figure 4 shows the spatial distribution of ICP electron density at different discharge powers. When the power was increased from 300W to 500W, the n_e of the electronegative core region increased from $2.73 \times 10^{11}cm^3$ to$4.01 \times 10^{11}cm^3$. The increase of n_e was approximately proportional to the power because increased power would cause more intense electron-neutral particle collision thus much more electron would be generated.

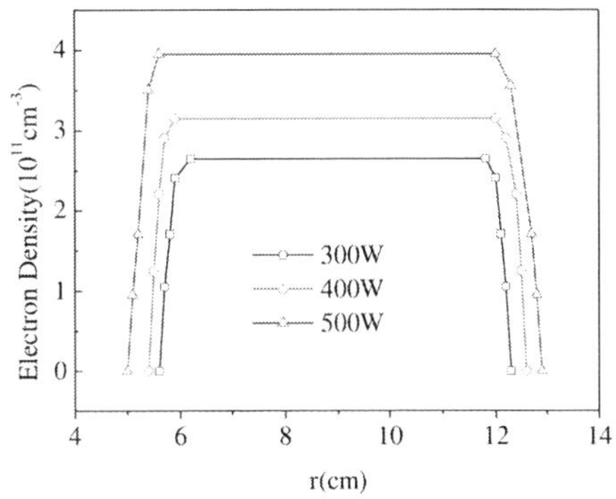

Figure 4. The n_e distribution of ICP varied from power versus radius in100Pa.

4. Introduction Calculation of RCS

4.1 Description of calculation

According to the shape and size of the S-shaped inlet and the annular plasma, the three-dimensional model was established in CATIA, as shown in Figure 5. The electron density in the edge halo decay fast and its effect on the stealth of inlet could be ignored, so only the electronegative core region of ICP was considered. The RCS numerical simulation was carried out by using Remcom's XFDTD electromagnetic simulation software. This software calculated the electromagnetic field based on the finite difference time domain method (FDTD) which directly discredited the Maxwell equations in time and space, alternately sampling the electric field and the magnetic field, so that the process can completely simulate the electromagnetic wave with time [10].

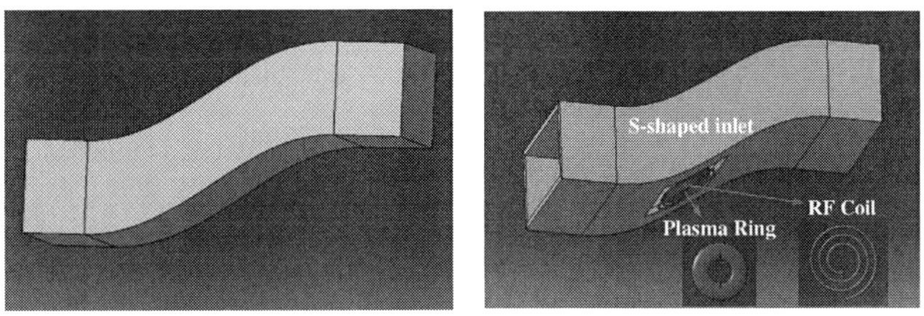

(a) Without ICP generator (b) Installed ICP generator

Figure 5. The three-dimensional model of S-shaped inlet

Process-related electromagnetic parameters of the simulation were set as follows:

The inlet and the RF antenna were set as good conductors. The plasma was set as a Drude medium and its relative permittivity can be expressed as:

$$\varepsilon_p = \varepsilon_\infty + \frac{\varepsilon_s - \varepsilon_\infty}{1 + j\omega_0\tau_0} \quad (3)$$

Where ε_∞ is the relative dielectric constant of the plasma when the collision frequency approaches infinity whose value was 1 for the plasma, ε_s was the static relative permittivity: $\varepsilon_s = 1 - (\omega_p/\nu_m)^2$, τ_0 was the relaxation time: $\tau_0 = 1/\nu_m$, and the value of the plasma conductivity σ could be define as:

$\sigma = \varepsilon_0 \omega_p^2 / \nu_m$.

The plasma oscillation frequency ω_p and the plasma collision frequency ν_m could be obtained from the electron density n_e of the plasma [11-12]:

$$\omega_p = \sqrt{(n_e e^2)/(\varepsilon_0 m_e)} \tag{4}$$

$$\nu_m = 6.3\times10^{-9} n_m \sqrt{\frac{T}{300}} \tag{5}$$

where n_m was the density of neutral particle which was decided by the gas state equation: $n_m = 2.415\times10^{14} P$, P was the pressure of the chamber and T was the gas temperature which was always defined as the indoor temperature: T=300K.

The RCS calculation was carried out in the azimuth ranged from -60° to 60° of the heading direction and at the direction of 0° angle of attack of the S-shaped inlet. The mesh size of the inlet duct region was 5mm and was 1mm in the plasma and RF antenna regions. The incident waves were sinusoidal plane waves, polarized in the $\varphi\varphi$ direction with 1-10GHz of the frequency, and the absorption boundary was set to 7 layers of PML. The convergence criterion was that the residual decreased by -30 dB.

4.2 RCS calculation results

Figure 6 showed the RCS of the S-shaped inlet changed with the azimuth angle with/without ICP generator in different discharge power when the EM wave frequency was 5GHz. It could be seen that, without ICP generator, RCS of ±60° range in the head presented a symmetrical distribution of tens of lobes and the main lobe appeared in the vicinity of azimuth 0° with the width of about 20° and the peak value could reach up to 10dBsm. With the azimuth away from 0°, the peak value of the lobes gradually decreased.

The RCS of the S-shaped inlet could be effectively reduced by installing the ICP generator. This attenuation was more pronounced near the main lobe. When the ICP discharge power increased to 300 W, the RCS decreased by about 10dBsm at 0° azimuth angle and the width of the main lobe decay from 20° down to about 10°. As the azimuth angle deviated from 0°, the attenuation effect decreased gradually. It could be concluded that the ICP generator had a good attenuation effect on the head-scatter of the inlet.

When the discharge power of ICP increased from 300W to 400W, the attenuation effect of RCS was improved obviously, while the power increased from 400W to 500W, the attenuation effect was weakened. That's because with the increase of power, the electron density increased, attenuation of electromagnetic waves by plasma would move to high-frequency area, and attenuation of low-frequency electromagnetic wave would be weakened [13]. So aiming at different frequency radar wave, the ICP generator should be adjusted to an appropriate discharge power to enhance the absorption.

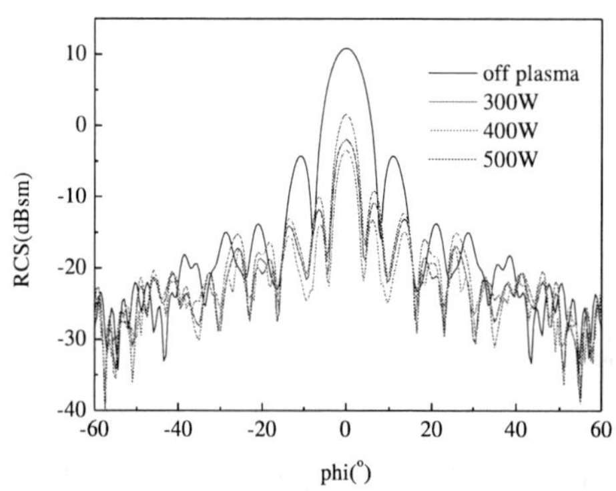

Figure 6. RCS of inlet versus azimuth

Figure 7 was the RCS of S-shaped inlet changed with the frequency of electromagnetic wave when the azimuth angle was 0° with / without ICP generator of different discharge power. It could be seen that the RCS attenuation effect of the ICP generator varied with the change of discharge power, and the attenuation peak of the RCS was near the plasma frequency. When ICP discharge power rose from 300W to 500W, the attenuation of the main frequency band from 4.5-5GHz increased to 5.5-6.5GHz, the attenuation peak would also been significantly improved.

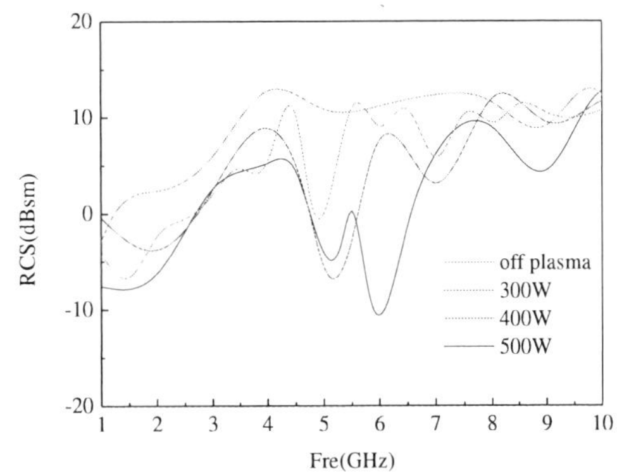

Figure 7. RCS of inlet versus frequency when the azimuth is 0°

It could be seen that the attenuation of RCS of inlet caused by ICP generator presented the characteristic of band, when the frequency of EM wave was much lower than the plasma frequency: $\omega << \omega_p$, the absorption of EM wave by ICP was mainly caused by the cut-off reflection which was much limited. And when the EM wave was much higher than the plasma frequency: $\omega >> \omega_p$, the electronic couldn't response to the change of the EM wave thus there couldn't be polarization phenomenon, leading to a lower attenuation effect.

5. Conclusion

The ring plasma with electron density ranging from$2.73\times10^{17}m^{-3}$ to $4.01\times10^{17}m^{-3}$could be obtained in the experiment, the attenuation of the forward RCS of inlet could be 11-25dB caused by the plasma ring and there is an optical discharge power to maximum the attenuation the RCS of inlet at a certain frequency radar wave. When the discharge power was increased, the attenuation effect the attenuation peak region would move to the high frequency region.

References

[1] Sang Jianhua. 2013, Aircraft stealth technology, Aviation Industry Press, Beijing (in Chinese).
[2] JI Jin-zu, WU Zhe, LIU Zhanhe. 2009, JOURNAL OF XIDIAN UNIVERSITY., 4: 746(in Chinese).
[3] P.Scheubert, U.Fantz, P.Awakowicz. 2001, Journal of Applied Physics., 90: 587.
[4] Michael A. Lieberman, Allan J. Lichtenberg. 2005, Principles of Plasma Discharges and Materials Processing, John Wiley & Sons Inc, New Jersey.
[5] Lee C C, Boekicher C. 1985, AIAA-85-3073 Colorado
[6] WEI Xiao-long, XU Hao-jun, LIN Min etl. 2015, SPECTROSCOPY AND SPECTRAL ANALSIS 36(4) (in Chinese).
[7] WEI Xiao-Long, XU Hao-Jun, LI Jian-Hai etl. 2015 Acta Phys.Sin 64:175201 (in Chinese).
[8] Heald M A, Wharton C B.1978, Plasma Diagnostics with Microwave. Krieger, New York.
[9] PeredaJA,VegasA,PrietoA 2002 IEEE Trans. Mi-crowave Theory and Techniques., 50: 1689.
[10] Ge Debiao, YAN Yubo. 2011, Finite-Difference Time-Domain Method for Electromagnetic Waves. XiDian Press, Xian(in Chinese).
[11] YUAN Zhongcai,SHI Jiaming. 2004, Nuclear Fusion and Plasma Physics.,24:157.
[12] Jin Baozi. 2004The Electromagnetic wave propagation in plasma.Science Press, Beijing(in Chinese).
[13] Lin Min, XU Haojun, WEI Xiaolong, et al. 2015, Plasma Science and Technology., 10: 847.

IOP Conf. Series: Materials Science and Engineering **751** (2020) 012070 doi:10.1088/1757-899X/751/1/012070

Method of Probe Radius Compensation for Optical Complex Surface Measurement

L J Chen[*] and Y D Kang

School of Mechanical Engineering & Automation, Northeastern University, Shenyang 110819, China

[*]E-mail: lijchen@mail.neu.edu.cn

Abstract. When surface measuring with a contact method, the data obtained by the LM50 laser interferometer is the data of the center of the probe actually, not the real measurement point. Therefore, based on the theory of NURBS and the theory of Delaunay triangulation, a method of "enveloping surface - triangulation" is proposed, and the normal vector of the measuring points of the surface is derived so that the radius of the probe is compensated accurately and quickly. After triangular segmentation, the normal vector of each triangle around a measuring point has an influence on the normal vector of the vertex, so it is necessary to calculate the weight function of each triangulation. In this paper, the influence of the edge length and the angle factor on the vertex normal vector is considered, and the calculation of the vertex normal vector is improved in the triangulation, and the programming and calculation are simplified on the premise of guaranteeing the accuracy.
Key words: contact measurement; the theory of NURBS; the theory of Delaunay triangulation; probe radius compensation; vertex normal vector

1. Introduction

A complex surface can be defined as a rotating invariant surface without an axis, with arbitrary shape, regular or irregular surface structure[1].Optical complex surfaces are widely used in new and traditional fields, such as communication and new energy automotive engineering, high-end equipment manufacturing, biomedicine and so on, so the surface shape accuracy of surfaces must be guaranteed. Compared with non-contact assessment, the measurement efficiency is not very high, but its evaluation accuracy can reach a high standard, which is widely used in the detection of various complex curved parts. The measuring device adopts the LM50 laser interferometer of German company, which can realize the measurement resolution of nanometer level in the range of 50mm, and the contact measurement method is used to evaluate the measurement.This type of measuring instrument is composed of probe, electrical box and data processing software, The instrument has its own measurement program. In order to better apply to the machine tool, the instrument also has a dynamic link library. C program can be used for secondary development to complete the measurement and evaluation of various complex parts. The surface is measured by contact measuring method[2-5], The coordinate value of the real measuring point of the surface obtained by LM50 laser interferometer is not the coordinate value of the real measuring point of the surface, but the value of the center of the probe is the actual measured data, so the radius of the probe should be compensated. In the case of complex curved surface, the method of measuring the radius of the measuring head has the micro-plane method, the micro-sphere method, the surface quasi-legal and the Delaunay triangulation method, and the method has to be further improved in terms of the accuracy of the radius compensation and the calculation efficiency. In this paper, the radius compensation method of probe based on surface fitting and Delaunay triangulation is proposed by using surface fitting method and Delaunay triangulation method, which can improve the surface measurement.

2. Radius compensation based on NURBS Surface Theory

2.1. Mathematical Model and partial derivative Vector of complex Surface

A NURBS surface is a double variable vector value function as shown below[6-10]:

$$S(u,v)=\frac{\sum_{i=0}^{n}\sum_{j=0}^{m}N_{i,p}(u)N_{j,q}(v)\omega_{i,j}P_{i,j}}{\sum_{i=0}^{n}\sum_{j=0}^{m}N_{i,p}(u)N_{j,q}(v)\omega_{i,j}},\quad 0\le u,v\le 1 \tag{1}$$

The control grid is formed by $\{P_{i,j}\}$, $\{\omega_{i,j}\}$ is a weighted factor, $\{N_{i,p}(u)\}$ and $\{N_{j,q}(v)\}$ is the NURBS basis function specified in the node vector U And V , the definition $\{N_{i,p}(u)\}$ is as follows:

$$\begin{cases} N_{i,0}(u)=\begin{cases}1, u_i\le u\le u_{i+1}\\ 0, else\end{cases} \\ N_{i,p}(u)=\dfrac{u-u_i}{u_{i+p}-u_i}N_{i,p-1}(u)+\dfrac{u_{i+p+1}-u}{u_{i+p+1}-u_{i+1}}N_{i+1,p-1}(u) \\ stipulate\dfrac{0}{0}=0 \end{cases} \tag{2}$$

The same principle is available $\{N_{j,q}(v)\}$.

where

$$U=\left\{\underbrace{a,\cdots,a}_{p+1},u_{p+1},\dots,u_{r-p-1},\underbrace{b,\cdots,b}_{p+1}\right\}$$

$$V=\left\{\underbrace{a,\cdots,a}_{q+1},v_{q+1},\dots,v_{s-q-1},\underbrace{b,\cdots,b}_{q+1}\right\}$$

here $r=n+p+1, s=m+q+1$.

$\{Q_{k,l}\}$ is $(n+1)\times(m+1)$ data points given $k=0,1,...,n$ and $l=0,1,\cdots,m$. Create a $\{p,q\}$ degree surface which interpolated to these given data points.

$$Q_{k,l}=S(\bar{u}_k,\bar{v}_l)=\sum_{i=0}^{n}\sum_{j=0}^{m}N_{i,p}(\bar{u}_k)N_{j,q}(\bar{v}_l)P_{i,j} \tag{3}$$

firstly, U and V can be respectively solved by a centripetal parameter method.

$$U=\left\{\underbrace{0,\cdots,0}_{p+1},u_{p+1},\dots,u_{r-p-1},\underbrace{1,\cdots,1}_{p+1}\right\}$$

$$V=\left\{\underbrace{0,\cdots,0}_{q+1},v_{q+1},\dots,v_{s-q-1},\underbrace{0,\cdots,0}_{q+1}\right\}$$

The second is the reverse seeking control point.(3) is a set of equations composed of $(n+1)\times(m+1)$ equations. $S(u,v)$ is a tensor product surface, $P_{i,j}$ can be operations more conveniently and more quickly by a series of interpolation operations. For fixed l, (3) is writen as

$$Q_{k,l}=\sum_{i=0}^{n}N_{i,p}\left(\bar{u}_k\right)\left(\sum_{j=0}^{m}N_{j,q}\left(\bar{v}_l\right)P_{i,j}\right)=\sum_{i=0}^{n}N_{i,p}\left(\bar{u}_k\right)R_{i,l} \tag{4}$$

In the formula (4)

$$R_{i,l}=\sum_{j=0}^{m}N_{j,q}\left(\bar{v}_l\right)P_{i,j} \tag{5}$$

(4) is Interpolation operation for data point. $Q_{k,l}\left(k=0,1,\ldots n\right)$. $R_{i,l}$ is the control point on the surface for $S(u,v)$ when $v=\bar{v}_l$. Now order i no change, let l change, (5) is equations of data points performing curve interpolation. $P_{i,0},\ldots,P_{i,m}$ is the control vertex to calculate. The form of its matrix can be written as

$$Q=AP \tag{6}$$

where

$$\begin{bmatrix} Q_0 \\ Q_1 \\ Q_2 \\ \vdots \\ Q_{n-2} \\ Q_{n-1} \\ Q_n \end{bmatrix}=\begin{bmatrix} 1 & & & & & \\ N_{0,1} & N_{0,1} & N_{0,1} & N_{0,1} & & \\ & N_{0,1} & N_{0,1} & N_{0,1} & & \\ & & \ddots & \ddots & \ddots & \\ & & N_{n-3,n-2} & N_{n-2,n-2} & N_{n-1,n-2} & \\ & & N_{n-3,n-1} & N_{n-2,n-1} & N_{n-1,n-1} & N_{n,n-1} \\ & & & & & 1 \end{bmatrix}\begin{bmatrix} P_0 \\ P_1 \\ P_2 \\ \vdots \\ P_{n-2} \\ P_{n-1} \\ P_n \end{bmatrix} \tag{7}$$

Using Gaussian elimination method to solve the system of equations. So find all the control points $P_{i,j}$. The algorithm of:

①With U and parameters $\bar{u}_k$ do interpolation operation of $m+1$ times: $l=0,1,2,\ldots m$, The curve equation interpolated at the point $Q_{0,l},\ldots,Q_{n,l}$ is formed in this order. As a result, $R_{i,l}$ can be obtained;

②With V and parameters $\bar{v}_l$ do interpolation operation of $n+1$ times: $i=0,1,2,\ldots n$, The curve equation interpolated at the point $R_{i,0},\ldots,R_{i,m}$ is formed in this order. As a result, $P_{i,j}$ can be obtained;
What needs to be required is the partial derivative vector of the NURBS surface. order

$$S(u,v)=\frac{\omega(u,v)S(u,v)}{\omega(u,v)}=\frac{A(u,v)}{\omega(u,v)} \tag{8}$$

where

$$A(u,v)=\sum_{i=0}^{n}\sum_{j=0}^{m}N_{i,p}(u)N_{j,q}(v)\omega_{i,j}P_{i,j} \tag{9}$$

thus

$$S_\alpha(u,v)=\frac{A_\alpha(u,v)\text{-}\omega_\alpha(u,v)S(u,v)}{\omega\ (u,v)} \tag{10}$$

The first-order partial guide vector is formed by the complex curved surface, In the equation. α Representative u or v.

2.2. Radius compensation Operation in Surface Measurement

The tangent vector of any point on a complex surface are all contained in the plane of $S_u(u,v),S_v(u,v)$ that is open. The normal vector at each point on the surface is the normal vector for the plane to be opened,The expression of the unit normal vector is[11]

$$\vec{n} = \frac{S_u(u,v) \times S_v(u,v)}{\left|S_u(u,v) \times S_v(u,v)\right|} \tag{11}$$

Radius compensation using NURBS interpolation method, that is the process by which the interpolation surface moves in the direction of the normal vector. (r is the radius of the interferometer probe). Hypothesis $P_i(x,y,z)$ is a measuring point, then the contact point $P_i^{'}(x,y,z)$ is

$$P_i^{'}(x,y,z) = P_i(x,y,z) \pm r\vec{n} \tag{12}$$

3. Delaunay Compensation of Segmentation Measurement Theory

3.1. Delaunay Triangulation based on Bowyer-Watson [1 2]

In order to segment the surface by Delaunay triangulation, the following two conditions must be satisfied: maximizing the minimum angle and empty circle, maximizing the minimum angle to ensure that the smallest angle in the triangle formed by triangulation is the largest, and empty circle ensuring the uniqueness of triangulation. The specific subdivision steps are as follows:

Step1:Form a maximum triangle so that it includes all the measuring points;

Step2:The points measured by each interferometer are inserted step by step, and each point is inserted into to find the triangle containing the points contained in the outer circle of the triangle.

Step3:Remove the searched triangles to form a cavity of polygons;

Step4: A new triangle is built by connecting the polygons cavity with the inserted measurement points, which are vertexed by the inserted measurement points.

3.2. Calculate the normal vector of each triangulation triangle

Calculate the normal vector of $\Delta a_1 a_2 a_3$,Set the coordinates of each vertex $a_1(x_1, y_1, z_1)$, $a_2(x_2, y_2, z_2)$, $a_3(x_3, y_3, z_3)$,The formula for calculating the normal vector $\vec{n}(n_x, n_y, n_z)$ of standard $\Delta a_1 a_2 a_3$ is as follows:

$$\vec{n} = \begin{bmatrix} n_x \\ n_y \\ n_z \end{bmatrix} = \begin{bmatrix} i & j & k \\ x_2 - x_1 & y_2 - y_1 & z_2 - z_1 \\ x_3 - x_1 & y_3 - y_1 & z_3 - z_1 \end{bmatrix} \tag{13}$$

3.3. Calculation of improved Vertex normal Vector

As shown in figure 1,After triangulation of the surface, the normal vector of each triangle around the measuring point a has an influence on the normal vector of the vertex, so it is necessary to calculate the weight function of each triangle surface. In this paper, the influence of edge length and angle on the normal vector of the vertex is considered synthetically. The larger the edge length is, the farther away the other vertices are from the center point. The smaller the influence of the triangular patch on the normal vector of the vertex, the greater the angle of the triangle at the apex of the triangle, and the greater the influence of the triangle on the normal vector of the vertex.

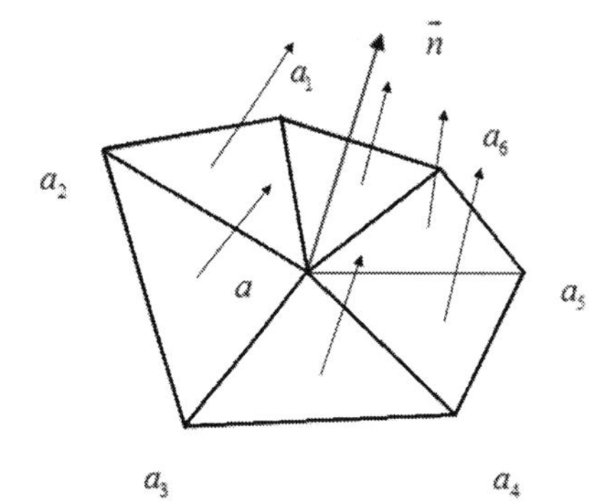

Figure 1 Vertex normal vector weighted average

The formula for calculating the weighted average of the normal vector $\vec{n}$ of the Vertex is as follows[13-14]:

$$\vec{n} = \frac{\sum_{i=1}^{k} \frac{\overrightarrow{N_i}}{R_i}}{\left\| \sum_{i=1}^{k} \frac{\overrightarrow{N_i}}{R_i} \right\|} \tag{14}$$

where k is the number of triangular slices around the point a, $\overrightarrow{N_i}$ is the normal vector of the i triangular patch around a point, R_i is the radius of the outer circle of the triangular patch. Compared with the previous model, it is simplified in calculation and programming.

4. Envelope Surface-triangulation Radius compensation method

The method of NURBS surface fitting is used to measure the accuracy and the compensation error is small, but the data processing is more complicated and the calculation amount is large, which is not conducive to the rapid measurement.Maximum error of triangulation The minimum error is close to the surface fitting method, and the average error is large, but the data processing is relatively simple and the calculation is small.Through analysis, the smaller the curvature of the center trajectory surface of the triangulation probe is, that is, the larger the curvature radius is, the higher the fitting accuracy is, the higher the calculated normal vector direction is to the normal vector direction of the real surface, and the higher the compensation accuracy is. On the contrary, the larger the curvature of the probe center trajectory surface, the lower the fitting accuracy, the greater the difference between the normal vector direction of the calculated measuring point and the real normal vector direction, and the lower the compensation accuracy.

In view of the above two methods, a "envelope surface triangulation" method based on NURBS theory and Delaunay segmentation theory is proposed. That is, after the design surface is given, the curvature at each interpolation point is obtained. When the maximum curvature of the surface exceeds the given value, the NURBS surface fitting method is adopted, and when the maximum curvature of the surface is less than the given value, the Delaunay segmentation method is adopted, so that the radius error can be compensated quickly and accurately.

5. Simulation and analysis

In order to verify the above method, the simulation of this method is carried out, Let theoretical surface ① be as follows:

$$z(x, y) = 30 - 30\sin\left(\frac{\pi x}{300}\right)\sin\left(\frac{\pi y}{300}\right) \quad 0 \le x, y \le 300$$

Distribution points in equal step size according to the surface parameter u, v, a series of point sets $P_{i,j}$ are obtained. The maximum curvature of the surface exceeds the curvature threshold, and the normal vector of each point in the point set is obtained, Each point on a complex surface moves upward in the direction of the normal vector 1.5mm(Probe radius size), the new point set is the measurement point, and then the new point set is inversely calculated by NURBS surface interpolation, and the new surface is obtained, and the normal vector at the top of the new surface is obtained as shown in figure 2. Move 1.5 mm down along the normal vector direction, and get a new point set $P_{i,j}^{'}$ after compensation. The index determined by deviation $e=\left|P_{i,j}-P_{i,j}^{'}\right|$ for accuracy is shown in Table 1, Using the micro-plane method to perform the radius compensation calculation, the average surface error average value is 8.63×10^{-3} mm.

Table1 Enveloping Surface - Triangulation Probe Radius Compensation accuracy analysis

Surface No	Surface①	Surface②
Maximum error $e_{\max}$ (mm)	2.93×10^{-4}	4.67×10^{-4}
Least error $e_{\min}$ (mm)	1.28×10^{-4}	1.65×10^{-4}
average error $\bar{e}$ (mm)	1.95×10^{-4}	2.14×10^{-4}

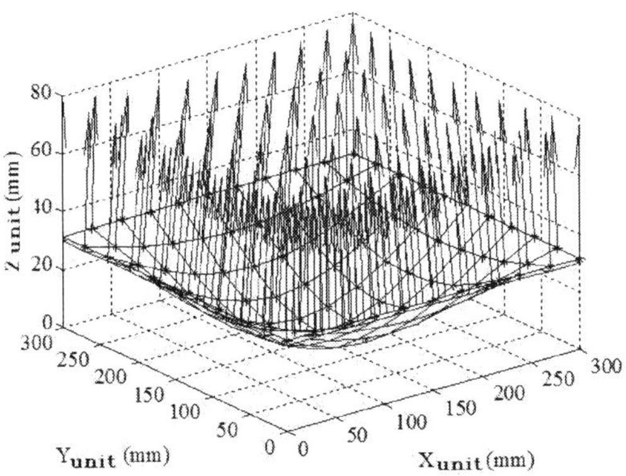

Figure 2 Radius Compensation simulation

The theoretical surface ② is:

$$z(x,y)=15\sin\left(\frac{\pi x}{300}\right)\sin\left(\frac{\pi y}{300}\right)\quad 0\le x,y\le 300$$

Distribution points in equal step size according to the surface parameter u, v, a series of point sets $P_{i,j}$ are obtained. By calculating the maximum curvature of the surface, the normal vector of each point is obtained, Moving 1.5mm upward along the normal vector at each point on the surface (Probe radius size), The new point is the measuring point, the triangular dividing surface is obtained by triangulating the set of new points, and the normal vector of each triangular dividing surface is obtained as shown in figure 3,the weighted average obtains the normal vector of each triangle vertex as shown in figure 4, Move 1.5 mm down along the normal vector direction, and get a new point set $P_{i,j}^{'}$ after compensation.The index determined by using deviation $e=\left|P_{i,j}-P_{i,j}^{'}\right|$ as precision is as shown in Table 1. From the experimental data, it can be seen that this method is feasible.

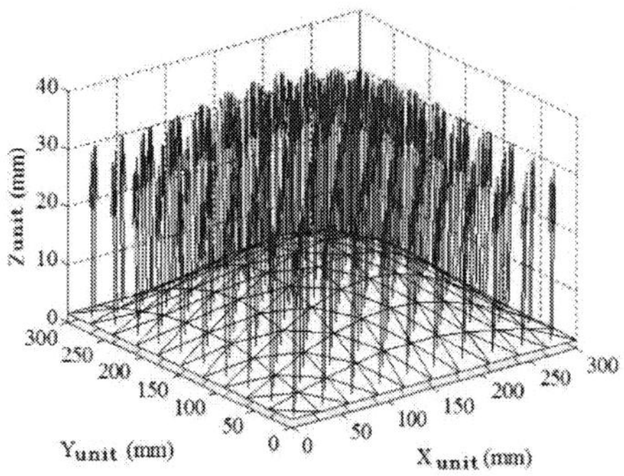

Figure 3 Triangular patch normal vector

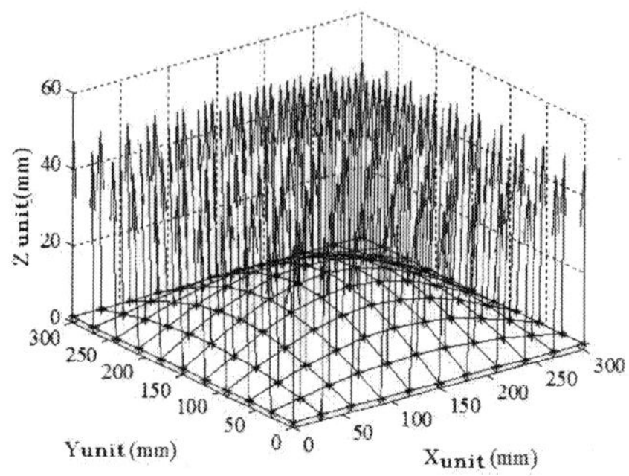

Figure 4 The normal vector of the weighted average posterior vertex

6. Conclusion

The "envelope surface-triangulation" method to realize the radius compensation of the measuring head is put forward, and the surface radius compensation problem for different size curvature is successfully solved.

The mathematical model of calculating the normal vector of the Vertex after triangulation is established, and the algorithm is improved, and the simplification of calculation and programming is realized.

By means of MATLAB simulation, the "envelope surface-triangulation" method is compared with the micro-plane method, and the precision is greatly improved.

Acknowledgments

The project is sponsored by the Natural Science Foundation of P. R. China (U1508206)

References

[1] Speck A,Zelzer B, Kannengieber M,et al. Inspection of freeform intraocular lens topography by phase measuring deflectometric methods[J]. Application. Optics,2013,52(18):4279-4286.

[2] Mayer J,Mir Y, Trochu F,et al. Touch probe radius compensation for coordinate measurement using kriging interpolation[J]. Proceedings of the Institution of Mechanical Engineers, Part B: Journal of Engineering Manufacture,1997,211(1):11-18.

[3] Dermani H S, Nategh M J. Upgrading the CNC system of hexapod machine tool by adding the five axis tool radius compensation[J]. Mechanical Engineering Department, Faculty of Engineering, Tarbiat Modares University,2016,16(7):93-98.

[4] Suleiman M O, Shivakumar R. An intelligent sampling method for inspecting free-form surfaces[J]. The International Journal of Advanced Manufacturing Technology,2009,40(11-12):1125-1136.

[5] Wang Yang. Prediction sampling method and probe compensation technology of CMM research [D]. Shenyang: School of Mechanical Engineering & Automation, Northeastern University,2016.

[6] Shi Fa-zhong. Computer-aided geometric design & Non-uniform rational B-spline [M]. Beijing: Higher Education Press,2013.

[7] Sekar M, Suresha B, Kantharaj I. Design of smooth ramp feedrate for machining complex NURBS paths [J]. Materials Science & Engineering Conference Series,2017,247(1):12-17.

[8] Kang P, Youn S K. Isogeometric topology optimization of shell structures using trimmed NURBS surfaces [J]. Finite Elements in Analysis and Design,2016,120:18-40.

[9] Coelho M,Roehl D, Bletzinger K U. Material model based on NURBS response surfaces[J]. Applied Mathematical Modelling,2017,51:574-586.

[10] Xie Wei-cheng, Zou Xiu-fen, Yang Jian-dong, et al. Iteration and optimization scheme for the reconstruction of 3D surfaces based on non-uniform rational B-splines[J]. Computer-Aided Design,2012,44(11): 1127 -1140.
[11] Wang Li-cheng, Huang Xin-da, Ding Han. Error analysis and compensation for touch trigger probe of on-machine measurement system[J]. China Mechanical Engineering,2012,23(15):1774-1778.
[12] Zhou Xue-mei, Li Ying-fei. Research on the generation algorithm based on Bowyer-Watson triangulation[J]. Computer Engineering and Application,2013,49(6):198-200.
[13] He Wang-jun, Liu-Jiping, Zhang Fuhao. GPU-based parallel computing method for topographic vertex normal vectors [J]. Journal of Liaoning Technical University,2017,36(7):734-738.
[14] Max N. Weights for computing vertex normals from facet normals[J]. Journal of Graphics Tools,1999,4(2):1-6

IOP Conf. Series: Materials Science and Engineering **751** (2020) 012071 doi:10.1088/1757-899X/751/1/012071

Analysis of Failure Effect of Cage Thickness on Spherical Roller Bearings

Fang Li [1,*],Wang Biao Qiu[1],Jia Bin Cai[1],Kai Wang[1] and Zheng Pan[1]

[1] School of Mechanical Engineering,Guizhou University,550025,Guizhou,China;

[*]E-mail: 2536099203@qq.com

Abstract. Spherical roller bearings are important parts of high-frequency crushers.Using the finite element analysis software to carry on the transient dynamics analysis to the bearing obtains the bearing deformation, the pressure and so on characteristis. Then, by contrasting and analysising, it is found that under pure radial load, radial load and axial load mixed force, influence of cage thickness on bearing performance. The results show that the thickness of the cage has little effect on the deformation of the bearing under pure radial load.Under the combined action of the radial load and the axial load, the radial deformation of the bearing decreases compared with the pure radial load.The deformation of the cage under the mixed load is increased than that under the radial load.

1.Introduction

The high-frequency crushers is an excavator attachment tool that crushes the target rock by the eccentric force generated by the eccentric component [1]. The Spherical roller bearing is an important part of the high-frequency crushers. It as a joint role in the working process.Its structural design has an important influence on the overall performance of the high-frequency crushers. Because self-aligning roller bearings bear a larger load in the use process, so easy to self-aligning roller bearings failure.To improve the performance of the bearing, the dynamic analysis of the bearing is performed by Ansys WorkBench finite element analysis software to find out the role played by the cage in the use of the bearing.The simulation result can be a product. Further design and optimization provide a basis for judgment.

At present, many scholars at home and abroad have carried out dynamic and static simulation analysis of bearings. The components of the bearing have a very important influence on the reliability of their use. However, there are relatively few comparative studies on this aspect. The quasi-static analysis of the optimized analytical model is carried out, and the load distribution of the double-row tapered roller bearing under the complex force and ultimate load is obtained respectively[2]. The axial clearance value under the maximum fatigue life curve, the literature uses the finite element analysis method and theoretical calculation to study the characteristics of the deep groove ball bearing cage and the steel ball during operation[3].The important influence on the cage and the relationship between the change of the number of steel balls and the bearing load analyzed the dynamics of the cylindrical roller bearing, and obtained the force of the cage and the rolling element[4]. The literature shows the position of the contact area of the bearing body during the use of the rolling element and the inner and outer rings of the bearing during the use of the dynamic finite element dynamic response analysis[5].The literature through the use of display dynamics finite element analysis the stress distribution at different speeds was studied[6]. It was found that the contact stress between the rolling element and the outer ring was positively correlated with the rotational speed.The force of the outer ring hardly changed with the rotational speed.The traditional Newton-Raphson iterative method is improved, the reliability of the solution result is effectively improved, and the solution result is verified by controlling the value of the contact angle[7].The literature has been subjected to the cylindrical roller bearing by finite element method[8].Force analysis shows that the stress is symmetrically distributed along the contact point to both sides of the bearing. The above analysis of bearings rarely involves the comparative analysis of double-row spherical roller bearings. In view of this,the cage bearings with different parameters are compared and analyzed.The characteristics of cage bearing with different parameters and the influence of cage on bearing force are found out It provides a reference for the further research and optimization of bearing.

2 .Bearing installation structure and research method

2.1 Analysis of the structure of spherical roller bearings

The self-aligning roller bearing, the helical gear and the eccentric block are arranged on the same shaft of the gear box. The bearing and the shaft are matched by the interference fit. The bearing acts as the supporting part of the gear and the eccentric block, and its performance directly affects the running efficiency of the crusher. During the working process, the hydraulic pump of the crusher drives the eccentric block and the bearing on the main shaft by driving the helical gear. Under the action of Centrifugal Force, the eccentric block drives the bucket teeth to impact the rock with a certain frequency.The force will have a certain impact on the bearing, and the arrangement of the parts on the shaft is shown in Figure 1.

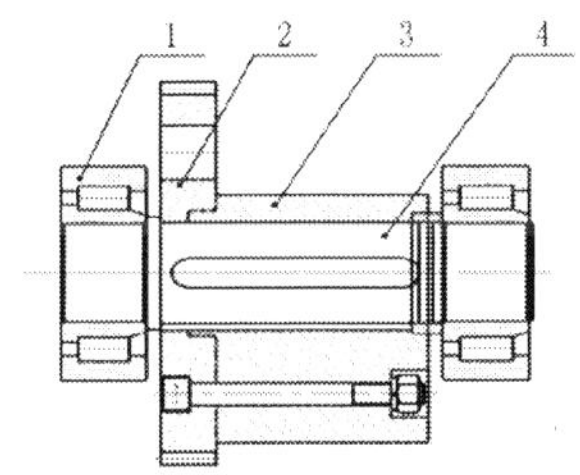

Figure 1. Part assembly drawing
1. Bearing 2. helical gear 3. Eccentric block 4. axis

According to the working principle of the high-frequency crushers, the formula for calculating the excited force generated by the eccentric block [1] is

$$F_j = 2m\left(\frac{2\pi n}{60^o}\right)^2 e \sin\frac{2\pi nt}{60^o} \tag{1}$$

Where F_j is the eccentric mass excitation force, N; m is the mass of eccentric block, kg; n is the eccentric block rotation speed, r/min

e——eccentric block eccentricity, mm

The excited force is generated by the eccentric mass centrifugal motion. When the center of gravity of the eccentric block is in the vertical direction, the excited force at this time is at the maximum value. After calculation, the maximum excited force is 303.3kN.

The unbalanced force is generated during the movement, and the eccentric block only produces the radial load. The force generated by the helical gear on the bearing can be decomposed into the axial load and the radial load. The calculation formulas are respectively

$$F_t = \frac{2T}{d} \tag{2}$$

$$F_r = \frac{F_t.tan\alpha_n}{cos\beta} \tag{3}$$

$$F_a = F_t.cos\beta \tag{4}$$

Where F_t is the circumferential force, N; F_r is the radial force, N; F_a is the axial force, N

IOP Conf. Series: Materials Science and Engineering **751** (2020) 012071 doi:10.1088/1757-899X/751/1/012071

β——node circle helix angle, degree

α_n——normal pressure angle, degree

Using the calculated values of radial load and axial load of the bearing, according to principle of moment balance, The acting force produced by the eccentric block and the helical gear on the bearing is superposed and calculated and it is found that the bearing close to the eccentric block is subjected to a large force and is subjected to Force analysis and theoretical calculation [9] show that the maximum radial force of the bearing is 273060N, and the maximum axial force is 54611N.

A high-frequency crushers of a factory uses a double-row spherical roller bearing of type 22322CA, and the cage material is cage. The specific material property parameters are shown in Table 1.

Table 1. Spherical roller bearing related parameters

material	density /kg.m^{-3}	Elastic Modulus /N.m^{-2}	Radial load /kN	Axial load /kN	Number of rollers	Poisson's ration
GCr15	7830	2.19E+11	273.06	54.611	30	0.3
Cage	8800	8.00E+10				0.35

2.2 Basic theory of transient dynamics

Transient dynamics can be used to analyze the linear and nonlinear dynamic responses of structures under steady, transient and harmonic loads.When the system is subjected to the impact load, the transient dynamic equation that the system needs to satisfy is [10]:

$$[K]\{u\} + [R]\{u'\} + [M]\{u''\} = \{P\} - \{I\} \tag{5}$$

Where [M] is the mass matrix; [C] is the damping matrix; [K] is the stiffness matrix; {u} is the displacement of each node; $\{u'\}$ is the velocity of each node; $\{u''\}$ is the Node acceleration; {P} is the external force of each node; {I} is the force within each node.

It can be seen that the amount of displacement deformation is directly related to structural stiffness, damping, velocity, and acceleration.

3. Finite element model establishment

The transient dynamics analysis is suitable for solving the dynamic response of the impact load. Because the calculation process requires a large amount of computational memory, the model is simplified under the premise of ensuring the reliability of the result. Ignore structures such as rounded corners [11].

In the course of Finite Element Analysis, the grid division plays an important role. The quality of the grid directly affects the reliability of the calculated results. Therefore, in the course of the grid division, Different Grid Division methods are adopted for different parts of the bearing The hexahedral Mesh method is adopted for the inner and Outer Rings of the bearing model, while the Tetrahedral Mesh method with good adaptability is adopted for the roller and cage of the bearing. The check result of grid quality is that the grid quality of the contact area of key parts such as roller is above 0.89, as shown in figure 2.

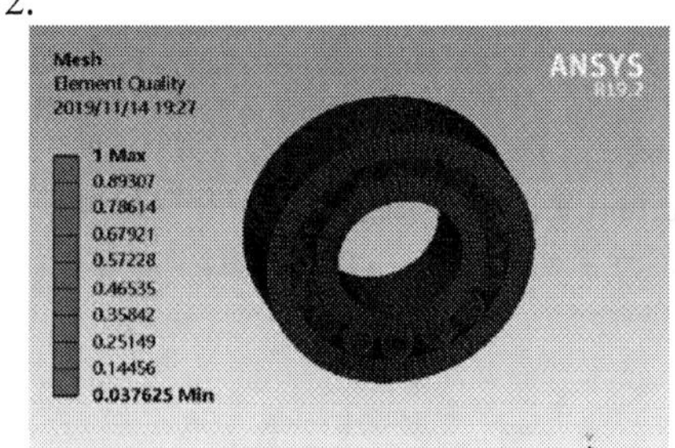

Figure 2. grid quality map

4 .Comparison of finite element analysis results

4.1 Deformation under pure radial load

Pure radial load is almost non-existent in the actual operation of the bearing. In order to compare and analyze the simulation results of spherical roller bearing under combined load and observe the deformation region of spherical roller bearing under single load, the spherical roller bearing is analyzed. By changing the thickness of cage, the working condition of self-aligning roller bearing under pure radial load is analyzed, and the total deformation of bearing and cage is compared and analyzed, The simulation results are shown in figure 3.

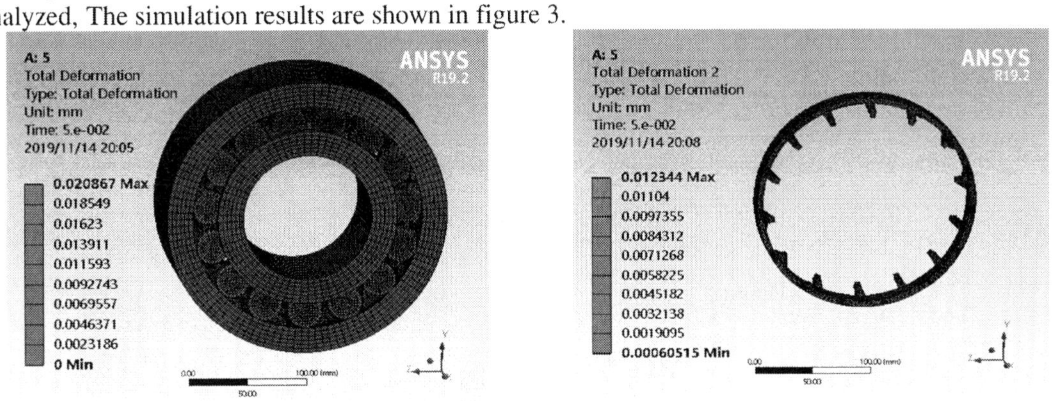

(a) 5mm cage simulation results

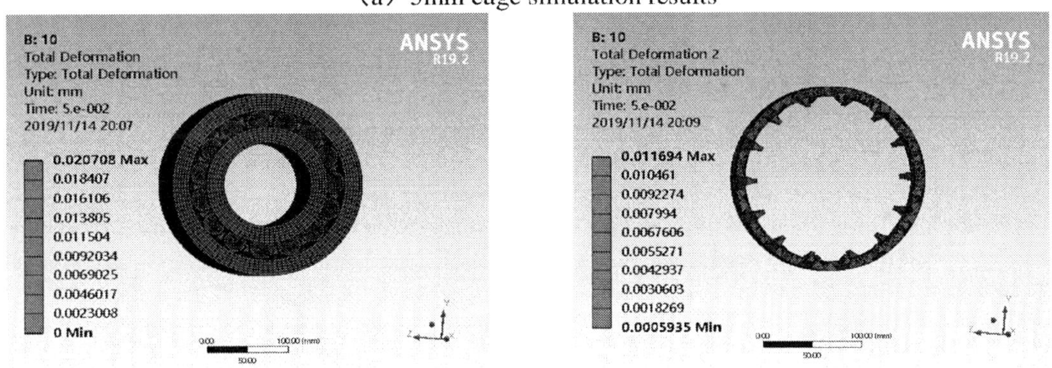

(b) 10mm cage simulation results

Figure 3. Deformation under pure radial load

It can be seen from the simulation results in the Figure 3. that the effects of changing the thickness of the bearing cage on the overall deformation of the bearing is almost unchanged.The total deformation of the bearing varies slightly with the thickness of the cage when the load is constant, which is mainly due to the role of the cage in keeping the roller position under pure radial load.Because there are different degrees of collision between the roller and the cage during the working process.When the load is constant, the thickness of the cage increases.Which increases its own bearing capacity, so the thick cage deformation is relatively small.Under load, the maximum load area of bearing and cage is invariable.

4.2 Deformation under Combined load

The force produced by the high-frequency crusher in the course of its operation is largely due to the centrifugal motion of the eccentric block, because of the combined action of helical gear and eccentric block, and the spherical roller bearing bears both radial and axial combined load.The existence of combined load leads to the eccentric load effect of self-aligning roller bearing, which has an important influence on the service life of the bearing.Therefore, it is very important to study the bearing under eccentric load.By changing the thickness of cage, the effect of cage under combined load is analyzed,the results are shown in Table 2.

Table 2. Bearing Total Deformation parameters under combined load

5mm Cage		10mm Cage	
Total Deformation/um	Cage Total Deformation /um	Total Deformation /um	Cage Total Deformation /um
20.134	12.03	19.958	11.575

As it can be seen from the simulation results in Table 2.When the thickness of the cage increases, the total deformation of the bearing and the deformation of the cage also decreases.This is because the axial load is much smaller than the radial load.The bearing performance is improved to some extent by increasing the thickness of cage.Compared to the pure radial load, the load on the cage is increased.This is because the axial load causes the bearing to appear the partial load, then squeezes the cage to cause.

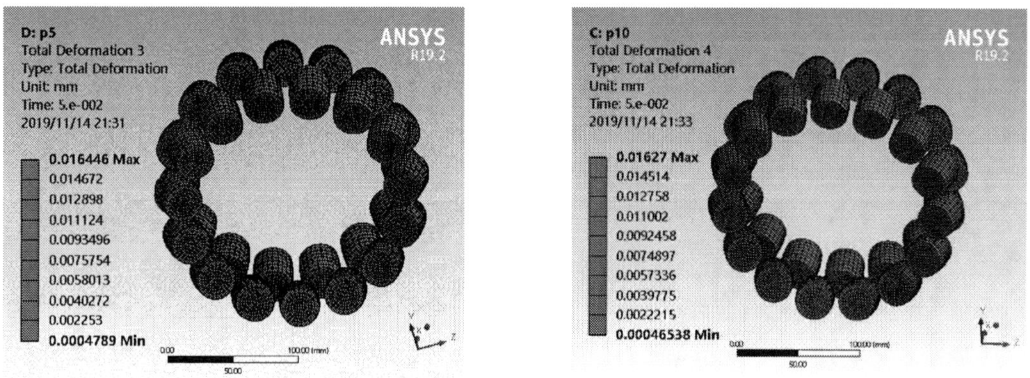

(a) 5mm cage Total Deformation of rollers (b) 10mm cage Total Deformation of rollers

Figure 4. Total Deformation of the roller

It can be seen from the simulation results in Figure 4.The maximum deformation of the roller occurs on the opposite side of the axial load application position, which is caused by the axial load. And the opposite side of the roller is subjected to extrusion, so the amount of deformation is larger than other rollers.Due to the existence of eccentric load,the roller of the upper half ring is subjected to different degrees of load.The maximum load region of roller is located directly below the radial load on the bearing, and the load of roller on both sides decreases gradually to both sides.In addition, through comparative analysis, it can be seen that the increase of cage thickness can improve the roller force environment.

4.3 Equivalent Stress under Combined load and pure radial load

Stress is another important value that needs to be referenced under bearing load. In order to further study and analyze the spherical roller bearing, the stress of the cage under different thickness and different loads is simulated and analyzed.Compare the simulation results, as shown in Table 3.

Table 3. Cage Equivalent Stress parameter

Pure radial load /N		Combined load /N	
5mm Cage	10mm Cage	5mm Cage	10mm Cage
Equivalent Stress/MPa	Equivalent Stress/MPa	Equivalent Stress/MPa	Equivalent Stress/ MPa

50.846	20.494	203.43	199.67

According to the values in the Table 3, under the single load or combined load, the average stress value of the cage is reduced with the increase of the thickness of the cage.Under combined load, the average stress of cage with the same thickness is greater than that of pure radial load.The simulation results further confirm that the damage of the bearing by the combined load is more serious,and the increase of the thickness of the cage can increase the service life of the cage.

5 .Conclusion

(1) The increase in the thickness of the cage under pure radial loading enables the bearing and cage itself to have better deformation resistance.

(2) Under the action of the combined load, the bearing is squeezed, on the one hand, the bearing roller on the opposite side of the force contact zone is subjected to a large force, and On the other hand, the roller of the upper half ring of the bearing is subjected to different degrees of load; The thickness is inversely related to the amount of deformation of the bearing.

(3) Under pure radial load or combined load, the load on the rollers at the middle position of the lower half ring of the bearing is maximum.And the load on the rollers on both sides is symmetric, and decreases gradually.

(4) The combined load has a more serious influence on the bearing failure than the single load.Through the analysis of the bearing deformation and the average stress,it can be seen that increase in bearing cage thickness increases the service performance and life of the bearing.

References

[1] Liu Wen. Design and simulation of 50 type high frequency breaker [D]. Guiyang: *Guizhou University*,**2017.**

[2] Li Zhen, Zheng Lin Zheng, Zhang Xu, et al. Quasi-static analysis of double row tapered roller bearings[J].*Journal of Harbin Engineering University*.**2017, (2)**:276-281

[3] Chan Wen Bing, Yao Ting Qiang, Xie Wei, et al. Finite element simulation analysis of dynamics of deep groove ball bearings[J].*Journal of Kunming University of Science and Technology(Natural Science)*.**2013, (5)**:45-51.

[4] Du Hui, Zhou Yan Wei, Deng Si Er, et al. Transient dynamic analysis of interaction force between high-speed cylindrical roller bearing parts[J].*Bearing*.**2005,(9)**:9-12,15.

[5] Zhang Gang, Liang Song, Zhang Yi Min. Dynamic Simulation Analysis of Rolling Bearings[J].*Machinery Design & Manufacture*. **2013,(9)**:32-34,38.

[6] Lin TengJiao, Rong Qi, Li Run Fang, et al. Finite Element Analysis of Dynamic Characteristics of Deep Groove Ball Bearings during Operation[J]. *Journal of Vibration and Shock*. **2009,(1)**:21-39.

[7] Sun Dao Yong, Lu Ji, Liu Wun. Dynamic Characteristics Analysis of High Speed Rolling Bearings with Axial Loading[J]. *Chinese scientific paper*. **2014,(8)**:920-922,926.

[8] Liu Ning, Zhang Gang, Gao Gang, et al. Finite Element Stress Analysis of Cylindrical Roller Bearing Based on ANSYS[J]. *Bearing*.**2006,(12)**:8-10.

[9] Kong Ling Jia, Wang Xiao Li, Wang Wen Zhong. Mechanical Design Second Edition [M]. *Beijing Institute of Technology Press*.**2013.8.**

[10] Kuang Rui, Zhou Hai Tao, Wang Di, et al. Research on anti-impact design of a shipboard engine control cabinet based on transient dynamics simulation analysis [J]. *Ship Electronic Process*. **2019, (8):** 194-198

[11] Cai Jia Bin, Liu Wen, Ding Cheng Bo, et al. Failure Analysis of Spherical Roller Bearings in High Frequency Breakers [J]. *Mechanical Design and Manufacturing*. **2018, (3):** 13-15.

Research on Construction Method, Risk Identification and Evaluation Technology of UHV AC Composite Cross Arm Pole and Tower

Chunhua Hu*, **Jiancheng Wan, Ming Jiang and Yujing Hao**

China Electric Power Research Institute Co., Ltd, Beijing, China

*E-mail: 2546132092@qq.com

Abstract. The research on construction method of UHV AC composite cross arm pole and tower mainly combines the material and structural characteristics of composite cross arm insulator, analyzes the danger sources which may cause damage to construction links such as packaging, transportation, storage, construction hauling, lifting and installation, and then this paper conducts evaluation, and proposes corresponding construction measures, thus forming the construction technical scheme of UHV AC composite cross arm pole and tower. This paper conducts analysis and assessment to the risk (danger source) of damage to composite cross arm insulator during the construction by utilizing the risk evaluation method, and then proposes corresponding construction full-protection measures and counter measures targeting the assessment results.

Key Words: Composite cross arm pole and tower; risk identification; risk grading

1. Introduction

Compared with routine iron tower, the requirements for technical conditions for composite cross arm pole and tower are higher, and it is required to adopt a series of special construction measures in tower assembly construction links such as packaging, transportation, assembly, lifting and installation, thus completing the installation construction of composite cross arm smoothly. In our country, although we have accumulated some experience, the research on existing construction technology of composite cross arm pole and tower, especially that the UHV AC circuit adopts new type of pole and tower construction technology of composite cross arm, is still neither systematic nor complete; in addition, the standard construction technical scheme of new type of composite cross arm insulator hasn't been formed, and the requirements for promotion and application of composite cross arm of UHV AC circuit cannot be met yet.

The research on construction technology of UHV AC composite cross arm pole and tower mainly combines the material and structural characteristics of composite cross arm insulator, analyzes the danger sources which may cause damage to the construction links such as packaging, transportation, storage, construction and hauling, lifting and installation, and then conducts evaluation. In addition, this paper proposes corresponding construction measures, forms a systematic and complete construction technical scheme of UHV AC composite cross arm pole and tower. On the premise of guaranteeing the construction safety, the scheme can enhance the construction quality and efficiency of composite cross arm pole and tower, thus providing technical support for UHV AC circuit adopting composite cross arm pole and tower as well as its promotion and application. [1-5]

2. Research on Construction Risk Evaluation Technology

2.1. Safety Evaluation Technology

The risk evaluation and risk control technology origins from the 1930s, and it develops along with the development demand of insurance industry of western countries. The insurance companies undertake various risks for customers, so they have to charge fees, and the amount of fees is decided by the risk level which an insurance company undertakes. Therefore, there is a problem about balancing of risk degree, and the process of balancing and determining risk degree is a process of safety assessment, so it is also called "Risk Assessment".

The safety evaluation and risk control technology was developed greatly in the second half of the 20th century, which benefited from the improvement and development of system safety engineering theory. The system safety theory was firstly applied into the military industry of the USA. In April 1962, the USA issued the first instructions about system safety i.e. System Safety Engineering of Ballistic Missile of Air Force, and it took this as the system safety requirements proposed for contractors related to the minuteman missle plan; and this was the first actual application of system safety theory. In 1969, the United States Department of Defense approved and issued the most typical system safety military standard Key Points of System Safety Outline (MIL-STD-822), which proposed safety requirements, procedures and targets covering the whole life cycle of the system with regard to the targets, plans and manners, including design, measures and assessment, in the aspect of safety of system completion. This Standard was revised into MIL-STD-822A in 1977. In 1984, it was revised into MIL-STD-822B, which generated huge influence on the world engineering safety and fireproof field, and was promoted to the aviation, spaceflight, nuclear industry, petroleum and chemical industry all over the world successively, and after constant development and improvement, a theory and method system of modern system safety engineering was formed. Therefore, it occupies an important position in the current safety science.

At the early stage of the 1980s, the safety system engineering is introduced into our country, and many research units, industrial management departments and some enterprises start research and actual application of safety assessment method. To incorporate the safety assessment work into the legal system and use it better in the actual work, in October 1996, the former Ministry of Labor Forces issued the Decree No. 3 - Rules on Supervision of Labor Safety and Hygiene of Construction Project (Engineering); in May 1999, the former State Economic and Trade Commission issued the Notice about Conducting Qualification Recognition to Labor Safety and Hygiene Pre-assessment Units of Construction Project (Engineering) (Guo Jing Mao An Quan [1999] No. 500); in June 2002, the State Administration of Work Safety (National Coal Mine Safety Administration) issued the Opinions on Enhancing Management of Safety Assessment Mechanism. On November 1, 2002, the Production Safety Law of the People's Republic of China was issued and implemented, which had a great promotion function to the safety assessment. With the issue of relevant matched regulations such as the Regulations on Safety Management of Dangerous Chemicals (Decree No. 344 of State Council), the safety assessment is implemented gradually. At present, the safety evaluation is expanded to 4 categories, i.e. safety pre-assessment, safety inspection & acceptance assessment, current safety status assessment and specialized safety assessment from original pre-assessment of labour safety and hygiene, covering the full life cycle of the engineering and system, which has obtained preliminary achievements.

The practice proves that the safety evaluation can not only enhance the intrinsic safety degree of enterprise and production equipment, but provide powerful technical support for decision-making, supervision and inspection of Work Safety Production, Supervision and Management Department at each level.

2.2. LEC Safety Risk Assessment Method

The LEC assessment method (Graham (Benjamin Graham, 1894-1976) Assessment Method) is a kind of semi-quantitative safety assessment method for danger sources in operation environments with potential dangerousness. It is used to assess the dangerousness and hazard when the operation personnel are operating in an environment with potential danger.

LEC method is a kind of semi-quantitative safety assessment method for danger sources in operation environments with potential dangerousness. This method is used to assess the personal casualty degree by multiplying the three factor index values related to the system risk, and these three factors are: L (possibility of accident occurrence), E (frequency degree of personnel exposed in dangerous environment) and C (consequences in case of occurrence of accident). Determine different scores for different grades of the three factors respectively, and then adopt the D, which is the product of the three scores, to assess the dangerousness of operation conditions, i.e. $D=L\times E\times C$.

Risk value $D=L\times E\times C$. The higher the value of D is, the higher the danger of the system will be, and it is necessary to add safety measures, or change the possibility of accident occurrence, or reduce the frequency degree of human body being exposed in the dangerous environment, or relieve the accident loss, till that the risk value is adjusted into the allowable scope.

It should be noted that the division to danger grade of LEC risk assessment method is to be judged relying on experience to some extent. During the application, it is required to consider the limit, and then conduct correction according to actual situation.

3. Calculation of Risk Value with LEC Safety Risk Assessment Method

The risk value adopts LEC method for quantitative calculation, and the inherent risk grade is determined according to the inherent risk value. The risk value $D=L\times E\times C$. See Table 1-Table 4 for the relationship among the value taking of risk factors L, E and C, risk value D and the risk grade:

Table 1. Possibilities of Accident or Risk Event Occurrence (L)

Score	Occurrence Possibility
10	Continuous
6	Work hours each day
3	Once each week
2	Once each month
1	Several times each year
0.5	Very rarely

Table 2. Frequency Degree of Risk Event Occurrence (E)

Score	Occurrence Possibility
10	Unable to be predicted completely
6	Quite likely
3	Likely, but not frequent
1	Less likely, completely accidental
0.5	Highly unlikely, allowed to assume
0.2	Extremely unlikely
0.1	Practically unlikely

Table 3. Consequences of Risk Event Occurrence (C)

Score	Consequences of Risk Event Occurrence
100	Catastrophe, unable to bear the loss
40	Disaster, almost unable to bear the loss
15	Extremely severe, extremely major loss
7	Severe loss
3	Major loss
1	General loss

Table 4. Table of Relationship between Risk Value D and Risk Grade

Risk Value D	Risk Degree	Risk Grade
>320	Extremely high risk. It is required to adopt measures to reduce risk grade, otherwise, the operation cannot be continued.	5
160～320	High risk. It is required to conduct rectification instantly.	4
70～160	Obvious risk. It is required to conduct rectification.	3
20～70	General risk. Please pay attention to this.	2
<20	Slight risk, but acceptable.	1

4. Division of Risk Grade of Construction Safety

The construction safety risk grade of power transmission and transformation engineering is divided into five grades in total:

Grade-I risk (slight risk): it refers to the construction operation that there is relatively low safety risk during the operation process, and if no control is conducted, there may be minor injury or event below this degree;

Grade-II risk (general risk): it refers to the construction operation that there is a certain safety risk during the operation process, and if no control is conducted, there may be minor personal injury accident;

Grade-III risk (obvious risk): it refers to the construction operation that there is relatively high safety risk during the operation process, and if no control is conducted, there may be severe personal injury or death accident, or there may be Grade-VII power grid event;

Grade-IV risk (high risk): it refers to the construction operation that there is very high safety risk during the operation process, and if no control is conducted, there may be personal death accident easily, or there may be Grade-VI power grid event;

Grade-V risk (extremely high risk): it refers to the construction operation that there is extremely high safety risk during the operation process, and if no control is conducted, there may be mass death or casualty accident, or there may be Grade-V power grid event.

5. Dynamic Risk Calculation

Calculation method of dynamic adjustment coefficient

Dynamic risk value $D1 = D/k$

Dynamic adjustment coefficient k:

$$k = \frac{\sum_{i=1}^{4} a_i \cdot b_j}{N} \tag{1}$$

In the formula: N = 4 (4 dimensions)

ai - it corresponds to the weighted value of importance assessment of the dimension Ui.

bj - it corresponds to the average value of various sub-item risk factors of the dimension Ui.

In the power transmission and transformation engineering, the relative importance of each dimension is considered as per the higher one temporarily, i.e. ai=1, and the dynamic risk adjustment coefficient of the Project adopts the simple assessment method, i.e.:

$$k = \frac{\sum_{i=1}^{4} b_j}{N} \tag{2}$$

Relationship between actual situation of dimension and risk value taking of risk factor at each dimension.

6. Research on Assessment Technology of Construction Safety and Quality Risk of Composite Cross Arm Pole and Tower

6.1. Packaging Requirements and Risk Analysis

- With regard to the composite insulator, prior to storage of the Using Unit, it is required to inspect and make sure that there is no obvious extrusion or damage trace on the external packaging of composite cross arm part, and the part shall not be stored unless the external packaging is proved to be qualified.
- The composite cross arm part shall be stored in a dry and ventilated warehouse, so as to avoid damage due to external force such as water logging and damage by rats. It is prohibited to put heavy objects or stack sharp objects with corner angles onto the composite cross arm accessories directly, so as to avoid deformation or mechanical damage of umbrella skirt.
- The stacking height of packing object shall meet requirements of the manufacturer, and the number of layers should not exceed 8.
- With regard to outdoor storage, besides meeting above ruled requirements, it is still necessary to adopt rainproof measures..

According to above packaging and storage requirements, analyze main risk sources in the packaging and storage, as shown in Table 5 below:

Table 5. Table of Analysis and Assessment of Danger Sources of Composite Cross Arm Packaging

No.	Risk Source	Risk Factor	Possible Risk	Risk Assessment D	Risk Grade
1	External packaging of composite cross arm part	There is packaging damage.	There may be part damage.	18	1
2	Storage of composite cross arm part	The environment is not suitable.	There may be part damage or mildew.	50	2
3	Storage of composite cross arm part	The packing doesn't meet requirements.	There may be part damage.	40	2
4	Outdoor storage of composite cross arm part	No rainproof measure is adopted.	There may be part damage or mildew.	50	2

6.2. Transportation Requirements and Risk Analysis

The transportation refers to transporting the composite cross arm part from the storage location to the installation site. Both the transportation and hauling shall be conducted when the package is intact.

- In case of adopting automobile for transportation, the stacking of packaging box (drum) of composite cross arm part shall meet stacking rules. In case of unloading, it is strictly prohibited to conduct dragging or throwing, so as to guarantee that the packaging box (drum) can fall to the ground stably.
- Both transportation and hauling shall be conducted when the packaging box (drum) is intact, so as to prevent the material from being damaged or deformed, especially that the umbrella skirt part shall not be stressed, and it is required to avoid the contact friction with the ground.

According to above transportation requirements, the analysis on main risk sources in the transportation is as shown in Table 6 below:

Table 6. Table of Analysis and Assessment of Transportation Danger Sources

No.	Risk Source	Risk Factor	Possible Risk	Risk Assessment D	Risk Grade
1	Automobile transportation	Stacking failing to meet requirements	Part damage	40	2
2	Automobile transportation	Vehicle injury and traffic accident	Personal casualty and property loss	54	2
3	Hauling	Falling, object crashing, and umbrella skirt wear	Personal injury and part damage	50	2
4	Transportation and lifting	General lifting collision	Lifting injury	63	2
5	Transportation and lifting	The weight reaches 95% of the rated load of lifting machinery, and the lifting machinery conducts lifting below the power transmission line or in the place relatively close to the charged boy in a special way.	Mechanical injury, lifting injury, and electric shock	240	4

6.3. Research on Construction Risk of UHV Composite Cross Arm Pole and Tower

The construction of composite cross arm is mainly divided into ground assembly and lifting. During the construction, there are mainly safety and quality risks, and after sorting its construction flow, the analysis on ground assembly and lifting risk is shown as below:

6.3.1. Analysis on Ground Assembly Risk. The ground assembly should adopt crane, and the crane shall be assembled according to the sequence ruled by the Manufacturer. When the Manufacturer has no rules, it is necessary to assemble the pressure lever on the tower body side, and then assemble it to the side of cross arm head successively; prior to connection of pressure lever, it is required to check the alignment; adopt support to connect two sections of pressure levers which have been assembled well, and then install the front support; two pieces of pulling rods shall be lifted independently after the cross arm is in place; the shielding ring on both sides of the cross arm head shall be installed at the accessory stage after the stringing is completed. According to above main construction process, the analysis on main risk sources at the ground assembly stage is as shown in Table 7 below:

Table 7. Table of Analysis and Assessment of Ground Assembly Danger Sources

No.	Risk Source	Risk Factor	Possible Risk	Risk Assessment D	Risk Grade
1	Lifting of crane	General lifting collision.	Lifting injury	63	2
2	Assembly of composite cross arm	The assembly is not conducted as per ruled sequence.	The assembly cannot be completed or there is member damage.	40	2
3		There is collision between it and the umbrella skirt of silicon rubber	Umbrella skirt damage	100	3
4		The umbrella contacts oily, acidic or alkaline substances.	Umbrella skirt damage	90	
5		No measure is adopted to isolate it from the ground in case of assembly.	Umbrella skirt damage	100	3

6.3.2. Analysis on Lifting Risk. Prior to lifting of composite cross arm, fasten the iron tower body bolt in place (except the part connected with the cross arm) firstly. It would be better to adopt the crane or other routine tower assembly methods for overall lifting. Prior to lifting, it is necessary to design the arrangement of lifting points or adopt auxiliary hanging bracket for lifting firstly; tie two pulling rods on the lifting belt, and it is not necessary to conduct reinforcement; keep two pressure levers at a design included angle at the root for auxiliary support and reinforcement. In addition, conduct lifting point binding according to the position and method ruled in the operation instructions. When the cross arm is 0.5m away from the ground, the lifting shall be paused temporarily, and it is necessary to adjust the pressure lever angle to make it close to the angle after installation. The Field Commander is responsible for inspecting stress situation of each part, and after making sure that there is no error, loosen the control rope slowly, and then continue to lift the cross arm. During the process of cross arm lifting, the Commander shall stand in a safe position in favour of observing cross arm status exactly, pay close attention to the distance away from the tower body, command and coordinate the lifting of cross arm and loosening of control rope, and adopt effective measures to make sure that the cross arm doesn't collide with the tower body. After the lifting is in place, it is required to install the pressure lever firstly, and finally install the pulling rod.

According to above main construction process, the analysis on main risk sources at the lifting stage is as shown in Table 8:

Table 8. Table of Analysis and Assessment of Lifting Danger Sources

No.	Risk Source	Risk Factor	Possible Risk	Risk Assessment D	Risk Grade
1	Lifting of Crane	General lifting collision	Lifting injury	63	2
		The weight reaches 95% of the rated load of lifting machinery, and the lifting machinery conducts lifting below the power transmission line or in the place relatively close to the charged boy in a special way.	Mechanical injury, lifting injury, and electric shock	240	4
2	Lifting of Suspending Pole	Temporary grounding	Electric shock	45	2
		Arrangement of pole system	Lifting injury	90	3
		Selection and setting of ground anchor pit	Object striking, lifting injury, and ground anchor pulling	120	3
		Installing pole upright	Machine and tool injury	45	2
		Lifting cross arm	Object striking, and falling from height	63	2
		Enhancing pole	Object striking	63	2
		Removing pole	Object striking	63	2
3	Lifting of Powered Winching (Utilizing Ground Wire Support)	Insufficient ground wire support strength	Object striking, and falling from height	150	3
4	Quality Risk	Colliding tower body during the lifting process	Cross arm structure and umbrella skirt damage	120	3
		Firm installation of cross arm	Non-uniform fastening, insufficient fastening torque, and non-tight joint of connection surface	70	2
		Pre-arch of cross arm	The pre-arch fails to meet design requirements.	45	2

7. Conclusions

IOP Conf. Series: Materials Science and Engineering **751** (2020) 012072 doi:10.1088/1757-899X/751/1/012072

The research on construction method of UHV AC composite cross arm pole and tower mainly combines the material and structural characteristics of composite cross arm insulator, analyses the danger sources which may cause damage to construction links such as packaging, transportation, storage, construction hauling, lifting and installation, and then this paper conducts evaluation, and proposes corresponding construction measures, thus forming the construction technical scheme of UHV AC composite cross arm pole and tower. This paper conducts analysis and assessment to the risk (danger source) of damage to composite cross arm insulator during the construction by utilizing the risk evaluation method, and then proposes corresponding construction full-protection measures and counter measures targeting the assessment results.

Acknowledgements

This Research was supported by Science and Technology Research Project of State Grid (Research on Key Technologies of New Type Composite Cross-Bar Insulator for UHV AC. GYW17201700029).

References

[1] Wang Qiuling, He Zhouwen, Zhang Zhuo, et al. Molding Process of Carbon Fiber Core Material of New Composite Cable [J]. Electric Power Construction, 2010, 31(3): 106-108.
[2] Cai Yuncai, Guo Jian, Wang Chuqi, et al. Research on the Performance and Application of Carbon Fiber Composite Core Conductor [J]. Agricultural Science & Technology Equipment, 2013, 9(9): 49-51.
[3] Li Lei, Chen Wei, Wan Jiancheng. Technical and Economic Analysis on Popularization and Application of Conductor with Reinforced Core of Carbon Fiber in China [J]. Electric Power Construction, 2012, 33(9): 79-82.
[4] Xu Jingcan. Study on Status and Correlative Factors of Mental Health among Electric Linemen in Hunan [D]. Changsha: Central South University, 2011.
[5] Tang Xujun. Construction Machinery and Equipment of Power Transmission Line [M]. Beijing: China Electric Power Press, 2009.

IOP Conf. Series: Materials Science and Engineering **751** (2020) 012073 doi:10.1088/1757-899X/751/1/012073

Application of Weak Signal Denoising Based on Improved Wavelet Threshold

Ning Zhang[a], Pengfei Lin[b] and Lei Xu[c]

College of Weaponry Engineering, Naval University of Engineering, Wuhan 430033, China;

Correspondence: [a]18602710800@163.com; [b,*]xiaoatai1991@163.com; [c]xldxulei@163.com

Abstract: In this Study, the application of wavelet threshold denoising in weak signal detection of underwater targets under complex sea conditions is discussed. Firstly, the basic principle and steps of wavelet threshold denoising method based on Mallat decomposition are analyzed. Then a new threshold function is proposed to overcome the shortcomings of traditional Donoho hard and soft thresholds. The denoising effects of wavelet and wavelet packet threshold denoising methods under hard threshold, soft threshold and new threshold functions are simulated, and the signal-to-noise ratio (SNR) and mean square error (MSE) after denoising are taken as evaluation indexes. The results show that the improved threshold denoising method outperforms the traditional threshold denoising methods.

1. Introduction

The background noise of marine environment has a serious impact on the detection of underwater target signal. The generation, processing and transmission of underwater target physical field signal are inevitably disturbed by noise. Especially in the detection of weak target signal, the noise elimination is the key issue [1].

For the wavelet transform has the ability to concentrate the signal energy on a few wavelet coefficients [2], the effect of white noise on the signal can be well removed by using the wavelet transform. There are three methods commonly used wavelet denoising: Mallat's denoising algorithm based on the wavelet transform modulus maxima technique [3]; Xu's spatial correlation denoising algorithm [4]; Donoho's threshold denoising algorithm[5]. Among them, the threshold denoising method was proposed by D.L. Donoho in 1995. For this method can get the best estimation in Besov space, and no other linear estimation can achieve the same estimation effect, the wavelet threshold denoising method has attracted wide attention of scholars at home and abroad. However, the soft threshold and hard threshold functions used in this method still have some problems to be improved.

The paper 1 proposed an improved wavelet threshold method and subsequently verified its de-noising performance through simulations and actual applications, using SNR and MSE as the evaluation indexes in the simulations [6]. The paper 2 proposed a new wavelet denoising algorithm, the proposed method overcomes the deviation between the estimated wavelet coefficient and the noisy signal coefficient, and has a better regulation and continuity [7].

Wavelet packet transform further decomposes the high frequency part of the signal on the basis of the wavelet transform and improves the processing ability of the signal. In this paper, the basic principle and implementation steps of wavelet threshold denoising method are analyzed. In view of the shortcomings of traditional soft and hard thresholding methods, a new threshold function is proposed, and the denoising effects of wavelet and wavelet packet threshold denoising methods are simulated and analyzed.

2. Mallat decomposition algorithm

2.1 The Wavelet Transform

The continuous signal $z(t)$ is expanded on the wavelet basis to generate the continuous wavelet transform (CWT). The expression of CWT is:

$$WT_z(a,b) = < z(t), \varphi_{a,b}(t) >= a^{-1/2} \int_R z(t)\varphi(\frac{t-b}{a})dt \quad (1)$$

The $\varphi(t)$ is the wavelet basis function, A family of wavelet functions can be obtained by scaling and translating the wavelet basis function. Expressed as:

$$\varphi_{a,b}(t) = a^{-1/2}\varphi(\frac{t-b}{a}) \quad (2)$$

2.2 The wavelet transform of the Mallat algorithm

Mallat decomposition algorithm is a fast wavelet transform algorithm based on multi-resolution analysis. Each decomposition produces high frequency detail component $d_j(k)$ and low frequency approximation component $c_j(k)$. The transformation process is equivalent to using a set of high-pass and low-pass filters to decompose time series signals step by step.

Firstly, the wavelet basis function which is closest to the signal is selected, then the level J of the wavelet decomposition is determined, $J \le \log(2gN)$ and the J is natural number. The multiresolution decomposition formulas are as follows:

$$c_{j,k} = \sum h_0(m-2k) \bullet c_{j-1,m} \quad (3)$$

$$d_{j,k} = \sum h_1(m-2k) \bullet c_{j-1,m} \quad (4)$$

$h_0(n) =< \phi_{j,0}(k), \phi_{j-1,n}(k) >$ is equivalent to a low-pass filter bank, and the approximation coefficient $c_j(k)$ is decomposed. $h_1(n) =< \varphi_{j,0}(k), \phi_{j-1,n}(k) >$ is equivalent to a high-pass filter bank, and the detail coefficients $d_j(k)$ is decomposed, ϕ and φ are the corresponding scaling function and wavelet basis function respectively.

2.3 Reconstruction of Wavelet

After signal $z(t)$ is processed by wavelet transform, it can be reconstructed to obtain useful signals:

$$x(t) = C_\varphi^{-1} \int_0^\infty \frac{da}{a} \int_{-\infty}^{\infty} WT_z(a,b) \frac{1}{\sqrt{a}} \varphi(\frac{t-b}{a}) db \quad (5)$$

The $C_\varphi = \int_R \frac{|\psi(w)|^2}{|w|} dw < \infty$, $\psi(w)$ is the Fourier transform of the basis function $\varphi(t)$.

The reconstruction of discrete wavelet can be expressed as:

$$x(t) = C_N(t) + \sum_{j=1}^{N} D_j(t) \quad (6)$$

In formula, $C_N(t)$ represents the reconstruction information from the approximation coefficients of layer N, and $D_j(t)$ represents the reconstruction information from the approximation detail coefficients of different scales.

3. The Selection of Threshold

3.1 Donoho threshold function

At present, the hard and soft thresholding denoising method proposed by Donoho is the most widely used. Donoho hard threshold processing is to compare the absolute value of signal transformation with the threshold value. The hard thresholding is carried out as follows:

$$\hat{w}_{j,k} = \begin{cases} w_{j,k}, & |w_{j,k}| \geq \lambda \\ 0, & |w_{j,k}| < \lambda \end{cases} \tag{7}$$

In the case of soft thresholding the coefficients are carried as follows:

$$\hat{w}_{j,k} = \begin{cases} sign(w_{j,k}) \cdot (|w_{j,k}| - \lambda), & |w_{j,k}| \geq \lambda \\ 0, & |w_{j,k}| < \lambda \end{cases} \tag{8}$$

Where sign(x) is +1 if x is positive and -1 if x is negative, $w_{j,k}$ is the coefficients, λ is the threshold. The $w_{j,k}$ is the wavelet coefficients obtained after thresholding processing.

The disadvantage of Donoho thresholding method is that it uses a fixed threshold to process the wavelet coefficients, which will distort the weak feature components contained in the original signal, thus causing the distortion of the reconstructed signal. In signal processing, many weak feature components are important. If the noise is removed indiscriminately, the reconstructed signal can not accurately reflect the characteristics of the signal.

3.2 Construction of New Threshold Function

The core of wavelet thresholding denoising method is the threshold processing of wavelet coefficients. Although Donoho's hard threshold and soft threshold functions have achieved good results, they also have some shortcomings. The hard thresholding function is discontinuous at $\pm\lambda$, and the signal reconstructed by $\hat{w}_k^{n,j}$ may produce some oscillations. Although the soft threshold function has good overall continuity, when the wavelet coefficients are large, there is always a constant deviation λ between $\hat{w}_k^{n,j}$ and $w_k^{n,j}$, which will bring inevitable errors to the reconstructed signal. But it is not necessarily the best to reduce the deviation to zero [8]. In order to overcome the shortcomings of hard and soft threshold function, a new threshold function is constructed in this paper.

$$\hat{w}_k^{n,j} = \begin{cases} sign(w_k^{n,j}) \cdot \left[(w_k^{n,j})^2 - \dfrac{\lambda^2}{1+\exp[(|w_k^{n,j}| - \lambda)/N]} \right]^{\frac{1}{2}}, & |w_k^{n,j}| \geq \lambda \\ 0, & |w_k^{n,j}| < \lambda \end{cases} \tag{9}$$

In the formula, N is a variable parameter and positive integer. The schematic diagram of the new threshold function at N=2 and N=12 is shown in Figure 1.

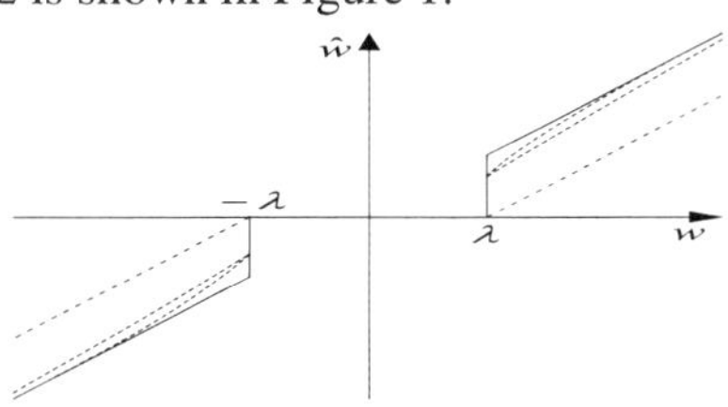

Figure 1. New threshold function diagram

In Figure 1, the solid line represents the hard threshold, the dot dash line represents the soft threshold, and the dotted line represents the new threshold function. When N=2, the dotted line is closer to the hard threshold line. Through mathematical analysis of the new threshold function, it is not difficult to get that it takes $\hat{w}_k^{n,j} = w_k^{n,j}$ as asymptote, with the increase of $w_k^{n,j}$, $\hat{w}_k^{n,j} \rightarrow w_k^{n,j}$, it overcomes the shortcomings of the constant deviation between the soft thresholds $\hat{w}_k^{n,j}$ and $w_k^{n,j}$. When $\left|w_k^{n,j}\right| \geq \lambda$, the new threshold function is high-order differentiable, which is convenient for various mathematical processing. The new threshold function is between the soft threshold and hard threshold, and closer to the hard thresholds, which is closer to the real coefficients, but not completely equal. At the same time, the smaller the N is, the faster the $\hat{w}_k^{n,j}$ converges to $w_k^{n,j}$. As long as the parameter N is adjusted properly, better denoising effect can be achieved. It can be seen that the new threshold function is a flexible choice between soft and hard threshold. The practical and effective threshold function can be obtained by selecting N.

4. Experiments simulation

4.1 The settings of experiment simulation

Four typical signals (Blocks, Bumps, Heavy sine and Doppler) used by Donoho are selected as experimental signal, and the simulation results of Doppler signal are given. Wavelet and wavelet packet threshold denoising methods are used to denoise noisy signal. Hard threshold, soft threshold and new threshold functions are selected for the threshold functions respectively, and the denoising effects of various methods are compared. The same assumption is made for both wavelet and wavelet packet denoising methods: assuming the signal-to-noise ratio of input signal is 2dB and the additive noise is zero-mean Gauss white noise, sym4 wavelet is selected as the wavelet basis, the scale of wavelet and wavelet packet decomposition is 4, and the threshold is fixed threshold (sqtwolog threshold), that is $\lambda = \sqrt{2\ln N}$, N is the length of the signal. In MATLAB, the selection of the optimal wavelet packet basis is realized by the entropy criteria and threshold parameters, and the selection of the entropy criteria depends on different signal analysis requirements.

The signal-to-noise ratio (SNR) and mean square error (MSE) after denoising are selected as the evaluation indexes of denoising effect.

$$SNR = 10\lg \frac{\sum s_0^2(n)}{\sum [s(n) - s_0(n)]^2} \tag{10}$$

$$MSE = \frac{1}{N}\sum [s(n) - s_0(n)]^2 \tag{11}$$

In the formula, $s_0(n)$ is the original signal and $s(n)$ is the denoised signal. Obviously, the bigger the SNR is and the smaller the MSE is, the better the denoising effect is.

The original Doppler signal and the noisy signal are shown in figure 2.

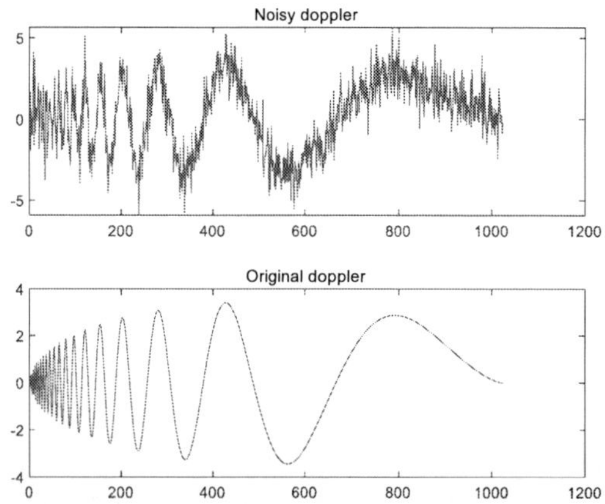

Figure 2. The original signal and the noisy signal

The results of wavelet denoising using three threshold functions are shown in Figure 3, and the corresponding results of wavelet packet denoising are shown in Figure 4.

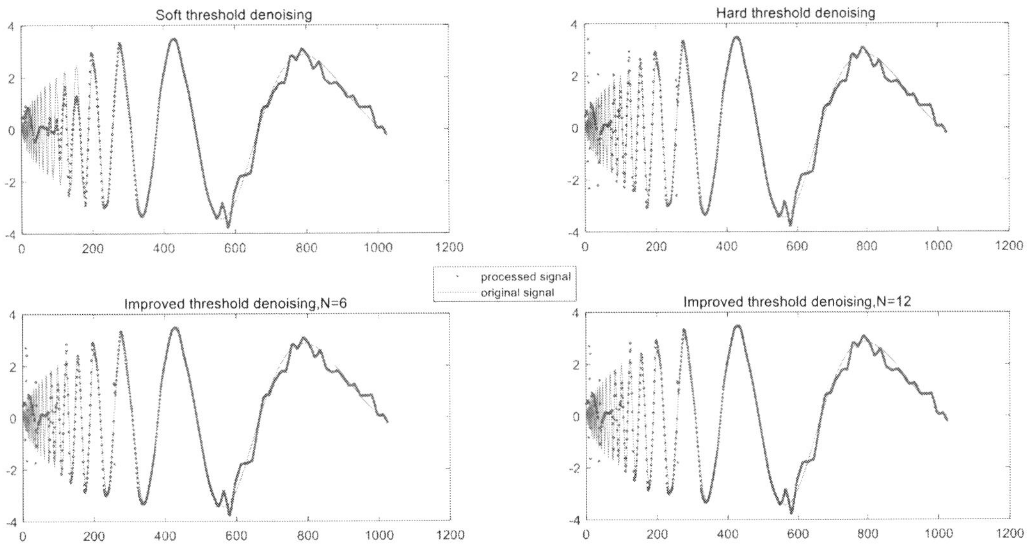

Figure 3. The result of wavelet denoising

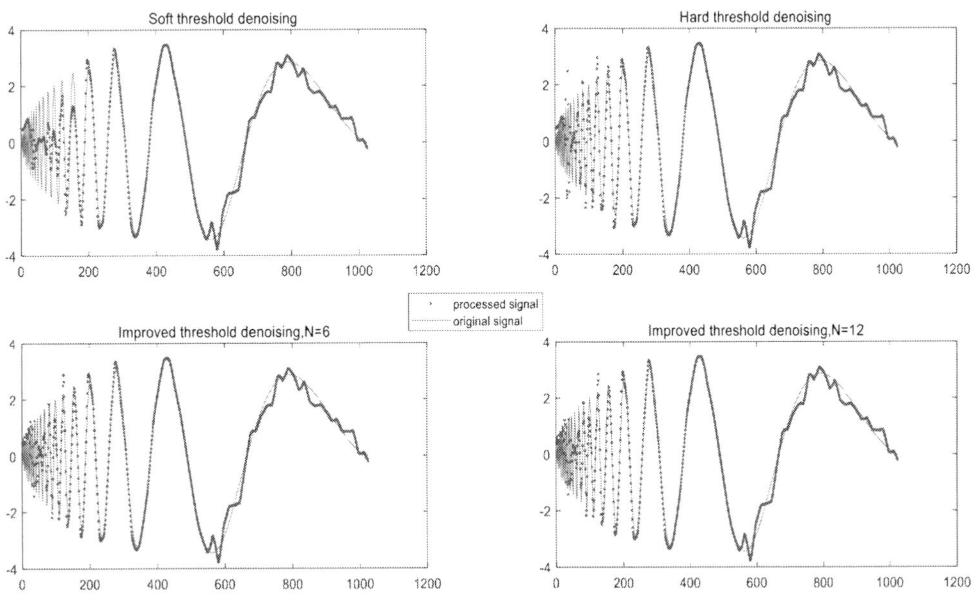

Figure 4. The result of denoising by wavelet packet

The comparison of SNR and MSE after denoising is shown in Table 1.

Table 1. Comparison of SNR and MSE for noisy doppler signals after denoising

Denoising algorithm		Hard threshold	Soft threshold	New threshold function	
				N=6	N=12
Wavelet denoising	SNR/dB	13.5859	12.8300	15.5328	15.5473
	MSE	0.1750	0.2083	0.1107	0.1103
Wavelet Packet Denoising	SNR/dB	14.0456	13.6101	15.3085	15.3231
	MSE	0.1574	0.1740	0.1382	0.1377

IOP Conf. Series: Materials Science and Engineering **751** (2020) 012073 doi:10.1088/1757-899X/751/1/012073

4.2 The analysis of the denoising performance

From figure 3, figure 4 and table 1, we can see that the denoising effect of the wavelet packet threshold denoising method is better than that of the wavelet threshold denoising method when the same threshold function is used, But the effect is not obvious . The reason is that the high-frequency part of the signal is also decomposed by multi-scale wavelet packet transform, which is not available in wavelet transform. At the same time, the simulation results also show that the denoising effect of using hard threshold is better than that of soft threshold, because there is always a constant deviation between $\hat{w}_k^{n,j}$ and $w_k^{n,j}$ in soft thresholding processing, which brings some errors to reconstructed signal. The new threshold function proposed in this paper has better denoising effect than the traditional soft and hard thresholding method, the SNR of the new threshold function is about 2 dB higher than that of hard threshold function, and about 3 dB higher than that of soft threshold function. As can be seen from Table 1, the value of parameter N has little influence on the denoising effect.

5. Conclusion

In this paper, the application of wavelet threshold denoising method in weak signal detection is studied. A new threshold function is proposed to overcome the shortcomings of traditional soft and hard thresholding method. The simulation results show that the threshold denoising using wavelet packet transform has advantages over using wavelet transform, and is more conducive to restoring the high frequency information of the original signal. The denoising effect of the new threshold function is better than that of the traditional soft and hard thresholding method. However, how to design the best threshold function according to the characteristics of different types of signal remains to be further studied.

References

[1] Lin Chunsheng, Gong Shenguang. Ship Physical Field[M]. Beijing: Weapon Industry Press, 2007.

[2] Wu Zhengguo, Xia Li, Yin Weimin. Modern Signal Processing[M]. Wuhan: Wuhan University Press, 2003.

[3] Mallat S, Hwang W L. Singularity detection and processing with wavelets[J]. IEEE Trans. on IT, 1992, 38(2): 617- 643.

[4] Xu Y S, Weaver J B, and Healy D M, et al. Wavelet transform domain filters: a spatially selective noise filtration technique [J]. IEEE Trans. on Image Processing, 1994, 3(6): 747-758.

[5] Donoho D L. De-noising by soft-thresholding [J]. IEEE Trans. on Information Theory, 1995, 41(3): 613-627.

[6] Wang Yaokui, Yu Xiang, Wang Heng, et al. Using wavelet threshold function to improve signal SNR for seismic exploration[J]. Earth Science Frontiers, 2017,24(3): 319-324.

[7] Li Mingli, Jiao Xinxin. Application of a new wavelet threshold denoising algorithm in engineering. Computer & Digital Engineering[J], 2019, 47(07):1627-1629.

[8] Liu Jun, Zhang Bingzhen. Weak Signal Detection Technology[M]. Beijing: Publishing House of Electronics Industry, 2005.

IOP Conf. Series: Materials Science and Engineering **751** (2020) 012074 doi:10.1088/1757-899X/751/1/012074

Design of Swinging Somatosensory Game Rehabilitation System

Lili Kou*, Chaoyu Xia, Zhihua Hu and Xiaojun Jiang

Shanghai Polytechnic University, Department of Automation, Jinhai Road, No. 2360, Pudong New area, Shanghai, China

*Corresponding author's e-mail: llkou@sspu.edu.cn

Abstract. This study designed an arm-swinging somatosensory game for physically uncoordinated personnel, which realized the movement attitude acquisition, data transmission, and motion-driven game character movement. The measurement results show that in the range of 0.3g to 0.5g, the probability distribution of the ordinary person is twice or more the accuracy of the limb dissonance in the face of the same action, which proves that the game has the ability to detect and measure the reaction speed of the person. It can be used as a standard for rehabilitation levels. In addition, the game calculates relative data of the acceleration and uses which to stimulate the movement of the game character, which relaxed the threshold. Therefore, it can stimulate participation interest and promote rehabilitation.

1. Introduction

The development of human social civilization is accompanied by the aging of society, which is due to the extension of life expectancy and the degradation of birth rate. The prolonged life expectancy and poor living habits increase the proportion of strokes and social disability [1-3]. The uncoordinated patient's recovery is the common aspiration of the patient's family and society. Rehabilitation studies have shown that exercising more on one side of the arm and artificial intervention can speed up recovery. But for most patients, repeated boring rehabilitation exercises and high rehabilitation massage costs slow down the pace of recovery. The somatosensory game is a new type of electronic game that is performed (operated) by detecting changes in the movement of the limbs. However, most of the somatosensory games retain the handle, and the operation and the somatosensory effect of the game handle are not direct [4-7]. Another type of somatosensory game is the way of visually collecting data. There is processing delay in visual collection, so it can't replace the touch screen game[8, 9]. At the same time, the visual processing price is often high and cannot be widely used. Mechanical sensors have fast sensing speed, high sensitivity and good stability, but they need to wear corresponding wearable equipment, but they are still used by researchers.

In this paper, the directional acceleration sensor is used as the detection limb motion collector. The Super Marie game interface is compiled by LabVIEW, and the saved data is recorded. The data is processed to measure the effect of hand rehabilitation.

2. The design of the game

The system needs to detect the upward, downward, forward and backward movement of the limb. After the data is accepted, the noise is removed and transmitted to the computer [10]. The LabVIEW PC game interface detects the data and drives the animated character movement according to the proportion of the game interface. At the same time, the detected data is saved, and the overall design is as follows

2.1. Limb motion detection method

We selected the commercially available MPU6050 six-axis attitude angle sensor to detect the hand swing motion data and fix the accelerometer on the glove. The sensor adopts the northeast sky coordinate system, which can detect the angles of the three directions (x, y, and z) of the hand. The acceleration and other data are calculated on the basis of the angle. Both angle and acceleration are important reference indicators for limb detection, but in this paper, the motion acceleration is mainly studied.

2.2. PC software data acceptance interface

The upper computer LabVIEW communicates with the serial interface instrument through VISA, and provides a set of independent standard I/O low-level functions that can be easily called. By calling these standard functions, VISA will automatically retrieve the corresponding interface driver routines based on the actual interface type, thus enabling data reception and transmission.

2.3. Game character

2.3.1.The design of the character. The game character selects the strict custom control type (Strict Type Def.), because the selected game character is a control that is not included in LabVIEW, and the control is required to be updated every time, after each subsequent call, the control can use the update. In the form, and the controls in each VI (the function of the Lab view) are consistent. In the LabVIEW startup interface, select the custom control, select the new Boolean in the custom control window, and then create a PNG format image of the Mary character. The image is used to replace the logical "true" state of the Boolean control; select the previously imported image and import it again into the clipboard to replace the logical "false" state of the Boolean control. Thus design of the image of Super Mario is finished.

2.3.2.Dynamic movement design of game characters. To implement the dynamic movement of the control (Super Mario character) during the running of the LabVIEW program, first create a Position property node for the control to be moved in the back panel, select All Elements, and then convert the Position node to the written state. At the input end of the node, a control cluster is generated. The cluster contains Left and Top variables. After the program is run, only the corresponding coordinates are input to the left and the Top.

2.3.3.Data acceptance. The serial communication module should be divided into several parts such as data collection and data extraction processing display, so multiple cycles are required at the same time, and the "producer and consumer" loop structure in LabVIEW satisfies the need to run multiple processes at the same time. And will not affect the execution speed. The main loop executes in parallel with the slave loop, the main loop performs data acquisition, and the data is saved from the loop. The main loop generates and passes data to the "element to queue" function, and controls the loop speed through the "wait (ms)" function, choosing to wait for 65ms. Returns the data in the queue from the loop and stores the data in an array.

The VISA serial communication parameter configuration method is as follows. First, use the VISA Resource name control parameter module to configure the serial port number, baud rate, data bit, stop bit, and check bit; then use the time delay function to match the VISA device clear function block. The input and output buffers of the device are cleared, then the VISA Read function reads the data in the device specified by the VISA Resource Name, and finally opens the session connection with the VISA resource, which takes up the system resources of the computer, so when the VISA program ends All open session channels must be closed. LabVIEW provides the VISA Close function for this purpose.

This function closes the communication process of the device specified by the VISA Resource Name parameter and releases the computer system resources occupied by the VISA Session.

The data sent by the sensor module to the upper computer is divided into three data packets, namely acceleration packet, angular velocity packet and angle packet, and three data packets are sequentially output. When the baud rate is 115200, 1 frame of data is output every 10 ms. The data is sent in hexadecimal form, with 11 data per packet and a total of 33 data. The data is configured in the VISA Read function so that each time 33 data is read from the device specified by the VISA Resource Name and stored in an array.

2.4. *Limb movement data transmission.*

Different steps in the game interface require Mary to jump different lengths or heights, corresponding to different accelerations. The acceleration data collected by the experiment is a series of irregular sinusoids, and the game extracts the maximum value of each acceleration data.

Each time the sensor module sends 33 hexadecimal data sequentially to the upper computer. Therefore, when parsing data, we firstly have to determine what kind of data packet the packet is, then determine the value of the second data of each packet of data; X (Y, Z) axis acceleration low byte is the 2nd (4th, 6th) bit of the array. X (Y, Z) axis acceleration high byte is the 3rd (5th, 7th) bit of the array; X (Y, Z) axis angle low byte is the 24th (26th, 28th) bit of the array, X (Y, Z) axis acceleration high byte is the 25th (27th, 29th) bit of the array.

Taking the X-axis acceleration as an example, for the first, extracting the zeroth bit and the first bit in the array of stored data, to determine whether it is 0X55 and 0X51. If yes, the data packet is an acceleration package, extract the second and third digits in the array, convert it into a hexadecimal integer string, and then concatenate the two strings with the high byte in the first low byte after the resulting character. The string is re-transferred to a numerical value. Thirdly, the X-axis acceleration is obtained by using the calculation formula. The other two-axis accelerations are calculated as the X-axis acceleration of the three-axis angle analysis method. Finally, the acceleration value is obtained by using the shift register. Compare with the previous time value until the maximum value is obtained. The game moving distance is calculated allows acceleration data, and then the Super Mario character start to move.

3. Data measurement and analysis

3.1.Measurement method

To drive Super Mario forward, backward, up or down, the person needs to swing his arm in the appropriate direction. Data records show that each person has his or her own unique motor response for the same distance. This kind of reaction is random, in order to accurately drive the movement of the characters, the personnel need to try repeatedly. The data obtained are divided into two categories, including longitudinal measurement data and lateral measurement data. Longitudinal data refers to the acceleration measured by different participants for the same distance. When the response is not appropriate, the game personnel need to self-correct. Lateral measurements are pointers to different distances that require different accelerations. When the response is not appropriate, the game player needs to constantly self-correct.

3.2.Measurement results and analysis

Figure 1 is the acceleration raw data measured within 200s, from which we extract the maximum value of each measurement. The max value is shown in Figure 2.

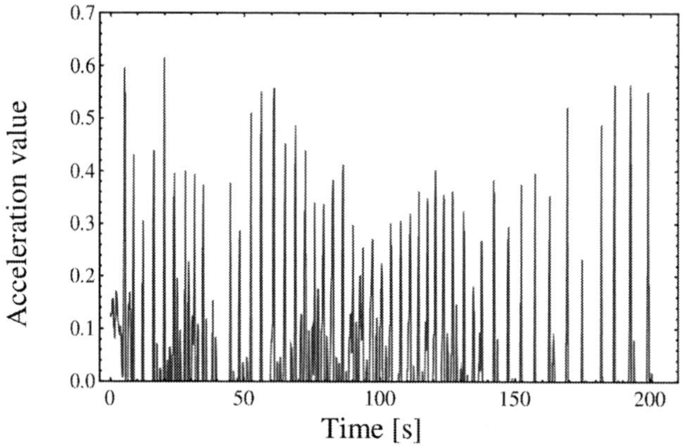

Figure 1. Acceleration raw data measured within 200s.

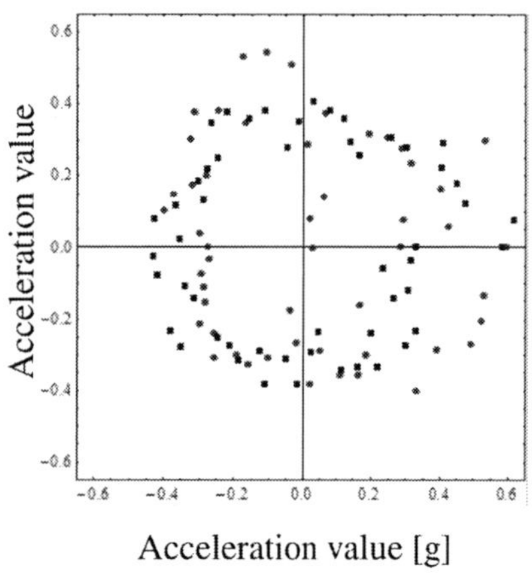

Figure 2. The acceleration maximum is evenly distributed from 0 to 360 degrees. In which, the black square and the blue dot is respectively the maximum acceleration of each limb movement.

Figure 2 is a typical example of the obtained longitudinal contrast data which is uniformly distributed on a plane of 0-360 *degrees* with the maximum value of the acceleration value of each swing arm as a radius. The black square point (data A) is the maximum acceleration measured by ordinary people, and the blue dot (data B) is the maximum acceleration measurement point of the limb uncoordinated person. It can be seen that the blue point is more dispersed. It shows that the acceleration of the limb uncoordinated person varies around a central value, but the standard deviation of acceleration is larger, and the speed of self-correction is smaller than that of normal people. Calculate the standard deviation σ from the data, the *equation* (1) is as follows, where $v_i$is the residual.

$$\sigma = \left(\frac{v_1^2+v_2^2+\cdots+v_n^2}{n-1}\right)^{1/2} \tag{1}$$

The standard deviation of data A is nearly twice the standard deviation of data B. In order to more clearly compare the dispersion of the values, we use the Normal distribution formula to plot the probability distribution curves of the two types of data. The equation (2) is as follows

$$y = \frac{1}{\sigma\sqrt{2\pi}} e^{-\frac{(v_i)^2}{2\sigma^2}} \tag{2}$$

The calculation results are shown in Figure 3. The black line A is the distribution curve of the ordinary person, and the blue line B is the distribution curve of the uncoordinated person. In the range of 0.3g-0.5g, when the area of the data A is marked as approximately 68% of the whole area, data B has only 33% of that, and this measurement is stable rather than accidental. It is proved that this game have the ability to assess the limb health condition of the game players.

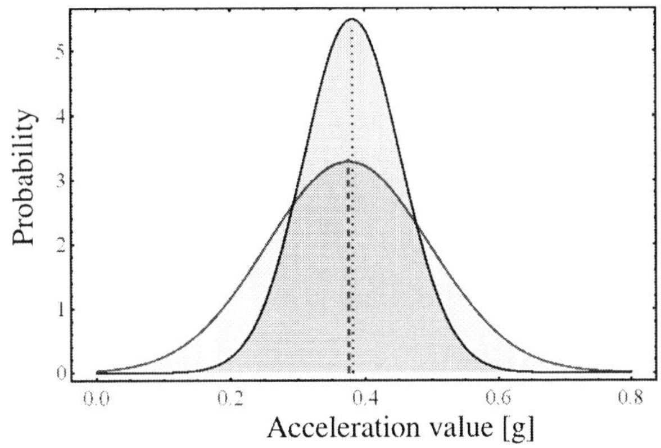

Figure 3. The normal distribution of the two data, in which A, B respectively presents the data obtained by normal player and the uncoordinated player.

In order to meet the needs of special groups, the game uses relative value instead of absolute value to drive the character. Therefore the game was more appropriate to the uncoordinated players, which will stimulate the interest of the uncoordinated players, and thus promote the long-term training and rehabilitation process.

4. CONCLUSION

This paper designed a game-based rehabilitation system that includes a game software interface compiled with Lab VIEW and hardware for collecting limb motion data. Use the acceleration sensor to detect the swing of the hand, send the motion parameters to the game interface, match the game action, and drive the game character to move. Comparing the detected hand movement parameters with the game demand parameters, it is found that each person has unique sports characteristics. The hand-poor patients have a larger range of motion than ordinary people, and the values are scattered. Within the same threshold range, the accuracy is less than 2/3 of the one of the normal person data and below. It shows that the system can be used to detect and assess the level of limb rehabilitation of patients. In addition, from the perspective of game experience, it is necessary to use the relative value of acceleration drive the character, so as to stimulate the interest of the game, and then promote limb rehabilitation.

Acknowledgments

Authors wishing to acknowledge financial support from the Shanghai Polytechnic University Enterprise Cooperation Automobile Electronic Joint Experiment Center with Grant Number A11NH190704.

References

[1] Hung Y H, Lai H Y, and Shih C H 2015 Study of multi-sensory stimulation for the design of hand rehabilitation equipment for stroke patients *J Chin Inst Eng* **32** 425

[2] Akhlaquor R and Adel A J 2013 Design and Development of a Bilateral Therapeutic Hand Device for Stroke Rehabilitation *Int J Adv Robot Syst* **10** 405

[3] Fabio V, Palmira B, Chiara M, Francesca G, Chiara M, Giovanni T, Simonetta S and Alberto L 2017 Feasibility and efficacy of a robotic device for hand rehabilitation in hemiplegic stroke patients: a randomized pilot controlled study *Clin Rehabil*, **31**, 351

[4] Tsoupikova D, Stoykov N S Corrigan M et.al, 2 015 Virtual Immersion for Post-Stroke Hand Rehabilitation Therapy *Ann Biomed Eng*, **43**, 467

[5] Leeanne M C and Thomas A M 2011 Frequency of discriminative sensory loss in the hand after stroke in a rehabilitation setting *J Rehabil Med* **43** 257

[6] Hallett M 2001 Plasticity of the human motor cortex and recovery from stroke *Brain Res Rev* **36** 169

[7] Smith P S, Dinse H R, Kalisch T, Johnson M and Walker-Batson D 2009 Effects of repetitive electrical stimulation to treat sensory loss in persons poststroke *Arch Phys Med* Rehabil **90** 2108

[8] Daud O A, Oboe R, Agostini M and Turolla A 2011 Performance evaluation of a VR-based hand and finger rehabilitation program *IEEE Inter Symp on Industrial Electronics* Gdansk, Poland, 934

[9] Hsu H W, Lee C L, Hsu M J, Wu H C, Lin R, Hsieh C L and Lin J H 2013 Effects of noxious versus innocuous thermal stimulation on lower extremity motor recovery 3 months after stroke *Arch Phys Med Rehabil* **94** 633

[10] De Diego C, Puig S, Navarro X, 2013 A sensorimotor stimulation program for rehabilitation of chronic stroke patients *Restor Neurol Neuros*, **31**, 361

IOP Conf. Series: Materials Science and Engineering **751** (2020) 012075 doi:10.1088/1757-899X/751/1/012075

Intelligent Sequencing of Working Steps Using Genetic Algorithm

Y P Zhu[1], H G Zhou[1,*], J F Liu[1] and G Z Tian[1]

[1]School of Mechanical Engineering, Jiangsu University of Science and Technology, Zhenjiang 212003, China

[*]hgzhou205@163.com

Abstract. There are many factors that need to be considered in the operation sequencing, and there are mutual relations among them, so it has become a hot topic worthy of study. In this paper, based on the parallel consideration of process constraint and minimum manufacturing cost, the concepts of workingsteps sequence matrix and process constraint matrix are proposed. The process constraint matrix is used to ensure that the workingsteps sequences conform to the precedence relation constraint, and the genetic algorithm is used to ensure the minimum manufacturing cost. Finally, the optimal solution of the workingsteps sequence is obtained. The connecting rod of marine diesel engine is taken as an example to verify.

1. Introduction

As a bridge between part design and manufacturing, process planning has been regarded as a research hotspot in the field of machining. First of all, for the more complex parts, there are many machining features and different machining methods corresponding to each feature, so the arrangement of processes has more flexibility. Secondly, the operation sequencing needs to meet all the precedence constraints and minimal manufacturing cost, which makes the operation sequencing becomes a problem has complex precedence relations.

More and more scholars use various algorithms to study the operation sequencing. Hu, Qiao, and Peng[1] proposed a novel modified ant colony optimization algorithm for the operation sequencing. Li et al.[2] presented an ant colony algorithm based cooperative optimization for machining scheme selection and operation sequencing of engine cylinder. A novel discrete particle swarm optimization (DPSO) approach is proposed to solve the operation sequencing problems in CAPP[3]. Dou, Li and Su[4] proposed a novel discrete particle swarm optimisation (DPSO) named feasible sequence oriented DPSO (FSDPSO) to solve the operation sequencing problems in CAPP. Nallakumarasamy et al.[5] used meta-heuristic simulated annealing technique (SAT) to generate the feasible sequences of operations based on the precedence cost matrix and reward-penalty matrix. Qiao, Hu and Zhang[6] conducted an algorithm to generate optimal machining sequence of the distributed holes in a machined part by using simulated annealing. Phung et al.[7] presented a method for optimizing operation sequence using modified clustering algorithm. Sundstrom et al.[8] showed how the sequence planner language (SPL) model can be converted into a constraint programming model for optimization. Fok, Thimm and Britton[9] presented the use of datum hierarchy trees, back propagation neural network and evolutionary algorithm in operation sequencing.

Among them, the most studied is the application of genetic algorithm in operation sequencing. Dou, Zhao and Su[10] developed an improved genetic algorithm (IGA) to minimize the total cost, which includes feasible operation sequence (FOS) permutation coding method, fragment crossover and fragment mutation, as well as a new elitist-based crossover strategy. A new Feasible Operation Sequence Oriented Genetic Algorithm (FOSOGA) was developed to solve the operation sequencing problem. In the FOSOGA, the crossover with adaptive crossover probability and the mutation with adaptive mutation probability were designed[11]. Liu and Qiao[12] provided feature constraint matrix, the operation constraint matrix and an iterative genetic algorithm to simplify process constraint aggregations. Su et al.[13] presented a modified GA that use an edge selection based chromosome encoding approach to make sure all the precedence constraints are met in every step. Su et al.[14] offered an edge selection strategy based on GA to improve GA's converging efficiency. An improved genetic algorithm to optimize worksteps sequence was proposed for mill/turn machining center[15].

However, these studies do not link and optimize the two requirements of the precedence constraints and minimal manufacturing cost in a unified way. At the same time, the genetic algorithm has the advantages of small correlation with the initial conditions, parallel search for global optimal solution and good operability. Therefore, this paper proposes a workingsteps sequence optimization algorithm with the workingsteps sequence matrix as the constraint means and the genetic algorithm as the optimization idea.

2. Intelligent sequencing based on genetic algorithm

Traditional process planning usually determines the sequence of processing operations first, then the machine and tools. Although such a process can ensure the requirements of process constraints, it is not easy to meet the requirements of minimal manufacturing cost. So this paper considers these two requirements in parallel. Using the matrix to restrict the clear precedence relation, using genetic algorithm to comprehensively consider the influence of tools, machines and fixtures, and finally determine the order. The complete flow chart of the method is shown in figure 1.

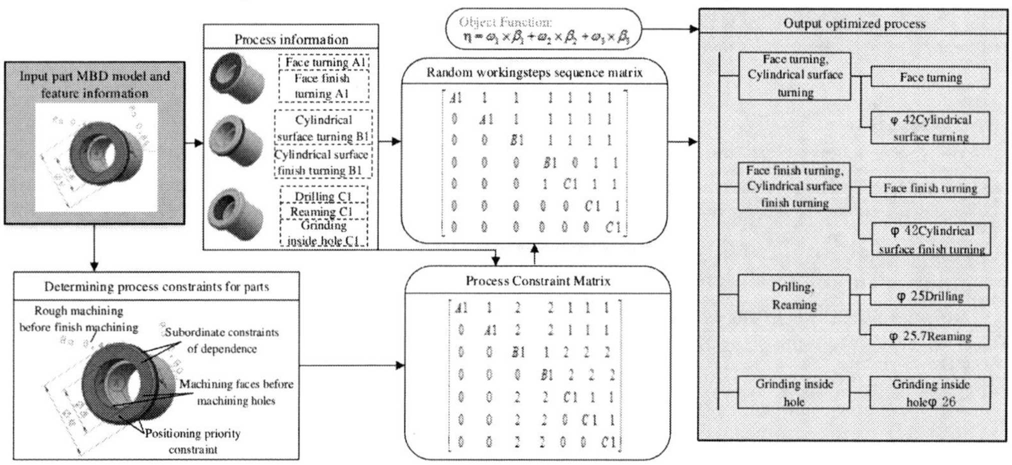

Figure 1. Complete flow chart of the method.

2.1. Objective function and gene coding

In order to meet the requirements of the minimal manufacturing cost, it is necessary to ensure the minimum replacement of process resources. Therefore, this paper aims to reduce the replacement of tools, machines and fixtures.

The change factor $\beta = \frac{real(the\ number\ of\ steps\ corresponding\ to\ the\ unchanged\ process\ resources)}{max(the\ number\ of\ steps\ corresponding\ to\ the\ unchanged\ process\ resources)}$ is given to quantitatively express the influence of each optimization objective. The denominator Max represents the maximum value of the number of working steps corresponding to the unchanged process resources, that is, the difference of the total number of working steps minus the number of corresponding process resources required for processing. The molecular Real represents the actual value of the number of working steps corresponding to the unchanged process resources, that is, the difference of the total number of working steps minus actual replacement times of corresponding process resources.

The change factors of tool, machine and fixture are set as β_1, β_2, and β_3 respectively. For comprehensive consideration, the objective function given is the weighted sum of three change factors, i.e. the objective function $\eta = \omega_1 \times \beta_1 + \omega_2 \times \beta_2 + \omega_3 \times \beta_3$, and each weight can be determined by the importance between the two elements by Analytical Hierarchy Process(AHP).

The natural digital chain gene coding is used to represent the workingsteps sequence which has many factors and mutual influence. In order to meet the minimum manufacturing cost constraint, it is necessary to consider the replacement times of tool, machine and fixture at the same time. Therefore, these three factors are directly displayed in each gene of chromosome, that is, gene is composed of four parts: working step number, machine number, fixture number and tool number. The examples of gene coding for working steps are shown in figure 2.

The examples of gene coding for working steps

Workingsteps	Features	Machines	Fixtures	Tools	**Genes**
Step1	A1	M01	F01	T01	01010101
Step2	A1	M01	F01	T01	02010101
Step3	B1	M01	F01	T01	03010101
Step4	B1	M01	F01	T01	04010101
Step5	C1	M02	F02	T02	05020202
Step6	C1	M02	F02	T03	06020203
Step7	C1	M03	F03	T04	07030304

Figure 2. The examples of gene coding for working steps.

2.2. Workingsteps sequence matrix and process constraint matrix

It is necessary to consider not only the influence of minimum manufacturing cost, but also the constraint of precedence relation and process experience between machining features to ensure manufacturability. The precedence constraints between features mainly include two aspects: process constraints and geometric constraints. Among them, process constraints mainly include benchmark constraint, location constraint, clamping constraint and non-destructive constraint. Geometric constraints mainly include the topological relation constraint and subordination relation constraint. Furthermore, the process experience constraints include: machining faces before machining holes, rough machining before finish machining and main features before secondary features.

The discrete constraints can't restrict the workingsteps sequence clearly and intuitively, so the concept of workingsteps sequence matrix is put forward. An N-order matrix is established to quantify the order relations among N working steps. The value of each element is determined by the order of the working steps corresponding to the numbers of row and column. For example, the element M_{ij} represents the sequence relation between step i and step j. When the working step i precedes the working step j, $M_{ij} = 1$. When the working step i is after the working step j, $M_{ij} = 0$. When there is no clear sequence relation between the two working steps, $M_{ij} = 2$. Since the main diagonal element only represents the working step itself, it is set as the machining feature identification to avoid self-contradictory.

Process constraint matrix can be regarded as a special workingsteps sequence matrix. It takes the constraint rules as the premise input, obtains the precedence relation table according to the constraints, and determines the values of corresponding elements according to precedence relation table and the concept of workingsteps sequence matrix. The process constraint matrix is used to determine the initial population and to screen the chromosomes after mutation. The chromosomes in the population are transformed into the corresponding workingsteps sequence matrix according to the order between the working steps, and then the chromosomes can be determined whether they meet the precedence constraints by comparing the element values with the process constraint matrix.

2.3. Chromosome selection and crossing

A set of initial population which meets the constraint requirements can be obtained by comparing the process constraint matrix. By calculating the objective function and combining with roulette selection method, the preliminary selection of workingsteps sequences can be completed. The chromosomes are crossed and mutated to obtain the global optimization results. The mutation operation is realized by the direct exchange of two random working steps, and the mutation probability is smaller. Compared with mutation operation, crossover operation has higher probability and more flexibility. In order to avoid repeatedly using the process constraint matrix for testing, Subtour Exchange Crossover is used for reference. As shown in figure 3, select a group of genes in parent 1, and then find these genes in parent 2. Keep the unselected genes in parent 1 unchanged, and replace the selected genes in parent 1 according to the sequence of these genes in parent 2 to get the offspring.

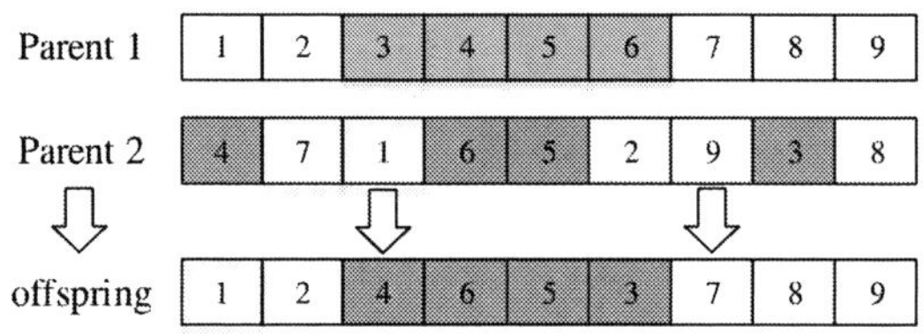

Figure 3. The example of chromosome crossing.

3. Case study

The marine diesel engine connecting rod is taken as the research object to prove the operability of the method in this paper. According to the coding rules and specific machine, fixture and tool for workingsteps, the gene coding of all the machining features of the connecting rod is given, as shown in figure 4.

Workingsteps	Features	Machines	Fixtures	Tools	Genes
Step1	Top surface	M01	F01	T01	01010101
Step2	Top surface	M02	F01	T02	02020102
Step3	Bottom surface	M01	F01	T02	03010102
Step4	Bottom surface	M02	F01	T02	04020102
Step5	Big end hole	M03	F02	T03	05030203
Step6	Big end hole	M03	F02	T04	06030204
Step7	Small end hole	M03	F02	T05	07030205
Step8	Small end hole	M03	F02	T06	08030206
Step9	Left hole	M04	F03	T07	09040307
Step10	Left hole	M04	F03	T08	10040308
Step11	Right hole	M04	F03	T09	11040309
Step12	Threaded hole	M04	F03	T010	12040310
Step13	Countersink hole	M04	F03	T011	13040311
Step14	Arc of radius 250	M02	F04	T012	14020412
Step15	Arc of radius 250	M02	F05	T013	15020513
Step16	Arc of radius 40	M02	F06	T014	16020614
Step17	Side profile	M02	F07	T015	17020715
Step18	Oil groove	M02	F08	T016	18020816

Figure 4. Gene coding of workingsteps of connecting rod.

Using the genetic algorithm, the workingsteps sequence with the minimum objective function can be found. Taking the connecting rod as an example, the optimal workingsteps sequence is shown in figure 5. Among them, the number of machine replacement is 4, the number of fixture replacement is 7, and the number of tool replacement is 15.

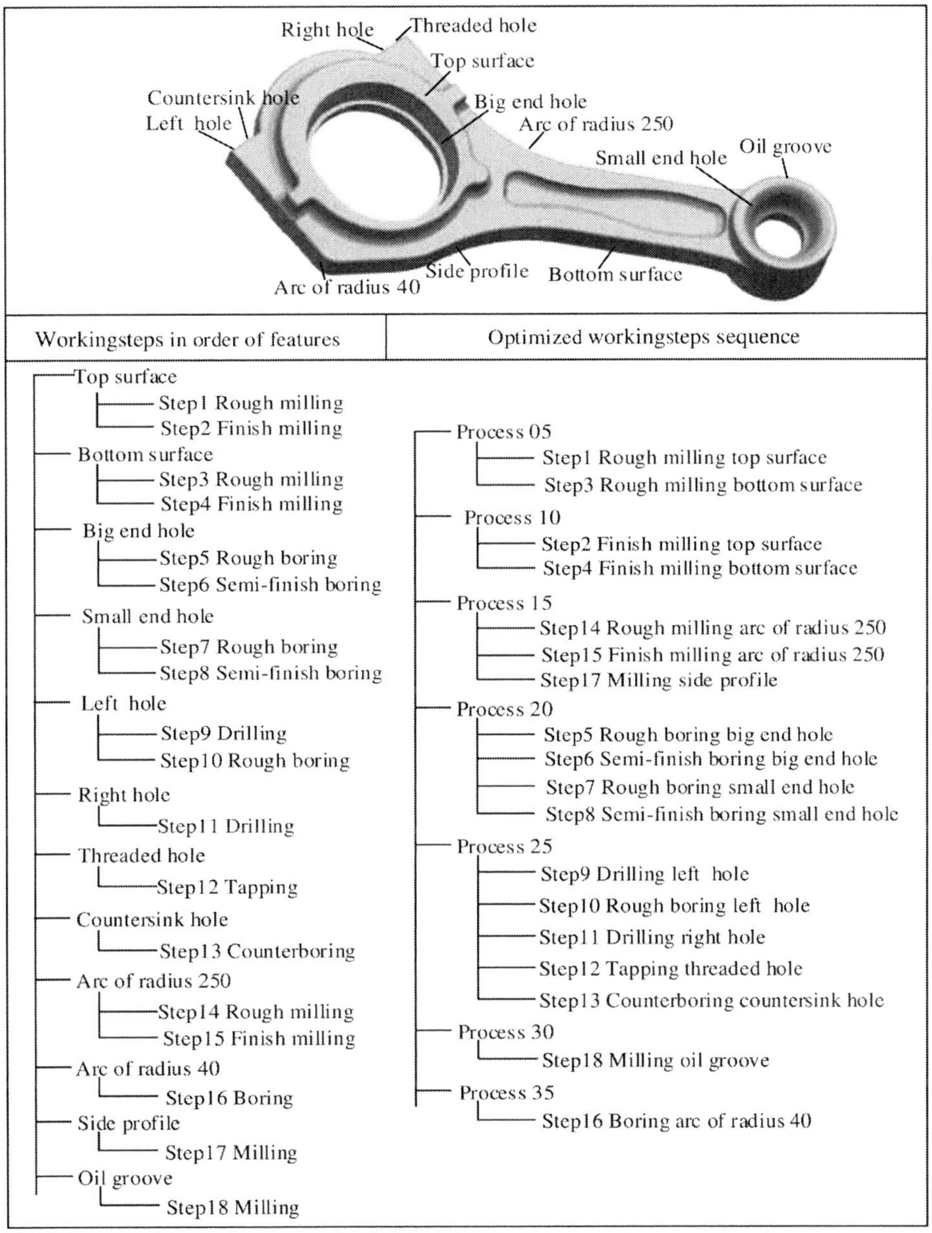

Figure 5. Optimal workingsteps sequence of connecting rod.

4. Conclusions

There are many factors that need to be considered in the operation sequencing, and there are mutual relations among them, so it has become a hot topic worthy of study. In this paper, based on the parallel consideration of process constraint and minimum manufacturing cost, the method of process constraint matrix combined with genetic algorithm is adopted to obtain the optimal solution of workingsteps sequence. And the connecting rod of marine diesel engine is taken as an example to verify. The result of the process sequence is basically consistent with the process sequence compiled by the factory based on many year experience.

References

[1] Hu Q, Qiao L, and Peng G 2017 An ant colony approach to operation sequencing optimization in process planning *Proceedings of the Institution of Mechanical Engineers, Part B: Journal of Engineering Manufacture* **231** 470-489

[2] Li A, Zhu J, Lu J, and Liu X 2016 Cooperative optimization of machining scheme selection and operation sequencing for engine cylinder *Journal of Tongji University* **44** 1084-1090

[3] Dou J, and Zhao X 2017 A feasible sequence oriented discrete particle swarm optimization for operation sequencing in CAPP *ASME 2017 International Design Engineering Technical Conferences and Computers and Information in Engineering Conference* (Cleveland, OH, United states)

[4] Dou J, Li J, and Su C 2018 A discrete particle swarm optimisation for operation sequencing in CAPP *International Journal of Production Research* **56** 3795-3814

[5] Nallakumarasamy G, Srinivasan P S S, Venkatesh Raja K, and Malayalamurthi R 2011 Optimization of operation sequencing in CAPP using simulated annealing technique (SAT) *International Journal of Advanced Manufacturing Technology* **54** 721-728

[6] Qiao L, Hu Q, and Zhang H 2012 Machining sequencing optimization based on simulated annealing algorithm *3rd international Conference on Manufacturing Science and Engineering* (Xiamen, China) pp 1632-1636

[7] Phung L X, Van Tran D, Hoang S V, and Truong S H 2017 Effective method of operation sequence optimization in CAPP based on modified clustering algorithm *Journal of Advanced Mechanical Design, Systems and Manufacturing* **11**

[8] Sundstrom N, Wigstrom O, Falkman P, and Lennartson B 2012 Optimization of operation sequences using constraint programming *14th IFAC Symposium on Information Control Problems in Manufacturing* (Bucharest, Romania) 1580-1585

[9] Fok S C, Thimm G, and Britton G 2010 Application of neural network and evolutionary algorithm in operation sequencing using datum hierarchy trees *International Journal of Machining and Machinability of Materials* **8** 114-130

[10] Dou J, Zhao X, and Su C 2018 An improved genetic algorithm for optimization ofoperation sequencing *15th IEEE International Conference on Mechatronics and Automation* (Changchun, China) pp 695-700

[11] Dou J, Li J, and Su C 2019 Optimization of operation sequencing based on feasible operation sequence oriented genetic algorithm *Computer Integrated Manufacturing Systems* **25** 1981-1990

[12] Liu L, and Qiao L 2012 Operation sequencing using genetic algorithm *9th International Conference on History of Mechanical Technology and Mechanical Design* (Tainan, Taiwan) pp 57-61

[13] Su Y, Chu X, Che D, and Chu D 2014 A modified genetic algorithm for precedence constrained operation sequencing problem in process planning *2014 IEEE International Conference on Industrial Engineering and Engineering Management* (Selangor, Malaysia) pp 84-88

[14] Su Y, Chu X, Chen D, and Sun X 2018 A genetic algorithm for operation sequencing in CAPP using edge selection based encoding strategy *Journal of Intelligent Manufacturing* **29** 313-332

[15] Su Y, Chu X, Sun X, and Tang C 2010 Worksteps sequencing and optimization of mill/turn machining center based on improved genetic algorithm *Nongye Jixie Xuebao/Transactions of the Chinese Society of Agricultural Machinery* **41** 216-222

IOP Conf. Series: Materials Science and Engineering **751** (2020) 012076 doi:10.1088/1757-899X/751/1/012076

Availability Assessment of Avionics Display System based on MBSA using Fault Dependent Matrix

Haiyong Dong[1,*], Zhiyuan Cao[2], Zhengjun Zhai[1], Qingfan Gu[3], Yanhong Lu[1], Guoqing Wang[1,4] and Miao Wang[4]

[1]School of Computer Science and Engineering, Northwestern Polytechnical University, Xi' an, 710072, China
[3]University of Michigan – Shanghai Jiao Tong University Joint Institute, Shanghai, 200240, China
[3]China National Aeronautical Radio Electronics Research Institute, Shanghai, 200241, China
[4]School of Aeronautics and Astronautics, Shanghai Jiao Tong University, Shanghai, 200240, China

[*]Corresponding e-mail: donghaiyong@mail.nwpu.edu.cn

Abstract. The availability of avionics display systems plays an important role in flight safety and requires a safety analysis during the design phase. The current safety analysis process is not carried out based on the system design model, so there may be inconsistencies. This paper proposes a model-based safety analysis method using a fault dependent matrix, which can be applied to commonly used system modelling tools such as Enterprise Architect. This article illustrates the utilization of the method by providing the elapsed flight time function, using MATLAB to generate the fault dependent matrix, and then obtaining the availability of the function. This method can be used to aid system safety design and improve system design efficiency.

1. Introduction

The avionics display system is an important system on the aircraft, providing information for pilots. To meet the airworthiness needs, the designation process and the development process of the display system need to follow the 4754 process [1]. During the system development process, safety analysis such as Preliminary System Safety Assessment (PSSA) and System Safety Assessment (SSA) is required, and there are multiple iterations in these processes. In the industry practice, tools such as EA are commonly used for system modelling while Relex is used for safety analysis [2-3]. This process is inefficient and may result in inconsistent system designation models and system safety models.

To solve the above problem, many researchers tried to change the system model directly to make it a system design model with safety properties, known as Model-based Safety Analysis (MBSA) [4]. Dong used Simfia to model the IMA network, and then wrote a program to analyse the availability of the exported model in the XML format [5]. Zhang once studied the safety of Head Up Display (HUD), and construct the safety model based on the system designation model to XML [6]. However, the safety analysis method of the above research is mainly the method of 4761, including Fault Tree Analysis (FTA), Dependence Diagram (DD), Markov Analysis (MA), Petri Net (PN), etc. [7]. Traditional analytical methods can hardly meet the practice needs of system safety analysis. Many researchers have expanded these methods. Due to October 2016, more than 847 safety analysis methods have been proposed and widely used in nuclear power, aviation, aerospace, etc. [8]. We proposed a safety analysis method based on a fault dependent matrix [9]. This method combines the advantages of FTA and PN methods. It can obtain the cut set using the top-down method and analyse the fault propagation process using the bottom-up method. This method has been expanded, but the model modelling process is always separated from the system design process and easily leads to inconsistencies [10-11].

This paper proposes an availability analysis method that satisfies MBSA based on a fault dependent matrix. The following structure of this paper is as follows. Section II introduces the availability analysis method based on MBSA using fault dependent matrix. Section III takes the availability of the display system as an example to explain the usage process of the method. The last Section summarizes this paper and proposes the future study.

2. Availability Assessment Method Based on MBSA using Fault Dependent Matrix

Safety analysis mainly includes availability and integrity analysis. The availability analysis is to analyse whether the system or function can provide the service, while the integrity refers to whether the service provided by the system is correct. The quantitative representation of the two is usually expressed by the indicator which represents the service is not available [7]. The safety analysis method based on the fault dependent matrix can be used to construct the safety model and evaluate the quantity probability of the availability of the specified function.

EA is a widely recognized graphical system modelling tool in the industry, its elements (Block, Port, etc.) support adding custom Tagged Value, and the system model can be exported to XML and other formats. Traditional safety analysis method does not perform the safety analysis process based on the system design model. This paper proposes adding safety related attributes directly to the system design model in EA and then perform the availability analysis. The method flow is shown in the figure below.

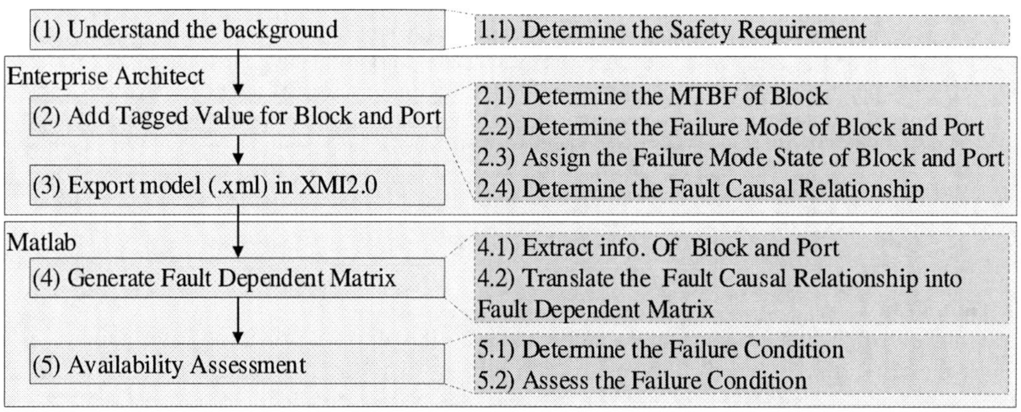

Figure 1. Availability assessment process based on MBSA using fault dependent matrix.

As shown in figure 1, the safety requirement is the goal of safety analysis, that is, to determine whether the system designation meets the safety requirement of the system. If so, then continue; if not, the system designation needs to be revised. Adding the safety information to the system design model in EA directly can ensure the consistency between the safety model and the system designation. The Tagged Value added to the Block and Port in the model is slightly different, as shown in the table 1.

Table 1. Tagged Value for safety model in EA.

Tagged Value	Description	Detail [a]	Block	Port
FailureMode	The Failure Mode of the component	Type=String; Default=Loss of the ability to do sth.;	☑[b]	☑
FailureModeState	The working status of the	Type=Enum;	☑	☑

	component	Values=Normal,Loss; Default=Normal;		
MTBF	The Mean Time Between Failure of the component	Type=Custom; Mask=dddddD; Template=___,___FH;	☑	☒[c]
FaultCausalRelationship	The Fault Causal Relationship among the Failure Mode	Type=String; Default=FM(device) AND/OR FM() LEAD FM(self);	☒	☑

[a] Each Tagged Value could be predefined its structure type before utilization [12].
[b] ☑ means the Tagged Value should be defined in the element.
[c] ☒ means the Tagged Value should not be defined in the element.

The fault causal relationship describes the causal relationship between failure modes. In this paper, all cause relationships are static fault causal relationships, which can be expressed as causality. The failure mode on the left side of LEAD can be device and port, and the self on the right side indicates the failure mode status of the port.

Suppose a port output is normal, if and only if the device and port inputs are normal, then the fault causal relationship will be expressed as FM(device) OR FM(inPort) LEAD FM(self), which can be expressed as fault dependent matrix as follows.

$$M_{FPM} = \begin{bmatrix} V_{FPM\_1} \\ V_{FPM\_2} \end{bmatrix} = \overbrace{\begin{bmatrix} -1 & 0 & 1 \\ 0 & -1 & 1 \end{bmatrix}}^{FM(device)\quad FM(inPort)\quad FM(self)} \tag{1}$$

Since the fault causal relationship is an OR relationship, it is split into multiple fault causal relationships. Where -1 indicates the source of the fault, 1 indicates the affected state of the fault, and 0 does not have meaning.

After obtaining the fault dependent matrix, the fault dependent matrix algorithm can be used to obtain the cut set of the failure condition. The cut set is composed of several device failure modes, and then the quantitative evaluation of the failure condition is obtained.

3. Case Study

3.1. Avionics Display Model

The current avionics display system of civil aircraft is based on the IMA architecture. The RDIU is responsible for collecting information such as sensors. The GPM with multiple hosted functions is responsible for processing information. The Switch is responsible for transferring data. The IDU is responsible for displaying the data in the form of graphic images for pilots. The examples in this paper have been simplified to illustrate the utilization of the method. The simple display system design architecture for displaying "Elapsed Flight Time" is shown in figure 2.

The *Elapsed Flight Time* displaying function collects the UTC Time provided by the flight management system and the wheel signal of the landing gear. After processing by GPM, the *Elapsed Flight Time* has been obtained and displayed to the pilot.

3.2. Assumptions

To simplify the system model, this paper takes several common assumptions.

- Failure will only occur on the devices and will not occur on external systems, cables, and ports;
- The failure mode state of the device can only be classified into two categories, one is Normal and the other is Loss;
- The devices are independent of each other, and the failure rate remains unchanged during use.

3.3. System Model with Safety Tagged Value

Add safety information to the model in figure 2, that is, assign the value to the Tagged Value of Blocks and Ports, as shown in table 2 and table 3.

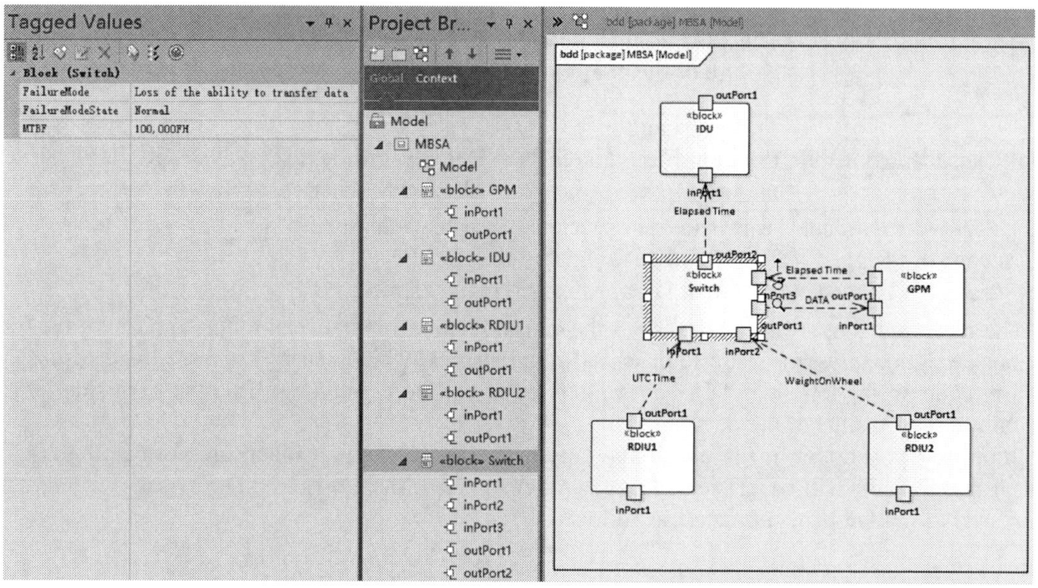

Figure 2. Physical architecture of display function "Elapsed Flight Time".

Table 2. Function and safety information of all Blocks in the model.

Component	Function	Failure Mode (Device Type)	FM No.	MTBF
GPM	Data Process	Loss of the ability to process data	1	50,000FH
IDU	Data Display	Loss of the ability to display information	2	16,000FH
RDIU1	Data Source collection	Loss of the ability to provide data	3	14,000FH
RDIU2	Data Source collection	Loss of the ability to provide data	4	14,000FH
Switch	Data Transfer	Loss of the ability to transfer data	5	100,000FH

All failure modes (FM) are numbered in table 2. The failure modes and fault causal relationship (FCR) are also numbered in the following table. These numbers are not stored in Tagged Value. All FMs and FCRs are numbered by MATLAB programs.

Table 3. Function and safety information of all Ports in the model.

Port	Failure Mode (Function Type)	FM No.	Fault Causal Relationship	FCR No.
GPM: inPort1	Loss of the ability to obtain source data from Switch	6	FM(Switch: outPort1) LEAD FM(self)	1
GPM: outPort1	Loss of the ability to provide source data to GPM	7	FM(device) OR FM(inPort1) LEAD FM(self)	2
IDU: inPort1	Loss of the ability to obtain processed data from Switch	8	FM(Switch: outPort2) LEAD FM(self)	3
IDU: outPort1	Loss of the ability to provide information (Elapsed Flight Time) for the crew	9	FM(device) OR FM(inPort1) LEAD FM(self)	4
RDIU1: inPort1	Loss of the ability to obtain source data (UTC Time)	10	/ [a]	/
RDIU1: outPort1	Loss of the ability to provide source data to Switch	11	FM(device) LEAD FM(self) [b]	5
RDIU2: inPort1	Loss of the ability to obtain source data (WeightOnWheel)	12	/	/

IOP Conf. Series: Materials Science and Engineering **751** (2020) 012076 doi:10.1088/1757-899X/751/1/012076

RDIU2: outPort1	Loss of the ability to provide source data to Switch	13	FM(device) LEAD FM(self)	6
Switch: inPort1	Loss of the ability to obtain source data from RDIU1	14	FM(RDIU1:outPort1) LEAD FM(self)	7
Switch: inPort2	Loss of the ability to obtain source data from RDIU2	15	FM(RDIU2:outPort1) LEAD FM(self)	8
Switch: inPort3	Loss of the ability to obtain processed data from GPM	16	FM(GPM: outPort1) LEAD FM(self)	9
Switch: outPort1	Loss of the ability to provide source data to GPM	17	FM(device) OR FM(inPort1) OR FM(inPort2) LEAD FM(self)	10
Switch: outPort2	Loss of the ability to provide processed data to IDU	18	FM(device) OR FM(inPort3) LEAD FM(self)	11

[a] Since external systems are assumed to keep working, the INPUT of both RDIUs could keep normal.

[b] Normally, the state of the output depends on the input, but since the input is always normal, it is simplified.

3.4. Fault Dependent Matrix Generation

MATLAB parses the XML file describing the safety model, obtains all the failure modes and fault causal relationships of the system, and then expresses the fault causal relationship one by one in the matrix. Take FCR#1 as an example. The fault causal relationship of GPM: inPort1 is FM(Switch:outPort1) LEAD FM(self), the FM number of Switch:outPort1 is 17, and the FM number of GPM:inPort1 is 6. Then the fault dependent matrix is shown as follows.

$$V_{FPM\_1} = \overbrace{\begin{bmatrix} \cdots & 1 & \cdots & -1 & \cdots \end{bmatrix}}^{\cdots \quad 6 \quad \cdots \quad 17 \quad \cdots} \tag{2}$$

Similarly, taking FCR#2 as an example, the fault causal relationship of GPM:outPort1 is FM(device) OR FM(inPort1) LEAD FM(self), the FM number of GPM is 1, and the FM number of GPM:inPort1 is 6. GPM: The FM number of outPort1 is 7, and the fault dependent matrix is shown as follows.

$$\begin{bmatrix} V_{FPM\_2} \\ V_{FPM\_3} \end{bmatrix} = \overbrace{\begin{bmatrix} -1 & \cdots & 0 & 1 & \cdots \\ 0 & \cdots & -1 & 1 & \cdots \end{bmatrix}}^{1 \quad \cdots \quad 6 \quad 7 \quad \cdots} \tag{3}$$

Since the rows of the same fault cause can be merged, the fault dependent matrix of the system is obtained as follows.

$$M_{FPM} = \begin{bmatrix}
0 & 0 & 0 & 0 & 0 & 1 & 0 & 0 & 0 & 0 & 0 & 0 & 0 & 0 & 0 & 0 & -1 & 0 \\
-1 & 0 & 0 & 0 & 0 & 0 & 1 & 0 & 0 & 0 & 0 & 0 & 0 & 0 & 0 & 0 & 0 & 0 \\
0 & 0 & 0 & 0 & 0 & -1 & 1 & 0 & 0 & 0 & 0 & 0 & 0 & 0 & 0 & 0 & 0 & 0 \\
0 & 0 & 0 & 0 & 0 & 0 & 0 & 1 & 0 & 0 & 0 & 0 & 0 & 0 & 0 & 0 & 0 & -1 \\
0 & -1 & 0 & 0 & 0 & 0 & 0 & 0 & 1 & 0 & 0 & 0 & 0 & 0 & 0 & 0 & 0 & 0 \\
0 & 0 & -1 & 0 & 0 & 0 & 0 & 0 & 0 & 0 & 1 & 0 & 0 & 0 & 0 & 0 & 0 & 0 \\
0 & 0 & 0 & -1 & 0 & 0 & 0 & 0 & 0 & 0 & 0 & 0 & 1 & 0 & 0 & 0 & 0 & 0 \\
0 & 0 & 0 & 0 & 0 & 0 & 0 & 0 & 0 & 0 & -1 & 0 & 0 & 1 & 0 & 0 & 0 & 0 \\
0 & 0 & 0 & 0 & 0 & 0 & 0 & 0 & 0 & 0 & 0 & 0 & -1 & 0 & 1 & 0 & 0 & 0 \\
0 & 0 & 0 & 0 & 0 & 0 & -1 & 0 & 0 & 0 & 0 & 0 & 0 & 0 & 0 & 1 & 0 & 0 \\
0 & 0 & 0 & 0 & -1 & 0 & 0 & 0 & 0 & 0 & 0 & 0 & 0 & 0 & 0 & 0 & 1 & 1 \\
0 & 0 & 0 & 0 & 0 & 0 & 0 & 0 & 0 & 0 & 0 & 0 & 0 & -1 & 0 & 0 & 1 & 0 \\
0 & 0 & 0 & 0 & 0 & 0 & 0 & 0 & 0 & 0 & 0 & 0 & 0 & 0 & -1 & 0 & 1 & 0 \\
0 & 0 & 0 & 0 & 0 & 0 & 0 & 0 & 0 & 0 & 0 & 0 & 0 & 0 & 0 & -1 & 0 & 1
\end{bmatrix} \tag{4}$$

There is a -1 and two 1 in the 11th line of equation (4), indicating that the loss of the switch can cause loss of both outport1 and outport2.

3.5. Analysis Results

As shown in table 2, the MTBF of the GPM is 50,000 FH, and the failure rate can be assumed to be constant 2E-5 / FH.

The function of this model is providing the "Elapsed Flight Time" for pilots. The failure mode FM#9 is the failure condition that needs to be analysed. Using the top-down analysis method, the cut set of failure modes that lead to the failure condition can be obtained, and the probability of failure condition is 2.354E-4 / HF. Based on the result, it is determined whether the safety requirement is met, and if it is not satisfied, the system designation needs to be improved.

4. Conclusions

This paper introduces an availability assessment method based on MBSA using fault dependent matrix, which ensures the safety model constructed could be consistent with the system designation model. Taking the availability analysis of providing *Elapsed Flight Time* for the pilot as an example, the EA model is built, the fault dependent matrix is generated by MATLAB program, and the availability is calculated, which can be to verify whether the safety requirements are met by the designed system.

However, this method is unable to deal with the availability analysis of the system design model with fuzzy fault causal relationship and time-related dynamic fault causal relationship, which will be further studied in the future.

Acknowledgments

The authors would like to thank Hao Rong for his technical assistance in the construction of the case model, Zilong He for his advice and assistance in proofreading the article.

References

[1] ARP4754A 2010 *Guidelines for Development of Civil Aircraft and Systems* (Pennsylvania: SAE International)
[2] Enterprise Architect. Available online: https://sparxsystems.cn/ (accessed on 29 August 2019)
[3] PTC Windchill Quality Solutions Software (Formerly Relex Software). Available online: http://www.relexsoftware.it/ (accessed on 29 August 2019)
[4] Gu Q, Wang G and Zhai, M 2014 *Proc. of the 14th AIAA Aviation Technology, Integration, and Operations Conference* (Atlanta: AIAA) p 2226
[5] Dong H, Gu Q, Wang G, Zhai Z, Lu Y and Wang M 2019 *Processes* **7** p 117
[6] Zhang F and Dong H 2019 *3rd Int. Conf. on Circuits, System and Simulation* (Nanjing: IEEE) pp 116-120
[7] ARP4761 1996 *Guidelines and Methods for Conducting the Safety Assessment Process on Civil Airborne Systems and Equipment* (Pennsylvania: SAE International)
[8] Everdij M and Blom H 2016 *Safety Methods Database. Version 1.1* (Netherlands: Netherlands Aerospace Centre NLR)
[9] Dong H, Gu Q, Wang G, Zhai Z, Lu Y and Wang M 2019 *Int. J. Performability Eng.* **15** p 2392-2399
[10] Dong H, Gu Q, Wang G, Zhai Z, Lu Y and Wang M 2019 *9th Int. Conf. on Electronics Information and Emergency Communication* (Beijing: IEEE) pp 1-5
[11] Dong H, Gu Q, Wang G, Zhai Z, Lu Y and Wang M 2019 *2nd Int. Conf. on Artificial Intelligence and Pattern Recognition* (Beijing: ACM Press) pp 196-200
[12] Predefined Structured Types. Available online: http://tool.uml.com.cn/ToolsEA/UserGuide/modeling_tools/predefinedtaggedvaluetypes.html (accessed on 29 August 2019)

IOP Conf. Series: Materials Science and Engineering **751** (2020) 012077 doi:10.1088/1757-899X/751/1/012077

Design of Image Recognition Algorithm for Critical Flight Parameters in Civil Aircraft Display System

Maodi Zhang*, **Yuanhui Zhou and Qingfan Gu**

China National Aeronautical Radio Electronics Research Institute, Shanghai, 200241, China

*Corresponding e-mail: zhang_maodi@careri.com

Abstract. In the current civil aircraft display system design, complex COTS devices such as CPU and GPU have been widely used. While improving product performance, it also greatly increases the difficulty of airworthiness certification. Taking the display system as an example, the data integrity of the critical flight parameters which is displayed on the display unit must reach 1.0E-7 or higher, and the development assurance level of each software and hardware module must reach level B or higher. However, according to industry experience, the data integrity of CPU and GPU can only reach the level of 1.0E-6. Therefore, for the critical flight parameters in the display system, it must be monitored to cover the entire data path from data processing to image generation to prove the data validity. Aiming at the numeric critical flight parameters of civil aircraft display system, this paper proposes a numeric image recognition technology realized by FPGA image processing function, which can be used to monitor the integrity of numeric critical flight parameters, and this technology has been performed and verified in real display unit.

1. Introduction

According to the current display system HMI design concept, the airspeed, barometric altitude which is from the flight control, and the engine parameters from the FADEC are generally in the form of numbers or letters, and will have a scrolling display effect. According to the safety assessment of display system, the display function of airspeed, barometric altitude and engine parameters shall be developed to a minimum of FDAL B or A, and the corresponding failure probability shall be 1E-7 or 1E-9 [1] [2]. In order to meet the safety requirements of these critical flight parameters (including DAL level and failure probability), each type of aircraft (such as A380, B787, etc.) will design corresponding monitoring means, it may be a cross monitoring between display unit [3], or a monitor of video processing channels inside a single display unit. The core of all the above monitoring methods is to obtain critical flight data from the current display screen, and compare and judge the data with the original data to ensure the integrity of the data.

The most direct and effective way to get data from the screen is through image recognition technology [4] [5]. Currently, the image recognition technology has been widely used in daily life, such as license plate recognition [6], face recognition [7] [8], etc. Therefore, this paper proposes an image recognition method based on pixel matching, which is used to identify the numeric critical flight parameters in the civil aircraft display unit, which can be used for the integrity monitoring of these data and improve the safety of the display system.

2. Design and Implementation

2.1. Technical Scheme

The technical scheme adopted in this paper is based on FPGA image processing and pattern recognition technology. It is based on the display characteristics of the numeric display parameters of the airborne cockpit display system, using the pixel normalization matching technology to identify the numeric critical parameters in the current display screen. Firstly, according to the display characteristics of the numeric flight parameters, the corresponding matching template needs to be developed. For example, the airspeed and barometric altitude are scrolling display, and the engine parameters are in direct change form. Each parameter needs to be templated according to the possible display state. Then, binarize the received image data to reduce the matched noise, and the corresponding region is intercepted according to the display position of each parameter. Finally, FPGA-based pixel block matching is used to compare the pixel values of the captured image with the display template line by line and select the number corresponding to the position with the largest similarity measure as the matching result.

In order to verify the results of image recognition in a real display unit, we designed a data comparison function in the FPGA. While identifying the critical flight parameters in the display, the FPGA also accepts the original data sent from the CPU via the SPI and compares the two data. If the result of the data comparison is within an acceptable threshold range, then the identified data is considered correct. The whole structure is shown in figure 1.

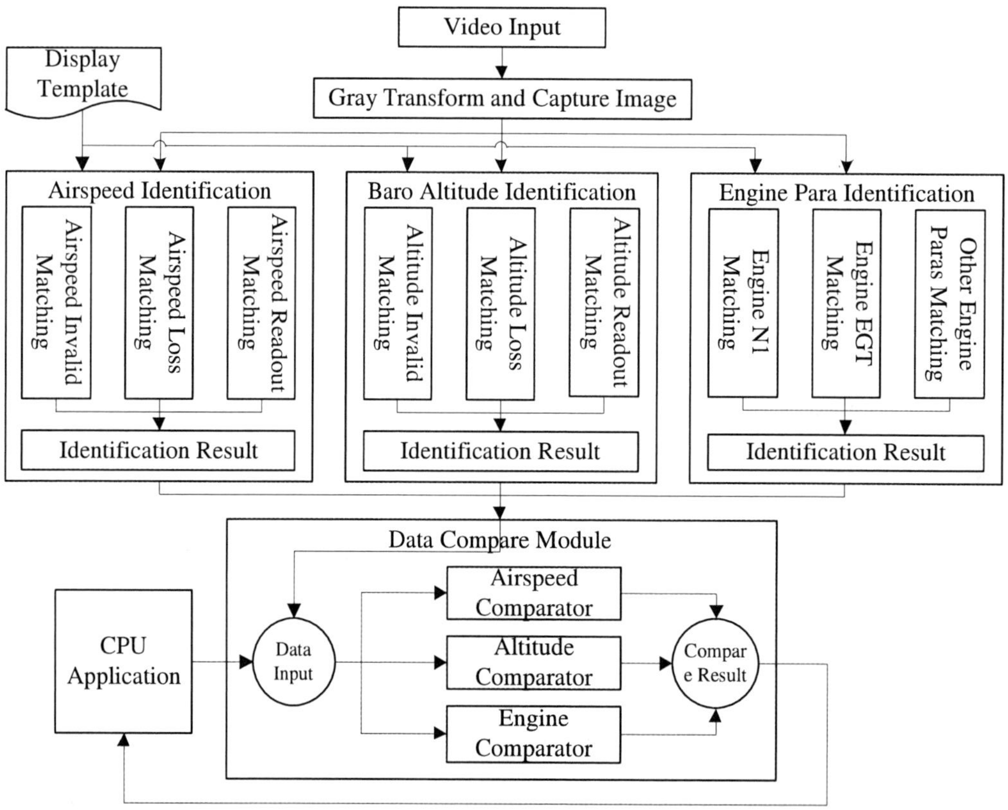

Figure 1. The verification architecture for the image recognition algorithm on display unit.

2.2. Implementation Details

When the airspeed, barometric altitude and engine critical parameters are displayed normally, the display number changes continuously from 0 to 9 according to the input. During the display of airspeed and barometric altitude, when the highest bit is 0, it is displayed as a grid line, and the barometric height may be negative. The display format and template is shown as figure 2.

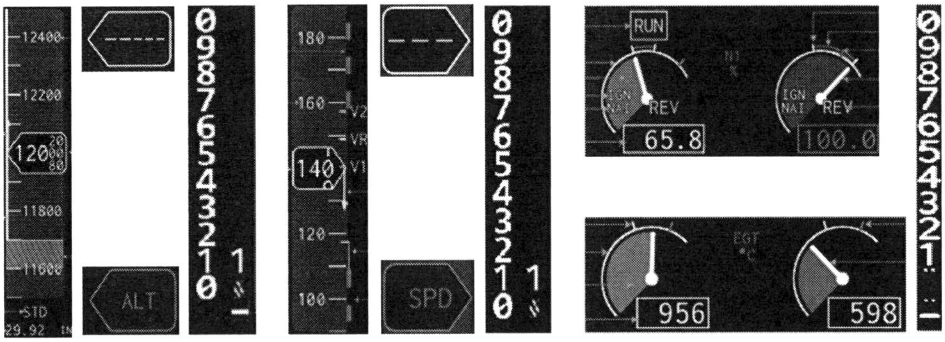

Baro Altitude Display Format and Template | Airspeed Display Format and Template | Engine Parameters Display Format and Template

Figure 2. Display format and template for numeric critical parameters.

Taking airspeed data as an example, after receiving the data input, the airspeed identification module intercepts the display areas of the digits, tens, and hundreds of digits, and then loads the respective display templates from the ROM. In the process of identifying each digit, it is necessary to instantiate one input data buffer RAM, one font template storage ROM, one matching module and one comparison positioning module. Then, the block image matching algorithm is used to match the input image data with the font template. The matching module adopts a parallel structure to improve the matching speed, and can complete the matching operation between the font template and the graphic data of 21 rows and 14 columns and each of the pattern shift registers inside the matching module has a tap number of 14 and a width of 8. During the matching operation, each unit registers one pixel of the graph, and shifting one unit to the right every other clock, to realize a point-by-point movement of a font template in an image, and then complete a real-time comparison between font template and image data. The arithmetic unit calculates the square of the error value between the image and the font template using (1), in which p_{ij}, q_{ij} represents the values of the corresponding pixels in the image block and the template. The main processing is shown as figure 3.

$$BM_{val} = \sum_{i,j} |p_{ij} - q_{ij}|^2 \tag{1}$$

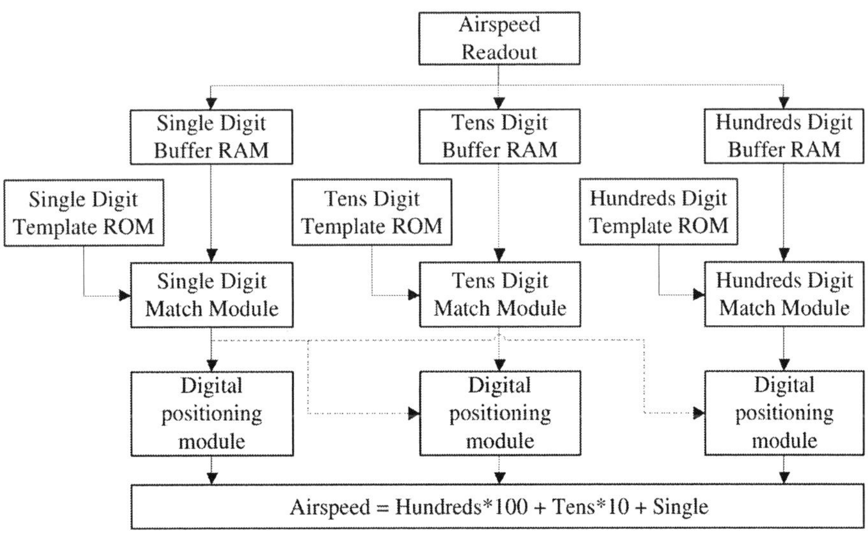

Figure 3. Airspeed identification process.

Since the digital display of airspeed and barometric altitude is scrolling, and it is possible that two consecutive numbers are incomplete, we have designed a digital positioning module. The digital positioning module is used to compare the calculation results of the matching metrics to determine the position of the best matching point. It has an internal input latching module which is used to latch the matching metrics to ensure data stability during operation. The comparison judging module compares the input data with the previous latched data and latches the smaller data for the next comparison. The address counting module is responsible for recording the number of matching metric values in the input latches participating in the comparison, so that the location of the font template search area corresponding to the metric value can be determined. When all the matching metrics are compared, what remains in the address latch is the position of the minimum matching metric in the entire sequence of matching metrics, thereby obtaining the position of the best matching point and converting it to the number of the corresponding data bit.

3. Experiments and Analysis

In order to verify the effectiveness of the technical method, we use simulation and real environment to implement verification activities.

3.1. Simulation Experiments

The simulation verification environment includes MATLAB and ModelSim. MATLAB simulation is mainly used for algorithm function verification in the early design stage. ModelSim simulation is mainly for FPGA code to verify that the FPGA code can meets the function and performance requirements.

The MATLAB simulation result is shown as figure 4:

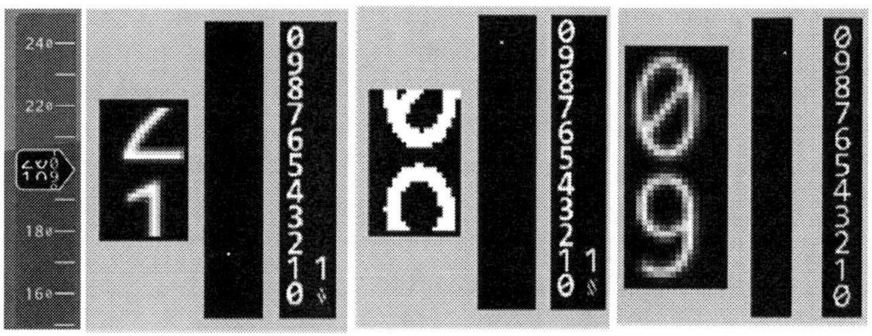

Figure 4. MATLAB simulation results.

On the left is part of the input video frame, we can see that the currently displayed airspeed number is between 199 and 200. The MATLAB simulation program gives the recognition result of each digit. In the recognition result of each digit, the left side represents the picture taken in the video frame, the right side represents the display font template, and the middle represents the position of the picture taken on the left side in the font template.

The ModelSim simulation result is shown as figure 5:

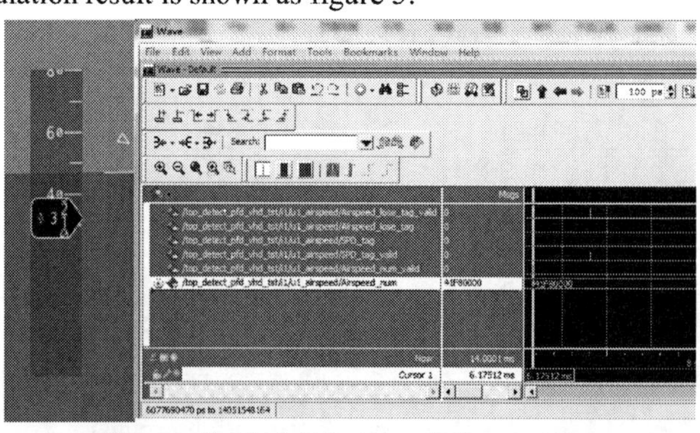

Figure 5. ModelSim simulation results.

Through the picture, we can see that the airspeed readout in the picture is 31, and the Modelsim simulation result of the image recognition algorithm is 41F80000, which is converted to a decimal number of 31, and this result is consistent with the airspeed displayed in the picture. Meanwhile, we can see that the calculation time of the airspeed recognition algorithm is about 6.18ms, considering the refresh frequency of the display frame is 60Hz, which is about 16.7ms. Therefore, the algorithm can complete data identification in the display time of one frame, and to realize the recognition of every display frame.

3.2. Experiments on Display Unit

The verification activities in the real display unit cannot be observed by us directly, so we designed a SPI protocol between FPGA and CPU application, which is used for the CPU application sent original data to FPGA, and the FPGA return the compared result to CPU. We use the CPU as the master of the SPI bus to control the SPI clock and data transfer. As the Slave end, the FPGA accepts the original data and returns the comparison result according to the SPI clock given by the CPU.

3.2.1. Send frame data definition. The send frame data includes: frame header, airspeed data, airspeed invalid or loss flag, barometric altitude data, barometric altitude invalid or loss flag, wherein each part is a 32-bit data. The send frame format is as table 1 (Take airspeed as an example):

Table 1. Send frame format definition.

Definition	Type	Length (32 bit)	Note
Frame Head	UINT 32	1	Frame header information, can be defined by a 32bit like 0xFFFFFFF0; Send by CPU
Airspeed_Num	Float 32	1	32bit float data, ranging from 30 to 450. Send by CPU
SPD_Loss_Tag	UINT 32	1	32bit integer data, send by CPU: 1) If the airspeed is valid, set as 0x00000000; 2) If the airspeed is invalid, set as 0x00000001; 3) If the airspeed is loss, set as 0x00000002;

3.2.2. Receive frame data definition. The received frame data includes: a frame header, an airspeed data comparison result, a barometric altitude data comparison result, an engine parameter comparison result, wherein each part is a 32-bit data. The received frame format is as table 2 (Take airspeed as an example):

Table 2. Receive frame format definition.

Definition	Type	Length (32 bit)	Note
Frame Head	UINT 32	1	Frame header information, can be defined by a 32bit like 0xFFFFFFF1; Send by CPU
Airspeed_Result	Float 32	1	32bit integer data, send by FPGA. 1) If the comparison data is normal, set as 0x00000000; 2) If the comparison data is error, set as 0x00000001; 3) If the comparison data is fault flag, set as 0x00000002;

The data comparison between original data and identified data need a reasonable threshold. We set the threshold for airspeed as 5, and the threshold for altitude as 20. During the FPGA operation, the SPI data and identified data can be captured through SignalTap. The result is shown as figure 6:

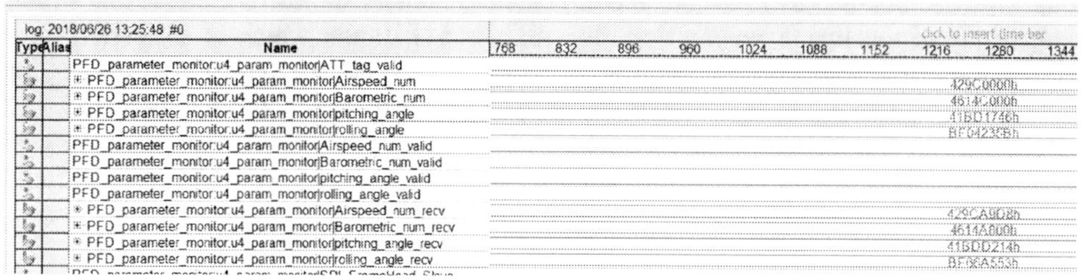

Figure 6. SignalTap result of experiment on display unit.

Table 3. Data compare result.

Parameter	Recognition Result	SPI Input
Airspeed	78 (429C0000)	78.33 (429CA9D8)
Altitude	9520 (4614C000)	9514 (4614A800)

According to the recognition result obtained by the capture and the SPI input data shown in table 3, the data of the two are basically consistent in consideration of the delay of the data transmission and the recognition accuracy. This also proves the effectiveness of this monitoring function.

4. Conclusion

This paper proposes an image recognition method based on FPGA pixel matching, which can be used to identify digital critical flight parameters in civil aircraft display systems. After tens of thousands of simulation tests, and dozens of consecutive tests in the display unit, the airspeed maximum recognition error does not exceed 1, and the barometric altitude maximum recognition error does not exceed 20. This error is acceptable for safety requirement. So this image recognition method is effective for critical flight parameters in civil aircraft display system.

References

[1] AC 25-11A 2007 Electronic Flight Deck Displays (Washington D.C.: FAA)
[2] Dong H, Gu Q, Wang G, Zhai Z, Lu Y and Wang M 2019 Processes 7 p 117
[3] TC 15/62 2016 Single Event Effects Mitigation Techniques Report (Washington D.C.: FAA)
[4] McAndrew A 2004 Introduction to Digital Image Processing with MATLAB (Boston: Victoria University of Technology)
[5] Seo J and Yoo C 2006 IEEE Trans. Signal Process 54 pp. 1537-1549
[6] Wang F, Zuo W, Lin L, Zhang D and Zhang L 2016 Proc. of the IEEE Conf. on Computer Vision and Pattern Recognition (Las Vegas: IEEE) pp. 1288-1296
[7] Boles W and Boashah B 1998 IEEE Trans. on Signal Processing 46 pp. 1185-1188
[8] Jain A, Bolle R and Pankanti S 2006 Biometrics: personal identification in networked society (New Jersey: Springer Science & Business Media)

IOP Conf. Series: Materials Science and Engineering **751** (2020) 012078 doi:10.1088/1757-899X/751/1/012078

Performance Simulation Analysis of Composite Thermal Management System for Hypersonic Vehicle

Yang Changbao[1,a], Zhou Hui[2,b] and Zhang Jingyang[2,*,c]

[1] Key Laboratory of Integrated Aviation Science and Technology for Aeronautical Mechatronic Systems, Jiangsu, Jiangsu Nanjing, 210016, China
[2] Nanjing University of Aeronautics and Astronautics, Jiangsu, Nanjing, 210016, China

Email:[a]1325542899@qq.com; [b]zhouhuihui0302@foxmail.com; [c,*]zjy@nuaa.cn

Abstract. A hybrid thermal management system model of hypersonic vehicle with liquid hydrogen as heat sink and thermoelectric conversion of Brayton cycle was established. The cooling effect, fuel flow demand and thermoelectric conversion characteristics of the thermal management system in flight envelope were studied by numerical simulation. The results show that the cooling fuel demand of refrigerant loop is about 0.22g/kW·s~0.24g/kW·s, that of aerothermal cooling fuel is about 0.11g/kW·s, and that of engine wall cooling fuel is about 0.08g/kW·s. The flow rate of engine wall cooling fuel can be saved by 25%~40% during thermoelectric conversion. Thermoelectric conversion cycle efficiency and generation efficiency can reach up to 38% and 30%. When Mach number is 3.7~4.2 under the condition of large aerodynamic heat load, the flow rate required for engine combustion is not enough to meet the cooling flow rate, and the maximum difference is 10%.

1. Introduction

The liquid hydrogen fuel, as it has the characteristics of high energy content and low density, is the most ideal booster for the hypersonic vehicle in recent time[1]. The boiling point of liquid hydrogen is 20K, and the upper limit temperature of hydrogen entering engine combustion is about 1033K[2]. Therefore, liquid hydrogen has a large heat capacity and can be used as heat sink of hypersonic vehicle thermal management system[3].Lei Yikun[4] proposed a scheme of integrated thermal/energy management system with combined turbine power plant as the core, and gave the system structure and overall working mode. However, the influence of fuel as heat sink on fuel system is not considered in the scheme. Gasner[5] et al. have simulated and evaluated the thermal management system of air-breathing hypersonic vehicle. Fuel is used to cool the airframe, engine and high temperature air. The result shows that the fuel radiator capacity can only meet the cooling requirements of hypersonic vehicle with Mach 5.5 during a specific design mission. With the development of hypersonic vehicle technology, especially the increase of flight Mach number, the aerodynamic thermal strength of aircraft body surface is getting higher and higher. Zhu Chunling[6] and others have raised an active and passive thermal protection structure for hypersonic vehicle based on the document[7]. C-C composite material structure is used as passive thermal protection measures in the outer layer of aircraft, and liquid hydrogen fuel is used as heat sink in the inner layer. The analysis shows that using liquid hydrogen fuel as heat sink can reduce fuel compensation loss. Aiming at the aerodynamic thermal cooling problem of hypersonic vehicle, Wang Peiguang[8] et al. proposed a thermal management scheme of "liquid hydrogen fuel as main heat sink and phase change heat storage material as auxiliary heat sink". The research shows that this scheme has good adaptability to high dynamic thermal environment at high Mach number. In order to improve fuel economy and engine cooling efficiency of aircraft, Glickstein[9] et al. put forward a comprehensive thermal management scheme for centralized management of fuel heat sink and engine cooling air by thermal cycle. Using fuel pre-cooled air to cool the engine can save the cooling air flow and increase the fuel calorific value. The results show that the specific fuel consumption is reduced by 0.4% and the fuel consumption rate is reduced by 1.6% under cruise condition. Balland[10] et al. proposed a heat/energy integrated thermal management system for MR2 aircraft with liquid hydrogen as heat sink. Cooling of the body is carried out by means of low-temperature fuel evaporation of the propellant tank, followed by cooling of the other hot end components, and finally injected into the combustion chamber. The numerical simulation results show that the scheme can meet the aerodynamic heat and engine cooling requirements of MR2 aircraft at Mach 5~8.Matthew[11] et al. proposed a scheme of Brayton cycle thermoelectric conversion technology, which can reduce the surface temperature of aircraft body while converting aerodynamic thermal energy into electrical energy. The numerical results show that when the heat exchanger efficiency reaches 0.95, the power generation efficiency of the system can reach 60%.Qin[12] et al. proposed a closed Brayton cycle heat/energy management system with scramjet wall heat load as the hot side and fuel as the heat sink to solve the problem of insufficient fuel heat sink of hypersonic vehicle. The numerical results show that the system can convert part of the heat into mechanical energy and electrical energy, and the maximum thermal cycle efficiency can reach 40%, and reduce the engine cooling demand for fuel flow.From the above research, it is not difficult to see that scholars mainly focus on the thermal management system scheme with fuel as heat sink and the thermoelectric conversion methods such as closed Brayton cycle, in order to solve the problem of efficient thermal/energy management of hypersonic vehicle. However, in the past studies, there is a lack of detailed discussion on the performance change rule of the heat management system scheme which combines cooling and thermoelectric conversion, and the feasibility study of the composite heat management system scheme is urgently needed in engineering.

This article is going to talk about the composite thermal management system for hypersonic aircraft, which uses liquid hydrogen fuel as coolant and establishes the idea of closed Brayton cycle. The numerical simulation method will be used to study the thermal management effect and thermoelectric conversion efficiency of the system under different flight modes and parameters in the flight envelope. The influence of thermoelectric conversion on the thermal management effect of the system and the fuel economy of the aircraft will be studied in order to provide a reference for the design of hypersonic vehicle engineering.

2. System Scheme and Model

2.1. System Solution

Figure 1 is the schematic diagram of the hypersonic vehicle thermal management system with liquid hydrogen fuel as heat sink[13]. The thermal management system uses cryogenic liquid hydrogen fuel as heat sink combined with closed Brayton cycle thermoelectric conversion scheme, which is mainly divided into four parts.In the first part, liquid hydrogen is used to cool the refrigerant loop. Driven by the pump, the refrigerant absorbs the thermal loads of the electronic cabin, hydraulic oil system and lubricating oil system, then enters the heat exchanger of the refrigerant loop, transfers heat to the cryogenic liquid hydrogen fuel, and releases the heat carrier back to the liquid collecting tank to complete a cycle. After absorbing the heat of the refrigerant loop, the fuel enters the combustion chamber of the engine. The second part depends on whether the thermoelectric conversion is carried out on the engine wall. In view of the stable temperature of the engine wall, which is benefit for the design of high efficient thermoelectric conversion, the thermoelectric conversion using the engine wall heat source is only considered in this scheme. In the first scheme, no thermoelectric conversion occurs, and the liquid hydrogen directly cools the engine wall and then enters the combustion chamber; in the second scheme, the thermal load on the engine wall is transformed into thermoelectric conversion before entering the combustion chamber[14]. Thermoelectric conversion adopts closed Brayton cycle. Low temperature and low pressure gas passes through compressor boost, then preheated by high temperature side fluid of heat exchanger, absorbs heat through engine wall, then enters turbine expansion to do work, exhaust gas is cooled by low temperature side fluid of heat exchanger, and then cooled by liquid hydrogen heat exchanger to the required inlet temperature of compressor. After that the gas will enter the compressor again, and complete the cycle.In the third part, liquid hydrogen enters the thermal protection system through the bypass connected with the pump, and directly cools the aerodynamic heat of the aircraft body. In the fourth part, when the fuel flow as heat sink can not generate enough thrust, the fuel in the fuel tank is directly transported to the engine by the bypass connected with the pump to compensate for the shortage of flow.

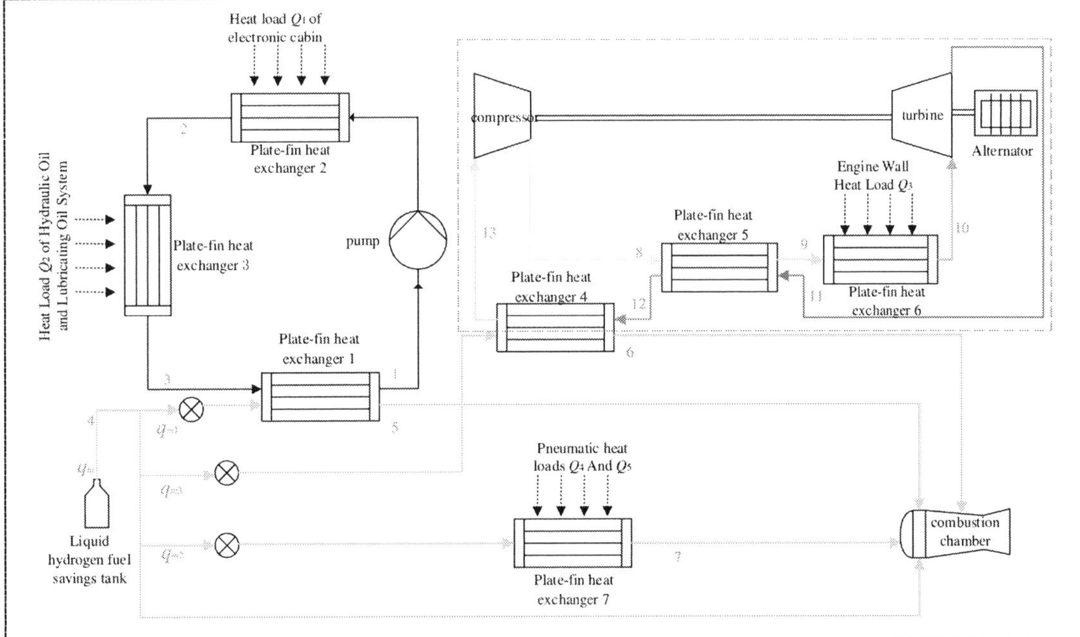

Figure 1. Schematic diagram of heat management scheme with liquid hydrogen fuel as heat sink

2.2. System Model

Figure 1 is a model of the combined heat management system of cooling and thermoelectric conversion. The parametric equation of refrigerant loop is shown in formulas (1) and (2).

$$T_2 = Q_1 / (q_m \cdot c_p) + T_1 \quad (1)$$

$$T_3 = Q_2 / (q_m \cdot c_p) + T_2 \quad (2)$$

Set the liquid hydrogen storage environment of the low temperature fuel tank: The temperature and pressure were -253 °C and 101 kPa. The parametric equation of the liquid hydrogen loop is shown in formulas (3) and (4):

$$T_5 = (T_3 - T_1) \times \varepsilon_1 + T_4 \quad (3)$$

$$T_7 = Q_4 / (q_{m2} \cdot c_{p1}) + T_4 \quad (4)$$

The parameters equation model of Brayton cycle thermoelectric conversion technology is as follows. The equations of compressor operating parameters are shown in formulas (5) and (6).

$$T_8 = T_{13} + T_{13} \times (\pi_k^{\frac{\kappa-1}{\kappa}} - 1) / \eta_k \quad (5)$$

$$W_c = q_{m4} \times (h_8 - h_{13}) \quad (6)$$

The working parameter equation of the intermediate heat exchanger is shown in formulas (7) and (8).

$$T_9 = (T_{11} - T_8)\times \varepsilon_1 + T_8 \tag{7}$$

$$T_{10} = Q_3 / (q_{m4} \cdot c_{p2}) + T_9 \tag{8}$$

The engineering parameter equation of turbine operation is shown in formulas (9) and (10).

$$T_{11} = T_{10} - T_{10}\times(1-\pi_t^{\frac{1-\kappa}{\kappa}})\times \eta_t \tag{9}$$

$$W_T = q_{m4}\times(h_{10} - h_{11}) \tag{10}$$

The calculation of cold-side heat exchangers is shown in formulas (11) and (12).

$$T_{12} = T_{11} - (T_{11} - T_8)\times \varepsilon_1 \tag{11}$$

$$T_{13} = T_{12} - (T_6 - T_4)\times \varepsilon_1 \tag{12}$$

$$Q = h_{10} - h_9 \tag{13}$$

$$\eta_r = (W_T - W_c)/Q \tag{14}$$

In the process of system calculation, Newton-Raphson algorithm is used to solve the equation to ensure that the convergence accuracy residuals are less than 10^{-6}.

2.3. Determination of System Parameters

2.3.1. Design of System Parameter. (1) fuel flow of hypersonic vehicle: In general, the mass flow rate of cryogenic liquid hydrogen fuel for thermal management will not exceed the fuel flow requirements of hypersonic vehicle flight.In this paper, the mass flow rate of liquid hydrogen fuel is calculated by using the hypersonic vehicle engine model[15], which provides the input condition for the heat sink flow rate of liquid hydrogen fuel for thermal management. The thrust variation of typical flight envelope, flight trajectory and typical flight trajectory of the hypersonic vehicle used in this paper[16] is shown in Figure 2. In this paper, the specific heat capacity of hydrogen is 14.3kJ/(kg·K) and the calorific value is 1.43×105kJ/kg. The fuel mass flow rate required for hypersonic vehicle engine combustion chamber is obtained by keeping the thrust variation in flight trajectory in Figure 2, as shown in Figure 3.

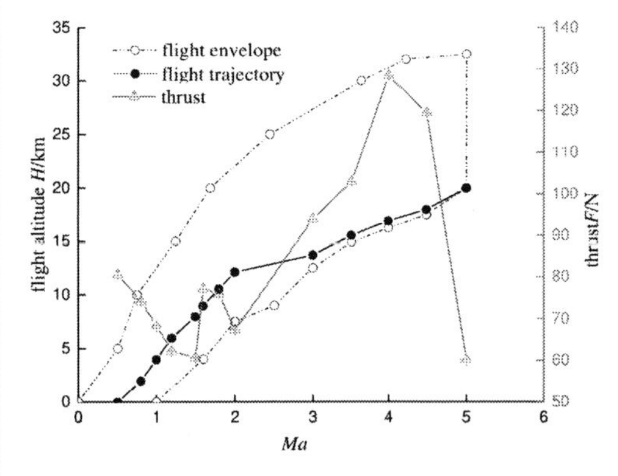

Figure 2. Thrust Variation under Typical Flight Envelope and Flight Trajectory

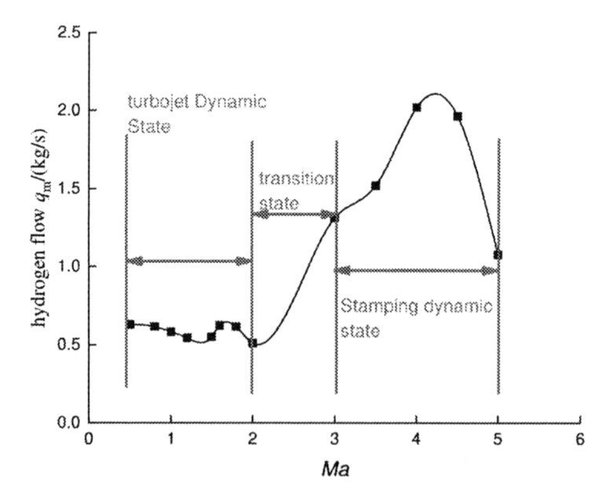

Figure 3. Mass Flow of Fuel Required in Engine Combustor

(2) Brayton cycle parameters of thermoelectric conversion: The fluid medium in the Brayton cycle is set to He, and the parameters of the Brayton cycle are designed as shown in Table 1[11]:

Table 1. Initial parameter

parameter	value
specific heat capacity c_{p3} of He	5.19kJ/(kg·K)
He adiabatic index k	1.66
compressor efficiency η_c	0.85
turbine efficiency η_t	0.9
heat exchanger efficiency ε_1	0.9
hot side heat load Q_3	100kW~800kW

2.3.2. Determination of System Parameters. Based on the above liquid hydrogen fuel flow rate, the cooling effect under different heat loads and the fuel consumption change of engine with or without thermoelectric conversion are studied in this paper, which can provide reference for the design of thermal management system under different aircraft design parameters and heat loads. Specific parameters range of electronic cabin and hydraulic oil, lubricating oil system, engine wall and aerodynamic heat load are selected according to existing literature and actual engineering data[17][18] as shown in Table 2.

Table 2. Change range of system heat load(kW)

system parameter	Electronic cabin heat load Q_1	Heat Load Q_2 of Hydraulic and Lubricating Oil System	Engine Wall Heat Load Q_3	Aerodynamic Heat Load Q_4 for Turbojet State	Pneumatic Thermal Load Q_5 in Ramjet State
Range of parameters	100~500	50~200	100~800	500~1800	2000~20000

3. Simulation Result

3.1. the Cooling of Coolant Loop

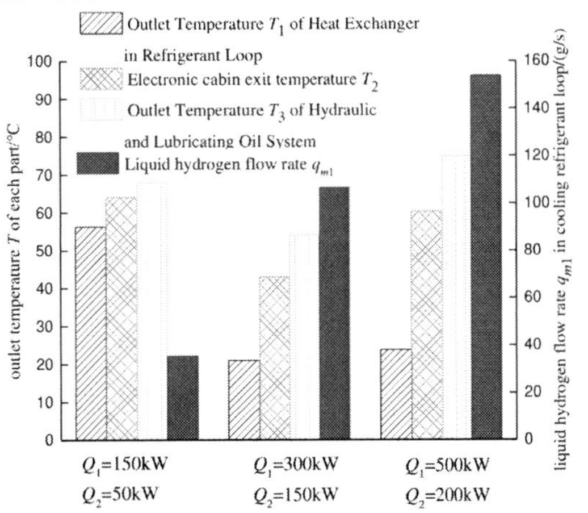

Figure 4. Temperature and flow rate of liquid hydrogen in refrigerant loop.

The refrigerant loop generally requires that the temperature of the electronic cabin, hydraulic oil and lubricating oil system should not exceed 80°C. Under this condition, the mass flow of liquid hydrogen fuel q_{m1} is calculated. Figure 4 shows the temperature of the refrigerant loop and the mass flow rate of liquid hydrogen fuel. While the mass flow rate of liquid hydrogen fuel in refrigerant loop remains unchanged at 3.34kg/s, the mass flow rate of liquid hydrogen fuel in refrigerant loop increases with the increase of heat load of electronic cabin, hydraulic oil and lubricating oil system. When the heat load of the electronic cabin is 100kW and that of the hydraulic oil and lubricating oil system is 50kW, the flow rate of liquid hydrogen fuel is 35.54g/s; when the heat load of the electronic cabin is 300kW and that of the hydraulic oil and lubricating oil system is 150kW, the flow rate of liquid hydrogen fuel is 106.63g/s; when the heat load of the electronic cabin is 500kW and that of the hydraulic oil and lubricating oil system is 200kW,the flow rate of liquid hydrogen fuel is 154.02g/s. When the total heat load of the electronic cabin and the hydraulic oil and lubricating oil system increases from 150kW to 700kW, the temperature of the electronic cabin, hydraulic oil and lubricating oil system does not exceed 80°C. The liquid hydrogen flow rate increases from 35.54g/s to 154.02g/s, and the cooling fuel demand is about 0.22g/kW·s~0.24g/kW·s.

3.2. the Cooling of Aerodynamic Heat

The aerodynamic heat load of hypersonic vehicle increases gradually with the increase of Mach number, so the aerodynamic heat load is smaller in the stage of turbojet power, larger in the stage of stamping power, and the total aerodynamic heat load varies with the area and shape of random body. In this paper, the total aerodynamic heat load in the stage of turbojet power is 500kW~1800kW,the total aerodynamic heat load in the stage of stamping power is 2000kW~20000kW, and the temperature of the body is generally not more than 400°C[18]. The mass flow rate q_{m2} of liquid hydrogen fuel for cooling pneumatic heat load is calculated under the condition of satisfying body temperature. Figure 5 shows the outlet temperature and flow rate of liquid hydrogen cooling aerodynamic heat in the turbojet power state, and Figure 6 shows the outlet temperature and flow rate of liquid hydrogen cooling aerodynamic heat in the stamping power state. It can be seen from the figures that with the increase of aerodynamic heat, the flow rate required for liquid hydrogen cooling will increase. With the increase of heat load, the flow rate of liquid hydrogen required for cooling under turbojet power state increases from 54.5g/s to 194.3g/s. With the increase of heat load, the flow rate of liquid hydrogen required for cooling under stamping power state increases from 215.63g/s to 2203.65g/s. When the body temperature of turbojet and stamping power stage does not exceed 400°C, the demand for cooling fuel is about the same. 0.11g/kW·s.

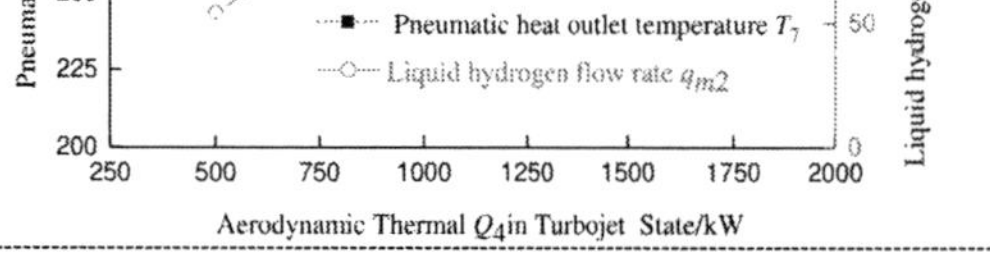

Figure 5. Aerodynamic heat outlet temperature and required flow rate of liquid hydrogen cooling under swirl injection

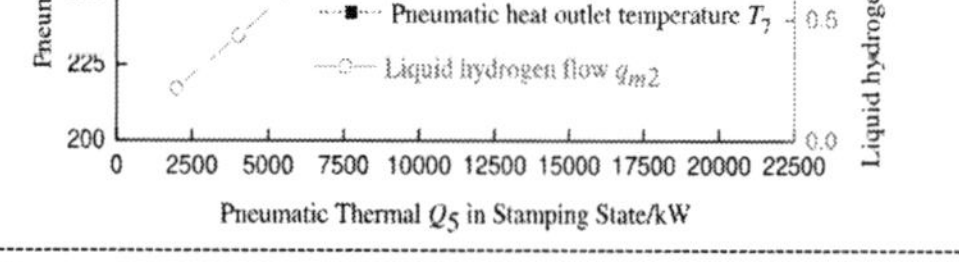

Figure 6. Pneumatic heat outlet temperature and required flow rate of liquid hydrogen cooling in stamping state

3.3. the Cooling of Engine Wall

Generally, the exit temperature of the engine wall should not exceed 600°C. Under this condition, the mass flow q_{m3} of liquid hydrogen fuel is calculated when the thermal load on the engine wall is converted into thermoelectric fuel or not. The simulation results are shown in Figure 7. The range of engine wall heat load varies from 100kW to 800kW. When the temperature of engine wall outlet does not exceed 600°C, the mass flow of liquid hydrogen fuel increases with the increase of engine wall heat load. When the engine wall heat load does not undergo thermoelectric conversion, more fuel is needed for cooling than when thermoelectric conversion is carried out. When the thermal load on the engine wall is not converted into thermoelectric power, the flow rate of liquid hydrogen required for cooling increases from 8.29g/s to 63.98g/s, and the demand for cooling fuel is about 0.08g/kW·s. When the engine wall heat load is converted into thermoelectric power, the liquid hydrogen flow rate for cooling increases from 5.89g/s to 37.9g/s, and the demand for cooling fuel is about 0.047g/kW·s~0.058g/kW·s, which is 25%-40% lower than that without thermoelectric conversion.

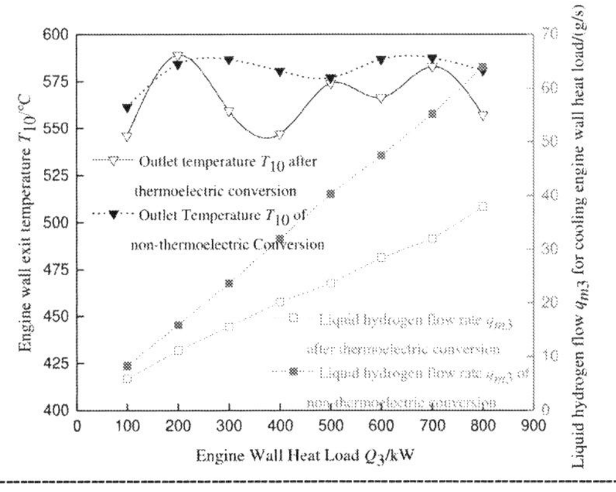

Figure 7. Outlet Temperature and Flow Rate of Liquid Hydrogen Cooled Engine Wall

According to the principle of Brayton cycle, the generation conditions under different heat loads are calculated. Formula (15) is generation power Wh and formula (16) is generation efficiency F of Bretton cycle.

$$W_h = W_T - W_c \tag{15}$$

$$\eta_f = (W_T - W_c) / W_T \tag{16}$$

As shown in Figure 8, with the increase of thermal load at the hot side (engine wall), the cycle thermal efficiency increases from 3% to 38%, and the generation efficiency increases from 17% to 30%. When the wall thermal load is about 200 kW, the increase of efficiency becomes smaller and smaller, resulting in more and more electric energy, from 17.5 kW to 244 kW.

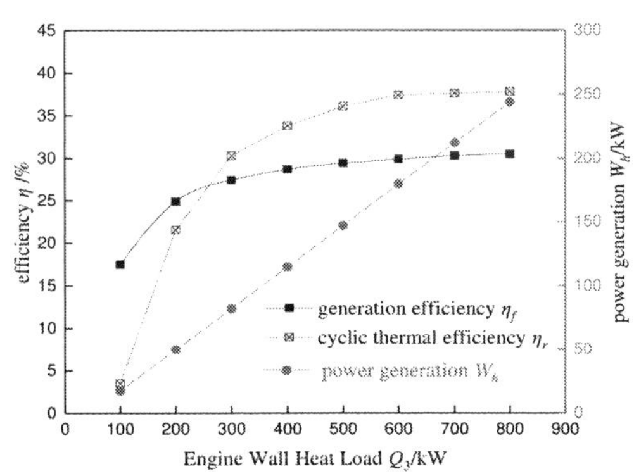

Figure 8. Generation efficiency and power of thermoelectric conversion on engine wall

3.4. Comparative Analysis of Total Cooling Flow Rate and Engine Combustion Flow Rate

In this section, the total cooling flow rate is compared with that required by engine to maintain thrust combustion, and the total cooling liquid hydrogen flow rate under different heat loads, such as refrigerant loop, engine wall and aerothermal load, and whether the combustion flow rate meets the total cooling flow requirement are discussed.The refrigerant loop in flight trajectory and the heat load on engine wall are generally stable, so it is considered as steady state. The maximum and minimum states in the previous study are selected as shown in Table 3, Working Conditions one and two respectively.

Table 3. Variation of Heat Load of Refrigerant Loop and Engine Wall

Working condition number	Heat load Q_1 of electronic cabin(kW)	Thermal Load Q_2 of Hydraulic and Lubricating Oil System(kW)	Engine Wall Heat Load Q_3(kW)
1	100	50	100
2	500	200	800

The aerodynamic heat load q_{m} in flight trajectory varies with Mach number Ma[18]. The larger and smaller aerodynamic heat loads are selected as shown in Figure 9, Working Conditions A and B respectively.The total cooling flow rate is discussed in three cases. In the first case, the heat load of the refrigerant loop and the engine wall is selected according to the smaller working condition one, which is 150 kW and 100 kW respectively. The aerodynamic heat load is considered under the smaller working condition A. In the second case, the heat loads of refrigerant loop and engine wall are selected according to the larger working condition two, 700 kW and 800 kW respectively, and the aerodynamic heat loads are considered according to the smaller working condition A. In the third case, the heat loads of refrigerant loop and engine wall are selected according to the larger working condition two, 700 kW and 800 kW respectively, and the aerodynamic heat load is considered according to the larger working condition B.The total flow rate $q_{m'}$ required for cooling of the refrigerant loop, the engine wall and the aerodynamic heat load in three cases is compared with the liquid hydrogen flow q_{m} required for combustion of the engine in the flight trajectory in Figure 4. The simulation results are shown in Figure 10. Within the range of the parameters studied in this paper, the changes of the refrigerant loop and the heat load on the engine wall have little effect on the total liquid hydrogen flow required for cooling. The liquid hydrogen fuel flow required for the first and second case can meet the cooling requirements for the whole thermal management system in the flight trajectories. The change of aerodynamic heat load has a great influence on the total cooling liquid hydrogen flow rate. In the second case, the aerodynamic heat load is smaller under working condition A, and the liquid hydrogen fuel flow required for combustion in flight trajectory can meet the cooling requirements of the whole thermal management system. In working condition B, the aerodynamic heat load is larger when the Mach number Ma in flight trajectory is 3.7~4.2. There are many fuels needed for cooling aerothermal load, and the flow rate of liquid hydrogen fuel required for combustion can not meet the cooling requirements of the system. The maximum difference is 10%.

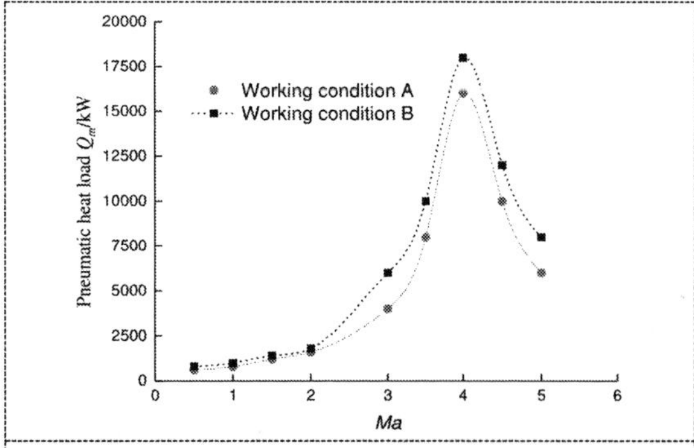

Figure 9. Variation of Aerodynamic Heat Load in Flight Trajectory

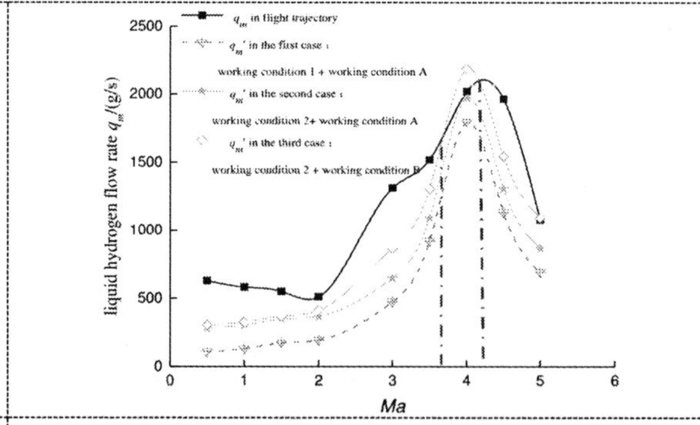

Figure 10. Liquid Hydrogen Flow Rate in Refrigerant Loop and Engine Wall Heat Load Chang

4. Conclusion

In this paper, a hybrid thermal management system for hypersonic vehicle is proposed, which uses liquid hydrogen fuel as the main heat sink and combines closed Brayton cycle thermoelectric conversion technology. Within the parameters of this study, the main conclusions are as follows:

(1) Under the condition that the refrigerant loop, engine wall and airframe do not exceed the prescribed temperature, the mass flow rate of liquid hydrogen fuel required for cooling each part is simulated and calculated. The demand for refrigerant loop cooling fuel is about 0.22g/kW·s~0.24g/kW·s, and the demand for aerodynamic heat load cooling fuel is about 0.11g/kW·s. The cooling fuel requirement is about 0.08g/kW·s when the engine wall heat load is not converted to thermoelectric power, and 0.047g/kkW·s~0.058g/kW·s when the thermoelectric conversion is carried out, which is 25%~40% lower than that when the thermoelectric conversion is not carried out.

(2) Discusses whether the flow rate of liquid hydrogen fuel for combustion meets the flow rate requirement for cooling under the change of heat load, such as refrigerant loop, engine wall and aerodynamic heat. The results show that the change of refrigerant loop and engine wall heat load has little effect on the total cooling liquid hydrogen flow rate, and the change of aerodynamic heat load has greater effect on the total cooling liquid hydrogen flow rate. When the Mach number is 3.7~4.2, the flow rate of liquid hydrogen fuel can not meet the cooling requirements of the system, and the maximum difference is 10%.

(3) The thermal load of the engine wall surface is thermoelectrically converted. The greater the thermal load on the engine wall, the higher the cycle heat efficiency. From 100 kW to 800 kW, the cycle thermal efficiency increases from 3% to 38%; the generated electric energy is increasing, increasing from 17.5kW. By 244kW, the power generation efficiency has increased from 17% to 30%, but the growth rate is getting smaller and smaller.

Acknowledgments

Thank you, Mr. Zhang Jingyang, for his great trust, support and guidance in the project topic, which has given me a lot of opportunities to learn and practice, and provided a solid foundation for the successful completion of my thesis. The authors gratefully acknowledge the financial supports for this project from the Aviation Foundation of China (grant No:20162852035)

Nomenclature

T_1 = inlet temperature of electronic cabin,°C
T_2 = outlet temperature of electronic cabin,°C
T_3 = outlet temperature of hydraulic oil and lubricating oil system,°C
T_4 = initial temperature of cryogenic liquid hydrogen fuel, °C
T_5 = inlet temperature of engine wall, °C
T_6 = outlet temperature after liquid hydrogen cooling,°C
T_7 = outlet temperature of pneumatic heat, °C
T_8 = outlet temperature of compressor,°C
T_9 = outlet temperature of plate-fin heat exchanger,°C
T_{10} = inlet temperature of turbine, °C
T_{11} = outlet temperature of turbine,°C
T_{12} = inlet temperature of cold-side heat exchanger, °C
T_{13} = inlet temperature of compressor, °C
q_m = flow of the cooling medium,kg/s
q_{m1} = flow rate required for liquid hydrogen cooling refrigerant loop,kg/s
q_{m2} = flow rate required for liquid hydrogen cooling aerodynamic heat, kg/s
q_{m3} = flow rate required for liquid hydrogen cooling engine wall,kg/s
q_{m4} = pipe medium flow rate, kg/s
c_p = specific heat capacity of cooling medium,kJ/(kg·K)
c_{pl} = specific heat capacity of liquid hydrogen, kJ/(kg·K)

c_{p2} = specific heat capacity of pipeline fluid,kJ/(kg·K)
Q_1 = thermal load of Electronic cabin,W
Q_2 = thermal load of hydraulic oil and lubricating oil system,W
Q_3 = engine wall heat load, W
Q_4 = aerodynamic heat load, W
Q = Brayton cycle system heat absorption, kJ/kg;
η_r = system cycle thermal efficiency
ε_1 = the efficiency of all heat exchangers in the system is the same
π_k = pressure ratio
η_k = compressor efficiency
π_t = expansion ratio
η_t = turbine efficiency
k = adiabatic index of pipeline gas
h_{13} = inlet enthalpy of compressor, kJ/kg
h_8 = outlet enthalpy of compressor, kJ/kg
h_{10} = inlet enthalpy of turbine,kJ/kg
h_{11} = outlet enthalpy of turbine,kJ/kg
Wc = work consumed by adiabatic compression of compressor, W
W_T = expansion output work of turbine,W

References

[1] Yan Mingsong, Tang Yifeng. Overview of Hypersonic Combat Platform [J]. Avionics Technology, 2016,47(1): 11-16.
[2] Anderson M S , Jackson L R . A carbon dioxide purge and thermal protection system for liquid hydrogen tanks of hypersonic airplanes.[J]. Adv.cryog.eng, 1967, 12(10):146-156.
[3] Kojima T , Taguchi H , Imamura S , et al. Conceptual Study on Heat Resistant and Cooling System of Hypersonic Airplanes[C]. 17th AIAA International Space Planes and Hypersonic Systems and Technologies Conference. 2011.
[4] Lei Yikun, Chang Haiping, Zhang Jingyang, et al. Simulation and analysis of integrated thermal/energy management system for aircraft [J]. 2014.
[5] Gasner J A , Foster R C , Fujimura C . Evaluation of thermal management for a Mach 5.5 hypersonic vehicle[J]. Aiaa Journal, 2013.
[6] Zhu Chunling, Ning Xianwen,Liu Yongji, et al. Integrated Thermal Management System for Hypersonic Unmanned Aerial Vehicles [C]. Papers of the First Conference on Modern Aerodynamics and Aerodynamics in China. 2006:351-355.
[7] Kwee Ngan Thio.Hypertension Plane Hot Sructure Thermal Protection System[D].Department of Aerospace Engineering California State University, 1997.
[8] Wang Peiguang, Liu Yongji, Wang Jun. Discussion on Integrated Thermal Management System for Hypersonic Vehicles [J]. China Engineering Science, 2007, 9 (2): 44-48.
[9] Glickstein M R , Spadaccini L J . Applications of Endothermic Reaction Technology to the High Speed Civil Transport[J]. 1998.
[10] Balland S , Villace V F , Steelant J . Thermal and Energy Management for Hypersonic Cruise Vehicles - Cycle Analysis[C]. Aiaa International Space Planes & Hypersonic Systems & Technologies Conference. 2013.
[11] Ma Zhe, Zhang Xingjuan, Yang Chunxin. Bretton thermoelectric conversion technology for thermal protection of hypersonic aircraft [J]. Tactical missile technology, 2014 (4): 20-25.
[12] Qin J , Bao W , Zhou W , et al. Thermal Management System Performance Analysis of Hypersonic Vehicle Based on Closed Brayton Cycle[C]. Aiaa/asme/sae/asee Joint Propulsion Conference & Exhibit. 2008.
[13] Villages V F, Steelant J. The Thermal Paradox of Hypersonic Cruisers[C].AIAA International Space Planes and Hypersonic Systems and Technologies Conference. 2013.

[14] Zhang Yaoli, Liang Dunhuang, Fang Chao, et al. Thermal characteristics of supercritical carbon dioxide Brayton cycle [C]. Papers of the 14th National Conference on Reactor Thermal Fluids and the Academic Annual Meeting of the Key Laboratory of Thermal and Hydraulic Technology for Central Nuclear Reactors in 2015. 2015.
[15] Zhao Yuqi. Modeling and Transition Performance Optimization of TBCC Engine [D]. Harbin University of Technology, 2014.
[16] Wang Weiwei, Li Dan, Zeng Jun. Power development of high-speed aircraft near space from horizontal takeoff and landing abroad [J]. Gas turbine test and research, 2014 (1): 57-62.
[17] Tang Hairong. Research on Engineering Estimation Method of Surface Heat Flux Density for Hypersonic Vehicles [D]. Nanjing University of Aeronautics and Astronautics, 2008.
[18] Zhu Jianyang. Aerodynamic and aerothermal numerical study of X43-like hypersonic vehicle [D]. Harbin University of Technology, 2008.

IOP Conf. Series: Materials Science and Engineering **751** (2020) 012079 doi:10.1088/1757-899X/751/1/012079

Research on Post-processing Microstructure and Property of Titanium Components with Selective Laser Melting (SLM)

Ming Qiang Chu[1, a], Shu Yan Zhang[1,2, b], Guan Qiao Su[1], Ren Gen Ding[1], Lei Wang[3] and Sanjooram Padde[1]

[1] Centre of Excellence for Advanced Materials, Songshan Lake, Dongguan 523808, China
[2] Materials laboratory of Song Shan Lake, Dongguan 523808, China
[3] Shanghai Aircraft Manufacturing Co., Ltd., China

Email address: [a] mingqiang.chu@ceamat.com; [b] shuyan.zhang@ceamat.com

Abstract. Additive manufacturing of titanium component holds promise to deliver benefit such as reduced cost, weight and carbon emissions during both manufacture and use. To capitalize on the benefits, it must be shown that the mechanical performance of parts produced by additive manufacturing can meet design requirement that are typically based on wrought material performance properties. Of particular concern for safety critical structures is the fatigue property of parts produced by additive manufacturing. Microstructure evolution, and its influence on mechanical properties of the alloy in the as-fabricated condition, has been documented by various researchers. However, fatigue crack propagation and the effects of the directional structure have not been sufficiently studied, imposing a barrier for this technology's potential extension to high-integrity applications. In this study, fatigue life (S-N) and fatigue crack growth (FCG) both parallel and perpendicular to the build directions was studied. The interaction between the directional as-fabricated SLM microstructure and FCG was investigated and compared to that of the hot isostatic pressing (HIP) specimens with and without the stress relief after fabricating with SLM.

1. Introduction

Additive Manufacturing (AM) is a process in which parts are built up by progressive consolidation of raw materials, such as powder or wire, in a layer-by-layer fashion. This is a different approach to traditional fabrication methods, such as machining from block or plate, where the final part geometry is produced by subtracting, or removing material. With this additive approach, parts of greater complexity can be economically produced. Furthermore, by optimizing the design of the part for a given boundary conditions, such as attachment interfaces with other parts and applied loads, using numerical techniques, while leveraging the complex geometric build capabilities of the AM process, parts with minimum weight can be achieved. An additional potential advantage of the AM process is reduction of carbon emissions during part manufacture compared to traditional processes like casting and machining. During the last decade, several additive manufacturing (AM) techniques for processing of complex metallic parts were developed and extensively investigated [1-6]. In recent years, the tensile strength values of additive manufactured laser powder bed (SLM) and electron beam melted (EBM) TiAl6V4 [3,4] have been well characterized for different heat treatments and surface conditions. In particular, Ti-6Al-4V has a great potential, since it shows a high specific strength, low density and high corrosion resistance at temperatures of up to 350 °C. To capitalize on the benefits, it must be shown that the mechanical performance of parts produced by additive manufacturing can meet design requirement that are typically based on wrought material performance properties. Of particular concern for safety critical structures is the fatigue property of parts produced by additive manufacturing.

The SLM technologies applied to Ti–6Al–4V due to the typical process conditions, i.e. layer by layer generative principle, short energy pulses resulting in highly localized melting and solidification and strong temperature gradients of this manufacturing process, which promotes the formation often acicular/lamellar alpha hcp phase (martensitic), a fine microstructure; the microstructure also exhibits strong directionality, i.e. columnar grain structure, consequently anisotropic structure. The strongly textured microstructure results in anisotropy of the mechanical properties [7-9]. Residual stresses are also expected due to this process and can be managed by appropriate post-fabrication heat treatment [10]. This further thermomechanical treatment adjusts the microstructure and reduces internal stresses, in comparison to 'as-cast' and wrought standards. Thus, the completely satisfied ductility and strength values can be optimal by the applied post-treatment. The post heat treatment for the SLM part can also be applied hot isostatic pressing, which is used in traditional powder metallurgy and foundry technology. HIP'ing seems to be very promising to restore Ti-6Al-4V properties due to the combined effect of high temperature and high pressure, and allows not only to adjust the microstructure, but also to fuse unmolten particles. This reduction of residual porosity is very important, since the pores within the sample can act as strong stress raisers and lead to failure, especially under fatigue loading. While many references, for example [8,9], discuss the link between process parameters and static mechanical properties of SLM Ti–6Al–4V, the fatigue strength characterization appears more recently in the literature [9-11]. In fact, fatigue crack propagation and the effects of the directional structure have not been sufficiently studied, imposing a barrier for this technology's potential extension to high-integrity applications. Fatigue testing is known to be expensive and time consuming as it is susceptible to a number of intrinsic and extrinsic factors that complicates the data generalization and exploitation. On the other hand, the fatigue behaviour of SLM Ti–6Al–4V is critical for sectors, in which high structural integrity is a paramount requirement for aerospace.

This study is aimed at Ti–6Al–4V materials presented and directed to the evaluation of the anisotropic fatigue behavior including the effect of post heat treatment on the fatigue life of Ti-6Al-4V obtained using the SLM process, While the directionality of static mechanical properties of SLM Ti-6Al-4V has been documented [11,12], little information is available of such influence on the high cycle fatigue behavior. Therefore, Fatigue life (S-N) and fatigue crack growth (FCG) both parallel and perpendicular to the build directions was studied. The interaction between the directional as-fabricated SLM microstructure and FCG was investigated and compared to that of the hipped specimens with and without the stress relief heat treatment (HT) after fabricating with SLM to understand whether the stress relief HT has effect on the fatigue property.

2. Experimental Procedure

Ti-6Al-4V alloy powder was produced by the gas atomization process and was spherical with a maximum particle size 53μm. Ti-6Al-4V alloy samples were fabricated on an EOS M280 machine including a laser unit delivering a continuous single mode laser power of 400W, which produces a laser beam with a wavelength of 1070 nm and an intensity distribution of Gaussian. The laser spot diameter was 100~500μm and the maximum scanning speed was 7m/s. In addition, the layer thickness can be selected between 20~100μm and the deposition was carried out on a 30mm thick Ti-6Al-4V alloy plate. During the SLM process, the processing chamber will be filled with argon in order to maintain the oxygen level during the process. A double optimization procedure is necessary in order to obtain a high quality SLM part. Firstly, the optimization of the SLM parameters for the minimization of inherent defects, and secondly, the application of further thermo mechanical treatment to minimize internal stresses and adjust the microstructure. These two stages of optimization are presented in this paper, which aims to comprehensively examine the factors influencing defects in Ti-6Al-4V processed by SLM. For this reason, more than 87 small Ti-6Al-4V cuboids were previously produced with the various scan parameters. The porosity of each test sample was quantitatively two- and three-dimensionally analyzed using optical microscope and X-ray computed tomography. The optimized process parameters were afterwards used in the second part of the program to produce cylindrical specimen for the mechanical testing.

After the deposition with SLM, the specimens underwent a stress relief heat treatment (HT) with exposure to 800 °C for 2 h in a vacuum furnace as specified in the optimization temperature of the stress relief heat treatment. The build configuration and SLM reference axis system on the build platform of the EOS machine is shown in Figure 1. The cylindrical specimens (Ø10 mm×70 mm) are orientated with their longitudinal axis perpendicular to the build platform (parallel to the build direction/ Z axis). The rectangular specimens have their longitudinal axis perpendicular to the build direction/Z axis, 10×10×70 mm^3 for horizontal samples. In this study, the terms vertical orientation and horizontal orientation are used to identify samples that are oriented parallel and perpendicular to the build direction/Z axis respectively. The diagonal samples to the build direction/Z axis also were tested as a comparison. The room temperature tensile property was tested according to ASTM E8/E8M. The tensile test was employed for the verification of mechanical property with SLM. Some of the samples were then hot isostatic pressed (HIP) to eliminate the porosities produced during the fabricating. Then an investigation using fatigue specimens fabricated using SLM were studied, fatigue crack propagation testing with force controlled constant amplitude axial, the loading condition was cyclic tension with R = 0.1 for the specimen as HT and HIP. In addition, each specimen was tested at different constant maximum stress levels, from 100 to 600 MPa, to failure in order to generate an approximate S–N curve for that given condition/orientation. Testing of the samples was carried out per ASTM E466 at a frequency of 20 Hz, a load ratio of R = 0.1 on an MTS load frame at room temperature in atmosphere. The fracture surfaces of the fatigue specimens were characterized with both optical and scanning electron microscopy (SEM) to provide further insight into the resulting fatigue test data. Then, the tensile properties and FCG fatigue resistance and S-N curve of all HIP treated and non-treated samples were tested, and compared with the wrought Ti-6Al-4V alloy. The microstructure, porosity and the received mechanical properties were analyzed and was comprehensively examined and discussed.

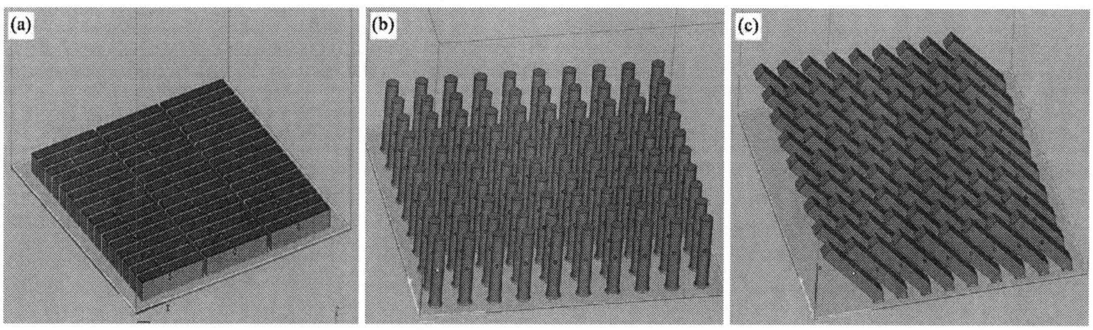

Figure 1. The specimen fabricated with SLM (a) Specimen building horizontal direction (b) Specimen building vertical direction (c) Specimen building diagonal direction

3. Results and Discussion

3.1. Characteristic of the Microstructures

Microstructures have been studied in Ti-6Al-4V alloy produced with SLM. Figure 2 shows the microstructure features of Ti-6Al-4V alloy fabricated by SLM in the X-Z and X-Y sectional directions. In general, the as-built condition exhibits fine martensitic acicular alpha morphology with elongated beta columnar grains that span across multiple build layers. It seems to mean that this unidirectional grain growth may also influence the anisotropy of properties of the SLM as-built alloy. Heat treatment, either stress relief or HIP, after SLM allows the fine acicular α to precipitate into a more lamellar $\alpha + \beta$

structure. These SLM microstructures are more similar to a weldment or casting than the equiaxed primary alpha grain structure in wrought Ti-6Al-4V mill products such as plate, sheet. As expected, this microstructure yields higher strengths and lower ductility than wrought material [3–10]. Figure 3 provides an overview of the microstructure for Ti-6Al-4V as built and as the post built-up heat treatments. It is assumed that the martensite was expected to be present (Fig. 3a). A significant coarsening of the lamellar microstructure can be observed for the hipped specimen. (Fig. 3c).

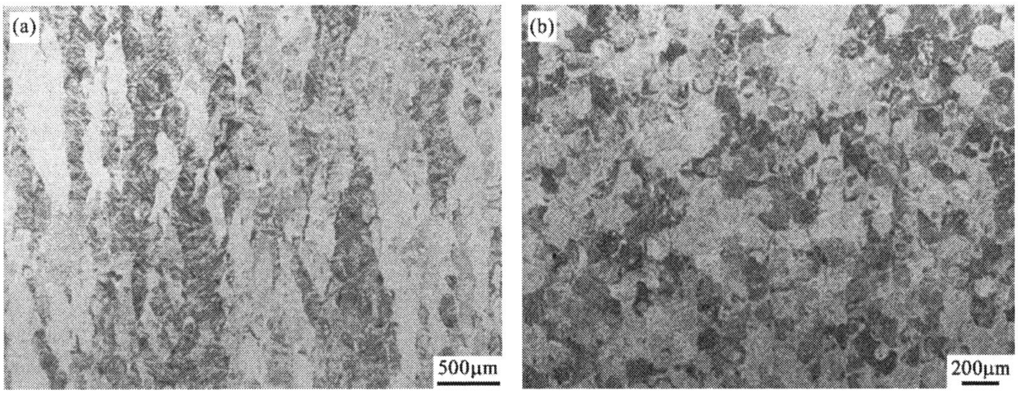

Figure 2. Microstructure of Ti-6Al-4V alloy fabricated by SLM: (a) is longitudinal section through columnar grains of the prior-β phase in the X-Z plane. (b) Cross-section through columnar grains of the prior-β phase in the X-Y plane

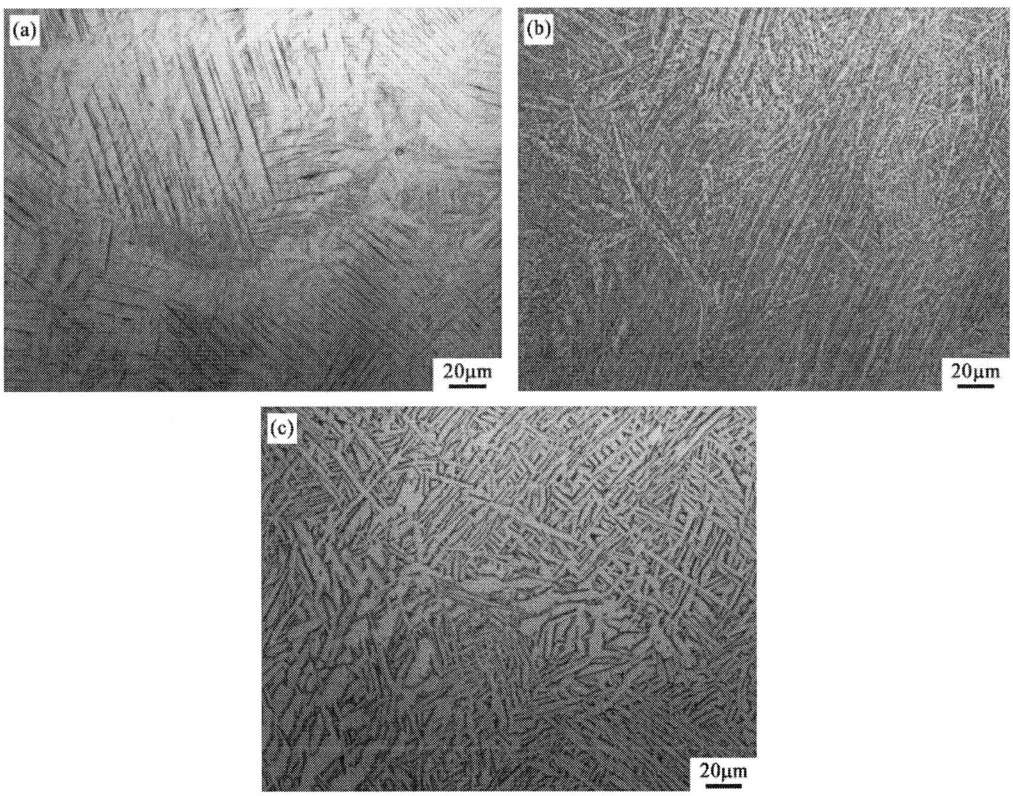

Figure 3. Microstructure of Ti-6Al-4V alloy (a) as fabricated (b) post heat treatment at 800 °C/2 h (c) Hipped after heat treatment at 800 °C/2 h

IOP Conf. Series: Materials Science and Engineering **751** (2020) 012079 doi:10.1088/1757-899X/751/1/012079

3.2. X-ray Computed Tomography Analysis

Ti–6Al–4V fabricated with SLM technology due to the strong temperature gradients of this manufacturing process, which promotes the formation of very fine pores. This residual porosity is very important, since the pores within the sample can act as strong stress raisers and lead to failure, especially under fatigue loading. Methods for detecting defects in AM parts can be divided into two categories: destructive testing and non-destructive testing. Destructive testing includes metallographic and scanning electron microscopy; non-destructive testing methods include ultrasonic testing, infrared detection, eddy current testing, and X-ray inspection. Among these methods, X-ray micro computed tomography (X-ray micro CT) has been widely employed to study defect and dimensions accuracy in additive manufacture parts because of its high precision spatial information of parts form almost any material [13]. Lopez E et al.[14] has compared micro CT with optical 3D scanner about different demonstrators and features by measuring parts made by electron beam melting (EBM) and SLM. Maskery I et al.[15] investigated the size and shape distribution of the pores inside the SLM alloy Al-Si10-Mg through micro CT, which prove than the X-ray CT has the ability of monitoring and detection for SLM process. The research of many scholars shows that X-ray scanning technology is a reasonable and feasible method to study the internal defects of metal additive products. Therefore, X-ray computed tomography was applied to measure the porosity distribution for various processing parameters in this study. A relatively large amount of porosity was observed as shown in Figure 4(a). Pore sizes of the annealed Ti–6Al–4V SLM specimen were typically found to be less than 60μm in diameter and randomly distributed. The hipped specimen after SLM was also measured with micro CT as a comparison shown in Figure 4(b). Obviously, the porosities were not detectable after hipping. Various types of defects inherent to the SLM have been well described by Kim et al[16]. and are very important for the mechanical properties of the material. Presently, the pores and other defects cannot be avoided completely, but they should be kept to a minimum at the stage of the powder-layer consolidation, as was clearly confirmed among other by Kruth et al.[17]. Detailed investigation of the influence of manufacturing process parameters on the defect as well as analysis of the surface roughness will be published elsewhere.

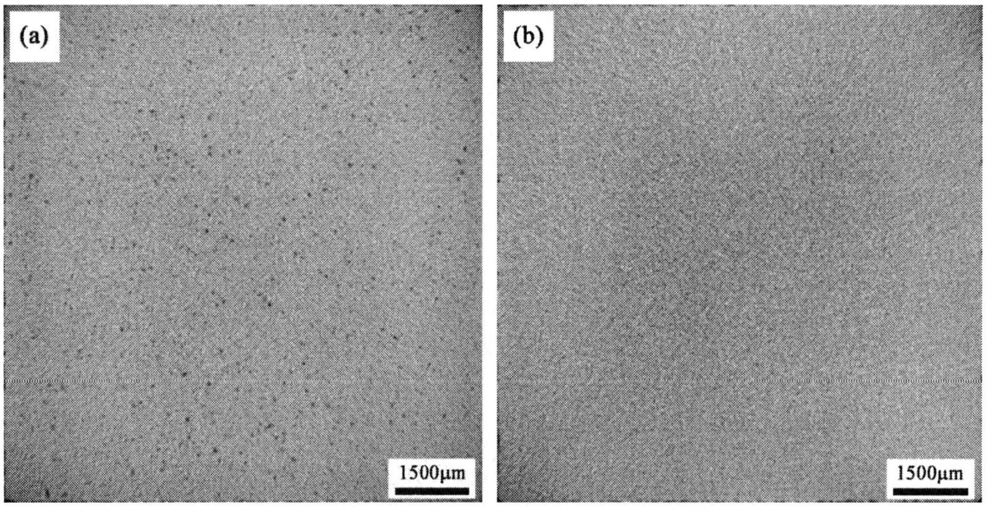

Figure 4. X-ray Computed Tomography image showing the pores in (a) Annealed sample (b) Hipped sample

3.3. Mechanical Testing

Microstructures have been studied in Ti-6Al-4V alloy produced with SLM. Figure 2 shows the microstructure features of Ti-6Al-4V alloy fabricated by SLM in the X-Z and X-Y sectional directions. In general, the as-built condition exhibits fine martensitic acicular alpha morphology with elongated beta columnar grains that span across multiple build layers. It seems to mean that this unidirectional grain growth may also influence the anisotropy of properties of the SLM as-built alloy. Heat treatment, either stress relief or HIP, after SLM allows the fine acicular α to precipitate into a more lamellar α + β structure. These SLM microstructures are more similar to a weldment or casting than the equiaxed primary alpha grain structure in wrought Ti-6Al-4V mill products such as plate, sheet. As expected, this microstructure yields higher strengths and lower ductility than wrought material [3–10]. Figure 3 provides an overview of the microstructure for Ti-6Al-4V as built and as the post built-up heat treatments. It is assumed that the martensite was expected to be present (Fig. 3a). A significant coarsening of the lamellar microstructure can be observed for the hipped specimen. (Fig. 3c).

3.3.1. Tensile. Figure 5 provides a summary of the mechanical properties obtained from this research on Ti-6Al-4V manufactured by SLM. These value are compared with accepted values for wrought Ti-6Al-4V tested in the annealed and solution treated and aged condition, dash line is AMS standard strength of annealed die forgings of Ti–6Al–4V at O~1400 ppm, Yield stress = 827 MPa, UTS = 896 MPa (El = 10%). The yield and tensile strengths in laser-based AM specimens are generally higher than those in annealed material and in the same range as age-hardened Ti–6Al–4V, which is likely due to the presence of a fine grained microstructure in laser-based AM specimens. An influence of specimen orientation with respect to the build direction on tensile properties was also observed as shown in Fig. 5. This was primarily attributed to the elongated, directionally oriented (vertically), grain structure across build layers. Heat treatment for the stress relief improves ductility but decreases strength. After hot isostatic pressing, the strength of specimen is decreased, however, plastic stability of the specimens is in a relatively good level, shown in table1, the reason for the good comprehensive performance with the hot isostatic pressing is that in the process of heat treatment, the test specimen under three directions to compressive stress, pore closure or disappear gradually, sample density increase. In addition, after the hot isostatic pressure, the microstructure is relatively coarser, resulting in a decrease in the strength level. However, the strength of specimen seems no much difference with and without the stress relief HT before hipping. Therefore, the best combination of strength and ductility for Ti-6Al-4V SLM specimens was obtained after hipping.

Table 1. Tensile properties for the microstructures obtained after different heating method.

Heating method	Sample Direction	YS/MPa	UTS/MPa	TE/%
	Horizontal	1039.9	1201.2	9.5
SLM	Veridical	1049.9	1194.8	10.2
	Diagonal	1070.4	1222.4	9.9
	Horizontal	944.6	1036.3	14.9
800°C/2h/FC	Veridical	964.9	1035.7	14.6
	Diagonal	967.3	1052.5	14.3
	Horizontal	866.8	961.1	16.1
SLM+SR*+HIP*	Veridical	851.4	950.4	15.3
	Diagonal	869.2	976	15.7
	Horizontal	859.8	968.7	16.7
SLM+HIP	Veridical	864.3	968.3	17.4
	Diagonal	870.4	979.7	15.8

SR: stress relief; HIP: hot isostatic pressing; UTS: Ultimate tensile strength; YS: yield strength; TE: total elongation.

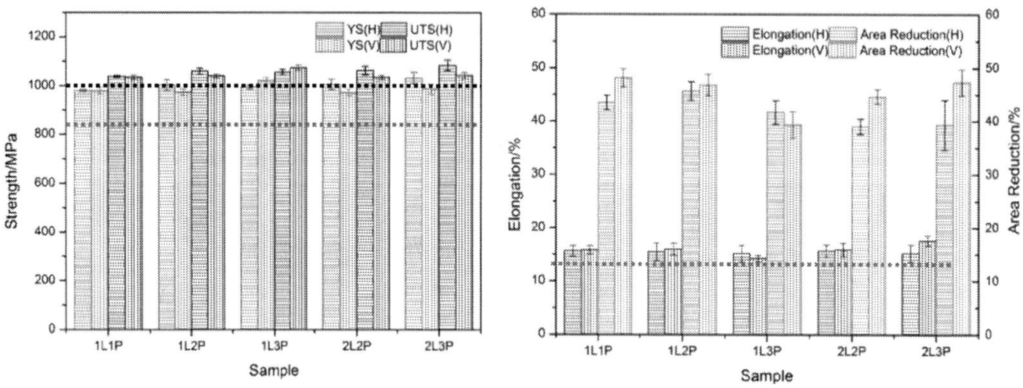

Figure 5. Tensile properties on specimen orientation with respect to the build direction

3.3.2. Fatigue. In addition to metallurgical and static performance evaluations, the primary goal of this study was to evaluate the fatigue performance of Ti–6Al–4V materials produced by the SLM additive manufacturing process as a function of build orientation. Furthermore, to minimize manufacturing costs and carbon emissions during manufacture, no post-fabrication heat treatment was desired, so it was also necessary to determine if residual stresses left in the part not removed by a subsequent heat treatment would be acceptable in terms of the resulting fatigue performance. Figure 6 is shown that fatigue crack propagation curves according to the crack plane oriented parallel (Z sample) and perpendicular to the build layers (X-Y sample). The trend in a linear-log plot is well behaved with fatigue crack growth of the Z specimens that are considerably different and slower than the other two specimen orientations. That is because that the microstructure exhibits strong directionality, i.e. columnar grain structure vertically, and crack plane oriented perpendicular to the columnar grain structure. Therefore, the fatigue crack growth path is slightly tortuous along the build layers. Experimental results also show that the impact of different heat treatment process on the fatigue performance as shown in Fig. 7, samples as fabricated and annealed heat treatment states fatigue life is much scattered, but the fatigue test performance of the samples after hot isostatic pressing treatment, the data divergence become smaller, and the repeatability of the sample in each direction is obvious, these results highlight a dependence of fatigue performance specimen porosities, while the analysis indicate that the hot isostatic pressing may eliminate the sample internal defects such as porosity, incomplete fusion, make samples of fatigue life and stability are greatly increased, the fracture surface as shown in Fig.8&9. Leuders al.[18] investigated an effect on the high cycle fatigue (HCF) behaviour of SLM-processed Ti-6Al-4V and revealed that minimization of porosity is much more important than the microstructure to avoid premature crack initiation under cyclic loading. The fatigue experiment in the HCF regime conducted by the aforesaid authors showed a multiple increase in durability of samples after HIP'ing. This was also confirmed in the work by Leuders et al.[19], they found that heat treatment and HIP'ing both increased fatigue life. HIP'ing improved the life by an order of magnitude over the as-deposited condition. Porosity was the primary cause for premature failure in the un-HIP'ed conditions, while microstructure and tensile residual stresses present in the as-built condition played a secondary role. It was also interesting to note that the fatigue life seems slightly better when specimen is directly HIP'ed after fabricated with SLM, which enables avoidance of any post-process stress relief heat treatment as well as the associated cost and the carbon emissions. It can conclude that fatigue life is improved by HIP'ing. In addition, Chan et al.[19,20] were able to show that when SLM specimens were machined, in the absence of porosity, the fatigue life became comparable with cast and wrought Ti-6Al-4V. Obviously, further research of HIP treatment on the SLM specimen with preliminary minimization of the porosity during the fabrication is certainly needed. To better understand these fatigue results, effect of the metallography, surface roughness characterization, residual stress measurements, tensile testing and fracture surface evaluations were also need to be considered to ensure the reliability of a SLM parts. Fatigue tests should be therefore with surfaces in the 'as built' and 'machined' conditions should be performed. A careful double optimization procedure is necessary to obtain a high quality material. Then the ductility and durability of SLM specimens produced with optimized process parameters and subsequently hot-isostatically pressed were equal to the values of the reference wrought material. As a result, the high quality production of a real, geometrically complex Ti-6Al4V part was enabled. While some data on the fatigue performance of AM titanium components is currently available, much more data is still needed due to the number of processing variables involved and inherent scatter associated with titanium fatigue testing in order to gain a more in-depth understanding and level of confidence in the process.

4. Conclusions

The following conclusions about Ti-6Al-4V specimens fabricated with SLM can be drawn from the results of this study:

1) The tensile strength, regardless of crack orientation with respect to the build direction, is higher than wrought and cast product forms. This increase is most likely attributed to the fine martensitic structure obtained from rapid solidification of the material in the as-deposited condition. Cracks orientated transversally to the build layers have higher ductility than those oriented parallel to the build layers due to the columnar grain structure vertically.

2) The fatigue crack growth of the Z specimens that are considerably different and slower than the Y-X specimen orientations. The fatigue life was found to be highly variable, or erratic, which was consistent with previous research in materials produced by this process.

3) The fatigue behavior of SLM Ti-6Al-4V is influenced by: (i) microstructure, because it introduces a directional effect; (ii) defects, such as porosity that being typically located between adjacent layers, affect most the direction Z, a HIP treatment often considered reduces that influence of defects but would lower the material strength.

4) Post-process heat treatment or HIP'ing should be tested to validate the assumption that fracture and crack growth properties would be increased by microstructural modification and removal of residual stresses.

5) In-situ neutron/synchrotron radiation combination needs to be done to understand the mechanism of the defects and deformation are caused by rapid solidification of molten pool under high temperature gradient.

5. Future Study

The relationships between microstructure, defects, post-processing and performance would be expected in AM parts such as those built in this study. The fatigue of SLM part can be comparable to wrought materials if defects, such as porosity, can be avoided and proper post-processing, like machining and stress relieving, are performed. The porosity present in the samples fabricated for this study. Figure 7 was likely a key contributor to the poor fatigue life observed relative to wrought material. These pores serve as stress concentrations and sites for fatigue crack initiation. The porosity was only identified qualitatively, and not quantitatively measured, in one small metallurgical sample that was built with equal dimensions in the x, y and z-directions. It is entirely possible, and highly likely, that the porosity levels could vary throughout with the volume of a sample and also be dependent on the sample aspect ratio, or build orientation. Such potential variations may have also contributed to the orientational dependence on fatigue performance observed. In the future, additional analysis should be done to further characterize the variation of porosity with respect to build orientation, location within a specimen and location of the specimen within the build chamber to better understand the resulting mechanical and fatigue performance dependence on these same variables.

Furthermore, during SLM, defects and residual stress are caused by rapid solidification of molten pool under high temperature gradient, to further investigate the influence of the processing parameters on the defect in SLM products, X-ray micro computed tomography (X-ray micro CT) has been widely employed to study defect and dimensions accuracy. However, the X-ray micro CT methods mainly analyze the mechanism based on the detection results after forming. So far, due to the lack of effective and intuitive means, the causes of the porosity, crack, deformation and so on are not clear, resulting in poor consistency and great contingency of the formed parts. Therefore, it is very necessary to do the research on manufacturing defects and deformation mechanism of laser additive based on in-situ neutron/synchrotron radiation combination, so as to obtain abundant information of melting behavior intuitively in situ. There are precedents for in-situ neutron/synchrotron radiation observation of welding fusion, phase change, stress change, heat treatment, mechanical stretching and other behaviors [21,22]. In situ neutron/synchrotron radiation based measurements of SLM process produce unique data for model validation and improved process understanding.

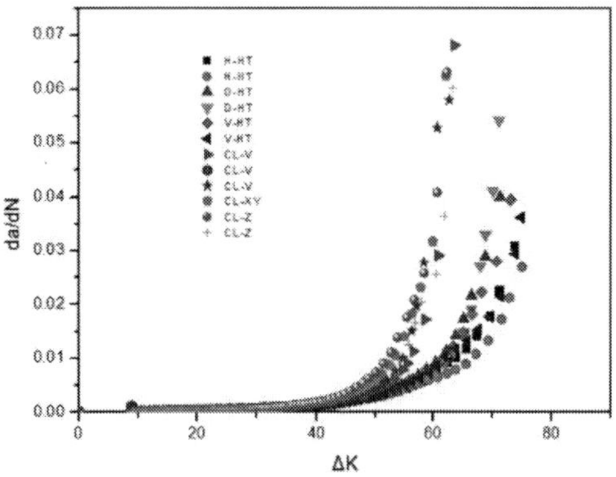

Figure 6. Fatigue crack propagation curves

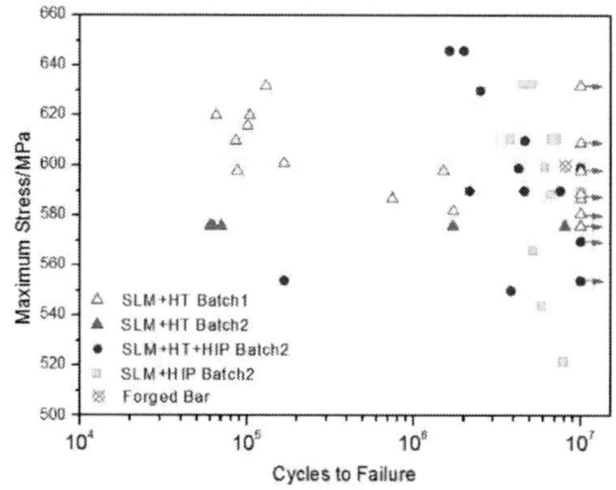

Figure 7. Fatigue life for different heat treatments

Figure 8. Typical fatigue fracture surfaces showing fatigue cracks initiated site

Figure 9. Typical fatigue fracture surfaces showing the fatigue striations

Acknowledgements
This project is supported by the Program for Guangdong Introducing Innovative and Entrepreneurial Teams（NO: 2016ZT06G025）and Guangdong Natural Science Foundation (NO: 2017B030306014)

References

[1] Murr L E, Martinez E, Gaytan S M, Ramirez D A, Machado B I, Shindo P W, Martinez J L, Medina F, Wooten J, Ciscel D, Ackelid U and Wicker R B 2011 *Metall. Mater. Trans. A* **42** 3491
[2] Li S J, Murr L E, Cheng X Y, Zhang Z B, Hao Y L, Yang R, Medina F and Wicker R B 2012 *Acta Mater.* **60** 793
[3] Singh S, Ramakrishna S and Singh R 2017 *J. Manuf. Process* **25** 185
[4] Alafaghani A, Qattawi A and Castañón M A G 2018 *Int. J. Adv. Manuf. Tech.* **99** 2491
[5] Heeling T, Cloots M and Wegener K 2017 *Addit. Manuf.* **14** 116

[6] Shah P, Racasan R and Bills P 2016 *Case Stud. Nondestruc. Test. Evaluat.* **6** 69
[7] Khorasani A M, Gibson I, Ghaderi A and Mohammed M I 2019 *Int. J. Adv. Manuf. Tech.* **101** 3183
[8] Thijs L, Verhaeghe F, Craeghs T, Humbeeck J V and Kruth J 2010 *Acta Mater.* **58** 3303
[9] Qiu C, Panwisawas C, Ward M, Basoalto H C, Brooks J W and Attallah M M 2015 *Acta Mater.* **96** 72
[10] Spierings A B, Starr T L and Wegener K 2013 *Rap. Prototyp. J.* **19** 88
[11] Edwards P and Ramulu M 2014 *Mater. Sci. Eng., A* **598** 327
[12] Chu M, Zhou X W, Sun X F and Wang L 2018 Evaluation of Mechanical Properties of Ti-6Al-4V Fabricated by Selective Laser Melting, 31st Congress of the International Council of the Aeronautical Sciences, ICAS 2018.
[13] Kim F H and Moylan S P 2018 Adv. Manuf. Ser. (NIST AMS) **100-16**
[14] Lopez E, Felgueiras T, Grunert C, Brückner F, Riede M, Seidel A, Marquardt A, Leyens C and Beyer E 2018 J. Laser Appl. **30** 032307
[15] Maskery I, Aboulkhair N, Corfield M, Tuck C, Clare A T, Leach R, Wildman R, Ashcroft I and Hague R 2016 *Mater. Charact.* **111** 193
[16] Zhang B, Li Y and Bai Q 2017 *Chin. J. Mech. Eng.* **30** 515
[17] Kruth J, Levy G, Klocke F and Childs T H C 2007 *CIRP Ann. Manuf. Technol.* **56** 730
[18] Leuders S, Thöne M, Riemer A, Niendorf T, Tröster T, Richard H A and Maier H J 2013 *Int. J. Fatigue* **48** 300
[19] Agius D, Kourousis K I and Wallbrink C 2018 *Metals* **8** 75
[20] Chan K S, Koike M, Mason R L and Okabe T 2013 *Metall. Mater. Trans. A* **44** 1010
[21] Leung C L A, Marussi S, Atwood R C, Towrie M, Withers P J and Lee P D 2018 *Nat. Commun* **9** 1355
[22] Calta N P, Wang J, Kiss A M, Martin A A, Depond P J, Guss G M, Thampy V, Fong A Y, Weker J N, Stone K H, Tassone C J, Kramer M J, Toney M F, Buuren A Van and Matthews M J 2018 *Rev. Sci. Instrum.* **89** 055101

Improving Clamping Accuracy of Thin-walled Workpiece in Turning Operation

Ammar H. Elsheikh[2,*], Jiajie Guo[1], Kun Bai[1], Kok-Meng Lee[1,3]

[1]The State Key Laboratory of Digital Manufacturing and Equipment Technology, School of Mechanical Science and Engineering, Huazhong University of Science and Technology, Wuhan 430074, China.
[2] Production Engineering and Mechanical Design Department, Tanta University, Tanta, 31527, Egypt.
[3]Woodruff Sch. of Mech. Eng. at Georgia Inst. of Tech., Atlanta, GA 30332, USA.

[*]Email: ammar_elsheikh@f-eng.tanta.edu.eg

Abstract. This study presents theoretical and experimental error analysis and avoidance which results from the fixation of a thin-walled metal disk workpiece on a duplex turning machine. The distance between the rotating workpiece surface and a fixed datum point is measured using a laser displacement sensor (LDS). The measured data are analyzed numerically to get the error in the fixation system in terms of Euler angles. Then the workpiece clamping points are axially adjusted according to the calculated angles. Moreover, the problem formulation and the solution procedures are presented in details. Finally, a case study has been used to verify the proposed approach. The results indicate that the proposed technique can be used to decrease the fixation error to a reasonable range.

1. Introduction

The machining of complex components such as thin disk-shaped workpieces has been one of the most challenging problems of manufacturing engineering. However, there has been little research about this problem. Research done to date has focused on thermal analysis, dynamic analysis and workpiece deflection during turning operations [1-4].

Currently, a strong need exists to improve the accuracy and surface integrity of such components [5]. Clamping cylindrical workpieces on the turning machine must achieve sufficient accuracy to prevent occurrence of shape and position deviation of the workpiece, hence dimensional and form accuracy of the machined components will be increased [6]. Compared to the rapid progress in the performance of machine tools, the advancement in the performance of chucks was very slow with little available research results [7]. Due to faster advance in the performance and accuracy of machine tools, clamping become the greatest weakness and a bottle neck in any machine processing system[8]. If clamping is not sufficiently accurate undue deviations in the shape and position of the work-pieces may ensue. Deviations in position between the machined and non-machined surfaces of the work-pieces are caused by deviations in the alignment between the axis of the workpiece and the axis of rotation of the chuck[9]. The poor form accuracies of the finished workpiece were mainly due to the inaccuracies in the conventional chucking, which allow the pre-existing errors to propagate to the workpiece in the subsequent machining operation[10]. Compared to the rapid progress in the performance of machine tools, the advancement in the performance of chucks was very slow with little available research results[7].

Byun and Liu develop a method for improving chucking accuracy [11]. All the major factors that affect the positioning accuracy and repeatability of a chucked workpiece have been identified by error budgeting and systematic measurements. Moreover, The effect of kinematic redundancy on chucking, especially, on the positioning accuracy of a cylindrical workpiece has been investigated[12].In general,chucking error may represent all sorts of errors caused by chucking operation in a broad sense; the chucking error of a workpiece conventionally represents the deviation in the position of the workpiece due to inaccurate chucking [9, 11].

To the best of the author's knowledge, all literature work was focused on study of cylindrical workpieces clamping and no previous work treat the chucking error of thin walled workpieces. This paper presents a novel study to improve chucking accuracy for such workpieces.

2. Objective and approach

The main aim of this study is to minimize the fixation error of a thin-walled circular workpiece. The paper was organized as follows. Firstly, it is important to formulate the relation between the workpiece orientation angles (Euler angles) and the axially displacements of the fixation points. Secondly, the axial displacement of the rotating workpiece surface is obtained using LDS based measurement system. Finally, a novel approach will be proposed to correlate the LDS measured data to the orientation angles in order to get the relative axially displacements of the fixation points which guarantee the coaxial fixation of the workpiece.

3. Mathematical formulation

For the purpose of the problem formulation the thin disk under study is considered to be a rigid body with a planar surface. The disk is rigidly attached to the inner ring of the bearing on a lathe machine through three points A, B and C on the dashed circle as shown in Fig. 1 (b). These three points define the fixation plane. A world coordinate system $O(X,Y,Z)$ is fixed at the bearing center to the disc at its center O, i.e., the dashed circle center, with Z axis normal to the fixation plane. Due to the fixation error, the geometric center of the disk O_P does not coincide with the rotation center O.

It is noted that location of O is unknown as the bearing has been embedded in the machine thus could not be measured directly. Let i, j and k denote three unit vectors pointing along the coordinate axis of the world frame O, and m, n and ω denote three unit vectors pointed along the rotation coordinate frame. The three unit vectors m, n and ω can be expressed in the world frame O as follows:

$$\begin{bmatrix} m & n & \omega \end{bmatrix} = \begin{bmatrix} i & j & k \end{bmatrix} \begin{bmatrix} c\theta & -s\theta & 0 \\ s\theta & c\theta & 0 \\ 0 & 0 & 1 \end{bmatrix} \begin{bmatrix} c\varphi & 0 & s\varphi \\ 0 & 1 & 0 \\ -s\varphi & 0 & -c\varphi \end{bmatrix} \tag{1}$$

where φ and θ are rotation angles of the rotation coordinate frame about Z and Y axes, respectively and, $c\varphi$ is a shorthand notation for $\cos(\varphi)$, $s\varphi$ for $\sin(\varphi)$, $c\theta$ for $\cos(\theta)$ and $s\theta$ for $\sin(\theta)$.

Then the actual rotational axis of the lathe is given by:

$$\omega = i\cos\theta\sin\varphi + j\sin\theta\sin\varphi + k\cos\varphi \tag{2}$$

If φ and θ are known, then the offsets of the clamps A, B and C are given as:

$$\begin{aligned} u_A &= Rs\varphi c(\theta),\ u_B = Rs\varphi c\left(\theta - \tfrac{2}{3}\pi\right), \\ u_C &= Rs\varphi c\left(\theta - \tfrac{4}{3}\pi\right) \end{aligned} \tag{3}$$

where u_A, u_B and u_C are the clamping offset at A, B and C respectively and R is the radius of the bearing inner ring which can be obtained from the specification sheet of the clamping system.

Defining another cylindrical coordinate at O with the same Z axis, any point on the fixation plane can be denoted as $(r_i, \theta_i, 0)$. As the LDS monitors the part motion at a fixed position, its measurement (along the direction of rotating axis) is obtained by:

$$u_i(t) = r_i\, s(\varphi) c(\theta - \Omega t_i) \tag{4}$$

Where $u_i(t)$ is a measured data point for a certain data index i at a certain time t, r_i is the measured data radius in mm and Ω is the angular velocity in rad/sec about ω axis. Equation (4) can be extracted using trigonometric relationships, and then the distance equation can be mathematically expressed as:

$$u_i(t_i) = r_i\, s(\varphi)\left[c(\theta)c(\Omega t_i) + s(\theta)s(\Omega t_i)\right] \tag{5}$$

At another measured date index, i.e. at $(i+1)$, (1) can be expressed as follows:

$$u_{i+1}(t_{i+1}) = r_{i+1}\, s(\varphi)\left[c(\theta)c(\Omega t_{i+1}) + s(\theta)s(\Omega t_{i+1})\right] \tag{6}$$

. In order to get the value of φ and θ, (5) and (6) should be solved simultaneously. However, since t_i is not equal to t_{i+1}, so the solution of these two equations might not be achieved. So that the most challenging problem faced the authors is to timing t_i and t_{i+1} to be the same. The methodology used to achieve that will be briefly described in the next section. For the moment, t_i and t_{i+1} are considered to be the same.

Subtracting (5) from (6) yields

$$\Delta u(t_i) = d_i\left[s(\varphi)c(\theta)c(\Omega t_i) + s(\varphi)s(\theta)s(\Omega t_i)\right] \tag{7}$$

where $\Delta u(t_i) = u_{i+1}(t_i) - u_i(t_i)$ and $d_i = r_{i+1} - r_i$

let $A = s(\varphi)c(\theta)$ and $B = s(\varphi)s(\theta)$ Equation (7) can be written as follows:

$$\Delta u(t_i)/d_i = c(\Omega t_i)A + s(\Omega t_i)B \tag{8}$$

The above equation can be put in the matrix form as:

$$\begin{bmatrix} c(\Omega_1) & s(\Omega t_1) \\ c(\Omega_2) & s(\Omega t_2) \\ . & . \\ . & . \\ c(\Omega t_i) & s(\Omega t_i) \end{bmatrix} \begin{bmatrix} A \\ B \end{bmatrix} = \begin{bmatrix} \Delta u(t_1)/d_i \\ \Delta u(t_2)/d_i \\ . \\ . \\ \Delta u(t_i)/d_i \end{bmatrix} \tag{9}$$

or equivalently,

$$\Psi \Pi = \Delta \tag{10}$$

then

$$\Pi = \left(\Psi^T \Psi\right)^{-1}\left(\Psi^T \Delta\right) \tag{11}$$

Once we get the value of Π, φ and θ can be found as follows:

$$\varphi = \sin^{-1}\left(\|\Pi\|\right) \tag{12}$$

$$\theta = a\tan 2\left(\frac{A}{\|\Pi\|}, \frac{B}{\|\Pi\|}\right) \tag{13}$$

4. Experimental setup

The experimental set-up to measure the deviation of the workpiece surface from a certain datum is shown in Fig.1(a). The workpiece is fixed on the lathe machine by the three clamps. These clamps are rigidly attached on the inner ring of the bearing of the machine. Deviation quantity is estimated utilizing LDS with no contact achieved. Commercial LDS (LK-HD500) of diffuse reflection type is used. The wave length of its light source is 650 nm with spot diameter at reference distance approximately 300 μm. More detailed specification can be found in [13].

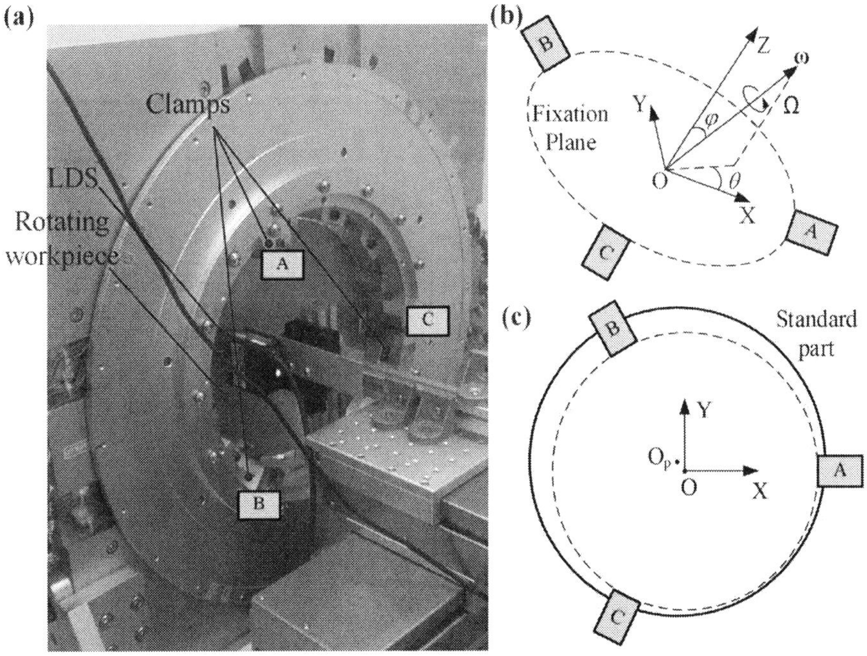

Figure 1. a) Experimental setup [14]; b,c) Coordinate frames

5. Illustrative example

The proposed model is illustrated by a thin-walled titanium disk workpiece fixed on the experimental setup which briefly discussed in the previous section. The disk dimensions are 300 mm outer diameter and 6.6 mm thickness. The workpiece surface roughness is limited to 0.05 mm .The workpiece is rigidly fixed on the machine clamp assembly at points A , B and C . LDS is used to measure the axial deviation of the workpiece surface from the vertical plan. The measured data (Fig.2 (a)) divided into three main portions. At the beginning, the motor should be still for a few seconds to detect the axial displacement of a reference point at a radial distance equal to r_{i+1} , which will be used as a datum point for the periodic measured data (Portion A). After that, the motor starts up with large fluctuation (Portion B). Then the motor rotates at approximately constant speed with a little fluctuation (Portion C). After complete the upper measurement sequence the motor is stopped and the disk is turned back to the same reference point. Then the LDS is moved radially a distance equal to d_i toward the center of the disk. A new data set measured starting from a new reference point which locates at r_i . These procedures are repeated until we get sufficient data sets.

To achieve adequate analysis of these data sets the following steps have been followed:

1) Data smoothing using Gaussian function, which is defined in terms below, in order to spread the data noise at each point over a predefined area.

$$g(x) = \frac{1}{\sigma\sqrt{2\pi}} e^{-\left[(x-\mu)^2 / 2\sigma^2\right]} \tag{14}$$

where σ is standard deviation, μ is mean and x is data point.

2) Determine the displacement value at the reference point by calculating the mean of the measured data within ‘Portion A’, see (Fig.2 (a)).

3) Locate the starting point of each periodic cycle (the point with the same displacement of the reference point).

4) Calculate the actual rotating speed of the motor Ω within 'Portion C'.

$$\Omega = {2\pi}/{T} \tag{15}$$

where T is the periodic time, which can be calculated by knowing of the LDS measuring frequency and the number of data inside one complete cycle.

5) Beginning from the reference point of the periodic cycle of two different measured data, calculate $\Delta u(\mathrm{t}_i)$.

6) Substituting Ω and $\Delta u(\mathrm{t}_i)$ into (9), hence calculate φ and θ from (12), (13).

7) Finally, calculate clamps offsets u_A , u_B and u_C using (3).

These steps have been applied to a workpiece in-hand. The measured data before and after data smoothing are shown in Fig.2 (b).

Fig. 2 (c) shows two different in-phase data sets (measured data at different radius) with different amplitudes and zero phase shift. The difference between two waves $\Delta u(\mathrm{t}_i)$ is also plotted in the same figure.

Fig.2 (d) shows the axial deviation of the disk surface from the vertical plane measured at the outer diameter of the disk, and hence this is considered to be the maximum deviation. It can be seen that deviation amplitude is approximately 1.2 mm. After the experiment was executed the deviation amplitude decreased from 1.2 mm to less than 0.2 mm. However, this amplitude (0.2 mm) is still larger than the target deviation (0.1 mm). Moreover, the insight into Fig.2 (d), the measured data reveals that the deviation fluctuates from 0.19 mm to 0.26 mm (the maximum deviation 0.07 mm) for 85% of each cycle (green rectangle). And for the remaining 15% of the cycle the deviation increased very quickly to reach 0.2 mm (red rectangle). This increasing may be resulted from some inherent flatness defects in the workpiece in this portion. Therefore, 0.2 mm is considered to be reasonable for the investigated disk. (0.2mm > 0.1mm+0.05mm, not reasonable yet).

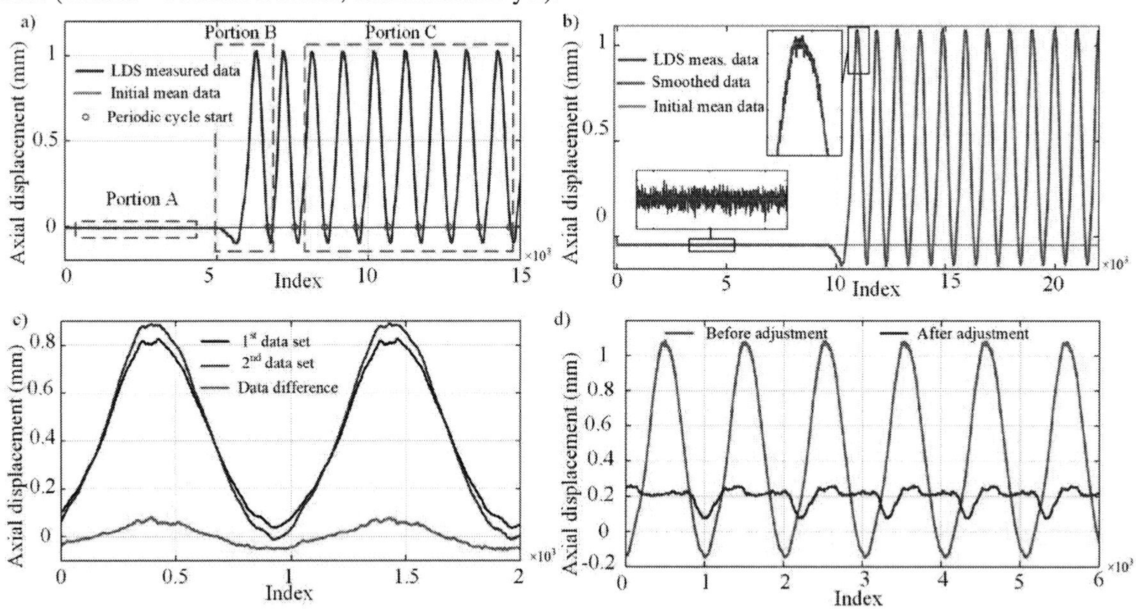

Figure 2. a) Portions of measured data; b) Data smoothing; c) In-phase data sets; d) Measured data before and after adjustment

IOP Conf. Series: Materials Science and Engineering **751** (2020) 012080 doi:10.1088/1757-899X/751/1/012080

6. Conclusion

Fixation system for thin disks on turning machine must fulfill a precise rigid fixation during machining operation. However, current fixtures systems do not guarantee coaxial fixation of the workpiece on the turning machine bearing, which may be result in workpiece damage. Hence, a novel approach is proposed to handle this problem. Experiments are executed to verify the proposed approach. The experimental results show high accuracy of the fixation system with a reasonable deviation.

Reference

[1] Cheung E, Yuan W and Hua M 1999 Physical Simulation of the Deflection in Turning of Thin Disk-Shaped Workpieces *The International Journal of Advanced Manufacturing Technology* **15** 863-8

[2] Guo J, Lee K-M, Wuguang L and Bo W 2015 Design Criteria Based on Modal Analysis for Vibration Sensing of Thin-Wall Plate Machining *Mechatronics, IEEE/ASME Transactions on* **20** 1406-17

[3] Elsheikh A H, Guo J and Lee K-M 2019 Thermal deflection and thermal stresses in a thin circular plate under an axisymmetric heat source *Journal of Thermal Stresses* **42** 361-73

[4] Elsheikh A H, Guo J, Huang Y, Ji J and Lee K-M 2018 Temperature field sensing of a thin-wall component during machining: Numerical and experimental investigations *International Journal of Heat and Mass Transfer* **126** 935-45

[5] Salman K h, Elsheikh A H, Ashham M, Ali M K A, Rashad M and Haiou Z 2019 Effect of cutting parameters on surface residual stresses in dry turning of AISI 1035 alloy *Journal of the Brazilian Society of Mechanical Sciences and Engineering* **41** 349

[6] Koepfer C 1996 Four considerations for selecting chucks and arbors *Modern Machine Shop* **69** 88-94

[7] Donmez M A, Blomquist D S, Hocken R J, Liu C R and Barash M M 1986 A general methodology for machine tool accuracy enhancement by error compensation *Precision Engineering* **8** 187-96

[8] Ema S and Marui E 1994 Chucking Performance of a Wedge-Type Power Chuck *Journal of Engineering for Industry* **116** 70-7

[9] Pahlitzsch G and Hellwig W 1968 *Advances in Machine Tool Design and Research 1967,* ed S A Tobias and F Koenigsberger: Pergamon) pp 97-118

[10] Byun J 2003 Methods for Improving Chucking Accuracy of a Cylindrical Workpiece in Finish Hard Turning. (United States -- Indiana: Purdue University)

[11] Byun J and Liu C R 2012 Methods for Improving Chucking Accuracy *Journal of Manufacturing Science and Engineering* **134** 051004-

[12] Byun J and Liu C R 2008 Improving Chucking Accuracy and Repeatability by Reducing Kinematic Redundancy. In: *3rd JSME/ASME International Conference on Materials and Processing,* (Evanston, Illinois, USA: ASME)

[13] Keyence http://www.keyence.co.uk/.

[14] Bai K, Qin J, Lee K-M and Hao B 2019 Design and chatter prediction analysis of a duplex face turning machine for manufacturing disk-like workpieces *International Journal of Machine Tools and Manufacture* **140** 12-9

The Numerical Simulation of Temperature Field in Friction Stir Welding of 7075 Aluminium Alloy

F J Ma*, D Y Fu, Y Liu, D P Yang, S F Zhang and Z H Sha

School of Mechanical Engineering, Dalian Jiaotong University, No. 794, Huanghe Road, Shahekou District, Dalian 116028, China

*The corresponding author's e-mail address: mafj@djtu.edu.cn

Abstract. Considering friction heat generation between rotating tool and workpiece and plastic deformation heat generation of material, a numerical simulation model of friction stir welding of 7075 aluminium alloy sheet was established. In this study, the temperature field of the welding process was analysed, and the effects of different stirring speeds and welding speeds on the temperature field were compared. The results show that the high temperature region is located below the shoulder and diffuses outward in a ring shape. After the welding is completed, the workpiece center temperature is high and the ambient temperature is low. The stirring speed of the tool has a large effect on the peak temperature during the welding process. The effect of the welding speed on the peak temperature is less than the stirring speed. The comparison shows that the welding temperature is about 478°C at a stirring speed of 800r/min and a welding speed of 120mm/min, which is a suitable welding parameter.

1. Introduction

Friction stir welding (FSW) technology is a new solid phase connection technology invented by the British Welding Research Institute (TWI) in 1991[1]. The frictional heat generated between the rotating tool and the workpiece to be welded during welding causes the material in the vicinity of the rotating tool to be thermoplastic. When the rotating tool moves forward, the dense solid phase connection joint is formed under the forging action of the shoulder.

FSW has been widely used because of its low soldering temperature, low post-weld residual stress, and high joint performance. In recent years, many scholars have analysed the temperature field of FSW welding process. Song et al. [2] established a three-dimensional heat source model based on the tribological principles and the shear rheology theory for FSW in a moving coordinate system. The final temperature field results are in agreement with the experimental results. Fu Ningning [3] obtained a FSW heat source model considering the frictional heat generation of the shoulder and the rotating tool. The results show that the temperature curve is sensitive to the horizontal location of weldment. Based on the heat generation theory of FSW, Wan Shengqiang [4] developed an adaptive moving heat source considering the friction head friction heat generation. He established a finite element model of 2219 aluminium alloy FSW process, and the welding temperature was 532.4°C.

The above research on the temperature field of FSW considers the frictional heat generation of the shoulder and the pin, ignoring the plastic deformation heat generation of the material. Chen [5] used fluid dynamics to simulate material flow to analyse heat production. But he ignored the frictional heat generation due to the potential sliding at the interface between tool and material.

In this study, the heat source model of FSW process is established by considering friction heat generation and material plastic deformation heat generation. A moving heat source is used to instead of the rotating tool to move evenly on the workpiece. The distribution law of the temperature field is obtained, and the effects of different speeds and welding speeds on the welding temperature field are compared.

2. Establishment of heat source model

2.1. Heat generation analysis of the welding process

During the FSW process, the friction between the rotating tool and the material generates heat to raise the temperature of the material to a plastic flow state, then the plastic state material flows backward to the stirring needle. In this process, the heat is also generated due to plastic deformation, so the welding heat is mainly composed of three parts: the heat generated by the friction between the head shoulder and material (shoulder heat generation), the heat generated by the friction between the stirring pin and material (pin heat generation), heat generated by the shearing of material. The location of the heat source is shown in Figure 1.

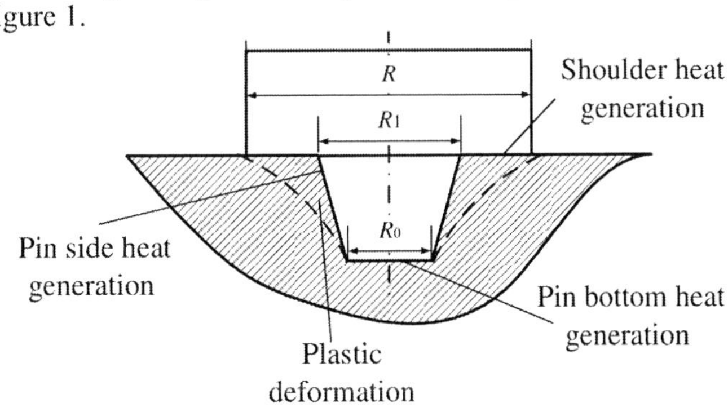

Figure 1 Heat source distribution

2.2. Frictional heat source model

According to the classical friction theory, the interfacial friction heat due to the relative movement between the tool and workpiece is given as follows [6]

$$Q = Q_1 + Q_2 + Q_3 = \frac{2}{3R^2}\mu\omega F\left(R^3 - R_1^3\right) + \frac{2}{3R^2 \sin\alpha}\mu\omega F\left(R_1^3 - R_0^3\right) + \frac{2}{3}\mu\omega F R_0 \tag{1}$$

In equation (1), Q_1, Q_2 and Q_3 are the heat generated by the shoulder, the pin side and the pin bottom, respectively. F is the shoulder pressure, ω is tool stirring speed, R is the radius of shoulder, R_1 is the radius of pin (bottom), R_0 is the radius of pin (upside), α is the cone angle of the pin.

2.3. Plastic deformation model

During the welding process, part of the deformation work is released into the surrounding environment by thermal energy, and the rest of the deformation work is used for plastic deformation of the material. The heat flux density of plastic deformation [7]

$$q_p = \bar{\sigma}\alpha_p\dot{\bar{\varepsilon}} \tag{2}$$

In equation (2), q_p is heat flux density, $\bar{\sigma}$ is equivalent stress, α_p is thermal conversion efficiency, usually between 0.9~0.95, $\dot{\bar{\varepsilon}}$ is equivalent strain rate.

The heat generated by plastic deformation can be expressed as

$$Q_v = \int_v \bar{\sigma}\alpha\dot{\bar{\varepsilon}}\,\mathrm{dV} \tag{3}$$

In equation (3), V is the volume of plastic material.

3. Establishment of finite element model

3.1. Material properties

In this study, the material to be welded is 7075 aluminium alloy, and the material density is 2810kg/m^3, Young's modulus is 71.7GPa and Poisson's ratio is 0.3. During the welding process, some material properties such as specific heat capacity and thermal conductivity are sensitive to the temperature. The specific parameters are shown in Table 1.

Table 1. The material parameters of 7075 aluminium alloy

Temperature/°C	Specific heat capacity/(W•m^{-1}°C^{-1})	Thermal conductivity/(J•kg^{-1}°C^{-1})
20	835.4	166.7
100	897	160.8
150	916.3	152.7
200	974	142.2
300	1012.5	135.7
400	1128	128.4
500	1205	114.8

3.2. Boundary conditions

During the FSW process, the ambient temperature and the convective heat transfer coefficient between the workpiece and air are known. This is consistent with the third type of boundary condition for heat conduction problems [8].

The boundary condition can be expressed as

$$-\lambda \frac{\partial t}{\partial n}\bigg|_s = h\left(t|_s - t_f\right) \tag{4}$$

In equation (4), subscript s indicates the boundary surface, $\partial t/\partial n|_s$ is the normal temperature gradient on the boundary surface of the object.

The temperature field of the workpiece is substantially symmetrical along the sides of the welded joint surface during the welding process. In order to simplify the calculation, the welded joint surface is treated as a symmetry plane. The bottom surface of the weldment that is in contact with the workbench is set as a heat conduction surface, and the remaining surfaces are convective heat transfer surfaces. The boundary conditions are shown in Figure 2.

3.3. Finite element model

The size of the workpiece is 100mm×80mm×6mm. The stirring pin is conical, the radius of shoulder is 20mm, the bottom radius of pin is 5mm, the upside radius of pin is 3mm, and the height of the pin is 5mm. An eight-node linear heat transfer hexahedral element is used to mesh the workpiece. From the centre to the side of the workpiece, the mesh density is gradually reduced.

In the heat source model, the shoulder heat generation is treated as a plane heat source that is axisymmetric with respect to the weld and the heat flux density increases as the shoulder radius increases. In order to facilitate the post-processing analysis, the starting point of the heat source is set to the coordinate origin, and the welding direction is the positive direction of the x-axis. The established finite element model is shown in Figure 3.

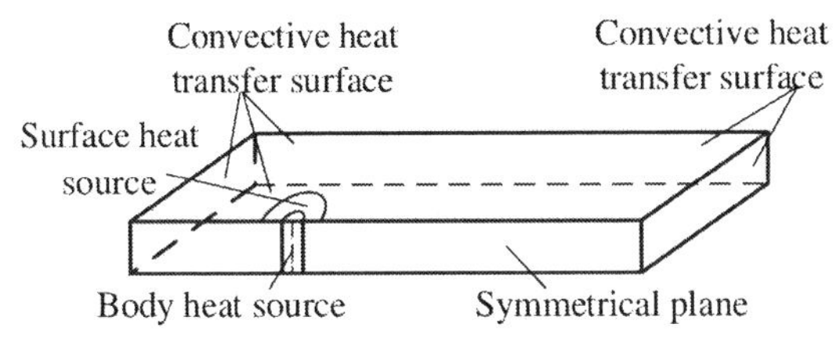

Figure 2. Boundary conditions.

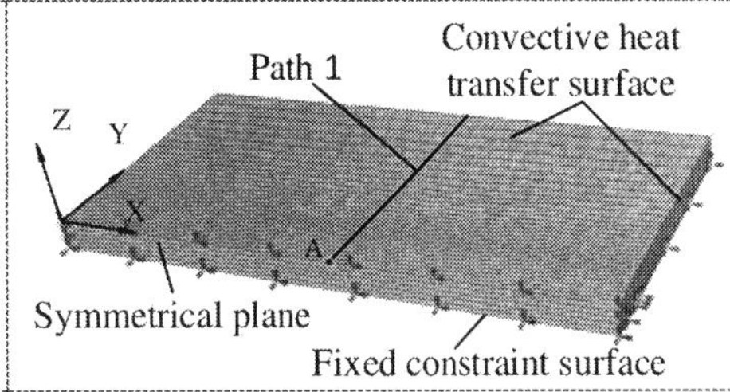

Figure 3. Welding finite element model.

4. Numerical simulation results and analysis

4.1. Overall temperature distribution of welding process

Under the condition that the plunge depth of shoulder is 0.1mm, the stirring speed is 800r/min, the welding speed is 120mm/min, the welding and the cooling time are 50s and 300s, respectively. The temperature field of 7075 aluminium alloy FSW was analysed by using the established finite element model. Temperature distribution at different stages are shown in Figure 4.

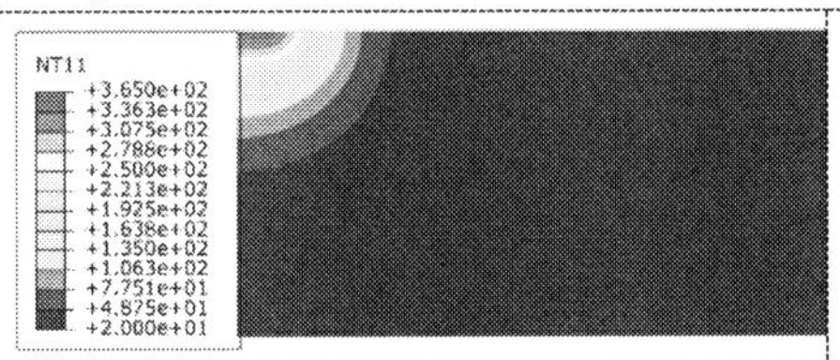

(a) t=2s

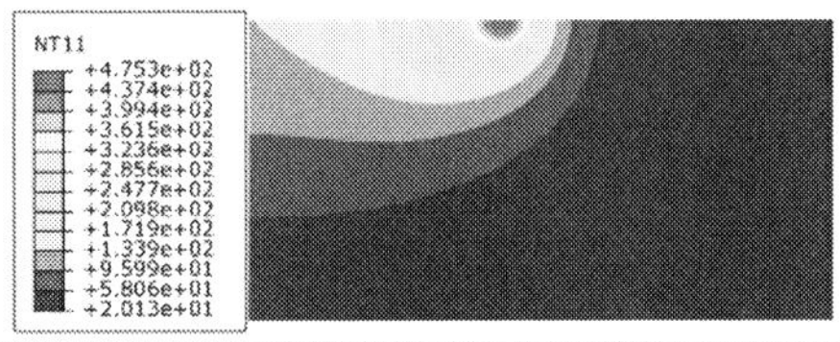

(b) t=30s

(c) t=50s

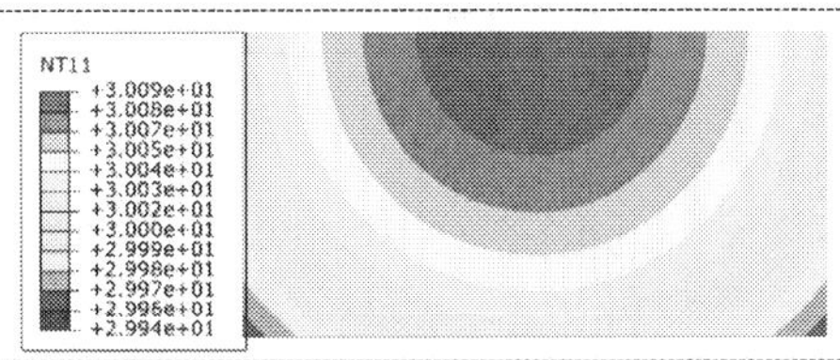

(d) t=350s

Figure 4. Temperature distribution at different times of the welding process.

According to Figure 4(a), the high temperature region is concentrated under the shoulder, which is ellipsoidal and spreads around. As shown in Figure 4(b) and Figure 4(c), as the welding progresses, the heat-affected zone expands and the workpiece temperature rises. At the end of the welding, the rotating tool pulled out quickly to allow the workpiece material to cool naturally. As shown in Figure 4(d), the convection heat exchange around the workpiece with the air and the middle part of the heat is superimposed so that the central temperature is high and the ambient temperature is low.

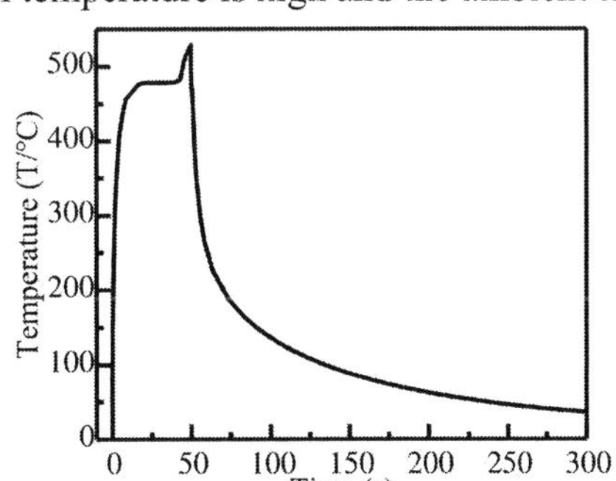

Figure 5. Welding process temperature graph.

Figure 5 shows the overall temperature distribution during the welding process. As the heat source moves, the welding temperature gradually rises from 20°C and stabilizes at around 478°C. The stabilization time lasts for about 30 seconds. This is the welding stabilization phase. The points at this stage are representative when performing temperature analysis. The feature point A (0, 50, 0) at the weld stabilization stage of the weld is selected, and the path 1 perpendicular to the weld direction is obtained from point A as shown in Figure 3. As the welding progressed, a temperature rise occurred near the end of the weldment and the peak temperature reached 530°C. At the end of the welding, the heat absorbed by the workpiece in a certain period of time is higher than the amount of heat lost. At the same time, the workpiece temperature accumulation is high when soldered to the end, and the combined effect of these factors causes the temperature to rise. After welding, the rotating tool leaves the workpiece, and the workpiece is rapidly cooled under convection and radiation heat dissipation.

4.2. Effect of stirring speed on temperature field of FSW

The single factor method was used for the simulation analysis in order to study the effect of the rotating tool speed on the welding temperature field. The temperature curve of the path 1 at different speeds is shown in Figure 6 when the plunge depth of the shoulder is 0.1 mm and the welding speed is 120 mm/min.

It can be seen from Figure 6 that the peak temperatures at five speeds are 398°C, 431°C, 478°C, 514°C, 546°C, and the welding temperature increases with the increase of the tool stirring speed. From equation (1), when the welding speed is constant, the higher the stirring speed, the more work the material rubs against the shoulder at the same time, and the higher the heat production. By equation (2) can be obtained the higher the speed, the faster the flow of the plastic material, the plastic deformation heat production increased. If the temperature is too high, the surface of the weld will be severely oxidized to reduce the quality of the weld. The temperature at 400 r/min is only 398°C, the material fails to reach the thermoplastic state, and it is easy to cause poor soldering.

4.3. Effect of welding speed on temperature field of FSW

At the stirring speed of 800r/min, the temperature curve of point A at the welding speeds of 180 mm/min, 150 mm/min, 120 mm/min, 90 mm/min and 60 mm/min is shown in Figure 7.

It can be seen from Figure 7 that the peak temperatures at the five welding speeds are 415°C, 458°C, 478°C, 518°C and 538°C, respectively. When the welding speed decreases, the peak temperature increases gradually. This occurs because the temperature accumulation time is so long that the heat input is increased to cause the temperature of the material to rise. If the soldering temperature is too high, void defects affecting the quality of the soldering will occur. When welding speed is 120 mm/min, the peak temperature can reach 478°C, which is 78% of the melting point of the aluminium alloy. This is consistent with the statement that during the FSW welding process, the melting point temperature of the joint generally does not exceed 80% of the melting point of the material to be welded [9].

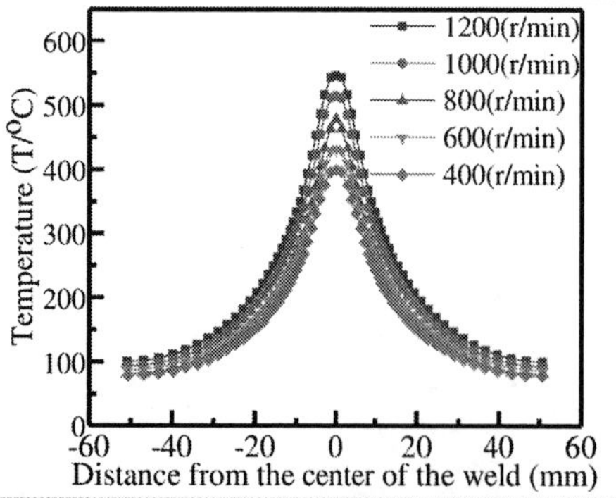

Figure 6. Temperature curve at different stirring speeds.

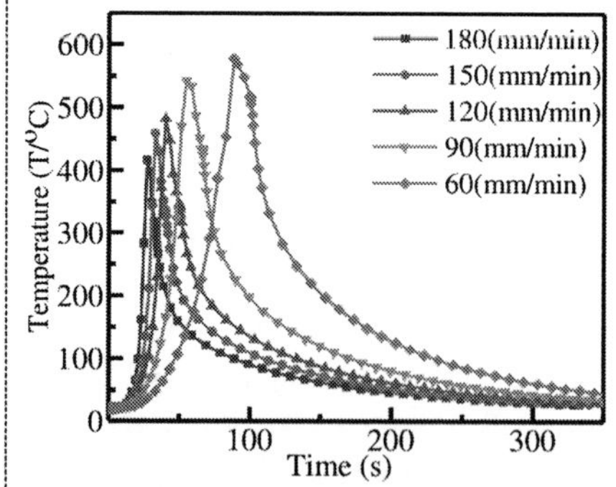

Figure 7. Temperature curve at different welding speeds.

5. Conclusion

(1) In this study, a heat source model of FSW is established, considering the friction heat generation and plastic deformation heat generation, and the temperature field of the welding process is analysed. The temperature of the workpiece gradually rises first, and then tends to be stable, and the high temperature area is located below the shoulder and spreads outward as a ring shape during the welding process.

(2) The stirring speed has a large effect on the peak temperature during the welding process. The increase of the stirring speed can cause the plastic deformation heat generation and the friction heat generation to be increased greatly, so the peak temperature of the welding changes greatly.

(3) The effect of the welding speed on the peak temperature is smaller than the stirring speed. The welding speed is low, and the heat accumulation time is long so that the temperature rises. So the peak temperature rises as the welding speed decreases. The welding temperature is about 478°C at a stirring speed of 800r/min and a welding speed of 120 mm/min, which is a suitable welding temperature.

Acknowledgments

The authors would like to acknowledge the financial support for this research from High-level Talents Innovation Support Project of Dalian (No. 2017RQ049).

References

[1] Thomas W M and Nicholas J C 1991 Friction stir butt welding. International patent application: No.PCT/GB92/02203 and G B patent application No.9125978.8.
[2] Song M and Kovacevic R 2003 Thermal modeling of friction stir welding in a moving coordinate system and its validation *International Journal of Machine Tools & Manufacture* **43**:605-15.
[3] FU NN, ZHU W and HUO QY 2016 Temperature field analysis of refilling friction stirring welded aluminium alloy plates *Journal of Jiangsu University of Science and Technology (Natural Science Edition)* **30**:435-8. (in chinese)
[4] WAN SQ, et al. Thermal mechanical coupling simulation of temperature and residual stress in friction stir welding of 2219 aluminium alloy *Hot Working Technology*, 2019. **48** 159-63. (in chinese)
[5] Chen G, Shi Q and Li Y 2013 Computational fluid dynamics on heat generation during friction stir welding of aluminium alloy *Computational Materials Science* **79** 540-6.
[6] Zhu JM 2013 Ultrasonic assisted friction stir welding process of 2024-T4 aluminium alloy sheet *Central South University*. (in chinese)
[7] Chen T 2012 Research the temperature field of friction stir welding numerical simulation *Northeastern University*. (in chinese)
[8] Zhang XM 2014 Heat transfer Beijing: China Building Industry Press.(in chinese)
[9] He DQ, Sun X and Zhu JM 2014 Numerical simulation on welding temperature field of 2024-T4 aluminium ultrasonic assisted friction stir *Hot Working Technology* **43** 162-168. (in chinese)

IOP Conf. Series: Materials Science and Engineering **751** (2020) 012082 doi:10.1088/1757-899X/751/1/012082

Study of Different Micro Milling Blades on Milling Titanium Alloy TC4

ZiChuan Zou[1,*], XianFeng Zhao[1], Lin He[2,3] and XueTing Jiang[1]

[1]School of Mechanical Engineering, Guizhou University, 550025, Guizhou, China;
[2]Liupanshui Normal College,553004, Guizhou, China;
[3]Guizhou Key Laboratory of Special Equipment and Manufacturing Technology, 550025, Guizhou, China

[*]E-mail address: zichuanzou@163.com

Abstract. With the development of micro-miniature, micro-milling technology has been widely used. Due to the difference in workpiece materials and processing technology, the selection of the edge structure of the micro-milling cutter during processing is particularly important. This paper analyzes the influence of micro-milling of titanium alloy TC4 on cutting force, cutting heat and tool stress distribution for micro-milling cutters with a diameter of 0.1mm based on Definite-3D, ANSYS and other finite element simulation software.The results show that the spiral-edge micro-milling cutter can effectively reduce the cutting force and cutting heat compared with the straight-blade micro-milling cutter. Considering the rigidity, strength, cutting performance and practicability of the tool, the traditional double-edged spiral-edge micro-milling is still the best choice. The research results of this paper can provide a theoretical basis for the design of micro-milling tools.

1. Introduction

With the continuous development of science and technology, the demand for micro-featured parts in the defense, military, aerospace, medical equipment and other manufacturing fields has increased significantly. Due to the high efficiency, high flexibility, high aspect ratio structure of micro-milling technology, micro-milling Tool technology has become a research hotspot in the field of micromachining [1-3]. Due to the small diameter and low rigidity of the micro-milling cutter, the overall rigidity of the tool is usually increased as much as possible in the tool design. Due to the different processing techniques and parts requirements, domestic and foreign scholars have developed a variety of micro-milling cutters with different blade structures. Asma Perveen [4] et al. used EDM to fabricate a micro-milling cutter with a V-groove and analyzed the effect of the shape of the cutting edge on the tool wear. It was found that the D-edge cutting tool has a triangular-to-square cutting tool. Better wear resistance. Uhlmann [5] et al. improved the rigidity of the tool by reducing the cutting length of the cutting edge and optimized the end mill with a diameter of 0.5 mm. Fang [6] et al. used finite element software to compare the structure of straight-blade triangle, D-shaped and pyramid-shaped micro-milling cutters with the traditional spiral end mill structure, and through a large number of experiments, from the tool stiffness, pointed out Cone D-shaped end mills are more suitable for micro-milling. Jiang et al.[7] optimized the spiral groove and edge shape of the end mill based on bionics, effectively suppressing the generation of serrated chips and reducing the fluctuation of cutting force. AURICH et al. [8] studied the design and sharpening of single-blade spiral micro-milling cutters, and proved that the micro-milling cutter with spiral groove structure can significantly reduce the workpiece burr. Li P et al.[9] used the analytical modeling and finite element method to geometrically optimize the commercial end mill with a diameter of 0.5mm, increase the tool neck angle and adopt the negative rake angle structure to improve the rigidity and strength of the tool. Small tool wear and extended tool life.

As a typical difficult-to-machine material, titanium alloy TC4 has been widely used in the industrial field due to its excellent comprehensive properties such as high temperature resistance and corrosion resistance[10-11], due to the high speed of the micro-milling spindle and the small diameter of the milling cutter. In the processing of difficult-to-machine materials such as titanium alloy TC4, the phenomenon that the blade is broken is more likely to occur, which is not conducive to the experiment. Therefore, many scholars usually use plastics such as copper and aluminum alloy to study [12-13]. There are few studies on micro-milling titanium alloy TC4 materials.

Although domestic and foreign scholars have carried out a lot of research on micro-milling technology, they mainly focus on the preparation of micro-milling cutters [14-17], structural optimization design [18-20], etc., for the milling performance of many different edge-type micro-milling cutters. The comparative study is not sufficient. Due to the complicated milling movement, the micro-milling scale is very small, modeling is difficult, and considering the long manufacturing cycle and complicated process of micro-milling cutter, this paper analyzes the milling of titanium alloy TC4 with different edge-type micro-milling cutters by using finite element software such as DEFORM and ANSYS. The influence law of cutting force and cutting heat is applied. The modal analysis of micro-milling cutters with different edge shapes is carried out. The static and dynamic characteristics of the micro-milling cutters are studied, which provides a reference for the optimization of the edge shape of the micro-milling titanium alloy TC4.

2. Establishment of finite element simulation model

Finite element is currently the common method for simulating the cutting process of the tool. The commonly used finite element software for the simulation cutting process is abaqus, Ls-dyna, Deform-3D and AdvantEdge. Deform-3D has the advantages of good robustness, high calculation accuracy and high reliability. Therefore, this article uses Deform-3D as the simulation software. The simulation uses side milling, the tool material is YG6, and its physical performance parameters are shown in Table 1.

Table 1. Performance parameters of YG6

Performance	Numerical value
Density/(g/cm3)	14.6-15.0
Bending strength/(MPa)	1450
Hardness/(HRA)	89.5
Impact toughness /(J/cm^2)	2.6

The workpiece material is Ti-6Al-4V, and its physical property parameters are shown in Table 2.

Table 2. Performance parameters of Ti-6Al-4V

Performance	Numerical value
Density/(g/cm3)	4.51
Poisson's ratio	0.3
Tensile strength/(MPa)	860
Breaking strength /(MPa)	967
Specific Heat Capacity /[J/(g·K)]	0.61
Coefficient of Linear Expansion /(K-1)	8.6×10-6
Elongation /(%)	8

IOP Conf. Series: Materials Science and Engineering **751** (2020) 012082 doi:10.1088/1757-899X/751/1/012082

Through the solidworks software, the tool cutting edge section is Δ shape (a, g), D shape (b, n), rectangle (c, i), regular pentagon (d, j), regular hexagon (e, k) And the traditional double-edged spiral (f, l) straight edge, spiral edge two types of micro-milling geometry model, as shown in Figure 1. The established milling cutter and workpiece geometry model are imported into Deform-3D for pre-processing. The workpiece size is 0.1mm×0.05mm×0.1mm, the bottom surface of the workpiece is set to full constraint, and the tool rotates around the Y axis and feeds along the Z axis. In order to improve the simulation efficiency, the cutting edge length is set to 0.2mm. Table 3 shows the machining process parameters used in the simulation experiment. The established micro-milling model is shown in Figure 2.

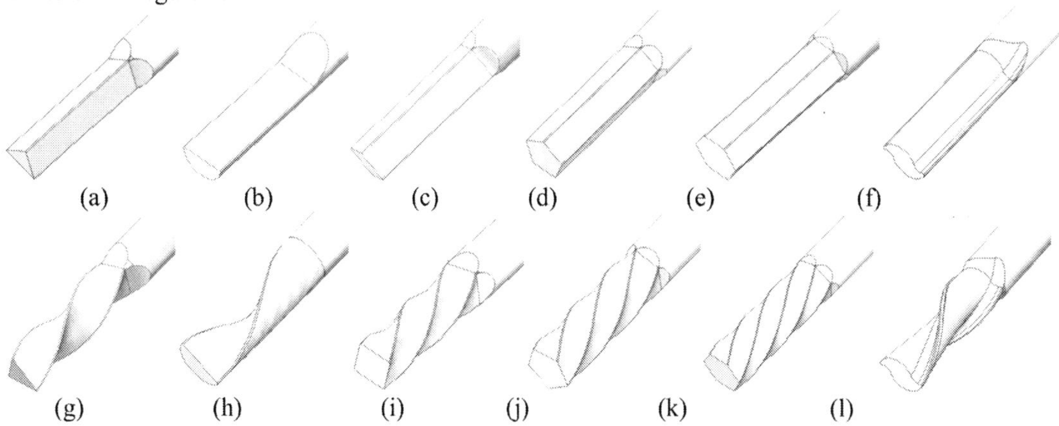

Figure 1. Different milling cutters

Table 3. Processing parameters

Process parameters	Numerical value
Spindle speed/(rpm)	40000, 45000, 50000, 55000, 60000
Feed per tooth/(mm/tooth)	0.005
Radial width of cut/(mm)	0.005
Axial width of cut/(mm)	0.05
Initial temperature/(°C)	20

Figure 2. Micro-milling model

3. Simulation result analysis

Single-factor experimental method was used to analyze the milling force and milling heat during the milling of titanium alloy TC4 under the same milling conditions for five different spindle speeds under the conditions of 40000, 45000, 50000, 55000 and 60,000 rpm. Influence the law. Due to the data processing, it is found that the milling force and the cutting heat change trend of the micro-milling cutters with different blade structures are basically the same under different speed conditions. Therefore, the data in the figure are taken as the autonomous axis rotation speed n=50000r, and the results are analyzed as follows.

3.1Analysis of cutting force

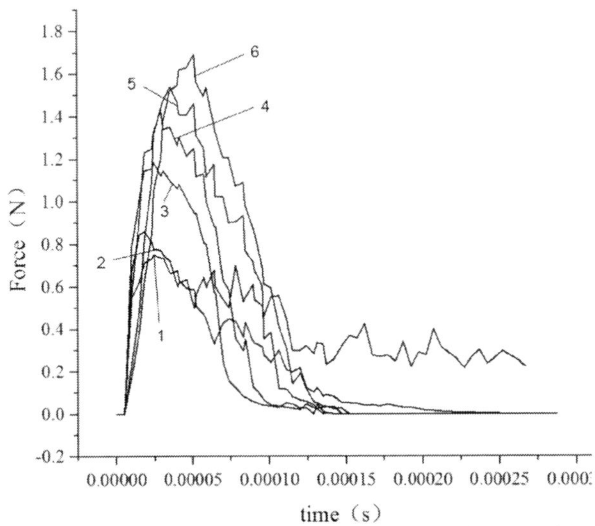

Figure 3. Comparison of milling force variation of straight edge micro-milling cutter
1.Traditional double-edged 2.D 3.△ 4.Rectangle 5.Positive pentagon 6.Hexagon

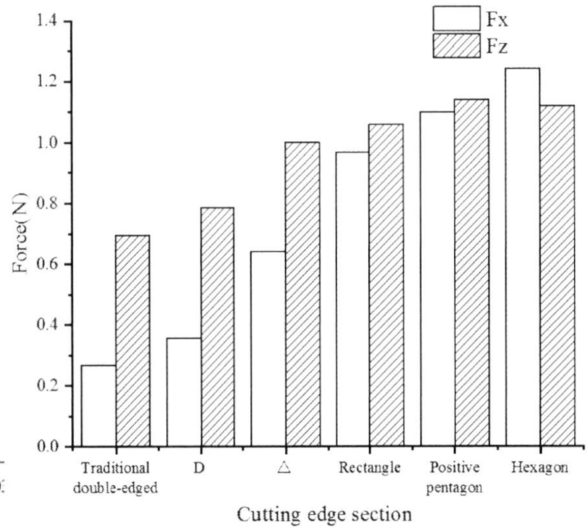

Figure 4. Comparison of cutter milling force of Straight edge type micro-milling

Figure 3. shows the milling force variation curve of a straight edge type micro-milling cutter with different cutting edge sections in one milling cycle. It can be seen that the milling force is the same as the milling cutter edge during the milling process. With the increase of the increase, the micro-milling cutter with the traditional double-edged spiral edge section can obtain smaller milling force than the micro-milling cutter of other sections; and the moment of the maximum milling force increases with the number of teeth of the milling cutter Gradually delayed, this is because the straight edge micro-milling cutter with △ shape, rectangular shape, regular pentagon shape and regular hexagonal cross section can be regarded as negative rake angle cutting during milling, and the corresponding rake angle is -30°, respectively. 45°, -54°, -60°, as the absolute value of the negative rake angle increases, the values of the radial force Fx and the tangential force Fz also gradually increase, and when the absolute value of the negative rake angle is greater than 54° The direction force Fx is gradually greater than the tangential force Fz, as shown in figure 4.

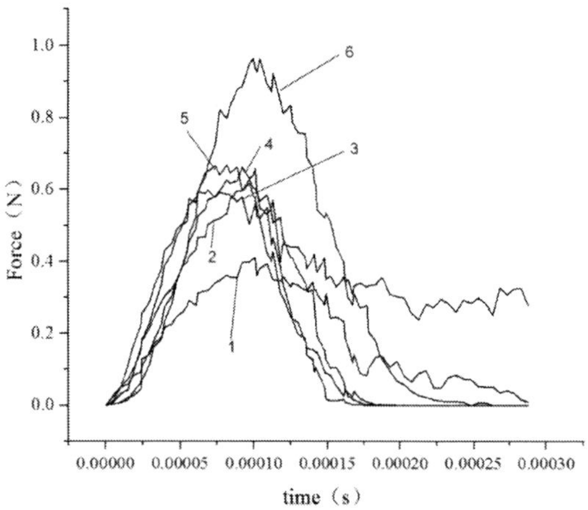

Figure 5. Comparison of milling force variation of spiral edge micro-milling cutter
1.Traditional double-edged 2.D 3.△ 4.hexagon 5.Rectangle 6.Positive pentagon

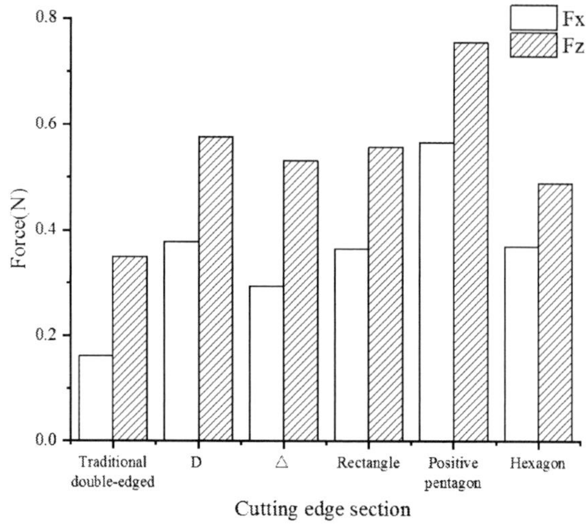

Figure 6. Comparison of cutter milling force of Spiral Edge Micro-milling

Figure 5. shows the milling force variation curve of a helical edge type micro-milling cutter in one milling cycle. Compared with Figure 3, it can be seen that the milling force generated by the spiral-edge micro-milling cutter is smaller than that of the straight-blade type micro-milling. The knife is small, the minimum milling force is only about 0.4N, and the milling force increases first and then decreases with the increase of the number of milling cutters. The spiral blade micro-milling cutter with the regular pentagon edge section is compared. The maximum milling force is generated, and the maximum milling force generally occurs during the half of the milling cycle, at which point the tool is rotated by 45°, which is the moment when the cutting edge is completely cut. The milling force component of the spiral edge type micro-milling cutter is shown in Figure 6. The difference compared with the straight edge type is that the radial force Fx is always smaller than the tangential force Fz as the absolute value of the negative rake angle increases.

3.2Analysis of cutting heat

During the metal cutting and forming process, the heat of cutting not only affects the service life of the tool, but also affects the stability of the machining process and destroys the surface quality of the workpiece. In general, the cutting heat is transmitted along with the cutting fluid and the chips. However, due to the particularity of the micro-milling process, the diameter of the micro-milling cutter is small, and the chip removal is difficult. The flow of the cutting fluid causes a certain impact on the tool, which leads to the tool. Breaking and other problems, so the cutting fluid is basically not allowed during processing. Therefore, it is especially important to select the right tool and match the correct processing parameters to effectively reduce the heat of cutting.

Through the analysis, the cutting temperature of the micro-milling cutter with different edge structures is changed with the shape of the cutting edge, as shown in Figure 7. It can be seen that: 1. The cutting temperature gradually increases with the increase of the number of edges, but the increasing range is gradually slower; 2. Under the same processing parameters, the traditional double-edged spiral micro-milling cutter can significantly reduce the cutting. Temperature, the temperature during cutting is 169.5 ° C, far lower than the processing temperature of other types of tools. Compared with the spiral-edged tool, the straight-blade tool produces a higher cutting temperature during the machining process. The main reason is that the straight-blade cutter is the cutting edge and cuts into the workpiece at the same time, and the spiral-edge cutter is progressively cut into the workpiece, reducing The unit cutting force, and the existence of the spiral groove improves the chip's chip and chip removal ability to a certain extent, thereby reducing the cutting temperature.

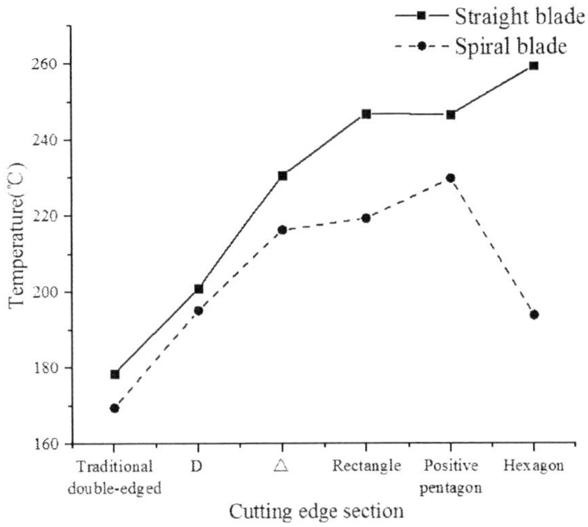

Figure 7. Milling temperature comparison of different edge type micro-milling

In general, the traditional double-edged spiral-edge structure micro-milling cutter can effectively reduce the cutting force and cutting temperature during cutting. Different tool edge shapes can be regarded as changes in the tool rake angle. As the rake angle increases, the cutting force and cutting temperature decrease continuously, but the larger the rake angle may reduce the overall rigidity of the tool, and the cutting edge is easier. damaged. Therefore, changes in tool stiffness must also be considered in tool design.

3.3 Stress analysis

Due to the small diameter and low structural strength of the micro-milling cutter, it is easy to break during the micro-milling process. Therefore, the stress analysis of the structure of the micro-milling cutter must be carried out to investigate the deformation and stress variation of the micro-milling cutters with different cutting edge shapes. The edge shape optimization of the micro-milling cutter can be used for reference. Referring to the conventional double-edged spiral micro end mill with a diameter of 0.1mm, the cutting edge structure and the cutting edge section are shown in Figure 8. The structural parameters of the micro milling cutter holder are shown in Figure 8. and Table 4, and different blade structures are established. A three-dimensional model of a complete micro-milling cutter.

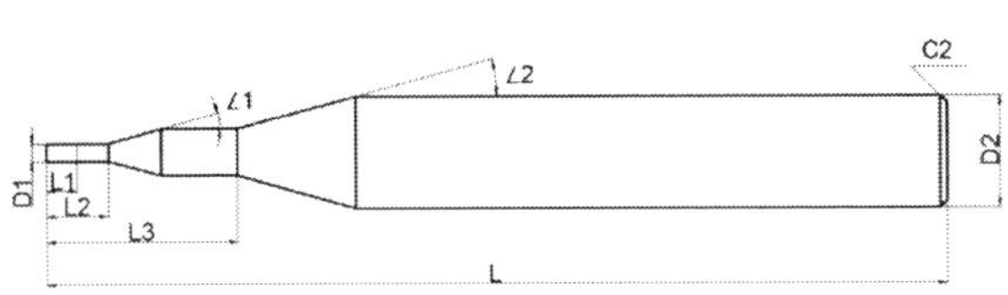

Figure 8. Micro-milling cutter structure diagram

Table 4. Structure parameters

D1 (mm)	D2 (mm)	L (mm)	L1 (mm)	L3 (mm)	∠1 (°)	∠2 (°)
0.1	4	50	0.2	10.5	15	15

The established micro-milling model is imported into the finite element software Ansys for static analysis. The tool overhang is set to 20mm. The overall structure of the tool is divided into three sections for mesh division. The head part (L2) has the smallest unit size and the grid is more Dense, can effectively improve simulation efficiency and accuracy. The surface loads are applied to the rake face of the micro-milling cutters with different edge structures to solve the displacement variation and the maximum stress value of the different milling cutters with different blade structures under the same external force load conditions.

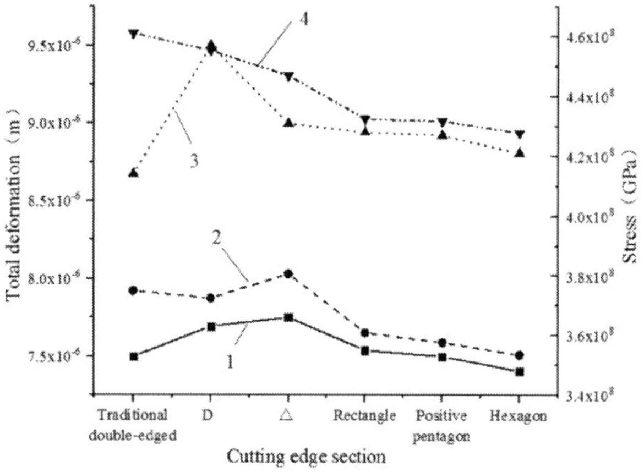

Figure 9. Comparison of deformation and maximum stress of micro-milling cutters with different cutting edge shapes
1.Spiral edge stress 2.Spiral blade deformation
3.Straight edge stress4.Straight blade deformation

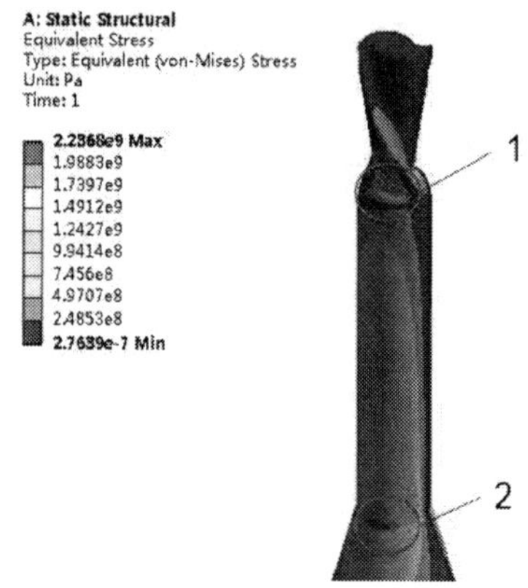

Figure 10. Micro-milling cutter stress distribution

As shown in Figure 9, the displacement variation and maximum stress of the micro-milling cutters with different cutting edge shapes can be seen. 1. The helical blade type micro-milling cutter has less deformation and stress than the straight-blade type. The maximum stress occurs at the end of the spiral groove of the cutting edge and the joint between the root of the cutter head and the neck, that is, the easy break point during the machining process, as indicated by the circle in Figure 10. 2. Compared with Fang et al. [6], the workpiece material is selected from brass, which has good plasticity. It is easier to process than titanium alloy TC4, and it adopts end milling mode, and its cutting amount is smaller. It is found through research that the stiffness and overall strength of the D-type tool are better than those of other edge-type structures under the same milling conditions, but it ignores the special structure of the D-type tool whose back angle is too small during the actual cutting process. It will increase the unnecessary contact between the flank and the machined surface, and the frictional force will increase, which will lead to the increase of cutting force and cutting heat, which may cause serious damage to the processed surface quality of the workpiece.

For the spiral cutter type micro-milling cutter, the stiffness and strength of the micro-milling cutter are not high. It is necessary to consider the cross-sectional shape of the cutting edge and the material removal rate of the cutter head. The main reason for the high rigidity of the multi-blade milling cutter may be due to the appropriate negative The front corner design retains a large amount of cutter head material, but this structure will inevitably lead to a decline in cutting performance, and considering the complexity of multi-blade micro-milling cutter manufacturing, the traditional double-edged spiral-edge micro-milling cutter is still the best Choice.

4.Conclusion

Based on the finite element software such as DEFORM and ANSYS, the influence law of different edge type micro-milling cutters for milling titanium alloy TC4 and the modal analysis of micro-milling cutters with different edge types were analyzed. The following conclusions were drawn:

(1) The spiral-edge micro-milling cutter has better performance than the straight-blade micro-milling cutter in reducing the cutting force and the cutting heat. The traditional double-edged spiral-edge micro-milling cutter can produce the minimum cutting force and cutting heat during the machining process.

(2) The cutting temperature gradually increases with the increase of the number of edges, but the increasing range is gradually slower, indicating that when the number of edges reaches a certain node, the effect of increasing the number of edges on the cutting temperature rise is gradually weakened.

(3) Straight edge type and spiral edge type micro-milling cutters under the static load, a stress concentration area will appear at the end of the cutting edge and the joint between the neck and the shank, which is also the actual operation of the micro-milling cutter. The location where the fracture is most likely to occur during processing.

(4) In the case of the same tool diameter, although the traditional double-edged spiral-edge micro-milling cutter retains less material, resulting in low stiffness and overall strength, considering the cutting performance and practicality, the traditional double-edged spiral Blade-type micro-milling cutters are still the best choice.

References

[1] Cheng X , Wang Z , Nakamoto K , et al. A study on the micro tooling for micro/nano milling[J]. *International Journal of Advanced Manufacturing Technology*, **2011, 53(5-8)**:523-533.

[2] Chen M J,Chen N,He N,et al.The Research Progress of Micromilling in Machining Mechanism[J].*Journal of Mechanical Engineering*,**2014,50(05)**:161-172.

[3] Yang Z J,Zhang Y B,Xu L Y.Study on On-site Preparation Technology of Micro-milling Tool[J].*Mechanical Science and Technology for Aerospace Engineering*, **2017, 36(09)**: 1396-1401.

[4] Perveen A , San W Y , Rahman M . Fabrication of different geometry cutting tools and their effect on the vertical micro-grinding of BK7 glass[J].*The International Journal of Advanced Manufacturing Technology*, **2012, 61(1-4)**:101-115.

[5] Uhlmann E , Schauer K . Dynamic Load and Strain Analysis for the Optimization of Micro End Mills[J]. *CIRP Annals - Manufacturing Technology*, **2005, 54(1)**:75-78.

[6] Fang F Z , Wu H , Liu X D , et al. Tool geometry study in micromachining[J]. *Journal of Micromechanics and Microengineering*, **2003, 13(5)**:726-731.

[7] Jiang Z X , Sun J , Li J F . Structural Design of Groove and Micro-Blade of the End Mill Based on Bionics[J]. *Key Engineering Materials*, **2016, 693:**1221-1227.

[8] Aurich J C , Reichenbach I G , Guido M. Schuler. Manufacture and application of ultra-small micro end mills[J]. *CIRP Annals - Manufacturing Technology*, **2012, 61(1)**:83-86.

[9] Li P, Oosterling J, Hoogstrate A, et al. Design of micro square endmills for hard milling applications[J].*International Journal of Advanced Manufacturing Technology* ,**2011, 57(9-12)**:859-870.

[10] Yang D,Liu Z Q,He M,et al.Experimental Investigation into Surface Integrity of Titanium TC4 after Low-Plasticity Burnishing[J].*Journal of South China University of Technology(Natural Science Edition)*,**2017,45(01)**:137-144.

[11] Zhao W,Wang S Z,Li L,et al.Evaluation of Cutting Tool Performance of End Mills for Titanium Alloy Components[J].*Journal of South China University of Technology(Natural Science Edition)*,**2015,43(09)**:121-127.

[12] Cai M , Gong Y D, Yu N,et al.Simulation and Experimental Study on the Milling Force of Aluminum Alloy 6061 in Micro-milling[J].*Journal of Northeastern University(Natural Science)*,**2018,39(01)**:76-81.

[13] Wen X L,Gong Y D,Cheng J,et al.Experimental Research on Force and Temperature Characteristics in Micro-grinding Al6061[J].*Journal of Mechanical Engineering*, **2014, 50(23)**:165-174.

[14] Suzuki H , Okada M , Fujii K , et al. Development of micro milling tool made of single crystalline diamond for ceramic cutting[J]. *CIRP Annals - Manufacturing Technology,* **2013, 62(1)**:59-62.

[15] Yan J , Uchida K , Yoshihara N , et al. Fabrication of micro end mills by wire EDM and some micro cutting tests[J]. *Journal of Micromechanics and Microengineering*, **2009, 19(2)**:025004.

[16] Zhang S J. Study on Micro-tools Fabrication by Focused Ion Beam Technology and Key Techniques[D].Tian Jin:*Tianjin University*,2009.

[17] Fonda P , Katahira K , Kobayashi Y , et al. WEDM condition parameter optimization for PCD

microtool geometry fabrication process and quality improvement[J]. *International Journal of Advanced Manufacturing Technology*, **2012, 63(9-12)**:1011-1019.

[18] Thanongsak Thepsonthi, Tuğrul Özel. Experimental and finite element simulation based investigations on micro-milling Ti-6Al-4V titanium alloy: Effects of cBN coating on tool wear[J]. *Journal of Materials Processing Technology*, **2013, 213(4):**532-542.

[19] Wu H.Research on Structural Optimization of PCD Micro End Tool [D].Nan Jing:*Nanjing University of Aeronautics and Astronautics*,2012.

[20] Heaney P J , Sumant A V , Torres C D , et al. Diamond coatings for micro end mills: Enabling the dry machining of aluminum at the micro-scale[J]. *Diamond and Related Materials*, **2008, 17(3)**:223-233

Experimental Evaluation for the Interference-fit Electromagnetic Rveting Joint with Headless Rivet

Danlong Song*, Xiangyang Zhang, Kai Liu, Yanchao Zhang and Zhenchao Yang

School of Mechanical and Precision Instrument Engineering, Xi'an University of Technology, Xi'an 710048, China

*E-mail: songdanlong@xaut.edu.cn

Abstract. Electromagnetic riveting (EMR) technology as a relatively new mechanical joining process has many advantages compared to traditional riveting methods. However, it needs to be investigated whether the traditional pressing riveting can be replaced by EMR process in aerospace manufacturing engineering. Therefore, in this paper, the differences of riveting quality between automatic riveting and handheld EMR were compared by riveting and mechanical experiments including the uniformity of interference fit, shear and pull-off strength and fatigue life of the joints respectively. It was observed that the interference fit of the riveting joint using EMR is more uniform along the axial direction of rivet than that of automatic riveting. In addition, the overhang of the rivet was difficult to control by handheld EMR, so the quality of rivets for EMR was poorer than that of automatic riveting. Thus, the shear and pull-off strength and fatigue life of EMR joints were affected by the quality of rivets, and a longer rivet is needed for the EMR.

Keywords: Interference fit, Electromagnetic riveting, Experiment, Mechanical evaluation;

1. Introduction

In advanced engineering fields, the reliability and safety of aircraft and vehicle is significantly important. Under service load, fatigue cracks are easily initiation and propagation in the structures, especially at the joints, more than 60% of failures occur at the joints [1]. A variety of fastening technologies are applied in manufacturing industry, such as welding, bolting, bonding and riveting. Riveting process which is a kind of interference-fit joint technology can improve the bearing capacity and sealing performance, which has proved that interference-fit joint of metal structure can enhance the fatigue life and strength in assembly of aircraft and automobile parts[2, 3]. It has been increasingly applied in many manufacturing industries[4, 5].

There are three kinds of riveting process such as pressing riveting, pneumatic riveting and electromagnetic riveting (EMR)[6]. It is proved that the quality of pressing riveting is very good. But the operating space should be open. The rivet deforms un-uniformly by pneumatic riveting. The structure skin may be scratched which will cause fatigue crack. As a new interference-fit riveting technology, the EMR, also known as stress-wave riveting, is an advanced electromagnetic rivet forming process, which act very fast [7, 8]. Usually, the rivet is formed within 1 *ms*, so the mechanical behavior of the material in high strain rate forming is entirely different with that of traditional pressing riveting process. The metal deforms by the mode of homogeneous slip in quasi-static riveting, while shear mode in EMR[9].

The residual stress distribution of EMR interference-fit joint was predicted by a theoretical model which provide scientific guidance for the process parameter design and riveting setup optimization in engineering application [10]. The pressing riveting and tension processes were investigated experimentally and numerically, and three kinds of failure modes including pull through, shank breaking, and head breaking were observed[11]. Newman et al.[12] explored microstructure in the rivet cross section of the riveted sample with countersunk rivets, and found that the cracks initiated and grew from the faying surface. The differences of riveting quality between pneumatic riveting and EMR were compared by experiments including interference quantity uniformity, shear and pull-off strength of the riveting, and the EMR quality was better [13, 14]. In addition, the microstructures and mechanical properties of the pressing riveting and EMR processes were analyzed by optical microscopy and tensile testing machine respectively [15]. But the geometric dimension analysis was ignored, and the rivet is one side formed by EMR.

Therefore, the mechanical property and geometric dimension of structures with headless rivet by EMR is not clear. It needs to be investigated whether the traditional pressing riveting process can be replaced by EMR technology in aerospace manufacturing engineering. So the contrastive research of EMR and pressing riveting is conducted in this paper. The aluminum alloy skin is fastened by headless rivet with EMR gun and automatic riveting machine which employs pressing riveting technology. Then the geometrical dimension, strength and fatigue life with various riveting voltages are investigated.

2. Riveting experiments of headless rivet

Based on the technical requirements, the material and dimensions of the rivet, the riveting test was conduct by the HH503 handheld EMR gun and G86 automatic riveting machine produced by Electroimpact and GEMCOR Inc. respectively shown in Figure 1. G86 automatic riveting machine which employs pressing riveting technology is widely used in aircraft assembly process. It is proved that the riveting quality of G86 is competent for engineering requirement. The EMR gun system, which consists of driver, gun body and handle shroud, is suitable for the rivets made of aluminum alloy and titanium alloy.

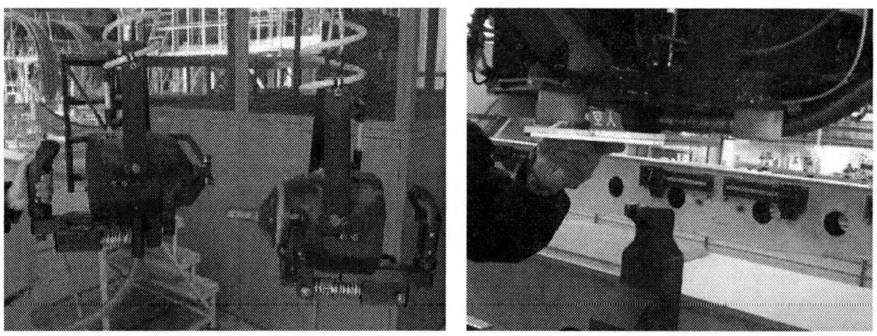

(a) Handheld EMR gun (b) Automatic riveting machine

Figure 1. Riveting machines used in the experiments

The EMR tests of headless rivet require two EMR guns, the riveting mould is installed on the primary gun to punch the rivet, and the riveting die is installed on the assistant gun to support the rivet. Based on the experiential parameters, the riveting voltage ranges are 170V ~ 180V and 220V ~ 250V for the headless rivets of Φ3/16" and Φ1/4" respectively. However, lower riveting voltage causes incomplete forming of rivets, the riveting mould is filled with gap. When the riveting voltage is too high, the riveting impact force will cause overlarge interference-fit percentage which may reduce the joint strength.

The material of headless rivets is 2117-T4 aluminum alloy, and the materials of the plates are 7055-T3 aluminum alloy. The automatic riveting specimens are riveted by G86 machine. The rivet holes for EMR process are drilled by G86 machine with drill-counterbore tools made of cemented carbide. The diameters are *Φ*4.90±0.02mm, *Φ*6.45±0.02mm for two types of rivets respectively. Then, the riveting process is implemented by handheld EMR guns. The specimens are shown in Figure 2.

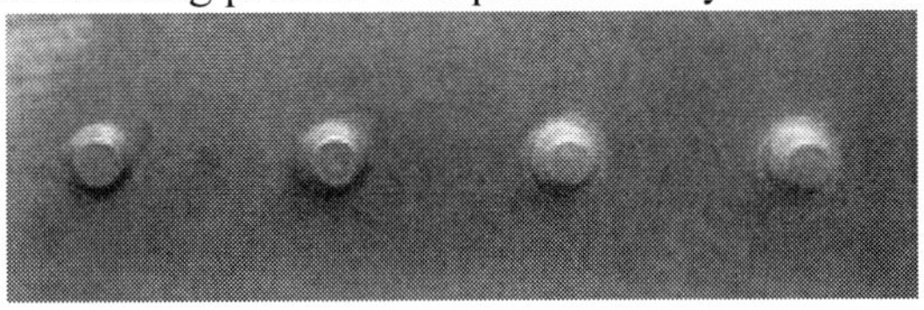

(a) Automatic riveting specimens

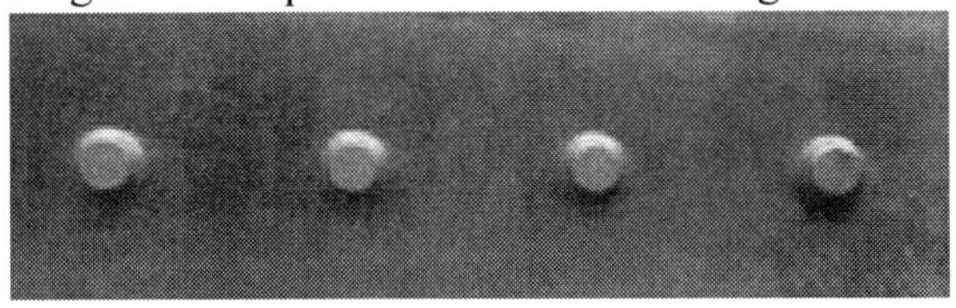

(b) Handheld EMR specimens

Figure 2. Specimens of two kinds of riveting tests

3. Geometrical analysis of the riveting joint

The geometrical shapes of the formed rivet are effected by many factors, such as riveting voltage, mould, material and dimension of headless rivets. The effects of riveting voltage on geometrical shapes, which include interference-fit percentage, the height and diameter of rivet head, are investigated by riveting experiments in details.

3.1. Interference-fit percentage

The interference-fit percentage, which affects the strength and fatigue life of the joint, is a very important parameter for the riveting joint in the field of aircraft assembly. The expansion percentage of the rivet hole during the EMR process is described by the interference-fit percentage which can be expressed by:

$$\Delta = \frac{D - D_0}{D_0} \times 100\% \tag{1}$$

where D_0 represents the initial hole diameter, and D represents the hole diameter after riveting which is equal to the rivet diameter after riveting.

The interference percentage is affected by riveting process. The riveting impact force is act on the head of the rivet during the riveting process, so the expansion of the rivet may not uniform along the axial direction of the rivet. Then, the specimens are decomposed by a saw to measure the diameter of the rivet at different location. The specimens and rivets cut by a saw are shown in Figure 3.

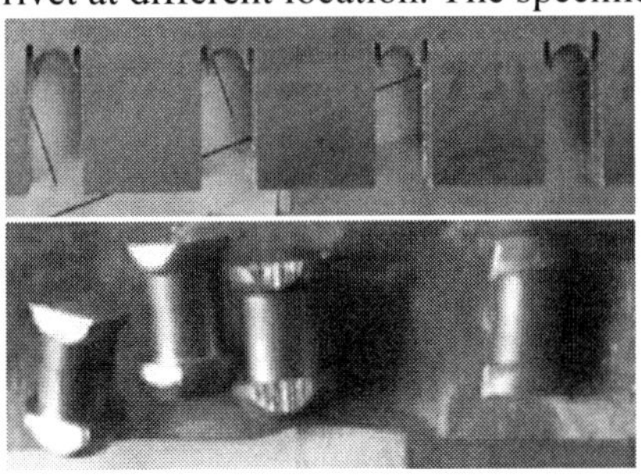

Figure 3. Specimens and rivets cut by a saw

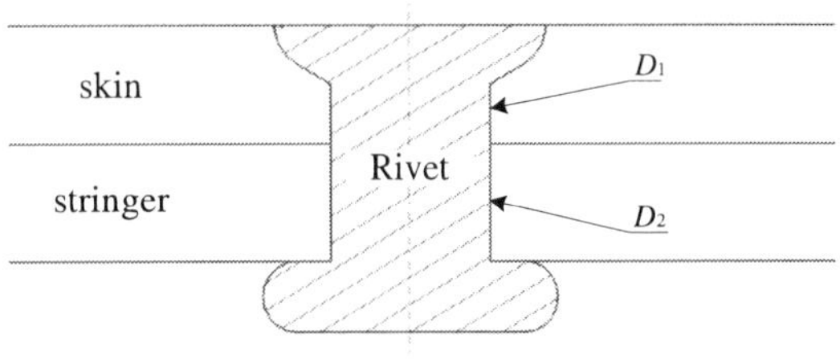

Figure 4. Location of interference-fit percentage

To explore interference percentages of different location in the joint clearly, D_1 and D_2, the interference-fit percentages of 2 locations along the axial direction of the headless rivet, shown in Figure 4. The specimens of EMR tests are disassembled by cutting machine. The diameters of each location are measured by a digimatic micrometer.

The interference-fit percentages of the headless rivets of *Φ*3/16" and *Φ*1/4" are shown in Figure 5 and Figure 6 respectively. The min and max curves are the allowable minimum and maximum interference-fit percentages respectively in engineering applications. The interference-fit percentages increase with the increasing of riveting voltage. It is proved that 180V and 210V are the optimal riveting voltages for*Φ*3/16" and *Φ*1/4" rivets respectively. However, the interference-fit percentages of Location D_2 are a little larger than those of Location D_1. Because the countersink of Location D_1 makes for the flow of metal material.

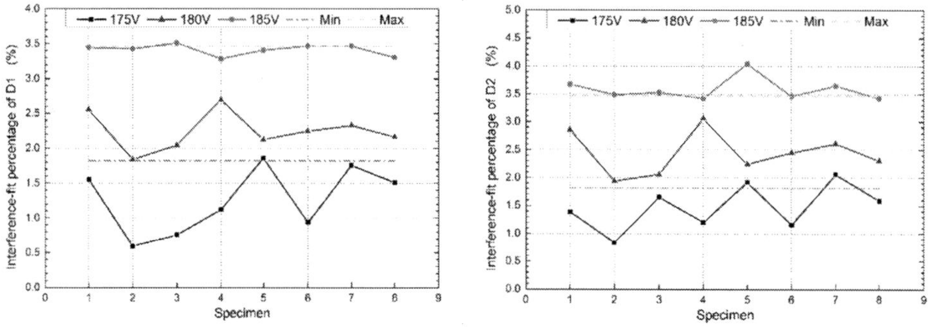

Figure 5. The interference-fit percentage of D_1 and D_2 for *Φ*3/16" rivets

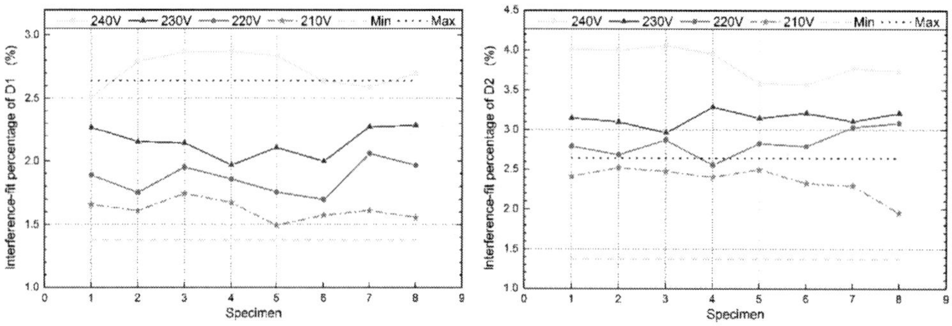

Figure 6. The interference-fit percentage of D_1 and D_2 for *Φ*1/4" rivets

In order to analyse the rivet deformation of EMR deeply, the average interference-fit percentages of EMR and automatic riveting are compared shown in Table 1. The results indicate that the average interference-fit percentages of EMR are larger than those of automatic riveting. Meanwhile, the interference-fit percentage along the axial direction of rivet by EMR is more uniform. The rivet completes plastic forming in one microsecond under the stress wave. From the point of the interference-fit percentage, EMR process is better than automatic riveting process.

Table 1. Comparison of the average interference-fit percentages

Specimens	*Φ*1/4" rivets	*Φ*3/16" rivets
Automatic Riveting	1.91%	2.30%
EMR	1.99%	2.34%

3.2. The height of rivet head

The height of rivet head for the headless rivets of *Φ*3/16" and *Φ*1/4" is shown in Figure 7. The Min curve is the allowable minimum height of rivet head in engineering applications. The height of rivet head decreases with the increasing of riveting voltage, because there is a positive correlation between the impact force for riveting and the riveting voltage. Besides, the height of rivet head meets the requirement of engineering applications in the range of testing voltages.

In addition, the height of rivet head conducted by EMR and automatic riveting are compared in Figure 8. It is obverted that the height of rivet head of EMR is about 7.8% smaller than that of automatic riveting with the same type of rivets. There is more material of the rivet for EMR flowing into the hole than that of automatic riveting. In addition, the overhang of the rivet was difficult to control by handheld EMR. But, the height of rivet head conducted by EMR could satisfy the engineering need.

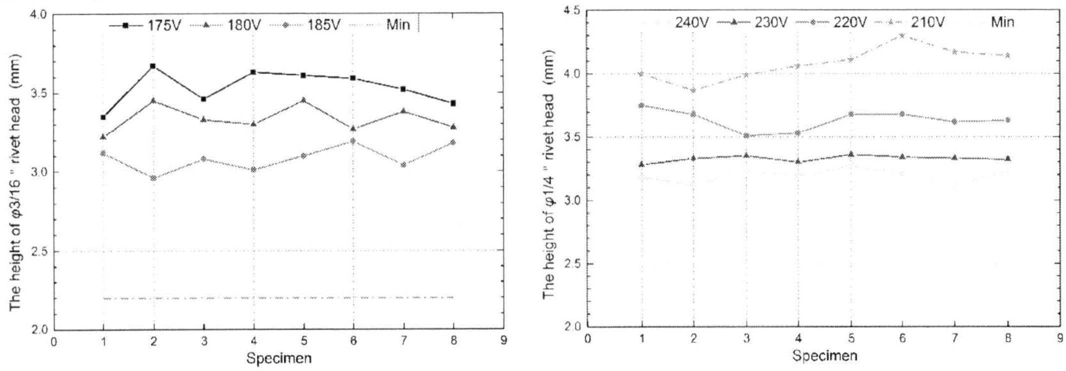

Figure 7. The height of rivet head for the Φ3/16" and Φ1/4" rivets

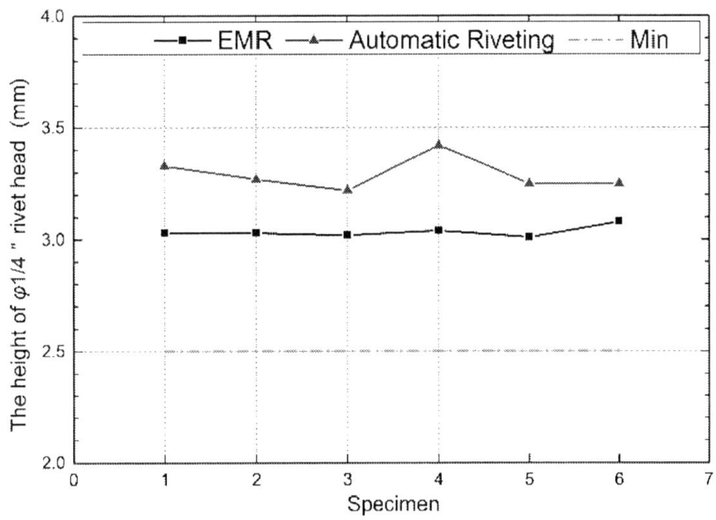

Figure 8. The height of rivet head of the EMR and automatic riveting

3.3. The diameter of rivet head

The diameters of rivet head for the headless rivets of Φ3/16" and Φ1/4" are shown in Figure 9. The min curve is the allowable minimum diameter of rivet head in engineering applications. The plastic forming of the headless rivet is under the condition without volume dilatation. The average diameters of rivet head conducted by EMR and automatic riveting are compared in Table 2. The average diameters of rivet head for EMR are the same as those for automatic riveting. However, the diameter of rivet head depends on the diameter of riveting die. The riveting voltage and riveting process have almost no effect on the diameter of rivet head.

Table 2. Comparison of the average diameters of rivet head

	Φ1/4" rivets	Φ3/16" rivets
Automatic Riveting (mm)	9.10	8.05
EMR (mm)	9.00	8.15

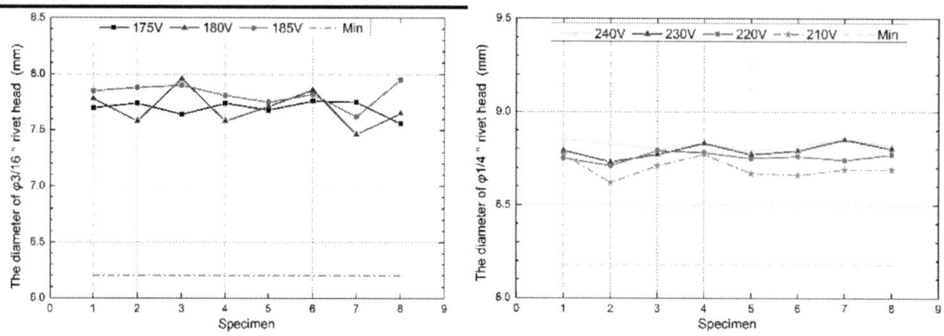

Figure 9. The diameters of rivet head for the Φ3/16" and Φ1/4" rivets

4. Mechanical performance of riveting joint

The geometrical shapes of EMR joint indicate the riveting quality directly. However, the mechanical performance and fatigue life of EMR joint still should be investigated by mechanical experiments in order to evaluate the engineering applicability of EMR process.

4.1. Strength of the riveting joint

The quasi-static mechanical performance experiments, such as tensile experiment and pull-out experiment, were conducted by INSTRON 2383 hydraulic universal material testing machine with the speed of 1mm/min. The specimens of tensile and pull-out experiment are shown in Figure 10. The load-displacement curves are record in real time.

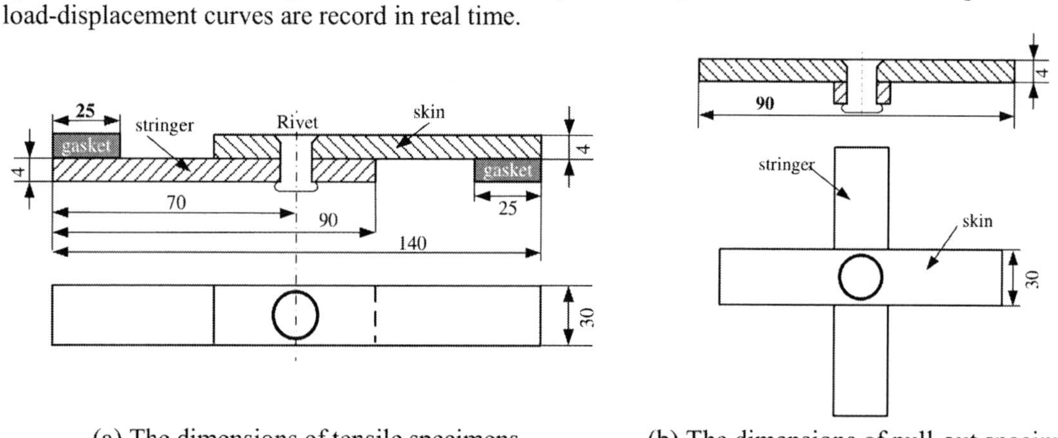

(a) The dimensions of tensile specimens (b) The dimensions of pull-out specimens

Figure 10. The dimensions of riveting specimens

The tensile experimental results of the riveting specimens is shown in Table 3. The average shear strength of the EMR specimens is 7.1% lower than that of the automatic riveting specimens. A same regularity for the pull-off strength is observed, which is shown in Table 4. The primary reason is that the height and diameter of rivet head of the EMR specimens are smaller than those of the automatic riveting specimens. But, the strength of EMR joint could satisfy the engineering need.

Table 3. Shear strength of riveting specimens

Specimens	1	2	3	4	Average
Automatic Riveting (MPa)	177.41	169.25	169.25	175.37	173.33
EMR (MPa)	157.01	167.21	157.01	163.13	161.09

Table 4. Pull-off strength of riveting specimens

Specimens	1	2	3	4	Average
Automatic Riveting (MPa)	12.4	13.0	12.6	12.8	12.7
EMR (MPa)	12.2	12.1	11.9	12.3	12.1

4.2. Fatigue life of the riveting joint

The specimens of fatigue experiment are shown in Figure 11. The fatigue experiments for the joint of EMR and automatic riveting processes were conducted by INSTRON 8801 hydraulic fatigue testing machine. The maximum stress is 140 MPa and the stress ratio is 0.1. The fatigue loading frequency is 15 Hz. The fatigue experiment should be stopped when visible cracks is appear or the number of loading cycles reaches 1.00×10^6.

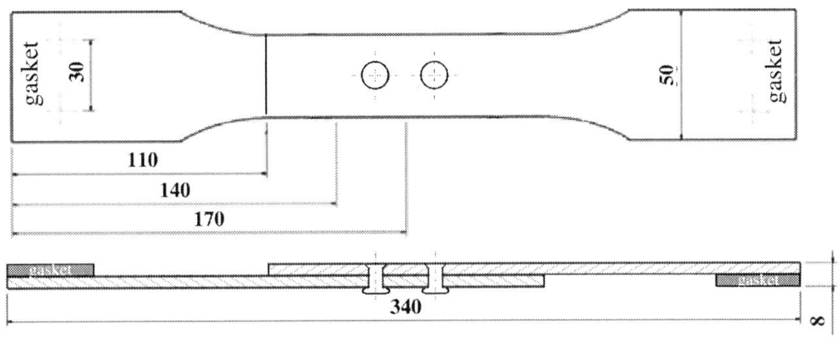

Figure 11. The dimensions of fatigue specimens

The fatigue life comparison of EMR and automatic riveting joints are shown in Table 5. It is observed that the average fatigue life of EMR joints is 34.61% smaller than that of automatic riveting joints, which is caused by that the smaller height and diameter of EMR rivet head than those of automatic riveting rivets. Therefore, the automatic riveting process is better than EMR process on the same condition. But, the fatigue of EMR joint could satisfy the engineering need.

Table 5. The numbers of loading cycles of fatigue specimens

Specimens	1	2	3	4	Average
Automatic Riveting (MPa)	8.39×10^5	1.00×10^6	9.82×10^5	9.47×10^5	9.42×10^5
EMR (MPa)	6.63×10^5	5.36×10^5	5.78×10^5	6.85×10^5	6.16×10^5

5. Conclusion

In allusion to the riveting process of aircraft assembly, a comparative investigation of EMR and automatic riveting processes was conducted in this paper. The geometrical dimensions and mechanical performance were discussed respectively for EMR and automatic riveting processes by riveting and mechanical experiments. It was proved that the riveting quality of EMR was lower than that of automatic riveting with the same type of rivet. But EMR as an advanced technology should be investigated deeply. Some results were in the following conclusions.

(1) The average interference-fit percentages of EMR are larger and more uniform than those of automatic riveting. But the height and diameter of EMR rivet head are smaller than those of automatic riveting rivets. Because the stress wave transmits very fast, and there is more material of the rivet for EMR flowing into the hole than that of automatic riveting.

(2) The average interference-fit percentages and diameters of rivet head increase with the riveting voltages. The optimal riveting voltages for *Φ*3/16" and *Φ*1/4" rivets are 180V and 210V are respectively.
(3) The shear and pull-off strength and fatigue life of EMR joint are lower than those of automatic riveting joint on the same condition, such as the length and overhang of rivet. So a longer rivet is needed for the EMR.

Acknowledgments
This work is supported by the Project funded by China Postdoctoral Science Foundation (Program No. 2018M643815XB), the Project Supported by Natural Science Basic Research Plan in Shaanxi Province of China (Program No. 2018JQ5025), the Scientific Research Program Funded by Shaanxi Provincial Education Department (Program No. 18JK0570) and the Young Talent fund of University Association for Science and Technology in Shaanxi, China (20190416). The authors would like to greatly acknowledge the editors and the anonymous referees for their insightful comments.

References
[1] Skorupa A and Skorupa M 2012 *Riveted lap joints in aircraft fuselage: design, analysis and properties* (Dordrecht, Heidelberg, New York, London: Springer)
[2] Zhang K, Hu J, Zou P, Cheng Y, Luo B and Cheng H 2019 Effect of secondary bending and bolt load on damage and strength of composite single-lap interference-fit bolted structures *J Compos Mater* **53** 4385-98
[3] Wei J, Jiao G, Jia P and Huang T 2013 The effect of interference fit size on the fatigue life of bolted joints in composite laminates *Compos Part B-Eng* **53** 62-68
[4] Chakherlou T N and Abazadeh B 2012 Investigating clamping force variations in Al2024-T3 interference fitted bolted joints under static and cyclic loading *Mater Design* **37** 128-36
[5] Zhang Q, Cao Z, Li H, Xiang C and Liu P 2017 Elastic fatigue enhancement mechanism of interference fit *Acta Aeronautica et Astronautica Sinica* **39** 235-43
[6] Li G, Jiang H, Zhang X and Cui J 2017 Mechanical properties and fatigue behavior of electromagnetic riveted lap joints influenced by shear loading *J Manuf Process* **26** 226-39
[7] Cao Z and Zuo Y 2019 Electromagnetic riveting technique and its applications *Chinese J Aeronaut*
[8] Psyk V, Risch D, Kinsey B, Tekkaya A and Kleiner M 2011 Electromagnetic forming - A review *J Mater Process Tech* 787-829
[9] Zhang X, Yu H P, Su H and Li C F 2016 Experimental evaluation on mechanical properties of a riveted structure with electromagnetic riveting *Int J Adv Manuf Tech* **83** 2071-82
[10] Zhang X, Jiang H, Luo T, Hu L, Li G and Cui J 2019 Theoretical and experimental investigation on interference fit in electromagnetic riveting *Int J Mech Sci* **156** 261-71
[11] Chen N, Luo H, Wan M and Chenot J 2014 Experimental and numerical studies on failure modes of riveted joints under tensile load *J Mater Process Tech* **214** 2049-58
[12] Newman J C and Ramakrishnan R 2016 Fatigue and crack-growth analyses of riveted lap-joints in a retired aircraft *Int J Fatigue* **82** 342-49
[13] Feng D and Cao Z 2012 Quality comparing analysis of electromagnetic riveting and pneumatic riveting *Forg & Stamp Tech* 123-26
[14] Skorupa M, Machniewicz T, Skorupa A, Schijve J and Korbel A 2015 Fatigue life prediction model for riveted lap joints *Eng Fail Anal* **53** 111-23
[15] Dong D, Sun L, Wang Q, Li G and Cui J 2019 Influence of electromagnetic riveting process on microstructures and mechanical properties of 2A10 and 6082 Al riveted structures *Arch Civ Mech Eng* **19** 1284-94

IOP Conf. Series: Materials Science and Engineering **751** (2020) 012084 doi:10.1088/1757-899X/751/1/012084

Research on Hull Assembly Process Planning Based on Rule Reasoning

Li Lei*, Liu Di, Wang Pengyu and Zhou Honggen

School of Mechanical Engineering, Jiangsu University of Science and Technology, Zhenjiang 212003,China

Corresponding author: lilei0064@sina.com

Abstract. Currently, hull assembly process is mainly dependent on experience, and lack of normative standard. In this paper, hull structure for assembly features is systematically studied, and an assembly sequence planning method based on Rule Reasoning is proposed. Firstly, assembly feature and sequence expression of hull structure is analyzed and the constraint relation of connection information is described. Secondly, the relationship between established structures and the constraint function is designed. Setting the double bottom of the ship as an example, hull assembly sequence is obtained based on geometric constraints and optimized by the fuzzy evaluation method and finally realized the computer aided evaluation of the hull assembly sequence.

1. Introduction

Hull assembly and welding are a core technology for the complicated systematic project of ship construction process, and greatly affect the efficiency and quality of ship building. However, hull assembly process is mainly dependent on the experience of technologists and lack of computer aided methods, which has seriously restricted the efficiency and quality of ship building.

With the development of computer information and intelligent technology, many scholars have carried out extensive research on the computer aided planning technology of ship assembly process. Cho set up a ship block assembly system, and realized the assembly sequence planning and scheduling by the technology of case-based reasoning [1] . Seo carried out the further research on the matching of similar instances [2]. Qu Shipeng combined the reasoning based on case with the reasoning based on geometric constraints and the optimization of assembly sequence was realized by genetic algorithm [3].Li Peiyong applied demolition method and analytic hierarchy process to optimize the assembly sequence.[4]. Wang Bo proposed a method of cut-set to generate assembly sequences, but it would bring combinatorial explosion problem for complex product combination [5]. Song Li-ping programmed the assembly sequence of the marine engine room with the interactive virtual assembly technology based on the demolition [6]; In addition, Wei Wei and Wang Jing pin used the ant colony algorithm to program the assembly sequence of the hull double floor [7-8]. However, because of the numerous of hull structures and the complexity of assembly process, modern intelligent algorithm is still difficult to meet the practical requirements. The method of case based reasoning and geometric reasoning own good practicability, However, they are obvious defects in the non-geometric information processing, which is difficult to meet the requirements of rationality and effectiveness of the hull assembly process planning.

In this paper, hull structure for assembly features are systematically studied, and an assembly sequence planning method based on rule reasoning is proposed, which realized the computer aided planning of hull assembly sequence.

2. Analysis of assembly feature and assembly sequence expression of hull structure

2.1. Structural analysis

The number of hull assembly structures is very large, but only a few of them will affect the planning process. In order to plan a better assembly sequence and reduce the difficulty of planning, the selection of the common structures is extremely essential. In practical applications, there are 8 types of common parts(A, B, C, K, L, P, S, W) according to the type of the shape and function[9].

Although the number of common part types is reduced through filtering, the difficulty of planning assembly sequence is not reduce because there may be many different shapes even the same type and function of the components. Therefore, based on the simplification principle, selecting several common components can greatly reduce the difficulty of the problem in practical production. By analyzing these ship hull parts, the assembly elements of different parts can be obtained as shown in Table 1:

Table 1. Function and assembly elements of hull Parts

Part type	Classification	Function	Assembly element
A	A1	T plate of web	Four edges of the parts
B	B1	Square elbow plate	Four edges of the parts
		....	
K	K1	Base plate	Four sides and two surfaces
K	K3	Base ware	Two surfaces
L	L1	Stiffening bar	Four edges of the parts

By analyzing the assembly process of the hull, there are mainly four kinds of connection relation between ship structural components: basic part, strutting piece, reinforcing member and envelope part. The relation diagram of structural parts assembly is shown in Figure 1:

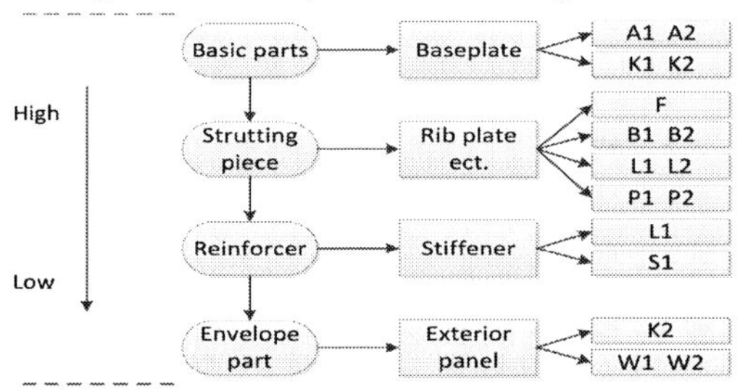

Figure 1. Relation diagram of structural parts assembly

2.2. Assembly sequence expression

The expression of assembly sequence should be concise and contain the main information of assembly. In this paper, four types of information (assembly code, initial information, terminal information and assembly type) are selected to express the assembly sequence. The assembly relationship can be expressed as: (Num, Sta, End, Lin). In which, Num is the order number of the assembly operation; Sta is the initial component information, including the initial component and its assembly elements; End is the information of the end component, including the end component and assembly elements; Lin is the type of connection.

For example: If the numbers of the structural are P3 and P4, the sequence is expressed as (2, (P3, 3), (P4, 2), (, 4). Its meaning is: in the whole assembly sequence, the assembly sequence number of P3 and P4 is 2 and the assembly elements 3 of P3 are assembled by elements 2 of P4. And the type of connection is fitting.

2.3. Constraint relation of connection information

The large number of rules and assembly elements may lead to the difficult of the assembly planning. Because not all the components of the assembly process will affect the assembly process, a lot of small components can be simplified such as reinforcing ribs, reinforcing plates. In addition, the method of rule based reasoning will incorporate all the possible connected components into the reasoning category. However, in practice, those components which have influence on the assembly planning are not necessarily in contact with each other, which will result in a dramatic increase in the number of reasoning results.

In this paper, a method of introducing the relation constraint as well as the connection information is proposed to plan the assembly sequence. This method is based on the connection relationship of the main components by means of information description. It can eliminate the connection relationship which may not exist in the space position and then reduce the difficulty of assembly sequence reasoning.

The expression of connection information constraint relation is:(C_L,P_M,P_N), in which, C_L is the connection number, P_M is the M_{th} component, P_N is the N_{th} component. In terms of the constraint information, the assembly sequence reasoning process can be carried out according to the assembly rules.

3. Inference of assembly sequence based on rules

The reasoning process of assembly sequence based on rules is shown in Figure 2:

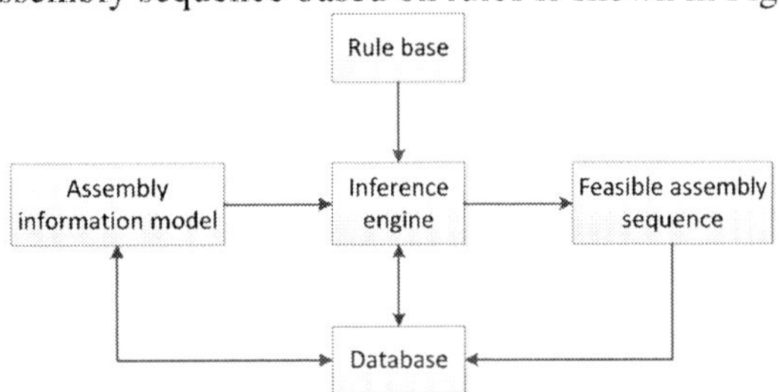

Figure 2. The reasoning process of assembly sequence planning

(1)Arrange the sequence of component installation according to the assembly priority of component, that means, installation of structural parts according to priority level, and then find the relevant structure in the same level using the associated graph model.

(2)The assembly process structure of same assembly level can be considered as the matching process of assembly rules. The connecting assembly component element and assembly type are determined by matching rules.

(3)After completing the assembly of component, the assembly of sub-assembly and unit can be carried out which is similar to the component layer. The only difference is that the assembly priority of the sub-assembly layer is determined according to the relevant components of the component parts.

By analogy, the assembly of the whole hull is realized and the assembly sequence is obtained by the process above.

Compared to the other intelligent algorithms, Rule based reasoning (RBR) for the assembly sequence planning technology is more practical. At the same time, it would obtain more feasibility of assembly sequence by the increasing of the assembly rules, although the RBR has a higher requirement for the actual assembly experience. In this paper, the method of constraint reasoning combined with the rule based reasoning is proposed to complete the effective planning of assembly sequence. This method introduces the assembly constraints such as assembly contact and interference matrix, which can help RBR to complete the whole process of assembly sequence planning.

3.1. Establishment of relationship between structures

In order to express the logical relations between the components, an adjacent interference matrix is established to represent the contact and assembly interference of each structure.

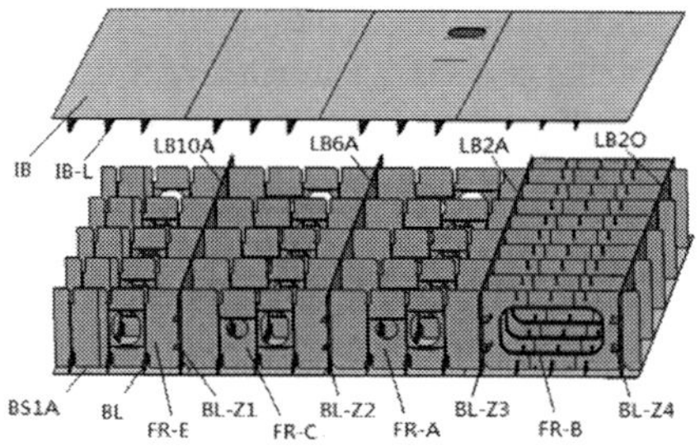

Figure 3. Structure of the double bottom

Taking the double bottom of the ship as an example (shown in Figure.3), the six directions $(x, y, z, -x, -y, -z)$ of the three-dimensional space of the components is represented by the Descartes coordinates. Design variables $L(p,q)$ and $G(p,q)$ are used to represent the assembly relationship, the variables are defined as follows:

$$L(p,q) = (L_1, L_2, L_3, \cdots, L_i), L_i \in \{0,1\} i = 1 \sim 10 \tag{1}$$

$$G(p,q) = (G_1, G_2, G_3, \cdots, G_j), G_j \in \{0,1\} j = 1 \sim 10 \tag{2}$$

In which, L_i represents the contact relation of any two structural elements in the i_{th} directions, $L_i = 1$ means that the component p and q have contact relationship in the i_{th} direction of p; $L_i = 0$ indicates that the component p and q do not have any contact in that direction. G_j indicates that a component has an assembly interference relationship in the j_{th} direction relative to another. $G_j = 1$ shows that the q in the direction of the relative i in the p movement is not affected by any of its interference, and $G_j = 1$ means that there is no interference relationship.

The relationship of contact and interference can be obtained in the process of establishing a 3D model. In the establishment of contact and interference matrix, the influence caused by the sequence of assembly can also be directly identified, that is the difference of the turn over number. Because the outer and mainland components are similar, only the outer component is selected to establish the contact and interference matrix, which are shown as follows:

$$L_{pq} \rightarrow \begin{array}{c} \\ BS-L \\ LB10A \\ FR-E \\ LB6A \\ FR-C \\ FR-A \\ LB2A \\ FR-B \\ LB2Q \end{array} \begin{array}{ccccccccc} BS-L & LB10A & FR-E & LB6A & FR-C & FR-A & LB2A & FR-B & LB2Q \\ (0,0,0,0,0,0) & (0,0,1,0,0,0) & (1,0,1,1,0,0) & (0,0,1,0,0,0) & (1,0,1,1,0,0) & (1,0,1,1,0,0) & (0,0,1,0,0,0) & (1,0,1,1,0,0) & (0,0,1,0,0,0) \\ (0,0,0,0,0,1) & (0,0,0,0,0,0) & (0,1,0,1,1,0) & (0,0,0,0,0,0) & (0,0,0,0,0,0) & (0,0,0,0,0,0) & (0,0,0,0,0,0) & (0,0,0,0,0,0) & (0,0,0,0,0,0) \\ (1,0,0,1,0,1) & (1,1,0,0,1,0) & (0,0,0,0,0,0) & (0,0,0,0,0,0) & (0,0,0,0,0,0) & (0,0,0,0,0,0) & (0,0,0,0,0,0) & (0,0,0,0,0,0) & (0,0,0,0,0,0) \\ (0,0,0,0,0,1) & (0,0,0,0,0,0) & (0,0,0,0,0,0) & (0,0,0,0,0,0) & (0,0,0,0,0,0) & (1,0,0,0,0,0) & (0,0,0,0,0,0) & (0,0,0,0,0,0) & (0,0,0,0,0,0) \\ (1,0,0,1,0,1) & (0,0,0,1,0,0) & (0,0,0,0,0,0) & (1,1,0,0,1,0) & (0,0,0,0,0,0) & (0,0,0,0,0,0) & (0,0,0,0,0,0) & (0,0,0,0,0,0) & (0,0,0,0,0,0) \\ (1,0,0,1,0,1) & (0,0,0,0,0,0) & (0,0,0,0,0,0) & (0,0,0,1,0,0) & (0,0,0,0,0,0) & (0,0,0,0,0,0) & (1,0,1,0,0,1) & (0,0,0,0,0,0) & (0,0,0,0,0,0) \\ (0,0,0,0,0,1) & (0,0,0,0,0,0) & (0,0,0,0,0,0) & (0,0,0,0,0,0) & (0,0,0,0,0,0) & (0,0,0,1,0,0) & (0,0,0,0,0,0) & (1,0,1,0,0,1) & (0,0,0,0,0,0) \\ (1,0,0,1,0,1) & (0,0,0,0,0,0) & (0,0,0,0,0,0) & (0,0,0,0,0,0) & (0,0,0,0,0,0) & (0,0,0,0,0,0) & (0,0,1,1,0,1) & (0,0,0,0,0,0) & (1,0,1,0,0,1) \\ (0,0,0,0,0,1) & (0,0,0,0,0,0) & (0,0,0,0,0,0) & (0,0,0,0,0,0) & (0,0,0,0,0,0) & (0,0,0,0,0,0) & (0,0,0,0,0,0) & (0,0,1,1,0,1) & (0,0,0,0,0,0) \end{array} \tag{3}$$

$$G_{pq} \rightarrow \begin{array}{c} \\ BS-L \\ LB10A \\ FR-E \\ LB6A \\ FR-C \\ FR-A \\ LB2A \\ FR-B \\ LB2Q \end{array} \begin{array}{ccccccccc} BS-L & LB10A & FR-E & LB6A & FR-C & FR-A & LB2A & FR-B & LB2Q \\ (1,1,1,1,1,1) & (1,1,1,1,1,0) & (0,1,1,0,1,0) & (1,1,1,1,1,0) & (0,1,1,0,1,0) & (1,1,1,1,1,0) & (0,1,1,0,1,0) & (0,1,1,0,1,0) & (1,1,1,1,1,0) \\ (1,1,0,1,1,1) & (1,1,1,1,1,1) & (0,0,1,1,0,1) & (1,1,1,1,1,1) & (1,1,1,0,1,1) & (1,1,1,1,1,1) & (1,1,1,1,1,1) & (1,1,1,1,1,1) & (1,1,1,1,1,1) \\ (0,1,0,0,1,1) & (1,0,1,0,0,1) & (1,1,1,1,1,1) & (1,1,1,1,1,1) & (1,1,1,1,1,1) & (1,1,1,1,1,1) & (1,1,1,1,1,1) & (1,1,1,1,1,1) & (1,1,1,1,1,1) \\ (1,1,0,1,1,1) & (1,1,1,1,1,1) & (1,1,1,1,1,1) & (1,1,1,1,1,1) & (0,0,1,1,0,1) & (1,1,1,0,1,1) & (1,1,1,1,1,1) & (1,1,1,1,1,1) & (1,1,1,1,1,1) \\ (0,1,0,0,1,1) & (0,1,1,1,1,1) & (1,1,1,1,1,1) & (1,0,1,0,0,1) & (1,1,1,1,1,1) & (1,1,1,1,1,1) & (1,1,1,1,1,1) & (1,1,1,1,1,1) & (1,1,1,1,1,1) \\ (0,1,0,0,1,1) & (1,1,1,1,1,1) & (1,1,1,1,1,1) & (0,1,1,1,1,1) & (1,1,1,1,1,1) & (1,1,1,1,1,1) & (1,1,1,0,1,1) & (1,1,1,1,1,1) & (1,1,1,1,1,1) \\ (1,1,0,1,1,1) & (1,1,1,1,1,1) & (1,1,1,1,1,1) & (1,1,1,1,1,1) & (1,1,1,1,1,1) & (0,1,1,1,1,1) & (1,1,1,1,1,1) & (1,1,0,0,1,0) & (1,1,1,1,1,1) \\ (0,1,0,0,1,1) & (1,1,1,1,1,1) & (1,1,1,1,1,1) & (1,1,1,1,1,1) & (1,1,1,1,1,1) & (1,1,1,1,1,1) & (0,1,0,1,1,0) & (1,1,1,1,1,1) & (1,1,0,0,1,0) \\ (1,1,0,1,1,1) & (1,1,1,1,1,1) & (1,1,1,1,1,1) & (1,1,1,1,1,1) & (1,1,1,1,1,1) & (1,1,1,1,1,1) & (1,1,1,1,1,1) & (0,1,0,1,1,0) & (1,1,1,1,1,1) \end{array} \tag{4}$$

3.2. Design of constraint function

Assembly constraint relation can guide the reasoning of assembly sequence. In the process of ship assembly, the planning of the block assembly sequence can be simplified as the management of the constraint relation of different structures. The space constraint relation of the assembly structure can be called rigid constraint. In this paper, the rigid constraint function is used to describe the geometric relations of the segmented structures. In order to realize the geometric feasibility of assembly sequence, the relations of the contact and the interference of the assembly structure must be guaranteed and any assembly sequence must meet the following two conditions:

(1)There is at least one connection relationship between the parts to be assembled and the previously installed structures.

(2) When any structural member is assembled, at least one direction does not interfere with other structural elements.

The above two conditions are necessary to determine the geometric feasibility of assembly sequences. The formula (1) and the formula (2) define the contact relation and interference relation between any two components, respectively.

In order to verify whether the assembly sequence is satisfied with the first constraint condition, formula (5) and (6) are defined to make relevant judgments, $(TL_i)_k$ in formula (5) is a six dimensional vector, which is used to represent the contact relationship between the assembly component and the assembled component in six different directions in space. $L_i(k,\bullet)$ represents the contact relationship between the component k and the installed structure, Symbol "$\vee$" represents the "or" operation of a vector, If at least one dimension vector value is equal to 1 in $(RTL)_k = 1$, then the component is contacted with the assembled part. Otherwise, there is no contact relationship and the assembly sequence is not reasonable.

$$(TL_i)_k = L_i(k,k-1) \vee L_i(k,k-2) \vee \cdots \vee L_i(k,1) \tag{5}$$

$$(RTL)_K = (TL_1)_k \vee (TL_2)_k \vee (TL_3)_k \vee (TL_4)_k \vee (TL_5)_k \vee (TL_6)_k \tag{6}$$

Similarly, formula (7) and (8) are given to verify whether the assembly sequence can satisfy the second constraint conditions, and i represent any direction in the Descartes coordinate. $(TG_i)_k$ is a six dimensional vector, which is used to indicate the presence of interference between the structural member and any other structural elements that have been assembled in the direction of movement to the specified position. $G_i(k,\bullet)$ indicates the interference of a structural member when compared to other structural members during assembly. Symbolic "$\wedge$" is used to find the "and" operation of the vector. $(RTG)_K = 1$ indicates that it can move in a certain direction to the specified position without interference with other installed structures. Otherwise, the structure can not be effectively assembled, that is, the assembly sequence is unreasonable.

$$(TG_i)_k = G_i(k,k-1) \wedge G_i(k,k-2) \wedge \cdots \wedge G_i(k,1) \tag{7}$$

$$(RTL)_K = (TG_1)_k \wedge (TG_2)_k \wedge (TG_3)_k \wedge (TG_4)_k \wedge (TG_5)_k \wedge (TG_6)_k \tag{8}$$

3.3. Hull assembly sequence reasoning based on geometric constraints

Taking the double floor as an example and it is divided into 10 components which can be seen as a new combination of components, according to the assembly rules, the Outsole component $BS-L$ and the floor or stringer should be assembled first and the floor $FR-B$ is selected. The assembly of the two components can be seen as an assembly sequence$(BS-L, FR-B)$. Then the assembly sequence $(BS-L, FR-B, LB2Q)$ can be analyzed as follows:

(1)Judging the contact relationship

Considered the $(BS-L, FR-B)$ as a new sub assembly, and the contact relationship between the new sub assembly and the component $LB2Q$ is judged by formula (3):

$$L(BS-L, LB2Q)=(0,0,1,0,0,0)$$

$$L(FR-B, LB2Q)=(1,0,1,0,0,1)$$

According to formula (5), the value of TL_i can be calculated as shown in Table 2.

Table 2. Truth table of contact relationship

	L_1	L_2	L_3	L_4	L_5	L_6
$L(BS-L, LB2Q)$	0	0	1	0	0	0
$L(FR-B, LB2Q)$	1	0	1	0	0	1
TL	1	0	1	0	0	0

Taking the "or" operation to TL according to formula (6), it can obtain $RTL=1$, which means there have contact relationship between $LB2Q$ and $(BS-L, FR-B)$.

(2)Determine the assembly direction of the new structure

This process is similar to the step of determining the relationship of structural parts. Firstly, it can obtain $G(BS-L, LB2Q)=(1,1,1,1,1,0)$ and $G(FR-B, LB2Q)=(1,1,0,0,1,0)$ according to the formula (4). Secondly, the value of TG can be calculated according to formula (7) as shown in Table 3.

Table 3. Truth table of interference relation

	G_1	G_2	G_3	G_4	G_5	G_6
$G(BS-L, LB2Q)$	1	1	1	1	1	0
$G(FR-B, LB2Q)$	1	1	0	0	1	0
TG	1	1	0	0	1	0

Continuing to carry on the "or" operation to TG according to the formula (8), it can obtain that $RTG=1$. According to table 3, $LB2Q$ own three assembly degrees of freedom, which means the component $LB2Q$ can be successfully installed on the sub assembly $(BS-L, FR-B)$ and it can meet the practical geometric conditions according to the sequence of $(BS-L, FR-B, LB2Q)$. Because the number of the assembly sequence is great, Only 5 representative assembly sequences are selected and shown in Table 4:

Table 4. Typical assembly sequence

Initial component	Assembly sequence number	Assembly sequence
$BS-L$	1	$BS-L \to LB10B \to FR-E \to LB6B \to FR-C \to FR-A \to LB2B \to FR-B \to LB2Q$
	2	$BS-L \to FR-A \to LB6B \to FR-C \to LB10B \to FR-E \to LB2B \to FR-B \to LB2Q$
	3	$BS-L \to FR-B \to LB2Q \to LB2B \to FR-A \to LB6B \to FR-C \to LB10B \to FR-E$
$IB-L$	4	$IB-L \to LB2B \to FR-B \to FR-A \to LB6B \to FR-C \to LB10B \to FR-E \to LB2Q$
	5	$IB-L \to LB10B \to FR-E \to LB6B \to FR-C \to FR-A \to LB2B \to FR-B \to LB2Q$

According to the above 5 feasible assembly sequences, the scheme can be optimized according to the fuzzy evaluation method. The index and the weight value of the evaluation model can be adjusted by the assembly process personnel according to the specific situation.

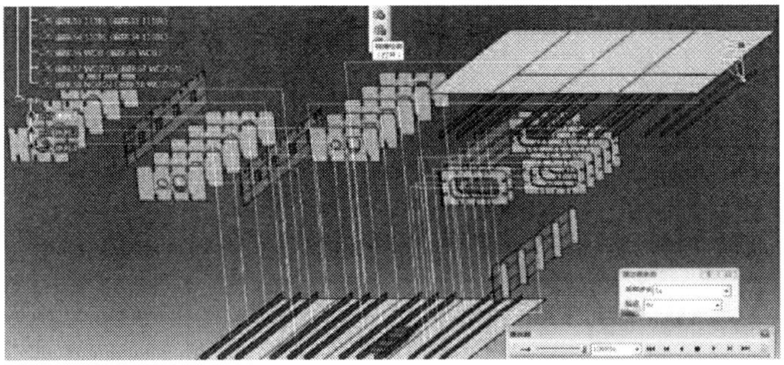

Figure 4. Dynamic interference inspection

In addition, the assembly and dis-assembly process simulation can be carried out through dynamic interference inspection (shown in Figure.4), so as to avoid the collision interference problem in the assembly process, and to optimize the final assembly method.

4. Conclusions

This paper aims at the problems of low assembly efficiency and unstable product quality, which are mainly caused by manual and experience operation. A method of assembly sequence planning based on rule reasoning is presented based on the research on the assembly features of ship hull structures. Setting the double bottom of the ship as an example, hull assembly sequence is obtained based on geometric constraints, which realized the computer aided of the hull assembly sequence evaluation.

Acknowledgement

This work was supported by the Special Funding Project for Key Technologies of Ship Intelligence Manufacturing from the MIIT of China under Grant MC-201704-Z02.

References

[1] Cho K K, Lee S H and Chung D S.1996 An Automatic Process-Planning System for Block Assembly in Shipbuilding J *CIRP Annals - Manufacturing Technology* vol **45** pp 41-44.

[2] Seo Y, Sheen D and Kim T.2007 Block assembly planning in shipbuilding using case-based reasoning J *Expert Systems with Applications* vol **32** pp 245-253.

[3] Shipeng Q, Zuhua J and Ningrong T. 2013 An integrated method for block assembly sequence planning in shipbuilding J *The International Journal of Advanced Manufacturing Technology* vol **69** pp 1123-1135.

[4] Li Peiyong,Cui Jin,Gao Feng .2015 Research on the assembly sequence of a ship block based on the disassembly interference matrix *J Journal of Ship Production and Design* vol **31** pp 230-240.

[5] Wang B. 2006 The research on naval virtual assembly technological optimization and simulation D Dalian University of Technology, Design and manufacture of ship and Marine structures.

[6] Song L P, Fan X M, Ma D Z. 2006 Research on interactive ship assembly sequence planning in virtual reality environment J *Journal of Computer Integrated Manufacturing Systems* vol **12** pp 862-867.

[7] Wei W, Guo C, Duan X D, et al.2014 Method of Assembly Sequence Planning Based on Ant Colony Genetic Hybrid Algorithm J *Journal of System Simulation* vol **26** pp 1684-1691.

[8] Wang J P, Hu Z Y, Lin J S ,et al. 2008 Assembly Sequence Planning Based on Genetic Algorithm J *Journal of Mechanical Engineering & Automation* vol **2** pp 9-11.

[9] Zhang Z Y, Jiang Z B. 2007 Approach on Automation Method of Hull Assembly Process Based on Rule Reasoning J *Journal of Shanghai Jiao tong University* vol **41** pp 31-35.

IOP Conf. Series: Materials Science and Engineering **751** (2020) 012085 doi:10.1088/1757-899X/751/1/012085

Construction Process of Continuous Unreeling and Setting of Leading String Based on Trajectory Theory

Haoting Wu[1], Yong Zhu[2], Qinghao Wang[2], Haichao Wang[2], Lei Zhao[5], Pingjun Wu[4], Ming Jiang[3] and Lei Yang[3,*]

[1] Communication construction branch of State Grid Corporation of China, Beijing, China.
[2] Beijing Electric Power Engineering Co., Ltd., Beijing, China
[3] China Electric Power Research Institute Co., Ltd, Beijing, China
[4] Beijing ZhongDian Golden Bridging & Technology Development Co., Ltd, Beijing , China
[5] State Grid Beijing Electric Power Company, Beijing, China

[*]Corresponding author: 379247828@qq.com

Abstract. The increasing construction of power transmission line promotes the development of construction technology of overhead power transmission line in our country into a brand-new era; while the paving of leading string of traditional overhead power transmission line is still to be completed by climbing the iron tower and manual operation of construction personnel. Therefore, it has been extremely urgent to implement research on the construction process of the power transmission line's leading string threading into pulley, thus reducing the frequency of operation at height of construction personnel. This paper aims to introduce the ejection construction technology of UAV based on trajectory theory to reconstruct the line tackle device, states the construction process flow of multi-rotor UAV utilizing ejection technology to conduct continuous unreeling and setting of leading string with ejection technology in detail, summarizes the full process of construction, conducts contrastive analysis with traditional construction process, and summarizes the social benefit and economic benefit of new type of construction process.

Key words: power transmission line, multi-rotor UAV, pulley reconstruction, leading string

1. Introduction

At present, during the stringing construction process of power transmission line in our country, the UAV is gradually taking the place of dynamic parachute and helicopter to conduct unreeling and setting of leading string by virtue of its small volume, light weight and good manipulation performance, and has been applied into multiple projects. The construction enterprises have accumulated abundant using experience. However, the paving of leading string still needs to be completed by climbing iron tower and manual operation of construction personnel. The construction personnel need operating at height, with high safety risk, many uncontrollable factors, high labor intensity and low construction efficiency.

The UHV power transmission line project applies UAV to conduct unreeling and setting of leading string widely. The UAV is adopted as traction equipment to cooperate with dedicated small-sized tension machine, so as to implement the tension unreeling and setting of “one traction and one tension”. During the process of unreeling and setting, the primary leading string is led by the UAV to pass through the top of tower at the unreeling and setting section base by base, and the personnel on the tower will place the primary leading string into the tackle wheel groove which faces upward via dedicated tool, thus completing the operation of each tower successively. After completion of unreeling and setting, through remote control on the ground, the UAV will cut off the primary leading string and then return to the take off site for landing. [1-7]

This paper proposes new assumption innovatively: utilize UAV to carry the ejection device, eject the leading string via pneumatic ejection, so as to make sure that the leading string can thread into the pulley exactly, so as to realize continuous unreeling and setting. The detailed content is shown as below:

- Reconstruct the receiving device of line tackle pulley, make sure that the leading string can enter the middle wheel groove exactly and under control.
- Research the construction scheme of threading leading string into the line tackle, verify the reliability, exactness and manoeuvrability of utilizing UAV leading string ejection device to eject leading string at height thus threading into the tackle.

2. Introduction to Unreeling and Setting of Leading String

2.1. Introduction to Construction Environment

This paper mainly introduces a construction for the transmission line with crossing barriers including railway, other power transmission lines, and trees. As shown in Figure 1, the line tackle is directly hung on the suspended composite insulator.

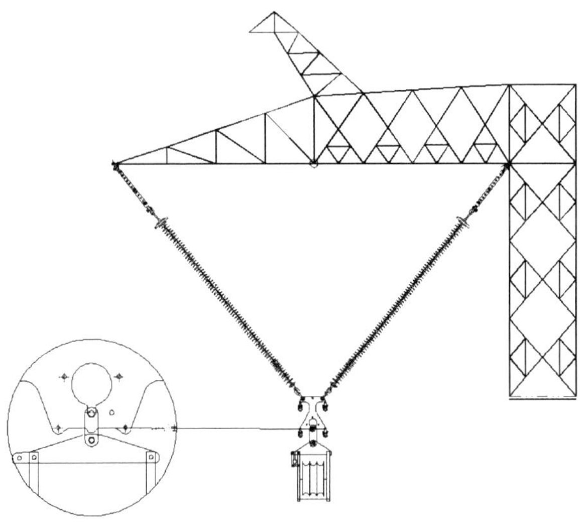

Figure 1. Schematic Diagram of Suspension of Straight-line Tower Single Tackle

2.2. Analysis on the Forces of Primary Leading String for UAV to Handle

At present, the selected rope projectile is 18.7mm in internal diameter, about 22mm in external diameter and about 198mm in length, and the weight of rope projectile is 140g. In case of researching the law of shooting off trajectory of rope projectile, to simplify the computation, we assume the following conditions:

- The shape and mass distribution of rope projectile are of axial symmetry;
- The gas thrust vector of rope projectile coincides with the center shaft of rope projectile;
- Under standard meteorological conditions, the influence of Coriolis inertial force and the change of acceleration of gravity along with latitude are not taken into consideration;

- Do not consider the influence of wind disturbance.

Based on above conditions and assumptions, we can obtain the longitudinal motion equation set of rope projectile:

$$\begin{cases} m\dfrac{dV}{dt} = P\cos\alpha - X - mg\sin\theta \\ m\dfrac{d\theta}{dt} = P\sin\alpha + Y - mg\cos\theta \\ J_{t1}\dfrac{dw_{t1}}{dt} = Mt1 \\ \dfrac{d\theta}{dt} = w_{t1} \\ \dfrac{dx}{dt} = V\cos\theta \\ \dfrac{dy}{dt} = V\sin\theta \\ \dfrac{dm}{dt} = -m \\ \alpha = v - \theta \end{cases} \tag{1}$$

Wherein

- X refers to resistance, X=CxqS, and Cx refers to non-dimensional proportional coefficient, which is called resistance coefficient.
- Y refers to rising force, Y=CyqS, Cy refers to non-dimensional proportional coefficient, called lift coefficient.

According to above theory, the relationship between flight speed and time of rope projector is simulated as shown in the Figure 2:

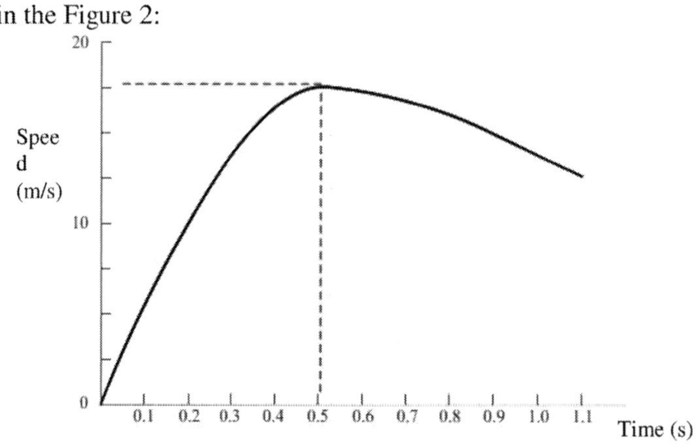

Figure 2. Relationship between Flight Speed and Time of Rope Projectile

From the Figure above, it can be known that after the rope projectile is ejected, the maximum initial flight speed in theory is 17.5m/s, and before 0.5s, the rope projectile is at the acceleration stage; at this time, the gas thrust is greater than the leading string and air resistance; however, after this, the resistance of traction rope is greater than the gas thrust (or the thrust is 0), and the rope projectile is at the deceleration stage.

2.3. Analysis on Stress of Utilizing UAV to Conduct Unreeling and Setting of Leading String

According to existing matured theory [8-9], the sag of leading string in the midpoint of span:

$$f = \frac{l^2 W}{8H_1 cos\varphi} = \frac{l^2 W}{8T cos^2\varphi} \quad (2)$$

Sag at any point of span:

$$f_x = \frac{x(l-x)W}{2H_1 cos\varphi} = \frac{x(l-x)W}{2T cos^2\varphi} = \frac{4x\left(1-\frac{x}{l}\right)f}{l} \quad (3)$$

Initial tension of overhead line (rope):

$$H_1 = \frac{l^2 W}{8f cos\varphi} = T cos\theta \quad (4)$$

As shown in Figure 3(a), when the suspension point A is higher:

$$\gamma_o = Y_B - \frac{W_0 X(l-X) + 2H_1 X sin\varphi}{2H_1 cos\varphi} \quad (5)$$

As shown in Figure 3(b), when the suspension point A is lower:

$$\gamma_o = Y_B - \frac{W_0 X(l-X) - 2H_1 X sin\varphi}{2H_1 cos\varphi} \quad (6)$$

In the formula:
T - Average tension of overhead line, N;
H_1 - Horizontal tension of overhead line, N;
Y_A - Height difference from suspension point A to ground (or object), m;
Y_B - Height difference from suspension point B to ground (or object), m;
L - Line span distance, m;
x - Horizontal distance from crossed object or near-ground span ground to the suspension point A, m;
X - Horizontal distance from crossed object or near-ground span ground to the suspension point B, m;
W - Dead weight of unit length of overhead line, N/m;
h - Height difference of suspension point, m;
f - Sag of midpoint line (rope) of span, m;
ϕ - Height difference angle of suspension point of conductor, degree;
fx - Sag of crossed object, m;
γ_0 - Vertical clearance from rope (line) to ground (or object), m

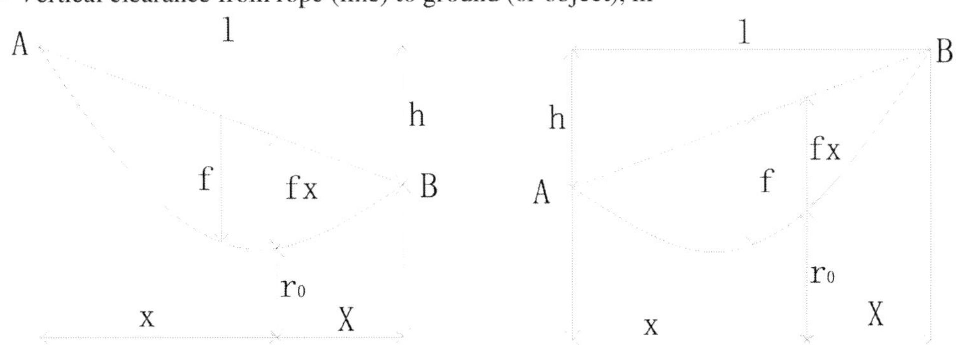

(a) When the suspension point A is higher (b) When the suspension point A is lower

Figure 3. Overhead Line Suspension Curve when Two Suspension Points are not at the Same Height

In case of conducting unreeling and setting of leading string with UAV, the construction cannot be conducted unless that the wind speed reaches fresh breeze, and the influence of wind pressure when the wind speed is 10m/s shall be taken into consideration. The wind pressure refers to the pressure of wind which the plane vertical to the air flow suffers from; the influence of wind pressure which the nylon thread suffers from within a short time period during the unreeling and setting can be neglected; according to the Bernoulli equation, we can obtain the relationship between wind and pressure, and the dynamic pressure of wind is:

$$wp = 0.5 \times \gamma_0 \times V^2 = 0.5 \times \gamma \times \frac{V^2}{g} \tag{7}$$

In the formula:

wp - Wind pressure, kN/m^2;

γ_0 - Air density, kg/m^3;

V - Wind speed, m/s; under the standard status (the atmospheric pressure is 1,013hpa, and the temperature is 15°C), the unit weight of air is $\gamma = 0.01225$kN/m^3, and the acceleration of gravity at the dimension of 45° is g=9.8m/s^2.

It should be pointed out that the air unit weight and the acceleration of gravity are changed along with the dimensionality and altitude. Generally speaking, the γ/g on the plateau is smaller than that on the plain, i.e. at the same temperature, the wind pressure on the plateau generated by the same wind speed is smaller than that on the plain;

If the length is one meter and the diameter is 2mm, then the stressed area for Dubond silk or Deenyma rope is calculated as below:

$$S = \pi \times r \times l = 3.14 \times 0.02 \times 10^{-3} \times 1 = 6.28 \times 10^{-5} \tag{8}$$

$$F = wp \times S = 0.5 \times 0.01225 \times 10 \times \frac{10}{9.8} \times 6.28 \times 10^{-5} = 3.925 \times 10^{-6} kN = 3.925 \times 10^{-3} N \tag{9}$$

Then, calculate the horizontal tension for the leading string using software and get results as shown in Figure 4.

T(N)	I(m)	YA(m)	YB(m)	x(m)	X(m)	W(N/m)
21.36	279	32	33	164	115	0.02195
H$_1$(N)	h(m)	φ (°)	f	fx(m)	higher(γ_0)	lower(γ_0)
21.36	1	0.00	10	9.69		22.90

(a) Calculation of Tension for Unreeling and Setting of 2mm Dubond Silk after Considering Wind Pressure

T(N)	I(m)	YA(m)	YB(m)	x(m)	X(m)	W(N/m)
52.47	279	32	33	164	115	0.053925
H$_1$(N)	h(m)	φ (°)	f	fx(m)	higher(γ_0)	lower(γ_0)
52.47	1	0.00	10	9.69		22.90

(b) Calculation of Tension for Unreeling and Setting of 2mm Dyneema Rope after Considering Wind Pressure

Figure 4. Analysis on Stress in case of Unreeling and Setting of Leading String

Based on the calculation of the traction force, the UAV selected as shown in Figure 5 has following technical data:

- Rising limit of altitude: in case of routine power unit: $\geq$ 2,000m
- Power battery: 22,000 mah
- Maximum flight speed: 10m/s
- Duration of flight: 40min
- Flight radius: 8km
- Full load flying weight: 15.0kg
- Wind resistance capability: grade V (wind speed 8.0-10.7m/s)
- Working temperature: -10 to +40 °C

Figure 5. Picture of Multi-rotor UAV

Characteristics of utilizing UAV to conduct unreeling and setting of leading string includes:

- The flying precision is relatively high, and it can realize the RTK positioning and automatic obstacle avoidance function, and is able to have a cruising flight to the pre-set position exactly, thus meeting the demands for distance of line tackle and safety of iron tower itself;
- It can complete hovering easily, and can realize the posture adjustment via stabilized gyro.

3. Reconstruction of Line Tackle which Utilizes UAV to Conduct Unreeling and Setting of Leading String

To realize the continuous operation construction of utilizing UAV to conduct unreeling and setting of leading string, it is necessary to reconstruct relevant equipment of stringing, make sure that the leading string can thread into the line tackle exactly and then enter the tackle wheel groove selectively; in addition, the pulley after reconstruction will not influence the subsequent tension stringing. The reconstruction scheme of line tackle is shown in Figure 6.

IOP Conf. Series: Materials Science and Engineering **751** (2020) 012085 doi:10.1088/1757-899X/751/1/012085

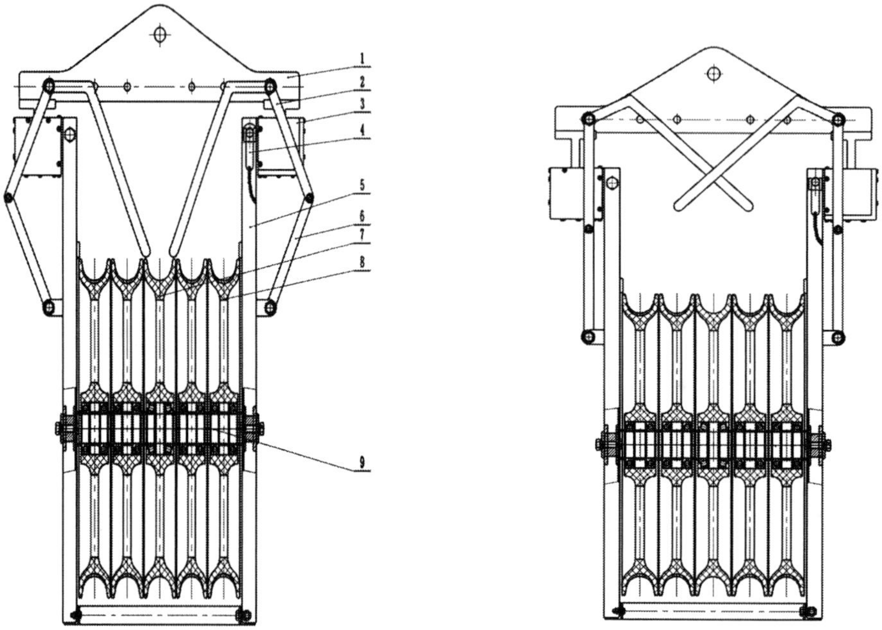

(a) Schematic Diagram of Line Tackle with the Lowest Position of Guide Plate

(b) Schematic Diagram of Line Tackle with the Highest Position of Guide Plate

1. Hanging Plate 2. Guide Plate 3. Self-adjustment Mechanism 4. Handle 5. Rack 6. Rack Connection Plate 7. Steel Wire Rope Wheel 8. Guide Wire Rope 9. Main Shaft

Figure 6. Schematic Diagram of Line Tackle

Prior to unreeling and setting of leading string with UAV, hang the line tackle below the cross arm of iron tower. At this time, for the steel wire rope wheel doesn't suffer from any stress, the line tackle is in the status in Figure 6(a).

When the UAV utilizes ejection device to eject the leading string to the triangular zone between two guide plates, the leading string will slide into the steel wire rope wheel groove automatically along with the guide plate.

With the unreeling and setting of leading string, the vertical load which the steel wire rope wheel suffers from will be increased gradually. The spring in the self-adjustment mechanism will be deformed due to stress; the rack of line tackle, steel wire rope wheel, guide rope wheel and main shaft will have displacement downwards as a whole; and the guide plate will be raised gradually.

After completion of unreeling and setting of leading string, the root of leading string is connected with traction rope for traction unreeling and setting. At this time, the vertical load which the steel wire rope wheel of line tackle is further increased, and the spring is further compressed, till that the spring pressure plate contacts the upper surface of sleeve at the roof plate of spring. The spring will not be compressed any longer, the overall position of rack of line tackle, steel wire rope, guide rope wheel and main shaft will reach the lowest point. Meanwhile the guide plate will be raised to the highest position, as shown in the Figure 6(b).

At this time, the distance between the lower edge of guide plate and the steel wire rope wheel & upper edge of guide rope wheel is relatively large, which will not influence the threading of machines and tools such as subsequent traction plate and splicing sleeve protection device into the line tackle.

4. Continuous Operation Construction of Utilizing UAV to Conduct Unreeling and Setting of Leading String Based on Trajectory Theory

4.1. Operating Procedure

The operating procedure for using UAV to conduct the unreeling of leading string is shown in Figure 7.

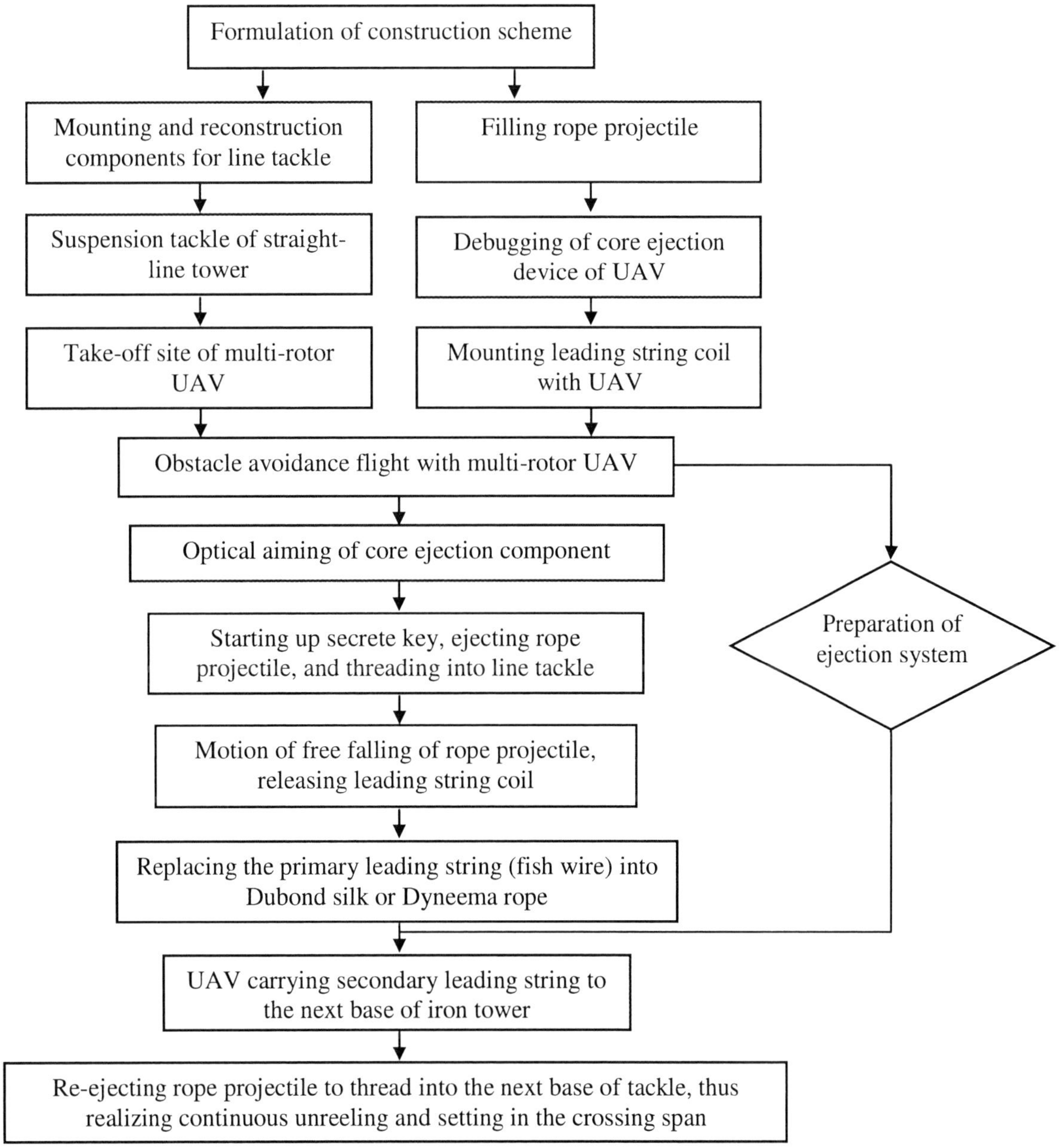

Figure 7. Operating procedure for using UAV to conduct the unreeling of leading string

4.2. Construction Preparation

- Conduct detailed investigation to the construction site and its environment, master the terrain, landform and crossing span situation on the site clearly, and sort the coordinate and elevation of tower position in the crossing span. Pay attention to the disturbance problem of electromagnetic wave.

IOP Conf. Series: Materials Science and Engineering **751** (2020) 012085 doi:10.1088/1757-899X/751/1/012085

- Investigate and implement situations and relevant agreements & treatment measures about crossed objects such as farmland, orchard, house, power line, railway, highway, etc. in the corridor channel, and gather the detailed technical constraints.
- To arrange the unreeling and setting of leading string of tension stringing reasonably, make full use of personnel and time, it is required to allocate personnel for the construction shift of unreeling and setting of leading string.
- Conduct technical training, learn relevant technical and process rules meticulously, master the operation characteristics and operation methods. Organize all construction personnel to conduct technical safety and technical disclosure meticulously.

4.3. Introduction to Continuous Operation Construction Flow of Leading String

*4.3.1. UAV Carrying Leading String for Flying and Ejection Device Control.*The UAV utilizes the GPS module which is mounted on itself to set target position, and then flies to the target position under the control of autonomous navigation program; in addition, it is allowed to be intervened by acting beyond the authority to guarantee the safety of flight during the flight process.

This system adopts portable ground control terminal to control the UAV to start operation by one key, and is able to observe the flight status of UAV in a real-time way; in addition, it is allowed to adopt multiple ground control terminals to control the same UAV. The multi-rotor UAV is provided with automatic obstacle avoidance functional module to enhance the safety of UAV flight, so as to avoid error during the planning of flight route or prevent the UAV from colliding the surrounding object due to wrong artificial flight control and intervention, thus guaranteeing that the UAV can conduct deceleration and hovering before the tackle.

The UAV utilizes the laser ranging and high-precision solid gyroscope to measure the distance from ejection system to the tackle and the angle of pitch exactly, and then work out the aim-taking system of aiming point accurately according to measurement parameters and trajectory theory. Research and develop double-spring damping type impulse absorption and release device, and then absorb the recoil impulse at the moment of ejection.

4.3.2. Construction Process of Threading Leading String into Tackle. Utilize the onboard leading string ejection device of multi-motor UAV; the core component for ejection takes the CO2 compressed gas which is released at a high speed as the power source of ejection of rope projectile, pushes the rope projectile to tract primary leading string (nylon thread) to move forward and separate away from the launching tube; after separation, the gas cylinder will release gas continuously to continue to push the rope projectile to fly forward and to drag the nylon thread to conduct motion of a free falling body after crossing the tackle span. After the projectile falls to the ground, separate the rope releaser away, so as to realize the unreeling and setting of leading string of tower base. Through the artificial traction, replace the nylon thread to φ2mm Dubond silk at the projectile falling point.

4.3.3. Utilizing Multi-motor UAV to Tract Secondary Leading String for Unreeling and Setting Level by Level. After the multi-motor UAV continues to fill rope projectile and rope releaser, connect it with the φ2mm Dubond silk at the falling point of rope projectile nearby the tower base, rise slowly and vertically, adjust the flight direction of UAV according to wind direction on the site, and then tract the φ2mm Dubond silk to fly to the next base of iron tower.

Control the tension of Dyneema leading string, adjust the flight height appropriately if necessary, make the Dyneema leading string pass the obstacle smooth, and in case of being close to the tackle of next base of iron tower, utilize the ejection construction technology of UAV based on the trajectory theory repeatedly, and at this time, the unreeling and setting of leading string in this crossing span is completed, and it is necessary to bind the end of Dyneema rope which has been unreeled and set well firmly in time.

The continuous work flow is shown in Figure 8.

Figure 8. Continuous Unreeling and Setting of Leading String with UAV

5. Conclusions

By utilizing the multi-motor UAV to conduct continuous unreeling and setting of leading string, the destruction rate of young crops and the compensation at the early stage can be reduced; in addition, it can reduce the operation at height, enhance efficiency, and shorten construction period. There are good economic and social benefits in case of utilizing small-sized UAV to conduct unreeling and setting of leading string, and this will be the development direction of construction where there are numerous crossed objects and construction in the forest zone. This method can not only simplify the construction steps, but reduce the dangerousness of construction, guarantee the safety and stability of power transmission line construction, with relatively large research and development prospect.

Acknowledge

This Research was supported by Science and Technology Research Project of State Grid (Research on the construction technology of launching guide rope through pulley based on ballistic theory, GCB17201800306).

References

[1] DL 5106-2017 Operation Code of Cross Power Transmission Line in Installing the Conductor.
[2] DL/T 875-2016 Basic Technical Requirements for Construction Machines and Tools of Overhead Power Transmission Line.
[3] DL/T 685-1999 Basic Requirements, Inspection Specification and Determination Methods of Conductor Stringing Sheave.
[4] DL/T 5343-2018 Construction Technology Guidance for Tension Stringing of 110kV -750kV Overhead Transmission Line.
[5] DL/T 371-2010 Stringing Blocks of Overhead Transmission Lines
[6] GB 50233-2014 Code for Construction and Acceptance of 110kV-750kV Overhead Transmission Line.
[7] GWGF018-2010-SD-XL Typical Construction Method for Unreeling and Setting of Primary Leading String with Remotely Controlled UAV Helicopter.
[8] Li Bozhi. Calculation Part of Stringing Project in Technical Manual of High-voltage Overhead Power Transmission Line Construction Technology [M]. Beijing: China Electric Power Press, 2008.
[9] Li Qinglin. Construction Manual of Overhead Power Transmission Line [M]. Beijing: China Electric Circuit Press, 2002.

IOP Conf. Series: Materials Science and Engineering **751** (2020) 012086 doi:10.1088/1757-899X/751/1/012086

Fatigue Behavior of Friction Stir Welded Lap Joints for Dissimilar AA7150-AA2524 Aluminum Alloy

A Chen*, W F Zang, D K Dong and Y Z Gong

Aircraft Strength Research Institute of China. Xi'an, China

*Email: andychen1986@163.com

Abstract. The detail fatigue rating (DFR) value of friction stir welded (FSW) lap joints for dissimilar AA7150-AA2524 aluminum alloy was experimentally investigated. The metallographic analysis was carried out on by optical microscope (OM). Fatigue fractography was observed under scanning electron microscope (SEM). The results indicate that DFR value of AA7150-AA2524 FSW lap joints is 191.54MPa. There is no distinct different boundary between HAZ and BM of the upper AA7150 sheet. The grains in the upper NZ of AA7150 are coarser than those in the bottom NZ of AA2524. Fracture morphologies shows that lots of hook defects in the FSW lap joint. Fractography indicates that fatigue crack is initiated from the AA2524 bottom surface around the weld center, where weak-bonding defects can be seen. The hook defects and weak-bonding defects have obvious influence on the fatigue property of friction stir welding lap joints, which seriously reduce the fatigue strength.

1. Introduction

As a typical solid-state joining technology, friction stir welding (FSW) is very suitable for joining aluminum alloys. FSW allows the materials to be welded well below the melting temperature of the materials. Many metallurgical reactions between the dissimilar materials at the elevated temperature can be avoided. Thus FSW is a prospective welding process which can also join the dissimilar materials having incompatibilities [1]. Lots of investigations have been studied the defects, microstructure, lap shear failure loads and fracture modes of the FSW for dissimilar materials joints [2,3].

In aerospace industry, 2xxx and 7xxx series aluminum alloys are two outstanding high-strength aluminum alloys which are usually used to fabricate structural components. For instance, AA2524 is used to skins of aircraft fuselage panels and airplane wings, and AA7150 is used extensively to stingers of civil airplane fuselage panels. In this case, dissimilar AA2524 and AA7150 lap joints are inevitably involved. Chen studied the fatigue performance of FSW butt joints for 2524-T3 aluminum alloys based on detail fatigue rating (DFR) [4]. Dubourg carried out process optimization and mechanical tests of friction stir lap welded of AA7075-T6 stringers on AA2024-T3 skin [5]. However, only a few researches for friction stir lap welded dissimilar joints of AA2524 and AA7150 aluminum alloys can be found.

In this present study, the fatigue property of FSW lap joints for dissimilar AA7150-AA2524 aluminum alloy was investigated. The DFR value of FSW lap joints was calculated based on the fatigue test results. The macro and microstructures of weld structure were examined by optical microscope.

2. Experiment

AA2524-T3 and AA7150-T77511 aluminum alloy were used to perform FSW experiments. The nominal chemical compositions of both alloys are listed in table 1. Dimension of the AA2524-T3 sheet was 400 mm × 320 mm ×1.8 mm, which was cleaned with emery paper to wipe off the oxidation layer prior to welding. Dimension of the AA7150-T77511 stringer formed into Z-section pultruded geometry was 28 mm ×2 mm × 400 mm. All welds were made in the lap joint configuration with AA7150-T77511 on top and AA2524-T3 on the bottom as shown in figure 1. The flange and web of stringer were cut in order to convenience of fatigue test. The measured process parameters were a rotational speed of 800 rpm and a travel speed of 200 mm/min. The lap joints were welded by China Friction Stir Welding Center.

Table 1. Chemical compositions of AA2524 and AA7150 (mass fraction, %).

Materials	Fe	Cu	Mn	Mg	Zn	Si	Al
AA2524	0.08	4.8	1.3	1.9	/	0.06	other
AA7150	0.08	2.11	/	2.33	6.42	0.05	other

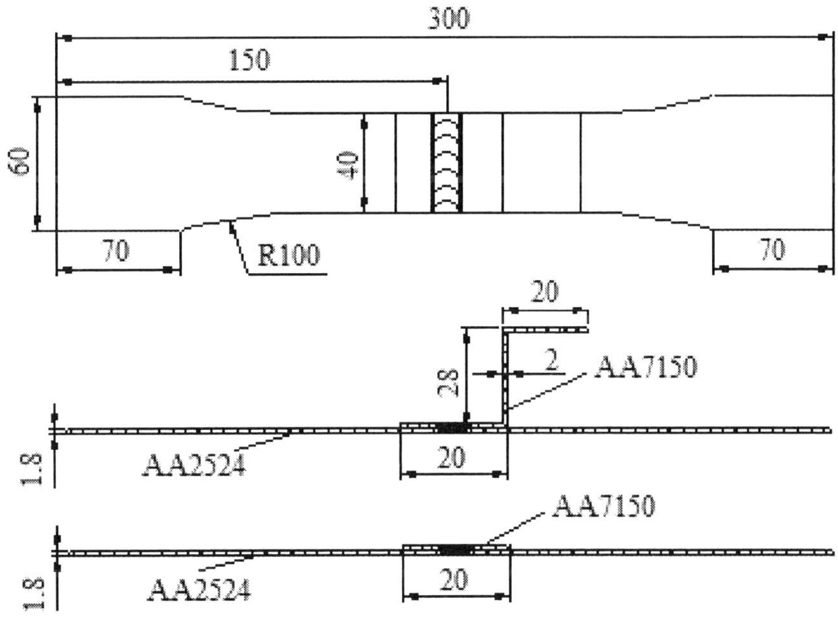

Figure 1. FSW lap joint specimen.

The cross-section surface of the weld was polished and etched with the Keller etchant (3mL HNO_3, 6mLHCL, 6mL HF, 150mL H_2O). The metallographic analysis was carried out on an optical microscope (OLYMPUS-GX51).

Fatigue tests (figure 2) were conducted under INSTRON-8801 servo-hydraulic material testing machine with a load cell of 100 kN. The specimens were loaded in uniaxial tension with an R ratio of 0.06. All tests were performed in load control mode at a frequency of 20 Hz. In all performed tests, only the number of cycles until failure is recorded. After the fatigue test, the fracture features of the FSW lap joint specimens were analyzed.

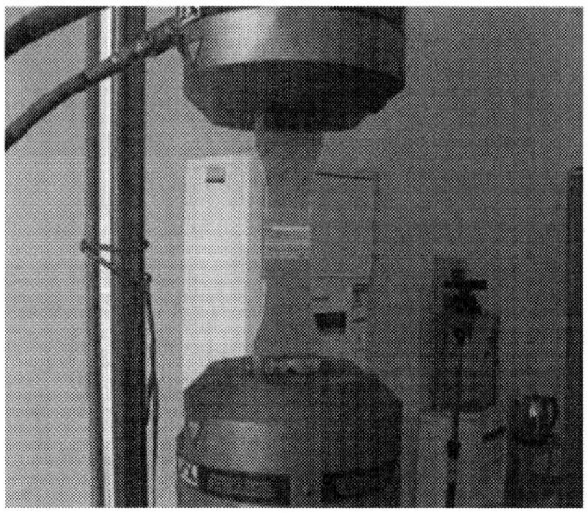

Figure 2. FSW lap joint test.

3. Results and Analysis

3.1. Fatigue behavior

The detail fatigue rating (DFR) means the maximum nominal of 10^5 cycles under constant amplitude loading with stress ratio 0.06, 95% confidence level and 95% confidence. Nowadays, this method is widely used for fatigue life analyses of aircraft structures because of its advantages of simplicity and convenience. The relationship between the DFR and other fatigue parameters is shown as follows [6].

$$DFR = \frac{(1-R)\sigma_{m0}}{0.94\sigma_{m0}/\sigma_{\max} S^{(5-\lg N)} - (0.47S^{(5-\lg N)} - 0.53) - R(0.47S^{(5-\lg N)} + 0.53)} \tag{1}$$

$$N_{95/95} = \frac{\hat{\beta}}{S_C \cdot S_R \cdot S_T} \tag{2}$$

$$\hat{\beta} = \left(\frac{1}{n} \sum_{i=1}^{n} N_i^{\alpha} \right)^{\frac{1}{\alpha}} \tag{3}$$

Where σ_{m0} is material life curve and the σ_m axis intersection value, S is a constant S-N of the slope of the curve parameter, $\hat{\beta}$ is the point estimation of characteristic fatigue life, S_C is the confidence level factor, S_R is the reliability level factor , S_T is the specimen factor, n is number of valid data, N_i is experiment life, α is the scale parameter, which determines the shape of the distribution density curve of Weibull distribution.

S_C with a 95% confidence level for n=6 is 1.15. S_C with a 95% confidence level is calculated as S_R=2.1. Specimen factor S_T is determined as 1.3.

For aluminum alloy: σ_{m0}=310MPa, S=2 and α=4.

The results of fatigue reliability assessment of AA7150-AA2524 FSW lap joint are shown in table 2. The DFR value of AA7150-AA2524 aluminum alloy with FSW lap joint is 191.54MPa.

Table 2. Test results of fatigue life.

σ_{max}(MPa)	Fatigue life (cycle)	$\hat{\beta}$	$N_{95/95}$	DFR(MPa)
192	367113 / 218166 / 368258 221223 / 267024 / 310275	310218	98811	191.54

3.2. Metallographic analysis

The cross-sections and microstructure of AA2524 and AA7150 aluminum alloy lap joint are shown in figure 3. Three typical regions can be distinguished on the cross-sections, corresponding to the nugget zone (NZ), the thermo-mechanical affected zone (TMAZ), the heat affected zone (HAZ) and base material (BM).There is no distinct different boundary between HAZ and BM of the upper AA7150 sheet. Material in the TMAZ undergoes both the thermal cycle and the mechanical stirring during the welding process. Therefore, the grains in TMAZ of AA7150 are strongly distorted. Microstructure in NZ is characterized by fine and equiaxed grains because of undergone complete dynamic recrystallization. The grains in the upper NZ of AA7150 are coarser than those in the bottom NZ of AA2524. The heat in FSW process mainly originates from the shoulder's friction with the surface of the top sheet. The BM adjacent to this HAZ of the bottom AA2524 sheet mainly consists of grains elongated along the rolling direction.

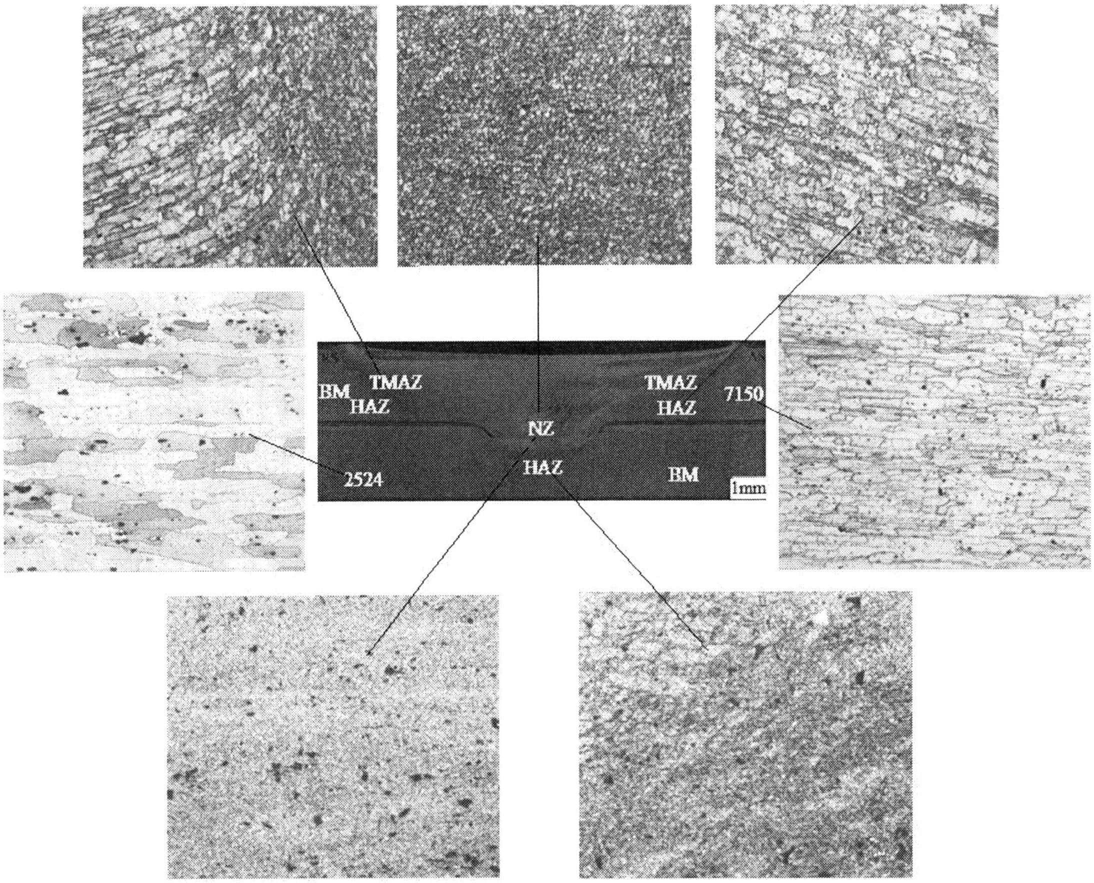

Figure 3. Macro and microstructures of AA7150-AA2524 lap joint.

3.3. Fracture morphologies

Figure 4 illustrates the fracture position of the AA7150-AA2524 FSW lap joint after fatigue test. Fatigue fracture morphologies represent that the crack initiates from the tip of the hook defect on advancing side (AS), propagates upwards along the NZ/TMAZ interface and finally fractures in the NZ. As shown in figure 4, a lot of hook defect presented in the FSW lap joint, which seriously reduced the joint properties of fatigue strength. The reason is that the hook defect provides a favorable orientation for crack propagation and causes stress concentration.

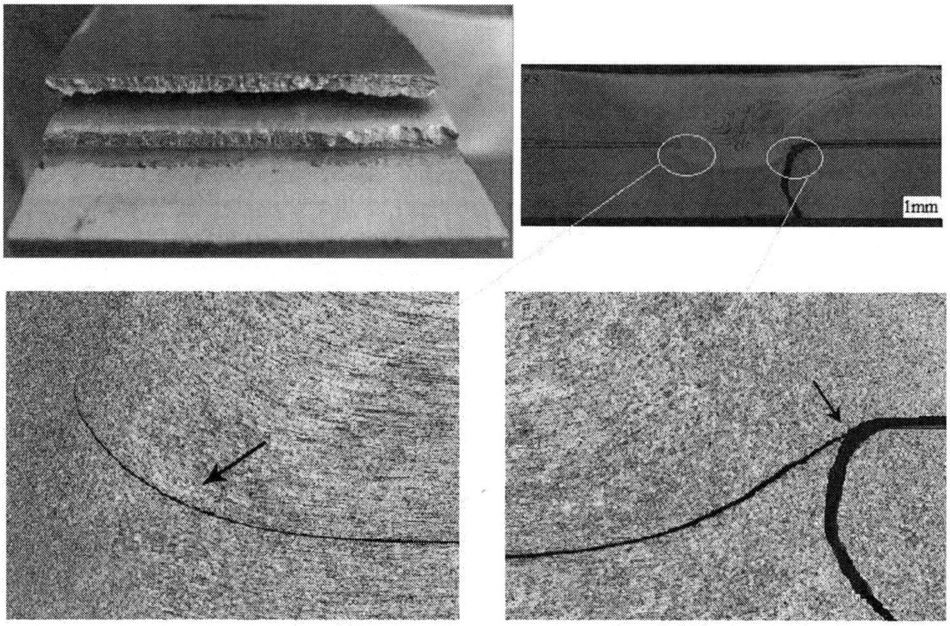

Figure 4. Fracture morphologies of the FSW lap joints.

3.4. Fatigue fractography

The fatigue morphologies of AA7150-AA2524 FSW lap joint specimens were analyzed by using SEM (Model: Quanta600). The following micromorphologies characteristic were observed on the fatigue fracture surface as shown in figure 5.

As seen in figure 5a, the SEM fractography consist of three parts: fatigue crack initiation area , fatigue crack propagation area and final fracture area. Figure 5a shows that the critical fatigue crack was initiated from the AA2524 surface. The detailed view of the crack initiation site is shown in figure 5b,where weak-bonding defects can be seen at the AA2524 bottom surface around the weld center. It also can be seen some inclusions inside the crack initiation area of specimens. The weak-bonding defects and inclusions have obvious influence on the fatigue property of friction stir welding lap joints., which shorten the fatigue lives. Fatigue striations (figure 5c) can be observed in the crack propagation area, associated with some secondary cracks at an intermediate stage of propagation extensively existed. Final fracture area (figure 5d) presents a ductile dimpled characteristic. There are many dimples and microscopic voids with different size in this area. A large number of primary voids are seen due to the broken constituent particles during FSW.

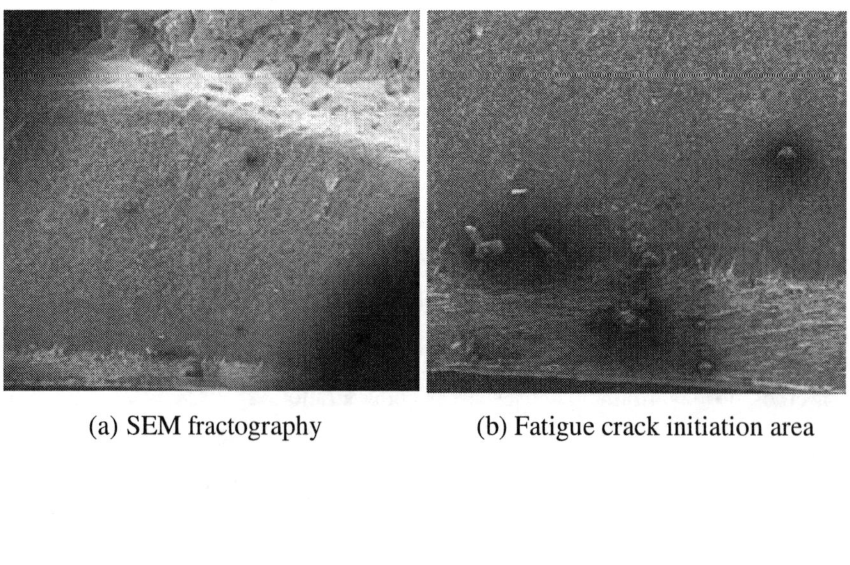

(a) SEM fractography (b) Fatigue crack initiation area

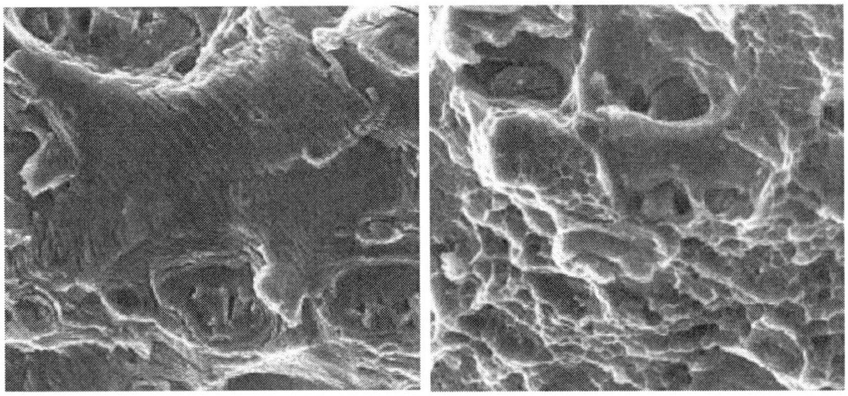

(c) Fatigue crack propagation area (d) Final fracture area

Figure 5. Fatigue fractography

4. Conclusions

The DFR value of FSW lap joints for dissimilar AA7150-AA2524 aluminum alloy is 191.54MPa. There is no distinct different boundary between HAZ and BM of the upper AA7150 sheet. Material in the TMAZ undergoes both the thermal cycle and the mechanical stirring during the welding process. The grains in the upper NZ of AA7150 are coarser than those in the bottom NZ of AA2524. The BM adjacent to this HAZ of the bottom AA2524 sheet mainly consists of grains elongated along the rolling direction. Fatigue fracture morphologies shows that the crack initiates from the tip of the hook defect on AS, propagates upwards along the NZ / TMAZ interface and finally fractures in the NZ. The hook defect provides a favorable orientation for crack propagation and causes stress concentration, which is the main reason for the FSW lap joints fracture. The critical fatigue crack was initiated from the AA2524 bottom surface around the weld center, where weak-bonding defects and inclusions can be seen. The hook defects and weak-bonding defects have obvious influence on the fatigue property of friction stir welding lap joints, which shorten the fatigue lives.

References

[1] Li Y, Murr L E and McClure J C 1999. *Mater Sci Eng A* **271** 213-223
[2] Y B Song, X Q Yang, L. Cui, X P Hou, Z K Shen, and Y Xu 2014. *Materials. Design* **55** 9-18
[3] M Azizieh, A Sadeghi Alavijeh, M Abbasi, Z. Balak and H S KimMechanical 2015. *Materials Chemistry and Physics* 1-10
[4] Chen A, Xu F, Dong D K and Li W Y 2014. *Electric Welding Machine* **44** 140-143
[5] Dubourg L, Merati A and Jahazi M.2010. *Mater Des* **31** 3324-3330
[6] Chen A, Li S S, Dong D K and Xu F. Studying 2016. *Mechanical Science and Technology for Aerospace Engineering* **35** 489-492

IOP Conf. Series: Materials Science and Engineering **751** (2020) 012087 doi:10.1088/1757-899X/751/1/012087

Effect of Load Ratio on the Mechanical Field around Crack Front of Dissimilar Metal Weld Joints

Jia Wenna[1], Zhao Lingyan[2,*], He Jinxuan[1] and Rehmat Bashir[1, 3]

[1]School of Mechanical Engineering, Xi'an University of Science and Technology, Xi'an 710054, China;
[2]School of Science, Xi'an University of Science and Technology, Xi'an 710054, China;
[3]University of Engineering and Technology, Lahore 54890

[*]E-mail address: gloomy2@foxmail.com

Abstract. Stress corrosion cracking (SCC) of the dissimilar metal weld (DMW) joints in nuclear power equipment has been paid more attention in recent years. Besides the geometric size, load level, multiaxial stress state and mechanical heterogeneity, the relaxation and redistribution of residual stress make it very difficult to directly study the influence of residual stresses on the actual welded components. To study the effect of load ratio on residual stress around the crack front in the weld zone of DMW joints, the distributions of residual stress and stress triaxiality under different load ratios are simulated and discussed. The analyses show that the stress becomes larger with the increasing load ratio. The stress is tensile stress near the crack front. While the stress becomes compressive stress when it is far away from the crack front. The external load rotio has great effect on the stress around the circumferential direction at the crack front. The higher load ratio are more likely to cause the crack propagation. In the region of crack angle θ=80-120 °, the load ratio has little effect on the stress triaxiality.

1. Introduction

In nuclear power equipments such as pressure vessels, there are many similar and dissimilar metal weld joints in the pipeline where the weld zones and geometric discontinuity zones, which are the victim of failure in the nuclear power plants[1]. As a weld filler metal, the high-temperature yield strength induced by the residual stress of Alloy 182 makes it more susceptible to SCC [2]. Under the extreme working conditions of high temperature and high pressure, all kinds of complex loads, the residual stresses introduced in the process of welding, forging and bending, SCC of the DMW joints in nuclear power equipment has been paid more attention in recent years [3]. However, the actual welding, due to the geometric size, load level, multiaxial stress state and mechanical heterogeneity is very complex [4, 5]. It is very difficult to directly study the influence of residual stresses on actual welded components, generally without considering the relaxation and redistribution of residual stresses. By the mechanical pre-compression method, residual stresses were introduced into to predict the propagation of creep crack, and the residual stresses measured by the experiments are in good agreement with the finite element simulation data [6]. The crack initiation and creep relaxation behavior of creep crack were also studied under the influence of residual stress [7, 8]. In the actual welding components, the material properties, specimen geometries and loading modes can affect the local residual stress field of the components. Therefore, it is necessary to study the effect of load ratio on the stress field around the crack front in the weld zone of DMW joints under the combined action of residual stresses and applied load [9, 10].

In this paper, the residual stress is introduced into the local crack front of a compact tension (CT) specimen in a DMW joint by the mechanical pre-compression method, the stress-strain fields at the crack front of notched CT specimen under different load ratios are simulated, and the influence of load ratio on the redistribution of residual stress around the crack front is also analyzed.

2. Finite element modelling

2.1. Specimen model

The finite element numerical simulation adopts a high restraint CT specimen, which is cut from the welded zone of a typical nozzle safe-end DMW joint. The geometric size of the CT specimen is shown in Figure 1. The width of the specimen is defined as W=20 mm and crack depth factor as a/W=0.5. The specimen notch radius is defined as r=0.25 mm and there is a groove with a width of 2.5 mm. P is a rigid body displacement load, which is used to generate residual stress by pre-compression in the initial stage. In this study, P is defined to be 0.3 mm. This is modeled by defining the distance between the loading point of rigid body displacement and the left side of the specimen d_c, and the external tensile load T, which is used to simulate the interaction of various complex loads and internal pressure in the FE formulation.

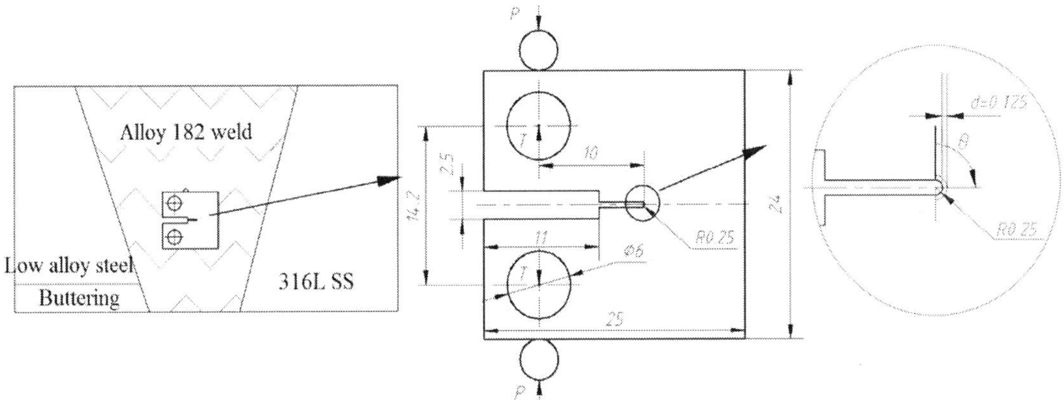

Figure 1. Geometry of a notched CT specimen (W=20 mm, a/W=0.5)

Due to good strength, toughness and corrosion resistance, many Nickel-based alloys are selected as weld filler metals of DMW joints, and their non linear relationship between stress and strain beyond yield is described by Ramberg-Osgood equation in this simulation:

$$\frac{\varepsilon}{\varepsilon_0} = \frac{\sigma}{\sigma_0} + \alpha\left(\frac{\sigma}{\sigma_0}\right)^n \tag{1}$$

Where ε is the strain, including elastic and plastic strain. σ is the total stress; ε_0 is the yield strain of the material, σ_0 is the yield stress of the material, and n is the strain hardening exponent of the material, α is the offset coefficient of the material[11].

$$n = \frac{1}{\kappa \ln(1390/\sigma_0)} \tag{2}$$

Where κ=0.163.

The mechanical properties data and the Ramberg-Osgood constitutive parameters of Alloy 182 at 343 °C are listed in Table 1.

Table 1. Mechanical properties of Alloy 182 at 343°C

Material	Young's modulus E (MPa)	Poison ratio ν	Yield stress σ_0 (MPa)	Hardening exponent n	Hardening parameter α
Alloy 182	193	0.288	517	4.779	1.0

2.2. Boundary conditions and loading

The symmetrical boundary condition is applied to the ligament area at the crack front and the full constraint is applied to avoid the rigid displacement at the right corner of the specimen. The displacement and rotation in the horizontal direction are limited to the reference point where the displacement load is applied, and only the movement in the upper and lower directions is allowed. The introduction of residual stress is realized by applying a certain compression displacement to the up and down rigid bodys.

In the numerical simulation, a variety of complex loads such as self-weight, internal pressure, tensile stress and bending moment are simplified to a tensile load T, which induces the stress intensity factor at the crack front is 30 Mpa·m$^{0.5}$. The redistribution of the stress field of the crack tip is studied by changing the load ratio R under the co-action of residual stresses and external load. The variable load process is divided into six stages, as shown in Figure 2. When the period repeats, the stress field distribution has no effect on the stress field distribution. Therefore, three load ratios are selected for the first two cycles and the comparative analysis is carried out. It is assumed that the values of load ratio R is 0.3, 0.5 and 0.7 respectively. The residual stress caused by the pre-compression method is applied at the unloaded stage.

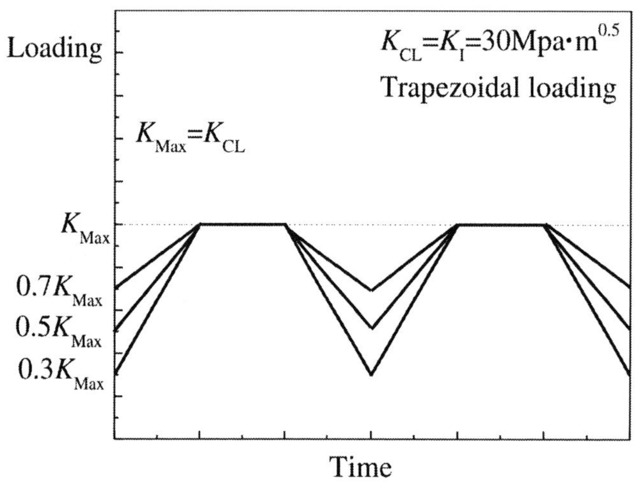

Figure 2. The loading changed with calculation and analysis steps

2.3. FE Model

To reduce the computation time, half of the actual CT specimen is selected to establish the finite element model. A symmetry constraint is set on the symmetry line of the CT specimen. Figure 3 gives the mesh of the notched CT specimen (global model and the area around the crack front), where X-axis is the crack growth direction and Y-axis is the direction perpendicular to the crack growth direction. 17260 4-node bilinear plane strain elements are adopted in the global model. A more refined mesh is adopted near the crack front, in order to obtain a more detailed and accurate data. The minimum size of the element is about 0.025 mm in the area around the crack front.

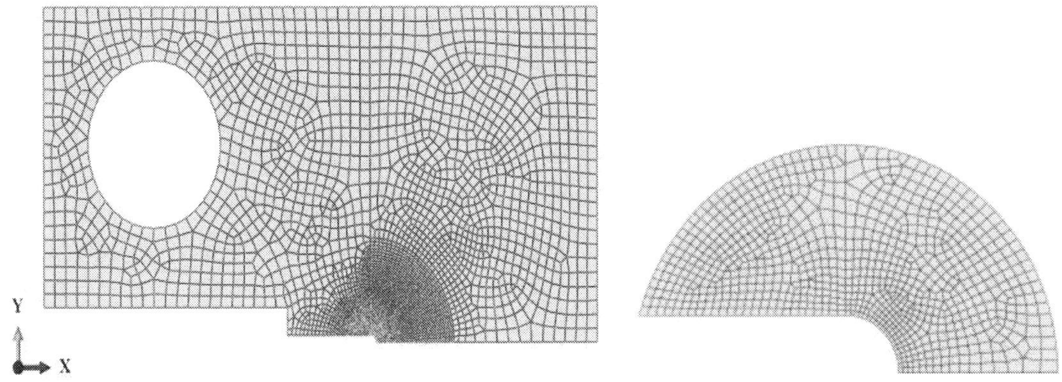

(a) Meshes in global model (b) Meshes refinement around the notch

Figure 3. Finite element model of the CT specimen

3. Results and discussion

3.1. Effect of load ratio on the mechanical field

A set of specimens with passivated circular radius r=0.25 mm were selected and subjected to the mechanical pre-compression. Before unloading, the compression displacement P=0.3 mm was applied at the point where the rigid body compression position dc=15 mm. Thus, the residual stresses introduced at the front end of the crack and the external tensile load T=274.98 N (K_{CL}=K_I=30 Mpa·m$^{0.5}$) are able to work together.

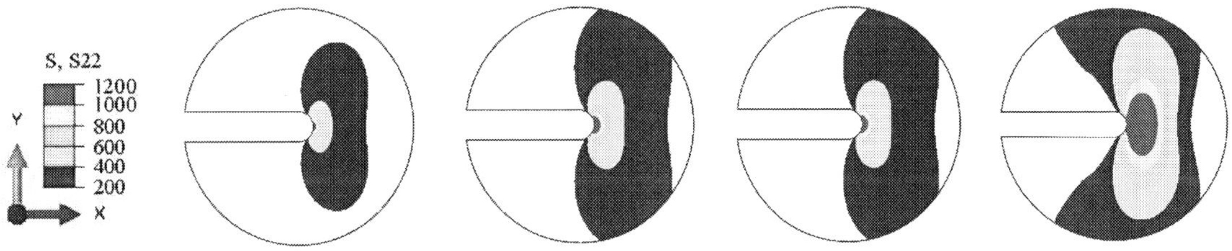

(a) R=0.3 (b) R=0.5 (c) R=0.7 (d) Constant loading

Figure 4. Residual stress distributions around the crack front

Figure 4 shows the residual stress distributions around the crack front of the CT specimens. Under the function together with residual stress and applied load, the high-stress zone around the crack front is much larger under constant loading. With the increase in the load ratio of the applied load, the stress becomes larger. The stress is tensile stress near the crack front. While the stress becomes compressive stress when it is far away from the crack front.

Figure 5(a) shows the residual stresses along the ligament length at the crack front with different load ratios of trapezoidal loading. The stress increases to the peak of a tensile stress, then drops gradually and finally becomes a compressive stress. At the same time, the larger of the load ratio, the higher the tensile stress peak is.

The residual stress at the characteristic distance (d=0.125 mm) ahead of the crack front is shown in Figure 5(b). The crack angle θ varies from 0° to 180°around the crack front. The maximum stress occurs when the crack angle θ=90°, which leads to the crack propagates at θ=90° firstly. The larger of the load ratio, the greater the stress around the circumferential direction is. The external load rotio has great effect on the stress around the circumferential direction at the crack front.

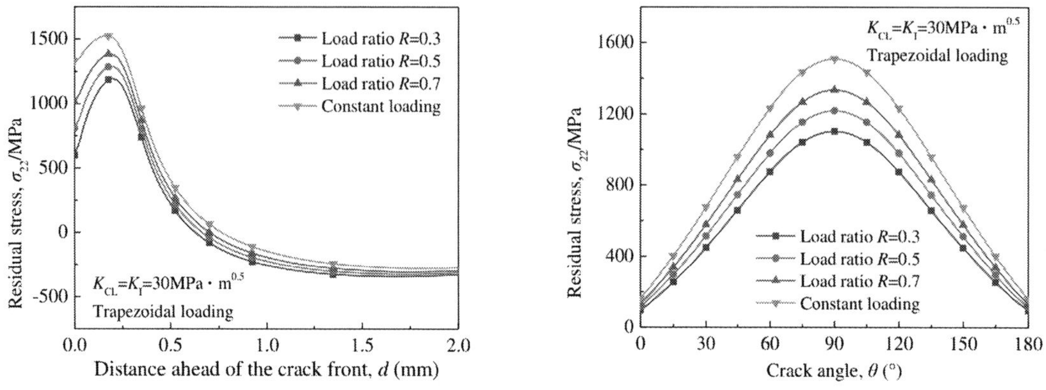

(a) (b)

Figure 5. Distributions of residual stress, (a) along the ligament length, (b) around the crack front

3.2. *Effect of load ratio on stress triaxiality*

Figure 6(a) shows the stress triaxiality along the ligament length at the crack front. The peak value appears at about 0.25 mm away from the notch tip and the stress triaxiality is negative, which is due to the fact that the stress triaxiality is mainly controlled by the residual tensile stresses. The value of stress triaxiality increases with the increasing load ratio, which indicates the higher load ratio are more likely to cause the crack propagation.

As shown in Figure 6(b), the peak of stress triaxiality at the characteristic distance (d=0.125 mm) ahead of the crack front occurs when the crack angle θ=90°. In the region of θ=80-120 °, the load ratio has little effect on the stress triaxiality, which can influence the crack morphology. The effect of external load on the stress along the ligament length at the crack front is more obvious than that around the circumferential direction.

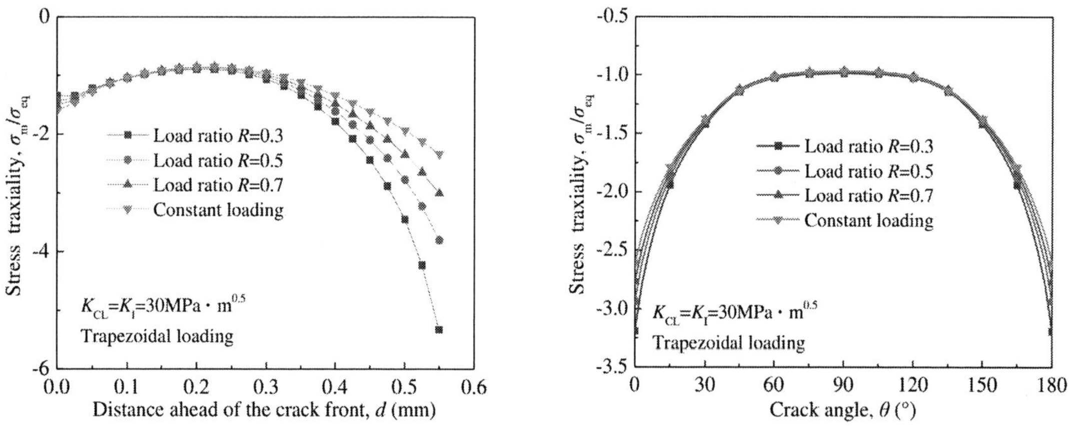

(a) (b)

Figure 6. Distributions of stress triaxiality, (a) along the ligament length, (b) around the crack front

4. Conclusions

Finite element analyses have been conducted for notched CT specimen with the combined action of residual stress and external load. The distributions of residual stress and stress triaxiality under different load ratios are simulated and analyzed. The main results are as follows:

- With the increase in the load ratio of the applied load, the stress becomes larger. The stress is tensile stress near the crack front. While the stress becomes compressive stress when it is far away from the crack front.
- The maximum stress occurs when the crack angle θ=90°, which leads to the crack propagates at θ=90° firstly. The external load rotio has great effect on the stress around the circumferential direction at the crack front.
- The higher load ratio are more likely to cause the crack propagation. In the region of crack angle θ=80-120 °, the load ratio has little effect on the stress triaxiality, which can influence the crack morphology. The effect of external load ratio on the stress along the ligament length at the crack front is more obvious than that around the circumferential direction.

Acknowledgments

This work is financially supported by the National Natural Science Foundation of China (Grant no. is 11502195).

References

[1] Li G and Congleton J 2000 *Corros. Sci.* **42** 1005
[2] Lim Y, Kim D, Kim S and Kim H 2019 *Nucl. Eng. Technol.* **51** 228
[3] Li G, Li G, Fang K and Peng J 2011 *Acta. Metall. Sin.* **47** 797
[4] Kim W, Park J, Lee H, Hong S, Kim Y and Kim S 2013 *Int. J. Pres. Ves. Pip.* **110** 66
[5] Sugiura R, Jr A, Suzuki K, Yokobori T, Suzuki K, Tabuchi M 2010 *Eng. Fract. Mech.* **77** 3053.
[6] Turski M, Bouchard P, Steuwer A and Withers P 2008 *Acta. Mater.* **56** 3598
[7] Chen L, Wang G, Tan J, Xuan F and Tu S 2013 *Eng. Fract. Mech.* **97** 80
[8] Song X, Wang G, Xuan F and Tu S 2015 *Eng. Fract. Mech.* **149** 45
[9] Chen B, Spindler M and Smith D 2010 *ASME 2010 PVP* K 357
[10] Bray D, Dennis R and Smith M 2010 *ASME 2010 PVP* K 387
[11] Xue H and Shi Y 1998 *Int. J. Pres Ves. Pip.* **75** 575.

IOP Conf. Series: Materials Science and Engineering **751** (2020) 012088 doi:10.1088/1757-899X/751/1/012088

Improvement of Microstructural and Mechanical Characteristics of Friction Stir Welded AA6061 Aluminium Alloy Joint

Taher A. Shehabeldeen[1, 2, 3], T. S. Mahmoud [2], Mahmoud Nemat-Alla[3], E.Y. El-Kady[2], Yajun Yin[1], Xiaoyuan Ji[1], Xu Shen[1] and Jianxin Zhou[1,*]

[1]State Key Laboratory of Materials Processing and Die & Mould Technology, Huazhong University of Science and Technology, Wuhan, 430047, China
[2]Mechanical Engineering Department, Faculty of Engineering at Shoubra, Benha University, Egypt.
[3]Department of Mechanical Engineering, Faculty of Engineering, Kafrelsheikh University, Kafrelsheikh, 33516, Egypt.

[*]E-mail: zhoujianxin@hust.edu.cn

Abstract: To improve mechanical characteristics of friction stir welded joint made of AA6061-T6, AA2024 was used as a filler material due to its high strength and compatibility with AA6061 aluminium alloy. Different filler strip thickness (1, 1.5, and 2 mm) were investigated. Homogenous plastic deformation of the mixture with soft and refined grains was observed in the stir zone due to dynamic recrystallization of the grains. Friction stir welded joints with filler strip showed high hardness and tensile strength compared with filler-free joints. Welded joint with 1 mm filler strip has the highest tensile strength among other joints due to the dynamic recrystallization of the refined grains as a result of heat concentration. In addition, using greater thickness of filler strips makes the joining process more difficult, causes joint defects and reduces the tensile strength of welded joints.

Key words: Friction Stir Welding; dissimilar alloys; AA6061 Aluminum Alloy; Mechanical Characteristics; Microstructure.

1. Introduction

Friction stir welding (FSW) was developed in Cambridge by TWI, UK in 1991 [1]. As a solid-state welding method, It has several advantages over traditional fusion welding techniques, such as high joint strength, lack of cracks, less distortion, clean and save power [2-4]. FSW is widely applied for various alloys in comparison with traditional welding techniques. FSW can be used for similar or dissimilar materials like aluminium alloys, titanium and copper alloys. As well as, a lot of application use FSW technique via various types of FSW machines like horizontal, vertical and longitudinal in the industrial range such as the floor panel of recent ships or train [5]. Nowadays, FSW has many worldwide applications, for example, fabrication of aluminium components, manufacturing of airframe structures for commercial and military aircraft, recent automotive manufacturing and space shuttle external tanks. The main idea of FSW is very simple, a revolving tool of non-consumable material has three main functions, heating of the workpiece, material stirring to produce the joint, and restraint of the soft metal under the shoulder of the tool and causes plastic deformation in the metal which leads to solid-state joining in the surrounding area [6]. The strength of FSW joints depends on many parameters such as tool rotation, traverse speed, tool tilt angle, plunge depth and flow of material. The flow of material through FSW has important role which controls the formation of joints without defects and the microstructural characteristics during the weld, therefore. Joining of dissimilar materials is difficult when compared to joining similar materials, due to the difference in chemical composition and mechanical characteristics of the base materials [7-9] . Lap joints which consisted of AA6061–AA 5052 alloys and AA6061 alloy above it, had high Mechanical strength [10] due to the change of joint strength using different thicknesses of plates. For dissimilar lap joining of AA2024 and AA7075 aluminium alloys, increasing in hook size was observed when the AA7075 alloy was located above the joint [11]. Enhanced mechanical properties are achieved when a stronger material is involved in the advancing (AS) side and softer material in the retreating side (RS) [12-14]. AA6061 and 2024 aluminium alloys were selected in this study because they are used in many fields such as aerospace, marine manufacturing and storage tanks [15]. Much effort has been made to improve the mechanical properties of FSW joints by computational [16, 17] and practical methods [18-20].

Herein, the microstructural and mechanical properties of friction stir welded AA6061-T6 aluminium alloy joint have been improved where strips of AA2024 with different thickness were added vertically in between the base metal of the joints. AA2024 strips were used as a filler material which has high strength and good compatibility with AA6061. Traditional FSW tool was used, mechanical and microstructural characteristics of all joints were investigated. The main objective of this work is to increase the strength of AA6061 aluminium alloy joints using high strength material such as AA2024 by FSW process.

2. Materials and Methods

In the present investigation, AA6061-T6 aluminium alloy was used as a base metal and the AA2024 aluminium alloy was used as a filler metal. The chemical compositions of AA6061-T6 and AA2024 aluminium alloys were presented in Table 1. Two plates (120 ×50×8) mm were prepared for the formation of a butt joint, while three AA2024 filler strips with the same length and different thickness (1, 1.5, 2) mm were inserted vertically between the base metal plates, as well as AA6061-T6 joint was welded without filler strip for comparison purpose. FSW was carried out using computer numerical control (CNC) milling machine. FSW was performed using H13 steel tool with 20 mm shoulder diameter and a cylindrical threaded pin with 6 mm diameter and 3 mm length.

Table 1. Chemical composition of AA6061 and AA2024 aluminum alloys (wt. %).

Alloy	Si	Fe	Cu	Mn	Mg	Zn	Cr	Ti	Al
AA6061	0.543	0.0009	0.237	0.00551	1.07	0.00549	0.221	0.047	97.86
AA2024	0.0516	0.131	6.35	0.475	1.85	0.0648	0.0001	0.0554	91.02

AA6061/AA2024/AA6061 butt joints were welded using tool rotational speed of 2000 rpm, welding speed of 25 mm/min, and zero title angle. welding was performed perpendicular to rolling direction. A schematic illustration of the FSW technique with filler strip was shown in Figure 1.

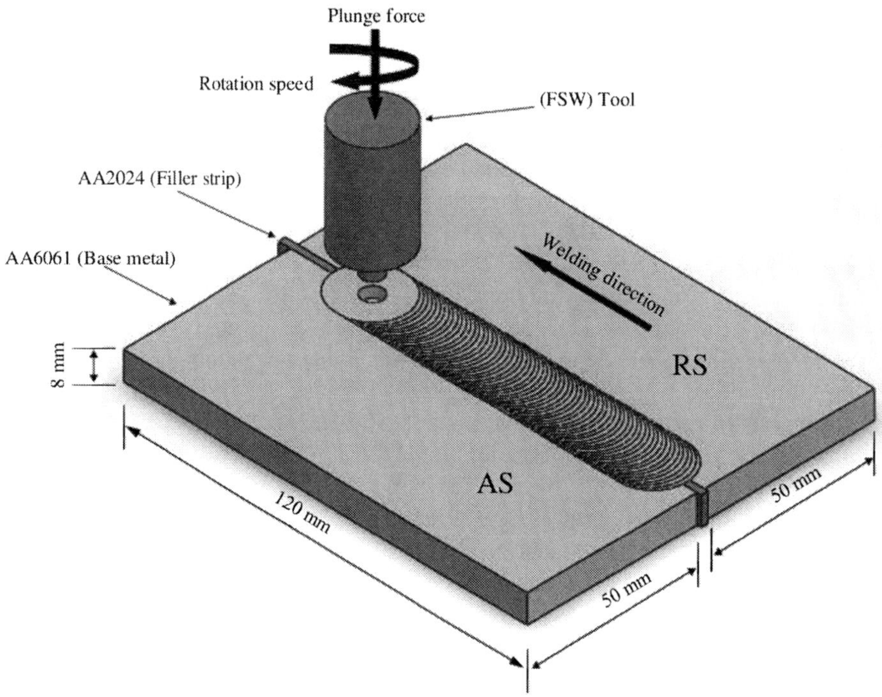

Figure 1. Schematic drawing of FSW technique with filler strip.

3. Results and discussion

3.1 Microstructural Investigations

Figure 2 shows a micrograph of the microstructure of AA6061 base alloy. The microstructure consists of coarser elongated α-Al grains in which the grains are oriented along the rolling direction, and black spots attributed to over etching during microscope investigation. Figure 3 shows typical micrographs of the microstructure at the center of the SZs developed using a different AA2024 strip thicknesses. SZs exhibited finer grains compared to the base alloy. This may be due to the concentration of heat that causes excessive fine grains in equiaxed shape in the SZ, where higher plastic deformation of the material occurs in this region due to dynamic recrystallization [21]. Highly refined grains may enhance the strength of the welded joint [22]. The results revealed that when using a strip with thickness of 1 mm or 1.5 mm, SZs show a lower average grain size when compared to SZs developed without a filler strip, while SZ developed using 2 mm strip shows the largest average grain size. Consequently, it is evidence of a direct relationship between grain size and strip thickness, as the heat generated from the tool shoulder is insufficient to soften and refine the filler material grains.

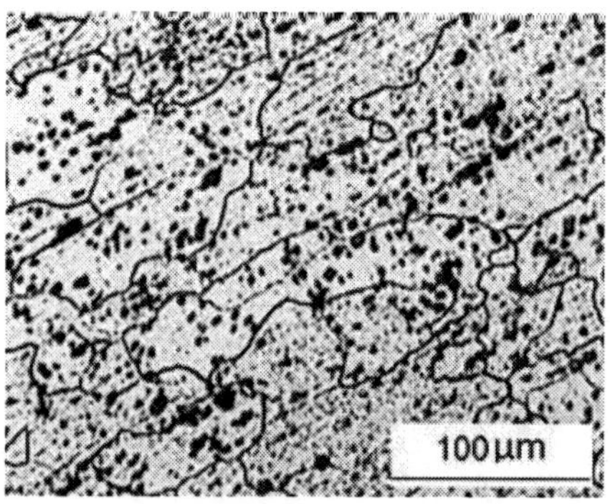

Figure 2. The microstructure of AA6061-T6 base alloy.

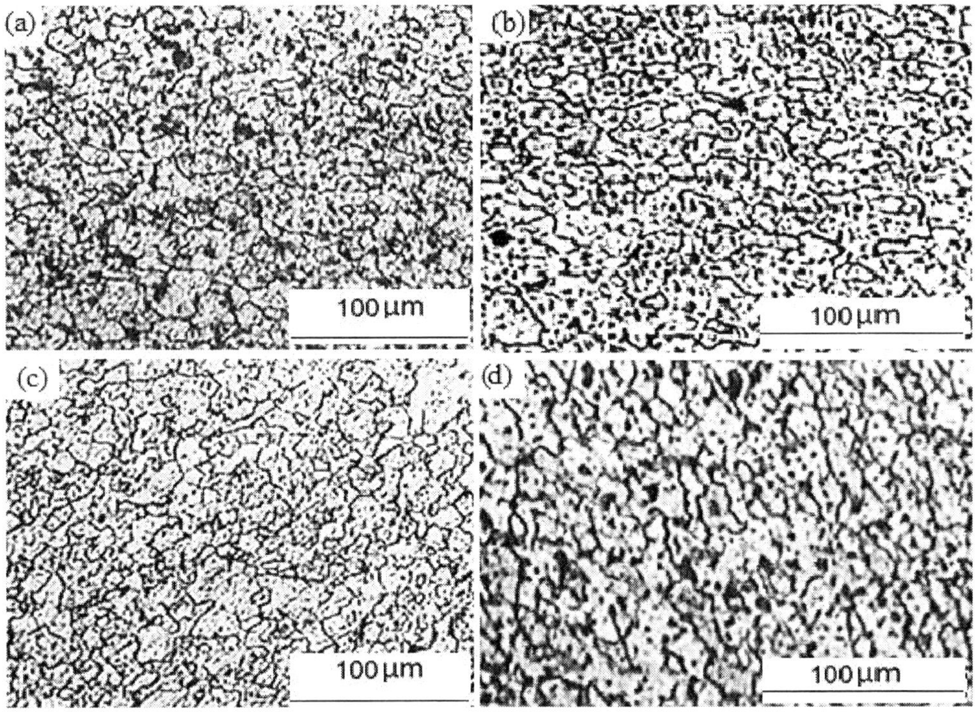

Figure 3. Microstructure at the center of the stirred zones of AA6061-T6 FS welded plates (a) without filler strip (b) with AA2024 filler strip of 1 mm (c) with 1.5 mm and (d) with 2 mm thickness.

3.2 Mechanical Characteristics

3.2.1 *Hardness of welded joints.* The Hardness of welded regions developed using AA2024 filler strips is higher than that developed without filler strip. The peak hardness of SZ at the joint without filler strip was about 49 VH. The SZ developed using 1 mm strip showed higher peak hardness, compared to that developed using 2mm and 1.5mm strips. The peak hardness of welded regions developed using 1, 1.5, and 2mm filler strip were 106, 94 and 101 VH, respectively, as shown in Figure 4 (a). Thus, a proportional relation between hardness and strip thickness was observed. The increased hardness of welded regions developed using AA2024 filler strips may be attributed to higher hardness of AA2024 alloy which contains Cu particles as well as interference between AA6061 and AA2024 grains with homogenous distribution which increases hardness of the joints [23].

3.2.2 *Tensile Characteristics.* Tensile test was performed for FS AA6061 specimens with and without AA2024 filler strips, and it was observed that all tensile failures occurred outside the SZ, which was attributed to the fact that the SZ strength was higher than other zones, moreover, specimens with 1 mm AA2024 filler strip has the highest tensile strength. A gradual decreasing in the tensile strength was detected for specimens with 1.5 and 2 mm filler strip, respectively, while a joint without filler strip has less tensile strength compared to other joints, as shown in Figure 4(b). Small thickness of the filler strip resulted in sufficient contact surface of the revolving pin between the workpiece and the filler strip, making an adequate blend and improved plastic flow. Results show that maximum ultimate tensile strength (UTS) value for the joint with 1 mm strip is 215 MPa, while the UTS for the as received base metal is 175 MPa, this may be attributed to deep dimples that enhance ductility and strength.

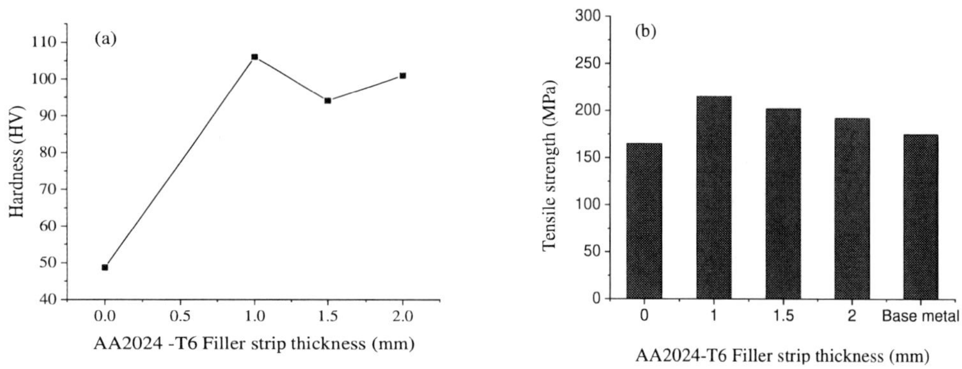

Figure 4. Mechanical properties of welded joints with different strip thickness (a) Hardness at SZ, (b) Ultimate tensile strength.

4. Conclusions

In order to improve the mechanical characteristics of AA6061-T6 joint, AA2024 strips were used as a filler material between the joint plates with a suitable process parameter. The following conclusions can be drawn:

- Homogenous plastic deformation of the mixture was observed in the stir zone, soft and refined grains were also detected due to the dynamic recrystallization of the grains as a result of heat concentration, especially for the joint with lower filler strip thickness.
- Increasing the thickness of AA2024 filler strip results in joint defects and reduces the tensile strength of welded joints.
- FSW joints with filler strip showed high hardness and tensile strength at the SZ compared to joints without AA2024 filler strip.

References

[1] W.M. Thomas, E.D.N., J.C. Needham, M.G. Murch, P. Templesmith, C.J. Dawes, G.B. Patent Application No. 9125978.8 (December 1991).

[2] Wan, L., et al., *Effect of self-support friction stir welding on microstructure and microhardness of 6082-T6 aluminum alloy joint.* Materials & Design, 2014. **55**: p. 197-203.

[3] Zhang, L., et al., *Friction stir welding of Al alloy thin plate by rotational tool without pin.* Journal of Materials Science & Technology, 2011. **27**(7): p. 647-652.

[4] Nandan, R., T. DebRoy, and H. Bhadeshia, *Recent advances in friction-stir welding–process, weldment structure and properties.* Progress in materials science, 2008. **53**(6): p. 980-1023.

[5] Zargar, O.A., *Friction Stir Welding Aluminum Alloy H20-H20 Conventional and Overlap Joints Mechanical Properties.* Manufacturing Science and Technology, 2014. **2**(2): p. 27-33.

[6] Cam, G. and S. Mistikoglu, *Recent developments in friction stir welding of Al-alloys.* Journal of Materials Engineering and Performance, 2014. **23**(6): p. 1936-1953.

[7] Heidarzadeh, A., et al., *Tensile behavior of friction stir welded AA 6061-T4 aluminum alloy joints.* Materials & Design, 2012. **37**: p. 166-173.

[8] Amancio-Filho, S., et al., *Preliminary study on the microstructure and mechanical properties of dissimilar friction stir welds in aircraft aluminium alloys 2024-T351 and 6056-T4.* Journal of materials processing technology, 2008. **206**(1-3): p. 132-142.

[9] Babu, S., et al., *Microstructure and mechanical properties of friction stir lap welded aluminum alloy AA2014.* Journal of Materials Science & Technology, 2012. **28**(5): p. 414-426.

[10] Soundararajan, V., E. Yarrapareddy, and R. Kovacevic, *Investigation of the friction stir lap welding of aluminum alloys AA 5182 and AA 6022.* Journal of Materials Engineering and Performance, 2007. **16**(4): p. 477-484.

IOP Conf. Series: Materials Science and Engineering **751** (2020) 012088 doi:10.1088/1757-899X/751/1/012088

[11] Song, Y., et al., *Defect features and mechanical properties of friction stir lap welded dissimilar AA2024–AA7075 aluminum alloy sheets.* Materials & Design, 2014. **55**: p. 9-18.
[12] Xue, P., et al., *Effect of friction stir welding parameters on the microstructure and mechanical properties of the dissimilar Al–Cu joints.* Materials science and engineering: A, 2011. **528**(13-14): p. 4683-4689.
[13] Padmanaban, G. and V. Balasubramanian, *Selection of FSW tool pin profile, shoulder diameter and material for joining AZ31B magnesium alloy–an experimental approach.* Materials & Design, 2009. **30**(7): p. 2647-2656.
[14] Kwon, Y., I. Shigematsu, and N. Saito, *Dissimilar friction stir welding between magnesium and aluminum alloys.* Materials Letters, 2008. **62**(23): p. 3827-3829.
[15] Sadeesh, P., et al., *Studies on friction stir welding of AA 2024 and AA 6061 dissimilar metals.* Procedia Engineering, 2014. **75**: p. 145-149.
[16] Shehabeldeen, T.A., et al., *Modeling of friction stir welding process using adaptive neuro-fuzzy inference system integrated with harris hawks optimizer.* 2019.
[17] Okuyucu, H., et al., *Artificial neural network application to the friction stir welding of aluminum plates.* 2007. **28**(1): p. 78-84.
[18] Ma, H., et al., *Gap-tolerance control for friction stir butt welding of 2A14 aluminium alloy.* 2019. **148**: p. 106915.
[19] Rajakumar, S., et al., *Influence of friction stir welding process and tool parameters on strength properties of AA7075-T6 aluminium alloy joints.* 2011. **32**(2): p. 535-549.
[20] Costa, M., et al., *Dissimilar friction stir lap welding of AA 5754-H22/AA 6082-T6 aluminium alloys: Influence of material properties and tool geometry on weld strength.* 2015. **87**: p. 721-731.
[21] Babu, S.R., et al., *Microstructural changes and mechanical properties of friction stir processed extruded AZ31B alloy.* Procedia Engineering, 2012. **38**: p. 2956-2966.
[22] El-Hafez, H.A., *Mechanical properties and welding power of friction stirred AA2024-T35 joints.* Journal of materials engineering and performance, 2011. **20**(6): p. 839-845.
[23] Kalaiselvan, K., I. Dinaharan, and N. Murugan, *Characterization of friction stir welded boron carbide particulate reinforced AA6061 aluminum alloy stir cast composite.* Materials & Design, 2014. **55**: p. 176-182.

AUTHOR INDEX

Institute of Physics
Dirac House, Temple Back
Bristol BS1 6BE UK

ISSN: 1757-8981
ISBN 978-1-7138-4700-7

CURRAN ASSOCIATES INC.
proceedings.com